# HVAC
## *Level Four*

### Trainee Guide

 Pearson

Boston   Columbus   Indianapolis   New York   San Francisco   Amsterdam
Cape Town  Dubai  London  Madrid  Milan  Munich  Paris  Montreal  Toronto  Delhi
Mexico City  Sao Paulo  Sydney  Hong Kong  Seoul  Singapore  Taipei  Tokyo

**NCCER**

*President:* Don Whyte
*Vice President:* Steve Greene
*Chief Operations Officer:* Katrina Kersch
*HVAC Project Manager:* Chris Wilson
*Senior Development Manager:* Mark Thomas
*Senior Production Manager:* Tim Davis

*Quality Assurance Coordinator:* Karyn Payne
*Desktop Publishing Coordinator:* James McKay
*Permissions Specialists:* Kelly Sadler
*Production Specialist:* Kelly Sadler
*Production Assistance:* Hannah Payne
*Editors:* Graham Hack, Debie Hicks

**Writing and development services provided by Topaz Publications, Liverpool, NY**

*Lead Writer/Project Manager:* Troy Staton
*Desktop Publisher:* Joanne Hart
*Art Director:* Alison Richmond

*Permissions Editor:* Andrea LaBarge
*Writers:* Troy Staton, Thomas Burke, Terry Egolf

**Pearson**

*Director of Alliance/Partnership Management:* Andrew Taylor
*Editorial Assistant:* Collin Lamothe
*Program Manager:* Alexandrina B. Wolf
*Assistant Content Producer:* Alma Dabral
*Digital Content Producer:* Jose Carchi
*Director of Marketing:* Leigh Ann Simms

*Senior Marketing Manager:* Brian Hoehl
*Composition:* NCCER
*Printer/Binder:* LSC Communications
*Cover Printer:* LSC Communications
*Text Fonts:* Palatino and Univers

17 2023

ISBN-13:   978-0-13-518506-3
ISBN-10:   0-13-518506-8

# Preface

## To the Trainee

Heating, ventilating, and air-conditioning (HVAC) systems technicians install and repair systems that regulate the temperature, humidity, and the total air quality in residential, commercial, industrial, and other buildings. Refrigeration systems, which are used to transport or store food, medicine, and other perishable items, are also a major part of the HVAC technician's job. Hydronics (water-based heating systems), solar panels, and other specialized heating, cooling, and refrigeration systems are also important facets of the HVAC industry.

As a technician, you must be able to install, maintain, and troubleshoot problems throughout the entire system. You must know how to follow drawings or other specifications to maintain any system. You will have to understand air movement, temperatures, and pressures. You may also need basic working knowledge of other crafts such as sheet metal, welding, basic pipefitting, and electrical practices.

Nearly all buildings and homes in the United States alone use forms of heating, cooling, and ventilation to maintain comfort. The increasing advancement of HVAC technology causes employers to recognize the importance of continuous education and keeping up to speed with the latest equipment and skills. Hence, technical school training or apprenticeship programs often provide an advantage and a higher qualification for employment.

NCCER's HVAC program has been designed by highly qualified subject matter experts. The four levels present an apprentice approach to the HVAC field, including theoretical and practical skills essential to your success as an HVAC technician. The US Department of Labor projects faster than average job growth in the HVAC industry.

We wish you the best as you begin an exciting and promising career. This newly revised HVAC curriculum will help you enter the workforce with the knowledge and skills needed to perform productively in either the residential or commercial market.

## New with *HVAC Level Four*

NCCER is proud to release *HVAC Level Four* with updates to the curriculum geared towards water and air quality, building systems design and commissioning, and specialty heating, cooling, refrigeration, and energy conservation methods.

In this edition, you will find a number of important updates to the content. In "Indoor Air Quality," new information has been added about volatile organic compounds, carbon monoxide detectors, heat pipes, and all EPA and health safety requirements. "Energy Conservation Systems" has updated training for required maintenance tasks on heat pipes and heat recovery ventilators (HRVs). The module "Heating and Cooling System Design" now covers the use of smartphone duct system sizing applications.

We wish you success as you progress through this training program. If you have any comments on how NCCER might improve upon this textbook, please complete the User Update form located at the back of each module and send it to us. We will always consider and respond to input from our customers.

We invite you to visit the NCCER website at **www.nccer.org** for information on the latest product releases and training, as well as online versions of the *Cornerstone* magazine and Pearson's NCCER product catalog.

Your feedback is welcome. You may email your comments to **curriculum@nccer.org** or send general comments and inquiries to **info@nccer.org**.

## NCCER Standardized Curricula

NCCER is a not-for-profit 501(c)(3) education foundation established in 1996 by the world's largest and most progressive construction companies and national construction associations. It was founded to address the severe workforce shortage facing the industry and to develop a standardized training process and curricula. Today, NCCER is supported by hundreds of leading construction and maintenance companies, manufacturers, and national associations. The NCCER Standardized Curricula was developed by NCCER in partnership with Pearson, the world's largest educational publisher.

Some features of the NCCER Standardized Curricula are as follows:

- An industry-proven record of success
- Curricula developed by the industry, for the industry
- National standardization providing portability of learned job skills and educational credits
- Compliance with the Office of Apprenticeship requirements for related classroom training (*CFR 29:29*)
- Well-illustrated, up-to-date, and practical information

NCCER also maintains the NCCER Registry, which provides transcripts, certificates, and wallet cards to individuals who have successfully completed a level of training within a craft in NCCER's Curricula. *Training programs must be delivered by an NCCER Accredited Training Sponsor in order to receive these credentials.*

# Special Features

In an effort to provide a comprehensive and user-friendly training resource, this curriculum showcases several informative features. Whether you are a visual or hands-on learner, these features are intended to enhance your knowledge of the construction industry as you progress in your training. Some of the features you may find in the curriculum are explained below.

## Introduction

This introductory page, found at the beginning of each module, lists the module Objectives, Performance Tasks, and Trade Terms. The Objectives list the knowledge you will acquire after successfully completing the module. The Performance Tasks give you an opportunity to apply your knowledge to real-world tasks. The Trade Terms are industry-specific vocabulary that you will learn as you study this module.

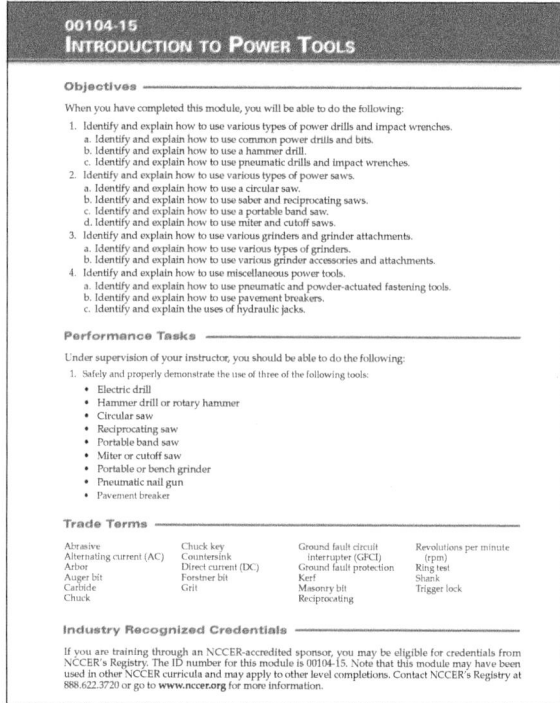

## Trade Features

Trade features present technical tips and professional practices based on real-life scenarios similar to those you might encounter on the job site.

### Bowline Trivia

Some people use this saying to help them remember how to tie a bowline: "The rabbit comes out of his hole, around a tree, and back into the hole."

## Figures and Tables

Photographs, drawings, diagrams, and tables are used throughout each module to illustrate important concepts and provide clarity for complex instructions. Text references to figures and tables are emphasized with *italic* type.

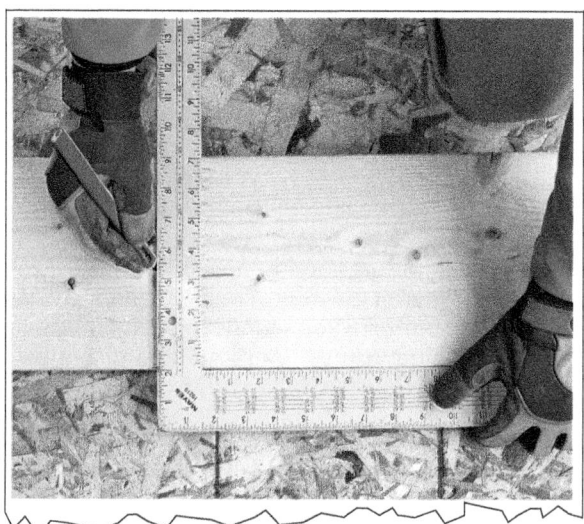

## Notes, Cautions, and Warnings

Safety features are set off from the main text in highlighted boxes and categorized according to the potential danger involved. Notes simply provide additional information. Cautions flag a hazardous issue that could cause damage to materials or equipment. Warnings stress a potentially dangerous situation that could result in injury or death to workers.

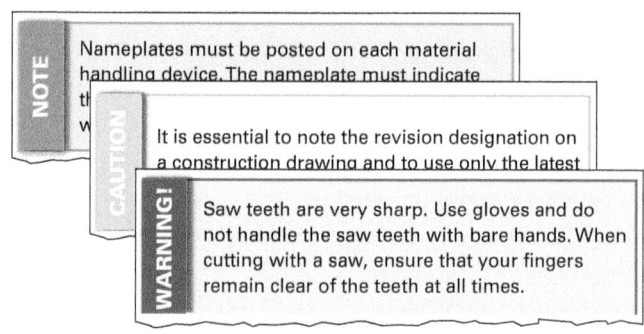

NOTE: Nameplates must be posted on each material handling device. The nameplate must indicate...

CAUTION: It is essential to note the revision designation on a construction drawing and to use only the latest...

WARNING! Saw teeth are very sharp. Use gloves and do not handle the saw teeth with bare hands. When cutting with a saw, ensure that your fingers remain clear of the teeth at all times.

## Case History

Case History features emphasize the importance of safety by citing examples of the costly (and often devastating) consequences of ignoring best practices or OSHA regulations.

**Case History**

**Requesting an Outage**

An electrical contractor requested an outage when asked to install two bolt-in, 240V breakers in panels in a data processing room. It was denied due to the 24/7 worldwide information processing hosted by the facility. The contractor agreed to proceed only if the client would sign a letter agreeing not to hold them responsible if an event occurred that damaged computers or resulted in loss of data. No member of upper management would accept liability for this possibility, and the outage was scheduled.

**The Bottom Line:** If you can communicate the liability associated with an electrical event, you can influence management's decision to work energized.

## Going Green

Going Green features present steps being taken within the construction industry to protect the environment and save energy, emphasizing choices that can be made on the job to preserve the health of the planet.

**GOING GREEN**

### Reducing Your Carbon Footprint

Many companies are taking part in the paperless movement. They reduce their environmental impact by reducing the amount of paper they use. Using email helps to reduce the amount of paper used,

## Did You Know

Did You Know features introduce historical tidbits or interesting and sometimes surprising facts about the trade.

**Did You Know?**

### Safety First

Safety training is required for all activities. Never operate tools, machinery, or equipment without prior training. Always refer to the manufacturer's instructions.

## Step-by-Step Instructions

Step-by-step instructions are used throughout to guide you through technical procedures and tasks from start to finish. These steps show you how to perform a task safely and efficiently.

Perform the following steps to erect this system area scaffold:

*Step 1* Gather and inspect all scaffold equipment for the scaffold arrangement.

*Step 2* Place appropriate mudsills in their approximate locations.

*Step 3* Attach the screw jacks to the mudsills.

## Trade Terms

Each module presents a list of Trade Terms that are discussed within the text and defined in the Glossary at the end of the module. These terms are presented in the text with bold, blue type upon their first occurrence. To make searches for key information easier, a comprehensive Glossary of Trade Terms from all modules is located at the back of this book.

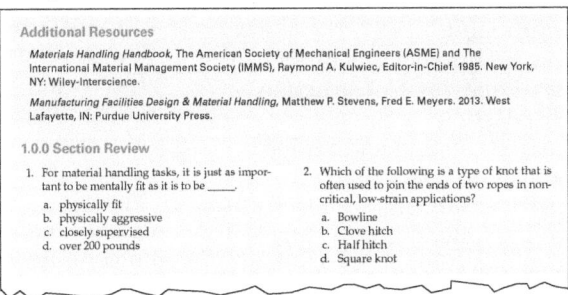

During a rigging operation, the load being lifted or moved must be connected to the apparatus, such as a crane, that will provide the power for movement. The connector—the link between the load and the apparatus—is often a sling made of synthetic, chain, or wire rope materials. This section focuses on three types of slings:

## Section Review

Each section of the module wraps up with a list of Additional Resources for further study and Section Review questions designed to test your knowledge of the Objectives for that section.

**Additional Resources**

*Materials Handling Handbook*, The American Society of Mechanical Engineers (ASME) and The International Material Management Society (IMMS), Raymond A. Kulwiec, Editor-in-Chief. 1985. New York, NY: Wiley-Interscience.

*Manufacturing Facilities Design & Material Handling*, Matthew P. Stevens, Fred E. Meyers. 2013. West Lafayette, IN: Purdue University Press.

**1.0.0 Section Review**

1. For material handling tasks, it is just as important to be mentally fit as it is to be _____.
   a. physically fit
   b. physically aggressive
   c. closely supervised
   d. over 200 pounds

2. Which of the following is a type of knot that is often used to join the ends of two ropes in non-critical, low-strain applications?
   a. Bowline
   b. Clove hitch
   c. Half hitch
   d. Square knot

## Review Questions

The end-of-module Review Questions can be used to measure and reinforce your knowledge of the module's content.

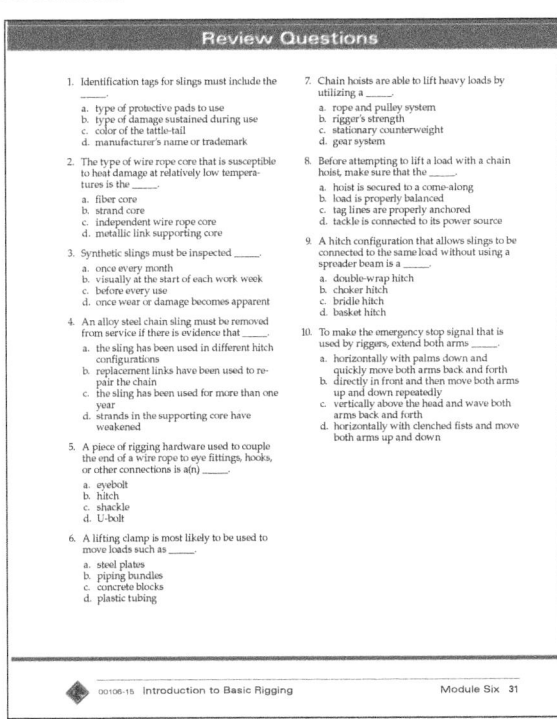

**Review Questions**

1. Identification tags for slings must include the _____.
   a. type of protective pads to use
   b. type of damage sustained during use
   c. color of the tattle-tail
   d. manufacturer's name or trademark

2. The type of wire rope core that is susceptible to heat damage at relatively low temperatures is the _____.
   a. fiber core
   b. strand core
   c. independent wire rope core
   d. metallic link supporting core

3. Synthetic slings must be inspected _____.
   a. once every month
   b. visually at the start of each work week
   c. before every use
   d. once wear or damage becomes apparent

4. An alloy steel chain sling must be removed from service if there is evidence that _____.
   a. the sling has been used in different hitch configurations
   b. replacement links have been used to repair the chain
   c. the sling has been used for more than one year
   d. strands in the supporting core have weakened

5. A piece of rigging hardware used to couple the end of a wire rope to eye fittings, hooks, or other connections is a(n) _____.
   a. eyebolt
   b. hitch
   c. shackle
   d. U-bolt

6. A lifting clamp is most likely to be used to move loads such as _____.
   a. steel plates
   b. piping bundles
   c. concrete blocks
   d. plastic tubing

7. Chain hoists are able to lift heavy loads by utilizing a _____.
   a. rope and pulley system
   b. rigger's strength
   c. stationary counterweight
   d. gear system

8. Before attempting to lift a load with a chain hoist, make sure that the _____.
   a. hoist is secured to a come-along
   b. load is properly balanced
   c. tag lines are properly anchored
   d. tackle is connected to its power source

9. A hitch configuration that allows slings to be connected to the same load without using a spreader beam is a _____.
   a. double-wrap hitch
   b. choker hitch
   c. bridle hitch
   d. basket hitch

10. To make the emergency stop signal that is used by riggers, extend both arms _____.
    a. horizontally with palms down and quickly move both arms back and forth
    b. directly in front and then move both arms up and down repeatedly
    c. vertically above the head and wave both arms back and forth
    d. horizontally with clenched fists and move both arms up and down

# NCCER Standardized Curricula

*NCCER's training programs comprise more than 80 construction, maintenance, pipeline, and utility areas and include skills assessments, safety training, and management education.*

Boilermaking
Cabinetmaking
Carpentry
Concrete Finishing
Construction Craft Laborer
Construction Technology
Core Curriculum: Introductory
 Craft Skills
Drywall
Electrical
Electronic Systems Technician
Heating, Ventilating, and Air
 Conditioning
Heavy Equipment Operations
Heavy Highway Construction
Hydroblasting
Industrial Coating and Lining
 Application Specialist
Industrial Maintenance Electrical
 and Instrumentation Technician
Industrial Maintenance Mechanic
Instrumentation
Ironworking
Manufactured Construction
 Technology
Masonry
Mechanical Insulating
Millwright
Mobile Crane Operations
Painting
Painting, Industrial
Pipefitting
Pipelayer
Plumbing
Reinforcing Ironwork
Rigging
Scaffolding
Sheet Metal
Signal Person
Site Layout
Sprinkler Fitting
Tower Crane Operator
Welding

## Maritime

Maritime Industry Fundamentals
Maritime Pipefitting
Maritime Structural Fitter

## Green/Sustainable Construction

Building Auditor
Fundamentals of Weatherization
Introduction to Weatherization
Sustainable Construction
 Supervisor
Weatherization Crew Chief
Weatherization Technician
Your Role in the Green
 Environment

## Energy

Alternative Energy
Introduction to the Power Industry
Introduction to Solar Photovoltaics
Power Generation Maintenance
 Electrician
Power Generation I&C
 Maintenance Technician
Power Generation Maintenance
 Mechanic
Power Line Worker
Power Line Worker: Distribution
Power Line Worker: Substation
Power Line Worker: Transmission
Solar Photovoltaic Systems Installer
Wind Energy
Wind Turbine Maintenance
 Technician

## Pipeline

Abnormal Operating Conditions,
 Control Center
Abnormal Operating Conditions,
 Field and Gas
Corrosion Control
Electrical and Instrumentation
Field and Control Center
 Operations
Introduction to the Pipeline
 Industry
Maintenance
Mechanical

## Safety

Field Safety
Safety Orientation
Safety Technology

## Supplemental Titles

Applied Construction Math
Tools for Success

## Management

Construction Workforce
 Development Professional
Fundamentals of Crew Leadership
Mentoring for Craft Professionals
Project Management
Project Supervision

## Spanish Titles

Acabado de concreto: nivel uno
 (*Concrete Finishing Level One*)
Aislamiento: nivel uno
 (*Insulating Level One*)
Albañilería: nivel uno
 (*Masonry Level One*)
Andamios (*Scaffolding*)
Carpintería: Formas para
 carpintería, nivel tres
 (*Carpentry: Carpentry Forms, Level Three*)
Currículo básico: habilidades
 introductorias del oficio
 (*Core Curriculum: Introductory Craft Skills*)
Electricidad: nivel uno
 (*Electrical Level One*)
Herrería: nivel uno
 (*Ironworking Level One*)
Herrería de refuerzo: nivel uno
 (*Reinforcing Ironwork Level One*)
Instalación de rociadores: nivel uno
 (*Sprinkler Fitting Level One*)
Instalación de tuberías: nivel uno
 (*Pipefitting Level One*)
Instrumentación: nivel uno, nivel
 dos, nivel tres, nivel cuatro
 (*Instrumentation Levels One through Four*)
Orientación de seguridad
 (*Safety Orientation*)
Paneles de yeso: nivel uno
 (*Drywall Level One*)
Seguridad de campo
 (*Field Safety*)

# Acknowledgments

This curriculum was revised as a result of the farsightedness and leadership of the following sponsors:

Builders Association of North Central Florida
CareerSafety Center
Center for Employment Training
Daiken
Duke Energy
Fort Scott Community College
Hubbard Construction

Industrial Management and Training Institute, Inc.
Lee Company
Lincoln Tech
Santa Fe College
Windham School District

This curriculum would not exist were it not for the dedication and unselfish energy of those volunteers who served on the Authoring Team. A sincere thanks is extended to the following:

Corey Driggs
Art Grant
Joseph Pietrzak
Norman Sparks

John Stronkowski
Tony Vazquez
Ted Watts

A sincere thanks is also extended to the dedication and assistance provided by the following technical advisors:

Senobio Aguilera          Lenny Joseph          Chris Sterrett

# NCCER Partners

American Council for Construction Education
American Fire Sprinkler Association
Associated Builders and Contractors, Inc.
Associated General Contractors of America
Association for Career and Technical Education
Association for Skilled and Technical Sciences
Construction Industry Institute
Construction Users Roundtable
Design Build Institute of America
GSSC – Gulf States Shipbuilders Consortium
ISN
Manufacturing Institute
Mason Contractors Association of America
Merit Contractors Association of Canada
NACE International
National Association of Women in Construction
National Insulation Association
National Technical Honor Society
National Utility Contractors Association
NAWIC Education Foundation
North American Crane Bureau
North American Technician Excellence
Pearson
Prov
SkillsUSA®
Steel Erectors Association of America
U.S. Army Corps of Engineers
University of Florida, M. E. Rinker Sr., School of Construction Management
Women Construction Owners & Executives, USA

**NCCER Business Partners**

# Contents

## Module One
### Water Treatment

Focuses on the methods and devices that are used to treat water in HVAC systems. It introduces the main characteristics of water and explains how they affect HVAC system performance, pointing out the problems that could occur if improperly treated water were used. It presents various types of filtration devices and water treatment equipment. It also covers how to use water analysis test kits and emphasizes the importance of following safety precautions when working with water treatment chemicals. It explores in detail how to identify and address specific water-related problems in various types of recirculating water systems. (Module ID 03308; 10 Hours)

## Module Two
### Indoor Air Quality

Provides trainees with guidance on how to maintain good indoor air quality (IAQ) and comply with established IAQ standards. It describes how the health and comfort of people are affected by contaminants contained in the air that circulates through a building. It covers in detail how to detect sources of building air contaminants. It explains how control of ventilation, temperature, humidity, and chemical and microbial contaminants helps maintain acceptable IAQ. It explores various methods and HVAC devices and equipment that are used to address IAQ problems. It also discusses liability issues associated with servicing HVAC systems. (Module ID 03403; 12.5 Hours)

## Module Three
### Energy Conservation Equipment

Describes energy conservation technologies and devices used in residential and commercial HVAC systems. The operation of several innovative devices used for energy recycling and reclamation is described. It also discusses methods being implemented for electric energy-demand reduction and the practical use of ice storage systems. (Module ID 03404; 7.5 Hours)

## Module Four
### Building Management Systems

Provides trainees with guidance related to building management systems. Developing the necessary skills to understand the applications, principles, and troubleshooting of building management systems are vital to the future success of trainees in the HVAC trade. (Module ID 03405; 12.5 Hours)

## Module Five
### System Air Balancing

Provides trainees with information and skills needed to balance air systems. Developing air balancing skills is vital to the future success of trainees in the HVACR trade. To that end, a portion of this module is devoted to hands-on practice and the successful completion of its required performance tasks. (Module ID 03402; 15 Hours)

## Module Six
## System Startup and Shutdown

Provides trainees with information and skills needed to start up and shut down commercial HVAC equipment, including boilers and chillers. Developing these skills is vital to the future success of trainees in the HVACR trade. To that end, a portion of this module is devoted to hands-on practice and the successful completion of its required Performance Tasks. (Module ID 03406; 15 Hours)

## Module Seven
## Construction Drawings and Specifications

Focuses on the interpretation of construction drawings and specifications associated with HVAC installations in new construction. It explores the many different types of drawings that HVAC technicians and installers work with. It covers the use of specifications and submittals for HVAC equipment installation. It also covers the performance of the takeoff process for HVAC equipment and materials. (Module ID 03401; 12.5 Hours)

## Module Eight
## Heating and Cooling System Design

Focuses on the selection of proper heating and cooling equipment along with proper design of air distribution and refrigerant piping systems. It emphasizes how a good understanding of the many factors that influence HVAC system design enables HVAC contractors to select the most effective equipment while minimizing installation costs. (Module ID 03407; 22.5 Hours)

## Module Nine
## Commercial/Industrial Refrigeration

Focuses on commercial and industrial applications of refrigeration. It covers the equipment, control systems, and refrigerants used for these purposes. It compares and contrasts the methods and components used in commercial and industrial applications with those used in comfort cooling systems. (Module ID 03408; 20 Hours)

## Module Ten
## Alternative and Specialized Heating and Cooling Systems

Provides trainees with guidance related to alternative and specialized heating and cooling systems. Developing the necessary skills to understand the applications, principles, and troubleshooting of these systems are vital to the future success of trainees in the HVACR trade. (Module ID 03409; 7.5 Hours)

## Module Eleven
## Fundamentals of Crew Leadership

Teaches skills needed to become an effective crew leader, as well as knowledge and abilities required to transition from craftworker to crew leader. The module also covers workforce diversity and organization, basic leadership skills, safety, and project control. (Module ID 46101; 22.5 Hours)

## Glossary

## Index

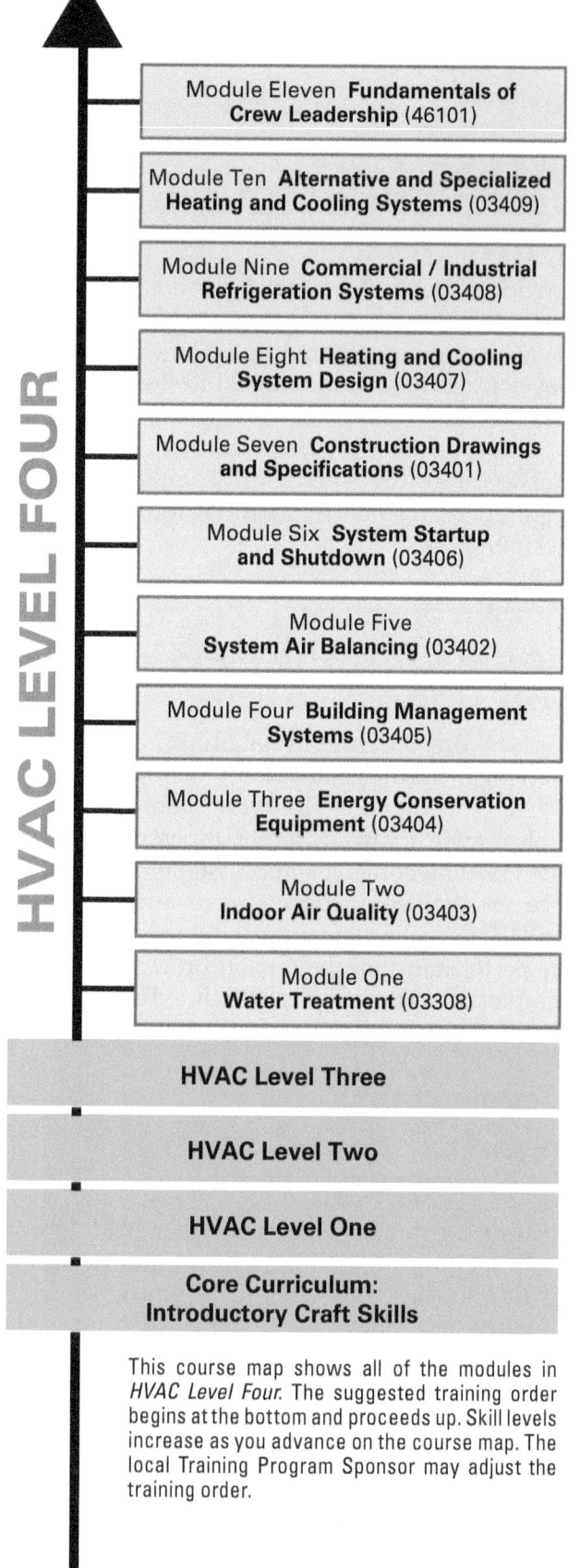

HVAC LEVEL FOUR

Module Eleven **Fundamentals of Crew Leadership** (46101)

Module Ten **Alternative and Specialized Heating and Cooling Systems** (03409)

Module Nine **Commercial / Industrial Refrigeration Systems** (03408)

Module Eight **Heating and Cooling System Design** (03407)

Module Seven **Construction Drawings and Specifications** (03401)

Module Six **System Startup and Shutdown** (03406)

Module Five **System Air Balancing** (03402)

Module Four **Building Management Systems** (03405)

Module Three **Energy Conservation Equipment** (03404)

Module Two **Indoor Air Quality** (03403)

Module One **Water Treatment** (03308)

**HVAC Level Three**

**HVAC Level Two**

**HVAC Level One**

**Core Curriculum: Introductory Craft Skills**

This course map shows all of the modules in *HVAC Level Four.* The suggested training order begins at the bottom and proceeds up. Skill levels increase as you advance on the course map. The local Training Program Sponsor may adjust the training order.

# Water Treatment

## OVERVIEW

This module focuses on the methods and devices used to treat water in HVAC systems. It introduces the main characteristics of water and describes how the use of untreated water can be harmful to the various types of HVAC equipment. It also discusses some of the common water treatment chemicals, devices, and methods that are used in the field. Water treatment of potable (drinking) water is a related but separate subject that is not covered here.

# Module 03308

Trainees with successful module completions may be eligible for credentialing through NCCER's National Registry. To learn more, go to **www.nccer.org** or contact us at **1.888.622.3720**. Our website has information on the latest product releases and training, as well as online versions of our *Cornerstone* magazine and Pearson's product catalog.

Your feedback is welcome. You may email your comments to **curriculum@nccer.org**, send general comments and inquiries to **info@nccer.org**, or fill in the User Update form at the back of this module.

This information is general in nature and intended for training purposes only. Actual performance of activities described in this manual requires compliance with all applicable operating, service, maintenance, and safety procedures under the direction of qualified personnel. References in this manual to patented or proprietary devices do not constitute a recommendation of their use.

*03308 V5*

## Objectives

When you have completed this module, you will be able to do the following:

1. Describe problems that the properties of water can cause in HVAC systems.
   a. Describe the properties of water that relate to water treatment in HVAC systems.
   b. Identify water quality problems that affect HVAC system performance.
   c. Describe how water test kits are used to collect samples for analysis.
   d. State the common safety precautions related to working with water treatment chemicals.
2. Identify types of mechanical water treatment devices and equipment.
   a. Identify types of filtration devices.
   b. Identify types of water treatment equipment.
3. Identify and describe how to address water-related problems that occur in specific types of hydronic and steam systems.
   a. Identify and describe how to treat water-related problems that occur in open recirculating water systems.
   b. Identify and describe how to treat water-related problems that occur in closed recirculating water systems.
   c. Identify and describe how to treat water-related problems that occur in steam systems.

## Performance Tasks

Under the supervision of your instructor, you should be able to do the following:

1. Use a water analysis test kit to test water.
2. Inspect a cooling tower or steam boiler and its related water piping system for signs of water treatment problems.

## Trade Terms

Alkalinity
Backwashing
Bleed-off
Colloidal substance
Concentration
Cycles of concentration
Dissolved solids
Electrolysis
Electrolyte
Fouling

Grains per gallon (gpg)
Hardness
Inhibitor
Milligrams per liter (mg/l)
Parts per million (ppm)
pH
Scale
Suspended solids
Total solids

## Industry-Recognized Credentials

# Contents

## Figures and Tables ─────────────

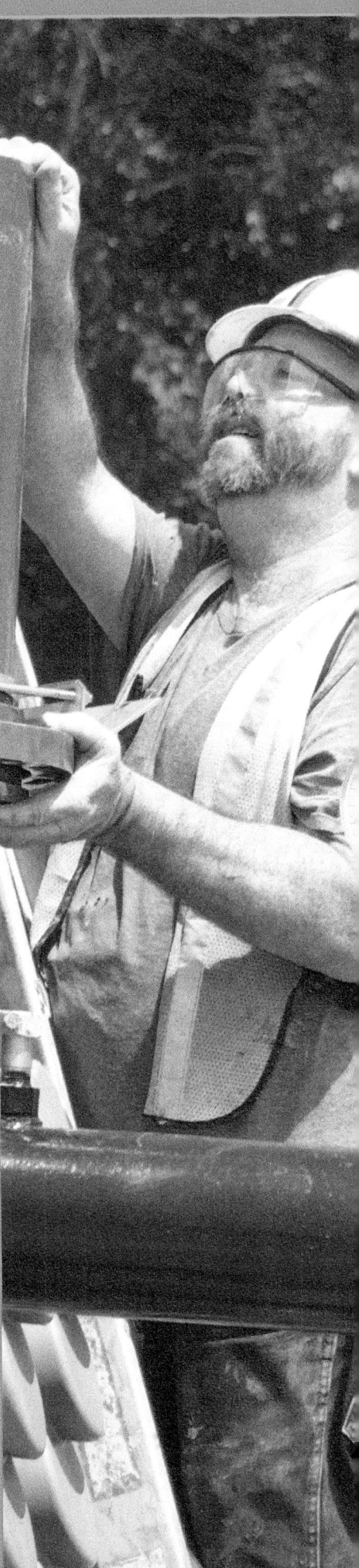

# SECTION ONE

## 1.0.0 INTRODUCTION TO WATER TREATMENT

## Objectives

Describe problems that the properties of water can cause in HVAC systems.

a. Describe the properties of water that relate to water treatment in HVAC systems.
b. Identify water quality problems that affect HVAC system performance.
c. Describe how water test kits are used to collect samples for analysis.
d. State the common safety precautions related to working with water treatment chemicals.

## Performance Task

1. Use a water analysis test kit to test water.

## Trade Terms

**Alkalinity**: A water quality parameter in which the pH is higher than 7. It is also a measure of the water's capacity to neutralize strong acids.

**Bleed-off**: A method used to help control corrosion and scaling in a cooling tower. A metered amount of water is consistently drained from the water circuit to help carry away minerals and other impurities that remain as water evaporates from the cooling tower.

**Concentration**: Indicates the strength or relative amount of an element present in a water solution.

**Cycles of concentration**: A measurement of the ratio of dissolved solids contained in a heating/cooling system's water to the quantity of dissolved solids contained in the related makeup water supply. The term *cycles* indicates when the concentration of an element in the water system has risen above the concentration contained in the makeup water. For instance, if the hardness in the system water is determined to be two times as great as the hardness in the makeup water, then the system water is said to have two cycles of hardness.

**Dissolved solids**: The dissolved amounts of substances such as calcium, magnesium, chloride, and sulfate contained in water. Dissolved solids can contribute to the corrosion and scale formation in a system.

**Electrolysis**: The process of changing the chemical composition of a material (called the electrolyte) by passing electrical current through it.

**Electrolyte**: A substance in which conduction of electricity is accompanied by chemical action.

**Fouling**: A term used for problems caused by suspended solid matter that accumulates and clogs nozzles and pipes, which restricts circulation or otherwise reduces the transfer of heat in a system.

**Grains per gallon (gpg)**: An alternate unit of measure sometimes used to describe the amounts of dissolved material in a sample of water. Grains per gallon can be converted to ppm by multiplying the value of gpg by 17.

**Hardness**: A measure of the amount of calcium and magnesium contained in water. It is one of the main factors affecting the formation of scale in a system. As hardness increases, the potential for scaling also increases.

**Inhibitor**: A chemical substance that reduces the rate of corrosion, scale formation, or slime production.

**Milligrams per liter (mg/l)**: Metric unit of measure used to specify exactly how much of a certain material or element is dissolved in a sample of water. One mg/l is equivalent to one ppm.

**Parts per million (ppm)**: Unit of measure used to specify exactly how much of a certain material or element is dissolved in a sample of water. For example, one ounce of a contaminant mixed with 7,500 gallons of water equals a concentration of about one ppm.

**pH**: A measure of alkalinity or acidity of water. The pH scale ranges from 0 (extremely acidic) to 14 (extremely alkaline), with the pH of 7 being neutral. Specifically, pH defines the relative concentration of hydrogen ions and hydroxide ions. As pH increases, the concentration of hydroxide ions increases.

**Scale**: A dense coating of mineral matter that precipitates and settles on internal surfaces of equipment as a result of falling or rising temperatures.

**Suspended solids**: The amount of visible, individual particles in water or those particles that give water a cloudy appearance. They can include silt, clay, decayed organisms, iron, manganese, sulfur, and microorganisms. Suspended solids can clog treatment devices or shield microorganisms from disinfection.

**Total solids**: The total amount of both dissolved and suspended solids contained in water.

Proper treatment of water used in HVAC systems is one of the most important factors in maintaining efficient system operation and preventing premature failure of the equipment. Water treatment is needed for all steam, hot-water, and chilled-water systems and components, regardless of use. Depending on the situation, water treatment may be done by chemical treatment, a physical means such as bleed-off, or a combination of the two. Correct water treatment involves making a detailed analysis of the available water, considering its intended use, then implementing the needed treatment program and/or equipment as required. Analysis of water with regard to its specific use and the development of a specific water treatment approach is normally done by a qualified water treatment specialist. HVAC technicians, however, need to be familiar with this topic to recognize problems related to water treatment or the lack thereof.

## 1.1.0 Characteristics of Water

Water without dissolved substances does not exist naturally. Pure water is obtained only through extensive and expensive treatment. Water is an effective solvent that dissolves gases from the air, minerals from the soil, and even materials from the piping and other components through which it flows. Water easily dissolves various minerals and organic chemicals. These impurities, and not the water itself, are the main cause of problems in HVAC systems. Even though the quality of a water supply makes it safe to drink, the water can still contain minerals and gases that may be extremely harmful if the water is used untreated in HVAC systems.

The different kinds of dissolved inorganic materials, including gases, and their relative amounts describe the characteristics of a specific water sample. The strength or relative amount of an element present in a water solution is commonly referred to as its concentration. The term parts per million (ppm) is used to express the concentration for a specific material or element that is dissolved in a water sample. For example, one ounce of a given contaminant mixed with 7,500 gallons of water equals a concentration of about one ppm. In comparison, one drop of a given material or element mixed in 60 quarts of water is about one ppm. Sometimes, an alternate unit of measure called grains per gallon (gpg) is used instead of ppm to express concentration. Grains per gallon can be converted to ppm by multiplying the value of gpg by 17. Concentrations in water are frequently expressed using the metric unit of measure, milligrams per liter (mg/l). One mg/l is equal to one ppm. This follows the fact that there are 1,000,000 mg in one liter.

The characteristics of water supplied to a given system can vary widely over time. This can occur when more than one source is being used or when the characteristics of a single supply source, such as a lake or river, have changed. Table 1 shows a hypothetical example of the many kinds of substances that can be found in an untreated water supply. Note that only dissolved solids are listed in the table because most water analyses omit dissolved gases.

Gases such as nitrogen have almost no effect on water use. Other gases such as oxygen, carbon dioxide, and hydrogen sulfide do affect water systems. Levels of oxygen and hydrogen sulfide must be measured when the water sample is collected. This is necessary because oxygen easily dissolves into and saturates water, according to the water's partial pressure in air and the water temperature.

The amount of carbon dioxide in water can either be measured directly or estimated from the water's pH and total alkalinity. Alkalinity is a measure of water's buffering capacity, which allows it to counteract the amount of pH reduction from the addition of acids. Water with low alkalinity will experience a greater reduction in pH from the addition of acid than water of higher alkalinity. Water above a pH of 7.0 is considered to be alkaline. Alkalinity is measured using two indicators: the phenolphthalein alkalinity (P-alkalinity) measures the strong alkali present; the M-alkalinity (methyl orange alkalinity) or total alkalinity measures all the alkalinity in the water. The term *pH* is a measure of alkalinity or acidity of water. The pH scale ranges from 0 (extremely acidic) to 14 (extremely alkaline), with a pH of 7 being neutral (*Figure 1*). Specifically, pH defines

**Table 1** Analysis of an Untreated Water Supply

| Substance | Concentration (ppm) |
| --- | --- |
| Silica | 37 |
| Iron | 1 |
| Calcium | 62 |
| Magnesium | 18 |
| Sodium | 44 |
| Bicarbonate | 202 |
| Sulfate | 135 |
| Chloride | 13 |
| Nitrate | 2 |
| Dissolved solids | 426 |
| Carbonate hardness | 165 |
| Noncarbonate hardness | 40 |

the relative concentration of hydrogen ions and hydroxide ions. As pH increases, the concentration of hydroxide ions increases.

The hardness of a water sample indicates the amount of dissolved calcium (limestone) salts and/or magnesium in the water. Both calcium and magnesium ions may result from several dissolved chemical compounds. For example, scale occurs as a calcium compound such as calcium bicarbonate, calcium chloride, or calcium sulfate.

The amount of calcium and magnesium in water is often reported as if it is all present as calcium carbonate ($CaCO_3$). This allows total hardness to be reported as one number. There are many hardness classification systems. *Table 2* shows the classification of hardness according to the American Society of Agricultural Engineers. As hardness increases, the amount of scaling usually increases.

Solid materials can exist in water in the dissolved state as suspended solids. Dissolved solids are not visible and consist mainly of calcium, magnesium, chloride, and sulfate. They may also include nitrates, silicates, and organic material. Dissolved solids can contribute to the corrosion and scale formation in a system. The greater the

**Table 2** Classification of Hardness

| Classification | Total Hardness as Calcium Carbonate (ppm) |
|---|---|
| Soft | 0.0 to 60 |
| Moderately hard | 61 to 120 |
| Hard | 121 to 189 |
| Very hard | Over 190 |

concentration of dissolved solids, the greater the potential for corrosion. Suspended solids may be visible as individual particles or they may give water a cloudy appearance. They can include silt, clay, decayed organisms, iron, manganese, sulfur, and microorganisms. Suspended solids can clog treatment devices or shield microorganisms from disinfection. Total solids refers to the amount of both dissolved and suspended solids in water.

The term cycles of concentration is frequently used when determining the total dissolved solids (TDS) in system water. It indicates when the concentration of an element in the water system has risen above the concentration contained in the makeup water. Cycles of concentration are determined by the ratio of dissolved solids in the existing system water to the quantity of dissolved solids contained in the makeup water. For instance, if the hardness in the system water is determined to be two times as great as the hardness in the makeup water, then the system water is said to have two cycles of hardness. If you know the amount of TDS in the makeup water, it can be multiplied by the cycles to find out how much TDS is being contributed by the makeup water supply. Note that this does not take into account the TDS contributed by water treatment compounds.

Bacteria, algae, fungi, and protozoa can also be present in water systems. Problems resulting from these organisms are usually limited to cooling systems, since these organisms tend to die in systems operating at higher temperatures.

### 1.2.0 Problems Related to Water Quality

Water problems in heating and cooling systems can reduce system efficiency and lead to premature system failure. Some symptoms that indicate potential water problems include the following:

- Greatly reduced cooling or heating ability of the equipment because insulating deposits have formed on the heat exchanger surfaces.
- Reduced water flow resulting from partial or complete blockage of pipelines, condenser tubes, or other water line orifices.

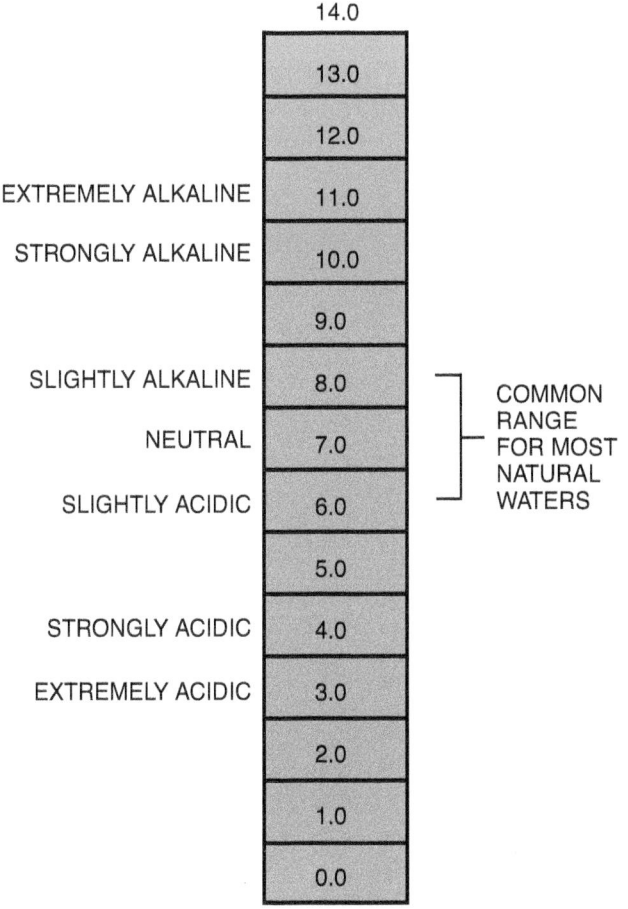

*Figure 1* The pH scale.

03308-13_F01.EPS

- Excessively rapid wear of moving parts such as pumps, shafts, and seals.

The most common problems that untreated water can cause in heating and cooling systems include the following:

- Corrosion
- Scale formation
- Biological growth
- Suspended solid matter

## 1.2.1 Corrosion

Corrosion is a state of deterioration or the wearing of metal surfaces within a system. It can occur both internally and externally to system units. External corrosion appears as rust or oxidation. Internal corrosion can be caused by acids, oxygen, and other gases present in the water. Left unchecked, air entrainment in water will cause carbon dioxide ($CO_2$) pitting corrosion that will eventually cause a failure of the pipe wall. Internal corrosion can also be caused by electrolysis. Electrolysis is the process of changing the chemical composition of a material by passing electrical current through it.

Corrosion occurs in the following forms:

- *Pitting corrosion* – A localized type of corrosion normally caused by the breakdown of the passive film at the metal surface, resulting in a blister or pit tubercle. This blister concentrates ions, which accelerates metal loss and creates the typical pinhole leak.
- *Grooving* – A uniform type of corrosion indicated by narrow grooves. Groove corrosion commonly appears along the edges of riveted joints.
- *Embrittlement (caustic cracking)* – A crystallizing or hardening process that causes fine cracks in metal, especially around riveted joints and holes. This can be a dangerous type of corrosion because it weakens the metal walls and joints in equipment, making them more vulnerable to stress under high operating pressures.
- *Galvanic corrosion* – An accelerated electrochemical corrosion produced when one metal is in contact with another dissimilar metal, and both are in contact with the same electrolyte (water). An electrolyte is a substance in which conduction of electricity is accompanied by a chemical action. The corrosion is caused by electrolytic action, which transfers the molecules from one metal to the other dissimilar metal. The rate at which the corrosion occurs is slower when the water has a low mineral content, which gives the water a lower electrical conductivity than water with a high mineral content.
- *Uniform corrosion* – A common corrosion caused by acid.
- *Air corrosion* – The carbon dioxide and oxygen compounds of air released by turbulence in piping systems can corrode pump impellers and other components.

## Cooling Tower Construction Materials

The type of material that cooling towers are constructed from can be a means of addressing some of the problems related to corrosion. Stainless steel and fiberglass are widely used in cooling tower construction because both of these materials are strong and are not susceptible to corrosion. Although fiberglass towers do not have the overall durability of stainless steel, they are valued in applications where weight and cost are significant factors.

03308-13_SA01.EPS

Corrosion control is typically done with simple mechanical filtration and the use of inexpensive chemical inhibitors such as sodium silicate or phosphate-silicate mixtures. An inhibitor is a chemical substance that reduces the rate of corrosion, scale formation, or slime production.

### 1.2.2 Scaling

Scaling is a coating of mineral matter that precipitates and settles out on internal surfaces of equipment as a result of rising and falling water temperatures. Even a small buildup of scale on a heat exchanger surface will reduce water flow and decrease the exchanger's ability to transfer heat. Likewise, as shown in *Table 3*, as scale thickness increases, a corresponding increase in energy consumption occurs. Scale particles can also speed up the wear of moving parts.

### 1.2.3 Biological Growth

Temperatures in a water system ranging between 40°F and 120°F increase the potential for bacterial growth. Most waters contain organisms capable of producing slime when conditions are favorable. These conditions include sufficient nutrients accompanied by proper temperatures and pH conditions. Algae use sunlight to convert bicarbonate or carbon dioxide into the energy needed for growth.

Algae, bacterial slimes, and fungi can interfere with the operation of a cooling system if they clog distribution passages and restrict water flow. A film composed of bacteria, other organisms, and slimes produced by microorganisms, acts as an insulator on heat transfer surfaces. Also, accumulations of organic matter can cause localized corrosion that results in equipment failure. Fungus buildup can also destroy wooden cooling towers because some fungi use wood as a nutrient and thus weaken the tower lumber.

### 1.2.4 Suspended Solid Matter

Suspended solid matter can include silt, sand, grease, bacteria, algae, oil, and corrosion prod-

ucts. The matter can accumulate in a system and clog nozzles and pipes, restricting circulation or otherwise reducing heat transfer. Problems resulting from suspended solids are often referred to as fouling.

### 1.3.0 Water Testing

Water systems are most commonly tested using commercially available test kits. Some test kits allow quick on-site analysis of the water (*Figure 2*). Typically, these kits test for pH, alkalinity, hardness, and other water properties, in addition to potential contaminants.

More comprehensive tests require that samples be sent to a test lab for analysis. These kits contain directions for gathering the water sample and include all the containers and packaging necessary to ship the samples back to the lab. Most also contain a data sheet (*Figure 3*) that provides the testing facility with important information about the system being tested. For the water samples submitted, the lab sends back a report showing the measured or calculated amount of each contaminant.

HVAC technicians or water treatment specialists often perform water testing using commercially available test kits. When more extensive or specialized testing is required, a water chemist should perform the tests.

When testing water using test kits, remember the following:

• Properly collecting, handling, and preserving water samples is crucial for an accurate water test.

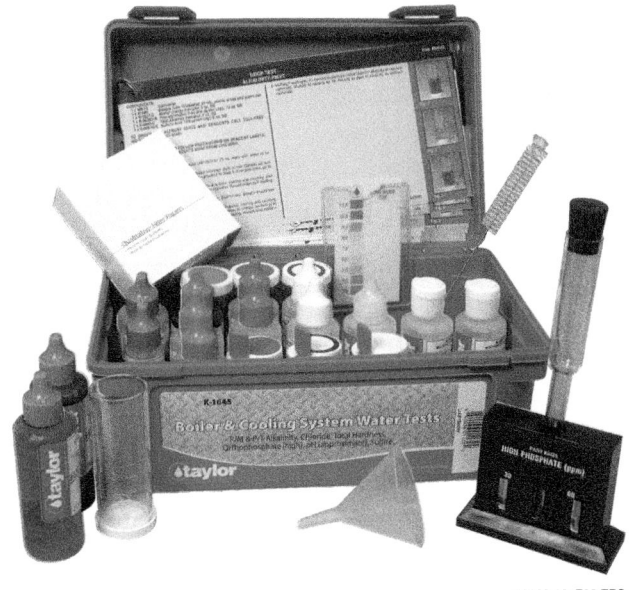

03308-13_F02.EPS

*Figure 2* Water analysis test kit.

**Table 3** Example Increases in Energy Consumption as a Function of Scale

| Scale Thickness (inches) | Increased Energy Consumption (%) |
|---|---|
| 1/32 | 8.5 |
| 1/16 | 12.4 |
| 1/8 | 25.0 |
| 1/4 | 40.0 |

DATE OF SAMPLING: _____

# WATER ANALYSIS DATA SHEET

STOP! This page must be filled out completely. Failure to do so will prevent an analysis and sample will be discarded.

|  | NAME | STREET | CITY | STATE |
|---|---|---|---|---|

JOB _____

CONTRACTOR/DEALER _____

WHOLESALER _____

REPORT SHOULD BE MAILED TO

COMPANY: _____ FAX NO: _____

ATTENTION: _____ E-MAIL: _____

_____

## WATER SAMPLE:

STOP! Make sure that you are sending a make-up (supply) water sample. Do not send recirculating or "system" sample by itself. Failure to send a make-up sample will prevent or delay your analysis report.

## WATER:

PROBLEM INVOLVED:   Lime Scale ☐   Corrosion ☐   Algae ☐   Slime ☐   Other ☐

SOURCE OF WATER:   City ☐   Private Well ☐

SOURCE OF SAMPLE:   Make-up ☐   Recirculating* ☐

## EQUIPMENT INVOLVED:

1. Evap. Condenser:   Make_____Tonnage_____Model # _____

Existing Bleed-off_____gal/hr.

2. Ice Machine:   Make_____Type: Cuber ☐   Flaker ☐   Drum ☐

Model # _____Capacity_____lbs. ice/day

3. Closed System:   Chilled ☐   Hot Water Boiler ☐   Both ☐

Size _____Tons_____Btu/h (output): Capacity_____gals.

4. Steam Heating
   Boiler   Make_____Btu/h _____

5. Evaporative
   (Swamp) Cooler:   Make_____ Size _____ cfm

## TREATMENT:

1. Chemical:   Brand _____ Type or Name_____

Part/Code No. _____ Results: Good ☐ Bad ☐

2. Feed Equip:   Chemical Feed Pump (size_____GPD) Drip Feed _____

TDS Monitor/Controller _____ Lockout Timer _____

3. Date System Last Cleaned: _____

### TEST RESULTS
(Field or Office Use)

pH:_____
Alk:_____ppm
Hard:_____ppm
Cl:_____ppm
SiO2:_____pprr
Cond:_____mmhos
_____ppm
_____ppm

**Nu-Calgon**
Wholesaler, Inc.

Calgon is a licensed trade name.

03308-13_F03.EPS

*Figure 3* Example of a water sample data sheet.

- Use only those containers provided or recommended by a testing laboratory.
- Carefully follow the laboratory and/or the test kit manufacturer's instructions.
- Usually, water must be sampled both before and after it goes through treatment equipment to ensure that a treatment device or program is working properly.
- The type of container needed depends on the test being performed. For microbiological testing, the container must be sterile and contain a chemical to deactivate any residual chlorine.
- Sampling containers used to collect water for chemical analysis often contain a fixing compound that prevents loss or breakdown of the specific chemical. Laboratories specially prepare containers for each category of contaminant. These containers should never be rinsed before use or filled to overflowing.

### 1.3.1 pH Testing

Water pH can be tested very simply using one of the following four methods:

- *Red and blue litmus paper* – If red litmus paper turns blue, the solution is basic, with a pH above 7.0. If blue litmus paper turns red, the solution is acidic, with a pH below 7.0.
- *pH test papers* – Immerse the test papers in the water, then remove them and compare their color with a color chart to determine the pH.
- *Dye* – Add dye to the water sample and compare the color with a color chart to determine the pH.
- *Electronic pH meter* – Use as directed by the manufacturer's instructions to measure the pH.

### 1.3.2 Alkalinity Testing

An alkalinity test, using a kit like the one shown in *Figure 2*, is conducted in the following (or a similar) manner. Note that the alkalinity test is a two-part test. One part is known as the PM test, and the other as the PT test.

> **WARNING!**
> Carefully read and follow all the precautions found on reagent bottles and test kits. Some reagents can be hazardous. Wear proper gloves, safety goggles, and any other PPE that may be recommended by the test kit manufacturer.

The PM test is completed using the following general technique. Note that the colors mentioned will be matched to a color chart in the kit to ensure accuracy.

- Rinse and fill a 25 mL sample tube with water from the source.
- Add the prescribed number of drops of indicator solution. Gently swirl the tube to mix the test solution with the water. The sample will likely turn to some shade of pink. If it does not, skip the next step.
- Add drops of a sulfuric acid solution, one at a time. Gently swirl the tube to mix after each drop, watching the color change that occurs. Counting the number of drops as you proceed, record the number of drops required to change the water back to a colorless liquid. The number of drops is recorded as the P value.
- Now that the liquid is colorless, add the prescribed number of drops of methyl orange indicator solution. Gently swirl the tube to mix. This time, the liquid should turn yellow.
- Again add drops of the sulfuric acid solution, one drop at a time, gently swirling after each drop while watching for color changes. Count the number of drops required to turn the solution orange. Add this number of drops to the number of drops of sulfuric acid solution used to determine the P value. Record this total as the M value.
- Now multiply the P value by 10 and record the reading as Pppm alkalinity as calcium carbonate. Next, multiply the M value by 10 and record as the Mppm alkalinity as calcium carbonate.

The second part of the test, the PT test, is done in a similar manner, as follows:

- Rinse and fill a 25 mL tube with the source water.
- Add the prescribed number of drops of indicator solution. Gently swirl the tube to mix the test solution with the water. The sample will likely turn to some shade of pink. If it does not, skip the next step.
- Add drops of a sulfuric acid solution, one at a time. Gently swirl the tube to mix after each drop, watching the color change that occurs. Counting the number of drops as you proceed, record the number of drops required to change the water back to a colorless liquid. The number of drops is recorded as the P value.
- Now that the liquid is colorless, add the prescribed number of drops of a total alkalinity indicator solution. Gently swirl the tube to mix. This time, the liquid should turn green.
- Begin adding the sulfuric acid solution one drop at a time, gently swirling the tube to mix after each drop. Count the number of drops required to turn the solution to a red color. Record the number of drops required as the T value.

- Multiply the P value by 10 and record it as Pppm alkalinity as calcium carbonate. Multiply the T value by 10 and record it as the Tppm alkalinity as calcium carbonate.

### 1.3.3 Hardness Testing

To determine the water hardness, follow these general instructions or a similar procedure:

- Rinse and fill a 25mL tube with the source water.
- Add the prescribed number of drops of a hardness buffer. Gently swirl the tube to mix.
- Add the prescribed amount of hardness indicator powder. Again swirl the tube gently to mix. Assuming some level of hardness is present, the mixture will turn red.
- Begin adding a hardness reagent, one drop at a time, while watching for the color to change to blue. Count to number of drops required to reach the proper shade of blue.
- Multiply the required number of drops by 10 and record it as Total Hardness ppm as calcium carbonate.

## 1.4.0 Chemical Safety Precautions

> **WARNING!**
>
> Always add chemicals to a solution tank that is already holding water. Adding chemicals first and then adding the water on top can produce a violent reaction and tremendous heat under the right circumstances.

HVAC technicians using chemicals in water treatment programs must be certain that the chemicals do not harm system components. This includes, but is not limited to, ferrous, nonferrous, plastic, natural rubber, and synthetic rubber items. Of greater concern is how safe a chemical is in terms of health, handling, and spills, and the impact of the chemical on the environment.

Unless proven otherwise, all chemicals should be treated as hazardous. Take the following precautions when working with chemicals:

- Wear appropriate personal safety equipment such as gloves, safety goggles, respirator, protective clothing, and/or aprons to protect your skin, eyes, and respiratory system from contact with chemicals. The exact type of equipment required depends on the potential hazards involved and the local and OSHA rules that apply to the job site.
- Ask for the related OSHA material safety data sheet (MSDS) or safety data sheet (SDS), and read and understand it.
- Do not mix or add chemicals other than as directed by the manufacturer or supplier.
- When using a mixture of chemicals, make sure that the chemical reactions in the system are not harmful to the user or the environment.
- Make sure to follow all federal, state, and local rules and regulations that govern the discharge into the environment and/or the proper disposal of all the chemicals used in a system.

## Additional Resources

*HVAC Water Chillers and Cooling Towers: Fundamentals, Application, and Operation.* Second Edition. Herbert W. Stanford III. Dekker Mechanical Engineering. Boca Raton, FL: CRC Press, Taylor & Francis Group

## 1.0.0 Section Review

1. The amount of calcium carbonate ($CaCO_3$) that is reported actually reflects both the calcium and magnesium in water.

   a. True
   b. False

2. Scaling can occur due to _____.

   a. electrochemical corrosion
   b. rising and falling water temperatures
   c. periodic blowdown
   d. cycles of concentration

3. Most comprehensive water analysis test kits contain a _____.

   a. hygrometer
   b. user's manual
   c. data sheet
   d. splash shield

4. Information about the hazards associated with a specific water treatment chemical is provided in the relevant _____.

   a. section of the Federal Code
   b. system operating parameters
   c. OSHA Incident Log
   d. MSDS (or SDS)

## SECTION TWO

### 2.0.0 MECHANICAL WATER TREATMENT EQUIPMENT

#### Objectives

Identify types of mechanical water treatment devices and equipment.
a. Identify types of filtration devices.
b. Identify types of water treatment equipment.

#### Trade Term

Backwashing: A procedure that reverses the direction of water flow through a multimedia-type filter by forcing the water into the bottom of the filter tank and out the top. Backwashing is performed on a regular basis to prevent accumulated particles from clogging the filter.

In addition to chemical water treatment methods, many mechanical or physical devices are used to filter or otherwise control water quality. These devices may be used alone or in conjunction with a chemical treatment program. Common mechanical water treatment equipment includes:

- Filtration equipment including strainers, filters, and centrifugal separators
- Evaporators
- Water softeners
- Deaerators
- Automatic chemical feeder systems
- Blowdown controllers and separators

### 2.1.0 Water Filtration Equipment

Mechanical filtration devices are used mainly to trap suspended solids circulating in a system. Strainers, filters, and separators are mechanical filtration devices commonly used to reduce suspended solids to an acceptable level.

#### 2.1.1 Strainers

Strainers (*Figure 4*) are closed containers with a cleanable screen element designed to remove foreign particles as small as 0.01" in diameter. Strainers are available in single or duplex units. They are usually cleaned manually, but some are self-cleaning.

Strainers can be made of cast iron, bronze, stainless steel, alloys, or plastic. Some strainers have magnetic inserts installed in them to catch microscopic iron or steel particles that may be present in the water being strained.

Strainers are, by design, simple to disassemble and clean or otherwise service. Duplex strainers can often be serviced without taking the system off-line or interrupting its operation. Cleaning generally consists of opening the vessel containing the basket, removing the basket, and back-washing it to remove any accumulated debris. An indicator, such as a lever, typically indicates which of the two baskets on a duplex strainer are in active use. For single basket strainers, the system usually must be shut down before the basket can be accessed. It is very important to ensure that the internal pressure has been relieved before opening the basket vessel. A gauge may be installed on the strainer assembly to monitor the pressure. A vent valve is also common, which is opened as a more positive means of ensuring that the vessel is not under pressure. Follow the manufacturer's instructions carefully.

> **WARNING!**
> Serious personal injury can be caused by liquid and chemical solution sprays from a strainer assembly that is opened under pressure. Wear the proper PPE, including chemical-resistant gloves, safety goggles and a face shield, when opening a strainer assembly. Ensure that the vessel is not under pressure.

#### 2.1.2 Cartridge Filters

Cartridge filters are closed units that use a cartridge-type filter element. They are commonly used to remove suspended particles ranging between 1 micrometer and 100 micrometers in size. They are typically used where the contamination levels are less than 100 ppm (0.01 percent by mass). Construction of filter cartridges varies widely, and they are made out of many different materials. *Figure 5* shows some common types of filter cartridges.

Filters are classified as depth-type or surface-type. Depth-type filters capture particles throughout the thickness of the filtering medium. These filters are typically made from yarns or resin-bonded fibers that normally increase in density toward their center. Large particles are captured at the surface, while smaller particles are captured deeper inside the media. Surface-type filters are thin-media filters made from pleated paper or similar material. These filters are designed to capture particles at or near the surface only. Surface-

NCCER – *HVAC Level Four* 03308

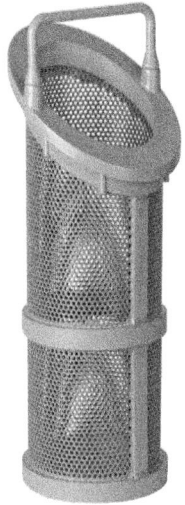

03308-13_F04.EPS

*Figure 4* Duplex strainer and typical basket.

type filters can normally handle higher flow rates because they have a lower resistance to water flow. The deeper media of a depth-type filter generally increases flow resistance.

Once a filter cartridge becomes plugged, as indicated by reduced water flow through the filter, the cartridge normally must be replaced with a new one. Some manufacturers provide guidelines for filter replacement by rating their cartridges according to the number of gallons they can treat. In practice, the frequency of replacement is determined mainly by the size and concentration of the suspended solids in the water that is being filtered.

### 2.1.3 Multimedia Filters

Multimedia filters consist of a filter tank containing three or four layers of different media. The different grain sizes and types of filtering media provide for depth filtration mainly to increase the capacity of the filter to hold suspended solids. This increases the length of time between backwashing. Backwashing is a procedure that reverses the direction of water flow through the filter by forcing the water into the bottom of the filter tank and out the top. Backwashing must be performed on a regular basis to prevent accumulated particles from clogging the filter.

## Inline Steam Separator

A typical steam separator, sometimes called an entrainment separator, is an inline device placed at the output of a boiler. The device can be designed to separate solids and water from the steam by centrifugal action and/or by passing the steam through a series of baffles that causes the steam to change direction at each baffle. Either type of action results in the separation of solids and water droplets from the steam. The drain at the bottom is routed back to the boiler.

03308-13_SA02.EPS

MULTI-CARTRIDGE FILTER VESSELS

PLEATED CARTRIDGE FILTERS

ABSORBENT FILTER CARTRIDGES

03308-13_F05.EPS

*Figure 5* Multi-cartridge filter vessels and filters.

In the multimedia filter shown in *Figure 6*, the coarsest filtering material is at the top with each successively lower layer being of a finer material. As shown, the filtering media used in this filter are bituminous coal, anthracite coal, sand, and garnet. Untreated water enters the top of the filter tank under pressure and flows down through each of the media layers. At each layer, smaller and smaller particles are trapped as the water works its way from the coarsest to the finest material.

### 2.1.4 Bag-Type Filters

Bag-type filters consist of a mesh or felt bag supported by a removable metal basket that Is enclosed in a tubular housing (*Figure 7*). The bag support basket is usually made from perforated stainless steel. In some units, the basket can be lined with fine wire mesh and used without a filter bag as a strainer. Some manufacturers make a two-stage bag filter consisting of an inner basket and bag that fits inside a larger outer basket and bag. This provides for a coarse filtering stage (inner bag) followed by a finer one (outer bag). Untreated water that enters the filter housing is routed to the top of the bag(s) and leaves the unit at the bottom, below the bag(s). Contaminants filtered out of the water as it passes through remain trapped inside the bag.

Filter bags are made from many kinds of materials such as cotton, nylon, polypropylene, and polyester. Bags made from felted material are common because of their finer pores and depth-filtering quality, which provide a high amount of dirt-loading. When plugged, these bags must be removed and replaced with new ones. Mesh bags are generally coarser, but are lower in cost because they tend to be reusable.

### 2.1.5 Centrifugal Separators

Centrifugal separators (*Figure 8*) are being used more often than in the past to keep closed and open water systems clean. Generally good for removing suspended particles greater than 45

## Duplex Multimedia Filter

Duplex multimedia filters provide two or more independent filter assemblies. The filter assemblies can be configured by their associated valve systems so that they filter the system water simultaneously, or they can be configured so that one is in the backwash mode while the other filters the system water.

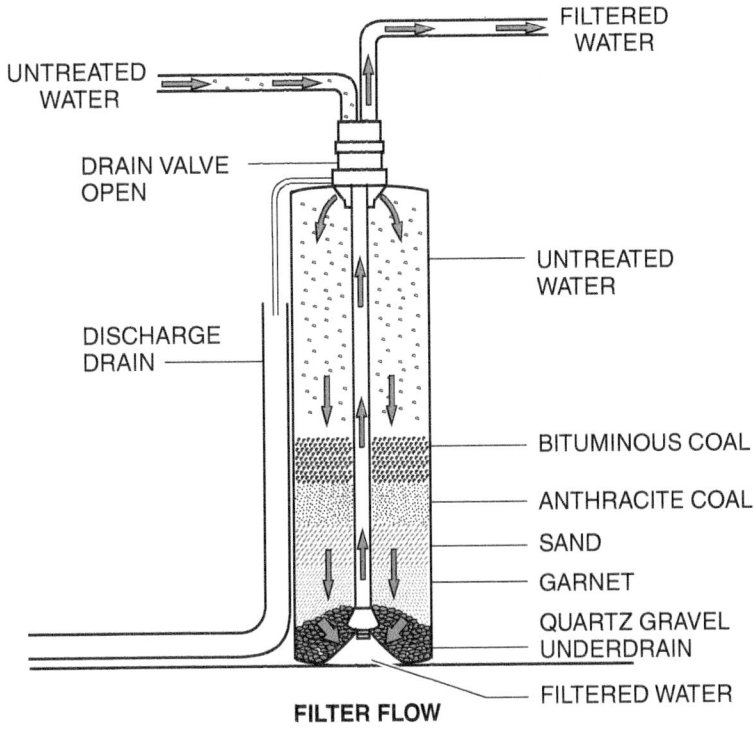

**FILTER FLOW**

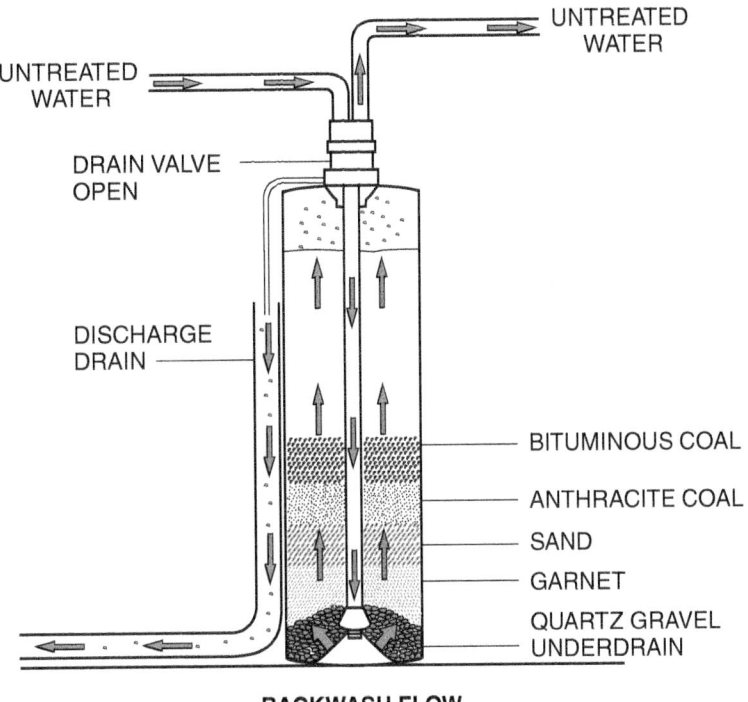

**BACKWASH FLOW**

03308-13_F06.EPS

*Figure 6* Typical multimedia filter operation.

micrometers in size, they tend to be inadequate at removing smaller particles. To increase their effectiveness, a coagulating compound is sometimes added to the water to make the particles larger, thus making them easier to remove. In the centrifugal separator, which is mounted vertically at a horizontal angle, the untreated water is drawn into the unit, where it is accelerated into a separation chamber. There, centrifugal force causes the heavier solid particles suspended in the water to be tossed to the edges of the separation chamber where the particles migrate into a collection chamber. The cleaned water is drawn to the low-pressure area of the separation cham-

BAG FILTER VESSELS

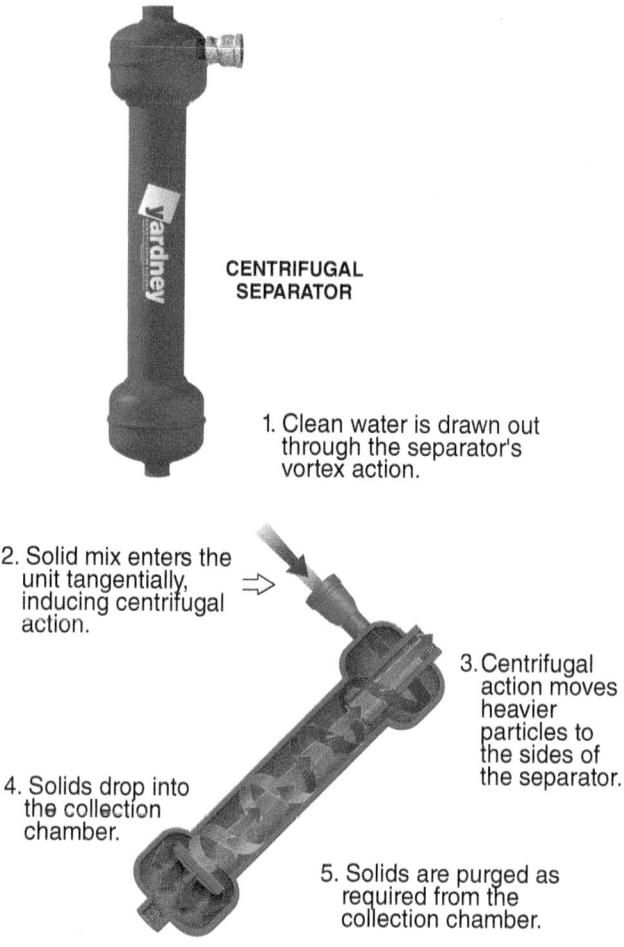

CENTRIFUGAL SEPARATOR

1. Clean water is drawn out through the separator's vortex action.

2. Solid mix enters the unit tangentially, inducing centrifugal action.

3. Centrifugal action moves heavier particles to the sides of the separator.

4. Solids drop into the collection chamber.

5. Solids are purged as required from the collection chamber.

CENTRIFUGAL SEPARATOR DIAGRAM

*Figure 8* Typical centrifugal separator.

FILTER BAGS AND STRAINERS

03308-13_F07.EPS

*Figure 7* Typical bag filter vessels and filter inserts.

ber and flows up through the outlet to the system. Solid particles removed from the water are either purged periodically or bled continuously from the separator unit.

## 2.2.0 Water Treatment Equipment

Water treatment equipment employs processes such as distillation, absorption, and precipitation to maintain water quality. Equipment commonly used for water treatment includes evaporators, water softeners, deaerators, automatic chemical feed systems, and blowdown controllers and separators.

### 2.2.1 Evaporators

Evaporators use a distillation process to remove concentrations of solids. Their use can also remove some organic contaminants and disinfect the water. Evaporators are inline, point-of-entry devices installed in the source water lines that supply makeup or feedwater to various pieces of water equipment. There are several types of evaporators. *Figure 9* shows one common type. It consists of an evaporation (distillation) chamber and an air-cooled condenser section. Untreated supply water enters the evaporation chamber and is heated to boiling. This vaporizes the water, causing all the solids to be retained in the evaporator chamber, where they are removed by a drain. The water vapor, free of solid particulate matter, flows into an air-cooled condenser, where it condenses back into water. This treated water is then collected below the condenser for distribution to the water system. The evaporator unit shown uses an electrical element to heat the water, but other heat sources are often used. Also, a water-cooled condenser is commonly used in some units, instead of an air-cooled condenser.

### 2.2.2 Water Softeners

Water softeners are used to remove the minerals from hard water that form scale and soap film. They are inline, point-of-entry devices installed in the source water lines that supply makeup or feedwater to the various kinds of water equipment. Many water softeners use a sand-like substance called zeolite that is saturated with sodium (salt) and arranged in a filter bed (*Figure 10*). The zeolite absorbs calcium and magnesium, the main components of hardness, more readily than it does sodium. As the supply of hard water filters through the zeolite, sodium is released and calcium and magnesium are absorbed. A distributor or baffle disperses the untreated water through the zeolite. This ensures that all the untreated water contacts the zeolite instead of channeling or passing through the softener unit without contact. As a result of this process, the water is softened by replacing calcium and magnesium bicarbonate with nonscale-forming sodium bicarbonate.

Through extended use, the sodium content of the zeolite bed gradually becomes depleted, thus preventing the unit from softening the water. To prevent this condition, the unit must be regenerated periodically by treating the unit with a strong dose of salt solution. This brine solution is generally mixed and applied from an associated brine tank connected to the softener.

### 2.2.3 Deaerators

Deaerators (*Figure 11*) are typically used in steam systems to remove air and other noncondensable gases from the system feedwater. Oxygen becomes less soluble as water temperature in-

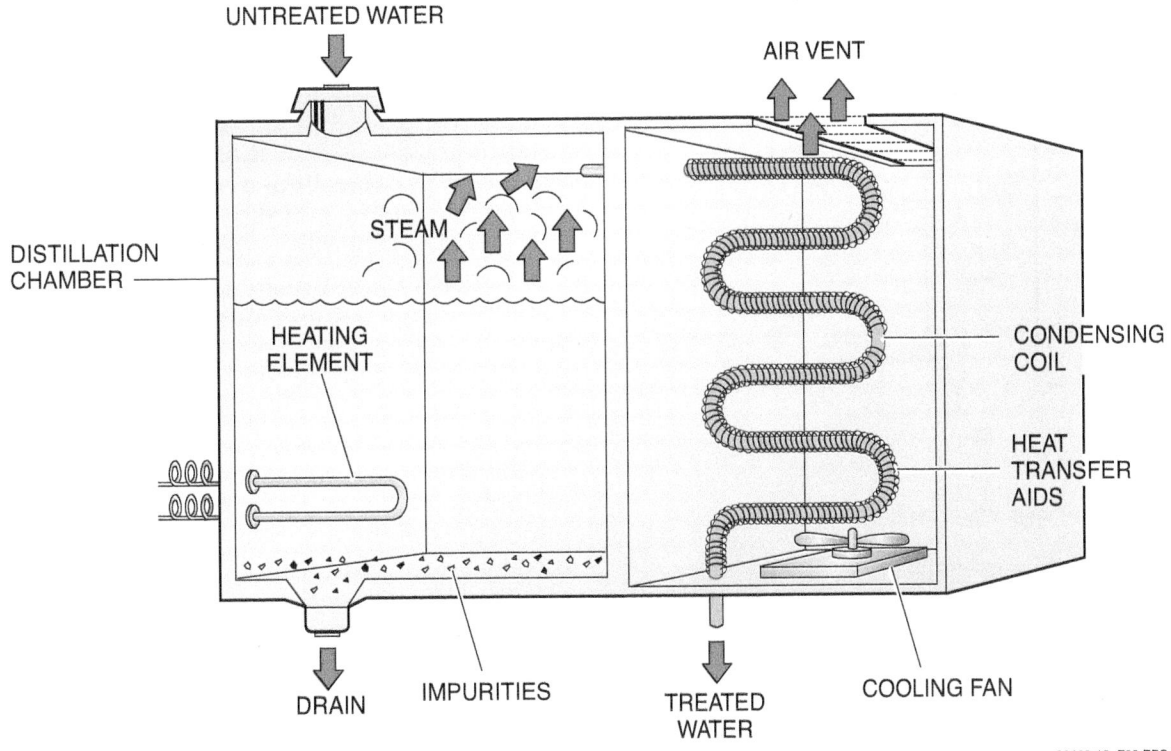

*Figure 9* Typical water treatment evaporator unit.

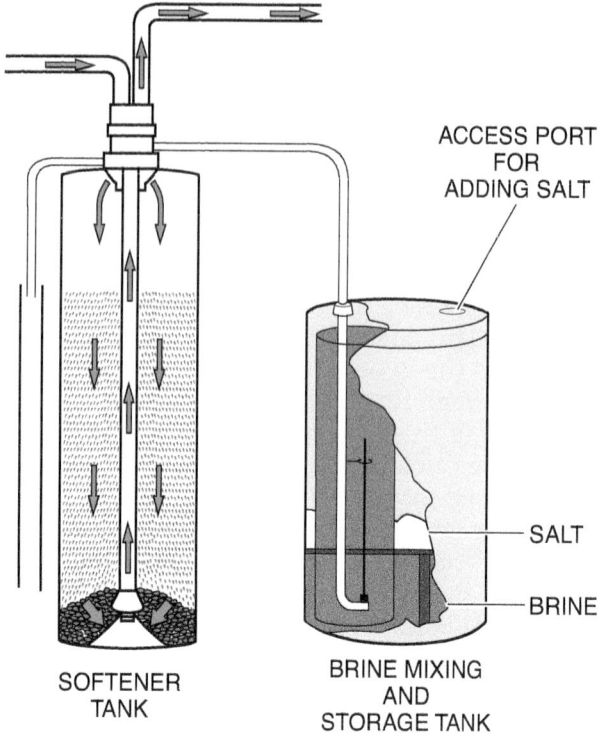

ACCESS PORT
FOR
ADDING SALT

SALT

BRINE

SOFTENER
TANK

BRINE MIXING
AND
STORAGE TANK

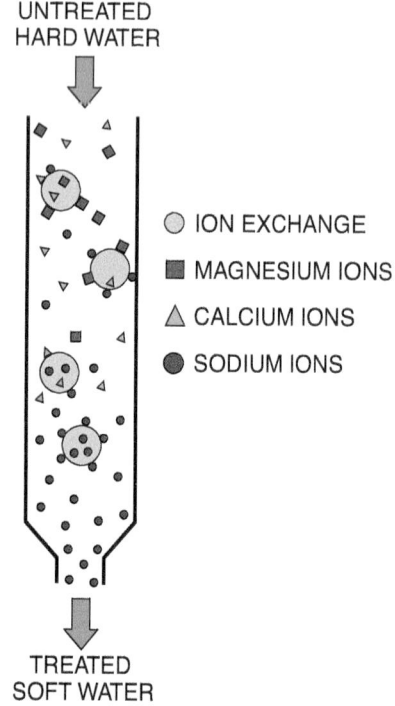

UNTREATED
HARD WATER

○ ION EXCHANGE

■ MAGNESIUM IONS

△ CALCIUM IONS

● SODIUM IONS

TREATED
SOFT WATER

**PRINCIPLE OF OPERATION**

03308-13_F10.EPS

*Figure 10* Typical water softener.

creases; therefore, removal of oxygen can be done by heating the incoming feedwater to its boiling point. As a result of the boiling, the oxygen and other gases are driven off and vented to the atmo-

sphere. Other impurities that cause hardness precipitate and settle to the bottom of the unit. There, a filtration system traps the solidified impurities.

There are several types of deaerators. One type uses system steam as the heat source. In this type, the incoming feedwater is discharged as a fine spray into the upper portion of a closed chamber. The steam enters and mixes with the sprayed water, subjecting the water to an intense scrubbing (cleaning) action. While the cleaning action continues, another traversing stream of steam heats the water to free the oxygen and other noncondensable gases. After contacting the feedwater, most of the steam condenses. Any remaining steam vapor, along with separated gases, enters a vented condenser where the steam condenses as it contacts tubes that contain cool, incoming feedwater. The liberated oxygen and other gases are vented to the atmosphere.

### 2.2.4 Automatic Chemical Feed Systems

On many steam and water systems, adding chemical treatment manually is impractical because inconsistent conditions and spikes in chemical levels result, which could be detrimental to system components. Automatic chemical feed controllers are often used to supply one or more biocides or other chemicals to the system. Simple controllers can provide pump control to add chemicals based solely on time and programmed dosage. Others have multiple inputs to monitor conductivity, pH, and other related water characteristics. In some cases, they also monitor the volume of makeup water added to the system, which also affects the amount of chemical treatments to be added and the amount of bleed-off that may be required. Most chemical feed controllers are capable of also controlling a cooling tower bleed-off solenoid valve to maintain proper conductivity and dissolved solid levels, further automating the water treatment and maintenance process.

Pumps used for this application are capable of very precise metering. Coupled with extremely sensitive and accurate controllers, the pumps help save on chemical costs and maintain consistent water conditions.

Chemical feed systems can also be connected to sophisticated building management systems, which allows the operator to monitor water conditions and chemical usage remotely. Consistent monitoring, especially on larger systems, can result in dramatic savings in chemical costs, and can minimize the environmental impact of biocide use.

Designing and programming the chemical feeder system requires that the water system specialist or other responsible parties understand the types of water problems to expect in the local

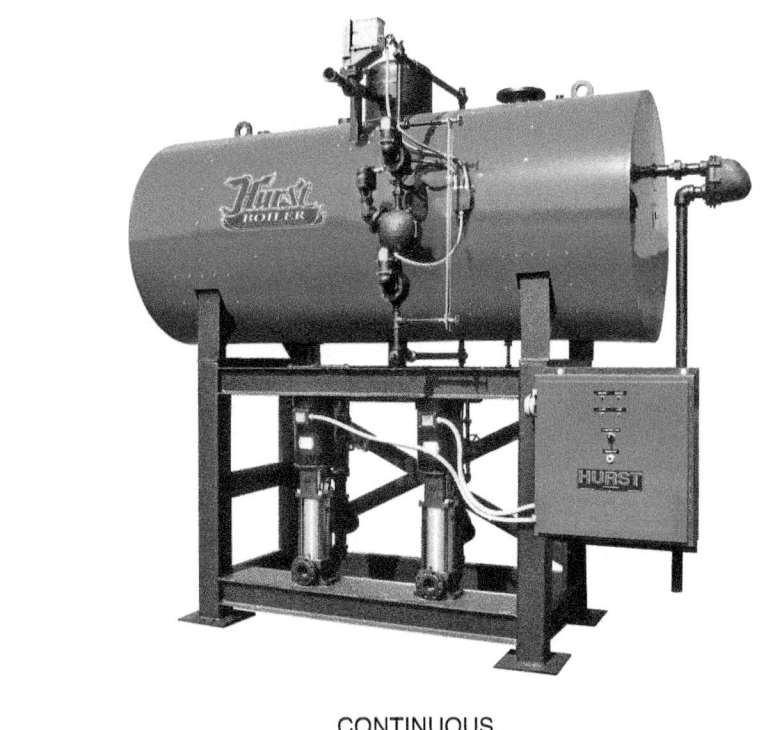

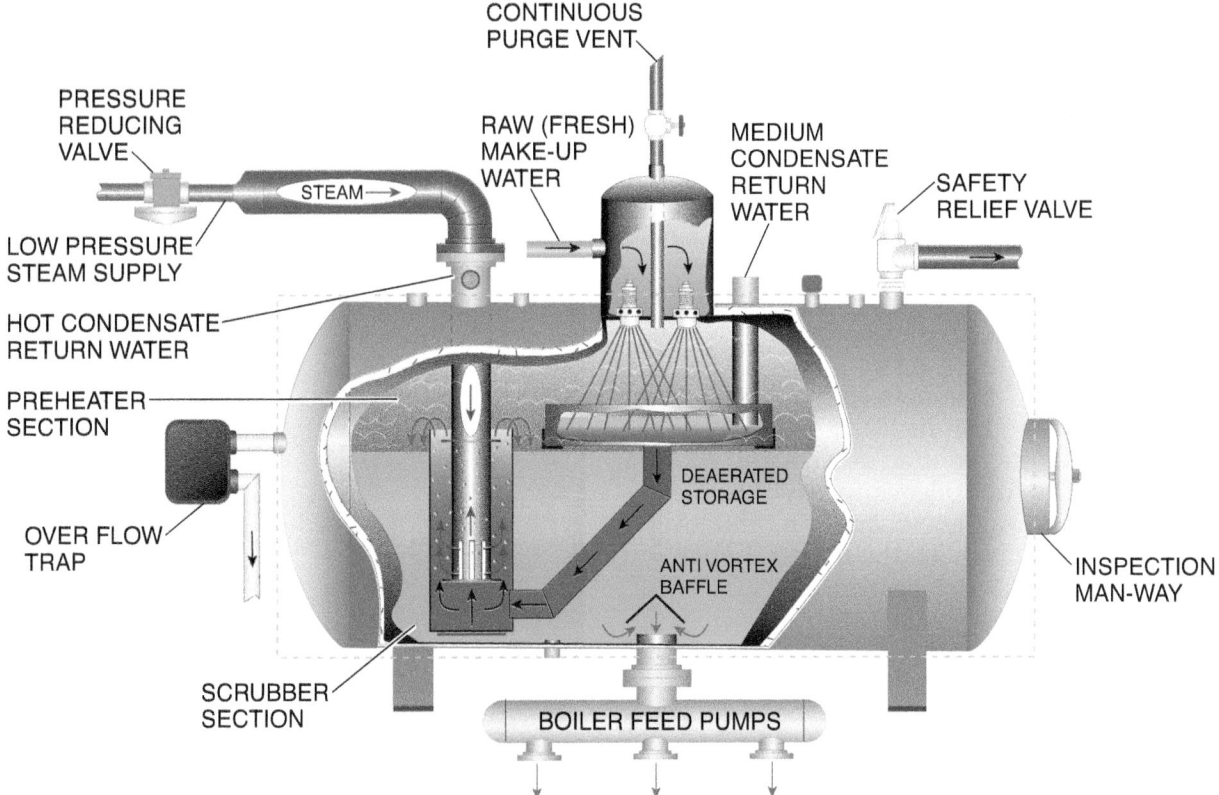

*Figure 11* Spray deaerator.

area and how to correct them. The chemicals used and their dosage levels will vary on the basis of a number of conditions, such as the following:

- The chemistry of the local water source
- Overall system design, including anticipated water temperatures and temperature changes

- The different materials of construction used in system components such as the piping, cooling towers, boilers, heat exchangers, and pumps
- The potential interaction between chosen chemicals

Because these parameters cause such a wide diversity in automatic chemical feeder systems, detailed coverage of these systems and their installation is beyond the scope of this module. More detailed information on automatic chemical feed systems can be found in documents regarding water treatment referenced at the end of this module.

### 2.2.5 Blowdown Controllers and Separators

Steam boilers require blowdown from both the surface and the bottom of the boiler to remove dissolved solids and other contaminants. Blowdown is the controlled draining of boiler water out of the boiler to maintain the desired concentration of suspended and dissolved solids in the boiler. Although this can be done manually, substantial cost savings can result from the use of automated blowdown control. Since the liquid and other material discharged during blowdown is at the same temperature as the boiler water, heat energy lost during the blowdown process can be significant. Steam boilers that have little or no condensate return typically need more blowdown due to the increased volume of contaminants entering with the makeup water. In addition, boiler operation and varying loads from one day to the next heavily impact the need for blowdown. As a result, when blowdown is accomplished manually, the operator cannot be certain how much blowdown is required to maintain the necessary TDS levels. An example of a packaged blowdown system is shown in *Figure 12*.

With automated blowdown, probes from the controller sense the conductivity of the water, which represents the volume of TDS and other contaminants. The controller compares this information to preset parameters, and a modulating valve is actuated to provide the necessary blowdown. Blowdown can be done from either the surface or the bottom of the boiler.

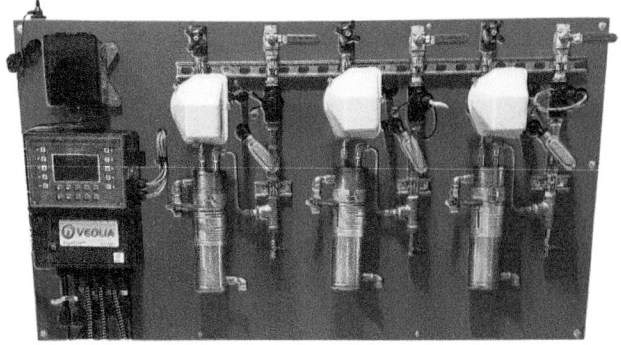

*Figure 12* Boiler blowdown control system.

03308-13_F13.EPS

*Figure 13* Blowdown separator.

Steam boilers may also be equipped with a blowdown separator (*Figure 13*). These devices are usually centrifugal-type separators that help reduce corrosion and scaling downstream by separating steam from water and collecting the dissolved solids. Blowdown separators also reduce the temperature of the water being discharged, which is often necessary to comply with environmental regulations and municipal codes.

---

# Typical Automatic Control and Feed System

In many cases, pre-engineered automatic systems for boilers, cooling, and process applications are available from manufacturers. These systems include sensors, controllers, and appropriate chemical metering pumps. Completely assembled systems make installation and startup relatively easy.

---

# Chemical-Free Water Treatment

Alternate technologies are available that provide water treatment and conditioning without the use of chemicals. Instead, electronic systems are used to reduce scaling, fouling, and biological growth. A typical example of this type of system employs a complex modulated signal field that causes the mineral and scale molecules in water to become agitated. The agitated molecules lose their adhesive properties and form a suspension that passes through system piping without adhering to the pipe surface, thus preventing and removing scale and biofilm.

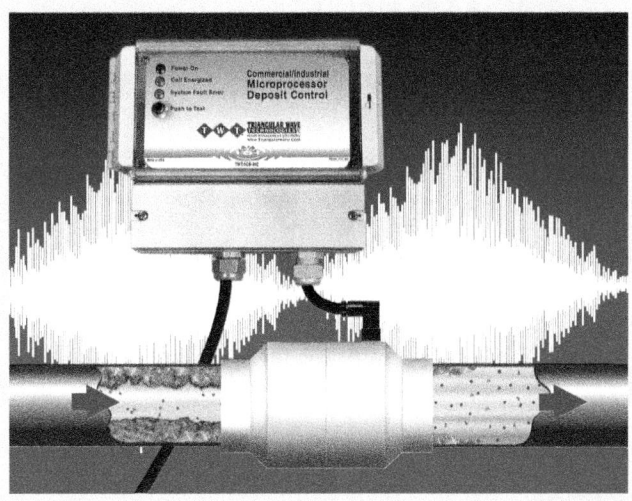

03308-13_SA03.EPS

## Additional Resources

*ASHRAE Handbook – HVAC Applications.* Latest Edition. Atlanta, GA: American Society of Heating, Refrigerating, and Air Conditioning Engineers, Inc.

*ASHRAE Handbook – HVAC Systems and Equipment.* Latest Edition. Atlanta, GA: American Society of Heating, Refrigerating, and Air Conditioning Engineers, Inc.

*Boilers Simplified.* Troy, MI: Business News Publishing Company.

*HVAC Water Chillers and Cooling Towers: Fundamentals, Application, and Operation.* Second Edition. Herbert W. Stanford III. Dekker Mechanical Engineering. Boca Raton, FL: CRC Press, Taylor & Francis Group.

*Water Treatment Specification Manual.* Troy, MI: Business News Publishing Company.

## 2.0.0 Section Review

1. A coagulating compound may be added to the water to increase the effectiveness of _____.
   a. duplex strainers
   b. bag-type filters
   c. cartridge filters
   d. centrifugal separators

2. Some automatic chemical feed controllers also monitor the _____.
   a. depth of zeolite in the filter bed
   b. age of silt deposits
   c. volume of makeup water added to the system
   d. type of algae in the recirculating water

## SECTION THREE

### 3.0.0 SYSTEM-SPECIFIC WATER TREATMENT PROBLEMS

#### Objectives

Identify and describe how to address water-related problems that occur in specific types of hydronic and steam systems.

   a. Identify and describe how to treat water-related problems that occur in open recirculating water systems.

   b. Identify and describe how to treat water-related problems that occur in closed recirculating water systems.

   c. Identify and describe how to treat water-related problems that occur in steam systems.

#### Performance Task

2. Inspect a cooling tower or stem boiler and its related water piping system for signs of water treatment problems.

#### Trade Terms

Colloidal substance: A jelly-like material made up of very small, insoluble, non-diffusible particles larger than molecules, but small enough to remain suspended in a fluid without settling to the bottom.

Water-related problems, such as scale and corrosion, adversely affect HVAC system performance. The specific types of problems that are most likely to occur depend on the type of system. Likewise, the specific steps for addressing water problems depend on whether the problems occur in open recirculating water systems, closed recirculating water systems, or steam systems.

### 3.1.0 Water Treatment in Open Recirculating Water Systems

Open recirculating systems use cooling towers (*Figure 14*) or evaporative condensers to cool the water leaving the cooling system condenser for reuse. Recirculating and spraying this water in the cooling tower allows it to come in contact with air. As a result, the composition of the water is drastically changed by evaporation, aeration, and other chemical and/or physical processes. Also, acidic gases and other contaminants in the air are absorbed into the water. Open towers are highly susceptible to the growth of algae, bacteria, and other living organisms, especially if the towers are located where the water surface is exposed to sunlight. The main reasons for water treatment in an open recirculating system are to prevent corrosion and scale and to eliminate algae, bacteria, and other living organisms.

#### 3.1.1 Corrosion and Scale Deposits

Cooling tower systems are exposed to many types of corrosion, from simple electrochemical corrosion to pitting caused by deposits or microorganisms. Microbiologically induced corrosion is one of the main types of corrosion in cooling towers. It results from improper control of biocides, which can lead to a high algae and/or bacterial count in tower water.

The buildup of solids also contributes to corrosion and scaling in an open recirculating water system. These solids and other impurities are caused by evaporation of the recirculating water in the cooling tower. As the system water recirculates through the tower, some of the water evaporates into the atmosphere. Any dissolved solids that were contained in the evaporated water remain behind in the recirculating system water. These solids consist of calcium and magnesium that were originally introduced into the system in the makeup water. Calcium carbonate and calcium sulfate are the main ingredients in the scale deposits. Additional solids and other impurities are introduced into the recirculating water each time fresh makeup water is added to the system to compensate for water lost through evaporation, controlled bleed-off, or other methods.

Chemical treatment of open recirculating systems to control corrosion and scale involves the use of blends of phosphates, phosphonate, molyb-

---

### Using Chromates for Corrosion Control

In the past, the use of chromates was common. Eventually, however, the EPA identified chromates as a suspected carcinogen. As a result, in January 1990, the EPA banned their use for cooling tower treatment in comfort cooling systems.

---

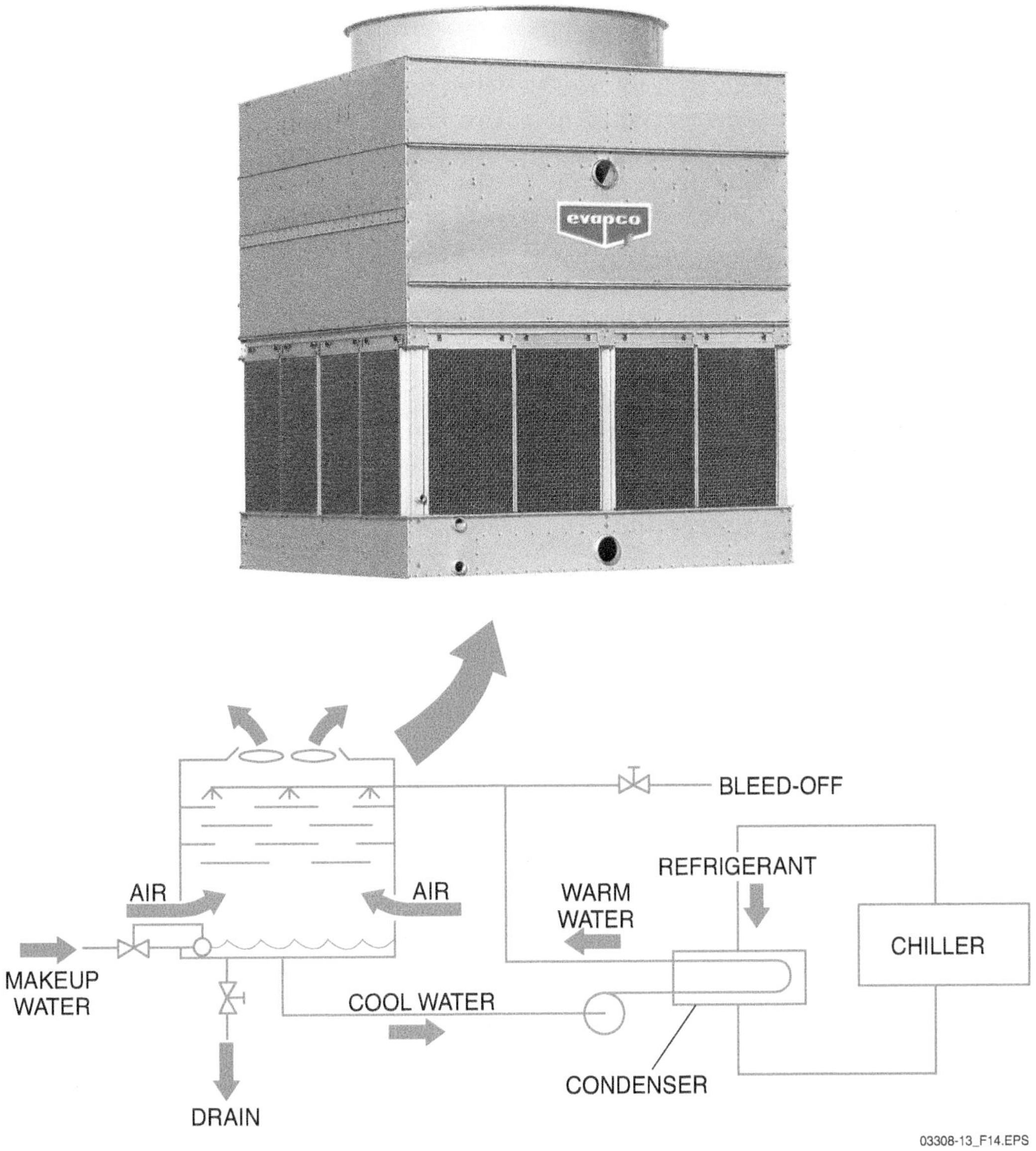

*Figure 14* Cooling tower.

03308-13_F14.EPS

date, zinc, silicate, and various polymers. Totyltri-azole or benzotriazole is added to these blends to protect copper and copper alloys from corrosion. Before using any chemical for water treatment, the water chemist or other responsible person must always follow the local, state, and federal environmental guidelines and regulations.

The range of pH in cooling tower water is typically maintained between 7.5 and 8.5. Sulfuric acid is generally used when pH control is needed because of high hardness or alkalinity in the makeup water. Sulfuric acid increases the solubility of calcium, thereby reducing the potential for scaling. If enough acidic gases are absorbed, the water can become very acidic with a correspondingly low pH. The pH of the recirculating water must be carefully regulated by adding caustic or alkali. Because acid is introduced into the recir-culating water from the air and not from makeup water, the correct amount of caustic or alkali that must be added to maintain the pH level cannot be calculated; it is usually determined by trial and error.

In addition to chemical treatments, corrosion and scaling are controlled by a procedure called bleed-off. In an open recirculating system, this procedure involves consistently draining a metered amount of water from the water circuit to help carry away minerals and other impurities that remain as water evaporates from the cooling tower. The bleed rate for most systems should be about 0.5 to 1 gallon per hour per ton of capacity.

### 3.1.2 Silt Deposits

Large quantities of airborne dirt enter an open system at the cooling tower and settle out as silt deposits. These deposits can cause corrosion, restrict water flow, and promote growth of microorganisms. Silt is commonly controlled by adding polymers that keep the silt in suspension while it flows through the system. This tends to cause the silt to accumulate in the tower basin where it can be manually cleaned during maintenance.

### 3.1.3 Biological Growth

The water contact with air that occurs in a cooling tower or evaporative condenser allows organic matter, bacteria, algae, and other airborne debris to be introduced into the system recirculating water. Algae are primitive plants that require sunlight for growth. They tend to grow in the perforated head pans of cooling towers and can eventually clog holes or nozzles in the head pan. They can also clog the system pipes and pumps. Slimes are jelly-like materials produced by the growth of bacteria. They tend to cling to many parts of the system. If they are allowed to grow, the flow of water through the nozzles can eventually be restricted. This blockage can cause high head pressure in the condenser, odor in air washers, and increased corrosion of all equipment and components. Microbiological activity can also cause pitting damage to equipment.

Control of algae and slimes can be difficult because they react differently to various chemicals. Also, they can build up a resistance to the chemical being used. Treatment with chemical algaecides or bactericides prevents or slows these growths. Regular shock treatments with chlorine or other chemicals are commonly used to control growth on the system. Typically, two different chemicals are used on an alternating basis to make sure that the algae or microorganisms do not build up a resistance to any one chemical.

Much larger dosages are needed if the material has been allowed to accumulate to any depth. The dosage used depends on the amount of water held in the system. Therefore, the volume of water normally held must be calculated. The volume can be calculated by multiplying the tonnage of the cooling tower by 6 and all other open systems by 4. The treatment frequency is usually determined by trial and error and the rate of growth of the algae or slime. Chemical treatment alone does not usually control algae or slime. For this reason, the equipment should also be thoroughly cleaned at regular intervals to avoid excessive growth.

### 3.1.4 Cooling Tower and Open Recirculating System Treatment

Regular maintenance is a must in any water system, but it cannot be stressed enough in open systems because of health and safety issues related to Legionnaires' disease. Poorly maintained cooling towers are known to support the growth of Legionella bacteria. The aerosol produced can potentially infect not only the people at the equipment site, but also the surrounding community. Legionnaires' disease can lead to pneumonia, and in some instances is fatal. If cooling towers are not properly maintained, a contractor could be held liable if someone contracts this disease.

The recommended maintenance actions and/or tests related to cooling towers/evaporative condensers and open recirculating water systems are summarized in *Table 4*.

After the installation or major repair of an open system, it is necessary to clean the internal water system of protective oils, films, grease, welding flux, dirt, and other debris. Cleaning is typically done according to the following guidelines:

*Step 1*   Fill, vent, and leak test the system.

*Step 2*   Add the chemical cleaning agent prescribed for the system in the amount and manner directed by the manufacturer.

*Step 3*   Operate all system pumps with their strainers installed for the time interval prescribed by the responsible water treatment specialist.

## Bleed-Off Water Control

Most people are becoming increasingly aware of the need to conserve the precious resource of clean water. Older cooling tower systems were often fitted with a simple manual valve to allow for bleed-off. In most cases, the valve was set by HVAC technicians and system operators to a flow rate far greater than required. In some cases, calculations were never made to determine the correct volume, nor was the volume of water flowing to bleed-off ever measured. As long as the objective of reducing impurities was reached, this method seemed sufficient.

With water conservation in mind, proper methods of monitoring the level of impurities are readily available and installed today, allowing for automated control of the bleed-off process. These monitors open and close a solenoid valve in the bleed-off line, opening it only when impurity levels indicate the need for dilution. As a result, literally millions of gallons of water across the country are saved each year.

**Table 4** Cooling Tower/Open Recirculating Water System Maintenance

| Maintenance Action/Test | Frequency |
| --- | --- |
| Test and record bacteriological quality of the system water. | Biweekly |
| Test and record biocide and inhibitor reserves, pH, and conductivity of the system water. | Biweekly |
| Check that dosing equipment containers are full, pumps are operating properly, and supply lines are not blocked. | Weekly |
| Check the bleed-off control equipment to make sure it is operating properly and the controller is in calibration. The solenoid valve should be manually operated to confirm that the flow of water to the drain is clear. | Weekly |
| Check the system for growths and deposits. There should be no algae growth in the towers or slimy feel to the fill pack, tower sump, or side walls. | Weekly |
| Check the operation of sump immersion heaters. There should be no visible corrosion on the outside of the heater, and the unit should activate when the setpoint on the thermostat is reached. | Monthly |
| Check the operation of sprays, fans, and drift eliminators. There should be no mechanical damage, and the components should be free of visible deposits. The bypass of aerosol droplets should be minimal when the fans are operating. The distribution system should have no deposits, with an even flow of water to all areas of the tower. | Monthly |
| Drain, clean, and disinfect cooling towers and associated pipe work in accordance with the method approved for the site. The chlorination period should be a minimum of five hours. Free chlorine residuals should be checked regularly. If possible, the tower pack should be removed for cleaning. Post-cleaning chlorination should be monitored to make sure that free chlorine residuals are maintained. | Semiannually |
| Review maintenance and water treatment program performance, including the quality of results obtained and the cost of system operation. | Annually |

*Step 4*  Drain the system; remove, clean, and replace all strainers; and then refill the system.

*Step 5*  Operate all system pumps for more than 30 minutes. During this time, temporarily open all dead ends, drain valves, and strainer flush valves.

*Step 6*  Drain and refill the system. Repeat Step 5. On a newly installed system, it may take as many as four flushes to clean the system adequately.

*Step 7*  Test the system water and certify its cleanliness using a procedure that measures the residual cleaning agent in the water. The final system water pH should be equal to ±0.3 pH units of that of the makeup water.

*Step 8*  After the system has been tested and certified to be free of cleaning agent, take samples of the system water for future use.

*Step 9*  Initiate the approved water treatment program specified for use during system operation.

Scale is a very common problem for systems that have been in operation for a while. If the amount of scale inhibitors prescribed for a water treatment program fails to control the scale buildup as predicted, the following factors should be checked:

*Step 1*  Check that the prescribed rate of bleed is being maintained.

*Step 2*  Check to see if the water can be made less scale-forming by using a more accurate dosage of the prescribed inhibitor chemicals. Take samples of recirculating and makeup water, and have them analyzed to check for the concentration of scale-forming salts. Look for changes in the concentration levels by comparing them with those previously used as the basis for the water treatment program.

*Step 3*  Because scale forms more rapidly at higher water temperatures, compare the temperature of the water in the sump with the ambient wet-bulb temperature. If the sump water temperature is more than 10°F above the wet-bulb temperature, possible causes are the following:
- Insufficient cooling tower capacity.
- A fan is not working properly due to loose belts, worn bearings, bent blades, or improper shrouding.

- A water-driven fan is operating at too low a speed due to plugged jets, improper water pressure at the fan, worn spindle, plugged louvers, plugged eliminators, or plugged packing in the cooling tower.
- The distribution and breakup of water is inadequate due to insufficient volume of water flow, plugged nozzles, slime-covered water distribution holes, or worn-out tower fill.
- Moist air from the discharge of the cooling tower is returning to the air intake.
- The air intake on an indoor tower is blocked.
- The air current path to the tower is restricted.

*Step 4* Because scale may be forming if the temperature rise of the water flowing through the system condenser is too high (more than 10°F), check to see if:

- Slime or debris is covering the pump screen.
- The water lines and/or condenser are undersized or restricted with corrosion or scale.
- There is a worn pump impeller or inadequate pump volume or pressure.
- There is a defective water valve.

## 3.2.0 Water Treatment in Closed Recirculating Water Systems

Hot-water, chilled-water, and dual-temperature systems are examples of closed recirculating systems. Closed recirculating systems are closed loop systems where water is continuously recirculated through sealed components and piping that prevent the water from having contact with air (*Figure 15*). As a result, acidic gases are not absorbed from the air. In a tight system, the water does not evaporate; therefore, dissolved minerals do not collect. No matter how tight, every closed system will sometimes need the addition of fresh or raw makeup water to compensate for leakage or some other cause of water loss. Dissolved oxygen and other impurities will be introduced into a closed system in this makeup water. Oxygen is one of the main agents of corrosion. Because it is a closed system, the purpose for water treatment is mainly to prevent corrosion and scale formation in the system.

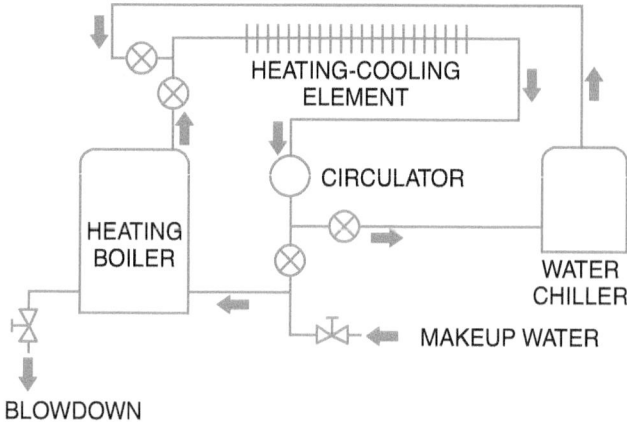

03308-13_F15.EPS

*Figure 15* Simplified dual-temperature closed water system.

### 3.2.1 Corrosion

Several chemicals can be used for corrosion control in a closed system. In the past, the use of chromates and hydrazine was common. However, the EPA has since identified both of those chemicals as suspected carcinogens; thus, they are subject to more stringent pollution control regulations. Some localities prohibit the use of chromates, while others allow their use only in closed systems. Sodium nitrite is commonly used as an alternative to chromates as an inhibitor in closed loop systems. Sodium nitrite has been proven nearly as effective as chromate for protecting steel and other ferrous metal components, but it gives little or no protection for components made of nonferrous metals. Other inhibitors must be included in the mixture to protect these components. One such nitrite-based inhibitor mixture includes borax as a pH buffer and sodium tolyltriazole as an inhibitor to protect copper, copper alloy, and other nonferrous metals. One drawback of using sodium nitrite is that it will not suppress bacterial growth at temperatures below 120°F.

## Acid Cleaning of a Tower or Condenser

Using a commercial acid cleaner usually has two benefits: most have inhibitors added to protect metal, and some contain a pH color indicator that shows when the acid is spent or neutralized. When the acid is spent during cleaning, more acid is added until the correct color Is maintained for 20 to 30 minutes of cleaning. Then, soda ash can be added to neutralize the acid, as indicated by the color, before the acid is flushed into drains.

### 3.2.2 Scale Deposits

In chilled-water systems, scale formation is normally not a problem even if the makeup water is fairly hard. Hot-water systems are more likely to have problems with scale formation due to the higher temperatures. Because a high pH promotes the buildup of scale, avoid using water with a high pH whenever possible.

### 3.3.0 Water Treatment in Steam Boilers and Systems

Steam boilers and steam systems are closed systems. The water used in a boiler must be kept clean for efficient operation. Never add dirty or rusty water to a boiler. Hard water may eventually interfere with the efficient operation of a boiler. For this reason, hard water should be chemically treated with water softeners before being used. Scale, corrosion, fouling, and foaming can all cause boiler problems. Scale deposits on heating surfaces increase boiler temperatures and lower operating efficiency. Corrosion can damage metal surfaces, resulting in metal fatigue and failure. Fouling clogs nozzles and pipes with solid materials, thereby restricting circulation and reducing the heat transfer efficiency. Foaming results in overheating and can cause water impurities to be carried along with the steam into the system. Foaming is a condition caused by high water surface tension due to a scum buildup from oil, grease, and/or sediment on the surface of the boiler water. Foaming prevents steam bubbles from breaking through the water surface and hinders steam production. The trapped bubbles rise and fill the steam space, resulting in impurities being entrained in the steam and then carried over into the steam system.

### 3.3.1 Corrosion

Corrosion erodes away the boiler metal and appears in different forms, including pitting, grooving, and embrittlement. Corrosion can be caused by acidic (low pH) feedwater. Another cause is the dissolved oxygen contained in the boiler feedwater. Oxygen, carbon dioxide, and other gases present in the feedwater are the main causes of corrosion. These gases must be eliminated by the use of a mechanical deaerator (deaerating heater) and/or scavenger chemicals such as sodium sulfite and hydrazine to remove traces of oxygen from the water.

Because makeup water is used, periodic blowdowns are also necessary to control chloride and hardness levels. To minimize corrosion, the boiler water pH should be maintained between 11 and 12 by an alkaline water treatment using soda ash, caustic soda, sodium silicate, or sodium phosphates. Note that the P-alkalinity should be maintained between 300 and 700 ppm. Maintaining high pH levels helps prevent corrosion, but it also promotes the formation of scale; therefore, scale inhibitors must be used in conjunction with high pH levels.

### 3.3.2 Scale Deposits

When water evaporates into steam in a steam boiler, the steam escapes to the system. Any calcium, magnesium, and other salts that were dissolved in the evaporated water remain behind in the water and are deposited on the boiler's tubes and other heat surfaces as scale. This condition can be aggravated (particularly in hard water areas) when steam or condensate losses are heavy, requiring the frequent addition of makeup water. The greater the amount of hardness in the original makeup water, the faster the boiler water reaches the point where scale can form. If a scaling condition becomes bad enough in firetube boilers, the tubes may bulge due to overheating. Tubes can burn out or possibly burst. Scale may also clog the insides of tubes. This can disrupt circulation and raise the boiler pressure, increasing the risk of an explosion.

Scale formation can best be avoided by removing all the calcium, magnesium, and other mineral salts from the feedwater before it enters the boiler. Chemicals commonly used to prevent scale formation are sodium carbonate and phosphate. Another water treatment for scale prevention involves the use of colloidal substances such as tannin or lignin. A colloidal substance is a jelly-like material made up of very small, insoluble, and non-diffusible particles that are larger than molecules, but small enough to remain suspended in a fluid without settling to the bottom. When added to the boiler water, these colloidal substances react with and absorb on their surface the calcium, magnesium, and other salts. The entire mass then forms a fluid sludge that is easily removed from the boiler by blowdown. Another common method is to add polymers that keep the sludge in suspension while they flow through the system.

### 3.3.3 Sludge Deposits

Sludge is a deposit formed by salts and other solids present in boiler water. It appears as lumps or thick masses of material that settle out in the low points of a boiler like a layer of mud. Some sludge in a system occurs naturally. Other sludge is produced on purpose by water treatment. Minerals that exist in water cannot be destroyed, but with the use of proper chemicals, the properties of these minerals can be changed to make them manageable by turning them into non-adhering sludge. Regardless of its source, if not controlled, sludge can insulate tubes, clog boiler circulation, and may bake out or harden, causing overheating and high operating temperatures. Sludge is removed by periodic blowdown of the boiler. In extreme cases, a thorough cleaning of the boiler with a chemical product may be necessary.

### 3.3.4 Foaming, Priming, and Carryover

Foaming, as discussed earlier, results in impurities from the boiler water becoming entrained in the steam and then carried over into the steam system. Once impurities enter the steam system, they can form deposits that damage heaters and other system components. Foaming is commonly controlled by filtration, antifoaming agents, and water treatment to remove solid impurities. Common antifoaming chemicals include organic materials such as polymerized esters, alcohols, and amides. Periodic surface blowdown is also done to help minimize foaming. Surface blowdown is covered later in this module.

Both priming and carryover also cause boiler water impurities to be entrained in the steam and then carried over into the steam system. Priming occurs when the boiler water level undergoes great changes and/or when violent discharges of bursting bubbles occur. Carryover is a condition in which water solids are entrained in the steam, even if there is no sign of foaming or priming. Carryover is usually indicated by drops of water appearing with the steam. Boiler water solids contained in the steam as a result of either priming or carryover disrupt operation of the equipment coming in contact with the steam and form deposits in terminals, valves, piping, and other components. In addition, any moisture carried over with the steam results in heat loss through the steam piping.

Controlling the type and amount of solids in boiler water is the main factor in the chemical control of carryover problems. Small amounts of organic matter and oil will form soap in the boiler and must not be allowed to enter. The amounts of solids and the alkalinity in the boiler water are also important. *Table 5* lists the maximum allowable total solids and suspended solids in boiler water recommended by the American Boiler Manufacturers Association.

**Table 5** Maximum Allowable Total Solids in Boiler Water

| Boiler Pressure (psi) | Total Solids in Boiler Water (ppm) | Suspended Solids (ppm) |
|---|---|---|
| 0 to 300 | 3,500 | 300 |
| 301 to 450 | 3,000 | 250 |
| 451 to 600 | 2,500 | 150 |
| 601 to 750 | 2,000 | 60 |
| 751 to 900 | 1,500 | 40 |

Solids and any slugs of water carried in steam due to carryover may be eliminated by use of a steam separator installed in the supply line. To help prevent carryover problems, do the following:

- Do not allow the boiler water level get too high.
- If possible, operate the boiler at an even rating.
- Arrange the distribution load so that the rate of steam flow is uniform.

### 3.3.5 Steam Boiler Water Treatment Guidelines

A boiler water treatment program can include pretreatment of the raw makeup water using filters, water softeners, and dealkalizers before it enters the boiler feedwater system. Chemicals can also be used to treat the boiler feedwater and boiler water and to condition any sludge to protect against corrosion and scaling in the boiler. In addition, boiler water treatment includes blowdown. The water that is removed by blowdown is then replaced with an equal amount of fresh, clean feedwater. Without blowdown, the concentration of suspended and dissolved solids in the boiler water can become excessive and result in the formation of sludge and scale. Priming, foaming, and/or carryover may also occur.

The amount of water removed by blowdown is normally determined by the concentration of total dissolved solids contained in the boiler water. Enough water must be removed to keep the TDS level below that at which priming, foaming, and/or carryover occurs in the boiler. Refer to *Table 5* for the limits for total and suspended solids rec-

## Steam System Water Hammer

In extreme cases of priming or carryover, water discharges from the boiler into the steam lines as a spray or as slugs, resulting in water hammer in the steam system.

ommended by the American Boiler Manufacturers Association for different boiler operating pressures.

Boilers are normally equipped with two kinds of blowdown valves: a bottom blowdown valve and a surface blowdown valve. The bottom blowdown valve is located at the bottom or lowest part of the boiler. It is used to manually purge the boiler of foreign matter by draining off some of the water from the bottom of the boiler to control the TDS in the water. This water contains sediment, scale, and other impurities that have settled out of the water as it is heated and have accumulated at the lowest point in the boiler.

The surface blowdown valve is connected to an internal collecting pipe that ends slightly below the boiler's normal operating water level. It is used to skim off impurities on the surface of the water inside the boiler. Surface blowdown usually involves a controlled and continuous draining of boiler water taken off the surface at the water level to skim off sediment, oil, and other impurities. The quantity of blowdown water removed from the boiler is controlled by the surface blowdown valve. The valve is adjusted to provide the flow rate needed to achieve the desired TDS level. Periodic adjustments of the surface blowdown valve must be made to increase or decrease the amount of blowdown, as determined by water test analysis. Note that some boilers can be equipped with more than one bottom and/or surface blowdown valve. Also, some boilers may not be equipped with a surface blowdown valve.

After a major repair, installation, or when the boiler is extremely dirty, skimming and blowdown may require that the boiler be completely drained and refilled with water one or more times. This process must be continued until the water discharged from the boiler runs clear. Always perform skimming, blowdown, and/or cleaning as directed in the boiler manufacturer's service instructions. General guidelines are given here.

## Wastewater Discharge Temperature

Typical plumbing codes require that the temperature of water entering a drainage system be 140°F or less.

Skimming is generally performed through the following steps:

*Step 1*  If it is not permanently connected to a drain, run a temporary connection from the boiler's skimming valve to a suitable drain.

*Step 2*  With the boiler empty and cool, slowly begin to add water. After a quantity of water has entered the boiler, never before, turn on the burners and adjust the flame so that the water being added is kept just below the boiling point. Boiling and turbulence must be avoided.

*Step 3*  Gradually raise the hot water level in the boiler to the point where the water just flows from the skimming valve, being careful not to raise it above this point.

*Step 4*  Continue to skim the boiler water until there is no trace of impurities. Water may be checked to make sure it is free from oil by drawing off a sample. If the sample is reasonably free from oil, it will not froth when heated to the boiling point.

Skimming will not usually clean the boiler of sediment that has accumulated at the bottom. After skimming, the boiler should be cleaned further by performing the following blowdown procedure:

*Step 1*  Check the water level in the boiler.

*Step 2*  Partially open the bottom blowdown valve. Once the water starts draining, fully open the valve.

*Step 3*  Remain at the blowdown valve. Monitor the gauge glass during blowdown to make sure that the water level is not lowered to a dangerously low point.

*Step 4*  When the desired amount of water has been drained from the boiler, close the blowdown valve.

> **NOTE**
>
> Only attempt blowdown at a light load. Ideally, temporarily suspend the boiler's heating process to halt water turbulence and allow the solids to settle out.

If an exceptional amount of dirt or sludge is present in a boiler, the boiler should be cleaned using an approved boiler cleaning compound according to the manufacturer's instructions. After cleaning, perform the blowdown procedure as needed to thoroughly flush all traces of the cleaning compound out of the boiler.

Record key boiler operating parameters and water conditions regularly. Changes in parameter values over time can reveal that the boiler operation is deteriorating and that corrective action should be taken. *Figure 16* shows an example of a form commonly used to record boiler operating and water quality parameters.

## BOILER WATER LOG SHEET

| RECOMMENDED READINGS | |
| --- | --- |
| ALKALINITY... 300-500 ppm | CONDUCTANCE:_____ mho |
| SULFITE... 30-60 ppm | CHLORIDE: _____ ppm |
| CONDENSATE pH... 8 to 8.5 | BOTTOM BLOWDOWN SCHEDULE: |

| DATE | ALKALINITY | SULFITE | CHLORIDE | CONDUCTANCE | CONDENSATE pH | TESTED BY |
| --- | --- | --- | --- | --- | --- | --- |
| | | | | | | |
| | | | | | | |
| | | | | | | |
| | | | | | | |
| | | | | | | |
| | | | | | | |
| | | | | | | |
| | | | | | | |
| | | | | | | |
| | | | | | | |
| | | | | | | |
| | | | | | | |
| | | | | | | |
| | | | | | | |
| | | | | | | |
| | | | | | | |
| | | | | | | |
| | | | | | | |
| | | | | | | |
| | | | | | | |
| | | | | | | |
| | | | | | | |
| | | | | | | |
| | | | | | | |
| | | | | | | |
| | | | | | | |
| | | | | | | |
| | | | | | | |
| | | | | | | |
| | | | | | | |
| | | | | | | |
| | | | | | | |
| | | | | | | |
| | | | | | | |

03308-13 F16.EPS

*Figure 16* Example of a boiler survey form.

## Additional Resources

*ASHRAE Handbook - HVAC Applications.* Latest Edition. Atlanta, GA: American Society of Heating, Refrigerating, and Air Conditioning Engineers, Inc.

*ASHRAE Handbook - HVAC Systems and Equipment.* Latest Edition. Atlanta, GA: American Society of Heating, Refrigerating, and Air Conditioning Engineers, Inc.

*Boilers Simplified.* Troy, MI: Business News Publishing Company.

*HVAC Water Chillers and Cooling Towers: Fundamentals, Application, and Operation.* Second Edition. Herbert W. Stanford III. Dekker Mechanical Engineering. Boca Raton, FL: CRC Press, Taylor & Francis Group.

*Water Treatment Specification Manual.* Troy, MI: Business News Publishing Company.

## 3.0.0 Section Review

1. Silt deposits in open recirculating water systems are commonly controlled by adding _____.

   a. algae
   b. phosphates
   c. makeup water
   d. polymers

2. Sodium nitrite is commonly used instead of chromates as an inhibitor in closed loop systems.

   a. True
   b. False

3. Corrosion-causing gases in the feedwater for a boiler are eliminated by using scavenger chemicals and/or _____.

   a. bag filters
   b. deaerating heaters
   c. colloidal inhibitors
   d. nonferrous cooling towers

# SUMMARY

Proper water treatment is an important aspect of HVAC equipment and system operation. It not only protects equipment, it also minimizes maintenance costs, saves energy, and prevents expensive repairs. These are common problems caused by untreated water in heating and cooling systems:

- Corrosion
- Scale formation
- Biological growth
- Suspended solid matter

Water treatment is site specific. The design of a water treatment program depends on the type of water system. The specific program is normally designed by a qualified water specialist. This involves making a detailed analysis of the available water, considering its intended use, then implementing the needed treatment program and/or equipment as required. Water treatment may be done by a physical means such as mechanical filtering, by chemical treatment, or both. After a water treatment program has been developed, its correct use and program maintenance are often the job of an HVAC technician.

Always remember that excessive use of chemicals to treat a water system can be just as harmful to the equipment as the use of insufficient chemicals.

1. Water with a pH of 9 is considered to be
_____.

   a. alkaline
   b. acidic
   c. neutral
   d. very acidic

2. As the hardness of water decreases, the potential for scaling _____.

   a. increases
   b. decreases
   c. remains the same
   d. fluctuates

3. Corrosion caused by electrolytic action is
_____.

   a. pitting corrosion
   b. grooving
   c. embrittlement
   d. galvanic corrosion

4. Problems resulting from suspended solids in water are often referred to as _____.

   a. fouling
   b. scaling
   c. caustic cracking
   d. electrolysis

5. When using a water analysis test kit with a sampling container that includes a fixing compound, you should _____.

   a. usually fill the container to overflowing
   b. always pretreat the fixing compound with an inhibitor
   c. never rinse the container before use
   d. sometimes apply a wax seal to the filled container

6. Unless proven otherwise, all water treatment chemicals should be considered hazardous.

   a. True
   b. False

7. Compared to a depth-type cartridge filter of the equivalent size, a surface-type cartridge filter _____.

   a. has a higher removal efficiency
   b. handles higher flow rates
   c. is usually less expensive
   d. contains less filter media

8. Centrifugal separators are used to remove
_____.

   a. water and impurities from steam
   b. dissolved particles from water
   c. suspended particles from water
   d. entrained gases from boiler blowdown

9. The zeolite filter bed in a typical water softener _____.

   a. distills minerals from hard water
   b. absorbs calcium and magnesium and releases sodium
   c. must be replaced when its sodium content is depleted
   d. is regenerated by backwashing

10. Deaerators are used to prevent corrosion in
_____.

   a. hot-water systems
   b. steam systems
   c. chilled-water systems
   d. water analysis kits

11. If the water recirculating through an open system cooling tower has a pH below 7.0, the most likely cause is absorption of _____.

   a. oxygen from the air
   b. airborne particles
   c. acidic gases from the makeup water
   d. acidic gases from the air

12. Biological growth is most likely found in
_____.

   a. hot-water systems
   b. water-to-water heat exchangers
   c. cooling towers
   d. steam systems

13. The use of chromates to prevent corrosion in closed recirculating systems is _____.

    a. permitted in all closed systems
    b. prohibited by EPA regulations
    c. not subject to any restrictions
    d. prohibited in some localities

14. Scale formation is more likely to be a problem in a closed hot-water system than in a closed chilled-water system.

    a. True
    b. False

15. To minimize corrosion in a steam boiler, the pH of the boiler water should be maintained between _____.

    a. 6.5 and 7.5
    b. 7.5 and 8.5
    c. 0.5 and 10.5
    d. 11 and 12

# Trade Terms Introduced in This Module

**Alkalinity:** A water quality parameter in which the pH is higher than 7. It is also a measure of the water's capacity to neutralize strong acids.

**Backwashing:** A procedure that reverses the direction of water flow through a multimedia-type filter by forcing the water into the bottom of the filter tank and out the top. Backwashing is performed on a regular basis to prevent accumulated particles from clogging the filter.

**Bleed-off:** A method used to help control corrosion and scaling in a cooling tower. A metered amount of water is consistently drained from the water circuit to help carry away minerals and other impurities that remain as water evaporates from the cooling tower.

**Colloidal substance:** A jelly-like material made up of very small, insoluble, nondiffusible particles larger than molecules, but small enough to remain suspended in a fluid without settling to the bottom.

**Concentration:** Indicates the strength or relative amount of an element present in a water solution.

**Cycles of concentration:** A measurement of the ratio of dissolved solids contained in a heating/cooling system's water to the quantity of dissolved solids contained in the related makeup water supply. The term *cycles* indicates when the concentration of an element in the water system has risen above the concentration contained in the makeup water. For instance, if the hardness in the system water is determined to be two times as great as the hardness in the makeup water, then the system water is said to have two cycles of hardness.

**Dissolved solids:** The dissolved amounts of substances such as calcium, magnesium, chloride, and sulfate contained in water. Dissolved solids can contribute to the corrosion and scale formation in a system.

**Electrolysis:** The process of changing the chemical composition of a material (called the electrolyte) by passing electrical current through it.

**Electrolyte:** A substance in which conduction of electricity is accompanied by chemical action.

**Fouling:** A term used for problems caused by suspended solid matter that accumulates and clogs nozzles and pipes, which restricts circulation or otherwise reduces the transfer of heat in a system.

**Grains per gallon (gpg):** An alternate unit of measure sometimes used to describe the amounts of dissolved material in a sample of water. Grains per gallon can be converted to ppm by multiplying the value of gpg by 17.

**Hardness:** A measure of the amount of calcium and magnesium contained in water. It is one of the main factors affecting the formation of scale in a system. As hardness increases, the potential for scaling also increases.

**Inhibitor:** A chemical substance that reduces the rate of corrosion, scale formation, or slime production.

**Milligrams per liter (mg/l):** Metric unit of measure used to specify exactly how much of a certain material or element is dissolved in a sample of water. One mg/l is equivalent to one ppm.

**Parts per million (ppm):** Unit of measure used to specify exactly how much of a certain material or element is dissolved in a sample of water. For example, one ounce of a contaminant mixed with 7,500 gallons of water equals a concentration of about one ppm.

**pH:** A measure of alkalinity or acidity of water. The pH scale ranges from 0 (extremely acidic) to 14 (extremely alkaline), with the pH of 7 being neutral. Specifically, pH defines the relative concentration of hydrogen ions and hydroxide ions. As pH increases, the concentration of hydroxide ions increases.

**Scale:** A dense coating of mineral matter that precipitates and settles on internal surfaces of equipment as a result of falling or rising temperatures.

**Suspended solids:** The amount of visible, individual particles in water or those particles that give water a cloudy appearance. They can include silt, clay, decayed organisms, iron, manganese, sulfur, and microorganisms. Suspended solids can clog treatment devices or shield microorganisms from disinfection.

**Total solids:** The total amount of both dissolved and suspended solids contained in water.

# Additional Resources

This module presents thorough resources for task training. The following resource material is suggested for further study.

*ASHRAE Handbook – HVAC Applications.* Latest Edition. Atlanta, GA: American Society of Heating, Refrigerating, and Air Conditioning Engineers, Inc.

*ASHRAE Handbook – HVAC Systems and Equipment.* Latest Edition. Atlanta, GA: American Society of Heating, Refrigerating, and Air Conditioning Engineers, Inc.

*Boilers Simplified.* Troy, MI: Business News Publishing Company.

*HVAC Water Chillers and Cooling Towers: Fundamentals, Application, and Operation.* Second Edition. Herbert W. Stanford III. Dekker Mechanical Engineering. Boca Raton, FL: CRC Press, Taylor & Francis Group.

*Water Treatment Specification Manual.* Troy, MI: Business News Publishing Company.

# Figure Credits

| Answer | Section Reference | Objective |
|---|---|---|
| **Section One** | | |
| 1.a | 1.1.0 | 1a |
| 2.b | 1.2.2 | 1b |
| 3.c | 1.3.0 | 1c |
| 4.d | 1.4.0 | 1d |
| **Section Two** | | |
| 1.d | 2.1.5 | 2a |
| 2.c | 2.2.4 | 2b |
| **Section Three** | | |
| 1.d | 3.1.2 | 3a |
| 2.a | 3.2.1 | 3b |
| 3.b | 3.3.1 | 3c |

# NCCER CURRICULA — USER UPDATE

NCCER makes every effort to keep its textbooks up-to-date and free of technical errors. We appreciate your help in this process. If you find an error, a typographical mistake, or an inaccuracy in NCCER's curricula, please fill out this form (or a photocopy), or complete the online form at **www.nccer.org/olf**. Be sure to include the exact module ID number, page number, a detailed description, and your recommended correction. Your input will be brought to the attention of the Authoring Team. Thank you for your assistance.

*Instructors* – If you have an idea for improving this textbook, or have found that additional materials were necessary to teach this module effectively, please let us know so that we may present your suggestions to the Authoring Team.

### NCCER Product Development and Revision
13614 Progress Blvd., Alachua, FL 32615

**Email:** curriculum@nccer.org
**Online:** www.nccer.org/olf

❏ Trainee Guide ❏ Lesson Plans ❏ Exam ❏ PowerPoints Other _____

Craft / Level: _____ Copyright Date: _____

Module ID Number / Title: _____

Section Number(s): _____

Description: _____

_____

_____

_____

Recommended Correction: _____

_____

_____

_____

Your Name: _____

Address: _____

_____

Email: _____ Phone: _____

# Indoor Air Quality

## OVERVIEW

Indoor air quality (IAQ) is one of the foremost concerns for HVAC designers, installers, and service personnel. Contaminants circulating in a building through the air can seriously affect the health and comfort of people breathing that air. Local and national government agencies require strict compliance with established IAQ standards. To meet these standards, HVAC technicians employ a variety of methods, systems, and equipment to help alleviate poor indoor air quality.

# Module 03403

Trainees with successful module completions may be eligible for credentialing through the NCCER Registry. To learn more, go to **www.nccer.org** or contact us at 1.888.622.3720. Our website has information on the latest product releases and training, as well as online versions of our *Cornerstone* magazine and Pearson's product catalog.

Your feedback is welcome. You may email your comments to **curriculum@nccer.org**, send general comments and inquiries to **info@nccer.org**, or fill in the User Update form at the back of this module.

This information is general in nature and intended for training purposes only. Actual performance of activities described in this manual requires compliance with all applicable operating, service, maintenance, and safety procedures under the direction of qualified personnel. References in this manual to patented or proprietary devices do not constitute a recommendation of their use.

*03403 V5*

## Objectives

When you have completed this module, you will be able to do the following:

1. Describe how indoor air quality (IAQ) affects humans.
   a. Define IAQ.
   b. Describe good IAQ and how the lack of it affects humans.
2. Identify sources of building air contaminants and describe how to detect such problems.
   a. Describe how building construction, materials, and equipment can affect IAQ.
   b. Describe how human occupancy can affect IAQ.
   c. Describe how external sources can affect IAQ.
   d. Explain how to conduct an IAQ survey.
   e. Describe the air sampling process.
3. Explain how acceptable IAQ can be achieved.
   a. Explain how the building design can affect IAQ.
   b. Explain how ventilation, temperature, and humidity control affect IAQ.
   c. Explain how to control chemical and microbial contaminants.
4. Identify IAQ-related HVAC equipment and describe specific activities used to address IAQ problems.
   a. Identify and describe HVAC equipment and devices used to improve IAQ and/or energy consumption.
   b. Explain how air distribution systems can contribute to poor IAQ and how these problems are addressed.
   c. Describe the liability that HVAC contractors may accept by servicing HVAC systems.

## Performance Tasks

Under the supervision of your instructor, you should be able to do the following:

1. Perform a building indoor air quality (IAQ) inspection/evaluation.
2. Make air measurements using at least one of the following devices:
   - $CO_2$ detector/sensor
   - CO detector/sensor
3. Use a manufacturer's humidifier capacity chart to find the humidifier capacity needed for various building types and sizes.

## Trade Terms

| | | | | |
|---|---|---|---|---|
| Arrestance efficiency | Desiccant | Friable | New building syndrome | Scintillation detectors |
| Biological contaminants | Dust-spot efficiency | Half-life | Off-gassing | Sick building syndrome (SBS) |
| Building-related illness (BRI) | Electret | High-efficiency particulate air (HEPA) filter | Ozone | Volatile organic compounds (VOC) |
| Chemically-inert | Environmental tobacco smoke (ETS) | Multiple chemical sensitivity (MCS) | Pontiac fever | |
| | Formaldehyde | | Radon | |

## Industry Recognized Credentials

If you are training through an NCCER-accredited sponsor, you may be eligible for credentials from NCCER's Registry. The ID number for this module is 03403. Note that this module may have been used in other NCCER curricula and may apply to other level completions. Contact NCCER's Registry at 888.622.3720 or go to **www.nccer.org** for more information.

# Contents

## Figures and Tables

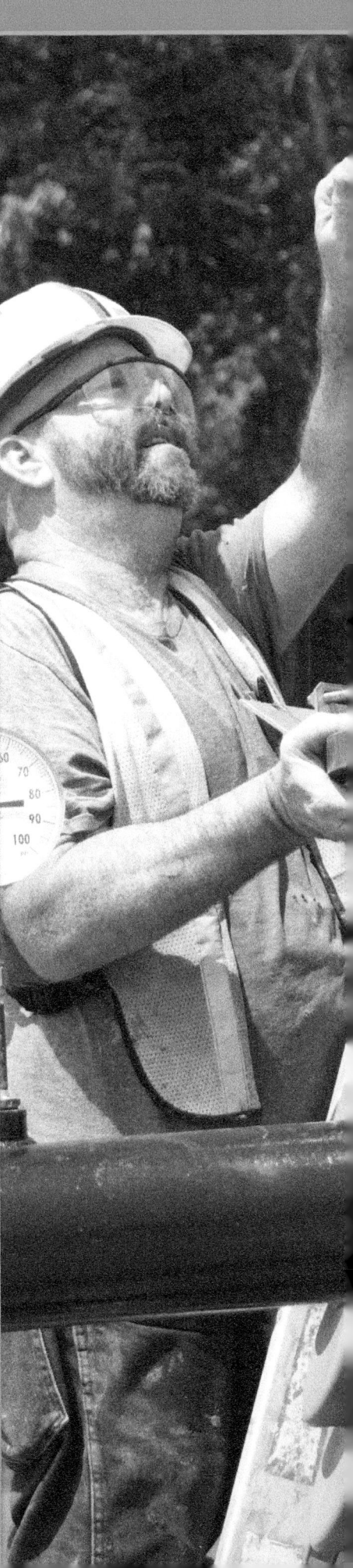

### 1.0.0 INDOOR AIR QUALITY

#### Objective

Describe how indoor air quality (IAQ) affects humans.

   a. Define IAQ.
   b. Describe good IAQ and how the lack of it affects humans.

#### Trade Terms

**Building-related illness (BRI):** A situation in which the symptoms of a specific illness can be traced directly to airborne building contaminants.

**Chemically-inert:** The property of a substance that does not readily react chemically with other substances, even under high temperatures and pressures, which usually accelerate chemical changes.

**Environmental tobacco smoke (ETS):** A combination of side-stream smoke from the burning end of a cigarette, cigar, or pipe and the exhaled mainstream smoke from the smoker.

**New building syndrome:** A condition that refers to indoor air quality problems in new buildings. The symptoms are the same as those for sick building syndrome.

**Radon:** A colorless, odorless, radioactive, and chemically-inert gas that is formed by the natural breakdown of uranium in soil and groundwater. Radon exposure over an extended period of time can increase the risk of lung cancer.

**Sick building syndrome (SBS):** A condition that exists when more than 20 percent of a building's occupants complain during a two-week period of a set of symptoms, including headaches, fatigue, nausea, eye irritation, and throat irritation, that are alleviated by leaving the building and are not known to be caused by any specific contaminant.

In the HVAC industry, awareness of indoor air quality (IAQ) as a health, economic, and environmental issue is very important. Poor IAQ can cause both long-term and short-term health effects. Aside from the health issues themselves, the economic impact includes the following:

- Direct medical cost for people whose health is affected by poor IAQ
- Loss of productivity from absence due to illness
- Decreased work efficiency
- Damage to equipment and materials due to exposure to indoor air pollutants

The challenge for HVAC technicians, building owners, and occupants is to increase their understanding of IAQ. This will help them to use a building in a manner that does not defeat the capabilities of the building HVAC equipment and/or compromise indoor air quality.

The causes of poor IAQ are varied. Recent energy conservation measures include tighter, well-insulated buildings, reduced capacities of HVAC systems, and HVAC system control schemes that minimize air movement in occupied spaces. These changes have been found to produce higher levels of contaminants in some buildings. Poor IAQ also originates with the types of materials used to construct, furnish, and maintain buildings today. Proper HVAC maintenance is always a key factor in the overall health of a building.

### 1.1.0 Indoor Environmental Quality Issues

Indoor air quality is a portion of a larger issue known as indoor environmental quality (IEQ). IEQ encompasses all elements of the indoor environment. The US EPA estimates that Americans spend roughly 90 percent of their time indoors. According to the Rocky Mountain Institute, productivity in the work environment can increase as much as 16 percent with the correct application of IEQ guidelines.

IEQ parameters include thermal comfort and fresh ventilation air, which are also IAQ attributes. Additionally, IEQ considers seating ergonomics, access to natural light, wall and ceiling colors and surface textures, and sound levels. Many of these are completely outside the HVACR realm, but are now part of the overall IEQ landscape, of which IAQ is an integral part.

Current building philosophy strongly leans toward the understanding that IEQ/IAQ problems are far easier and less expensive to prevent in the construction phase than they are to resolve after the fact. Many issues, such as noise from vibration, can be virtually impossible to address in existing buildings.

The principal impacts to the IAQ portion of the IEQ concerns include more zoning, more sensors for undesirable gases, larger volumes of fresh, filtered, outside air, and better control over air motion in occupied spaces. These areas of technology will continue to take on a larger role in the design, installation, and maintenance of HVACR systems.

The subject of indoor air quality is a complex one. Some debate exists as to what good IAQ is and how to best achieve it. For these reasons, technicians must make an effort to keep current on this subject by reading trade newspapers and journals. *ASHRAE Standard 62.1, Ventilation for Acceptable Indoor Air Quality*, specifies minimum ventilation rates and indoor air quality standards acceptable for human occupants while minimizing the potential for adverse health effects. To maintain a balance between IAQ and energy consumption, this standard incorporates both a ventilation-rate procedure and an air-quality procedure for ventilation design. The ASHRAE standard defines acceptable IAQ as, "air in which there are no known contaminants at harmful concentrations as determined by cognizant authorities, and with which a substantial majority (80 percent or more) of the people exposed do not express dissatisfaction."

To evaluate a building and its systems with regard to IAQ, it is necessary to know the acceptable levels of contaminants. There is much debate here, as well, as to the acceptable level for each

## Indoor Air Quality in Schools

The discussion of indoor air quality generally focuses on problems that occur in homes and commercial office buildings. However, studies show that about one-half of our nation's 130,000 public and private grade schools have problems linked to indoor air quality. This can potentially affect 20 percent of the United States population (about 56 million people) who spend their days in elementary and secondary schools. The concern is that students, especially younger children, are at greater risk from poor air quality because of the hours they spend in school facilities and because they are especially susceptible to pollutants.

## Certificate of Occupancy

In most municipalities throughout the United States, local building officials are required to inspect all residential and commercial buildings and issue a Certificate of Occupancy before allowing anyone new to move in. The certificate is proof that a space is suitable for occupancy based on a current inspection. Buildings must show compliance with prevailing IAQ requirements as one of the criteria for obtaining the certificate.

### GOING GREEN

## What is LEED?

LEED stands for Leadership in Energy and Environmental Design. It is an initiative started by the US Green Building Council (USGBC) with the goal of encouraging and accelerating global adoption of sustainable construction standards through a Green Building Rating System™. The rating system addresses six categories. Note that indoor environmental quality (IEQ) is one of the eight categories:

- Location and Transportation (LT)
- Sustainable sites (SS)
- Water efficiency (WE)
- Energy and atmosphere (EA)
- Materials and resources (MR)
- Indoor environmental quality (IEQ)
- Innovation in design (ID)
- Regional priority (RP)

LEED is a voluntary program in which the building owner chooses to participate. While most technicians will not have much, if any, say in whether a building is LEED-certified or not, they will have to maintain the systems to LEED standards in order for the building to retain its LEED rating. In other words, the LEED certification is for the life of the building, not just for the year of commissioning.

The USGBC is not a government entity, but a private, not-for-profit organization that is attempting to motivate and move market forces to a higher level. This initiative is strong enough that ASHRAE created their *Standard 189.1, Standard for the Design of High-Performance Green Buildings*, which many localities will likely adopt as code. In 2015, ASHRAE, along with the American Institute of Architects (AIA), the International Code Council (ICC), the USGBC, and the Illuminating Engineering Society (IES) signed an agreement to develop a new version of the International Green Construction Code (IGCC), which the consortium plans to release in 2018.

contaminant. Current standards can vary widely. For these reasons, this module does not provide specific contaminant control levels. Technicians should obtain copies of the current federal, state, and local standards for their specific location and use them for reference.

### 1.2.1 Long-Term and Short-Term Effects of Poor IAQ

Health effects resulting from poor IAQ may not show up until years after exposure. They may also occur after long or repeated periods of exposure. These long-term effects, which can include respiratory diseases and cancer, can be severely crippling or fatal. Long-term health effects are associated with indoor air pollutants such as asbestos, radon, and environmental tobacco smoke (ETS). These three pollutants have received a great deal of press coverage over recent years.

Radon is a colorless, odorless, chemically-inert, and radioactive gas. It results from the radioactive decay of naturally occurring uranium and thorium in soil and groundwater. Some organizations use the term *thoron* for radon from thorium. It is the largest source of ionizing radiation exposure (*Figure 1*).

ETS is a combination of smoke from the burning of tobacco products and the exhaled smoke from the smoker. With the introduction of electronic smoking devices (e-cigarettes) and the legalization of marijuana in some states, *ASHRAE Standard 62.1* now includes the airborne products of these activities in ETS.

Fixed short-term or immediate effects of poor IAQ may appear after a single high-dose exposure or as a result of long-term exposure to lower levels. Symptoms can include irritation of the eyes, nose, and throat, and headaches, dizziness, and fatigue. These conditions are normally treatable, often by simply avoiding exposure to the source of pollution, if it is identifiable. When a health professional can trace symptoms of a specific illness directly to airborne building contaminants it is referred to as a building-related illness (BRI).

There are also situations in which building occupants experience symptoms that do not fit the pattern of a particular illness. The symptoms are difficult to trace to any one source. These conditions can be temporary, but some buildings have long-term problems. Many IAQ professionals refer to this as sick building syndrome (SBS) or new building syndrome.

Sick building syndrome exists when more than 20 percent of a building's occupants complain during a two-week period of a set of symptoms—including headaches, fatigue, nausea, eye irritation, and throat irritation—that are alleviated by leaving the building, and which are not known to be caused by any specific contaminants. Some causes of these symptoms may include inadequate ventilation, chemical and biological contamination, and other non-pollutant factors such as temperature, humidity, and lighting.

New building syndrome refers to IAQ problems that present the same symptoms in occupants of new buildings.

> **NOTE**
>
> The 20-percent qualifier for SBS is an arbitrary limit based on ASHRAE's early work to define structural air quality. Today, many use it as the minimum limit for initiating remedial action. However, any time there are a number of complaints indicating the possibility of SBS, especially in large organizations, investigation should be considered to determine if there actually are building-related health problems.

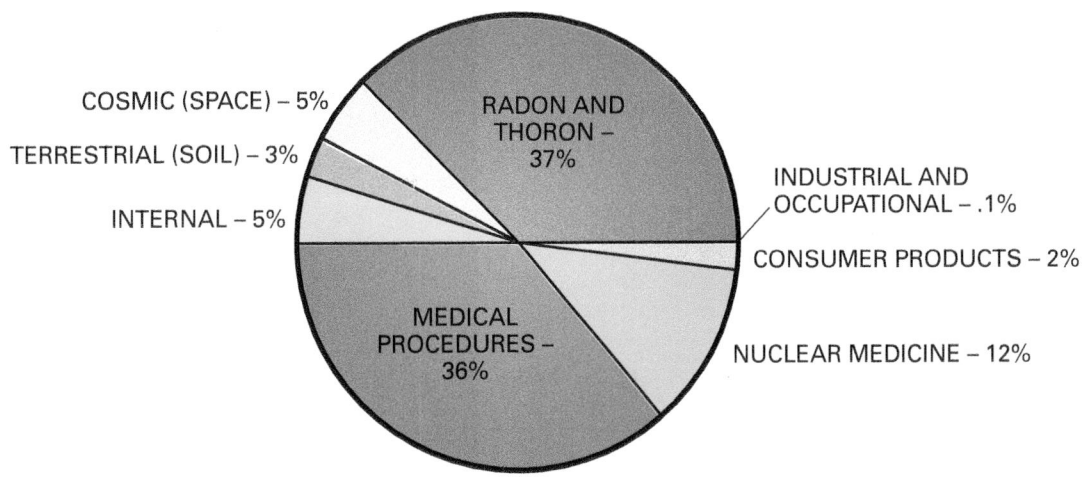

*Figure 1* Radiation sources.

## Additional Resources

*ASHRAE Standard 62.1, Ventilation for Acceptable Indoor Air Quality*. Current edition. Atlanta, GA: American Society of Heating, Refrigerating and Air-Conditioning Engineers (ASHRAE).

*ASHRAE Standard 189.1, Standard for the Design of High-Performance Green Buildings*. Current edition. Atlanta, GA: American Society of Heating, Refrigerating and Air-Conditioning Engineers (ASHRAE).

*Building Air Quality, a Guide for Building Owners and Facility Managers*. Current edition. Washington, DC: U.S. Environmental Protection Agency.

*Indoor Air Quality*. Current Edition. Chantilly, VA: Sheet Metal and Air Conditioning Contractors National Association (SMACNA).

*Indoor Air Quality in the Building Environment*. Ed Bas. 1993. Troy, MI: Business News Publishing Company.

## 1.0.0 Section Review

1. IAQ addresses _____.

   a. seating ergonomics
   b. wall and ceiling colors
   c. thermal comfort and fresh ventilation air
   d. surface colors and textures

2. Indoor air quality is considered acceptable if at least what percent or more of the people exposed to the air do *not* express dissatisfaction?

   a. 80
   b. 70
   c. 50
   d. 20

### 2.0.0 SOURCES OF BUILDING CONTAMINANTS

#### Objective

Identify sources of building air contaminants and describe how to detect such problems.

  a. Describe how building construction, materials, and equipment can affect IAQ.

  b. Describe how human occupancy can affect IAQ.

  c. Describe how external sources can affect IAQ.

  d. Explain how to conduct an IAQ survey.

  e. Describe the air sampling process.

#### Performance Tasks

1. Perform a building indoor air quality (IAQ) inspection/evaluation.
2. Make air measurements using at least one of the following devices:
   - $CO_2$ detector/sensor
   - CO detector/sensor

#### Trade Terms

**Biological contaminants:** Airborne agents such as bacteria, fungi, viruses, algae, insect parts, pollen, and dust. Sources include wet or moist walls, duct, duct liner, fiberboard, carpet, and furniture. Other sources include poorly maintained humidifiers, dehumidifiers, cooling towers, condensate drain pans, evaporative coolers, showers, and drinking fountains. Also given the terms microbial or microbiological contaminants.

**Electret:** A material or object that has the property of having both a positive and a negative electrical pole and generating an electrical field between them. The electrical equivalent of a magnet.

**Formaldehyde:** A colorless, pungent byproduct of synthetic and natural biological processes that can cause irritation of the eyes and upper air passages. A known human carcinogen.

**Friable:** The condition in which brittle materials can easily fragment, releasing particulates into the air.

**Half-life:** The time required for half of a number of atoms to undergo radioactive decay.

**Multiple-chemical sensitivity (MCS):** A condition found in some individuals who believe they are vulnerable to exposure to certain chemicals and/or combinations of chemicals. Currently, there is some debate as to whether or not MCS really exists.

**Off-gassing:** The process by which furniture and other materials release chemicals and other volatile organic compounds (VOC) into the air.

**Ozone:** An unstable form of oxygen that has a sharp, pungent odor like chlorine bleach, formed in nature in the presence of electric discharges or exposure to ultraviolet light. It is irritating to the mucous membranes and lungs in animals and to plant tissues. It is a strong oxidizer and forms a hazardous air pollutant near ground level.

**Pontiac fever:** A mild form of Legionnaires' disease.

**Scintillation detectors:** Radiation detectors that contain a substance which emits light when it absorbs a fast-moving nuclear particle (proton, neutron, or alpha). The sparkling effect caused by these emissions is called scintillation. The detector includes features that amplify the light flashes and display the rate of detections.

**Volatile organic compounds (VOC):** A wide variety of compounds and chemicals found in such things as solvents, paints, and adhesives, which easily evaporate at room temperature.

Poor IAQ can result from a building's construction or by pollutants released from sources located inside and/or outside of the building. Sources of indoor air pollution can include the following:

- Building construction
- Building materials and furnishings
- HVAC and other building equipment
- Cleaning compounds and pesticides
- Human occupancy
- Contaminant sources located outside the building

### 2.1.0 Building Construction

Fresh outdoor air can enter a building by infiltration, natural ventilation, and mechanical ventilation. Infiltrating air enters the building through openings such as joints, cracks in walls, and unsealed gaps around windows and doors. Natural ventilation air enters through open windows, doors, and passive, purpose-built air vents. Air movement during infiltration and natural ventilation results from air temperature and pressure differences between the indoor and outdoor environments.

Mechanical ventilation involves the use of fans vented to the outdoors that remove air from certain rooms, such as kitchens and bathrooms. Mechanical ventilation can also be an air handling system that uses fans and dampers to continuously remove indoor air and distribute filtered and conditioned outdoor air to the rooms in a building. The speed at which outdoor air replaces indoor air is called the exchange rate. When there is little infiltration, natural ventilation, or mechanical ventilation, the air exchange rate is low, and the levels of air pollutants originating inside the building can increase.

Older buildings normally allow more than enough outdoor air to enter by infiltration and natural ventilation to provide good IAQ. Newer buildings have much tighter construction (*Figure 2*). They are sometimes so tight that it creates problems. Without enough inward leakage of outdoor air by natural means or by mechanical ventilation, the indoor air can become unhealthy from internal pollutants. To the building's occupants, this is like living in an airtight box, where the absence of ventilation causes contaminants to accumulate.

### 2.1.1 Building Materials and Furnishings

Building materials and furnishings are the source of many contaminants. They can release volatile organic compounds (VOC) and/or have high particulate shed rates. VOCs include a wide variety of compounds and chemicals found in paints, adhesives, sealants, furniture, carpeting, and vinyl wall coverings. VOCs vaporize readily at

*Figure 2* Modern tightly constructed residence.

## Specifications and Standards

Codes, standards, and specifications published by associations such as ASHRAE undergo constant review by industry experts. These documents are updated periodically or as deemed necessary and are then republished. When using these documents, always make sure you have the current edition; this can be verified by a web search. The only exception to this rule would be when a project specifically refers to another edition of the code or standard.

normal air pressure and room temperature. Vaporization of volatile compounds out of solids is known as off-gassing.

A common VOC is formaldehyde, which manufacturers regularly use in particle board, plywood, and some foam insulation production. It is a colorless, pungent byproduct of hydrocarbons that can cause irritation of the eyes and upper air passages. In high concentrations, it may cause asthma attacks. Heat and higher humidity levels can sometimes increase the rate at which formaldehyde off-gasses.

> **WARNING!**
>
> Structures built before 1978 may contain lead-based paints. While not normally an airborne pollutant, any maintenance or demolition that produces dust from lead-based painted materials could create airborne particulate levels of lead that exceed OSHA standards. Before beginning such work, workers or occupational health professionals should test painted surfaces for lead. During demolition, workers should wear respirators that are NIOSH-approved for lead exposure. Occupied areas adjacent to the demolition work should have polyethylene barriers and positive pressure ventilation if it is known or suspected that lead paint is involved.

New building materials give off a great deal of moisture and VOCs. The level of VOCs from these materials may remain very high in a new building unless diluted by ventilation during and after construction. New carpet, padding, and adhesives are major sources of VOCs. When practical, installers should air out new carpeting for 24 hours before installation to reduce the volume of VOC emissions. The building should not be occupied for 72 hours following the installation.

# Carpet and Rug Institute Green Label Testing Program

The Carpet and Rug Institute (CRI) is a trade association for the carpet, rug, and flooring industry. In 1992, CRI, in association with the EPA, established a labeling program to identify products that have low VOC emissions. The CRI green and white logo displayed on carpet samples in showrooms informs the consumer that the product has been tested by an independent laboratory and has met the criteria for very low VOC emissions. To ensure that the product consistently complies, the program requires the retesting of various products quarterly.

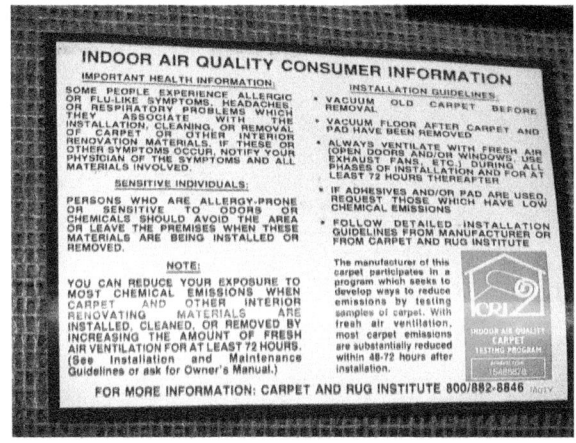

## Flooring Installation

The adhesives used to fasten floor materials can contribute to poor IAQ, especially when applied without adequate ventilation. All adhesives require that the subflooring materials have a very low moisture content, yet tight schedules associated with new construction often tempt contractors to apply flooring materials to wet floors. This can cause long-term IAQ risks including mold growth and the off-gassing of VOCs resulting from the decomposition of flooring components. Many construction specifications now require testing to ensure acceptable moisture content of concrete floors prior to installing sealant or flooring material.

Unlike the materials used in the past to furnish buildings, especially business offices, the fabrics and porous, soft materials used today act as incubators for microorganisms. Carpeting, wallpaper, acoustical tiles, and upholstered furniture in newer buildings have replaced the tile floors, painted walls, hard ceilings, and metal furniture found in older buildings. Depending on the temperature and humidity levels maintained in a building, these newer materials provide the ideal environment for mold, mildew, fungi, and bacteria.

Asbestos is another potential contaminant. This naturally occurring mineral was a common component of fireproofing materials, thermal insulation, floor tiles, and coverings for structural members in buildings constructed from 1930 to the mid-1970s. Because asbestos is a known carcinogen, workers must treat asbestos-containing materials with great caution.

Where asbestos occurs, it is not normally a problem as long as it remains encapsulated. However, if the surface of the material is deteriorating or subject to abrasion, asbestos fibers can become airborne. Brittle, fibrous materials that readily crumble and release airborne particles or fibers are considered to be friable. The EPA recommends that building owners leave undamaged asbestos materials alone as long as the materials are not likely to be disturbed. Qualified contractors must be employed to control any activities that may disturb asbestos and to perform asbestos removal and cleanup.

Some types of office equipment, especially photocopiers, can generate pollutants. These devices are sources of irritants such as ozone. While ozone is a naturally occurring gas in the atmosphere, when it exceeds even very low levels within buildings or at ground level, it can become hazardous to animals and plants.

Some of the health standards assigned by different organizations regarding ozone exposure are as follows:

- OSHA requires that people not be exposed to an average concentration exceeding 0.1 part per million (ppm) for eight hours.
- NIOSH recommends an absolute exposure limit of 0.1 ppm; 5 ppm is immediately dangerous to life or health (IDLH).
- EPA recommends a maximum average concentration of 0.07 ppm for eight hours.

# Mold Litigation and Legislation

Health hazards related to mold contamination are currently a hot topic of discussion and proposed government regulation. This is the result of a growing number of costly lawsuits pertaining to toxic mold. For example, a Texas homeowner won a $32-million judgment against her insurance company for failing to act properly and promptly to remove toxic mold contamination from her home. She claimed that as a result she and her daughter suffered severe health problems from exposure to the mold.

Due to the increased concern and litigation associated with toxic mold, many state and federal departments conducted studies of exposure to indoor toxic mold and its health effects. In 2005, Congress passed the Toxic Mold Safety and Protection Act. This law primarily addresses the prevention, inspection, and abatement of toxic mold in public housing and in dwellings with mortgages insured by federal programs. The Act requires the EPA to issue guidelines that define acceptable versus hazardous threshold limits for toxic mold. It also mandates that the EPA set standards for, and license, those who inspect and clean up mold sites.

**Figure Credit:** © iStock.com/cmannphoto

## 2.1.2 HVAC and Other Building Equipment

HVAC and refrigeration equipment, combustion (fuel-burning) equipment, and office equipment all have the potential to contribute to the pollution of indoor air, resulting in poor IAQ. Widespread use of total comfort heating/cooling systems with forced-air distribution is another cause of increased air pollution. These systems provide a comfortable building environment but can also distribute pollutants if they are poorly maintained.

Chilled-water cooling coils, evaporator or cooling coil condensate drip pans, humidifiers and dehumidifiers, cooling towers, and evaporative coolers and condensers are all moisture reservoirs that can become breeding grounds for biological contaminants. Mainly microbiological particulates, these contaminants include bacteria, fungi, viruses, algae, insect parts, pollen, and dust. The sources of these pollutants include the outdoors for pollen, and human and animal building occupants for viruses, bacteria, hair, and skin flakes.

Fiberglass duct board and insulation liners can trap moisture and contaminants and become a breeding ground for biological contaminants. Shutting down HVAC systems on weekends to save energy, coupled with water spills, leaks, and dripping plants in the building, can result in the building itself acting as an incubator. Rust or discoloration inside metal ductwork can be a sign of excessive moisture resulting from faulty humidifier operation.

Some highly publicized cases of Legionnaires' disease and Pontiac fever have been attributed to poorly maintained HVAC systems that allowed the incubation and distribution of disease-causing microorganisms. The bacteria responsible for Legionnaires' disease, a form of pneumonia, can live at temperatures between 78°F (25°C) and 124°F (51°C). The ideal growth temperature range is 90°F to 108°F (32°C to 42°C). Outside of this range, the bacterium lies dormant or dies; survival time depends on the extremes of temperature.

Problems can arise in a building from the mixed atmosphere of gases that may exist near, or migrate from, combustion equipment. All products of combustion can be dangerous and may compound pre-existing health problems. Furnaces, boilers, space heaters, wood stoves, gas stoves, and fireplaces can produce pollutants such as carbon monoxide (CO), oxides of nitrogen ($NO_x$), sulfur dioxide ($SO_2$), and airborne particles.

CO is a toxic gas that results from incomplete combustion. It is colorless, tasteless, and nonirritating. High levels of CO are extremely harmful. CO can slowly build up in the bloodstream where it binds with hemoglobin and displaces oxygen in the blood. Eventually the blood carries so little oxygen that it can no longer support life. Death from CO poisoning can happen suddenly. Its victims are often overcome and helpless before they realize they are in danger. Signs of CO poisoning include headaches, dizziness, fatigue, and nausea. Victims often think they have the flu or a common cold because the symptoms are similar. *Table 1* shows symptoms of CO poisoning at different exposure concentrations. Nitric oxide (NO) and nitrogen dioxide ($NO_2$) are two common airborne oxide gases of nitrogen, together symbolized as $NO_x$. All combustion processes can produce $NO_x$. Oxides of nitrogen form acids in the earth's lower atmosphere, where they contribute to acid rain. $NO_x$ and hydrocarbons react with sunlight to produce smog.

NO and $NO_2$ can also displace oxygen in the blood. $NO_2$ can irritate the skin and the mucous membranes in the eyes, nose, and throat. High levels of $NO_2$ may result in burning and pain in the chest, coughing, and/or shortness of breath. Oxides of nitrogen released from incomplete combustion can lodge in the lungs and irritate or damage lung tissue. They may also cause cancer.

Poorly installed or neglected chimneys and flues, as well as furnaces with cracked heat exchangers, are sources of pollutants as well. Negative pressures in tight buildings can cause back-drafting of a combustion appliance and the distribution of combustion byproducts throughout the building. In a warm-air furnace, a cracked heat exchanger can cause a buildup of toxic gases, including CO. The furnace blower may then distribute these gases into the conditioned space, causing sickness or death.

**Table 1** CO Levels and Related Symptoms

| Concentration | Exposure Times and Effects |
|---|---|
| 9 ppm | The maximum allowable concentration for short-term exposure in the living area according to ASHRAE |
| 35 ppm | The maximum allowable concentration for continuous exposure in any 8-hour period according to federal law |
| 200 ppm | Slight headaches, fatigue, dizziness, nausea after two to three hours |
| 400 ppm | Frontal headaches within one to two hours, life-threatening after 3 hours; also, the maximum ppm in flue gas (on a free-air basis) according to the EPA |
| 800 ppm | Dizziness, nausea, and convulsions within 45 minutes; unconsciousness within 2 hours; death within 2 to 3 hours |
| 1,600 ppm | Headache, dizziness, and nausea within 20 minutes; death within 1 to 2 hours |
| 3,200 ppm | Headache, dizziness, and nausea within 5 to 10 minutes; death within 1 hour |
| 6,400 ppm | Headache, dizziness, and nausea within 1 to 2 minutes; death within 30 minutes |
| 12,800 ppm | Death within 1 to 3 minutes |

**Did You Know?**

# Residential Carbon Monoxide Detectors

According to the National Conference of State Legislatures (NCSL) website, there are over 10,000 deaths in the United States due to carbon monoxide (CO) poisoning every year, and over 500 people require medical treatment for CO exposure. To reduce these statistics, many states have collaborated in enacting legislation requiring CO monitors in residences.

As 2017 began, 27 states had passed laws requiring CO detector installation in all homes; limited them to residential structures containing fossil-fuel heating systems; or required them to be installed only upon a change of property ownership. Eleven other states mandate detector installation when local codes adopt the International Residential Code.

States are also beginning to pass legislation requiring carbon monoxide monitoring in schools, hotels, and motels. The NCSL website, **http://www.ncsl.org/**, includes a current list of all states having CO monitor laws and regulations.

Furnace heat exchangers should be visually inspected to make sure they do not have cracks or pinhole leaks caused by corrosion. Be aware that a visual inspection may not detect tiny cracks that expand as the furnace heats up. A better method is to test and compare combustion gas readings taken before and after the furnace blower has turned on. A cracked heat exchanger should be suspected if there is a change of $O_2$ concentration in the flue gases greater than 0.5 percent or a change in the CO level greater than 25 ppm.

Any blockages in a chimney can cause inefficient combustion and produce dangerous levels of gases. Birds' nests, chimney deterioration, soot buildup, and other natural causes can create flue blockages. High levels of moisture in a chimney can cause the lining materials to decompose and create restrictions.

The routine and proper maintenance of combustion equipment such as furnaces, flues, and chimneys is the best way to prevent exposure to CO and $NO_x$. At a minimum, maintenance workers should inspect, clean, and adjust the equipment annually. Any needed repairs should be made promptly.

### 2.1.3 Cleaning Compounds and Pesticides

Cleaning compounds and pesticides are also sources of building contaminants. Pesticides are sources of many organic compounds whose effects range from minor irritation to potential cancer. Long-term damage to the liver and the central nervous system is possible in extreme cases of exposure. The safest way to protect against exposure from chemicals and pesticide pollutants is to always read the SDS and/or labels that clearly describe the specific health hazards related to the product. The product should be used only as directed; if in doubt, contact the EPA for more information.

Chemicals or pesticides used or stored in a building may cause some occupants to suffer from multiple chemical sensitivity (MCS). MCS is a medical condition found in some individuals who are vulnerable to exposure to certain chemicals and combinations of chemicals. There is some question as to whether MCS is a true physiological condition or simply a response to the perceived presence of certain chemicals.

### 2.2.0 Human Occupancy

Most people spend about 90 percent of their time indoors. As they breathe, humans consume oxygen and exhale carbon dioxide ($CO_2$). In a given area, a $CO_2$ concentration of 400 ppm is generally considered excellent air; 600 ppm, good air; 800 ppm, adequate air; and 1,000 ppm is minimally acceptable air. Experience has shown that $CO_2$ concentration levels above 1,000 ppm contribute to poor IAQ. At concentrations of about 1,200 ppm, people tend to get drowsy, have headaches, and/or function at lower activity levels. For this reason, the sensing and control of $CO_2$ concentrations in high occupancy areas of a building is important.

While $CO_2$ concentrations are widely used as an indicator of IAQ, low $CO_2$ levels do not necessarily mean there is no IAQ problem. There may be other contaminants in the air. Humans themselves are a source of contamination, emitting bioaerosols from the digestive process, perspiration, and exhalation (*Figure 3*). Their skin sheds, creating airborne particles. When they sneeze or cough, bacteria and viruses are sent flying into the air. Personal care products are yet another source of air contamination. Activities such as smoking and cooking also add to poor IAQ.

# Indoor and Outdoor CO₂ Levels

To dilute $CO_2$ levels and remove other indoor air contaminants, building mechanical ventilation systems are used to bring in adequate amounts of outdoor air. This approach works well as long as the outdoor air is less contaminated than the indoor air. Despite a continuing debate about $CO_2$ concentrations and their control and benefits, there is a general industry acceptance for comparing the indoor $CO_2$ level to the related outdoor $CO_2$ level. Normally, when air pollution levels increase outdoors, there is an associated rise in the outdoor $CO_2$ level. It is therefore possible on particularly smoggy days to bring more contaminated air into a building than already exists indoors. For this reason, it is good practice to sense both the indoor and outdoor $CO_2$ concentration levels to ensure that the outdoor air brought into a building has a lower $CO_2$ concentration level than the indoor level by at least 300 ppm.

When the outdoor air is heavily contaminated, many systems are configured to operate so that the indoor air is routed through air-cleaning devices such as ultraviolet lamps, electronic air cleaners, or activated charcoal and HEPA air filters for recirculation through the building.

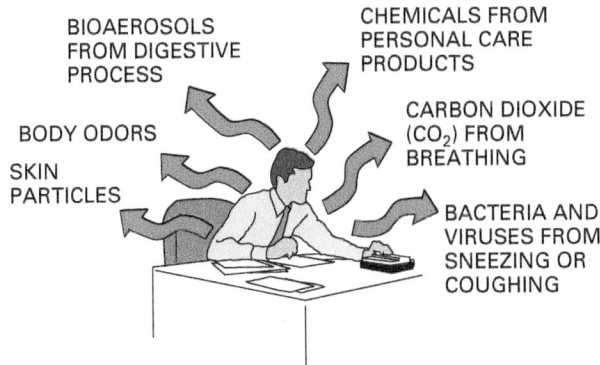

BIOAEROSOLS FROM DIGESTIVE PROCESS

CHEMICALS FROM PERSONAL CARE PRODUCTS

BODY ODORS

SKIN PARTICLES

CARBON DIOXIDE ($CO_2$) FROM BREATHING

BACTERIA AND VIRUSES FROM SNEEZING OR COUGHING

*Figure 3* Human sources of air contamination.

## 2.3.0 Contaminant Sources Located Outside a Building

Contaminants from outside a building, especially in urban areas, can be a major cause of poor indoor air quality (*Figure 4*). Contaminants may come from aboveground urban or industrial air pollution sources or from belowground sources that include pesticides, fertilizers, and soil gases such as radon.

Outdoor contaminant sources include the following:

- Building ventilation air intakes located too close to exhaust gas, loading docks, or kitchen/bathroom exhausts
- Short-circuited HVAC system exhaust
- Exhaust from vehicles
- Exhaust and other airborne discharges from neighboring manufacturing plants
- Urban smog
- Radon
- Pesticides or fertilizers
- Moisture or standing water that promotes microbial growth
- Neighboring building construction or demolition

Most of these sources of contaminants are manmade and cannot be completely eliminated. They can develop long after a building has been constructed. Many can be avoided by attention to the design of outside-air intakes and their position within the building structure. For example, outside-air intakes that are located well away from any exhaust openings and are positioned as high on the building structure as possible can eliminate some of these problems. Consistent attention to building pressure is also important; negative pressures encourage outside air to enter through every crack and crevice.

### 2.3.1 Radon Contamination and Testing

Of all the sources of outdoor pollution, radon is the least understood. The most common source of radon is uranium in soil and rocks. Another common source of radon is well water. As naturally occurring uranium and thorium atoms decay, they release radon. Because it is inert, radon does not combine with other materials in the environment; therefore, it diffuses freely from its sources into the atmosphere.

Radon enters buildings through cracks in concrete, wall and floor joints, hollow concrete block walls, or openings such as those for sewer pipes and sump pumps (*Figure 5*). Radon from well water is released into the air in a home when water is used for cooking, bathing, and other activities. The level of radon can vary depending on a building's construction and location. Generally,

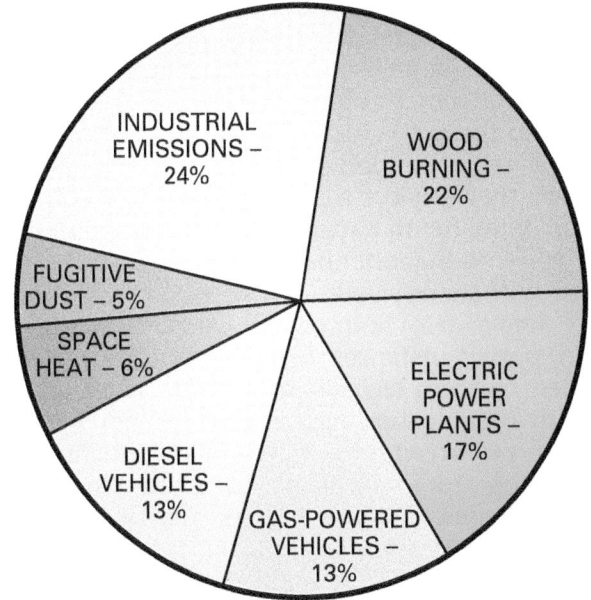

INDUSTRIAL EMISSIONS – 24%

WOOD BURNING – 22%

FUGITIVE DUST – 5%

SPACE HEAT – 6%

DIESEL VEHICLES – 13%

GAS-POWERED VEHICLES – 13%

ELECTRIC POWER PLANTS – 17%

*Figure 4* Sources of outdoor air pollutants.

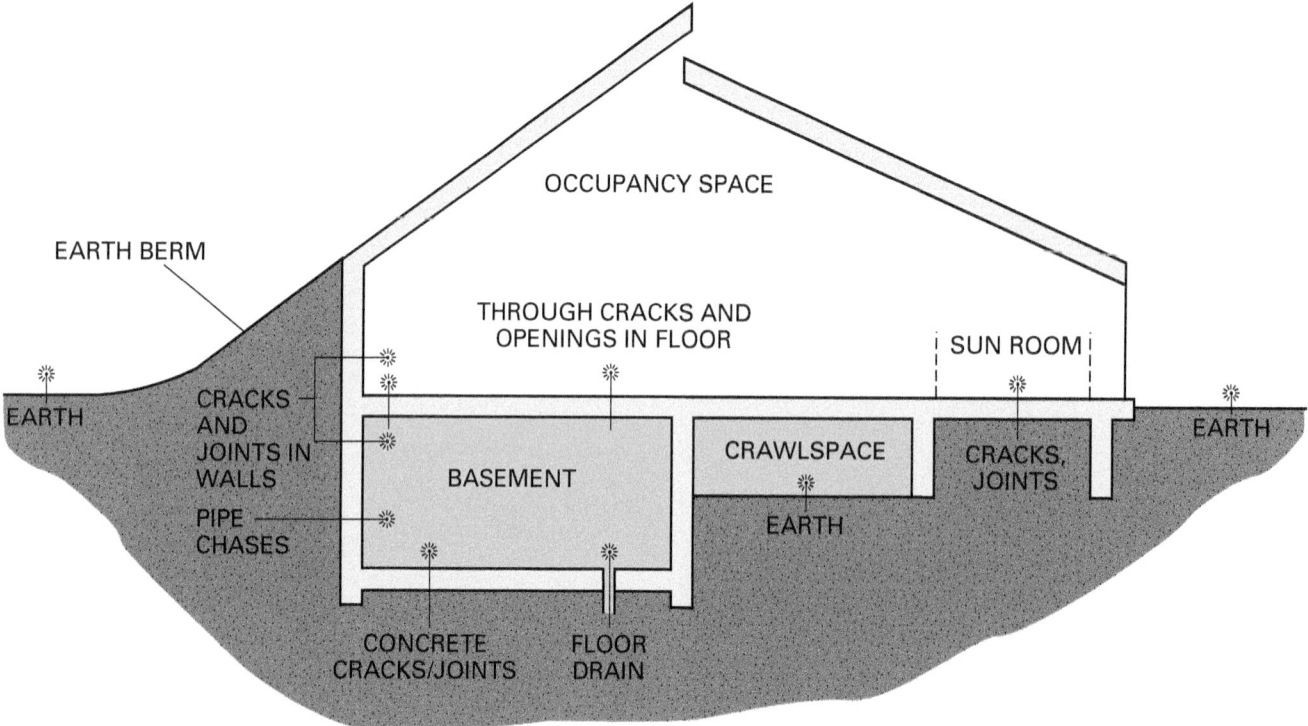

*Figure 5* Passage of radon gas into a building.

the fewer the cracks and openings in a building's foundation, the better the chances are of preventing radon entry. However, sealing the various parts of a building foundation is not completely effective by itself. Some openings in the building shell may not be accessible, and new openings can develop with time.

Radon produces radioactive decay products, called daughters, which emit elevated levels of alpha radiation. The daughter elements are chemically active metal elements, which allows them to become attached to tobacco smoke and dust particles in the air. When inhaled, these smoke and dust particles can lodge in the respiratory system where they subject the lung tissue to radiation. Each of the radon daughters has a relatively short half-life, so that after being deposited in the lung they will successively go through their radioactive decays in an hour or less. Currently, there are no reported instances of radon-related problems traced to short-term exposure. However, major health organizations agree that extended exposure to radon can increase the risk of lung cancer.

Testing for radon is normally done using electronic radon monitors and/or test kits. Inexpensive passive test kits (*Figure 6*) are available for short-term testing in residences. These devices do not require a power source to operate. Charcoal canisters and charcoal liquid scintillation detectors are the most commonly used passive types of detectors for short-term testing.

Approved test kits must have passed the EPA's testing program or be state-certified. Some of these test kits measure radon levels over two to three days; others measure it over one to three months. Follow the manufacturer's instructions for using the kit.

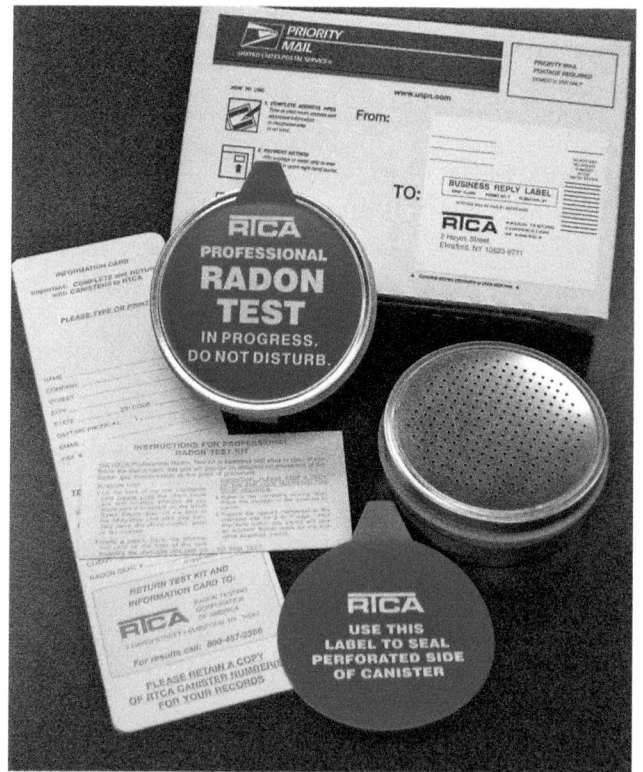

*Figure 6* Passive radon test kit.

The typical procedures for short-term testing with a passive test kit are as follows:

- The kit's detector is normally placed at a low level in the building where the highest concentration of radon is likely to be present. Place it at least 20 inches above the floor in a location where it will not be disturbed, away from drafts, high temperatures, high humidity, and exterior walls.
- If conducting a test lasting only two or three days, close all windows and outside doors at least 12 hours before beginning the test. Keep the windows and doors closed as much as possible during the entire test period.
- Heating and air conditioning system fans that recirculate air may be operated, but do not operate fans or other machines that bring outdoor air in during the test.
- Leave the test kit in place for the specified time.
- When the testing is finished, reseal the package and promptly send it for analysis to the lab specified on the package. The results will usually arrive within a few weeks.

Long-term testing is performed in much the same way as short-term testing, but it runs for more than 90 days. Alpha particle detectors (alpha track), electret ion chamber detectors, or continuous monitor detectors are commonly used for long-term radon tests.

Active testing devices, which require a power source to operate, are widely used by professional testers. These test devices are generally considered to be more reliable than passive devices. Continuous radon monitors, such as the one shown in *Figure 7*, are active sampling electronic devices that continuously measure and record radon decay products in the air. They may be used for both short- and long-term testing. After

a predetermined time has elapsed, the detector provides a measurement of the radon level in picocuries per liter (pCi/L). The monitor shown in *Figure 7* not only provides a digital display of the measured radon value; it also stores a report that can be printed out later using either an optional

## Airborne Activity Units for Radon

Early in the twentieth century, the unit for radioactivity (nuclear disintegrations per second) was the curie (Ci), named after the famous French physicist Marie Curie. One curie equals $3.7 \times 10^{10}$ disintegrations/second, based on the radioactivity of a gram of a type of radium. The modern radioactivity unit in the metric system is the becquerel (Bq), named for another pioneer in nuclear science, Antoine Henri Becquerel. One becquerel is equal to 1 disintegration per second.

Since radon is a free gas, standards concern themselves with its activity per unit volume of air. Thus, older airborne radioactivity standards given in picocuries per liter (pCi/L) can be converted to becquerels per cubic meter ($Bq/m^3$) using the following equation:

$$1 \text{ pCi/L} = 37 \text{ Bg/m}^3$$

Many home inspectors still report their findings in pCi/L, while scientists are more likely to use the modern unit of $Bq/m^3$.

*Figure 7* Continuous radon monitor.

thermal printer or PC software that is provided by the monitor's manufacturer.

There is some debate regarding the acceptable radon level. The Radon Abatement Act of 1988 established a national goal of achieving indoor radon levels that are no greater than outdoor levels. The EPA suggests a level not to exceed 4 pCi/L. Some countries are using higher threshold levels of about 11 pCi/L. The World Health Organization recommends lowering the reference level to 2.7 pCi/L. To date, the argument remains open as to whether the estimated health benefits would justify the costs associated with achieving a lower threshold level.

The most effective measures for reducing radon are the ones that limit soil gas entry into the building. One common method uses sub-slab depressurization (*Figure 8*). This method uses a radon mitigation exhaust system to reduce the pressure below the floor slab so that the air between the building substructure and the soil tends to flow out of, rather than into, the building.

Positive basement pressurization is a similar method used with some success that keeps the basement indoor air pressure slightly higher than the soil gas pressure. This prevents radon gas as well as other outdoor contaminants from entering the building, because the radon gas can only flow into an area of negative pressure.

## 2.4.0 Elements of a Building IAQ Survey

It is not easy to evaluate the performance of a building and its systems for IAQ problems. These problems can be complicated because of personal preferences, the complexity of the building design, and the fact that the evaluation standards and methods may be inconclusive. Ideally, the IAQ inspection team should include, at a minimum, an HVAC engineer and an industrial hygienist.

To successfully solve IAQ problems, an organized approach must be followed. The elements of this approach should include the following steps:

- Problem description
- Site visit and building walk-through
- Building HVAC and ventilation system inspection
- Air sampling and testing for specific contaminants
- Interpretation of test results and corrective actions

## Sub-Slab Depressurization System Monitor

A simple U-tube manometer mounted to the sub-slab depressurization system vent pipe in a basement is shown here. This method is widely used to monitor the operation of a sub-slab depressurization system fan. When the liquid levels in the legs of the manometer are uneven, a pressure differential exists, indicating the correct operation of the fan and system. Conversely, if the manometer liquid levels are even, the fan or system is not operating properly and needs repair.

### 2.4.1 Problem Description

To determine the scope of an IAQ problem, perform initial interviews with the building staff and/or the individual(s) who requested an investigation. The occupants' symptoms, as well as the location and duration of those symptoms, can be obtained via interviews or the use of questionnaires. *Appendix A* is an example of a form that can be used to conduct occupant interviews. This information should help to define and categorize the complaints. It may also help to determine whether the problem is localized to a particular part of the building, and identify any other relevant circumstances such as weather, time of day, day of week, building occupancy levels, or activities that improve the condition or make it worse.

### 2.4.2 Site Visit and Building Walk-Through

A site visit and building walk-through are needed to perform a preliminary evaluation of the overall condition of the building and its operating systems. A set of building plans and specifications is useful during this walk-through, as are any plans or drawings of major upgrades or changes. These drawings can be valuable for answering questions and pinpointing potential problem areas.

To be sure that no potential contamination source or component of the building is missed, use some type of formal checklist to document the results of the evaluation. An official EPA-produced guide entitled *Building Air Quality Guide: A Guide for Building Owners and Facility Managers* (1991) is the basis for the EPA's IAQ program. A 2002 update to the guide is titled *Indoor Air Quality Building Education and Assessment Model (I-BEAM)*.

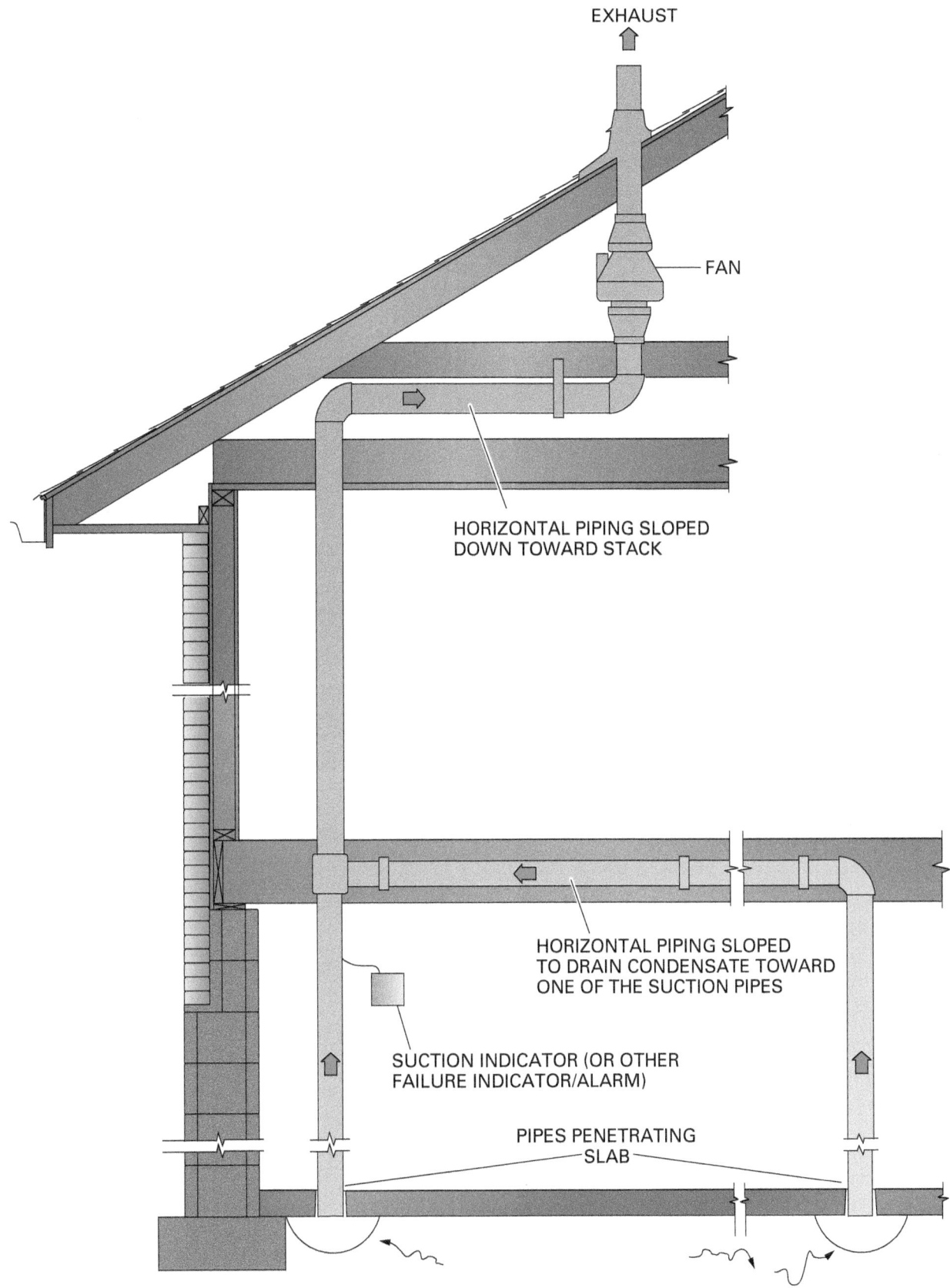

EXHAUST

FAN

HORIZONTAL PIPING SLOPED
DOWN TOWARD STACK

HORIZONTAL PIPING SLOPED
TO DRAIN CONDENSATE TOWARD
ONE OF THE SUCTION PIPES

SUCTION INDICATOR (OR OTHER
FAILURE INDICATOR/ALARM)

PIPES PENETRATING
SLAB

*Figure 8* Typical sub-slab depressurization (SSD) system.

The I-BEAM materials come on a CD, providing the latest information about IAQ problems and how to prevent or correct them. It contains training modules, checklists, references, and links to useful websites. Because this is a US government resource, it can be reproduced without permission. *Appendix B* is a reproduction of the *HVAC Checklist – Short Form* building IAQ evaluation available at the EPA's website.

Normally, a building's structure should be evaluated to determine if the following conditions exist:

- Tight building construction
- Low ventilation rates
- Rooms properly configured for their intended use
- Adequate separation between activities to control temperature, air movement, odors, liquids, noise, light, or vibration
- Recent addition of partitions or fire walls
- Reduced natural ventilation due to sealed windows or doors
- Windows or doors have been added, allowing outside pollutants to enter
- Asbestos, formaldehyde, and/or other harmful substances in building materials
- Insulation or recladding added to walls
- Recent use of caulks or sealants
- New openings for doors, ducts, and pipes that allow the transfer of dust, dirt, vapors, and/or odors between occupied spaces
- New furnishings or carpeting

### 2.4.3 Building HVAC Equipment and Ventilation System Inspection

After the structure has been evaluated, inspect the HVAC equipment, ventilation, and other building systems for potential sources of contamination. Check any building comfort-zoning scheme to make sure that system control of any one zone does not affect any other zone. Pay attention to the number of occupants in the various zones so that outside ventilation air is properly distributed without using excess energy. If the building uses a variable air volume (VAV) air distribution system, make sure that its layout and control arrangement allow for adequate ventilation to occupied areas when operating at minimum capacities. All systems should supply at least the minimum quantity of outside air required by the current issue of *ASHRAE Standard 62.1* or local codes, whichever is greater.

Even if a building has an adequate ventilation system, the system may not supply enough air to dilute some pollutants. This can happen if the system is shut down too often or for too long, such as during evenings and weekends. Other causes are poor air distribution or mixing, installation flaws, and incorrect system balance. Occupant intervention, such as home-made cardboard diffusers attached to ceiling vents to divert air, boxes or supplies piled near the vents, or office partitions that block airflow may also be the cause of inadequate ventilation.

> **WARNING!**
> A full system inspection may expose the inspector to contaminants such as mold and dust. Be sure to wear company-prescribed PPE, including gloves and respiratory protection, when performing this work.

Start by examining the ventilation system at the outside air intake. Air intakes are sometimes placed too close to building exhausts or sources of pollution, such as cooling towers, loading docks, and garbage bins. Temporary sources of pollutants often find their way to intake areas long after construction is complete. If necessary, use a smoke tube to check for airflow problems. Make sure that exhaust air is not being short-circuited and reintroduced into the building's air supply. Check for areas of negative pressure in the building that allow for migration of ground-level contaminants such as vehicle exhaust. Check the HVAC filtration system for efficiency, fit, condition, and the replacement schedule. Also check for the following:

- Dirty humidifier reservoirs
- Poorly draining condensate pans and trays
- Improperly pitched drain pans
- Missing or improperly designed traps
- Torn insulation
- Damp internal insulation
- Rusting internal surfaces
- Mold on internal surfaces
- Improperly maintained dampers, actuators, or linkages
- Incorrectly wired or inoperative fans and blowers
- Materials stored inside the air handling equipment

## 2.5.0 Air Sampling and Testing for Specific Contaminants

Testing can be time consuming and expensive, especially when testing for specific sources of contamination. Taking air samples and testing for various contaminants are tasks that are usually done over an extended period of time (a week, a month, or longer) in order to get enough samples to yield reliable results.

### 2.5.1 Gas Detectors and Analyzers

Gas detectors and analyzers provide an accurate way to detect and measure the presence of gaseous contaminants in the air. These devices can be either mechanical or electronic. Electronic instruments simplify testing and are more accurate. They can also perform automatic sampling and calculations. Some models can produce hardcopy reports of the date, time, and test results. Others can be connected to computers so that the test results can be transferred to and stored in a remote computer.

Electronic detectors and analyzers are available for use as stationary wall-mounted units or portable test instruments. Some units detect and/or measure only one specific kind of gas. Other units can detect and measure several gases. *Figure 9* shows some common gas detectors and analyzers.

Depending on the model, a gas detector may contain one or more sensors. Each sensor can detect a different gas. Typically, a sensing element includes three coated electrodes and a small quantity of an acid solution enclosed in a sealed plastic capsule or body. The three electrodes are related to the sensing, counting, and reference functions of the instrument. In use, the gases being sampled diffuse through a small opening on the sensor instrument's face or probe for application to the electrochemical sensor. There, a small current is generated that is proportional to the level of the gas being measured. Depending on the detector, the current sets off an alarm and is converted into a digital signal that represents the gas concentration for display on the device. On some detectors, the digital signal can also be used to drive a printer, or for display on a digital device.

The calibration of an electronic gas detector or analyzer should be done according to the manufacturer's schedule and procedure. Most units should be calibrated every six months using certified concentrations of test gases. The average life of a typical sensor is about two years. This type of sensor should be replaced at the interval specified by the manufacturer.

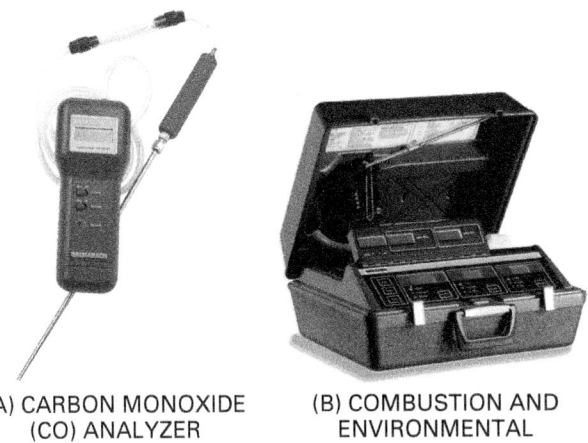

(A) CARBON MONOXIDE (CO) ANALYZER

(B) COMBUSTION AND ENVIRONMENTAL GAS ANALYZER

**PORTABLE DETECTORS**

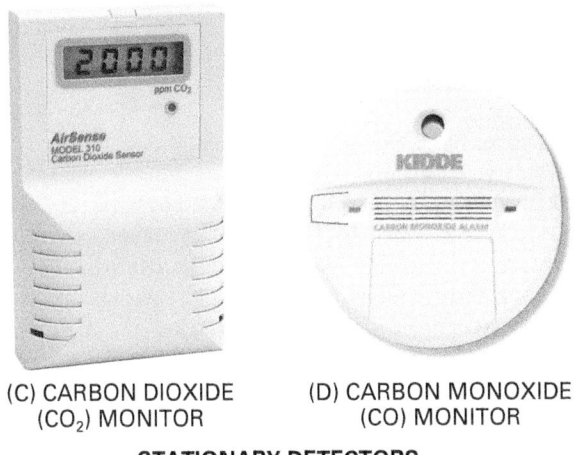

(C) CARBON DIOXIDE (CO$_2$) MONITOR

(D) CARBON MONOXIDE (CO) MONITOR

**STATIONARY DETECTORS**

*Figure 9* Gas detectors and analyzers.

The level of CO$_2$ is widely used as an indicator of suitability for human occupancy. Levels higher than 1,000 ppm can indicate ventilation problems. CO$_2$ sensors are commonly used for monitoring in non-industrial buildings. Some models can be used as ventilation controllers in demand-based ventilation control systems. When used as a ventilation controller, the CO$_2$ sensor determines the need for ventilation based on the CO$_2$ concentration. The sensor/controller then modulates the position of the building dampers to maintain acceptable ventilation. If a space is unoccupied, the CO$_2$ controller will set the air intake volume at a minimum setting that allows established ventilation rates to be maintained while reducing over-ventilation and saving energy.

The portable CO$_2$ detector/IAQ monitor shown in *Figure 10* detects real-time temperature and relative humidity (RH) readings as well as CO$_2$ readings. It has an audible alarm that sounds when measured CO$_2$ concentrations exceed a preset alarm threshold. It can also log data for

*Figure 10* Portable $CO_2$ detector/IAQ monitor.

downloading to a computer via a USB connector when the manufacturer's proprietary desktop software or mobile application is installed. Due to the variations in models and brands as well as the complexity of the instruments, the user must follow the manufacturer's operating manual carefully.

A CO detector is both a safety device and an IAQ device. Early detection of high CO is almost impossible without a CO detector. Stationary CO detectors are made for use in automated systems. The detectors are installed in strategic places throughout a building and will normally activate a contact closure and sound an alarm when a high level of CO is detected.

Portable CO detectors (*Figure 11*) are used either for testing HVAC combustion equipment, testing occupied areas, or locating the source of a CO leak. Most of these types of detectors provide a digital display of the CO level in seconds, rather than the minutes or even hours that are sometimes required for stationary CO alarms to respond. In many cases, a separate electronic sampling pump is used to draw a sample for the detector from a remote location, such as a confined space or a high ceiling. The detector shown here is very easy to use and can be operated without using a separate sampling pump. To ensure accuracy, it is important to allow it to stabilize at the temperature of the tested environment. A different type of detector is needed to test high-temperature airstreams, such as flue gases. The

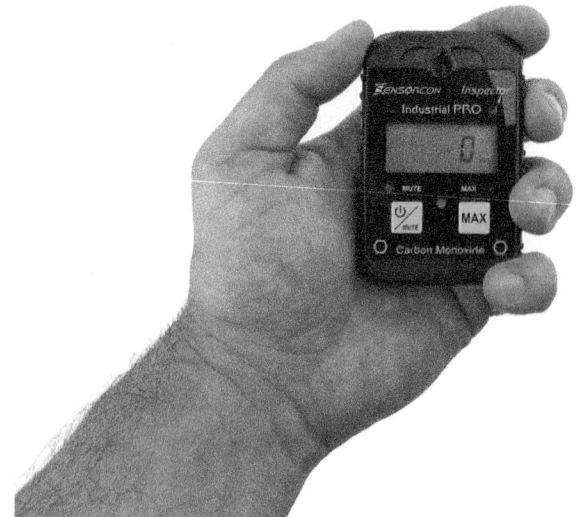

*Figure 11* Portable CO detector.

model pictured here is not used for testing hot combustion exhaust.

Volatile organic compound (VOC) detector/ sensors are often used to indicate non-occupant-related short-term changes in air contaminant levels. VOC sensors measure and react to a broad range of compounds. VOC sensors use an interactive, chemical-based oxidizing element. When this element is exposed to various compounds in the air, the sensor will vary its electrical resistance and provide an electrical output. This output is not specific to any one gas but reflects the total effect of a wide variety of compounds in the air. This type of sensor has no way of telling a harmful gas from a harmless gas; it can only indicate a change in the concentration.

VOC sensors are often tuned to the building space in which they are operating. Each of the individual sensors in the building is adjusted so that it provides a low output signal when the air in the space being monitored is considered to have good air quality. The sensor will then provide a higher signal output when there are more contaminants. The more contaminants, the higher the output signal is. Typically, the VOC sensor provides a one-in-five or one-in-ten scale output signal that represents the relative level of contamination. As a control, the sensor output can be used to activate an alarm. It can also be used to regulate building ventilation based on the actual level of pollutants sensed. This may or may not conflict with the established building ventilation scheme.

Many specialized detectors/analyzers are designed to detect and measure gases other than $CO_2$, CO, and VOCs. Some of the more common specialized detectors are identified as follows:

# Integrated Economizer and Demand-Control Ventilation

To save energy and still maintain good indoor air quality, some HVAC systems use demand-control ventilation (DCV) integrated with an economizer unit. The economizer determines whether free outside air can be used for cooling instead of running the system compressor. The DCV $CO_2$ sensor is located in the indoor occupied space. When the sensor detects an increase in space occupancy, it commands the system dampers to increase the amount of ventilation to the space.

Some systems also have an air quality sensor located outdoors that is used to determine whether the outdoor air is clean enough to bring indoors. If, according to the sensors, both the indoor air and the outdoor air are of poor quality, an alarm signal is generated which alerts the building maintenance personnel and notifies an automated building management system.

- *Oxygen detector* – Monitors the level of $O_2$ in the area. The normal level is 21 percent.
- *Hydrogen detector* – Monitors the $H_2$ in the area. Hydrogen is dangerous because of its flammability. It has a lower explosive limit in air of 4 percent. This is the level at which the air-gas mixture explodes.
- *Combustible gas detector* – Monitors the LP (propane, butane) and methane (natural) gas in the area. It is used to check for leaks.
- *Air pollution detector* – Monitors various gases that cause air pollution and endanger lives. It typically detects CO, $H_2$, alcohol, gasoline fumes, cigarette smoke, and exhaust fumes. Also referred to as an IAQ detector.
- *Refrigerant gas detector* – Provides area monitoring and early warning of refrigerant leaks.
- *Ammonia detector* – Provides area monitoring and early warning of ammonia ($NH_3$). Ammonia is poisonous and is dangerous even at very low levels.

## 2.5.2 Air Sampling

Basic air sampling typically involves measuring carbon dioxide ($CO_2$) levels inside a building as well as outside. Sampling usually begins in the morning, with samples taken at all sample locations, including outdoors. The $CO_2$ levels are recorded in ppm according to the specific location and time of day. The morning measurements are then compared to those taken throughout the day from all the sample locations.

The number of sampling periods is determined by the types and duration of activities that occur in the building, including lunch breaks. Samples should be taken at the end of the work day, just before everyone starts leaving the building, because the increased traffic usually results in more air exchange than during the main part of the day. To get a complete picture of the building air pattern, $CO_2$ measurements should also be made at night when the building is typically unoccupied.

The building temperature and humidity should be checked at various times and locations throughout the day and night. The airflow at the building air outlets and return grilles should also be checked. This will help determine whether the air outlets are working and whether the airflow is directed properly.

## 2.5.3 Testing for Microbial Contaminants

Microbial contaminants include bacteria, fungi, viruses, algae, insect parts, pollen, and dust. The sources of such contaminants include wet or moist walls, duct, duct liner, fiberboard, carpet, and furniture. Testing for specific microbial contaminants must be performed by qualified personnel such as microbiologists or industrial hygienists.

Even though sampling for airborne microorganisms is common, the results are often inconclusive. Airborne spores are often only present in large quantities for short periods of time. Spores released into the air depend on the current growing conditions. If growing conditions are intermittent, as is often the case in HVAC systems, then the release of the spores is also intermittent. Significant levels of airborne microbial contamination will be found only if sampling occurs during the time that spores are being released into the air.

Surface sampling is a non-airborne method used to test for the presence of microbial contamination. Also referred to as microbial sampling, it provides a historical reference of previous growth and indicates the potential for future growth. Surface sampling uses sampling strips, each with a pad on one end that contains a growth media for a wide range of organisms. The strip is activated, and one square inch of HVAC duct or other building surface is wiped with the pad. Two strips, each containing a different growth medium, are used at each location. After taking the sample, the strips are placed in sterile envelopes and returned to the manufacturer's laboratory for incubation and analysis. A report is returned with the results stated in colony counts per square inch. The results are also rated on a multistep severity index ranging from very low to severe.

To reduce or eliminate occupants' complaints resulting from high levels of surface microbial contamination, clean and sanitize building HVAC system components or building surfaces.

### 2.5.4 Interpreting Test Results and Taking Corrective Actions

Corrective actions are determined by what has been found during the site/building walk-through, HVAC and ventilation system inspections, and/or the air sampling and testing phases of evaluation. Depending on the complexity of the problem, it may be necessary to call in expert help, such as that of an industrial hygienist.

### Additional Resources

*ASHRAE Standard 62.1, Ventilation for Acceptable Indoor Air Quality.* Current edition. Atlanta, GA: American Society of Heating, Refrigerating and Air-Conditioning Engineers (ASHRAE).

*Building Air Quality, a Guide for Building Owners and Facility Managers.* Current edition. Washington, DC: U.S. Environmental Protection Agency.

*Indoor Air Quality.* Current Edition. Chantilly, VA: Sheet Metal and Air Conditioning Contractors National Association (SMACNA).

*Indoor Air Quality in the Building Environment.* Ed Bas. 1993. Troy, MI: Business News Publishing Company.

### 2.0.0 Section Review

1. Off-gassing is associated with which of the following contaminants?

    a. Legionella pneumophilia bacteria
    b. Aspergillus mold spores
    c. Environmental tobacco smoke (ETS)
    d. Volatile organic compounds (VOCs)

2. Occupants of a building affect IAQ by consuming oxygen and exhaling _____.

    a. ammonia ($NH_3$)
    b. nitric oxide (NO)
    c. carbon dioxide ($CO_2$)
    d. carbon monoxide (CO)

3. Which of the following is a source of radioactive air pollution?

    a. Ozone from a photocopier
    b. Radon in soil and rocks
    c. Formaldehyde in particle board
    d. Asbestos in building insulation

4. Ideally, an IAQ inspection team should include, at a minimum, an HVAC engineer and _____.

    a. an industrial hygienist
    b. the building's owner
    c. an occupants' representative
    d. an EPA officer

5. To obtain a complete picture of the building's air quality, $CO_2$ samples should include those taken _____.

    a. at the front door
    b. on the roof
    c. at night when it is unoccupied
    d. only once per day

## 3.0.0 ACHIEVING ACCEPTABLE INDOOR AIR QUALITY

### Objective

Explain how acceptable IAQ can be achieved.
  a. Explain how the building design can affect IAQ.
  b. Explain how ventilation, temperature, and humidity control affect IAQ.
  c. Explain how to control chemical and microbial contaminants.

### Performance Task

  1. Perform a building indoor air quality (IAQ) inspection/evaluation.

Acceptable indoor air quality can be achieved through awareness and control of the following related areas:

- Initial building design
- Ventilation control
- Thermal comfort control
- Control of chemical and microbial contaminants
- Scheduled building and equipment maintenance

### 3.1.0 Initial Building Design

The methods used to conserve energy in a building can impact indoor air quality. Ideally, they will improve the indoor air and result in better comfort and productivity. In some cases, however, they can degrade the indoor conditions and result in discomfort, sickness, and lost productivity. As buildings become more energy efficient, they become less forgiving environmentally. Energy conservation is necessary, but it must be done in a manner that also provides for acceptable IAQ.

Owners, architects, and engineers can achieve good indoor air quality, beginning with the design of a new building, with the implementation of the following guidelines:

- If possible, locate the building on a hill rather than in a valley. Sites should avoid major highways and parking lots. Locate upwind from a power plant, chemical plant, or other industrial pollution source.

- Incorporate HVAC equipment and mechanical ventilation systems that best meet the needs of the occupants in terms of comfort and performance, while reducing energy consumption and costs.
- Use building materials and furnishings that will keep indoor air pollution to a minimum.
- Consider outside air duct locations, prevailing wind conditions, and neighboring pollutant sources in order to get proper ventilation damper orientation and adequate filtration systems incorporated into the HVAC equipment.
- Ensure that combustion appliances are properly vented and receive enough supply air.
- Provide proper drainage, and seal foundations.
- Use radon-resistant construction techniques.
- In landscaping, avoid shrubs and trees that produce heavy pollen and can aggravate allergies, such as olive, acacia, oak, pine, and maple trees.
- Provide adequate HVAC system access, lighting, and work platforms.

### 3.2.0 Controlling Ventilation and Thermal Comfort

Controlling the flow, temperature, and humidity of the air that circulates through a building is essential for maintaining good indoor air quality. It also affects the health and comfort of the building's occupants.

#### 3.2.1 Ventilation Control

Ventilation control can correct or prevent poor indoor air quality. Both outdoor air and recirculated air is available for this purpose.

The use of additional outdoor ventilation is currently one of the best methods of correcting and preventing problems related to poor IAQ. Adequate ventilation pertains more to the level of $CO_2$ in a building than the level of oxygen in the air. When indoor air is stale or stuffy, or when drowsiness sets in, it is not a shortage of oxygen causing the problem; it is an excess of $CO_2$. If $CO_2$ or other pollutants accumulate inside a building, ventilation control is used to bring in more outside air for dilution. Even if a specific contaminant such as formaldehyde is identified as the cause of a problem, dilution can still be the most practical way of reducing exposure.

Outdoor ventilation air should be adequately distributed to all areas of a building during the entire time it is occupied. In many buildings that have their heating setpoint reduced at night, a morning warm-up program keeps the outdoor air dampers closed until a specific temperature is reached. Then the system and dampers return to their normal ventilation program.

The ventilation requirements of residential and commercial buildings are constantly changing in response to the latest IAQ standards established by ASHRAE. *ASHRAE Standard 62.1, Addendum n,* recommends minimum outdoor air ventilation rates for occupied spaces. These rates are not a one-size-fits-all determination. Factors such as space usage, floor area, number of occupants and occupant density, level of human activities, and environmental factors all contribute to the recommended ventilation rates. ASHRAE uses a series of equations to calculate minimum recommended air flows and outside air exchanges.

Proper balancing of the air supply and exhaust system may be all that is needed to achieve adequate airflow or air quality in buildings where new walls, partitions, or room dividers have been erected. Proper balancing may also cure problems in which insufficient amounts of outdoor makeup air have created negative pressure areas in a building. These are areas that allow untreated air and/or contaminants to infiltrate from outside. They also allow for the migration of odors or contaminants between areas within a building.

Some other correctable causes of poor airflow or low-quality indoor air related to ventilation problems are as follows:

- Closed or obstructed air outlets or diffusers that prevent adequate airflow to the supplied area.
- Outdoor dampers that are operating improperly. If mechanical ventilation is on, check the outdoor air dampers to make sure they are open. If the building is lightly occupied, make sure the outdoor dampers do not close beyond the minimum position. These dampers may have been set to the closed position deliberately to save energy, or they may have been closed automatically by a faulty control device.
- Supply or exhaust fans that are inoperative; blowers or fans rotating in the wrong direction.
- Filters in the air handling units are dirty.

Proper ventilation does not always guarantee good IAQ. While $CO_2$ levels are often used as the basis for controlling building ventilation, this does not necessarily mean that the building is free of IAQ problems based on this information alone. Suppose the outside air has more $CO_2$ than might be expected. Outdoor concentrations of 400 ppm $CO_2$ have regularly been recorded in large urban areas. This can create a problem if outside air is drawn into a building to reduce its $CO_2$ level, and the building is located in a busy traffic area with high $CO_2$. Another example can be found in supermarkets, which need extra ventilation during peak customer periods, but cannot lower the ventilation too much at night even if the $CO_2$ level is low, because chemicals and cleaners used at night can be absorbed into the food if the fumes are not diluted and dispersed.

There is some controversy as to whether building air should be recirculated to make up part of a building's demand for ventilation. Recirculated air is sometimes considered less healthy than outside air. However, recirculated filtered air is used in most ultraclean environments, such as manufacturing clean rooms. In some cases, recirculated air is actually better for the occupants than outdoor air, especially when a building is located in a highly polluted urban area. Air that has passed through high-efficiency or ultra-high-efficiency filters and through a charcoal-based adsorption filter bank can be far cleaner than air from outdoors.

Recirculation substitutes clean recirculated air for some portion of the outdoor air normally used to ventilate a building or space. For instance, assume the level of outdoor air required in a building is 15 cfm per person. To meet this requirement, a reduced outdoor airflow volume of 5 cfm/person might be used along with 10 cfm/person of filtered recirculated air.

Filtration methods can vary widely. Most use multistage filtering that obtains filter efficiencies needed to achieve good IAQ. One advantage of using recirculated air is that the size of building boilers, chillers, and other mechanical equipment can be smaller than the sizes needed to accommodate the use of higher outdoor airflow levels. This is possible because recirculated air has already been processed through the HVAC system.

## Air Pressure Relationships Within Healthcare Facilities

Most states require hospitals, health clinics, and other healthcare facilities to control and maintain certain air pressure relationships among the different rooms or activity areas in the facility. This is done to protect patients occupying these areas from exposure to odors and contamination from germs emanating from other areas. For example, laboratories handling infectious materials are maintained at a negative pressure relative to other rooms. Operating rooms and patient rooms are usually maintained at a positive pressure in relation to their surrounding areas.

## 3.2.2 Thermal Comfort Control

The human body is not always sensitive enough to tell the difference between slight thermal discomfort and actual IAQ problems. People may complain that the room is stuffy or the air quality is poor rather than feeling it is too warm. Generally, IAQ problems are judged to be worse when the ambient room temperature is above 75°F (24°C). Stuffiness can be the result of temperature stratification, which occurs when air distribution is inadequate. When air is improperly distributed, poor mixing occurs. Thermostats may not detect the changing room conditions until the temperature in the room becomes uncomfortable.

Humidity control is also related to thermal comfort. In summer, the human body can tolerate higher temperatures if the humidity level is lower than 60 percent RH. In winter, people are comfortable at lower temperatures when the humidity level is greater than 30 percent RH.

## 3.3.0 Controlling Contaminants

Chemical contaminants can be generated in a building during routine activities such as painting and cleaning. Microbial contaminants can grow as a result of something as simple as water draining from a humidifier. In either case, the contaminants must be controlled so that indoor air quality does not deteriorate, and the health of human occupants is not affected.

## 3.3.1 Controlling Chemical Contaminants

IAQ problems related to chemical contaminants can be derived from the organic gases commonly emitted from building materials as well as cleaning and maintenance products. They can also be derived from pesticides used to kill building pests or those used on lawns and gardens that then drift or are tracked into a building. Exposure to chemical contaminants can usually be eliminated or adequately controlled if the following guidelines are observed:

- Use local exhaust systems where needed to trap and remove contaminants generated by specific processes or equipment such as office machines. Exhaust room air outdoors from areas where solvents are used.
- Areas being remodeled, painted, or carpeted should be temporarily isolated from other occupied areas in the building. This includes temporarily isolating any related HVAC systems, if possible. As scheduling permits, perform this

# Thermal Comfort

A building's HVAC system must operate to maintain the indoor dry-bulb temperature and relative humidity within a comfort zone and provide adequate air circulation. Shown here are the parameters for a generally accepted comfort zone plotted on a portion of a psychrometric chart. The area bounded by the comfort zone represents the temperature and humidity conditions of air that will satisfy 80 percent of the occupants most of the time. Comfortable dry-bulb temperatures typically range from 68°F to 78°F (20°C to 26°C), and a comfortable relative humidity range is from 30 to 60 percent.

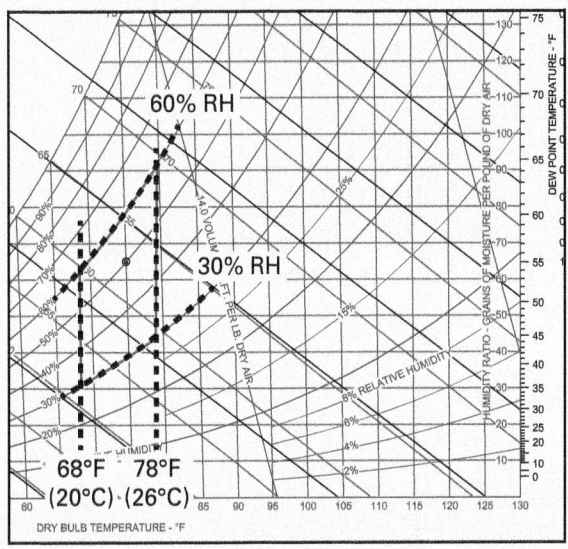

**Figure Credit:** Courtesy of Hands Down Software, www.handsdownsoftware.com

type of work during evenings and weekends. Also, supply the maximum amount of ventilation to the areas on a 24-hour basis to help eliminate any contaminants.

- Apply pesticides and disinfectants only when a building is unoccupied, and then thoroughly ventilate the building before it is reoccupied.
- Apply paints, paint strippers, and other solvents, wood preservatives, aerosol sprays, and cleaning products strictly according to the manufacturer's directions, and only in the recommended quantities.
- Make sure that outside air intakes or openings are not located close to places where motor vehicle or other emissions collect.

# Construction Dust Containment

Airborne dust generated during a remodeling job can be unhealthy for the occupants of a building. This is true especially for concrete dust containing silica, which can contribute to a respiratory condition called silicosis. The OSHA standards have recently changed and are more restrictive concerning the containment of silica. IAQ standards require that reasonable precautions be taken to isolate construction dust to the work area only.

Because of increased concern, new and better products have become commercially available that make it easier to isolate work areas. Two such products are reinforced vinyl dust doors used to temporarily block doorways and arches, and portable wall systems used to partition off the work area. It is recommended that any heating/cooling forced-air distribution system grilles and registers in the work area be sealed off so that construction dust is not circulated through the system to the rest of the building.

For more information about OSHA's latest regulations, search the term *silica dust control* at **www.osha.gov**.

---

### 3.3.2 Controlling Microbial Contaminants

Indoor air quality problems related to microbial contaminants are due to wet or moist sources in building materials, furnishings, and/or equipment. These contaminants can usually be eliminated or controlled by observing the following guidelines:

- Maintain relative humidity between 30 and 60 percent in all occupied spaces of the building. Although low humidity negatively affects comfort, higher humidity levels promote microbial growth.
- During the summer, make sure cooling coils are operating at low enough temperatures to properly dehumidify the conditioned air.
- Install and use fans in kitchens and bathrooms.
- Regularly check that humidifiers, filters, and sump pumps are clean. Water draining from this equipment can stagnate and promote the growth of microbial contaminants.

- Promptly detect and repair all water leaks. Eliminate or clean areas where water collects.
- Prevent and/or correct any causes of stagnant water accumulation around cooling coils and air handling units.
- If contamination has occurred in the plenum or ductwork downstream from a heat exchanger, add filtering downstream to better filter the air before it is introduced into occupied areas.
- Clean and disinfect surfaces where the accumulation of moisture has caused microbial growth, such as in drain pans and cooling coils. Use only biocide agents approved for this purpose.
- Replace, rather than disinfect, any porous water-damaged furnishings, including carpets, upholstery, drywall, and ceiling tiles.

## Additional Resources

*ASHRAE Standard 62.1, Ventilation for Acceptable Indoor Air Quality.* Current edition. Atlanta, GA: American Society of Heating, Refrigerating and Air-Conditioning Engineers (ASHRAE).

*Building Air Quality, a Guide for Building Owners and Facility Managers.* Current edition. Washington, DC: U.S. Environmental Protection Agency.

*Indoor Air Quality.* Current Edition. Chantilly, VA: Sheet Metal and Air Conditioning Contractors National Association (SMACNA).

*Indoor Air Quality in the Building Environment.* Ed Bas. 1993. Troy, MI: Business News Publishing Company.

## 3.0.0 Section Review

1. In the initial design of a building, energy conservation must be achieved in a way that also provides _____.

   a. environmental improvements
   b. acceptable indoor air quality
   c. a low intake/exhaust ratio
   d. maximum return on investment

2. One of the best methods currently used to correct and prevent problems caused by poor IAQ is _____.

   a. additional outdoor ventilation
   b. interior oxygen supplementation
   c. carbon dioxide recirculation
   d. ozone purge and regeneration

3. When applying a pesticide in a building you should _____.

   a. cover all the exhaust vents for the area being treated
   b. never isolate the HVAC system for the area being treated
   c. thoroughly ventilate the building before it is reoccupied
   d. keep the building's humidity level between 60 and 80 percent

### 4.0.0 IAQ EQUIPMENT AND SOLUTIONS

## Objective

Identify IAQ-related HVAC equipment and describe specific activities used to address IAQ problems.

a. Identify and describe HVAC equipment and devices used to improve IAQ and/or energy consumption.
b. Explain how air distribution systems can contribute to poor IAQ and how these problems are addressed.
c. Describe the liability that HVAC contractors may accept by servicing HVAC systems.

## Performance Task

3. Use a manufacturer's humidifier capacity chart to find the humidifier capacity needed for various building types and sizes.

## Trade Terms

**Arrestance efficiency:** The percentage of dust that is removed by an air filter. It is based on a test where a known amount of synthetic dust is passed through the filter at a controlled rate, then the weight of the concentration of dust in the air leaving the filter is measured.

**Desiccant:** A substance that has a high affinity to water vapor, creating and maintaining a state of dryness near it. Desiccants for HVAC purposes have low or no chemical reactivity.

**Dust spot efficiency:** The percentage of dust that is removed by an air filter. It is the number that is normally referenced in the manufacturer's literature, filter labeling, and specifications. The atmospheric dust spot efficiency of a filter is based on a test where atmospheric dust is passed through a filter, then the discoloration effect of the cleaned air is compared with that of the incoming air.

**High-efficiency particulate air (HEPA) filter:** An extended media, dry-type filter mounted in a rigid frame. It has a minimum efficiency of 99.97 percent for 0.3-micron particles when a clean filter is tested at its rated airflow capacity.

This section focuses on the equipment, systems, and activities that are used to address IAQ problems. It describes energy-efficient equipment and devices that improve IAQ while also minimizing energy consumption. It also discusses the proper cleaning and maintenance of air distribution systems, including various methods of duct cleaning. In addition, it describes the liability that HVAC contractors accept when agreeing to service HVAC systems.

### 4.1.0 IAQ and Energy Efficient Systems and Equipment

A wide range of equipment and systems is used to assure good IAQ in an energy-efficient way. Among the methods that can be used to improve indoor air quality are automated building management systems, improved air handling units, unit ventilators, and air filtration equipment.

## HVAC Equipment Design

With current concerns about indoor air quality, manufacturers of new HVAC equipment should be constructing their equipment so that maintenance personnel have easy access to inspect and clean the heat exchanging components, drip pans, and similar items that are likely to collect dirt and other forms of contamination. If you find that this is not the case, you or your employer should notify the manufacturer about your IAQ maintenance concerns. One way to get the manufacturer's attention in this regard is to put a note on the invoice if new equipment is delivered to the job site that does not meet these criteria. Express your IAQ maintenance concerns about the equipment and suggest recommendations for correcting the problem.

### 4.1.1 Automated Building Management Systems

To meet IAQ standards, most manufacturers of HVAC equipment and control systems have developed automated systems that control and monitor the ventilation air in a building. These systems communicate directly with the building's air handling and VAV systems (*Figure 12*). Modern applications work from desktop or handheld devices, and data resides in the Cloud to allow access and monitoring from any location. Many

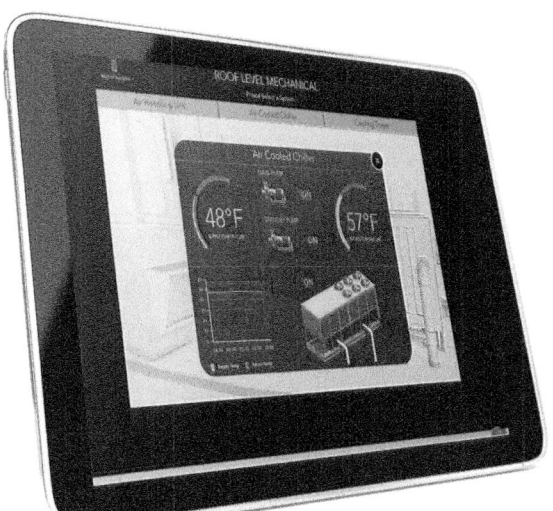

Figure 12 Automated building management system.

software developers are offering programs that permit controlling all aspects of intelligent buildings utilizing the Internet of Things concept.

The methods vary, but most automated systems involve the use of specialized computer hardware and software that make the air handling units and VAV systems operate more efficiently. These systems tie together building heating, cooling, and ventilation equipment in order to provide good IAQ without wasting energy. Through software management of the individual zones, the ventilation, temperature, humidity, and other desired zone parameters are monitored and controlled so that the best operating scheme is selected for each zone in the building. Each zone can communicate its need for heating, cooling, and ventilation, and a central controller then allocates resources or establishes priorities to satisfy zone requirements.

Normally, control of the outdoor and exhaust air dampers is provided in these systems so that the dampers are constantly modulated to maintain the ventilation airflow needed to satisfy IAQ requirements for the building. These systems also incorporate special building purge modes to cover temporary IAQ problems. Purge modes can allow for the maximum circulation of ventilation air in the building over extended time periods. The purge mode is typically used to purge the building air prior to occupying a new building. It may also be used to dilute increased levels of odors or chemical vapors that occur when activities such as painting or carpet cleaning are being performed.

> **NOTE**
>
> Building automation systems are presented in more detail in NCCER Module 03405, *Building Management Systems.*

### 4.1.2 Air Handling Units

Newer air handling units respond to the need for improved operation and efficiency by providing more fresh air and better service access, humidity control, and filtration. Many units are modular, like the one shown in *Figure 13*. This allows the unit to be customized to meet the specific IAQ and energy needs required by each customer. Adding, removing, or changing the components in the unit can be accomplished without the need for major modifications. Modular construction provides a hedge against any modifications that may be needed in the future in response to revisions in air quality and energy conservation standards and codes.

### 4.1.3 Unit Ventilators

Unit ventilators (*Figure 14*) provide ventilation and temperature control for individual rooms in a building. Unit ventilators have been used for years in offices, schools, and similar buildings. Newer unit ventilators have been vastly improved to provide better indoor air ventilation accompanied by

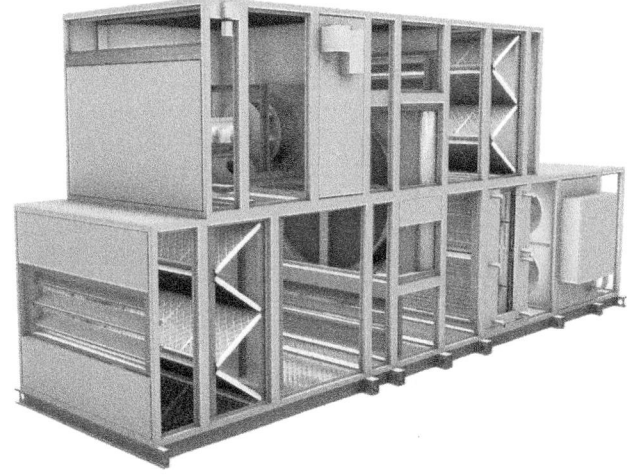

Figure 13 Modular air handler.

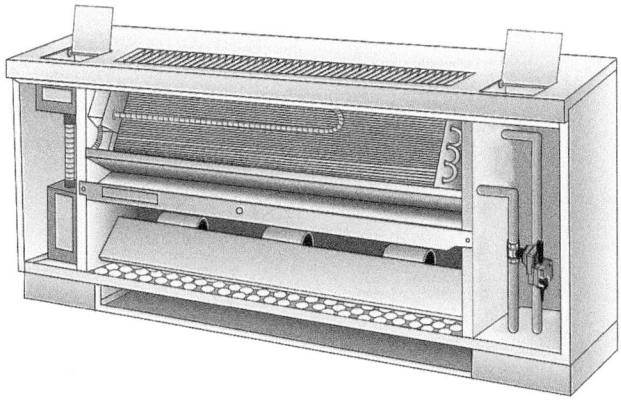

Figure 14 Unit ventilator.

energy conservation. The units can usually be controlled either from a local control panel in the unit or room or by digital control signals applied from a remote automated building management system. Unit ventilators come in a variety of sizes. A typical room unit is able to provide between 750 and 1,500 cfm of outside air to the conditioned space. Unit ventilators can be equipped with heat exchangers.

### 4.1.4 Air Filtration Equipment

Normal air contains varying amounts of natural and manmade foreign materials. Dirt and pollens contained in outdoor air enter a building as a result of infiltration. Indoor air is recirculated in a building many times, picking up dust, dirt, smoke, and other contaminants. This is especially true in airtight buildings. Airborne bacteria and mold spores are common both indoors and out.

*Figure 15* shows the relative sizes of some common particles that contaminate air. As shown, these particles have diameters that range in size from smaller than 0.001 micron to larger than 10 microns. About 99 percent of airborne particles are less than 1 micron in diameter. The remaining 1 percent consists of larger, heavier particles such as dust, lint, and pollen. Several types of air filters can remove contaminants from the air, making the air cleaner and healthier to breath. Mechanical filters, adsorption filters, and electronic air filters are in common use.

> **NOTE**
> A micron is a unit of length equal to one-millionth of a meter, or about one 25,400th of an inch. Engineers and scientists use the lower-cased Greek letter mu (μ) and meter unit (m) for its unit symbol (μm). The metric system uses the proper term *micrometer* for this unit of length.

Some filters can inadvertently increase the level of microbes. Once trapped in the filter, microbes can grow on the filter material. Unless filters are replaced frequently, or incorporate a safe and effective antimicrobial agent, they can become a major source of IAQ problems.

*ASHRAE Standard 52.2, Method of Testing General Ventilation Air-Cleaning Devices for Removal Efficiency by Particle Size* (2017), defines different methods for testing and rating air filters. This standard is a combination of what were

## Wet Air Filters

Wet air filters in an HVAC system can be a breeding ground for biological contaminants. Wet filters must be removed and replaced with dry new ones. Before installing the new filters, it is important to first determine why the filters were wet and correct the cause of the problem.

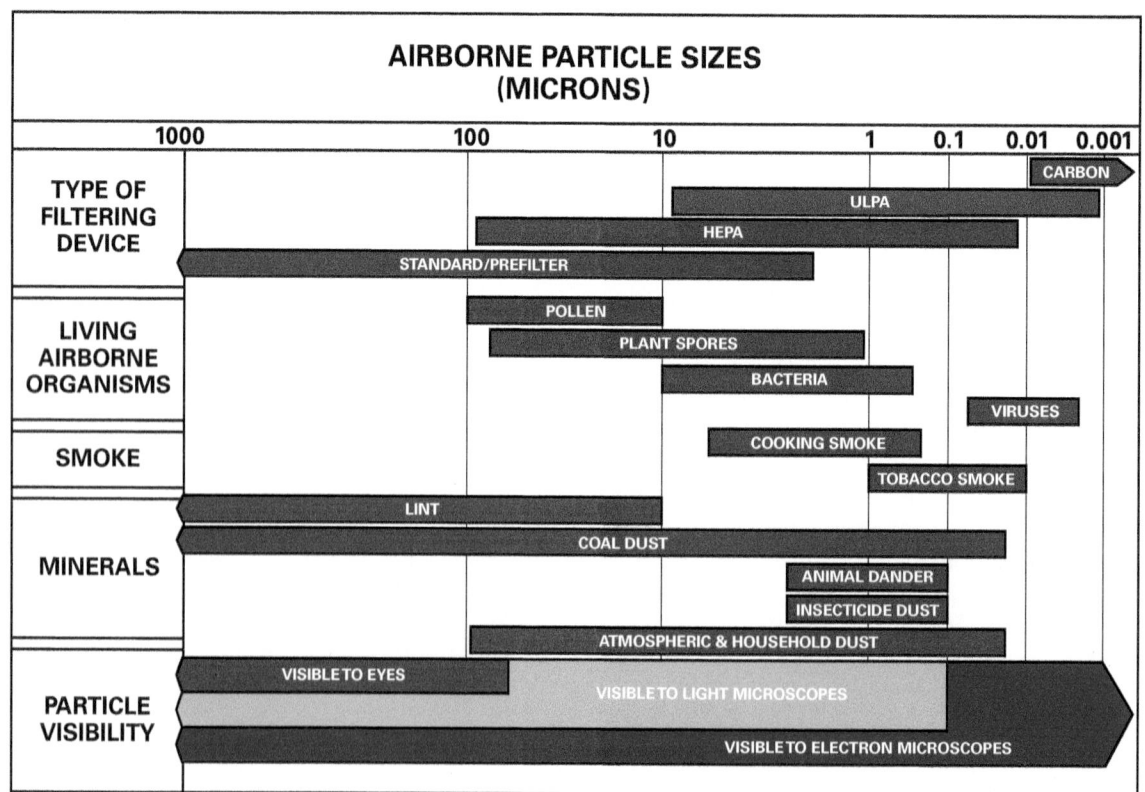

*Figure 15* Particle size in microns.

previously two separate standards. One was *Standard 52.1, Gravimetric and Dust-Spot Procedures ...* (1992), which rated filters on the basis of an overall percentage of dust spot efficiency and arrestance efficiency. The other was the original *Standard 52.2, Method of Testing General Ventilation Air-Cleaning Devices for Removal Efficiency by Particle Size* (1999), which rated filters on the basis of particle size, using the minimum efficiency reporting value (MERV) system. MERV ratings range from 1 to 20, with 20 being the highest. *Table 2* shows a comparison of *ASHRAE Standards 52.1* and *52.2*.

ASHRAE is continually updating their *Standard 52.2* to address filter performance in terms of the filter's ability to remove particles from the airstream, the total dust-holding capacity with arrestance (weight efficiency) for the filter, and the filter's resistance to airflow. Recent revisions have adjusted the MERV table ranges, addressed filter types and operating conditions allowed, and refined the instrumentation specifications.

Various kinds of mechanical air filters (*Figure 16*) are available for removing airborne particles. Most of these filters are disposable, but the metal wire screen and open-cell foam types of conventional filters can be washed and reused. *Table 3* reviews the performance characteristics of the different types of mechanical filters.

Adsorption filters remove gaseous vapors. The most common adsorption filter is the activated charcoal filter. This filter blocks materials with high molecular weights and allows materials with low molecular weights to pass through. When this type of filter becomes loaded, it must be replaced or regenerated to prevent off-gassing

> **NOTE**
>
> *Table 2* includes some abbreviations that may be unfamiliar:
>
> **ULPA:** Ultra Low Penetration Air filters remove at least 99.999% of particles larger than 0.1 μm from air.
>
> **DOP:** Dispersed Oil Particulate identifies dispersed aerosols used for testing a high-efficiency particulate air (HEPA) filter.

**Table 2** Filter Applications Guidelines

| MERV Std 52.1 | Average ASHRAE Dust Spot Efficiency Std 52.1 | Average ASHRAE Arrestance Std 52.1 | Particle Size Ranges | Typical Applications | Typical Filter Type |
|---|---|---|---|---|---|
| 1–4 | < 20% | 60 to 80% | > 10.0 μm | Residential—Minimum<br>Light Commercial—Minimum<br>Equipment Protection—Minimum | Permanent / Self Charging (passive)<br>Washable / Metal,<br>Foam / Synthetics<br>Disposable Panels<br>Fiberglass / Synthetics |
| 5–8 | < 20 to 35% | 80–95% | 3.0–10.0 μm | Industrial Workplaces/ Commercial<br>Residential—Better<br>Paint Booth—Finishing | Pleated Filters<br>Extended Surface Filters<br>Media Panel Filters |
| 9–12 | 40 to 75% | > 95 to 98% | 1.0–3.0 μm | Residential—Superior<br>Industrial Workplaces—Better<br>Commercial Buildings—Better | Non-Supported / Bag<br>Rigid Box<br>Rigid Cell / Cartridge |
| 13–16 | 80 to 95%+ | > 98 to 99% | 0.30–1.0 μm | Smoke Removal<br>General Surgery<br>Hospitals & Health Care<br>Commercial<br>Buildings—Superior | Rigid Cell / Cartridge<br>Rigid Box<br>Non-Supported / Bag |
| 17–20[1] | 99.972<br>99.992<br>99.9992 | N/A | ≤ 0.30 μm | Hospital Surgery Suites<br>Clean Rooms<br>Hazardous Biological Contaminants<br>Nuclear Material | HEPA<br>ULPA |

*Note:* This table is intended to be a general guide to filter use and does not address specific applications or individual filter performance in a given application. Refer to manufacturer test results for additional information:
(1) Reserved for future classifications
(2) DOP Efficiency
*Source:* National Air Filtration Association.

**Table 3** Mechanical Filter Performance Characteristics

| Filter Type | Removes | Efficiency | Typical Applications | Notes |
|---|---|---|---|---|
| Conventional Filter | Particles ≥10 microns (or as small as 5 microns if filter is coated with a tackifier) | Low | Used as prefilters for higher-efficiency filters. | A tackifier is a material that makes the filter medium sticky, which helps it retain and hold dust particles. |
| Extended Surface (Pleated Panel) Filter | Particles of 5 to 10 microns | Relatively low, but higher than conventional filters | Trapping mold, spores, dust, and most pollen. Often used in computer electronic equipment rooms and other areas where higher levels of cleanliness are needed. | Based on current IAQ requirements, this type of filter is generally considered to be the minimum standard for new installations. |
| Bag Filter and Box Filter | Particles of 0.30 to 3.0 microns | Medium to high efficiency; dust spot efficiencies of 30 to 95 percent | Used where tiny airborne particles cannot be tolerated, e.g., pharmaceutical process rooms. | Has a large holding capacity for particles. |
| High-Efficiency Particulate Air (HEPA) Filter | All particles in the 0.3 micron range; some can remove microscopic particles and microorganisms as small as 0.12 microns | Highest efficiency | Used to filter supply air for surgical rooms and to prevent process contamination during critical manufacturing processes. Also used in offices and residences to protect occupants who have asthma or allergies. | |

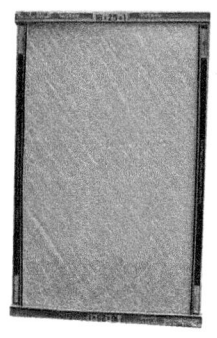  

(A) CONVENTIONAL FIBERGLASS FILTER    (B) ELECTROSTATIC PERMANENT FILTER    (C) MINI-PLEATED FILTER

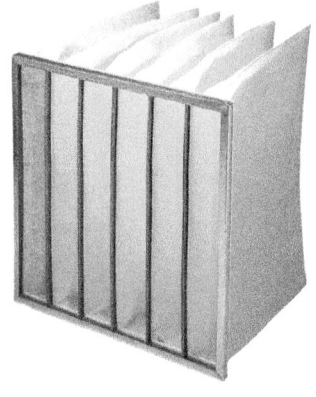 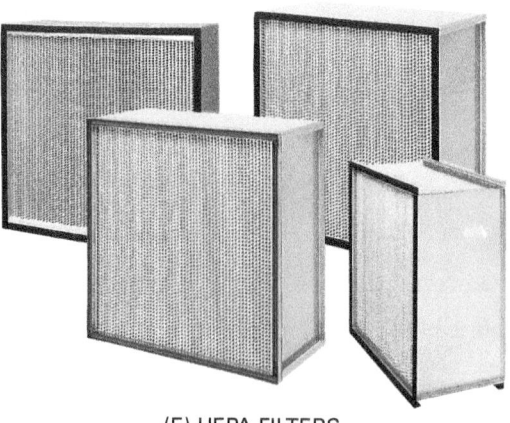

(D) BAG FILTER    (E) HEPA FILTERS

*Figure 16* Mechanical filters.

of previously adsorbed materials. Another type of gas filter uses porous pellets impregnated with active chemicals, such as potassium permanganate. These chemicals react with the contaminants and remove them or make them less bothersome or harmful. Maintenance consists of regenerating or replacing the chemicals.

Electronic air cleaners (*Figure 17*) outperform many of the mechanical air filters in trapping airborne particles and odors. They can be stand-alone units or can be mounted in the A/C system. Most electronic air cleaners contain a prefilter,

an ionizer, and a collector (*Figure 18*) to remove ionized particles. A charcoal filter for removing odors is an optional final section of the air cleaner.

### 4.1.5 Humidifiers and Dehumidifiers

Improper humidity levels can be a cause of sick building syndrome. Excess humidity promotes the growth of mold and mildew in ductwork, walls, and other interior spaces. *Figure 19* shows the relationship between humidity and common contaminants. As a rule of thumb, the RH in a building should be maintained at about 30

*Figure 17* Commercial electronic air cleaner.

## Disposing of Air Filters

Used air filters must be disposed of properly in accordance with the prevailing laws. This can mean disposal in a landfill, by incineration, or by recycling. Some air filters must be handled and disposed of as hazardous waste. Typically, filters used to trap hazardous waste are found in certain areas of hospitals, biomedical facilities, or in industrial processing plants. Do not attempt to dispose of filters used to capture hazardous waste unless you are equipped and licensed to do so.

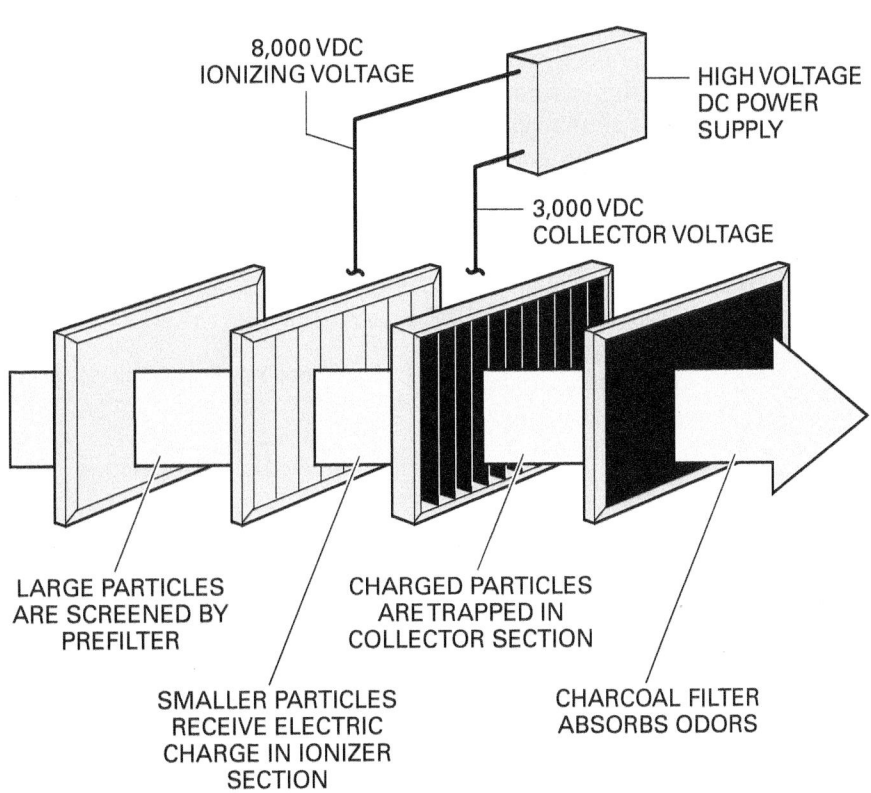

*Figure 18* Electronic air cleaner filtration stages.

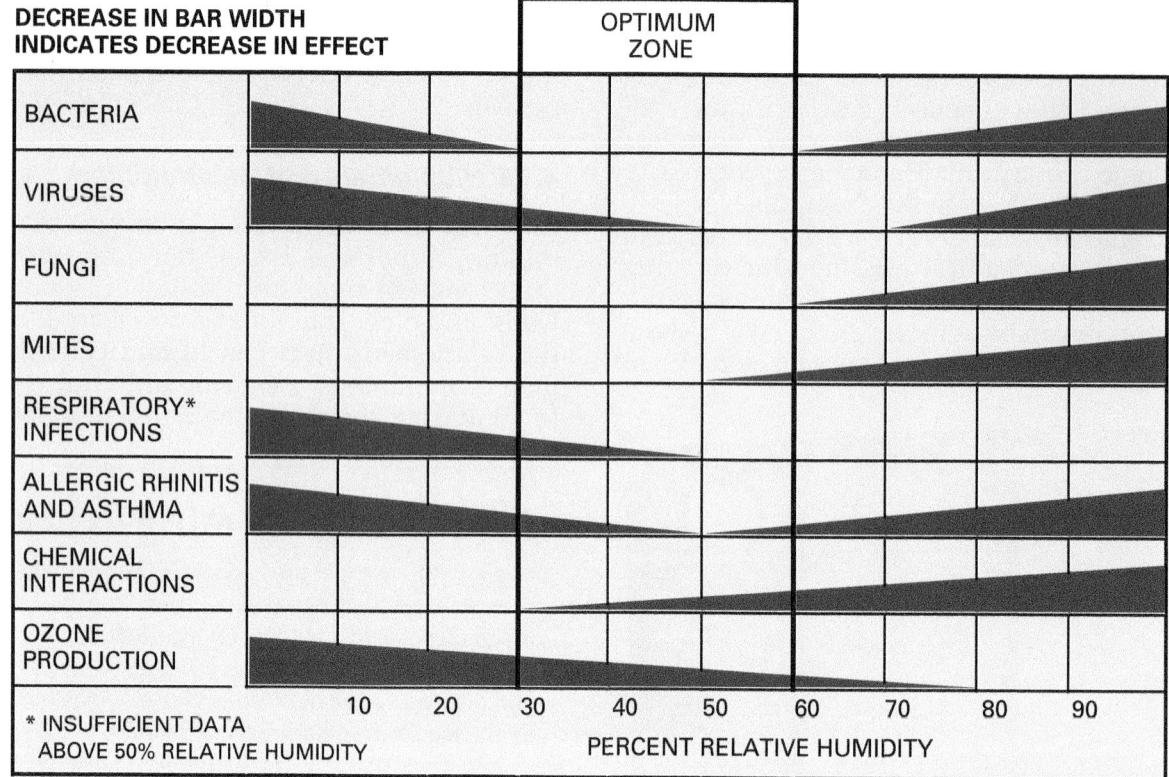

*Figure 19* Relative humidity range for health.

percent in the winter and 60 percent in the summer. From a system standpoint, RH levels over 30 percent are not practical in extremely cold weather because condensation on windows and other cold exterior surfaces can cause damage. This depends a lot on the thermal performance of the windows used. Similarly, an RH much below 40 percent cannot be achieved with most cooling equipment in the summer. Control of humidity involves the use of humidifiers when humidity levels are too low, and dehumidifiers when the humidity is too high.

Humidifiers are used to add humidity to a building or conditioned space. This is done by introducing water vapor into a building's conditioned air at a certain rate. Humidifiers can be portable units or they may be mounted in the HVAC system. The most popular types are wetted element, atomizing, ultrasonic, infrared, and steam models. Operation is controlled by a humidistat located in the conditioned space or unit.

Other than a failure of the humidifier or its control circuit, uncomfortable RH levels in a building can be caused by an incorrect humidistat setting relative to the outdoor temperature. Symptoms of excessive RH are condensation on windows and inside exterior walls. Too low an RH causes dry, itchy skin, static electricity shocks, clothing static cling, sinus problems, a chilly feeling, sickly pets and plants, and loose furniture joints. *Table 4* lists

the recommended indoor RH levels for various outdoor winter temperatures.

Another cause of too much or too little humidity can be a poorly sized humidifier. Humidifier capacities are normally rated in gallons of water per day. The capacity depends on the volume of the building or area in square feet ($ft^2$) or square meters ($m^2$). It also depends on the air-tightness of the building's construction. *Figure 20* shows a typical graph used for the selection of residential humidifiers. Similar graphs and/or charts are available for commercial and industrial humidifiers.

**Table 4** Recommended Indoor Winter Relative Humidity

| At Outdoor Temperature | | Recommended Indoor RH (%) |
|---|---|---|
| °F | °C | |
| –20– | 29 | 15 |
| –10 | 23 | 20 |
| 0 | 18 | 25 |
| 10 | 12 | 30 |
| 20 | 7 | 35 |
| 30 | 1 | 40 |

*Note:* Based on an indoor temperature of 72°F

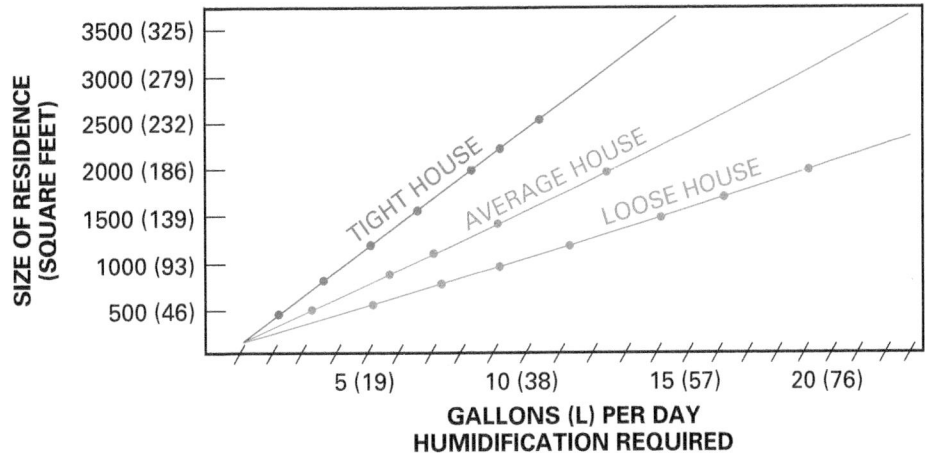

| LOOSE | AVERAGE | TIGHT |
|-------|---------|-------|
| No weatherstripping<br>No infiltration barrier<br>No vapor barrier<br>No fireplace damper<br>Undampered exhausts<br>Ductwork untaped or<br>  in unconditioned space<br>Indoor combustion air | Weatherstripping<br>Vapor barrier<br>Fireplace dampered<br>Dampered exhausts<br>Ductwork taped or in<br>  conditioned space<br>Indoor combustion air | Weatherstripping<br>Infiltration barrier<br>Seams/Penetrations<br>  sealed<br>Fireplace dampered<br>Dampered exhausts<br>Ductwork taped or in<br>  conditioned space<br>Outdoor combustion air |

*Figure 20* Humidifier capacity chart.

Using the capacity chart in *Figure 20*, note the difference in the size of humidifier that would be required for three buildings of the same size but having dissimilar air-tightness. A 2,000 ft$^2$ (186 m$^2$) building that is tightly constructed would require a humidifier capable of providing 8.5 gallons (32.2 L) per day. A similarly-sized building constructed with average air-tightness would require a humidifier capable of providing 12.5 gallons (32.5 L) per day. A loosely-constructed building would require a humidifier capable of providing over 19 gallons (71.9 L) per day. As you can see, a surprising volume of water may be required.

Dehumidifiers remove humidity from a building or conditioned space. Dehumidification of air occurs normally in a conventional cooling system. The system cooling coil normally removes both sensible heat and moisture (latent heat) from the entering air, which is a mixture of water vapor and dry gases. Both lose sensible heat during contact with the first part of the cooling coil, which functions as a dry cooling coil. Moisture is removed only in the part of the coil that is below the dew point of the entering air. When the coil starts to remove moisture, the cooling surfaces carry both the sensible and latent heat loads.

Portable dehumidifiers operate in the same way. These units are controlled by an adjustable humidistat that turns the unit on and off at pre-selected moisture levels. The capacity of a portable unit is normally rated in pints or liters per 24 hours. Again, the required capacity depends on the volume or area of the building. It also depends on the building's condition (wet, very damp, or moderately damp) without dehumidification during warm and humid outdoor weather conditions. *Table 5* provides sample guidelines for selecting portable dehumidifiers for residential or small commercial use. Similar charts are available for larger commercial and industrial dehumidifiers.

## Humidifier Restrictions

Many local authorities are beginning to ban the use of humidifiers for some applications because of bacterial growth. Before recommending or installing a humidifier, check with your local code administrator for verification that use of a humidifier is permitted for your location.

**Table 5** Dehumidifier Capacity Guide in Pints (Liters) per 24 Hours

| Conditions Without Dehumidification | Area in Square Feet (Square Meters) | | | | | |
|---|---|---|---|---|---|---|
| | 500 (46) | 1,000 (93) | 1,500 (139) | 2,000 (186) | 2,500 (232) | 3,000 (325) |
| Moderately damp – Space feels damp and has musty odor in humid weather. | 10 (4.7) | 14 (6.6) | 18 (8.5) | 22 (10.4) | 26 (12.3) | 30 (14.2) |
| Very damp – Space always feels damp and has musty odor. Damp spots show on walls and floors. | 12 (5.7) | 17 (8.0) | 22 (10.4) | 27 (12.8) | 32 (15.1) | 37 (17.5) |
| Wet – Space feels and smells wet. Walls and floors sweat, or seepage is present. | 14 (6.6) | 20 (9.5) | 26 (12.3) | 32 (15.1) | 38 (18.0) | 44 (20.8) |

Other dehumidifying equipment uses a liquid or solid desiccant, a material that has a high capacity for absorbing moisture. These units either collect the water on the surface of the desiccant or chemically combine with the water. One such piece of equipment is an air-to-air heat exchanger wheel (*Figure 21*). During the cooling season, the exhaust airstream recharges (dries out) a desiccant-coated wheel, causing the wheel to cool down. As the wheel rotates into the incoming airstream from outdoors, it absorbs the moisture and dehumidifies the outside air before delivering it to the air handler and cooling coils. In the heating season, some of the moisture leaving the building is transferred to the incoming airstream, raising the humidity of the incoming airstream. The wheel also transfers sensible heat between the two airstreams. Note that not all heat exchanger wheels are designed to exchange moisture. Many are designed solely for a sensible heat exchange.

Another accessory used to dehumidify an airstream is called a heat pipe (*Figure 22*). Use of a heat pipe in an HVAC system allows more of the system cooling coil capacity to go towards latent heat cooling by precooling the air before it gets to the cooling coil. Precooled air means less sensible cooling is required at the coil, allowing more capability for latent cooling (dehumidification).

> **NOTE**
>
> NCCER Module 03404, *Energy Conservation Equipment*, discusses the construction and operation of heat wheels and heat pipes more thoroughly.

### 4.1.6 Ultraviolet Light Air Purification Systems

Ultraviolet (UV) light air purification equipment can be used in HVAC air distribution systems to help prevent the growth of bacteria and other microorganisms known to cause indoor air problems and musty, mold-related odors.

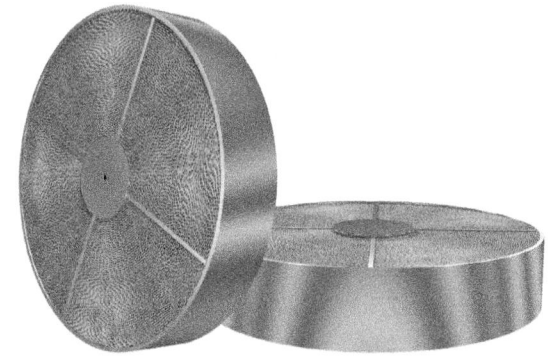

*Figure 21* Air-to-air heat exchanger wheel.

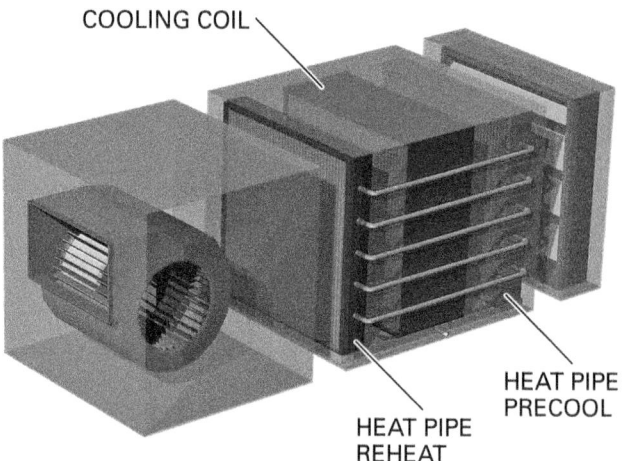

*Figure 22* Air-to-air, wrap-around heat pipe.

There are many manufacturers and designs of UV air purification equipment. However, the principle of operation is the same. C-band UV light (UV-C) energy with wavelengths in the 240- to 280-nanometer range destroys microorganisms by penetrating their cell wall. High-energy UV photons damage the protein structure of the cell and chemically alter the DNA. Once this occurs, the organism dies or cannot reproduce.

Germicidal effectiveness (microbe-killing power) is directly related to the UV dose applied, which is a function of time and intensity. HVAC system air purification by UV-C light is done in

# Ozone Generators

Advertising for many brands and models of ozone generators markets these devices as effective ways to improve indoor air quality. However, an EPA study documented in their publication entitled *Ozone Generators as Air Cleaners* states that available scientific evidence shows that at concentrations that do not exceed public health standards, ozone has little potential to remove indoor air contaminants. ASHRAE's Environmental Health Committee (EHC) published a similarly critical report, *Emerging Technologies without Clinical Evidence of Efficacy (Air Cleaners and Ionizers)*.

Furthermore, because ozone can be harmful to health at high concentrations, no agency of the federal government has approved the use of ozone generators in occupied spaces. An ASHRAE EHC report, *Ozone and Indoor Chemistry*, recommends minimizing the use of air cleaning technologies that produce ozone. It should also be pointed out that some authorities list ozone as a Class 2 or Class 3 carcinogen.

Possibly the main reason that ozone generators continue to be marketed as air cleaners is that government agencies do not have clear authority to control ozone emissions from air cleaning devices. To date, only California has established an outright ban on residential ozone generators marketed as air purifiers. Attempts by federal and other state agencies to address the problem have not been effective.

one of two ways: purification of a fixed object or purification of the moving air stream.

In fixed-object purification, the HVAC discharge side evaporator/indoor coil and drain pan are continuously irradiated with light rays generated by stationary quartz UV-C lamps or probes (emitters). The UV-C rays destroy bacteria and viruses present on the fixed object. The time required to destroy microorganisms on fixed objects depends on a number of things, including the distance the UV-C emitter is mounted from the fixed object, the size and intensity or killing power of the UV-C emitter, and the temperature of the air and UV-C emitter.

In UV purification of the moving air stream, the air in a duct system is illuminated as it moves past a stationary UV-C emitter. Achieving air purification using this method is much more difficult because of the short time (dwell time) during which the air moving past the emitter is illuminated. Typically, the air moves past the UV-C emitter at a speed of about 600 ft/min (3 m/s) or faster, spending only about 20 milliseconds in front of the probe/emitter. The intensity or killing power of the UV-C emitter, how fast the UV ray intensity decreases with distance as the air moves away from the emitter, and how far into the airstream the UV rays penetrate, determine the UV purification efficiency on the moving air. Because of the short dwell time, purification of the air stream normally requires the use of multiple UV light sources

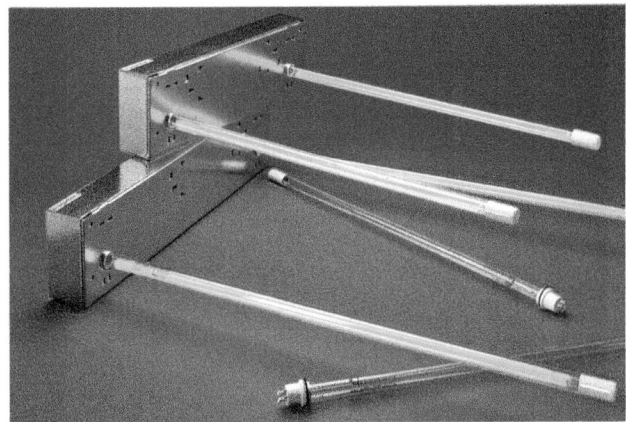

*Figure 23* UV-C air purification unit.

and reflectors that are capable of producing much stronger UV light rays than needed for a fixed object. For this reason, fixed object air purification systems are more widely used.

Properly installed UV-C sources inside air handlers can measurably reduce maintenance costs. They inhibit the growth of mold and bacterial films that can reduce air flow and heat transfer at coil surfaces.

*Figure 23* shows examples of a typical UV-C air purification unit. They are designed to protect coils, drain pans, and humidifiers from mold and bacterial growth while killing some airborne microorganisms. Each consists of a housing, power supply, and emitters. The components are

# UV Lights and Plastic Drain Pans

Exposure to UV light rays can cause some plastic drain pans to deteriorate; therefore, plastic drain pans exposed to UV light must be made of UV-resistant plastic.

incorporated into one assembly that is mounted outside the equipment at the cooling coil ductwork with the emitters protruding into the center of the coil and air stream.

## 4.2.0 IAQ and Forced-Air Duct Systems

In the United States, there are an estimated 60 million homes with forced-air heating and cooling systems. Studies have shown that these systems can lose up to 40 percent of the conditioned air through air duct leaks. Translated into wasted energy, it represents an annual fuel usage equivalent to that used by 13 million automobiles. Not only do the leaking ducts waste energy and contribute to poor comfort, they also can adversely affect indoor air quality and create health hazards.

### 4.2.1 Supply and Return Duct Leaks

In an ideal duct system (*Figure 24*), both the supply and return ducts are leak free. During operation, the system fan causes a pressure differential between the supply duct and the return duct. Positive (high) pressure in the supply duct causes the conditioned air to flow into the conditioned space. The negative (low) pressure in the return duct causes the air in the conditioned space to be drawn into the return duct. The pressure inside the structure itself is essentially neutral. Under these circumstances, the conditioned air inside the structure is circulated with little or no loss.

For the purpose of discussion, assume that the return duct is free of leaks, but the supply duct is leaking into an unconditioned space such as an attic or crawl space (*Figure 25*). Under these conditions, a slightly negative pressure is created in the conditioned space when the system fan

runs. The air lost through supply air leaks in the unconditioned space causes this slight negative pressure in the building. This air typically escapes the building through ceiling plenum, attic, or crawl space ventilation openings or structural defects. The now-negative pressure in the structure causes air to enter from the outside through cracks and small openings. Not only does this waste energy, but it can also cause additional unfiltered airborne contaminants to be drawn into the structure.

Similarly, assume that the supply duct is leak-free but the return duct, located in an attic or crawl space, is drawing in unconditioned air through leaks (*Figure 26*). Under these conditions, a slightly positive pressure is created in the structure because the quantity of air delivered by the supply ducts is greater than the amount of air being drawn from the conditioned space. The increased amount of air delivered by the supply duct is a result of the outside air that entered through leaks.

Leaks in the return duct can bring in all kinds of contaminants from inside and/or outside the structure. In actual practice, most structures have both supply and return duct leaks. In some cases, these two sets of leaks are relatively equal and cancel each other out. In most cases, however, one duct system will leak more than the other, causing a positive or negative pressure. In either case, contaminants from outside the structure can be brought into the structure.

## Moisture in Air Ducts

If moisture and dirt are present in air ducts, biological contaminants can grow and disperse throughout the building. Mold contamination in unlined sheet metal ducts can be successfully treated using an EPA-registered biocide. However, if fiberglass-lined sheet metal ducts or ducts made of fiberglass duct board become wet and contaminated with mold, cleaning is not sufficient to prevent regrowth, and there are no EPA-registered biocides for the treatment of porous duct materials. The EPA, National Air Duct Cleaners Association (NADCA), and the North American Insulation Manufacturers Association (NAIMA) all recommend the replacement of wet or moldy fiberglass material.

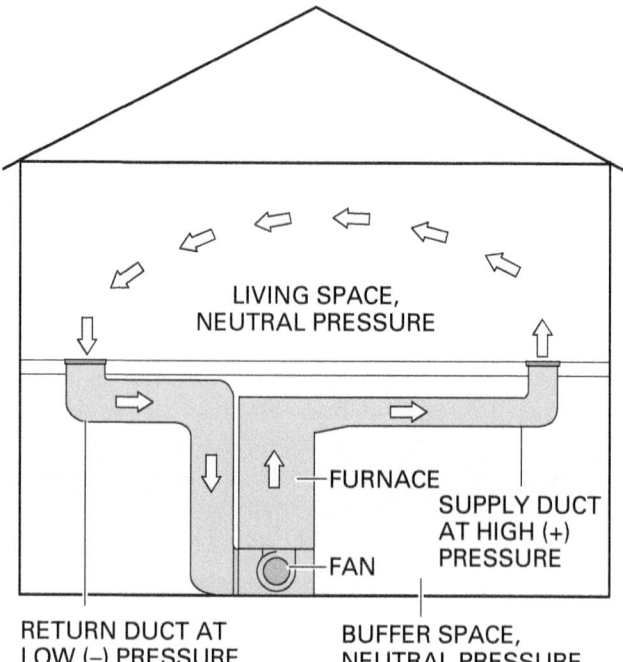

Figure 24 Simplified ideal air duct system.

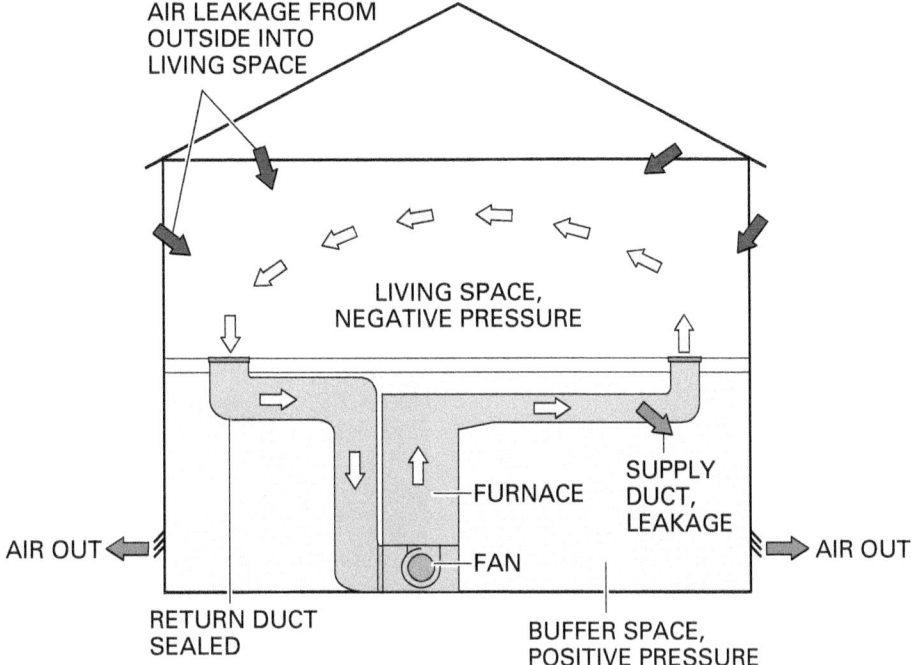

SUPPLY DUCT LEAKAGE OUTSIDE THE CONDITIONED SPACE
CREATES A NEGATIVE PRESSURE INSIDE THE SPACE.

*Figure 25* Simplified air duct system with leaks in the supply duct.

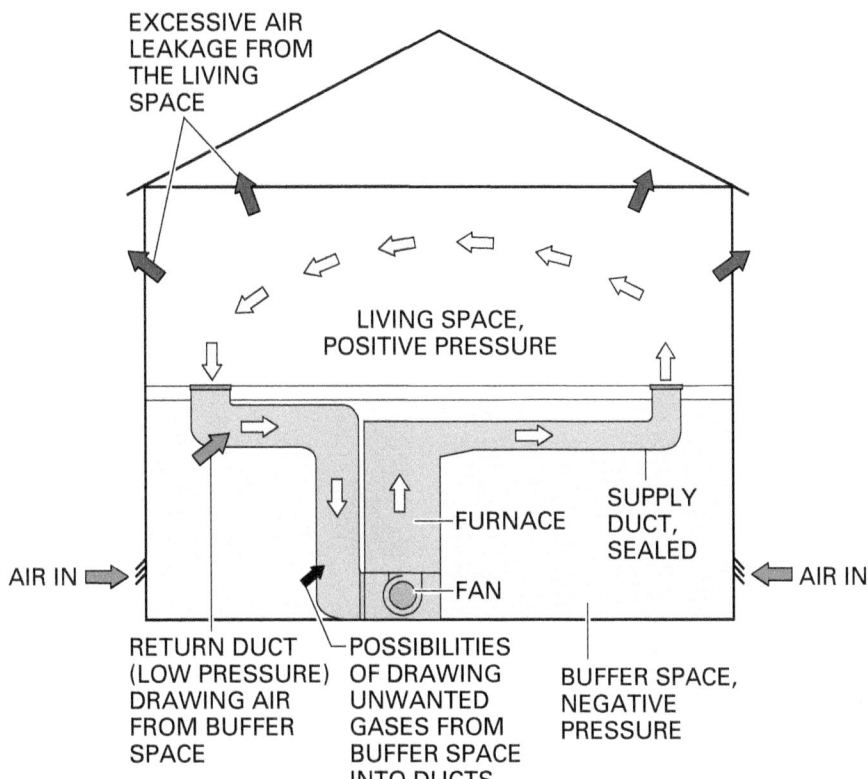

*Figure 26* Simplified air duct system with leaks in the return duct.

In addition to the IAQ problems that can be brought about by the introduction of contaminated outside air, leaking ducts can create hazardous conditions within the structure. If leaking return ducts draw in air from a garage or basement where a fuel-burning furnace is located, flue gases can be drawn in and circulated through the structure. This increases the potential for carbon monoxide poisoning. Leaking return ducts in a basement can also distribute radon gas that has infiltrated the area.

### 4.2.2 Sealing Air Duct Leaks

The best solution to duct leakage problems is to properly seal all ducts during installation. Unfortunately, many commonly accepted installation practices lead to duct systems that leak. Some states and localities realizing the scope of the problem have implemented construction practices and building codes to address the issue. However, this does not solve the problem of the millions of existing air duct systems that leak. Some duct leaks can be successfully sealed manually using caulks, mastic, or duct tape, assuming the leaks are accessible. The use of duct tape is not recommended, however. It tends to lose its adhesiveness after a few years and the tape will eventually loosen or fall off. Unfortunately, many leaks are not easily accessible, so manual sealing is not always an option.

New technologies have recently been developed that focus on sealing ducts from the inside. One such technology, called aerosol sealing (*Figure 27*), injects small dry-adhesive particles into a pressurized duct system. The air conditioning coil, fan, and furnace components are blocked off to isolate these components from the duct system. All of the registers or grilles are also carefully blocked. A fan (part of a sealing machine) temporarily connected to the supply or return plenum through a plastic connector tube is used to propel the particles through the duct system.

With the duct system plugged and under pressure, the only place air can escape is at the locations of the duct leaks. The sealant is deposited at those points. Over time, typically two to three hours, enough sealant builds up to stop the leaks. The process and resulting reduction in leakage can be monitored in real time by the technician using a laptop computer and sealing process software. The manufacturer for the process claims that leaks up to about $\frac{5}{8}$" (1.6 cm) in diameter can be successfully sealed and that the process can seal 70 to 90 percent of the existing leaks in a duct system.

### 4.2.3 Duct Cleaning

Duct cleaning generally refers to cleaning the various components of forced-air systems. This includes the supply and return ducts and registers, grilles and diffusers, heating and cooling coils, condensate drain pans, fan motor and fan housing, and the air handling unit.

Increased emphasis has been placed on the use of duct cleaning as a means of improving indoor air quality. There has been some controversy about its effectiveness and the methods for performing this task. There is also some question as

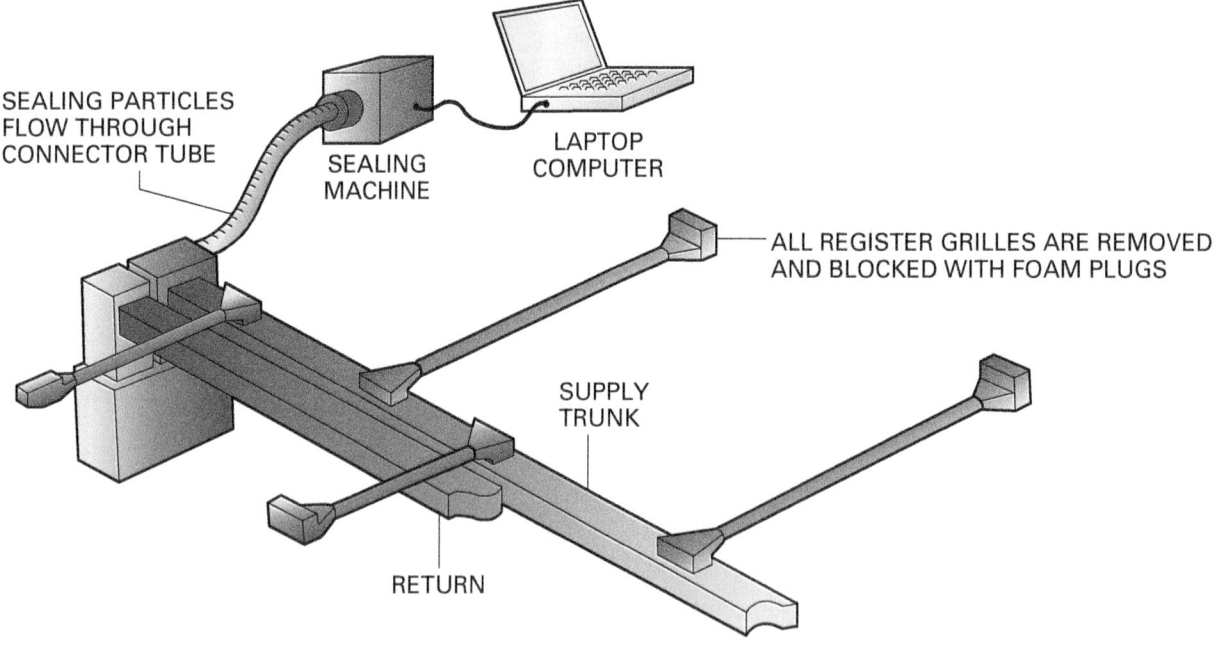

*Figure 27* Aerosol duct-sealing system.

SEALING PARTICLES FLOW THROUGH CONNECTOR TUBE

SEALING MACHINE

LAPTOP COMPUTER

ALL REGISTER GRILLES ARE REMOVED AND BLOCKED WITH FOAM PLUGS

SUPPLY TRUNK

RETURN

to when duct cleaning should be done and how the job can be validated. The EPA recommends that air ducts be cleaned only on an as-needed basis. In general, most HVAC systems should be inspected for cleanliness once or twice a year. Blowers should be cleaned regularly as part of scheduled maintenance, or more often if needed. The evaporator coil requires special attention, since fouling with dust and biological growths affects both its thermal performance as well as the system airflow. Some things to check during periodic maintenance include the following:

- Inspect and clean the indoor-coil fins. Straighten any that are bent.
- Check the tubing and fin connections for signs of oil, indicating leaks. Use a refrigerant leak detector if leaks are suspected.
- Clean the condensate tray or pan and check flow through the drain opening.
- Check for static pressure drop across the indoor coil.

Duct cleaning alone does not solve IAQ problems. Dirty air distribution systems are most often an effect, not the cause, of poor indoor air quality. The cause of a dirty system must be determined and corrected before cleaning. If HVAC systems are well maintained, frequent duct cleaning should not be needed. However, when duct cleaning is done properly, it can help to reduce the threat of indoor air pollution. Air filtration is not only designed to prevent contaminants from reaching building occupants, it is just as important to prevent airborne contaminants from accumulating in ducts and HVAC equipment.

In 1989, the National Air Duct Cleaners Association (NADCA) was formed by members of the duct cleaning industry. This organization adopted a standard in 1992 entitled *NADCA Standard 1992-01, Mechanical Cleaning of Non-Porous Air Conveyance System Components*. This document is now published by NADCA under the title *ACR 2013, Assessment, Cleaning, and Restoration of HVAC Systems*. Always refer to the latest edition of such specifications, as they are often revised.

Common duct cleaning equipment includes portable and/or truck-mounted HEPA vacuuming equipment and mechanical tools, such as rotary brushes and air whips, to dislodge dirt and debris in the ductwork (*Figure 28*).

*Figure 29* shows typical fiber-optic or video equipment used to inspect and document the conditions within the ductwork and other components before and after cleaning. This equipment can include borescopes and video cameras and monitors.

Follow NADCA standards and other NADCA-published guidelines when cleaning and accessing air conveyance systems. Before beginning cleaning, the operating system must be turned off and locked out using approved lockout/tagout procedures. Drop cloths should be used to protect furnishings in occupied areas.

## As-Built Drawings

When cleaning ducts in a large commercial system, it is good practice to obtain a set of air system as-built drawings. These can be used to identify obstructions such as coils, turning vanes, dampers, and similar devices within the duct system. The drawings can also be used to plan the locations for access points in the ductwork and to divide the ductwork system into workable sections for cleaning. Typically, cleaning sections should be no more than about 25 feet (8 m) in length. The drawings will also show sections of flexible or lined ductwork, certain types of which cannot be cleaned.

## Biocides and Chemical Treatments

Some duct cleaning services recommend applying chemical biocides to the inside of ductwork and other system components, Others also suggest applying chemical treatments (sealants or encapsulants) to cover the inside surfaces of air ducts and equipment housings. They believe that these treatments help control mold growth or prevent the release of dirt particle or fibers from ducts. These practices have not, however, been thoroughly researched and proven effective.

The EPA recommends that disinfectants, sanitizers, or other antimicrobial products not be applied to HVAC system if the products do not include directions for HVAC use (specifically, solid smooth, metal surfaces). No biocide products registered with the EPA are suitable for use inside of fiberglass-lined ductwork. The EPA also recommends that if biocides or chemical treatments are used, these substances should only be applied after the system has been properly cleaned of all visible dust or debris.

HEPA VACUUM COLLECTOR

POWER BRUSH                                        POWER WHIP

*Figure 28* Duct cleaning equipment.

# Isolating Cleaning Zones

The ductwork section being cleaned must be sealed off (isolated) from the adjacent sections to prevent loosened debris from contaminating the cleaned section or escaping past the vacuuming device into the downstream section. This is typically accomplished using inflatable balloons/bladders inserted into the ductwork at each side of the zone. When these balloons or bladders are inflated, they conform to the interior shape of the ductwork, sealing off the section being cleaned.

Common duct cleaning methods include contact vacuuming, air washing, and power brushing. Each of these methods follow what is sometimes referred to as the push-pull method described in *ACR 2013*. Mechanical agitation tools dislodge particulates, debris, and surface contamination from the duct (*Figure 30*), while vacuums, which must be operated continuously during the cleaning process, extract the dislodged material from the duct.

> **WARNING!**
>
> Duct cleaning will raise contaminants that may have settled in the ductwork. Be sure to wear company-prescribed PPE, including respiratory protection, when performing duct cleaning. Any products used must conform to current EPA requirements and local codes.

Contact vacuuming involves cleaning the interior duct surfaces by way of existing openings and outlets or, when necessary, through openings cut into the ducts (*Figure 31*). It uses portable vacuum equipment. Typically, the vacuum hose is about 2 inches in diameter. The vacuum unit should use HEPA filtering if it is exhausting into an occupied space. Starting at the return side of the system, the vacuum cleaner head is inserted into the section of the duct to be cleaned at the opening furthest upstream, and then the vacuum cleaner is turned on. Vacuuming proceeds downstream slowly enough to allow the vacuum to pick up all dirt and dust particles.

Inspection of each duct section is performed to determine whether the duct is clean. When cleanliness is confirmed, the vacuum cleaner head is removed from the duct and inserted through the next opening, where the process continues. This method is generally used for cleaning smaller sections of ductwork that are relatively easy to access.

In the air washing method (*Figure 32*), a vacuum collection unit with an 8- to 12-inch (20–30 cm) diameter hose is connected to the downstream end of the duct section through a suitable opening. The vacuum unit should use HEPA filtering if it is exhausting into an occupied space. The isolated section of duct being cleaned should be subjected to a minimum of 1" (2.5 mbar) negative air pressure to draw out loosened materials. Take care not to collapse the duct with an excessive negative pressure.

Compressed air is then introduced into the duct through a hose equipped with a skipper nozzle. This nozzle is propelled by the compressed air along the inside of the duct. For the air washing method to be effective, the compressed air source should be able to produce between

*Figure 29* Duct inspection equipment.

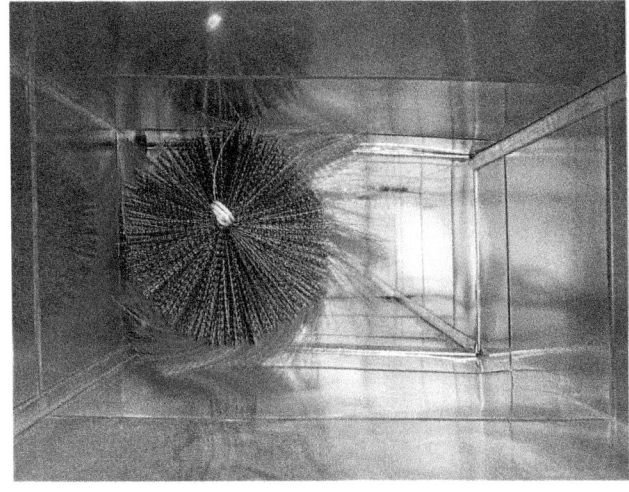

*Figure 30* Mechanical agitation during duct cleaning.

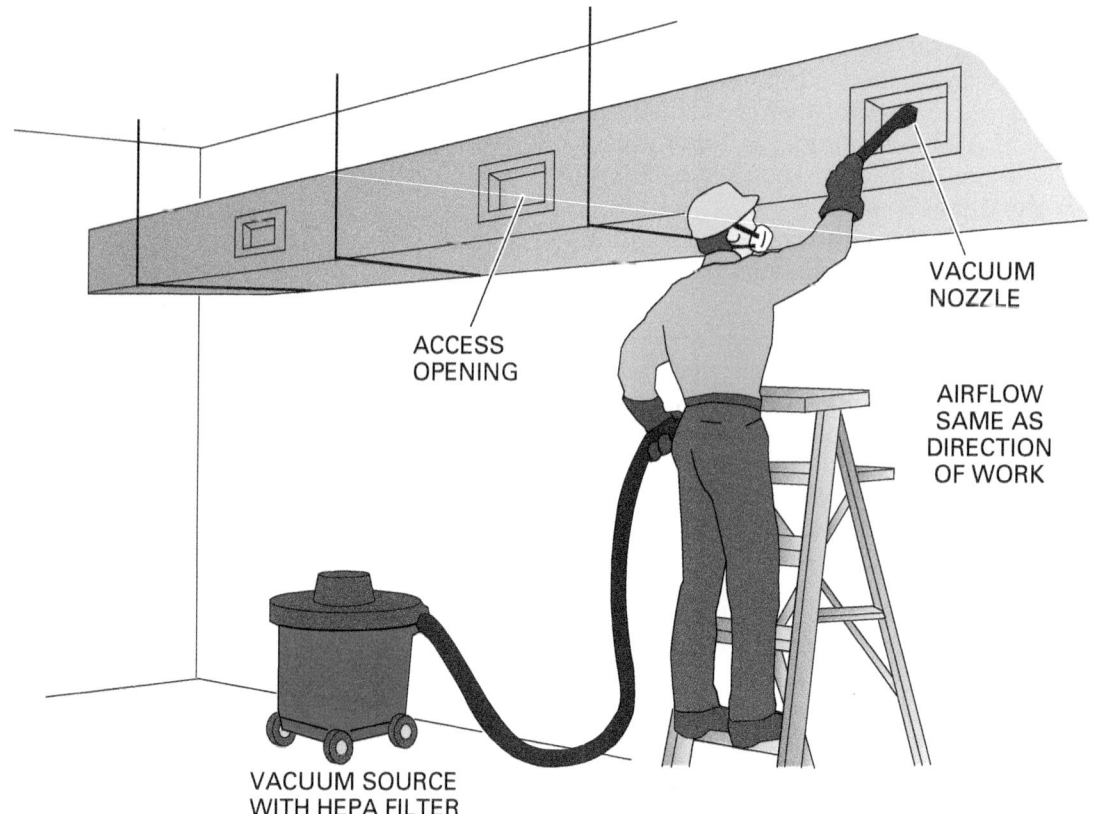

*Figure 31* Contact vacuuming duct cleaning method.

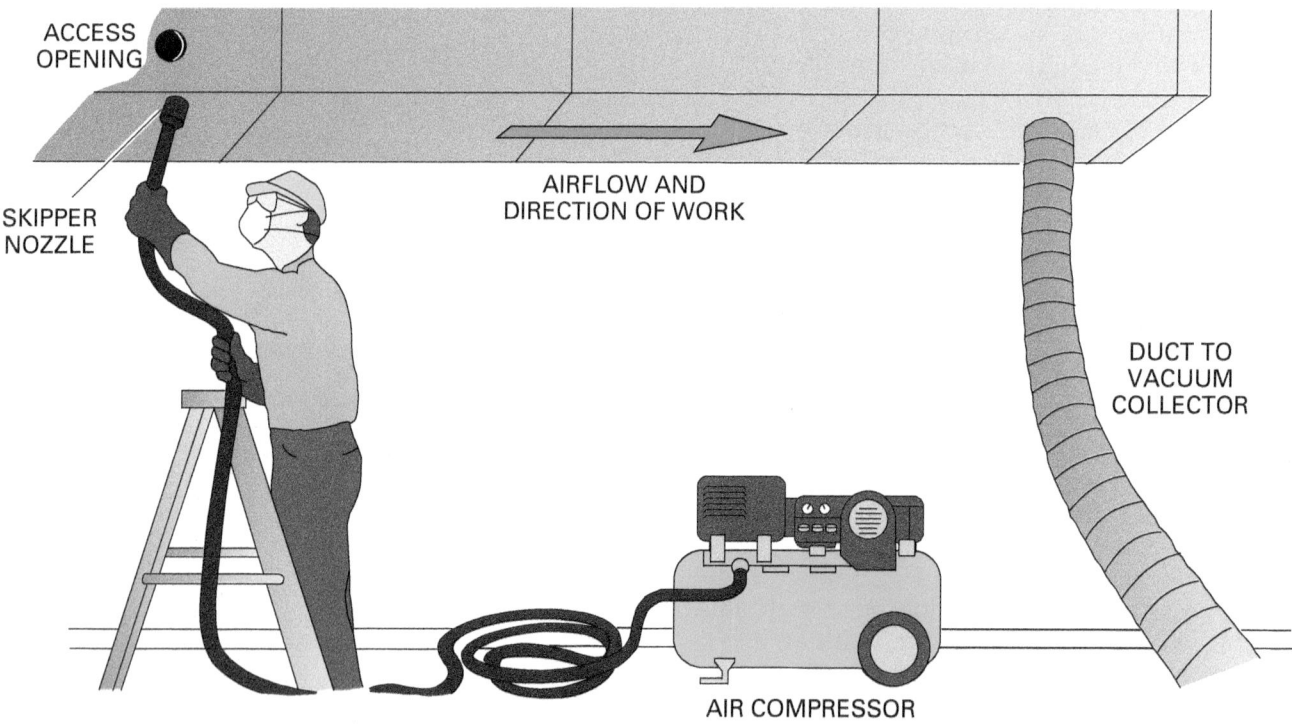

*Figure 32* Air washing duct cleaning method.

160 and 200 psi (1,100–1,380 kPa) air pressure and should have a 20-gallon (76 L) receiver tank. This method is most effective in cleaning ductwork interior dimensions no larger than 24" × 24" (61 cm × 61 cm). Inspection of each duct section is performed upon completion.

In the power brushing method, a powerful vacuum is connected to the duct in the same way. Pneumatic or electric brushes and air whips are used to dislodge dirt and dust particles, which become airborne and are then drawn into the vacuum (*Figure 33*). Compressed air is used to wash away any remaining dirt or debris. Power brushing can be used with all types of ducts and fibrous glass surfaces if the bristles are not too stiff and the brush is not allowed to remain in one place too long.

Power brushing usually requires larger access openings in the duct to allow for manipulating the equipment. The rotary brush is inserted into the duct section at the opening farthest upstream from the vacuum. The brush is moved downstream to dislodge dirt and dust particles. Inspection of each duct section and related components is performed to determine if the duct is clean. When the section of duct is clean, the brush is removed from the duct and inserted through the next opening, where the process continues.

Immediately after completing the duct cleaning, the entire system must be inspected to verify that it is clean before it is restarted. *ACR 2013*

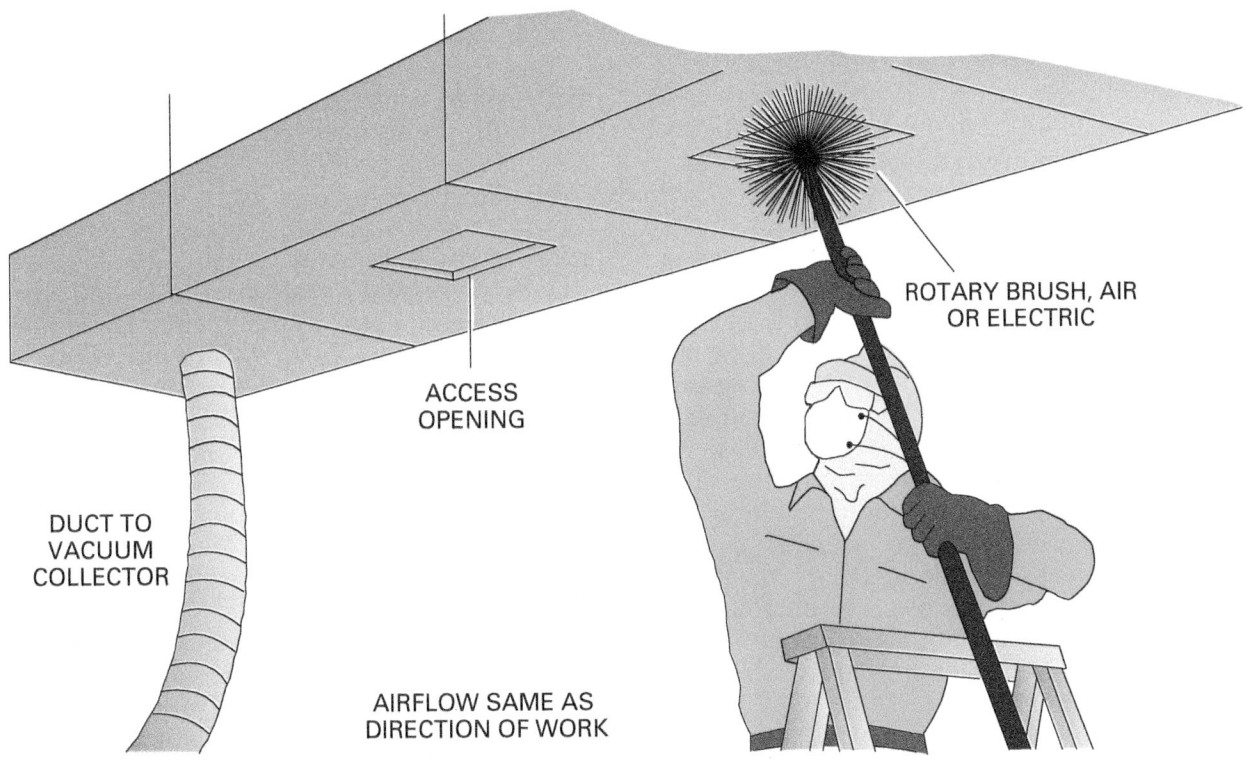

*Figure 33* Power brushing duct cleaning method.

ROTARY BRUSH, AIR OR ELECTRIC

ACCESS OPENING

DUCT TO VACUUM COLLECTOR

AIRFLOW SAME AS DIRECTION OF WORK

## Fiberglass Insulated Components

Care should be taken to prevent damaging fiberglass (also known as fibrous glass) insulated components during HVAC system duct cleaning. Potential damage that fiberglass insulation materials may be subject to includes delaminating, friable material, fungal growth, or damp, wet material. Fibrous glass insulated materials that are identified as damaged before or after system cleaning should be replaced. If fiberglass insulation material must be replaced, all replacement materials and repair work must conform to applicable industry standards and codes.

Although modern fiberglass duct liner and duct board products have surfaces that are resistant to the kind of abuse that occurs during duct cleaning, NAIMA recommends only contact vacuuming, air washing, and power brushing as suitable for use in cleaning insulated ductwork. NAIMA has published a manual that includes guidelines for cleaning insulated ductwork, *Cleaning Fibrous Glass Insulated Air Duct Systems—Recommended Practices.*

describes three methods of performing cleanliness verification: visual inspection, surface comparison testing, and the NADCA vacuum test.

Normally, visual inspection is the first method used. It involves making a visual inspection of the porous and non-porous HVAC system components, aided by remote photography/videography or fiber optics as necessary. If the surfaces of the components are free from non-adhered substances and debris, the components are considered clean and no further cleanliness verification is necessary.

If the visual inspection is inconclusive or disputed, surface comparison testing may be used to verify cleanliness. A specific contact vacuuming procedure is performed on a selected test area of the cleaned system. Then, the visible characteristics of the surface of the test area are compared with the visible characteristics of the surface prior to the contact vacuuming. The HVAC component surface is considered clean if there is no significant visible difference between the before and after surface characteristics at the location of the test.

If surface comparison testing is inconclusive or disputed, a NADCA vacuum test may be used to make a final cleanliness determination. This test is used for evaluating particulate levels of non-porous HVAC component surfaces; it does not apply to porous system components. During this procedure, a vacuum cassette with a filter media of mixed cellulose ester (MCE) is attached to an air sampling pump. The open face of the filter cassette is passed over two openings within the NADCA vacuum test template, which is applied to the component's airside surface. Generally, NADCA vacuum test samples are obtained by a qualified individual designated by the owner or owner's agent, and then sent to an accredited laboratory for analysis.

### 4.3.0 HVAC Contractor Liability

Today, many situations can lead to IAQ lawsuits by building owners, tenants, their employees, or others who use the building. For example, if a tenant discovers an IAQ problem in a building, the tenant may sue the owner, the HVAC contractor, or both, claiming that the building or office space is not environmentally safe. If people in a building become ill from an IAQ problem, they may sue the owner and/or HVAC contractor for damages, claiming they suffer from sick building syndrome. For these reasons, HVAC contractors must take steps to protect themselves from unjustifiable lawsuits that result from conditions or situations beyond their control.

HVAC contractors are fully aware of indoor air quality issues associated with HVAC systems, but many of their customers are not, including the owners of commercial buildings. The HVAC contractor should make customers aware and educate them about indoor air quality issues, including all local code requirements and changes. It is recommended that this information be presented in a formal and well-documented manner in case the information becomes relevant in a lawsuit.

The contractor should recommend periodic scheduled maintenance of all the HVAC equipment and should also recommend building walk-through inspections. If a problem or condition that affects IAQ is detected during such routine maintenance, the HVAC contractor should immediately inform the customer about the problem in writing and suggest recommendations for correcting it. The customer should be informed of the possible consequences if the repairs are not made and should be encouraged to make the repairs as soon as possible.

When retrofitting existing systems, the HVAC contractor should make sure that the customer is aware of and incorporates all the upgrades necessary to the system so that it will meet all current local IAQ codes and requirements. For example, equipment for supplying fresh outdoor air may need to be added to a system that originally did not have this capability.

## GOING GREEN

### Responsiveness

Heightened awareness of IAQ issues among the general public requires an urgent level of response to building occupants when a potential IAQ issue is raised. Although statistically less than 10 percent of all IAQ complaints begin with an actual contaminant, there is no way of knowing this until the problem has been fully researched. Problems are containable when they are seriously addressed. If the occupants feel that their complaints are being ignored, the issue will likely escalate. It could also manifest itself in other management issues for the people using the space. Building managers should address all IAQ complaints as real, make their responses public and transparent, and handle such complaints urgently to avoid further escalation of an IAQ situation.

## Additional Resources

*ACR, Assessment, Cleaning, and Restoration of HVAC Systems.* Current edition. Washington, DC: National Air Duct Cleaners Association.

*ASHRAE Standard 52.2, Method of Testing General Ventilation Air-Cleaning Devices for Removal Efficiency by Particle Size.* Current edition. Atlanta, GA: American Society of Heating, Refrigerating and Air-Conditioning Engineers (ASHRAE).

*ASHRAE Standard 62.1, Ventilation for Acceptable Indoor Air Quality.* Current edition. Atlanta, GA: American Society of Heating, Refrigerating and Air-Conditioning Engineers (ASHRAE).

*Building Air Quality, a Guide for Building Owners and Facility Managers.* Current edition. Washington, DC: U.S. Environmental Protection Agency.

*Cleaning Fibrous Glass Insulated Air Duct Systems—Recommended Practices.* Current edition. Alexandria, VA: North American Insulation Manufacturers Association (NAIMA).

*Indoor Air Quality.* Current edition. Chantilly, VA: Sheet Metal and Air Conditioning Contractors National Association (SMACNA).

*Indoor Air Quality in the Building Environment.* Ed Bas. 1993. Troy, MI: Business News Publishing Company.

## 4.0.0 Section Review

1. Which of the following is used to control and monitor the ventilation air in a building?

   a. Sub-slab depressurization system
   b. Total intake recirculation system
   c. Infrared exhaust-regeneration system
   d. Automated building management system

2. Which of the following conditions would likely result in crawlspace air being drawn into the HVAC system?

   a. Crawl space supply duct joint leak
   b. Loose-fitting grille cover
   c. Crawl space return duct joint leak
   d. Loose-fitting return air filter leak

3. Because of IAQ lawsuits brought by building owners, tenants, and others, HVAC contractors should _____.

   a. avoid addressing IAQ issues with their customers to minimize their liability
   b. educate building owners and customers of code requirements and changes pertaining to IAQ
   c. have their lawyers approve all contracts involving IAQ work
   d. direct their customers to contact equipment manufacturers for recommended maintenance

## SUMMARY

Most people are aware that outdoor air pollution can damage their health but do not realize that the effects of poor indoor air quality can be just as harmful. EPA studies show that indoor levels of many pollutants may be two to five times higher than outdoor levels. These pollutants are dangerous because most people spend as much as 90 percent of their time indoors.

Over the past several decades, exposure to indoor air pollutants has increased because of the following factors:

- Construction of tighter buildings
- Reduced ventilation rates to save energy
- Use of synthetic building materials and furnishings
- Increased use of cleaning products, pesticides, and personal care products

Major sources of pollution must be either eliminated or diluted using ventilation. Air systems must circulate reasonable quantities of fresh and filtered air. They must be maintained with proper filtration systems and kept free of accumulations of dust, dirt, and debris. Buildings, HVAC systems, and other systems must not be allowed to become breeding places for microbial contaminants.

Indoor air quality is affected by the following:

- Initial building design
- Ventilation control
- Thermal comfort control
- Control of chemical contaminants
- Control of microbial contaminants
- Building and equipment maintenance

1. Minimum ventilation rates and indoor air quality standards acceptable for human occupants are specified in _____.
   a. *ASHRAE Standard 52.2*
   b. *ASHRAE Standard 90.1*
   c. *ASHRAE Standard 62.1*
   d. *ASHRAE Standard 189.1*

2. Good IAQ requires that cognizant authorities have determined that the air contains _____.
   a. no more than 19.5 percent pure oxygen
   b. no known contaminants at harmful concentrations
   c. no known carcinogens at any level
   d. at least 75 percent relative humidity

3. Which of the following statements about the effects of poor IAQ on human health is true?
   a. Roughly 20 percent of the known indoor air pollutants account for over 80 percent of the cases of building-related illnesses in human occupants.
   b. Long-term health effects are nearly always the result of a single, high-dose exposure to indoor air pollutants.
   c. Health effects resulting from a person's exposure to poor IAQ may not show up until years after the person was exposed.
   d. Respiratory diseases and cancer are among the most common short-term effects of poor IAQ.

4. Buildings that have high air exchange rates _____.
   a. are more susceptible to poor indoor air quality
   b. are less susceptible to poor indoor air quality
   c. have low levels of dissolved oxygen
   d. have high levels of carbon dioxide ($CO_2$)

5. Building materials and furnishings can be a source of pollution caused by _____.
   a. ozone
   b. radon
   c. volatile organic compounds
   d. nitrogen oxide ($NO_2$)

6. Which of the following are indoor air pollutants that human occupants are responsible for?
   a. Bacteria and viruses
   b. Carbon monoxide
   c. Ozone and radon
   d. Carbon tetrachloride

7. Which of the following is an example of an aboveground source of contaminants outside a building that can cause poor indoor air quality?
   a. Polluted ground water
   b. Asbestos insulation
   c. Radon gas
   d. Vehicle exhaust

8. The first step in conducting an IAQ survey is usually _____.
   a. making a site visit and do a building walk-through
   b. taking air samples and test for specific contaminants
   c. conducting interviews to determine the scope of the problem
   d. inspecting the building's HVAC equipment and ventilation system

9. Basic air sampling typically involves measuring carbon dioxide ($CO_2$) levels _____.
   a. being exhausted outside a building
   b. at the air handler inside the building
   c. at air intake points inside a building
   d. both inside and outside a building

10. Air sampling usually begins _____.
   a. in the morning
   b. at midday
   c. in the afternoon
   d. in the evening

11. Which of the following design factors for a new building can improve its IAQ?

   a. If possible, locate the building in a protected valley.
   b. Design HVAC systems with every possible IAQ enhancement device, regardless of expense.
   c. Use building materials and furnishings that will keep indoor air pollution to a minimum.
   d. When landscaping, encourage the use of acacias, oaks, pines, and similar aesthetic plants.

12. When stale or stuffy indoor air causes the occupants of a building to become drowsy, the cause of the problem is typically _____.

   a. excess carbon dioxide
   b. insufficient oxygen
   c. high radon levels
   d. low relative humidity

13. Thermal comfort is related to temperature and _____.

   a. $CO_2$ pressure
   b. ozone levels
   c. off-gassing
   d. humidity

14. IAQ problems related to chemical contaminants can be derived from organic gases emitted from building materials, as well as _____.

   a. stagnant water around cooling coils
   b. cleaning and maintenance products
   c. accumulated mildew on ceiling tiles
   d. electrostatically charged filter media

15. The best way to eliminate or control an IAQ problem caused by microbial contaminants is to _____.

   a. maintain the building humidity between 30 and 60 percent
   b. dilute the indoor air with increased outdoor air ventilation
   c. close any system dampers to decrease outdoor air ventilation
   d. relocate the equipment that is the source of the contaminants

16. The minimum efficiency filters that should be used in new installations are _____.

   a. HEPA
   b. fiberglass and open-cell
   c. pleated panel
   d. bag and box

17. Five widely used types of humidifiers are wetted element, ultrasonic, infrared, steam, and _____.

   a. ionizing
   b. atomizing
   c. ultraviolet
   d. dry element

18. How can a leaking return duct that draws air in from a building's basement where a fuel-burning furnace is located cause IAQ problems?

   a. It can create a positive pressure that encourages the growth and distribution of microbial contaminants throughout the building.
   b. It can disrupt the air recirculation cycle for the entire HVAC system and expose building occupants to airborne particulates, mold spores, and radiation poisoning.
   c. It can draw in and circulate flue gases, possibly causing carbon monoxide poisoning to occupants of the building.
   d. It can discharge heat into the air outside the building and also waste energy by causing incomplete combustion of the fuel in the furnace.

19. Common duct cleaning methods include contact vacuuming, power brushing, and _____.

   a. aerosol steaming
   b. air washing
   c. dry scrubbing
   d. infrared flooding

20. It is recommended that HVAC contractors present customers with well-documented information about IAQ issues, including local code requirements and changes, because _____.

   a. the customer cannot obtain this information from any other reliable sources
   b. the International Building Code requires that contractors educate their customers
   c. the costs associated with correcting any IAQ problems are calculated on this basis
   d. the information could become relevant if the customer subsequently files a lawsuit

**Arrestance efficiency:** The percentage of dust that is removed by an air filter. It is based on a test where a known amount of synthetic dust is passed through the filter at a controlled rate, then the weight of the concentration of dust in the air leaving the filter is measured.

**Biological contaminants:** Airborne agents such as bacteria, fungi, viruses, algae, insect parts, pollen, and dust. Sources include wet or moist walls, duct, duct liner, fiberboard, carpet, and furniture. Other sources include poorly maintained humidifiers, dehumidifiers, cooling towers, condensate drain pans, evaporative coolers, showers, and drinking fountains. Also given the terms microbial or microbiological contaminants.

**Building-related illness (BRI):** A situation in which the symptoms of a specific illness can be traced directly to airborne building contaminants.

**Chemically-inert:** The property of a substance that does not readily react chemically with other substances, even under high temperatures and pressures, which usually accelerate chemical changes.

**Desiccant:** A substance that has a high affinity to water vapor, creating and maintaining a state of dryness near it. Desiccants for HVAC purposes have low or no chemical reactivity.

**Dust spot efficiency:** The percentage of dust that is removed by an air filter. It is the number that is normally referenced in the manufacturer's literature, filter labeling, and specifications. The atmospheric dust spot efficiency of a filter is based on a test where atmospheric dust is passed through a filter, then the discoloration effect of the cleaned air is compared with that of the incoming air.

**Electret:** A material or object that has the property of having both a positive and a negative electrical pole and generating an electrical field between them. The electrical equivalent of a magnet.

**Environmental tobacco smoke (ETS):** A combination of side-stream smoke from the burning end of a cigarette, cigar, or pipe and the exhaled mainstream smoke from the smoker.

**Formaldehyde:** A colorless, pungent byproduct of synthetic and natural biological processes that can cause irritation of the eyes and upper air passages. A known human carcinogen.

**Friable:** The condition in which brittle materials can easily fragment, releasing particulates into the air.

**Half-life:** The time required for half of a number of atoms to undergo radioactive decay.

**High-efficiency particulate air (HEPA) filter:** An extended media, dry-type filter mounted in a rigid frame. It has a minimum efficiency of 99.97 percent for 0.3-micron particles when a clean filter is tested at its rated airflow capacity.

**Multiple-chemical sensitivity (MCS):** A condition found in some individuals who believe they are vulnerable to exposure to certain chemicals and/or combinations of chemicals. Currently, there is some debate as to whether or not MCS really exists.

**New building syndrome:** A condition that refers to indoor air quality problems in new buildings. The symptoms are the same as those for sick building syndrome.

**Off-gassing:** The process by which furniture and other materials release chemicals and other volatile organic compounds (VOC) into the air.

**Ozone:** An unstable form of oxygen that has a sharp, pungent odor like chlorine bleach, formed in nature in the presence of electric discharges or exposure to ultraviolet light. It is irritating to the mucous membranes and lungs in animals and to plant tissues. It is a strong oxidizer and forms a hazardous air pollutant near ground level.

**Pontiac fever:** A mild form of Legionnaires' disease.

**Radon:** A colorless, odorless, radioactive, and chemically-inert gas that is formed by the natural breakdown of uranium in soil and groundwater. Radon exposure over an extended period of time can increase the risk of lung cancer.

**Scintillation detectors:** Radiation detectors that contain a substance which emits light when it absorbs a fast-moving nuclear particle (proton, neutron, or alpha). The sparkling effect caused by these emissions is called scintillation. The detector includes features that amplify the light flashes and display the rate of detections.

**Sick building syndrome (SBS):** A condition that exists when more than 20 percent of a building's occupants complain during a two-week period of a set of symptoms, including headaches, fatigue, nausea, eye irritation, and throat irritation, that are alleviated by leaving the building and are not known to be caused by any specific contaminant.

**Volatile organic compounds (VOC):** A wide variety of compounds and chemicals found in such things as solvents, paints, and adhesives, which easily evaporate at room temperature.

## INDOOR AIR QUALITY INTERVIEW FORM

---

**IBEAM: OCCUPANT COMPLAINT RECORDS**
IAQ Interview Form

page # 1 of 2

### D3:  Indoor Air Quality Interview Form

This form is used in conjunction with the Occupant Complaint Form. If the occupant called in the complaint and has not filled out the Occupant Complaint Form, ask him/her to do so as needed. Assure the occupant that all information provided will be kept confidential.

Building Name _____Address_____File #_____
Occupant Name_____
Room Number/Location_____Phone_____
Interviewed by _____ Date _____

**Symptom Patterns**

What kind of symptoms or discomfort are you experiencing?

Are you aware of other people with similar symptoms or concerns? Yes_____No_____. If yes, what are their names and locations?

Do you have any health conditions that may make you particularly susceptible to environmental problems?

| contact lenses | chronic cardiovascular disease | undergoing chemotherapy or radiation therapy |
| € allergies | chronic respiratory disease | |
| | chronic neurological problems | immune system suppressed by disease or other causes |

**Spatial Patterns**

Where are you when you experience symptoms or discomfort? _____

Where do you spend most of your time in the building? _____

Where else in the building do you frequent? _____

**Timing Patterns**

When did the symptoms/problems start? _____
When are symptoms/problems generally worse?

| Beginning of week | Morning | Afternoon | Spring | Summer |
| End of week | Particular times of day | | Fall | Winter |
| Particular days of week | | | Particular months | |

_____ _____ _____        _____ _____ _____        ___ _____ _____

**Figure Credit:** U.S. Environmental Protection Agency

Other pattern or no pattern (explain)

_____

_____

When do symptoms/problems go away?

After you leave the building ____yes____no: If yes, how long does it take to go away? _____

After you leave the space _____yes____no: If yes, how long does it take to go away? _____

Have you noticed any particular events, activities that you or others engage in, weather conditions, temperature or humidity conditions, odors, or other things that tend to occur around the same time or before your symptoms?

**Additional Information**

Do you have any observations about the building conditions that might need attention or might

help explain your symptoms?

Do you have other comments?

**Figure Credit:** U.S. Environmental Protection Agency

# HVAC Checklist - Short Form

*Page 1 of 4*

Building Name: _____  Address: _____

Completed by: _____  Date: _____  File Number: _____

*Sections 2, 4 and 6 and Appendix B discuss the relationships between the HVAC system and indoor air quality.*

**MECHANICAL ROOM**

■ Clean and dry?_____  Stored refuse or chemicals?_____

■ Describe items in need of attention _____

**MAJOR MECHANICAL EQUIPMENT**

■ Preventive maintenance (PM) plan in use?_____

Control System

■ Type_____

■ System operation _____

■ Date of last calibration _____

Boilers

■ Rated Btu input _____ Condition _____

■ Combustion air: is there at least one square inch free area per 2,000 Btu input? _____

■ Fuel or combustion odors _____

Cooling Tower

■ Clean?  no leaks or overflow?_____  Slime or algae growth?_____

■ Eliminator performance_____

■ Biocide treatment working?  (list type of biocide)_____

■ Spill containment plan implemented?_____  Dirt separator working?_____

Chillers

■ Refrigerant leaks?_____

■ Evidence of condensation problems?_____

■ Waste oil and refrigerant properly stored and disposed of?_____

**191**  *Indoor Air Quality Forms*

**Figure Credit:** U.S. Environmental Protection Agency

Building Name: _____  Address: _____

Completed by: _____  Date: _____  File Number: _____

**AIR HANDLING UNIT**

■ Unit identification _____  Area served _____

Outdoor Air Intake, Mixing Plenum, and Damper

■ Outdoor air intake location _____

■ Nearby contaminant sources? (describe) _____

■ Bird screen in place and unobstructed? _____

■ Design total cfm_____  outdoor air (O.A.) cfm _____  date last tested and balanced _____

■ Minimum % O.A. (damper setting)_____  Minimum cfm O.A. $\dfrac{\text{(total cfm} \times \text{minimum \% O.A.)}}{100} =$ _____

■ Current O.A. damper setting (date, time, and HVAC operating mode) _____

■ Damper control sequence (describe) _____

■ Condition of dampers and controls (note date) _____

Fans

■ Control sequence _____

■ Condition (note date) _____

■ Indicated temperatures        supply air_____   mixed air_____   return air_____   outdoor air_____

■ Actual temperatures           supply air_____   mixed air_____   return air_____   outdoor air_____

Coils

■ Heating fluid discharge temperature_____  ΔT_____  cooling fluid discharge temperature_____  ΔT_____

■ Controls (describe)_____

■ Condition (note date) _____

Humidifier

■ Type_____  if biocide is used, note type _____

■ Condition (no overflow, drains trapped, all nozzles working?)_____

■ No slime, visible growth, or mineral deposits? _____

*Indoor Air Quality Forms* **192**

**Figure Credit:** U.S. Environmental Protection Agency

Building Name: _____   Address: _____

Completed by: _____   Date: _____   File Number: _____

## DISTRIBUTION SYSTEM

| Zone/ Room | System Type | Supply Air | | Return Air | | Power Exhaust | | |
|---|---|---|---|---|---|---|---|---|
| | | ducted/ unducted | cfm* | ducted/ unducted | cfm* | cfm* | control | serves (e.g. toilet) |
| | | | | | | | | |
| | | | | | | | | |
| | | | | | | | | |
| | | | | | | | | |
| | | | | | | | | |

Condition of distribution system and terminal equipment (note locations of problems)

- Adequate access for maintenance? _____
- Ducts and coils clean and obstructed? _____
- Air paths unobstructed?   supply ____ return ____ transfer ____ exhaust ____ make-up ____
- Note locations of blocked air paths, diffusers, or grilles _____
- Any unintentional openings into plenums? _____
- Controls operating properly? _____
- Air volume correct? _____
- Drain pans clean? Any visible growth or odors? _____

### Filters

| Location | Type/Rating | Size | Date Last Changed | Condition (give date) |
|---|---|---|---|---|
| | | | | |
| | | | | |
| | | | | |
| | | | | |
| | | | | |

**193** *Indoor Air Quality Forms*

**Figure Credit:** U.S. Environmental Protection Agency

# HVAC Checklist - Short Form

Building Name: _____   Address: _____

Completed by: _____   Date: _____   File Number: _____

## OCCUPIED SPACE

Thermostat types _____

| Zone/ Room | Thermostat Location | What Does Thermostat Control? (e.g., radiator, AHU-3) | Setpoints | | Measured Temperature | Day/ Time |
|---|---|---|---|---|---|---|
| | | | Summer | Winter | | |
| | | | | | | |
| | | | | | | |
| | | | | | | |
| | | | | | | |
| | | | | | | |

Humidistats/Dehumidistats type _____

| Zone/ Room | Humidistat/ Dehumidistat Location | What Does It Control? | Setpoints (%RH) | Measured Temperature | Day/ Time |
|---|---|---|---|---|---|
| | | | | | |
| | | | | | |
| | | | | | |
| | | | | | |
| | | | | | |

■ Potential problems (note location) _____

■ Thermal comfort or air circulation (drafts, obstructed airflow, stagnant air, overcrowding, poor thermostat location)

_____

_____

■ Malfunctioning equipment _____

■ Major sources of odors or contaminants (e.g., poor sanitation, incompatible uses of space)

_____

*Indoor Air Quality Forms* **194**

**Figure Credit:** U.S. Environmental Protection Agency

# Additional Resources

This module presents thorough resources for task training. The following reference material is recommended for further study.

*ACR, Assessment, Cleaning, and Restoration of HVAC Systems*. Current edition. Washington, DC: National Air Duct Cleaners Association.

*ASHRAE Standard 52.2, Method of Testing General Ventilation Air-Cleaning Devices for Removal Efficiency by Particle Size*. Current edition. Atlanta, GA: American Society of Heating, Refrigerating and Air-Conditioning Engineers (ASHRAE).

*ASHRAE Standard 62.1, Ventilation for Acceptable Indoor Air Quality*. Current edition. Atlanta, GA: American Society of Heating, Refrigerating and Air-Conditioning Engineers (ASHRAE).

*Building Air Quality, a Guide for Building Owners and Facility Managers*. Current edition. Washington, DC: U.S. Environmental Protection Agency.

*Cleaning Fibrous Glass Insulated Air Duct Systems—Recommended Practices*. Current edition. Alexandria, VA: North American Insulation Manufacturers Association (NAIMA).

*Indoor Air Quality*. Current edition. Chantilly, VA: Sheet Metal and Air Conditioning Contractors National Association (SMACNA).

*Indoor Air Quality in the Building Environment*. Ed Bas. 1993. Troy, MI: Business News Publishing Company.

# Figure Credits

# Section Review Answer Key

| Answer | Section Reference | Objective |
|---|---|---|
| **Section One** | | |
| 1. c | 1.1.0 | 1a |
| 2. a | 1.2.0 | 1b |
| **Section Two** | | |
| 1. d | 2.1.1 | 2a |
| 2. c | 2.2.0 | 2b |
| 3. b | 2.3.1 | 2c |
| 4. a | 2.4.0 | 2d |
| 5. c | 2.5.2 | 2e |
| **Section Three** | | |
| 1. b | 3.1.0 | 3a |
| 2. a | 3.2.1 | 3b |
| 3. c | 3.3.1 | 3c |
| **Section Four** | | |
| 1. d | 4.1.1 | 4a |
| 2. c | 4.2.1 | 4b |
| 3. b | 4.3.0 | 4c |

# NCCER CURRICULA — USER UPDATE

NCCER makes every effort to keep its textbooks up-to-date and free of technical errors. We appreciate your help in this process. If you find an error, a typographical mistake, or an inaccuracy in NCCER's curricula, please fill out this form (or a photocopy), or complete the online form at **www.nccer.org/olf**. Be sure to include the exact module ID number, page number, a detailed description, and your recommended correction. Your input will be brought to the attention of the Authoring Team. Thank you for your assistance.

*Instructors* – If you have an idea for improving this textbook, or have found that additional materials were necessary to teach this module effectively, please let us know so that we may present your suggestions to the Authoring Team.

**NCCER Product Development and Revision**

13614 Progress Blvd., Alachua, FL 32615

**Email:** curriculum@nccer.org
**Online:** www.nccer.org/olf

❏ Trainee Guide    ❏ Lesson Plans    ❏ Exam    ❏ PowerPoints    Other _____

Craft / Level: _____    Copyright Date: _____

Module ID Number / Title: _____

Section Number(s): _____

Description: _____

_____

_____

_____

_____

Recommended Correction: _____

_____

_____

_____

_____

Your Name: _____

Address: _____

_____

Email: _____    Phone: _____

# Energy Conservation Equipment

## OVERVIEW

Because HVACR systems consume a great deal of energy, they are primary targets in the battle for energy conservation. To conserve energy, HVACR manufacturers have developed innovative devices that make use of heat that might otherwise be wasted. They have also introduced novel ways to minimize the use of mechanical cooling when outdoor air can be substituted. As technologies evolve, HVACR installers and service technicians need to be aware of the latest devices and how they contribute to energy conservation.

## Module 03404

*03404 V5*

## Objectives

When you have completed this module, you will be able to do the following:

1. Identify and describe the operation of various energy recycling and reclamation systems.
   a. Identify and describe the operation of energy- and heat-recovery ventilators.
   b. Identify and describe the operation of fixed- and rotary-plate air-to-air heat exchangers.
   c. Identify and describe the operation of condenser heat-recovery systems.
   d. Identify and describe the operation of coil energy-recovery loops.
   e. Identify and describe the operation of thermosiphon heat exchangers.
   f. Identify and describe the operation of twin-tower enthalpy recovery loops.
   g. Identify and describe the operation of flue-gas heat-recovery systems.
   h. Identify and describe the operation of steam heat-recovery systems.
2. Identify and describe the operation of electric energy-demand reduction and ice storage systems.
   a. Identify and describe the operation of electric energy-demand reduction systems.
   b. Identify and describe the operation of ice storage systems.

## Performance Tasks

This is a knowledge-based module; there are no Performance Tasks.

## Trade Terms

Enthalpy
Monel®
Runaround loop
Sensible heat

Sorbent
Total heat
Zeolite

## Industry-Recognized Credentials

If you are training through an NCCER-accredited sponsor, you may be eligible for credentials from NCCER's Registry. The ID number for this module is 03404. Note that this module may have been used in other NCCER curricula and may apply to other level completions. Contact NCCER's Registry at 888.622.3720 or go to **www.nccer.org** for more information.

# Contents

# Figures

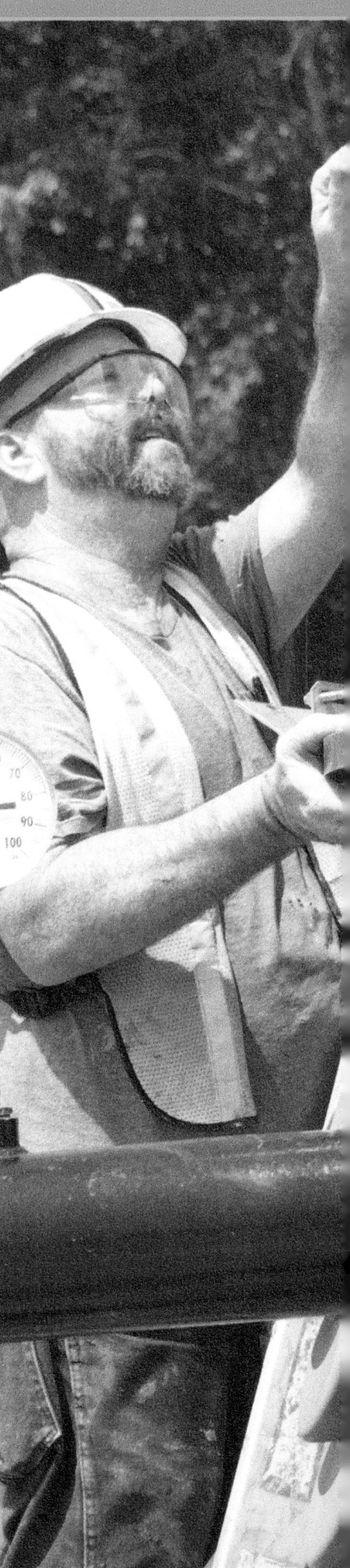

## 1.0.0 Energy Recycling and Reclamation

### Objective

Identify and describe the operation of various energy recycling and reclamation systems.

   a.  Identify and describe the operation of energy- and heat-recovery ventilators.

   b.  Identify and describe the operation of fixed- and rotary-plate air-to-air heat exchangers.

   c.  Identify and describe the operation of condenser heat-recovery systems.

   d.  Identify and describe the operation of coil energy-recovery loops.

   e.  Identify and describe the operation of thermosiphon heat exchangers.

   f.  Identify and describe the operation of twin-tower enthalpy recovery loops.

   g.  Identify and describe the operation of flue-gas heat-recovery systems.

   h.  Identify and describe the operation of steam heat-recovery systems.

### Trade Terms

**Enthalpy:** When used in HVACR technology, the total amount of thermal energy contained by a fluid; the sum of its thermal energy based on heat capacity, and its latent heats of fusion and vaporization, as applicable.

**Monel®:** An alloy made of nickel, copper, iron, manganese, silicon, and carbon that is very resistant to corrosion.

**Runaround loop:** A closed-loop energy recovery system in which finned-tube water coils are installed in the supply and exhaust airstreams and connected by counterflow piping.

**Sensible heat:** The heat exchange that simply raises or lowers the temperature of a fluid without causing or resulting from a phase change (condensation or evaporation). Compare with the term *a*.

**Sorbent:** A substance whose main purpose is to absorb something else by forming a mixture of the two substances.

**Total heat:** The heat exchange that not only raises or lowers the temperature of a fluid, but also causes or results from a phase change (condensation or evaporation).

**Zeolite:** An aluminosilicate compound that has a high affinity for water and water vapor, which can easily be dehydrated and rehydrated by heating and cooling. A compound commonly used in certain humidity-control systems.

The higher cost of energy, the need to conserve energy, and government-mandated efficiency standards are all factors that have caused an increase in the use of heat recovery and/or energy-saving devices in HVAC systems. Heat recovery devices save energy through the capture and reuse of heat that would otherwise be wasted. Other devices change the operation of the system in such a way that heating and cooling efficiencies are increased.

In addition to their heat- or energy-saving function, many of these devices are designed to help improve indoor air quality as well. The installation of one or more of these energy-saving devices in a new system often allows the selection of lower-capacity primary heating and/or cooling equipment because of the resulting improvement in system efficiency.

There are non-automated as well as automated energy management systems. They control the overall operation of a building's HVAC systems with the goal of greater efficiency and less wasted energy. The focus of this module is on some of the more common components, or groups of components, used in HVAC systems to help conserve energy.

Heat recovery (reclaim) equipment captures and uses heat that would otherwise be wasted. There are many kinds of heat recovery devices and processes in use, including the following:

- Energy- and heat-recovery ventilators
- Fixed-plate and rotary air-to-air heat exchangers
- Condenser-heat recovery
- Coil energy-recovery loops
- Heat-pipe and thermosiphon heat exchangers
- Twin-tower enthalpy recovery loops

### 1.1.0 Energy- and Heat-Recovery Ventilators

To maintain good indoor air quality, the American Society of Heating, Refrigeration, and Air Conditioning Engineers (ASHRAE) standards recommend that a building's indoor air be exchanged for fresh outdoor air at a rate of 0.35 air changes per hour. An alternative ASHRAE recommendation calls for an exchange rate of 15 cfm (0.4 cmm) per person, 20 cfm (0.6 cmm) per bathroom, and 25 cfm (0.7 cmm) per kitchen. Recovery ventilators are one type of HVAC equipment that can be used to bring a controlled amount of outside air into a building while helping to conserve energy. There are two types of recovery ventilators: energy recovery ventilators (ERVs) and heat recovery ventilators (HRVs). ERVs supply fresh air and recover energy from both heating and cooling operations; that is, they transfer both sensible heat and latent heat (heat contained in water vapor). HRVs supply fresh air and recover heat energy during the heating season. They are similar in construction and operation to the ERV unit, but they are designed to transfer only sensible heat between the fresh incoming air and the stale exhaust air.

According to the US Department of Energy, most recovery ventilators can recover about 60 to 80 percent of the energy from the exiting air, delivering the energy to the incoming fresh air. Typically, an ERV/HRV improves the indoor air quality by exchanging the air about every three hours.

Air from the occupied space is passed through the ERV or HRV unit and exhausted outside (*Figure 1*). At the same time, fresh air is brought in from the outside and sent through the unit. When the two airstreams pass through the heat exchanger core, a great deal of the heat is transferred from the warmer air to the cooler air. The core design allows this heat transfer between the entering and leaving airstreams to occur without directly mixing the two. The result is a constant stream of fresh air delivered to the living space that has been tempered to some degree, reducing the load on the building's HVAC systems.

ERVs and HRVs may use any one of several types of heat exchangers. Fixed-plate heat exchangers and enthalpy wheel (rotary wheel) heat exchangers are two of the most widely used. The main difference between an ERV and HRV is the way the heat exchanger core works. In the HRV, the core material can only transfer sensible heat. This is why these units are used mainly in colder climates. The core material in an ERV is capable of transferring both sensible and latent heat, allowing it to transfer heat in the winter and remove moisture from the air during the summer. This makes the use of ERVs popular in humid climates, such as in the southeastern United States.

When an ERV or HRV is installed, it is critical to balance the air distribution system to make sure that the amounts of incoming and outgoing air are equal.

Some commercial building air conditioning systems use an ERV unit in conjunction with a rooftop-package air conditioning unit. As shown in *Figure 2*, this can be done using a standalone ERV unit. Alternatively, the ERV is fastened directly over the outdoor intake of the rooftop unit. This way, the outdoor air first enters and is preconditioned by the ERV instead of entering the rooftop unit directly.

One of the benefits of a standalone ERV is that it allows an economizer to be used with the rooftop unit. This is because the ERV mounts on a separate roof curb rather than on the outdoor intake of the rooftop unit. During economizer operation, the rooftop unit typically controls the standalone ERV so that its exhaust fan continues to operate, but the supply fan and recovery wheel are shut down.

Another benefit of a standalone unit is that it eliminates the need for an exhaust fan, which is normally used to exhaust air from the building bathrooms, conference rooms, and similar areas. This is because the building exhaust air ductwork is connected to the ERV, and the ERV is used in place of an exhaust fan. Because an ERV fastened to the outdoor intake of the rooftop unit can only draw its exhaust air from the return duct, the use of a separate exhaust fan is still required.

When performing periodic or corrective maintenance on ERV and HRV equipment, always refer to the manufacturer's manual and instructions. ERV

## Fume Hood Exhaust Heat Recovery

Laboratories often require 100-percent makeup air for workstation environments where fume hood exhaust must be prevented from becoming re-entrained into the building's ventilation system. The makeup air generally needs to be preheated or precooled.

At its 20,000-square feet (1,860-square meter) facility for chemical research in Branford, Connecticut, Neurogen Corporation had Tri-Stack® heat recovery systems installed to help cut energy costs. In these systems, heat recovery coils filled with a solution of glycol and water extract heat from a workstation fume hood exhaust before it is discharged above the building's roof line. These heat recovery systems increased the intake air temperature by 10°F (5.6°C), reducing the need to use natural gas to preheat (or precool) the makeup air. As a result, a 30 percent reduction in energy costs was achieved.

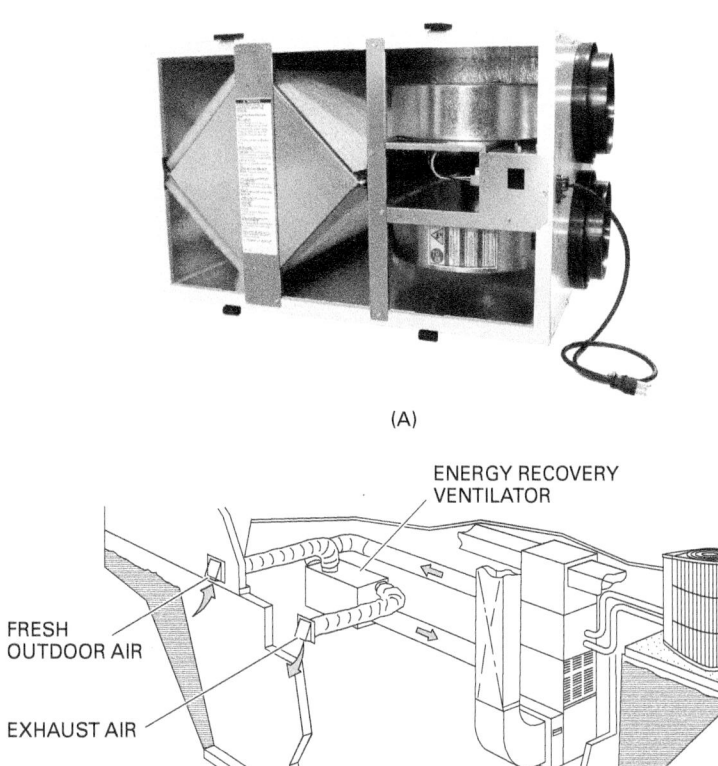

(A)

(B)

*Figure 1* Energy recovery ventilator.

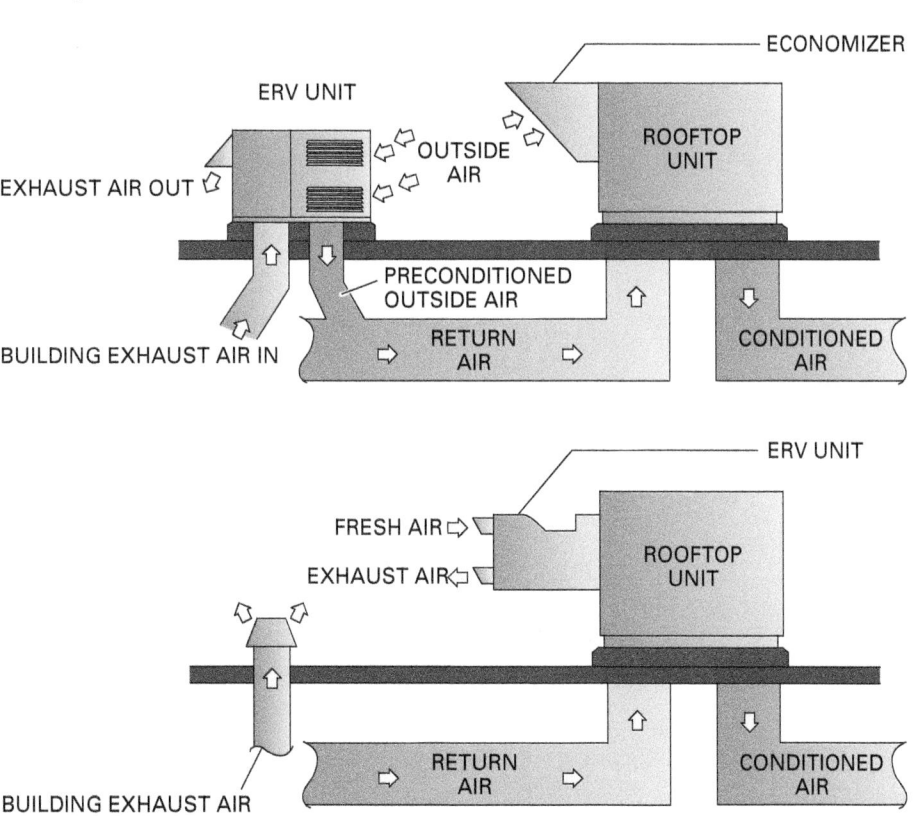

*Figure 2* ERV commercial application.

# Air-Side Economizers

An air-side economizer uses outdoor airflow as the medium to accomplish lower-cost cooling. It controls the amount of outside air brought into a conditioned space. A basic economizer system is shown here. It consists of a damper actuator assembly and a related economizer control module. Control signals applied to the economizer control module come from the thermostat located in the conditioned space, an enthalpy sensor located in the outdoor air duct, and a discharge air sensor located on the discharge side of the system evaporator coil. Changes in the air dry-bulb temperature and/or humidity determine the system changeover from cooling using compressor operation (mechanical cooling) to cooling using outside air (free cooling).

Some economizer systems use an optional second enthalpy sensor located in the return duct. When two sensors are used, the applicable term is *differential enthalpy*. The differential enthalpy economizer permits the use of outdoor air for cooling, but generally monitors and uses the air with the lowest enthalpy, regardless of whether it is suitable for total free cooling. It will use mechanical cooling unless outdoor air enthalpy is sufficiently low to manage the load.

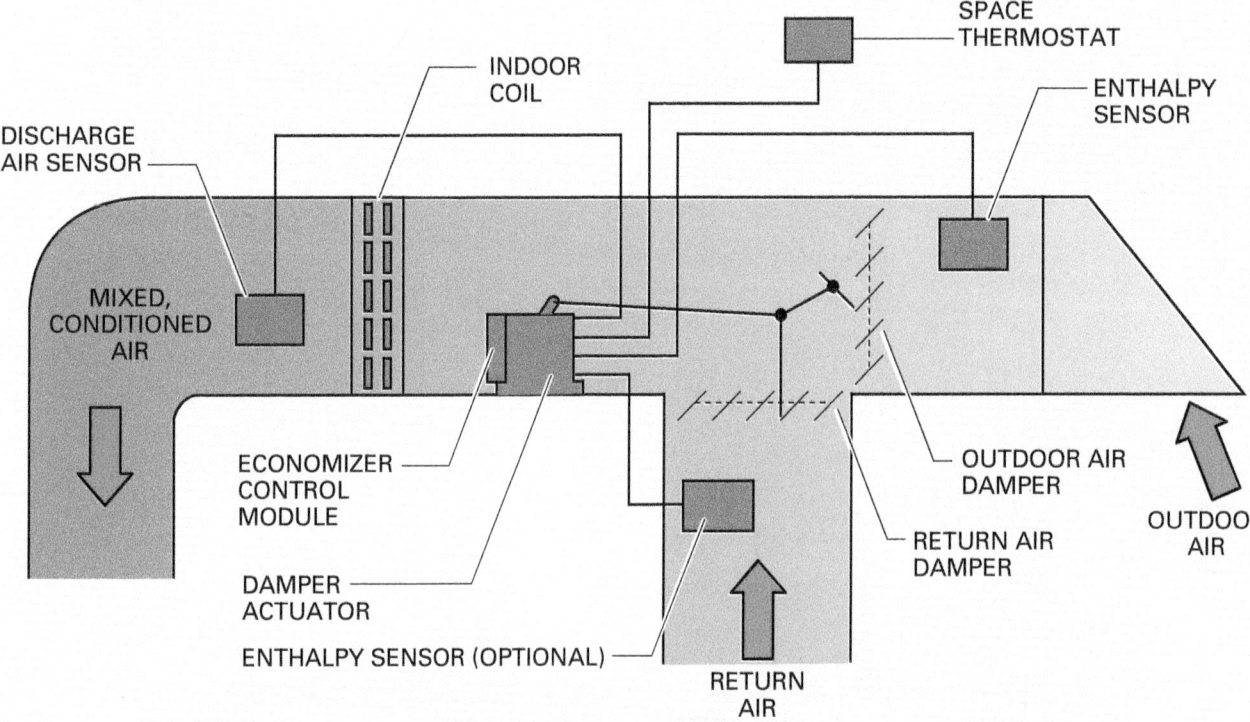

| MODE OF OPERATION | OUTDOOR AIR DAMPER | RETURN AIR DAMPER |
|---|---|---|
| OFF | CLOSED | WIDE OPEN |
| FAN ONLY & MECHANICAL COOLING | OPENS TO MINIMUM POSITION FOR VENTILATION | MODULATES TO COMPLEMENT OUTDOOR AIR DAMPER |
| FREE COOLING | MODULATES TO PROVIDE THE PROPER MIXED AIR TEMPERATURE | MODULATES TO COMPLEMENT OUTDOOR AIR DAMPER |
| HEATING | OPENS TO MINIMUM POSITION FOR VENTILATION | MODULATES TO COMPLEMENT OUTDOOR AIR DAMPER |

and HRV units typically require periodic inspection and maintenance as follows:

- Inspect and clean the intake and exhaust screens. These can be a source of particulate pollution which can reduce system flow rates.
- Wipe down and inspect the entire unit.
- Inspect and service the blower and wheel motors, belts, and drive mechanisms.
- Replace filters as specified by the manufacturer or building engineering specifications.
- Inspect and clean the exchanger cores, typically annually. Some cores may require washing or another permitted method of cleaning.
- After completing any maintenance, secure accesses to the equipment and ductwork, then confirm proper operation.

### 1.2.0 Fixed-Plate and Rotary Air-to-Air Heat Exchangers

Air-to-air heat exchangers are among the most common devices used to recover heat by transferring the heat between the supply and exhaust airstreams. There are air-to-air heat recovery devices that reclaim only sensible heat as well as those that reclaim total heat.

#### 1.2.1 Fixed-Plate Heat Exchangers

Fixed-plate heat exchangers are commonly used in ERV units (*Figure 3*). In colder weather, the ERV saves energy by transferring from 70 to 80 percent of the warmth contained in the heating system exhaust air to the cold ventilation air that is entering the building. In the summer, cooled indoor air flowing through the ERV unit is used to cool the warmer, incoming fresh air as it passes through.

During both the heating and cooling modes of ERV operation, the unit acts to improve the quality of the building air by allowing the exchange of stale indoor air for fresh outdoor air. Outdoor and indoor (return) airstreams are drawn by the ERV unit fan(s) through the ERV unit heat exchanger core and are then discharged from the ERV exhaust and supply air ducts.

The heat exchanger core is a fixed-plate air-to-air heat exchanger that contains no moving parts. It consists of alternately layered aluminum plates, separated and sealed, that form exhaust and supply airstream passages. Heat is transferred directly from the warmer air through the separating plates into the cooler air. This is done without mixing the two airstreams. The direction of the supply and exhaust airflow through the exchanger can be parallel, counter-flow, or cross-flow (*Figure 4*).

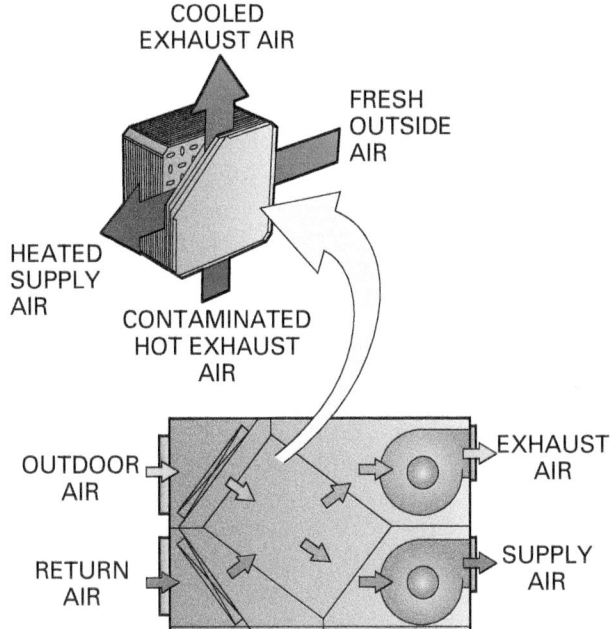

*Figure 3* Cross-flow ERV unit with a fixed-plate heat exchanger.

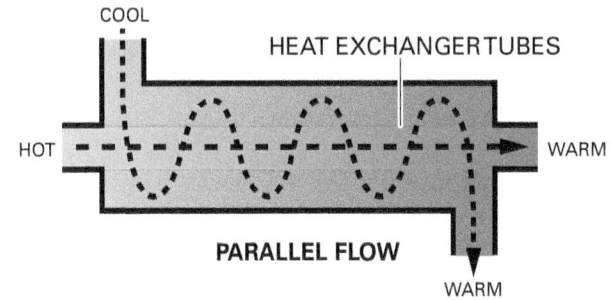

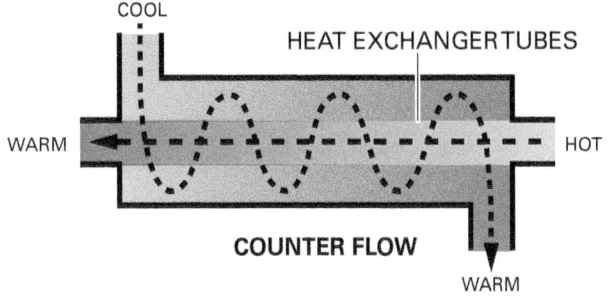

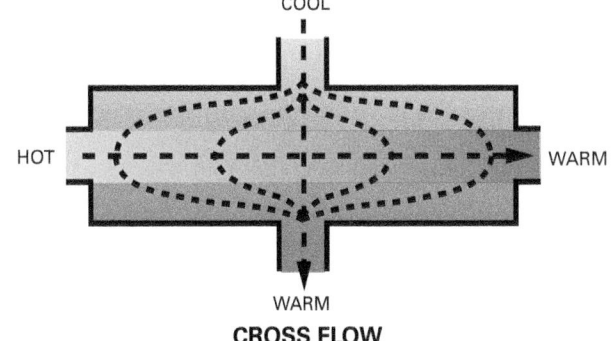

*Figure 4* Examples of heat exchange methods.

Fixed-plate heat exchangers used in ERVs achieve sensible heat recovery as well as latent heat recovery. In the winter, the result is that less facility energy is needed to heat the preheated fresh ventilation air than would be needed to heat cooler air that entered the building solely through natural infiltration or ventilation. Similarly, in the summer, less facility energy is needed to cool the precooled ventilation air than would be needed to cool warmer air that entered the building through natural infiltration or ventilation.

Because the fixed-plate heat exchanger in an ERV transfers some of the moisture from the exhaust air to the usually less-humid incoming winter air, the humidity of the building air during the winter tends to remain more constant. In the summer, the ERV fixed-plate heat exchanger transfers some of the water vapor contained in the more humid incoming air to the drier exhaust air leaving the building, thus providing for dehumidification of the incoming outside air. Because of the capability to transfer both sensible and latent heat, ERVs are recommended for use in most of the United States and some parts of Canada.

Some manufacturers offer an ERV with a fixed-plate heat exchanger that is coated with a desiccant so it can absorb moisture. While they are not able to handle extreme moisture, these heat exchangers provide effective energy and enthalpy transfer within reasonable limits.

Fixed-plate heat exchangers used in HRVs (which transfer only sensible heat) can accumulate frost as part of their normal operation when outdoor temperatures drop below freezing. The frost reduces and eventually blocks airflow through the heat exchanger, preventing a continuous supply of outdoor air. Some fixed-plate HRVs are designed to periodically defrost the heat exchanger for more efficient operation. Different manufacturers use different methods to defrost their equipment.

### 1.2.2 Rotary (Wheel) Heat Exchangers

A rotary air-to-air heat exchanger, also called an enthalpy wheel or heat wheel, consists of a motor-driven revolving cylinder containing heat transfer media through which the airstreams pass (*Figure 5*). The supply and exhaust airstreams flow through half of the heat exchanger in a counterflow pattern. To minimize the mixing of the two airstreams (cross-contamination), a partition or purge section separates the wheel sections. These barriers prevent exhaust air trapped within the heat-transfer media from being carried over to the supply side.

Rotary heat exchangers are available that can recover either sensible heat only or total heat, depending on the type of heat transfer media used in the wheel. Types of wheel media commonly used in sensible heat recovery units include aluminum, copper, stainless steel, or Monel®. Monel® is an alloy made of nickel, copper, iron, manganese, silicon, and carbon that is very resistant to corrosion. Media used in total heat recovery units include several kinds of metal or synthetic materials that are treated with a desiccant, such as lithium chloride or other mineral compound (zeolite).

As shown in *Figure 6*, sensible heat is transferred as the media picks up and stores heat from the warmer airstream and releases it to the cooler one. Latent heat is transferred as the media condenses moisture from the airstream with the higher humidity. It does this either because the media temperature is below its dew point, or by means of absorption (liquid desiccants) or adsorption (solid desiccants) with a simultaneous release of heat. Latent heat is also transferred by the release of moisture through evaporation (and heat pickup) into the airstream with the lower humidity ratio. Thus, moist air is dehumidified while the drier air is humidified. The transfer of sensible and latent heat occurs simultaneously.

The temperature below which frost will begin to accumulate (the frost threshold temperature) on the surface of a typical enthalpy wheel that moves both sensible and latent heat is generally 20°F to 30°F (–7°C to –1°C) lower than the frost threshold for a typical plate-type sensible heat exchanger. Therefore, enthalpy wheels can be used without frost protection in many regions where frost control is required for plate-type heat exchangers.

The capacity of rotary heat exchangers can be controlled by varying the speed of the wheel rotation via its drive motor. Another method commonly used is a supply-air bypass control. This method uses an air bypass damper controlled by a supply-air discharge sensor. The bypass control determines the proportion of supply air allowed to flow through and bypass the heat exchanger wheel.

Rotary heat exchangers should be maintained as directed by the manufacturer. The wheel media should be cleaned when lint, dust, or other foreign materials accumulate. Media with liquid desiccants for total heat recovery must not be wetted with cleaning agents.

Enthalpy heat exchangers are often used in ERVs. However, the enthalpy ERV may not be cost-effective in all applications because of the high initial cost and the increased maintenance

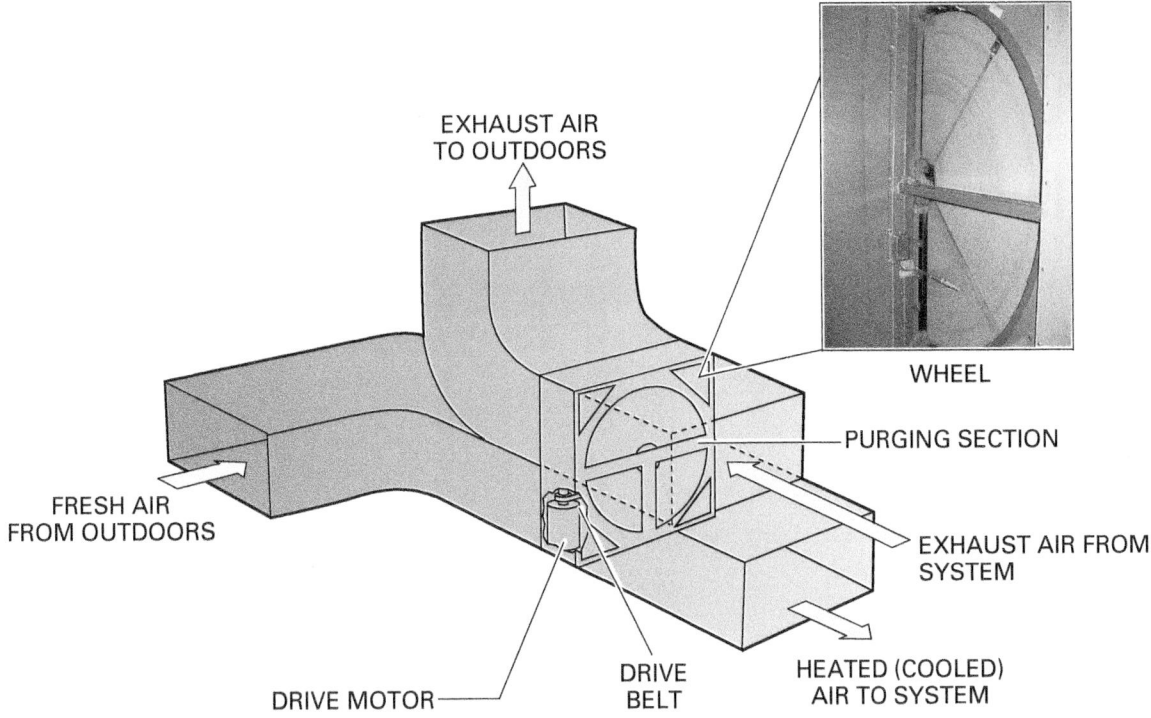

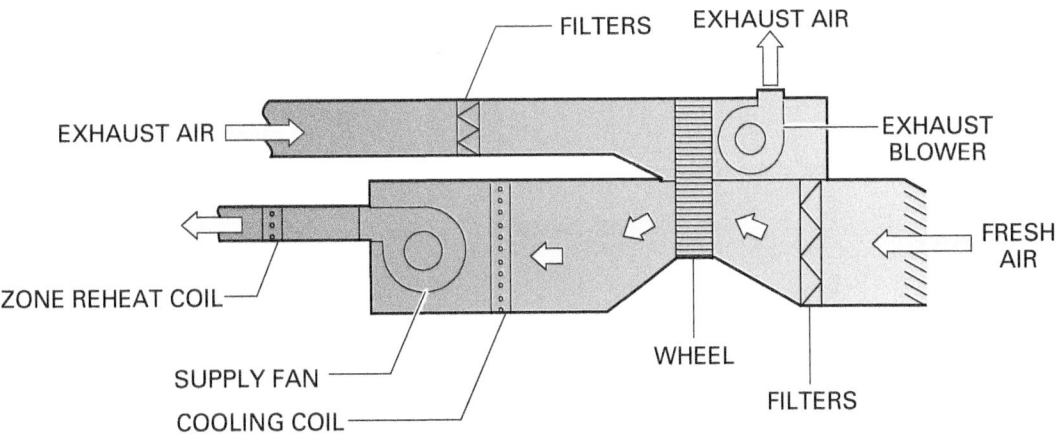

*Figure 5* Rotary air-to-air heat exchanger (heat wheel).

requirements. A cost-effective application would be in a building where the sensible heat load is relatively low in contrast to the latent heat load, such as in theaters and office buildings in areas with high relative humidity. In these environments, the cost of the enthalpy wheel ERV is a trade-off against the cost of higher-capacity cooling equipment that would otherwise be needed.

## 1.3.0 Condenser Heat-Recovery Systems

The use of rejected condenser heat in an HVAC system is not uncommon. A second condenser is used to extract heat from the hot refrigerant gas before it enters the condenser where it would

normally be rejected to the outside air or a water source. This recovered heat is then transferred to air or water that is used for another productive purpose.

### 1.3.1 Air Conditioning/Refrigeration System Condenser Heat Recovery

A dual-condenser refrigeration system that could be used in industrial buildings, or even supermarkets, is shown in *Figure 7*. When the building thermostat calls for heat, a refrigerant hot-gas diverting valve opens and routes the refrigeration system discharge gas through the heat recovery condenser. This air-cooled condenser is in the

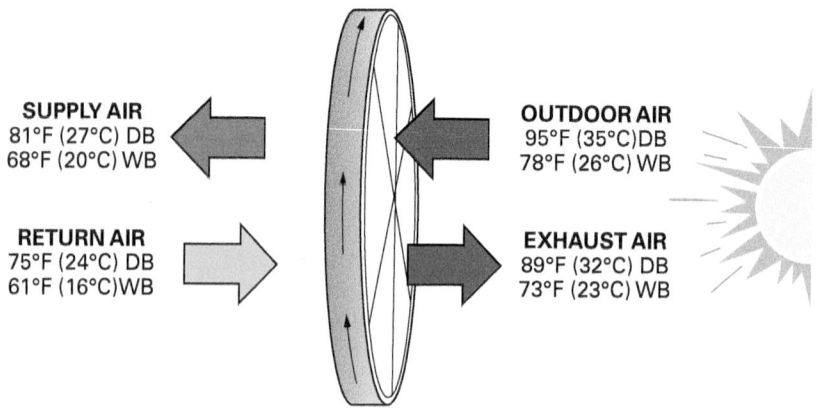

**SUMMER CONDITIONS**

SUPPLY AIR
81°F (27°C) DB
68°F (20°C) WB

OUTDOOR AIR
95°F (35°C) DB
78°F (26°C) WB

RETURN AIR
75°F (24°C) DB
61°F (16°C) WB

EXHAUST AIR
89°F (32°C) DB
73°F (23°C) WB

**WINTER CONDITIONS**

SUPPLY AIR
53°F (12°C) DB
40°F (4°C) WB

OUTDOOR AIR
7°F (−14°C) DB
6°F (−14°C) WB

RETURN AIR
72°F (22°C) DB
54°F (12°C) WB

EXHAUST AIR
27°F (−3°C) DB
20°F (−7°C) WB

*Figure 6* Heat wheel operation.

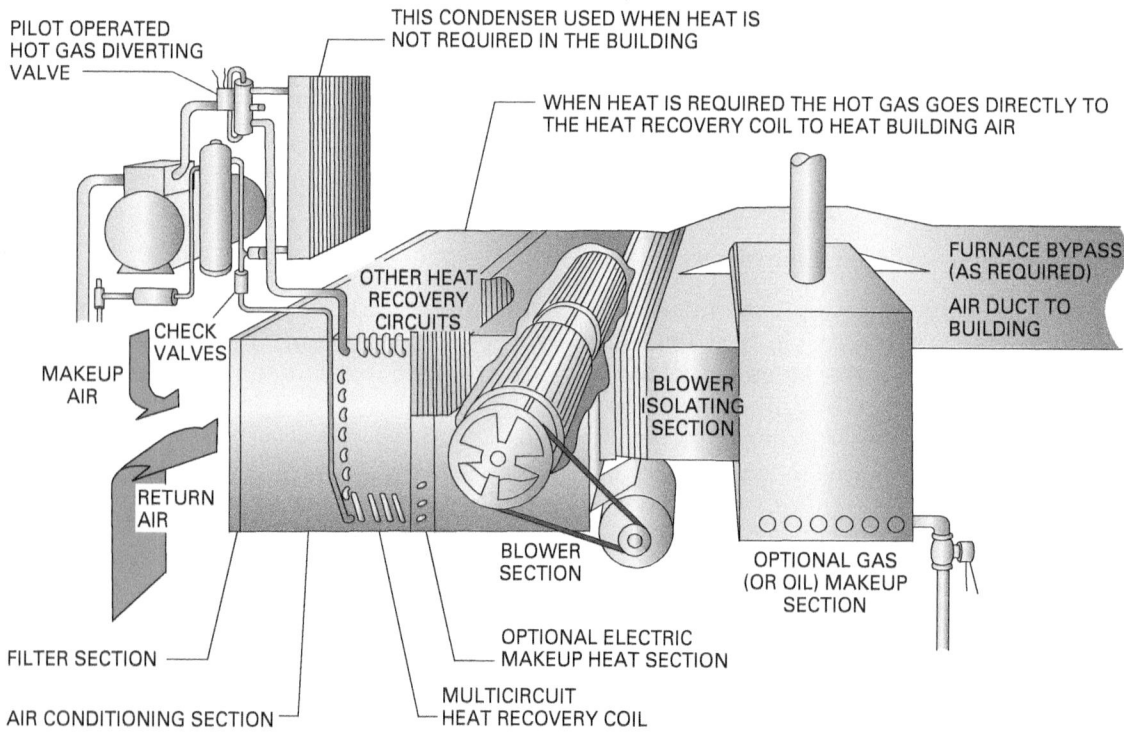

PILOT OPERATED HOT GAS DIVERTING VALVE

THIS CONDENSER USED WHEN HEAT IS NOT REQUIRED IN THE BUILDING

WHEN HEAT IS REQUIRED THE HOT GAS GOES DIRECTLY TO THE HEAT RECOVERY COIL TO HEAT BUILDING AIR

OTHER HEAT RECOVERY CIRCUITS

CHECK VALVES

MAKEUP AIR

RETURN AIR

FURNACE BYPASS (AS REQUIRED)

AIR DUCT TO BUILDING

BLOWER ISOLATING SECTION

BLOWER SECTION

OPTIONAL GAS (OR OIL) MAKEUP SECTION

FILTER SECTION

AIR CONDITIONING SECTION

MULTICIRCUIT HEAT RECOVERY COIL

OPTIONAL ELECTRIC MAKEUP HEAT SECTION

*Figure 7* Dual-condenser system.

building's air handling unit ductwork. As building air flows through the recovery condenser, the heat in the refrigerant gas is rejected to the cooler air. In turn, the heated air is circulated by the air distribution system blower(s) for subsequent dispersal through the building.

When the heating demand is greater than can be supplied by the recovery condenser, the difference is made up by either gas or electric heaters. When no building heat is required, the hot-gas diverting valve directs the gas to circulate through the refrigerant circuit's primary air-cooled condenser, where the heat is rejected to the atmosphere. This type of system can also be used to preheat water for domestic or laundry use.

A practical example of the application of this type of system is a factory in which refrigeration is used as part of an industrial process or to chill edible products in a food processing plant. The condenser heat from the industrial process can be used to heat office areas in the factory or to provide heat for another industrial process.

### 1.3.2 Chilled-Water System Heat Recovery

A chiller equipped with a heat recovery condenser is referred to as a chiller with a double-bundle condenser. Double-bundle condensers are typically formed by two independent water circuits enclosed in the same condenser shell with a common refrigerant chamber. The hot compressor discharge gas can be directed to the heat recovery condenser, the cooling condenser, or both (*Figure 8*).

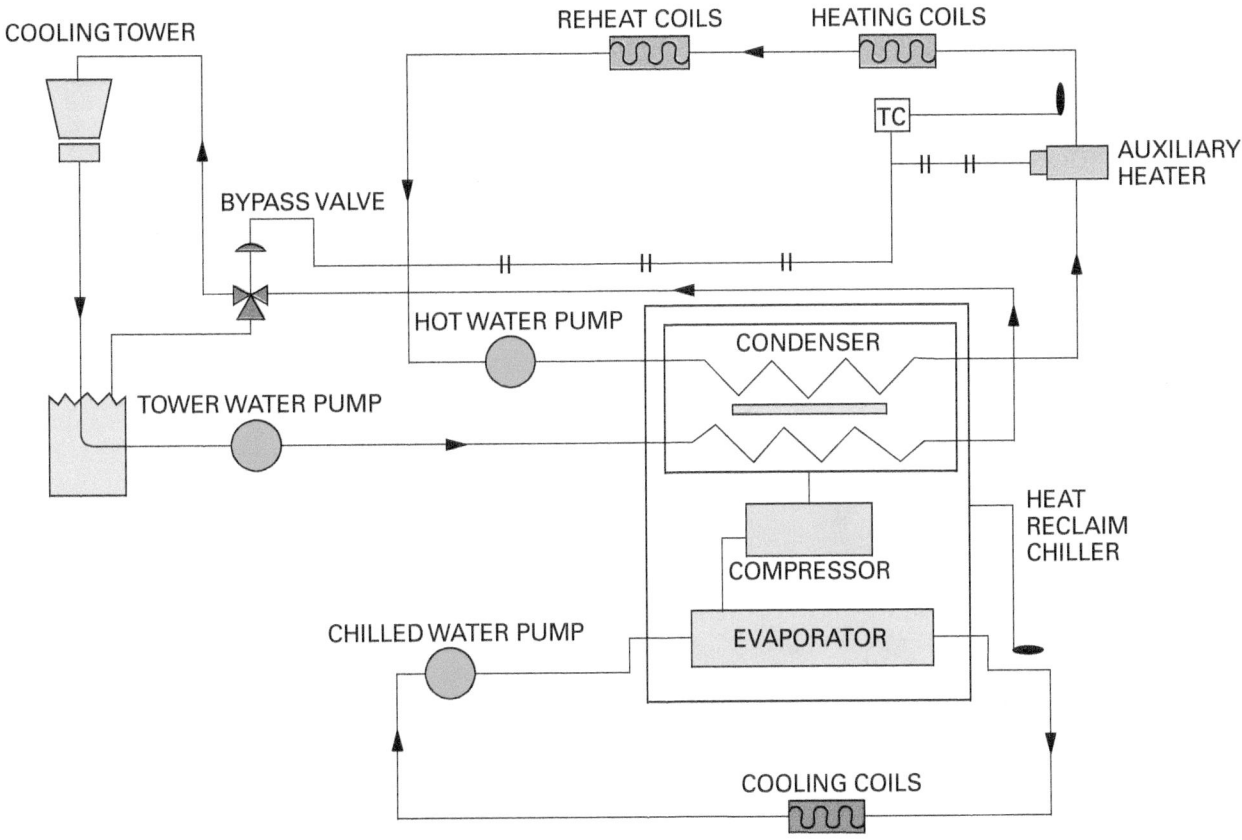

*Figure 8* Double-bundle heat reclaim system.

## Heat Conversion

The ability of a refrigeration system to move heat from one place to another makes all refrigeration systems, in effect, heat pumps. In some commercial refrigeration applications, the heat removed can be used for other useful purposes.

For example, in a large meat-packing plant, hundreds of beef carcasses must be cooled down. The evaporator coils in the meat lockers absorb tremendous amounts of heat from the meat as it cools. Instead of rejecting that heat to the outdoors in the usual manner, it can be transferred via heat exchangers for use in other industrial processes, to heat water, or to heat an office complex.

The heat recovery portion of the double-bundle condenser is piped into the building heating circuit and can supply all the building heating needs up to the total heating rejection capacity of the chiller bundle. When the available heat generated from the chiller system exceeds the building load, the surplus heat is rejected into the atmosphere by the cooling tower. In other words, the chiller simply returns to normal operation and disables the heat recovery process. When the building heating load is greater than can be supplied by the chiller system, the difference is made up by an auxiliary heater. The system is controlled by a heated water temperature controller that acts to control the tower bypass valve and auxiliary heater.

### 1.3.3 Swimming Pool Heat-Recovery Systems

Indoor swimming pools provide an excellent opportunity for heat recovery. Specialized high-capacity dehumidification systems, such as the one shown in *Figure 9*, are specifically designed for the high-moisture environments found in swimming pool structures, as well as in some commercial and industrial environments. This system uses heat pump technology to cool and dehumidify air

# Evaporative Pre-Coolers

When a compressor has to work harder, it consumes more energy. High head pressure caused by a high ambient temperature increases the load on the compressor and can make it work harder. If the compressor can be made to work less under the same ambient conditions, power consumption will drop. In hot, dry, desert areas, evaporative pre-coolers can be used to cool the air before it enters the condenser coil. Cooler air entering the condenser causes head pressure to drop. As a result, the compressor consumes less power.

An evaporative pre-cooler is nothing more than an evaporative cooler. It does not contain a fan but relies on the air being drawn in through the condenser coil for its airflow. The pre-cooler is positioned in front of the condenser coil. Water is applied from the top to wet the pad. The air being drawn across the condenser coil first passes through the pre-cooler where it is cooled.

Like all evaporative coolers, pre-coolers are less effective as the relative humidity increases, and they can consume a great deal of water. In addition, they have the same maintenance issues that all evaporative coolers have.

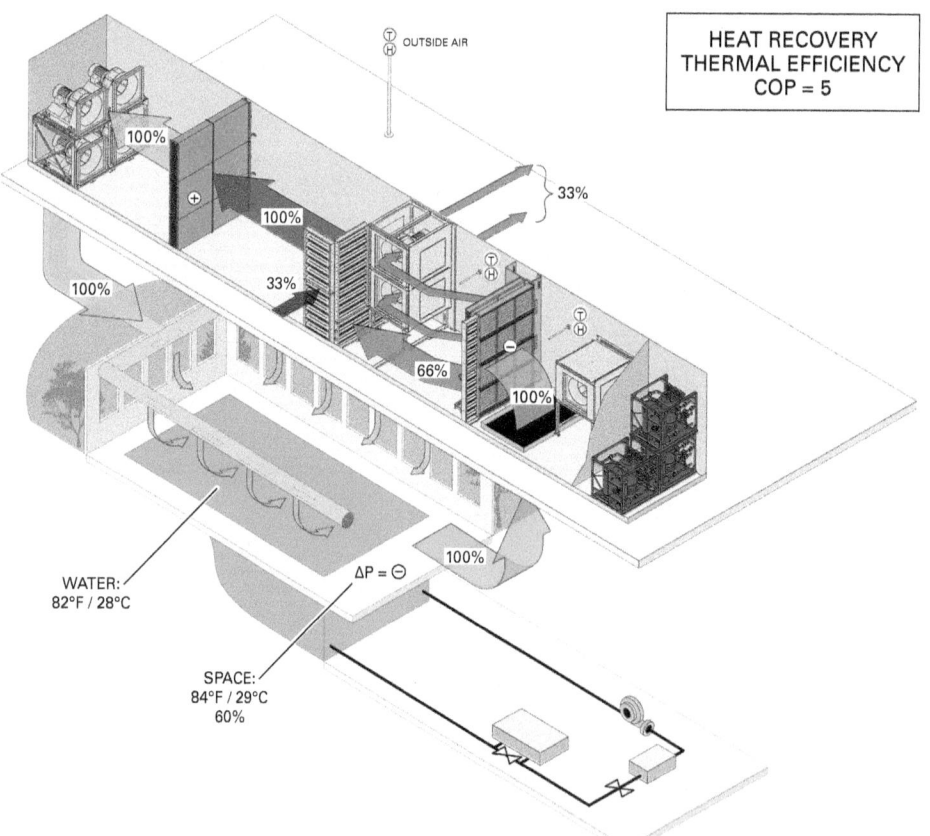

Figure 9 Swimming pool heat-recovery system installation.

from the pool enclosure while the heat from the warm, humid air in the pool enclosure is recovered for reuse. The recovered heat can be used to heat the structure or the pool water.

As shown in *Figure 10*, the hot, high-pressure gas leaving the compressor can be routed to the condenser/reheat coil, pool water condenser, or auxiliary condenser, as needed. A microprocessor control activates the solenoid valves based on demand. Liquid refrigerant leaving the condensers is stored in the receiver. As the refrigerant passes through the expansion valve, it is expanded to the operating pressure and temperature of the evaporator so that it can absorb heat from the air in the pool area. *Figure 11* shows the physical arrangement of the unit.

## Controlling Humidity

You may have noticed that paper tends to curl up in hot, humid weather. This fact led to the invention of modern air conditioning. The air conditioning system designed by Dr. Willis Carrier in 1902 was developed specifically to control heat and humidity in a paper-manufacturing facility. The technology was later applied to other industries, including textile manufacturing. These applications led to the recognition of the potential value of this technology, and eventually to its widespread use in comfort air conditioning.

## London Aquatics Centre

A major goal for London's 2012 Summer Olympics was to be the first truly sustainable Olympic Games. To that end, a concerted effort was made to promote energy conservation at the London Aquatics Centre, one of the main venues for the Games. The Centre was, and still is, the largest swimming venue in the United Kingdom. It accommodated 2,500 spectators; two temporary stands provided 15,000 additional seats during the 2012 Olympics. Achieving the desired comfort conditions while minimizing energy consumption in this venue was uniquely challenging.

The conventional energy-efficient approach would have been to install cogeneration, or combined heat and power (CHP), to power the building and use waste heat for heating the pool. In this case, that was not possible because the building is connected to the district heating network. Furthermore, the large volume of the main pool hall, which measures 180,000 cubic meters (6,357,000 ft$^3$), made it necessary to come up with some innovative ways of controlling the indoor environment and maximizing the building's energy-saving potential.

**Figure Credit:** Ben Sutherland, BBC

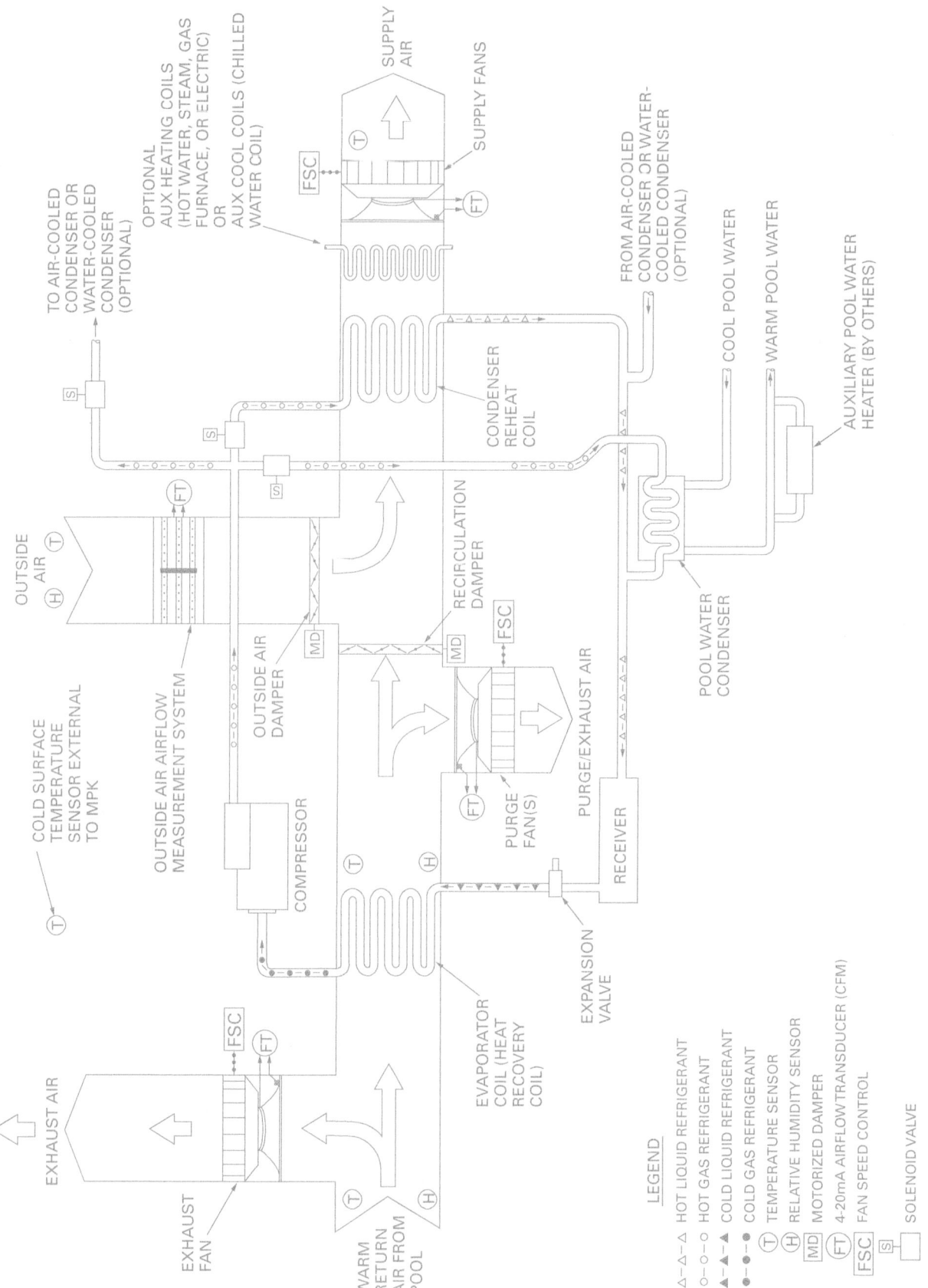

*Figure 10* Swimming pool heat-recovery system schematic.

NCCER – *HVAC Level Four*   03404

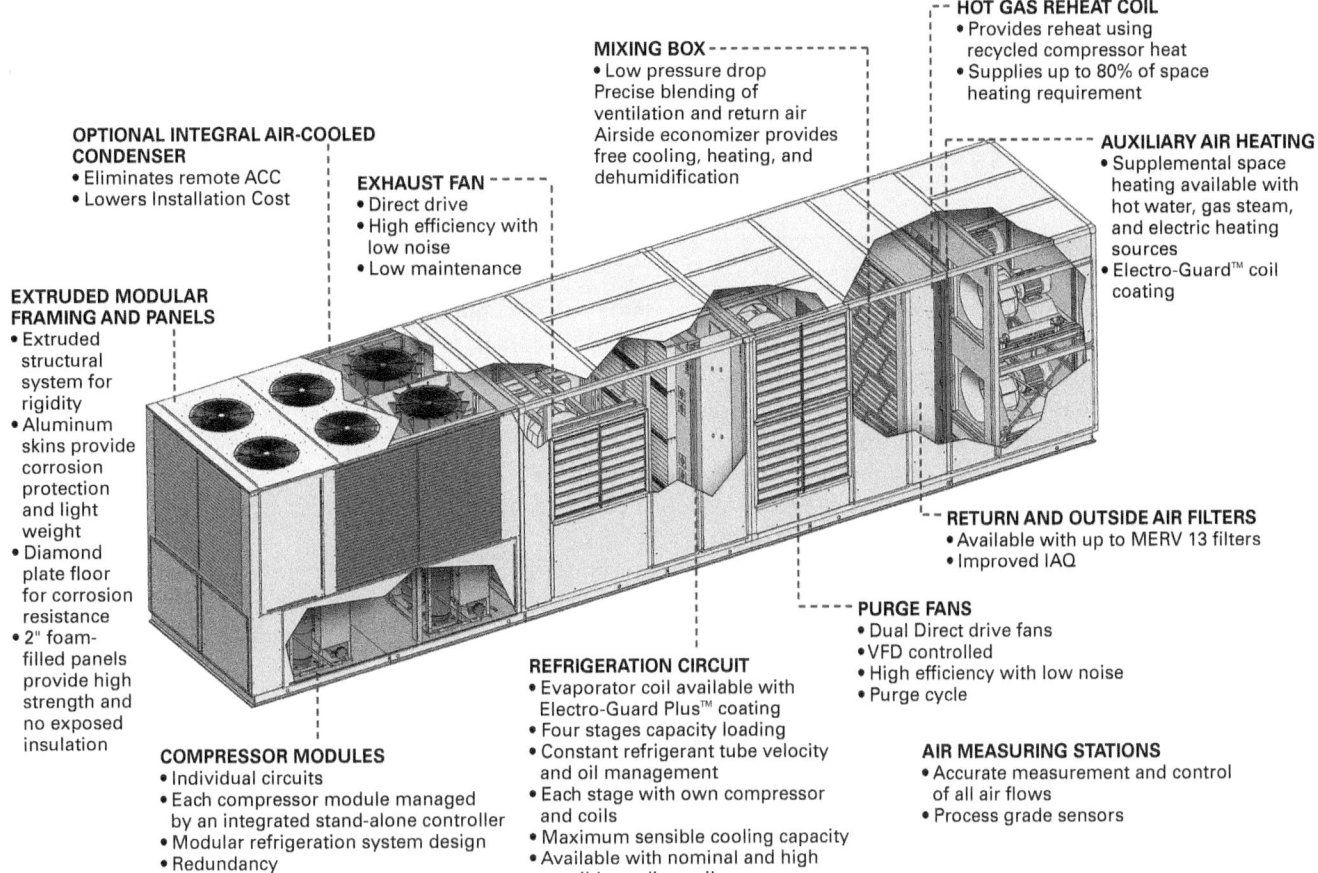

**OPTIONAL INTEGRAL AIR-COOLED CONDENSER**
• Eliminates remote ACC
• Lowers Installation Cost

**EXTRUDED MODULAR FRAMING AND PANELS**
• Extruded structural system for rigidity
• Aluminum skins provide corrosion protection and light weight
• Diamond plate floor for corrosion resistance
• 2" foam-filled panels provide high strength and no exposed insulation

**EXHAUST FAN**
• Direct drive
• High efficiency with low noise
• Low maintenance

**MIXING BOX**
• Low pressure drop Precise blending of ventilation and return air Airside economizer provides free cooling, heating, and dehumidification

**HOT GAS REHEAT COIL**
• Provides reheat using recycled compressor heat
• Supplies up to 80% of space heating requirement

**AUXILIARY AIR HEATING**
• Supplemental space heating available with hot water, gas steam, and electric heating sources
• Electro-Guard™ coil coating

**RETURN AND OUTSIDE AIR FILTERS**
• Available with up to MERV 13 filters
• Improved IAQ

**COMPRESSOR MODULES**
• Individual circuits
• Each compressor module managed by an integrated stand-alone controller
• Modular refrigeration system design
• Redundancy

**REFRIGERATION CIRCUIT**
• Evaporator coil available with Electro-Guard Plus™ coating
• Four stages capacity loading
• Constant refrigerant tube velocity and oil management
• Each stage with own compressor and coils
• Maximum sensible cooling capacity
• Available with nominal and high sensible cooling coils

**PURGE FANS**
• Dual Direct drive fans
• VFD controlled
• High efficiency with low noise
• Purge cycle

**AIR MEASURING STATIONS**
• Accurate measurement and control of all air flows
• Process grade sensors

*Figure 11* Swimming pool heat-recovery unit.

Other environments in which this dehumidification and reheat technology can be used include museums, printing facilities, warehouses, plywood manufacturing facilities, and water treatment plants. These are all applications in which humidity control is critical.

### 1.4.0 Coil Energy-Recovery Loops

In some commercial buildings or factories, stale or contaminated air must be exhausted and fresh air brought in. To prevent energy from being lost in the exhaust air and to condition the incoming air, a coil energy-recovery loop can be used. The typical coil energy-recovery loop (runaround loop) consists of two finned-tube water coils, a pump, a thermostatically controlled three-way valve, and related system piping (*Figure 12*).

The coils are connected in a closed loop by the piping through which water or another heat transfer fluid, such as glycol, is pumped. One coil is installed in the exhaust duct. The other coil is installed in the incoming air (supply) duct so that the incoming air that flows over the coil is preheated (or pre-cooled). Installation of the coils in these locations gives the greatest temperature difference between the outside air supply and

exhaust airstreams; therefore, maximum energy recovery is achieved.

Sensible heat is transferred between the exhaust and incoming supply airstreams without any cross-contamination. In comfort air conditioning systems, the heat transfer can be reversed. In winter, the supply air is preheated when it is cooler than the exhaust air; in summer, the supply air is precooled when it is warmer than the exhaust air. The recovery efficiency for runaround loops averages between 40 and 65 percent. The recovery efficiency of the cooling cycle is somewhat less than that for heating because the temperature difference between the airstreams is not as great.

When a glycol solution is used as the intermediate heat transfer fluid in a runaround loop system, there is some protection against freezing. However, moisture must not be allowed to freeze in the exhaust coil air passages, either. The dual-purpose, three-way temperature control valve prevents the exhaust coil from freezing. This valve is controlled to maintain the water entering the exhaust coil at a temperature over 30°F (–1°C). Bypassing some of the warmer water (or glycol solution) around the supply air coil helps to control the desired condition.

# London Aquatics Centre – How They Did It

Some of the techniques that were employed to make the London Aquatics Centre sustainable included the following:

- The main pool was used as a heat sink for heating systems and the condenser water system.
- Sensible heat recovery was employed in the pool ventilation systems.
  - Four high-performance specialized air handling units (AHUs) supplied conditioned air through the pool surround ventilation grilles, and air was extracted through air slots that were integrated within overflow water channels. Most air distribution occurred within service ducts on each side of the pool tanks. The AHUs had double-pass plate heat exchangers, and the AHU control adjusted the fresh air ratio into the air system to control the humidity level in the space. This arrangement was able to achieve 84 percent sensible heat recovery in the pool ventilation systems.
  - The low-velocity warm air supply louvers provided comfortable conditions in the pool surrounds, while the self-balanced, low-level air extraction service limited the buildup and spread of moisture and pollutants. Because the heat recovery was dedicated to the pool surround, the total air change over the pool hall volume was only approximately one air change per hour.
- To keep both the athletes and the spectators comfortable, the main pool was created with a system of microclimates—for poolside, spectator seating, and the roof void—that enabled heating and cooling to be provided efficiently in the different areas of the building.
  - Poolside: The pool and surroundings were kept at 25°C to 28°C (77°F to 82°F) and 60 percent relative humidity.
  - Spectator seating: Cooler fresh air was supplied in the areas where spectators were seated. A separate ventilation system was turned on to provide supplemental natural ventilation when events were taking place. The system consisted of perforated plate grilles at each seat. Slightly cooler air at low velocity was supplied through the grilles. Air was extracted at the top of the raked seating area. Two 390 kW water-cooled ammonia chillers provided cooling energy whenever fresh cooling air was required to cool the supply air. Each of the chillers was connected to a pair of dry air coolers located on the south end of the building and enclosed by a green wall system. The condensing water circuits were connected with the pool heating circuit for diverting waste heat for pool water heating, which reduced the overall heating demand.
  - Roof void: The main pool hall was enclosed by a two-way spanning steel roof, and there was a ceiling on the lower surface of the roof trusses that created a substantial roof void zone, which was insulated. To ensure air movement in the roof space, circulation fans were located at the walkways and operated on a cyclical basis or whenever the roof void temperature fell below a setpoint. Sensors monitored the roof temperature and humidity to control void temperature so that it was always above dew point.

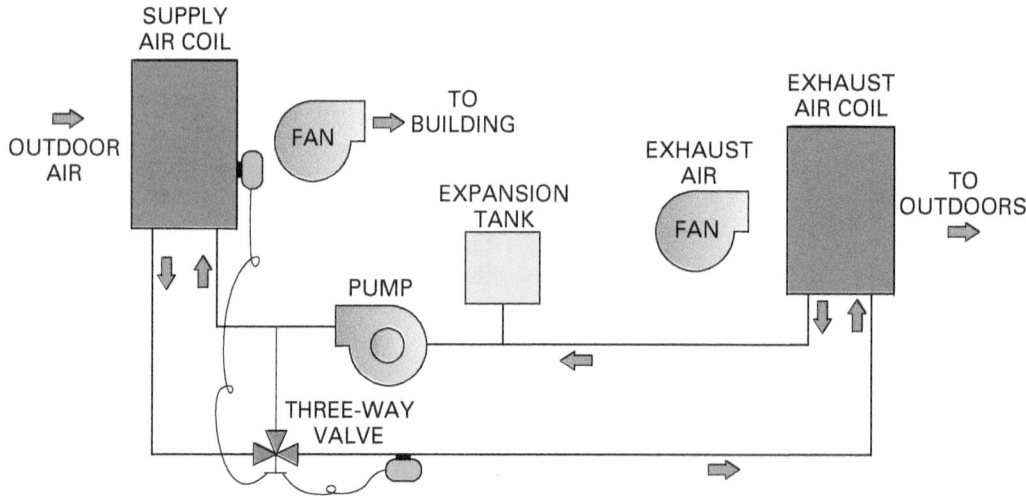

*Figure 12* Coil energy-recovery loop system.

## 1.5.0 Thermosiphon Heat Exchangers

Thermosiphon heat exchangers are closed systems that consist of an evaporator, condenser, interconnecting piping, and a two-phase (liquid and vapor) heat transfer fluid (refrigerant). They are passive devices that require no outside energy input other than the thermal energy added to and removed from the fluid during operation.

These elements may be enclosed in a single-shell tube containing a wicking lining (heat-pipe); in the tube without a wick lining (sealed-tube thermosiphon); or they may be physically separated (coil-loop thermosiphon). In each type, the natural convection-driven circulation of the two-phase refrigerant acts to transfer heat between two airstreams.

Because part of the system contains vapor and part contains liquid, the pressure in a thermosiphon is determined by the liquid temperature at the liquid-vapor interface. If the surrounding air causes a temperature difference between the liquid and vapor regions, the resulting pressure difference causes the vapor to flow from the warmer region (evaporator) to the cooler region (condenser). This flow is maintained by condensation in the cooler region and evaporation in the warmer region.

Depending on the mounting orientation of thermosiphon exchangers, the transfer of heat can be in both directions (bidirectional) or in one direction (unidirectional). When the heat transfer is unidirectional, the evaporator and condenser must be located so that the condensate can return to the evaporator by gravity, since no pumps are used in thermosiphon systems.

### 1.5.1 Heat-Pipe Heat Exchangers

In a conventional air conditioning system (*Figure 13*), dehumidification of the air occurs at the system's cooling coil. The coil normally removes both sensible and latent heat from the entering air, which is a mixture of water vapor and noncondensable gases. Both lose sensible heat during contact with the first part of the cooling coil. Latent heat of water vaporization is removed only in the part of the coil that is below the dew point of the entering air.

When the coil starts to remove moisture, thus dehumidifying the air, the cooling surfaces receive both the sensible and latent heat load. An air conditioner needs to operate longer to remove large amounts of moisture in a hot, humid environment; therefore, it consumes more energy. The thermostat setpoint may be lowered by the user to allow the system to run longer, which removes

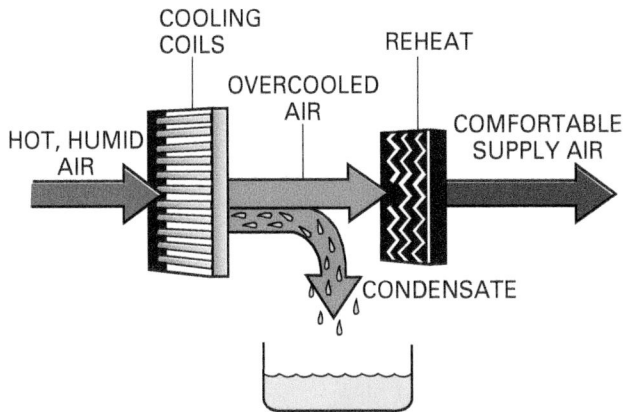

*Figure 13* Dehumidification in a conventional cooling system.

more moisture. However, this may result in the conditioned air being too cool for human comfort.

To remedy this condition, the air can be reheated before it is delivered to the conditioned space (*Figure 13*). Reheating may also be needed to decrease the relative humidity of the overcooled air. This reheating, which is often done using an electric reheat coil installed in the system, consumes a significant amount of electrical energy.

A heat-pipe exchanger is one type of thermosiphon device that can address these problems. It increases the dehumidification capacity of a system and reduces its energy consumption. It does this by pre-cooling the incoming air before it gets to the system cooling coil (*Figure 14*). The heat exchanger is an assembly formed by a bank of individual closed copper tubes with aluminum fins that are not interconnected. Each of these tubes is usually lined with a capillary wick, sealed at both ends, evacuated to a vacuum level, and charged with a refrigerant.

The individual heat pipes are assembled into finned banks inside a heat exchanger enclosure. When the exchanger is mounted in the system ductwork, the ends of the heat pipes in the hot duct act as an evaporator, while the ends of the pipes in the nearby cold duct act as a condenser. Heat-pipe heat exchangers are mainly sensible heat transfer devices, but condensation on the fins allows some transfer of latent heat.

During operation, the evaporator side of the exchanger is exposed to the incoming warm airstream. The evaporation of the refrigerant there precools the incoming air before it enters the active cooling coil. The warm refrigerant vapor rises up the pipe to the cooler condenser end at the downstream side of the duct cooling coil. The refrigerant vapor inside the tube transfers its heat to the cooler airstream and condenses. The cooled supply airstream warms to a more comfortable temperature.

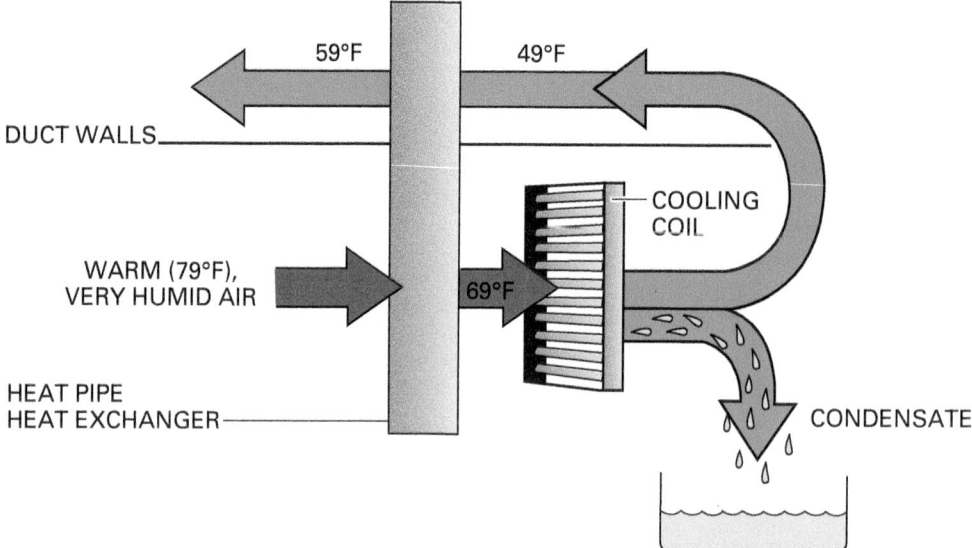

59°F

49°F

DUCT WALLS

WARM (79°F),
VERY HUMID AIR

69°F

COOLING
COIL

CONDENSATE

HEAT PIPE
HEAT EXCHANGER

*Figure 14* Heat-pipe heat exchanger.

After condensing, the liquid refrigerant in the condenser end of each heat pipe flows downhill to its evaporator end by gravity and/or the capillary wick. This closed-loop evaporation/condensation process in the heat-pipe heat exchanger continues as long as there is enough temperature difference between the two airstreams to drive the process. The amount of heat transferred by a heat-pipe heat exchanger can be controlled by changing the slope or tilt of the heat pipes between the two heat exchangers. This tilt control is normally done by a temperature-controlled actuator that rotates the exchanger around the center of its base.

Electronic humidistats, electronic motor speed controls, and variable-speed evaporator blower motors can be used to control the airflow over the evaporator coil so that dehumidification can take place without overcooling the room. Lower airflow over the evaporator coil allows the coil to extract much more moisture from the air than at higher airflows. Two-speed compressors and unloaders can also be used to control the performance of the evaporator to help prevent overcooling.

Heat-pipe exchangers require relatively little maintenance. The pipes themselves are sealed and should remain leak-free. Manufacturers recommend the following routine maintenance:

• Check for refrigerant leaks annually or during shutdown periods.

NOTE

Since heat pipe banks consist of individually sealed tubes, a leak in a single tube should not affect the others and will have minimal effect on the overall operation of the heat exchanger.

• Clean the heat pipes annually to ensure proper flow.
• Check and clean filters according to the manufacturer's or system's specified schedule.
• Check the air flow damper operation, if applicable.
• Inspect and clean the condensate drain pan.
• Test the performance of the heat exchanger annually. This requires measuring temperatures of supply and exhaust air at inlets and outlets, air flow rates, and operation of the tilt mechanism, if applicable.

### 1.5.2 Sealed-Tube Thermosiphons

Sealed-tube thermosiphon exchangers (*Figure 15*) are similar in construction to heat-pipe exchangers. Unlike the heat pipe, sealed-tube thermosiphons have no wick; they rely only on gravity to return the condensate to the evaporator end. Heat transfer will not take place if all the liquid resides at the cold end of the tube.

The sealed-tube thermosiphon exchanger works best when the exhaust and supply ducts are adjacent to each other. The evaporator and condenser regions are at opposite ends of a bundle of thermosiphon tubes. When the exchanger is mounted in the system ductwork, the ends of the tubes in the hot duct act as an evaporator, while the ends of the tubes in the cold duct act as a condenser.

### 1.5.3 Coil-Loop Thermosiphons

Thermosiphon loops (*Figure 16*) are used when the supply and exhaust air ducts are not mounted next to each other. A single closed loop consists of two coils interconnected by vapor and condensate return piping. The loop is charged with refrigerant in its saturation state, so that part of the loop is filled with liquid and part with vapor. The pressure in the loop depends on the type of refrigerant used and the fluid temperature at the liquid-vapor interface.

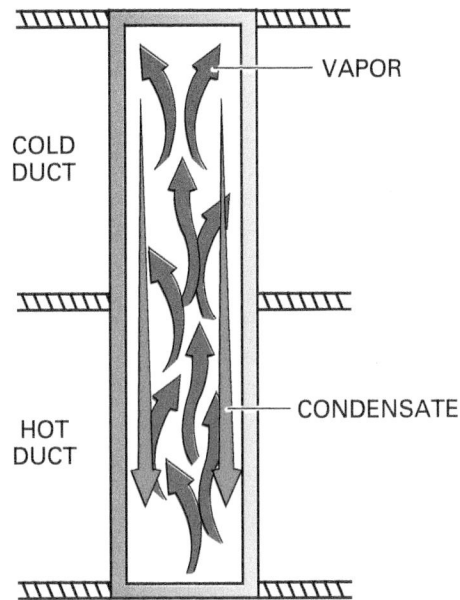

**UNIDIRECTIONAL SEALED TUBE THERMOSIPHON**

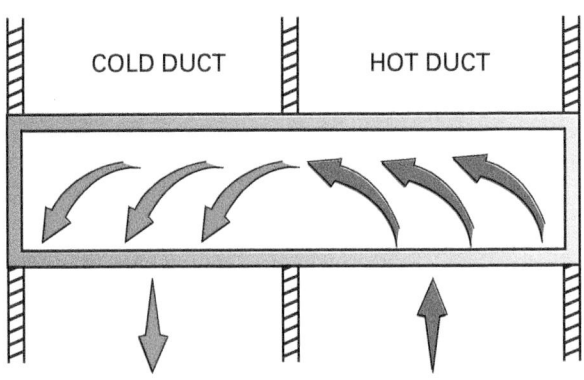

**BIDIRECTIONAL SEALED TUBE THERMOSIPHON**

*Figure 15* Sealed-tube thermosiphon recovery system.

Loops may be installed for unidirectional and bidirectional flow. Unidirectional loops are normally more efficient because the coil and loop charge can be selected to best satisfy only one function, rather than two functions (evaporation and condensation). When necessary, several coil-loop thermosiphons can be mounted in the supply and exhaust ducts to achieve a recovery effectiveness greater than that obtained with a single loop.

The routing of the interconnecting tubing must be considered because ambient conditions surrounding this piping can interfere with their successful operation. This is not an issue with heat-pipe and sealed-tube thermosiphons, because these flows operate over relatively short distances.

Note that the coil in the cold duct must be at a higher elevation than the coil in the hot duct (as shown in *Figure 16*) so that the fluid will flow by natural circulation.

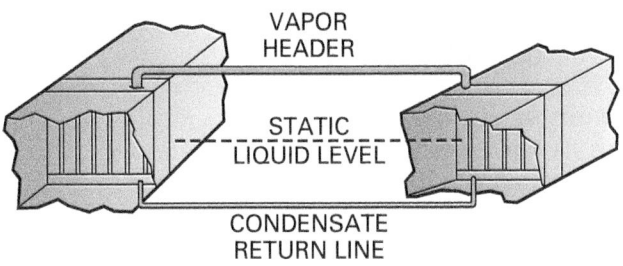

**BIDIRECTIONAL COIL LOOP**

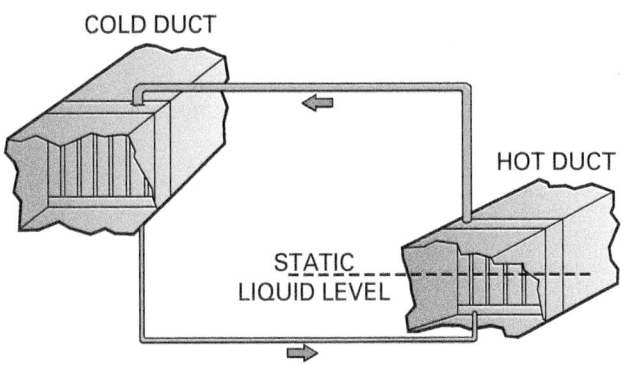

**UNIDIRECTIONAL COIL LOOP**

*Figure 16* Coil-loop thermosiphon system.

## 1.6.0 Twin-Tower Enthalpy Recovery Loops

A twin-tower enthalpy system is an air-to-liquid, liquid-to-air, heat recovery system. It consists of one or more towers (contactor towers) used to process the outdoor supply air, and one or more towers that process the building exhaust air (*Figure 17*). The towers may be oriented vertically or horizontally. Pumps continuously circulate the sorbent solution, typically lithium chloride and water, between the two towers. The fact that the two towers do not need to be adjacent to or even near each other allows them to be placed where it's most convenient—even on different floors of a structure.

In the towers, the sorbent solution is sprayed over the contact surfaces in the counter-flow (vertical tower) or cross-flow (horizontal tower) direction, where it comes into contact with the related airstream. Spraying the sorbent solution into the airstream enhances this contact. The air passes through demisters to remove droplets of the sorbent, which is highly corrosive.

A lithium chloride solution is strongly hydroscopic, especially at higher concentrations. The salt solution performs dehumidification when cool and concentrated and becomes more dilute as it absorbs water vapor. The solution itself acts as the transfer medium for latent and sensible heat between the two towers. When the solution is heated, either in an airstream or in the solution heater, the water evaporates. If this occurs in an airstream, humidification takes place. Loss of water increases the concentration of the sorbent solution. The system automatically adjusts the concentration of the sorbent solution using a balancing valve to obtain the optimum difference in sorbent concentrations in the two parts of the system.

Twin-tower enthalpy systems are used mainly for comfort air conditioning. The sorbent solution is also an effective antifreeze, allowing the system to operate in winter air temperatures as low as –40°F (–40°C). In the summer, they can operate with supply temperatures as high as 115°F (46°C). The solution transfers total heat as well as water vapor. Recovery efficiencies in the 60- to 70-percent range are typical.

When using the twin-tower system in colder climates, over-dilution of the sorbent solution can occur as the solution becomes saturated. This results in uneven supply air temperatures and humidity levels. To remedy this condition, a thermostatically controlled heater is often used to maintain constant-temperature supply air, regardless of the outdoor air temperature. The heater's control thermostat senses the air temperature leaving the supply tower and turns the heater on and off as needed.

## 1.7.0 Flue-Gas Heat Recovery System

The sensible heat available in the flue products of steam/hot water boilers can be reclaimed instead of wasted. This can be done by inserting a heat reclaimer (heat exchanger) in the flue between the boiler exhaust and the stack.

*Figure 18* shows a steam boiler with a stack heat exchanger that is used to preheat the boiler's feedwater supply. In this system, water from the boiler feed pump first enters the reheat tank. A portion of this water is pumped to the stack reclaimer heat exchanger, where it absorbs heat from the

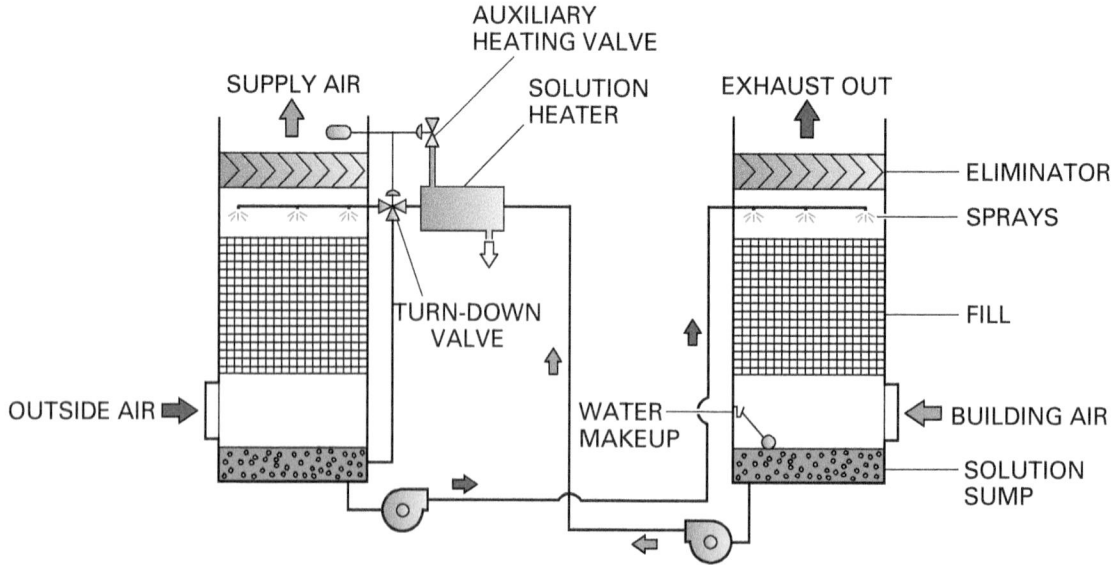

*Figure 17* Twin-tower enthalpy recovery loop.

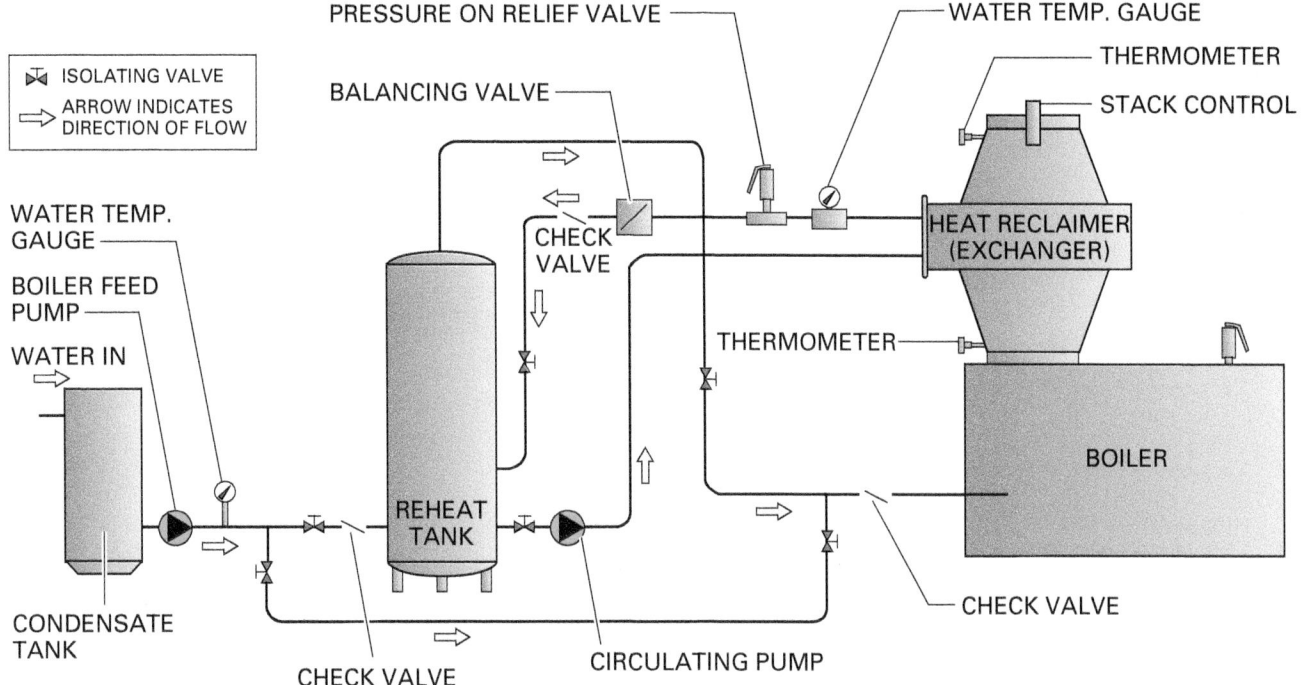

*Figure 18* Flue-gas heat recovery system.

exhaust flue. The now-hot water flows back to the reheat tank, heating the incoming feedwater from the feedwater pump. The warmed feedwater then enters the boiler.

By preheating the feedwater using waste heat, less fuel is needed to turn it to steam. This reduces the boiler burn time and can result in a typical increase in system efficiency of 10 percent or more. An automatic thermostat is normally used to control the feedwater circulating pump. A similar use involving the heat exchanger is to heat water used for the building's domestic hot water supply.

A major consideration when using a stack heat exchanger is that it must offer negligible resistance to the flow of the flue gases in the stack. A heat exchanger that offers too much resistance to flow can adversely affect the system by reducing both the flow and the temperature of the flue gases. This can cause moisture condensation in the flue and stack, leading to corrosion. A lower flue gas temperature can also adversely affect the flow of the flue gases out of the stack, posing a possible safety hazard.

Consistent cleaning and maintenance of the heat exchanger are required in order to ensure that it does not become clogged with combustion particulate byproducts. These particulates can block the flue and degrade heat transfer. Because of this requirement, it is very important that the heat exchanger be accessible to maintenance personnel.

### 1.8.0 Heat Recovery in Steam Systems

Heat recovery in steam systems is commonly done using direct heat recovery devices, such as heat exchangers/converters to heat air, fluid, or a process. In medium-pressure and high-pressure process steam systems, heat recovery can also be done using the heat in the system's liquid condensate to vaporize or flash some of the liquid to steam at a lower pressure. This lower-pressure steam can then be used for comfort heating.

### 1.8.1 Flash-Steam (Flash-Tank) Heat Recovery

Hot condensate in high-pressure steam systems is usually at or just below saturation temperature for the lower-pressure condensate system. This can cause some of the condensate to flash back into steam, causing pressure surges and steam binding in the condensate system. Engineers avoid these problems by directing hot condensate and steam trap discharges to flash tanks (*Figure 19*). In these tanks, flash steam can vent directly to the atmosphere or flow into a low-pressure steam main for reuse.

The heat content of flash steam is the same as that of live steam at the same pressure and temperature. Use of the flash steam in a low-pressure steam main allows this heat to be used rather than wasted. Low-pressure steam is commonly used for space heating and for heating or preheating water, oil, and other liquids.

---

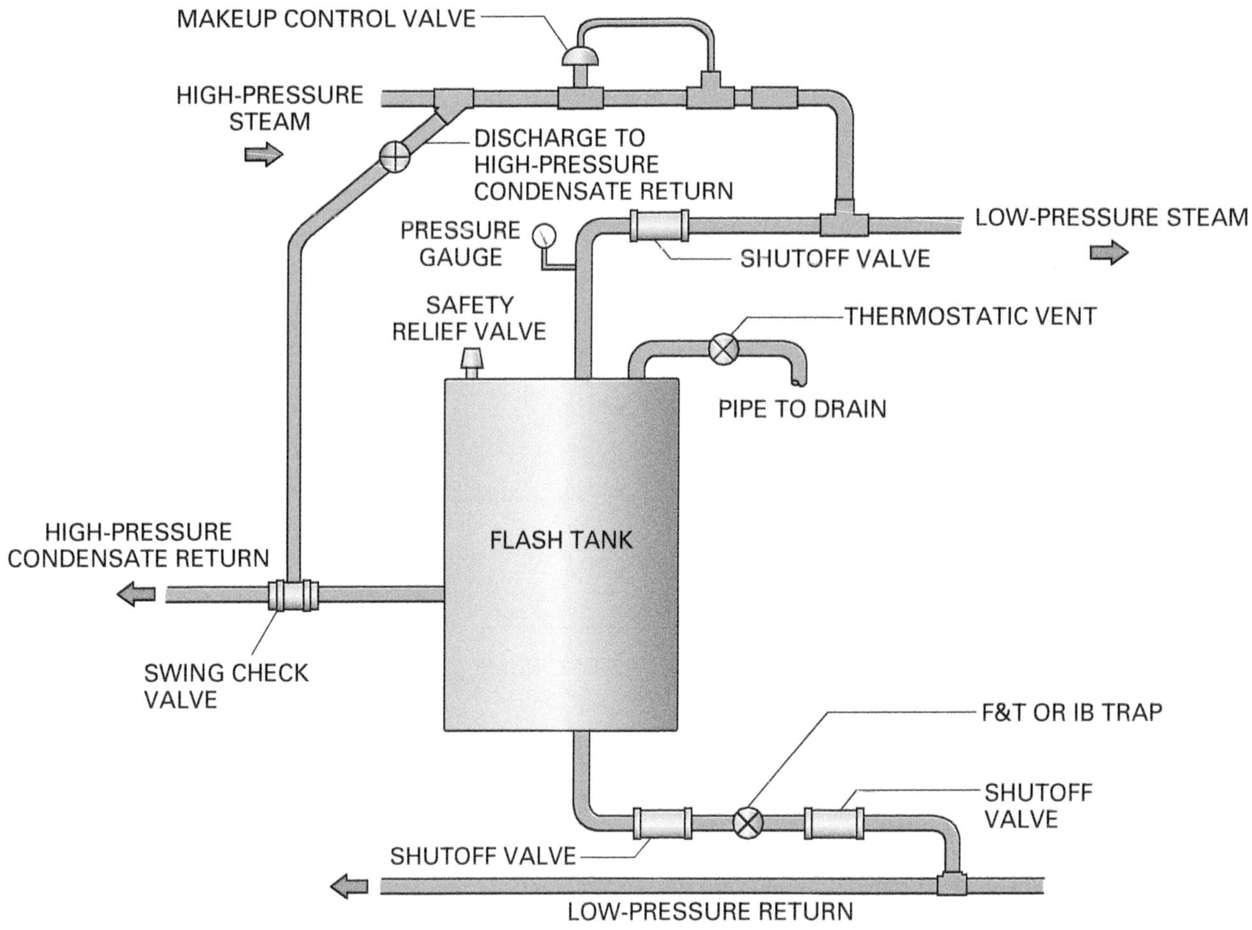

MAKEUP CONTROL VALVE

HIGH-PRESSURE
STEAM

DISCHARGE TO
HIGH-PRESSURE
CONDENSATE RETURN

LOW-PRESSURE STEAM

PRESSURE
GAUGE

SHUTOFF VALVE

THERMOSTATIC VENT

SAFETY
RELIEF VALVE

PIPE TO DRAIN

HIGH-PRESSURE
CONDENSATE RETURN

FLASH TANK

SWING CHECK
VALVE

F&T OR IB TRAP

SHUTOFF
VALVE

SHUTOFF VALVE

LOW-PRESSURE RETURN

*Figure 19* Steam system flash-tank schematic.

A typical flash-tank installation includes a high-pressure condensate inlet pipe, a low-pressure condensate discharge pipe, and an outlet connection to a low-pressure steam supply line. Flash steam and liquid water enter the tank via the high-pressure condensate line(s). Vertical flash tanks provide a better separation of steam and liquid condensate. The low-pressure flash steam enters the low-pressure steam header to serve the connected loads downstream. The liquid condensate in the tank enters the low-pressure condensate header via a float or thermostatic trap. The flash steam produced in the flash tank is often less than the amount of low-pressure steam that is needed. Therefore, a makeup reducer valve in the high-pressure steam line is used to supply any additional steam required.

For proper flash-tank operation, the condensate lines should pitch towards the tank. If more than one condensate line feeds the tank, a check valve should be installed in each line to prevent backflow. The top of the tank should have a thermostatic air vent to release any air that accumulates in the tank. The bottom of the tank should have an inverted bucket or float-and-thermostatic type steam trap.

The demand steam load on the flash tank should always be greater than the amount of flash steam available from the tank. If it is not, the low-pressure system could become over-pressurized. A safety relief valve must be installed at the top of the flash tank to protect the low-pressure line from over-pressurization.

### 1.8.2 Blowdown Water and Heat Recovery System

Blowdown water and heat recovery systems are used to recover heat from the boiler blowdown water and use it to preheat the boiler makeup water. Continuous or intermittent boiler surface blowdown, also known as skimming, is needed to purge the floating solids from a steam boiler system. This results in a constant heated loss.

Automatic control of the surface blowdown process is a part of a blowdown water and heat recovery system. A control valve in the unit senses the flow of makeup water and positions itself to maintain the desired ratio of blowdown flow to makeup water. As a result, the concentration of dissolved solids within the boiler is maintained automatically.

The control valve also provides for efficient heat recovery (about 90 percent) because the hot blowdown water flows through the heat exchanger only when there is a corresponding flow of cold makeup water. As skimmed water exits, makeup water enters to replace it. *Figure 20* shows a typical blowdown water and heat recovery system.

Some systems may include a flash-tank-like vessel for receiving the blowdown water. Flash steam can be condensed and returned to the boiler feedwater system while the concentrated blowdown water drains to the wastewater system. Codes in many jurisdictions require the blowdown water to be cooled to comply with environmental discharge rules before it is dumped into the municipal sewer system. Note the cold makeup water connection to the heat exchanger in *Figure 20*. This connection provides the means of cooling the blowdown water.

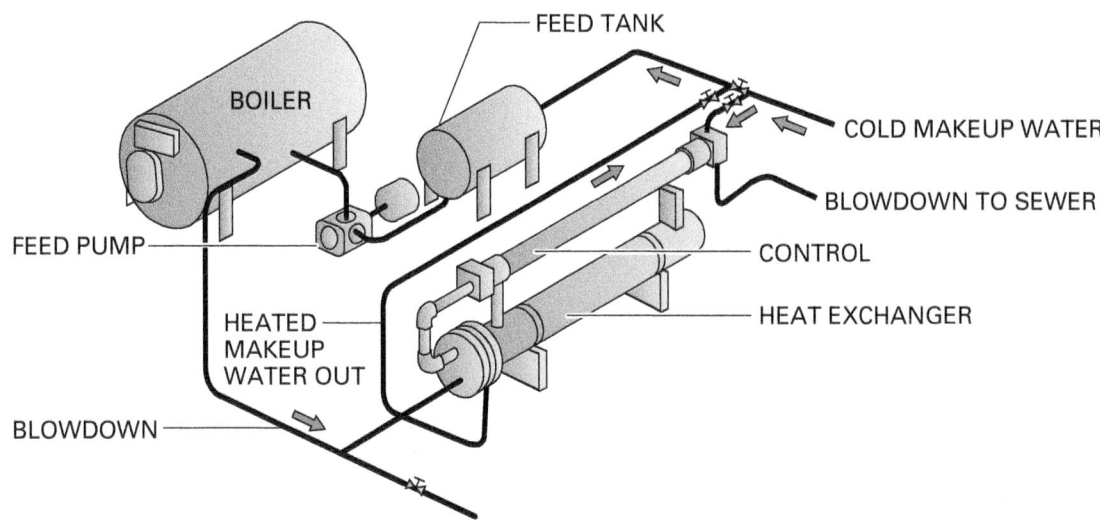

*Figure 20* Blowdown water and heat recovery system.

# Water-Side Economizers

Water-side economizers are often used in buildings that are cooled with chilled water. This type of economizer promotes energy conservation by using low-temperature cooling tower water to either pre-cool the entering supply air or to supplement mechanical cooling. If the cooling water is cold enough, it can be used to provide all of the system cooling. It has been found that the costs of a typical chilled-water plant are reduced by up to 70 percent during operation of a water-side economizer. Water-side economizers are best suited for climates where the wet bulb temperature is lower than 55°F (13°C) for 3,000 hours or more. These conditions apply to most of the United States, except for areas in the extreme Southwest and portions of the Southeast.

The economizer consists of a water coil installed upstream of the cooling coil. Cooling water flow is controlled by two valves, one at the input to the economizer coil and the other in a bypass loop to the condenser. One method of flow control keeps a constant water flow through the unit. In this mode, the two valves are controlled for complementary operation, where one valve is driven open while the other is driven closed.

Another method provides for variable system water flow through the unit. In this mode, the valve in the bypass loop is an on-off valve that is closed when the economizer is operating. Water flow through the economizer coil is adjusted by the valve in its input line. This varies the amount of water flow through the economizer coil in response to the system cooling load. As the cooling load increases, the valve opens more, increasing water flow through the coil. If the economizer valve is fully open and the economizer is unable to satisfy the system cooling load, the system controller turns on the system compressor to supplement the economizer cooling. When the unit is in the heating mode, both the economizer and bypass valves are closed.

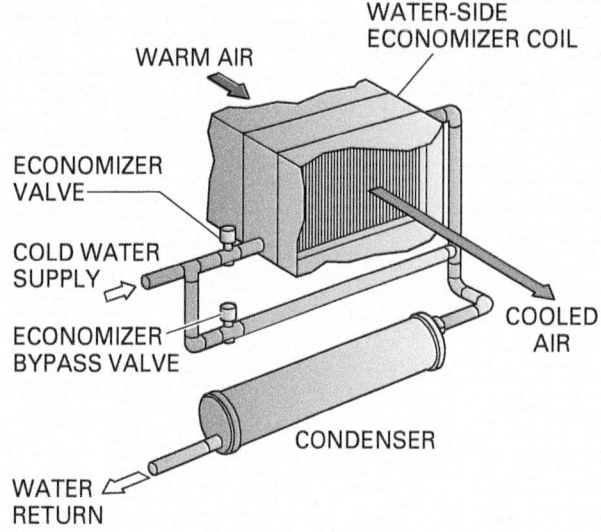

## Additional Resources

*ASHRAE Handbook—HVAC Systems and Equipment.* Atlanta, GA: American Society of Heating, Refrigerating, and Air Conditioning Engineers (ASHRAE), Inc.

*ASHRAE Handbook—HVAC Applications.* Atlanta, GA: American Society of Heating, Refrigerating, and Air Conditioning Engineers (ASHRAE), Inc.

## 1.0.0 Section Review

1. According to the US Department of Energy, what percent of the energy from the exiting air can most recovery ventilators recover?

   a. 20 to 30 percent
   b. 30 to 50 percent
   c. 45 to 60 percent
   d. 60 to 80 percent

2. The capacity of a rotary heat exchanger can be controlled by varying the _____.

   a. width of the purge section
   b. flow of air through the cooling coil
   c. flow of air through a return air bypass damper
   d. speed of the wheel rotation

3. To cool and dehumidify air from the pool enclosure, a swimming pool heat recovery system uses _____.

   a. a double-bundle coil
   b. heat pump technology
   c. flash evaporation cycles
   d. a specialized desiccant

4. The heat transfer fluid in a typical runaround loop system is either water or _____.

   a. glycol
   b. Monel®
   c. lithium chloride
   d. sorbitol

5. When a sealed-tube thermosiphon is mounted in system ductwork, the ends of the tubes in the cold duct act as a(n) _____.

   a. compressor
   b. flash tank
   c. condenser
   d. evaporator

6. The sorbent solution used for total heat recovery in a twin-tower enthalpy recovery loop is also an effective _____.

   a. algaecide
   b. antifreeze
   c. corrosion inhibitor
   d. desiccant

7. The main purpose of a flue-gas heat recovery system is to _____.

   a. preheat the boiler feedwater to reduce fuel consumption
   b. reduce flue temperatures to reduce stack corrosion
   c. produce flash steam for auxiliary steam systems
   d. deaerate the feedwater to reduce system corrosion

8. In a blowdown water and heat recovery system, the hot blowdown water flows through the heat exchanger only when _____.

   a. the steam system flash tank is fully charged
   b. the blowdown and makeup water temperatures are equal
   c. there is a corresponding flow of cold makeup water
   d. the system is in the off-peaking cooling mode of operation

---

 03404 **Energy Conservation Equipment**

## SECTION TWO

### 2.0.0 ENERGY-DEMAND REDUCTION AND ICE STORAGE SYSTEMS

#### Objective

Identify and describe the operation of electric energy-demand reduction and ice storage systems.

a. Identify and describe the operation of electric energy-demand reduction systems.
b. Identify and describe the operation of ice storage systems.

Approximately 40 percent of the energy consumption in the United States is in the electric power sector, and this continues to increase. Electric power demands vary with alternating periods of high (peak) and low demand, and the cost to produce electricity varies enormously according to the season and time of day. To maintain a reliable supply of electricity, electric utilities build new power-generating plants and maintain peaking units, such as combustion turbines, that are put into service only as needed to meet peak demands. Ultimately, the costs of building new plants and maintaining idle capacity is passed along to customers in the price they pay for electricity. In addition, with every fossil-fuel power plant that comes on line, carbon dioxide and other air pollution emissions increase.

Utilities and their customers can work together to address these issues by making use of electric energy-demand reduction systems. Facilities may also employ thermal energy storage systems, such as ice storage, that enable them to reduce energy use during the hours when electricity costs are normally the highest.

### 2.1.0 Electric Energy-Demand Reduction Systems

Electric utilities use various means to encourage customers to reduce consumption. For instance, they may perform free energy audits for customers, coupled with rebates and/or low interest loans for the purchase of energy-efficient equipment. Increasingly, utilities are finding that demand-side management (DSM) programs are a cost-effective way of addressing consumption demand challenges.

Basically, DSM programs decrease demand for energy by providing technologies and incentives that encourage customers to use less energy during peak hours, or to move the time of energy use to off-peak hours. These programs offer resources that can provide cost-effective substitutes for additional power plants. That is, DSM programs can provide the same types of services that a combustion turbine does, but at lower cost. These programs are often low-cost alternatives to the construction and operation of base-load coal and nuclear plants. Moreover, DSM programs help utilities reduce demands at critical (peak) times when it is most expensive to generate electricity.

### 2.1.1 Demand-Side Management Approaches

Three commonly used types of DSM are time-of-use (TOU) pricing, interruptible rates, and direct load control programs. All offer incentives that encourage customers to reduce or curtail their consumption of electricity during peak demand periods.

TOU pricing is generally aimed at large commercial and industrial users of electricity. It allows these customers to purchase their power at a lower rate during off-peak times. Typically, off-peak times are from midnight to 5:00 a.m., when demand is low. By encouraging large customers to use power during these times, the demand during peak usage hours is reduced.

---

### GOING GREEN

## Market Transformation Programs

Rather than focusing on individual customers, utilities sometimes run market transformation programs geared to influence the actions of contractors, builders, retailers, distributors, and even manufacturers. Here is one example:

Rather than offer a rebate to each customer who bought an energy-efficient refrigerator in 1994, a group of about 25 utilities promoted the Super-Efficient Refrigerator Program (SERP). They pooled about $30 million and ran a competition to select a manufacturer who could develop a refrigerator that was at least 30 percent more efficient than the 1993 federal standard. Whirlpool won the competition and its new, highly efficient refrigerators were featured in stores in the service areas of the participating utilities.

Applications of market transformation also have potential for the future. With the current rollout of smart grid technology, it may be that new appliances such as refrigerators and electric water heaters will be designed with controlling devices already built in for direct load control.

---

Interruptible rates are applied for customers who agree to service interruption during emergency or high-demand situations. In return, they receive a reduced rate, a credit on their bill, or other incentive. For instance, with the customer's agreement, some HVAC systems are equipped with power reduction features. One way to implement power reduction is for the utility to cycle equipment off during peak demand periods. The utility accomplishes this by sending a demand reduction signal to a device attached to the customer's HVAC equipment that interrupts power for a short time. The demand reduction signal can be sent by radio control (*Figure 21*), over phone lines, or through a modem connected to a building's computer-controlled energy management system.

The duration of the off cycle is long enough to reduce the utility's peak load but not long enough to significantly affect the indoor comfort of the user. If the utility has a large enough customer base participating in the demand reduction program, the duration of the equipment off-time cycle for each customer is reduced to the point that little individual comfort is lost while the peak demand is significantly lowered.

Direct load control programs in smart grid applications are perhaps the fastest growing form of DSM. These programs use digital (smart) equipment and technologies that allow customers' equipment or appliances to be directly controlled by the utility through switches or meter controls. The two most common targets of these programs are central air conditioning and electric water heaters, which make up 70 percent of the residential peak load nationwide. Electric clothes dryers are also high on the list of power consumption.

A smart grid is simply a digital grid that uses the same digital technology used in industries, such as telecommunications. The various components of a smart grid—hardware, software, consumer energy displays and controllers, etc.—create an advanced metering infrastructure (AMI) that enables two-way communication between a utility and its customers. Smart grids provide continuously updated information about a customer's energy use transmitted through electronic devices such as computers, smart phones, or energy management programs.

Smart meters (digital meters) are important components of smart grid systems. A smart meter records the amount of electric energy consumed in a given period (usually an hour or less) and regularly communicates that information to the utility. Information can also be sent from the utility back to the meter. Thus, the meter enables nearly real-time two-way communication

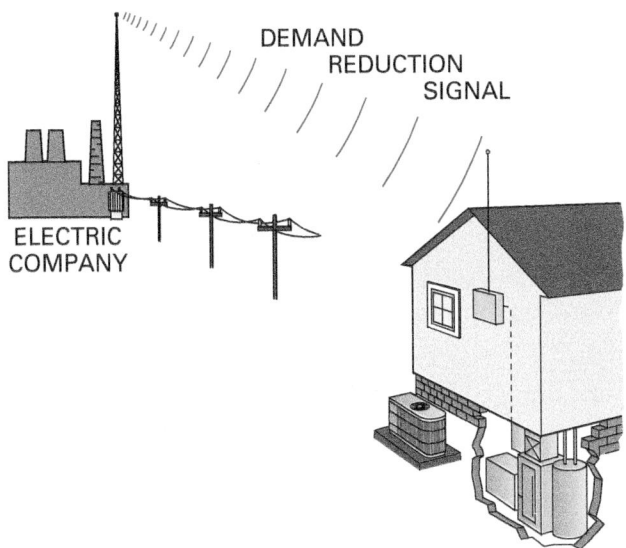

Figure 21  Utility demand reduction system.

### Electric Vehicles

Increasing numbers of electric vehicles and plug-in hybrid vehicles are being driven in the United States. As this trend continues, the economies of scale in regions with significant numbers of these types of vehicles could create a market for direct load control for electric vehicle charging stations.

**Figure Credit:** © Deanpictures/Dreamstime.com

between the meter and the central system. Smart meters can also gather data for remote reporting.

This two-way communication is critical for smart grid/smart meter applications. Traditional automatic meter reading does not qualify as smart technology because it is not capable of two-way communication. This older technology also has a much more limited range.

To date, there have been significant regional variations in the adoption of DSM and smart grid technologies. The majority of DSM utilities are in seven states—California, Florida, Massachusetts, North Carolina, New York, Washington, and Wisconsin. The leading utilities are concentrated along the east coast (especially the Northeast), west coast, and upper Midwest. In the Southeast, the Florida utilities, Duke Power (in North and South Carolina), and Georgia Power are among the leaders.

## 2.2.0 Ice Storage Systems

For many HVACR applications, peak cooling loads and peak utility demand tend to occur at the same time, making it difficult to apply demand reduction strategies such as interruptible rates or TOU pricing. An alternative for reducing HVAC energy consumption during peak periods is the use of thermal energy storage systems that shift cooling energy to non-peak times.

When cooling demand is low, thermal energy storage systems chill and/or freeze a storage media, usually ice. The stored energy in the ice is used later to meet air conditioning loads. This reduces operating costs by cooling with cheaper off-peak energy and reducing or even eliminating on-peak demands.

The basic equipment in a thermal storage system includes a tank containing the storage medium, a chiller or refrigeration system, and interconnecting piping, pumps, valves, and controls. *Figure 22* shows a typical arrangement for a thermal storage system.

Ice storage is often the system of choice for HVAC thermal energy storage because it can fit into relatively small spaces. Ice storage systems use the latent heat absorption capacity of ice at 32°F (0°C), the freezing point of water. The equipment must provide chilling fluid at temperatures of 15°F to 26°F (–9°C to –3°C), which is below the normal operating range of conventional stock chillers used for comfort cooling. Therefore, special ice-making equipment or standard chillers that are suitable for low-temperature applications are employed. The heat transfer fluid may be the refrigerant itself or a secondary coolant such as glycol with water or another type of antifreeze solution.

There are several available technologies for ice storage. The one that is currently most common for commercial HVAC applications is internal melt ice-on-coil. In this technique, ice forms on submerged pipes or tubes through which a refrigerant or a secondary fluid is circulated. The use of a secondary fluid, such as a glycol solution from a chiller, is the most popular method by far. The chiller operates at one temperature during occupied hours, and then resets the supply temperature to a value below freezing during the ice-building process. Ice-cooling occurs by circulating warm coolant through the pipes, which melts the ice from the inside and absorbs the latent heat of fusion from the coolant in the process.

Operating strategies are generally classified as either full-storage or partial-storage. This refers to the amount of cooling load that is transferred from on-peak to off-peak. In full-storage operation, the entire on-peak cooling load is shifted to off-peak

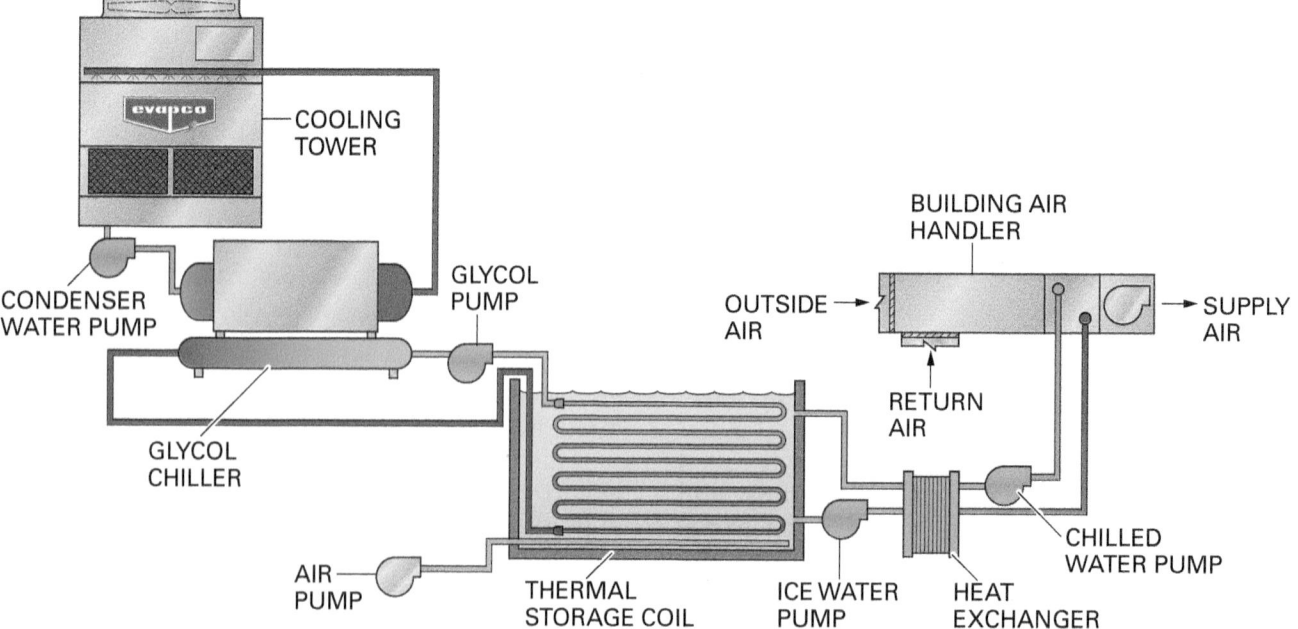

*Figure 22* Typical HVAC thermal storage system.

hours, and the system usually operates at full capacity to charge the ice storage during all non-peak hours. During on-peak hours, all cooling loads are met from storage, and the chiller does not operate. A full-storage system requires a relatively large chiller and storage capacities. It is generally used in applications in which on-peak demand charges are high or the on-peak period is short.

In the more widely used partial-storage operation, the chiller capacity is less than the design load. The chiller meets part of the on-peak cooling load, and storage meets the rest. This approach minimizes the required chiller and storage capacities. It is used most often in applications where the peak cooling load is much higher than the average load.

The partial-load strategy can be subdivided into demand-limiting and load-leveling. A demand-limiting partial-storage system operates the chiller at reduced capacity during on-peak hours. In a load-leveling system, the chiller typically runs at full capacity for the normal 24-hour cycle. When the load is less than the chiller output, the excess is used to charge storage. When the load exceeds the chiller capacity, the additional requirement is discharged from storage. Variations on the full- and partial-storage strategies can be achieved by operating multiple chillers.

Control strategies for partial-storage systems include chiller-priority control, storage-priority control, or a combination of the two. The chiller-priority strategy uses the chiller to directly meet as much of the load as possible. Cooling is supplied from storage only when the load exceeds chiller capacity. The storage-priority strategy meets as much of the load as possible from stored cooling; the chiller is used only when the daily load exceeds total stored cooling capacity. Some systems use combinations of these strategies. For example, chiller-priority may be used during off-peak hours and storage-priority during on-peak hours.

## 2.2.1 Ice Storage System Operation

The basic operation of one manufacturer's ice storage system, referred to as IceBank® energy storage or off-peak cooling, is illustrated in *Figure 23*. This example is a load-leveling partial-storage system that uses the internal melt ice-on-coil technique. IceBank® energy storage systems can also be configured for full-storage applications.

During off-peak hours, the cooling system chiller is used to make ice in large tanks (*Figure 24*). These are insulated, water-filled polyethylene tanks, each containing a spiral-wound plastic tube heat exchanger surrounded by water. These tanks may be installed indoors, outdoors, or even underground.

During the various modes of system operation, the automatic diverting valve controls the routing of the 25-percent glycol solution through or bypassing the building cooling coil as needed. The temperature modulating valve controls the downstream temperature of the glycol solution by mixing the flows from the chiller and ice tank.

When making ice, the cooling system chiller is used to produce a chilled glycol solution output at a below-freezing temperature, typically 25°F (−4°C). This solution is circulated through the ice tank heat exchangers instead following the normal path through the building air handler coil in the conditioned space. The below-freezing solution circulating through the tank heat exchangers causes the water in the tanks to freeze. This cycle of operation is called the charge cycle (*Figure 25*). It enables the chiller to be operated under the increased load needed to make ice during the less expensive off-peak power hours.

As the day progresses and the building cooling load increases, the chiller is operated to where the chilled glycol solution output is circulated serially, first through the heat exchangers in the ice storage tanks, then through the air handler coil in the conditioned space. The temperature of

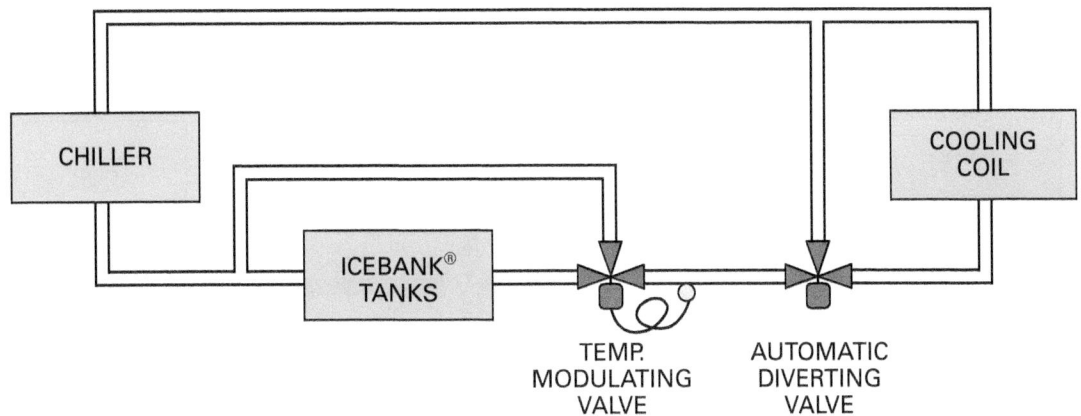

*Figure 23* IceBank® partial-load system general schematic.

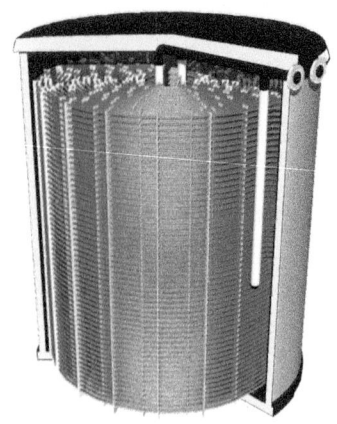

Figure 24 IceBank® energy storage tanks.

the solution produced by the chiller for input to the tank heat exchangers is higher than normal, typically 52°F (11°C). In some designs, the chiller will not run at all during this period; cooling is provided exclusively by the ice. This mode of operation is called the discharge cycle (*Figure 26*).

As the warm solution circulates through the ice tank heat exchangers, the ice in the tanks cools it to a temperature of about 34°F (1°C). The cold solution output from the tanks is then mixed with some of the warm 52°F (11°C) solution to produce a solution of about 44°F (7°C). This is the normal design temperature range for the cooling solution input to the building air handler coil. It reduces the amount of energy needed to run the chiller during the building's peak cooling load interval that is coincident with the more expensive peak demand power daytime interval.

At times when the building's actual cooling load is equal to or lower than the chiller's capacity, the ice tanks are completely bypassed, and the chiller glycol solution output is circulated at 44°F (7°C) through the air handler cooling coil in the normal fashion. This mode of operation is called the bypass cycle (*Figure 27*).

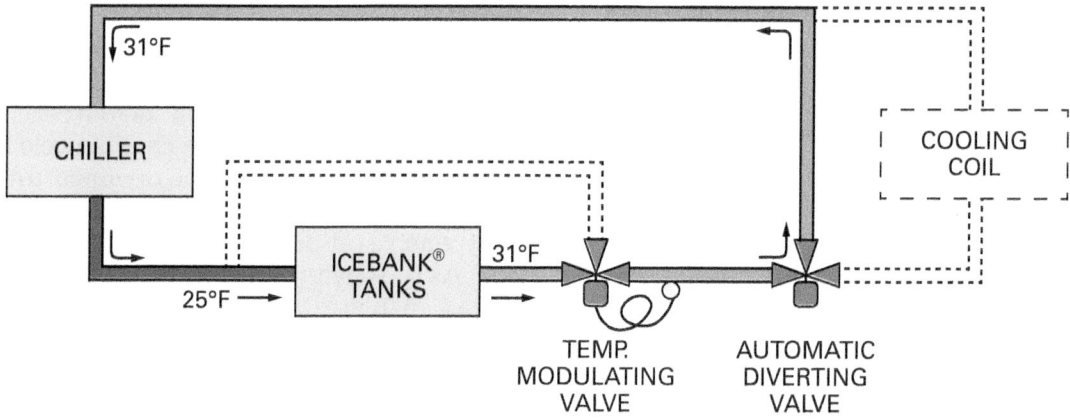

Figure 25 Flow path and temperatures for the charge cycle.

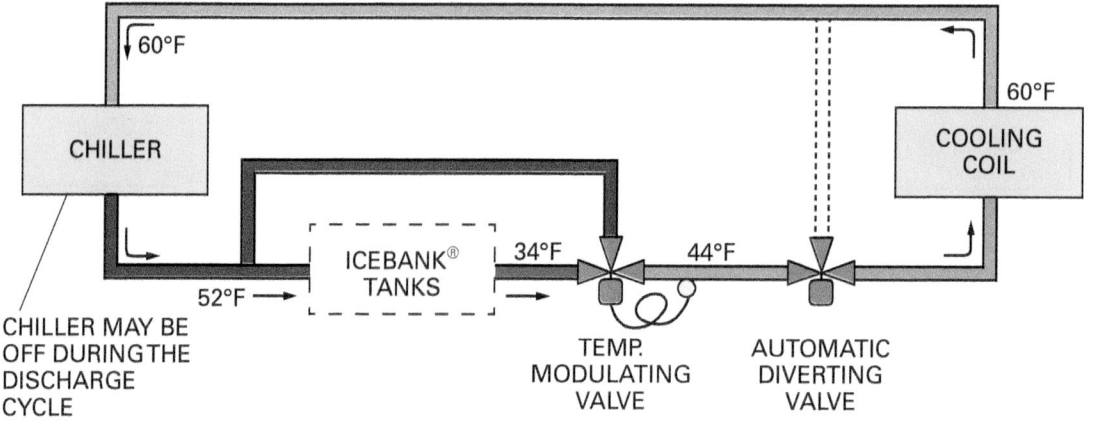

Figure 26 Flow path and temperatures for the discharge cycle.

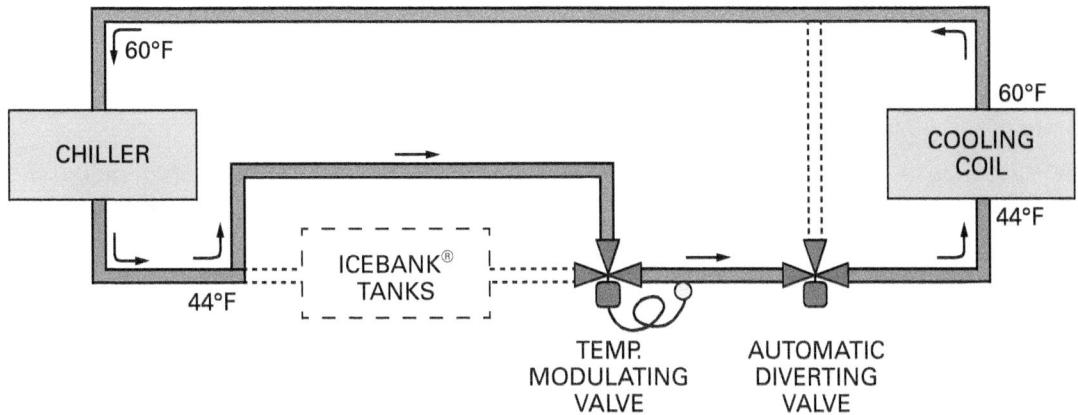

*Figure 27* Flow path and temperatures for the bypass cycle.

# Redding's Ice Bear Program

For the past decade, Redding, California has been developing an intelligent distributed-energy storage system, targeting small and mid-sized commercial buildings in the community. Redding Electrical Utility (REU) is in charge of developing this thermal energy storage program. The heart of the program is the Ice Bear® thermal storage unit, developed and distributed by Ice Energy, Inc.

The Ice Bear looks like a standard rooftop air conditioning enclosure. It acts like a thermal battery in reverse — it is a heat sink, containing a 450-gallon (119 L) storage tank of water. A small compressor turns the water to ice during off-peak hours. A small water pump circulates coolant water during operation. The Ice Bear complements 4- to 20-ton standard vapor-compression units for both commercial and residential applications.

The Ice Bear system connects to an existing air conditioning unit via an additional cooling coil installed in the supply air duct from the main system's air handler. Supply and return cooling water piping connects the cooling coil to the Ice Bear unit. During off-peak hours, the Ice Bear operates in its Ice Charge mode, where its compressor freezes the contents of its water tank.

As temperatures rise during the daytime, the Ice Bear's control system monitors a variety of parameters. At a programmed set of conditions, it shifts into Ice Cooling mode. Controls shut down the main air conditioning unit compressor and begin circulating ice-chilled water through the added cooling coil. An Ice Bear unit is capable of shifting a minimum of 7.2 kW of energy over 6 hours to off-peak service rates.

REU's program includes incentives for installing the Ice Bear systems and provides an economical source for the units themselves. It also offers engineering and site selection assistance for customers. The utility surveyed all of Redding's commercial and industrial buildings to identify candidates for the program.

The Ice Bear concept is scalable to any size HVAC system compatible with the design. The result is less expensive, more efficient cooling systems for customers while reducing costs and increasing efficiency for the utility itself. This keeps electrical costs lower for all its customers.

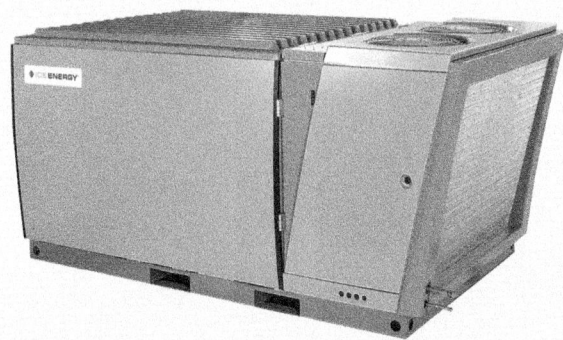

**Figure Credit:** Ice Energy

### Additional Resources

*ASHRAE Handbook—HVAC Systems and Equipment.* Atlanta, GA: American Society of Heating, Refrigerating, and Air Conditioning Engineers (ASHRAE), Inc.

*ASHRAE Handbook—HVAC Applications.* Atlanta, GA: American Society of Heating, Refrigerating, and Air Conditioning Engineers (ASHRAE), Inc.

### 2.0.0 Section Review

1. In an electric energy-demand reduction system, after the device that is attached to a customer's HVAC equipment receives a demand reduction signal, power to the HVAC equipment is _____.

   a. replenished
   b. demodulated
   c. stored
   d. interrupted

2. The routing of the glycol solution through the cooling coil during operation of an IceBank® energy storage system is controlled by the _____.

   a. temperature modulating valve
   b. ice storage tank level sensor
   c. automatic diverting valve
   d. off-peak chiller cycling modem

## SUMMARY

Facilities can reap substantial financial as well as environmental benefits by using energy recycling and reclamation equipment along with electric energy-demand reduction programs and ice storage systems.

Heat recovery devices save considerable amounts of energy through the capture and re-use of heat that would otherwise be wasted. Heat recovery systems commonly use air-to-air, water-to-air, water-to-water, and steam-to-water/air heat exchangers and/or coils to transfer heat from one part of an HVAC system for use in another part. This heat recovery can be achieved in process-to-process, process-to-comfort, or comfort-to-comfort applications.

Electric utility demand reduction systems include variable pricing structures and demand-side management (DSM) programs. The application of smart grid technologies in DSM programs engages consumers and utilities in two-way communications that empower consumers to use electricity in a more cost-efficient manner. Operating HVAC ice storage systems enables facilities to make ice during a utility's lower-cost off-peak hours, and then use the stored ice to help cool the facility during the utility's more expensive peak hours.

1. The two types of recovery ventilators are _____.

   a. MERV and UVC
   b. ERV and MERV
   c. ERV and HRV
   d. HRV and UVC

2. One advantage of using a standalone ERV in conjunction with a rooftop-package air conditioning unit is that it _____.

   a. maximizes recovery wheel efficiency
   b. eliminates the need for an exhaust fan
   c. reduces demand on the economizer
   d. increases airflow through the supply duct

3. A fixed-plate heat exchanger in an ERV can recover _____.

   a. total heat
   b. sensible heat only
   c. both sensible and latent heat
   d. latent heat only

4. One type of heat-recovery device in which the hot discharge gas from a system compressor is used to heat building air is called a _____.

   e. recovery condenser
   a. water-cooled condenser
   b. double-bundle evaporator system
   c. dual-evaporator system

5. When the available heat generated from the chiller system with a double-bundle condenser exceeds the building load, the surplus heat is _____.

   a. routed to and stored in the auxiliary heating system
   b. used to prevent the exhaust coil from freezing
   c. rejected into the atmosphere by the cooling tower
   d. directed to the refrigeration system to be cooled down

6. In a runaround loop system, the supply air is _____.

   a. precooled when it is warmer than the exhaust air
   b. charged with glycol in the exhaust coil
   c. preheated when it is warmer than the exhaust air
   d. bypassed into the exhaust coil air passages

7. A heat-pipe heat exchanger _____.

   a. transfers only latent heat
   b. increases a system's dehumidification capacity
   c. equalizes the supply and exhaust airstream temperatures
   d. requires the use of an electric reheat coil

8. The amount of heat transferred by a heat-pipe heat exchanger can be controlled by _____.

   a. trimming the capillary wick
   b. repositioning the cooling coil
   c. varying the speed of the motor
   d. changing the tilt of the unit

9. In a unidirectional coil-loop thermosiphon system, the coil in the cold duct must be _____.

   a. higher than the coil in the hot duct
   b. downstream of the coil in the hot duct
   c. at the same level as the coil in the hot duct
   d. lower than the coil in the hot duct

10. A twin-tower enthalpy recovery loop system is a(n) _____.

    a. air-to-liquid system
    b. air-to-liquid and liquid-to-air system
    c. liquid-to-air system
    d. air-to-air system

11. In a flue-gas heat-recovery system, a heat exchanger is inserted between the _____.

    a. reheat tank and the circulating pump
    b. boiler and the boiler feed pump
    c. reheat and condensate tanks
    d. boiler exhaust and the stack

12. To provide the best quality steam in a flash-steam heat-recovery system, the flash tank should be _____.

    a. purged periodically
    b. mounted vertically
    c. bypassed during blowdown
    d. discharged to an economizer

13. A control valve that responds to the flow of makeup water in a blowdown water and heat recovery system is used to provide efficient heat recovery and to _____.

    a. maintain the desired concentration of dissolved solids in the boiler
    b. keep the system's condensate lines pitched toward the flash tank
    c. discharge makeup water into a municipal sewer system
    d. precool the flash steam before it is vented to the atmosphere

14. The more customers participating in an electric utility's demand reduction program, the greater the likelihood that the duration of the equipment off-time cycle for each customer will _____.

    a. level out during the summer
    b. fluctuate during off-peak demand
    c. decrease along with peak demand
    d. increase when peak demand goes up

15. A method used to reduce peak utility electrical usage in HVAC equipment is _____.

    a. heat pump defrost
    b. flash-tank storage
    c. chiller barrel bypass
    d. ice storage

# Trade Terms Introduced in This Module

**Enthalpy:** When used in HVACR technology, the total amount of thermal energy contained by a fluid; the sum of its thermal energy based on heat capacity, and its latent heats of fusion and vaporization, as applicable.

**Monel®:** An alloy made of nickel, copper, iron, manganese, silicon, and carbon that is very resistant to corrosion.

**Runaround loop:** A closed-loop energy recovery system in which finned-tube water coils are installed in the supply and exhaust airstreams and connected by counterflow piping.

**Sensible heat:** The heat exchange that simply raises or lowers the temperature of a fluid without causing or resulting from a phase change (condensation or evaporation). Compare with the term *total heat*.

**Sorbent:** A substance whose main purpose is to absorb something else by forming a mixture of the two substances.

**Total heat:** The heat exchange that not only raises or lowers the temperature of a fluid, but also causes or results from a phase change (condensation or evaporation).

**Zeolite:** An aluminosilicate compound that has a high affinity for water and water vapor, which can easily be dehydrated and rehydrated by heating and cooling. A compound commonly used in certain humidity-control systems.

# Additional Resources

This module presents thorough resources for task training. The following resource material is suggested for further study.

*ASHRAE Handbook—HVAC Systems and Equipment.* Atlanta, GA: American Society of Heating, Refrigerating, and Air Conditioning Engineers (ASHRAE).

*ASHRAE Handbook—HVAC Applications.* Atlanta, GA: American Society of Heating, Refrigerating, and Air Conditioning Engineers (ASHRAE).

# Figure Credits

© Pavlo Vakhrushev/Dreamstime.com, Module opener

Courtesy of RenewAire LLC, Figure 1A

Engineered Air, Figure 5 (photo)

Airxchange, Inc., Figure 6

Courtesy of Hillphoenix, Figure 7

PoolPak LLC, Figures 9–11

Ben Sutherland, BBC, SA02

© Deanpictures/Dreamstime.com, SA04

Courtesy of Evapco, Inc., Figure 22

Ice Energy, SA05

CALMAC Corp., Figure 24A and B

| Answer | Section Reference | Objective |
|---|---|---|
| **Section One** | | |
| 1. d | 1.1.0 | 1a |
| 2. d | 1.2.2 | 1b |
| 3. b | 1.3.3 | 1c |
| 4. a | 1.4.0 | 1d |
| 5. c | 1.5.2 | 1e |
| 6. b | 1.6.0 | 1f |
| 7. a | 1.7.0 | 1g |
| 8. c | 1.8.2 | 1h |
| **Section Two** | | |
| 1. d | 2.1.1 | 2a |
| 2. c | 2.2.1 | 2b |

# NCCER CURRICULA — USER UPDATE

NCCER makes every effort to keep its textbooks up-to-date and free of technical errors. We appreciate your help in this process. If you find an error, a typographical mistake, or an inaccuracy in NCCER's curricula, please fill out this form (or a photocopy), or complete the online form at **www.nccer.org/olf**. Be sure to include the exact module ID number, page number, a detailed description, and your recommended correction. Your input will be brought to the attention of the Authoring Team. Thank you for your assistance.

*Instructors* – If you have an idea for improving this textbook, or have found that additional materials were necessary to teach this module effectively, please let us know so that we may present your suggestions to the Authoring Team.

**NCCER Product Development and Revision**

13614 Progress Blvd., Alachua, FL 32615

**Email:**  curriculum@nccer.org
**Online:**  www.nccer.org/olf

❏ Trainee Guide    ❏ Lesson Plans    ❏ Exam    ❏ PowerPoints    Other _____

Craft / Level: _____    Copyright Date: _____

Module ID Number / Title: _____

Section Number(s): _____

Description: _____

_____

_____

_____

_____

Recommended Correction: _____

_____

_____

_____

_____

Your Name: _____

Address: _____

_____

Email: _____    Phone: _____

# Building Management Systems

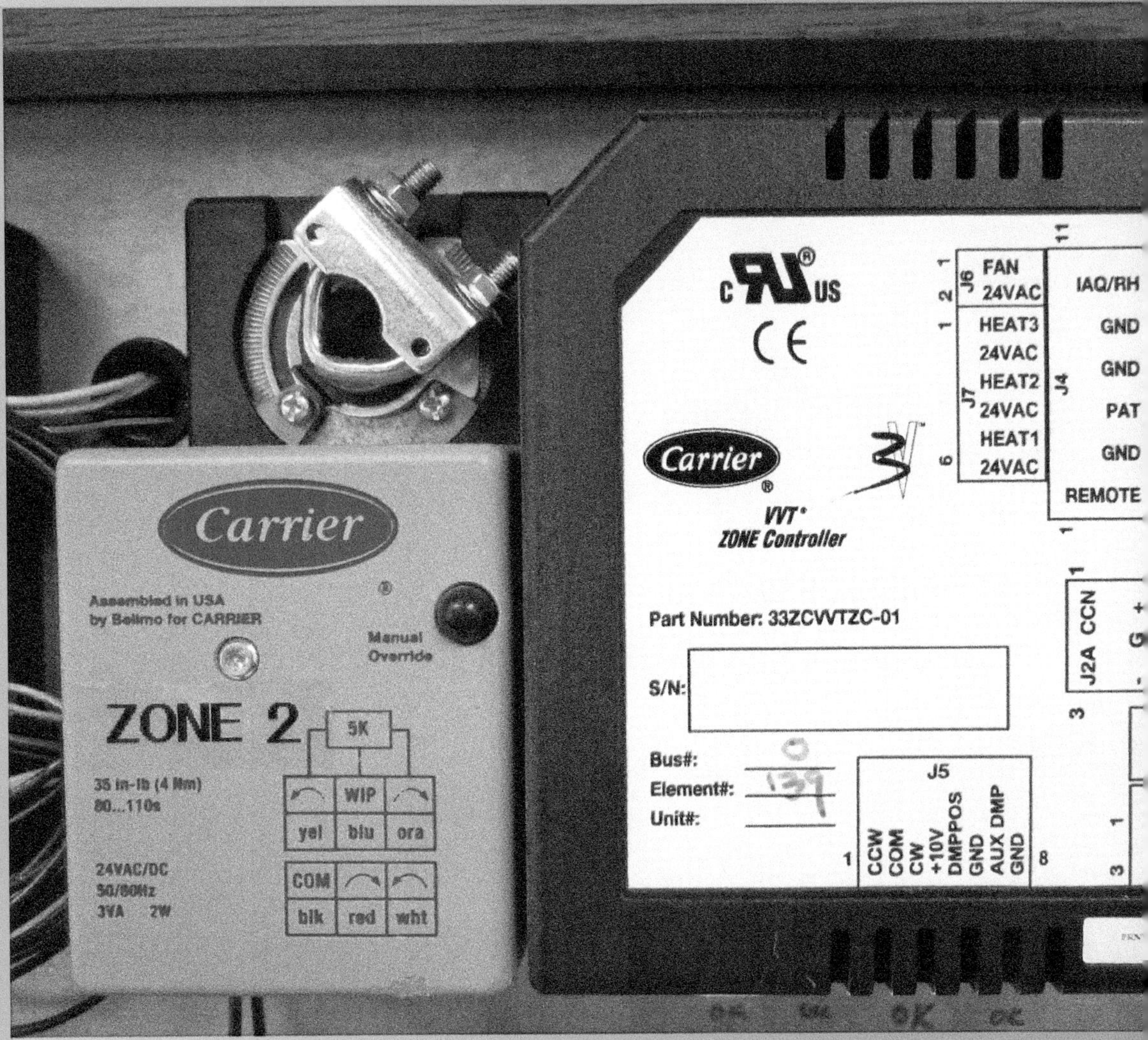

## OVERVIEW

Building management systems take the use of electronic controls and the internet to a higher level by integrating the control of all building systems, including HVAC, lighting, security, and fire control. All the building systems can be monitored and controlled from a central location, either local or hundreds of miles away. Technicians who install and service these systems must be able to interact with the systems. They must be able to monitor performance, diagnose problems, and establish or change operating parameters.

# Module 03405

Trainees with successful module completions may be eligible for credentialing through NCCER's National Registry. To learn more, go to **www.nccer.org** or contact us at **1.888.622.3720**. Our website has information on the latest product releases and training, as well as online versions of our *Cornerstone* magazine and Pearson's product catalog.

Your feedback is welcome. You may email your comments to **curriculum@nccer.org,** send general comments and inquiries to **info@nccer.org**, or fill in the User Update form at the back of this module.

This information is general in nature and intended for training purposes only. Actual performance of activities described in this manual requires compliance with all applicable operating, service, maintenance, and safety procedures under the direction of qualified personnel. References in this manual to patented or proprietary devices do not constitute a recommendation of their use.

*03405 V5*

## Objectives

When you have completed this module, you will be able to do the following:

1. Identify and describe the operation of basic digital controllers.
   a. Identify the four primary control point classifications.
   b. Describe analog and discrete input and output devices.
   c. Describe closed control loops and the related algorithms.
2. Describe the architecture of a building management system.
   a. Describe a DDC peer-to-peer network.
   b. Describe the functions of a packaged unit digital controller.
   c. Describe BMS control of an applied VAV system.
3. Describe various user-related tasks that can be achieved through a building management system.
   a. Describe the ways in which users interface with and access the system.
   b. Identify various tasks that are not related to temperature control that can be accomplished through the system.
4. Describe various building management system control strategies.
   a. Describe occupied building temperature control strategies.
   b. Describe unoccupied building temperature control strategies.
   c. Describe other building control strategies that are not related to temperature control.
5. Define the concept of interoperability and describe the various related protocols.
   a. Describe the four primary protocols in use.
   b. Define and describe web browser system integration.
   c. Project the course of interoperability in building management.

## Performance Task

Under the supervision of your instructor, you should be able to do the following:

1. Interpret operating data received through building management system software.

## Trade Terms

Algorithm
Application-specific controller
Baud rate
Bit
Building management system (BMS)
Bus
Control point
Data collection
Digital controller

Direct digital control (DDC)
Ethernet
Firmware
Gateway
Hypertext transfer protocol (http)
Internet protocol (IP)
Internet protocol address (IP address)
Interoperability

Load-shedding
Local area network (LAN)
Local interface device
Network
Product-integrated controller (PIC)
Product-specific controller
Protocol
Tenant billing
Wide area network (WAN)

## Industry-Recognized Credentials

If you are training through an NCCER-accredited sponsor, you may be eligible for credentials from NCCER's Registry. The ID number for this module is 03405. Note that this module may have been used in other NCCER curricula and may apply to other level completions. Contact NCCER's Registry at 888.622.3720 or go to **www.nccer.org** for more information.

# Contents

## Figures and Tables

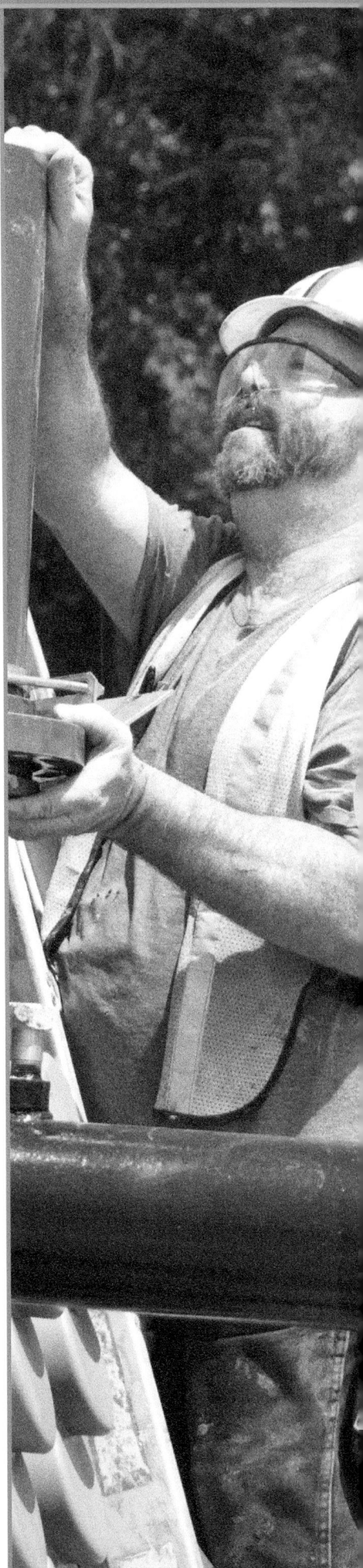

# SECTION ONE

## 1.0.0 BASIC DIGITAL CONTROLLERS

### Objectives

Identify and describe the operation of basic digital controllers.

    a. Identify the four primary control point classifications.

    b. Describe analog and discrete input and output devices.

    c. Describe closed control loops and the related algorithms.

### Trade Terms

**Algorithm**: A mathematical equation consisting of a series of logic statements used in a computer or microprocessor to solve a specific kind of problem. In HVAC applications, algorithms are typically used in microprocessor-controlled equipment to control a wide range of control function operations based on the status of various system sensor input signals.

**Building management system (BMS)**: A centralized, computer-controlled system for managing the various systems in a building. Also known as a building automation systems (BAS).

**Control point**: The name for each input and output device wired to a digital controller.

**Digital controller**: A digital device that uses an input module, a microprocessor, and an output module to perform control functions.

**Direct digital control (DDC)**: The use of a digital controller is usually referred to as direct digital control or DDC system.

**Product-integrated controller (PIC)**: A digital controller installed by a manufacturer on a product at the factory.

**Product-specific controller**: A digital controller designed to control specific equipment that may be installed by the equipment manufacturer. The controllers for chillers or boilers are examples of product-specific controllers. Also known as a product-integrated controller.

Most of the previous coverage of electronic controls has involved individual electronic control devices. Examples of such devices are programmable electronic thermostats and electronic modules that control furnaces and heat pumps.

In this module, the focus is on systems that use computer technology to control all building functions, including the HVAC system, lighting, smoke control, and building security access. The composite of these systems is known by many names, including building management system (BMS), building automation system (BAS), facility management system, and direct digital control (DDC) system.

Many building management systems are being designed as part of a process called integrated building design. Integrated building design is a collaborative decision-making process that uses a project design team from a project's inception through its design and construction phases.

The integrated building design process is, in turn, often incorporated into an overall larger commercial design trend called green buildings.

A green building is a sustainable structure that is designed, built, and operated in a manner that efficiently uses resources while being friendly to the environment. The resources used include, but are not limited to, occupant health, employee productivity, building materials, building systems, building operation, water use, and energy use.

Thus, in order to minimize resource use in many modern green building designs, the building management system must be able to integrate control strategies across multiple building systems.

The heart of any building management system is a series of digital controllers (*Figure 1*) used to monitor and control various building functions. The use of a digital controller is usually referred to as direct digital control (DDC). The term *direct*

## Application of a Building Management System

This is an example of a modern office building in which the building HVAC systems, lighting, smoke control, and other building systems are all under the control of a building management system.

03405-13_SA01.EPS

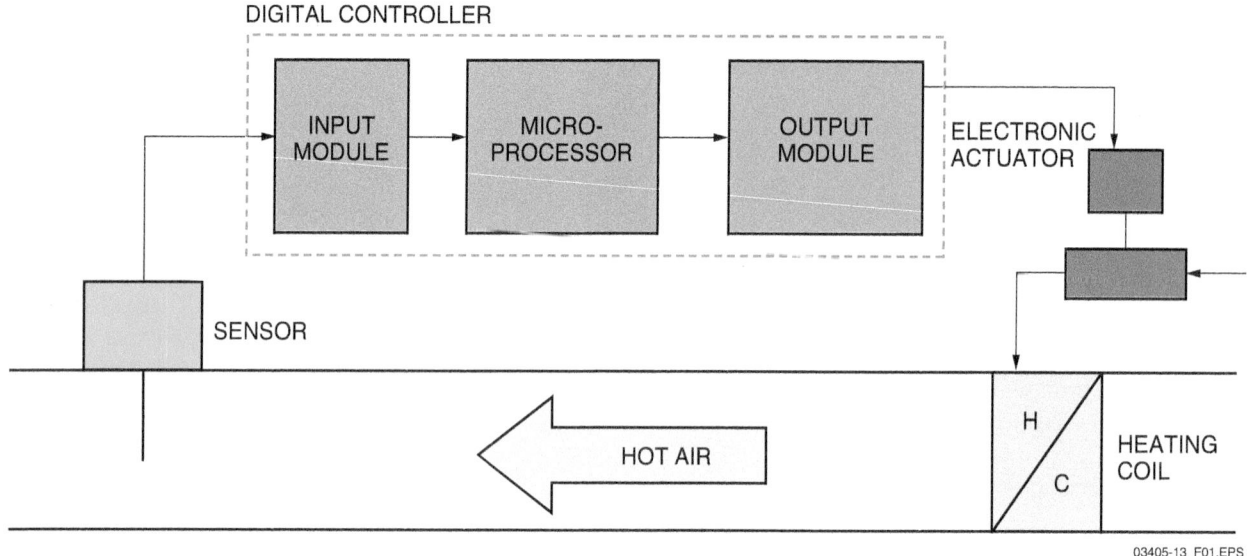

DIGITAL CONTROLLER

INPUT MODULE → MICRO-PROCESSOR → OUTPUT MODULE

ELECTRONIC ACTUATOR

SENSOR

HOT AIR

H / C HEATING COIL

03405-13_F01.EPS

*Figure 1* Typical digital controller.

refers to the direct use of the digital controller to perform control logic normally accomplished with electronic or pneumatic controllers. The digital controller consists of three major components: an input module, a microprocessor, and an output module.

Digital controllers fall into two basic groups: product-specific controllers, also known as product-integrated controllers (PIC), and general-purpose controllers. Product-specific controllers are digital controllers designed to control specific equipment and may be installed by the equipment manufacturer. The controllers for chillers or boilers are examples of product-specific controllers. General-purpose controllers, on the other hand, are designed to control a variety of equipment and are typically field installed. Such controllers would be used to control generic, non-complex devices with simple operating schemes, such as exhaust fans, pumps, and similar equipment.

### 1.1.0 Control Point Classification

*Figure 1* shows a typical digital controller. A sensor located in a hot-air duct is wired to the input module. The sensor is part of a low-voltage circuit in the input module. The input module converts resistance changes in the sensor into temperature value and transmits this value to the microprocessor. A control routine, called an algorithm, compares the duct temperature against a setpoint and sends instructions through the output module to the hot-water valve actuator. The valve opens or closes accordingly. Each input and output device wired to a controller is called a control point, or simply, a point.

Control points are either analog or binary. Binary points are also referred to as digital or discrete points. An explanation of each point type follows:

- *Analog input (AI)* – A sensor used to measure variables that have a range of values, such as temperature or pressure. The sensor signal varies over a predefined range of values. A duct temperature sensor is an example. An analog signal can vary significantly.
- *Discrete input (DI)* – A sensor used to measure the status of binary devices, such as the position of a pressure switch. The sensor is binary, or discrete, because the switch can be either open or closed; only two possible conditions exist.
- *Analog output (AO)* – An output signal used to modulate the position of a device, such as a valve or damper. The output signal from the controller to the AO point varies typically from 4 to 20mA or from 0 to 10VDC.
- *Discrete output (DO)* – An output signal used to change the position of a device from one state to another. Examples of discrete outputs are the starting and stopping of pump or fan motors. Again, there are only two possible states or conditions.

### 1.2.0 Input and Output Devices

Analog or discrete devices are wired to the input or output (I/O) modules in accordance with the types of points involved (inputs or outputs). The points on each module are grouped differently by manufacturer; the most common hardware components are available in blocks of 4, 8, or 16 points.

These points may be predetermined to be of one type, such as DO, or they may be universal, with their type being established during the installation and programming of the digital controller. The following describes typical examples of each type of input and output device.

### 1.2.1 Analog Input Devices

Temperature sensors (*Figure 2*) are the most common analog input (AI) devices used in HVAC control systems. Temperature sensors fall into one of three basic types: thermistor, resistance temperature detector (RTD), or transmitter.

A thermistor is a small bead of material that changes resistance in proportion to changes in temperature. The thermistor senses temperature at a single point. Thermistors can be used to sense the temperature of fluid within a duct or pipe or the temperature of air in a room or outdoors.

An RTD uses a long resistance-sensing element as opposed to a single point. Therefore, RTDs are commonly used to sense an average temperature. Measuring mixed air temperature in an air handler is one use of an RTD.

When the medium requires a sophisticated sensing element, or when there is a long distance between the sensor and the input module, a transmitter is used. The transmitter is a device that senses the specific range of a variable (like temperature) and sends (transmits) a linear current (4–20mA) or voltage (0–10VDC) signal back to the controller. The transmitter translates the variable into a current or voltage signal. However, do not confuse the term *transmitter* used here to mean that the device is wireless.

### 1.2.2 Discrete Input Devices

Discrete input devices are typically dry-contact closures from switches or relays used for feedback information from equipment. A dry contact is energized from an external source, rather than the voltage being provided by the device itself. Most common HVACR switches are considered

*Figure 2* Examples of thermistors.

to be dry. Such devices include auxiliary starter contacts, differential pressure switches (*Figure 3*), current sensing switches, and flow switches.

Another type of discrete input device is a meter that can send a pulsed signal to the input module. The frequency and duration of the pulses represents power consumption or flow rates. Each meter has a pulse conversion rate that is programmed into the processor to interpret the pulse signal correctly.

### 1.2.3 Analog Output Devices

Analog output devices are actuators that modulate valves or dampers. Most actuators accept industry standard output control signals of 4–20mA or 0–10VDC (*Figure 4*).

## The Use of Points

In the controls industry, each project is measured by the number of points to be installed. For example, a job is referred to as a 100-point job or a 1,000-point job, indicating the number of input or output devices in a system. This fixes the magnitude of labor required to install such a system.

03405-13_F03.EPS

*Figure 3* Differential pressure switch.

The transducer is another common analog output device. Transducers, such as the one shown in *Figure 5*, convert a digital controller output signal to a pressure range. The use of transducers is common in large or retrofit applications where valves or dampers are equipped with pneumatic actuators. To control a pneumatic actuator, a current-to-pressure (I/P) transducer converts a 4–20mA output signal into a 3–15 psig air signal. This approach, using electronics to control, pneumatic power to position a valve, can result in extremely precise control.

It is important to note that transducers can also serve as input devices. Just as they can convert signals to a pressure range, they can also convert a pressure value into a digital signal. The signal is then fed to the controller. One popular style of transducer is one that measures and converts current values into digital signals.

### 1.2.4 Discrete Output Devices

Discrete output (DO) devices are typically low-voltage (24VDC) control relays with contacts wired into motor control circuits to start or stop the motors. Pulsed or stepped actuators can also be used as discrete output devices if the devices are designed to receive a pulse-width modulating signal. This is called floating-point control. One DO point is used to pulse the device open and another is used to pulse the device closed.

### 1.3.0 Closed Control Loops

The basis for almost all digital control is a closed control loop. *Figure 6* shows such a loop being used to control the duct temperature in a warm-air heating application. A sensor located in the warm air measures duct temperature. The input module monitors the resistance change of the sensor and transmits duct temperature to the microprocessor.

The microprocessor compares the duct temperature, for example 95°F, to a programmed setpoint, perhaps 110°F, and calculates a milliamp signal to send to the hot-water valve actuator. The output module sends the milliamp signal to the valve actuator, and the valve opens to allow more hot water through the valve. A greater difference between the setpoint and the actual value causes a higher signal. For example, a setpoint 6°F or more higher than the actual value may result in a 20mA signal. As the values come closer together, the signal weakens to begin modulating the valve position. Of course, this is just one possible example of a programmed response. As the valve opens, the coil transfers the heat to the air, raising its temperature. As the air continues to warm, the signal begins to change.

The automatic feedback created by this loop makes the process a closed control loop. The sensor automatically measures the impact of any change ordered by the microprocessor.

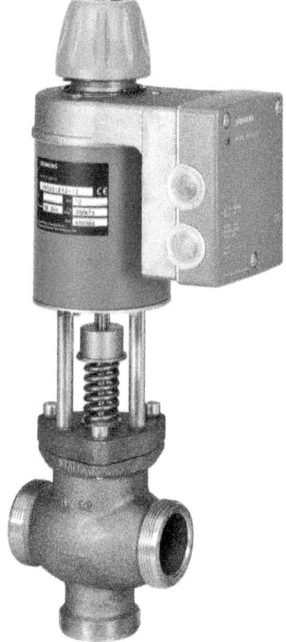

03405-13_F04.EPS

*Figure 4* Valve with actuator.

03405-13_F05.EPS

*Figure 5* I/P transducer.

## Control Signals

Most control signals to actuators are specified as 0 to 10VDC instead of 4–20mA. The reason is that it is easy to measure the voltage output signal being produced by a digital controller. A simple VOM can be used to measure the voltage output signal.

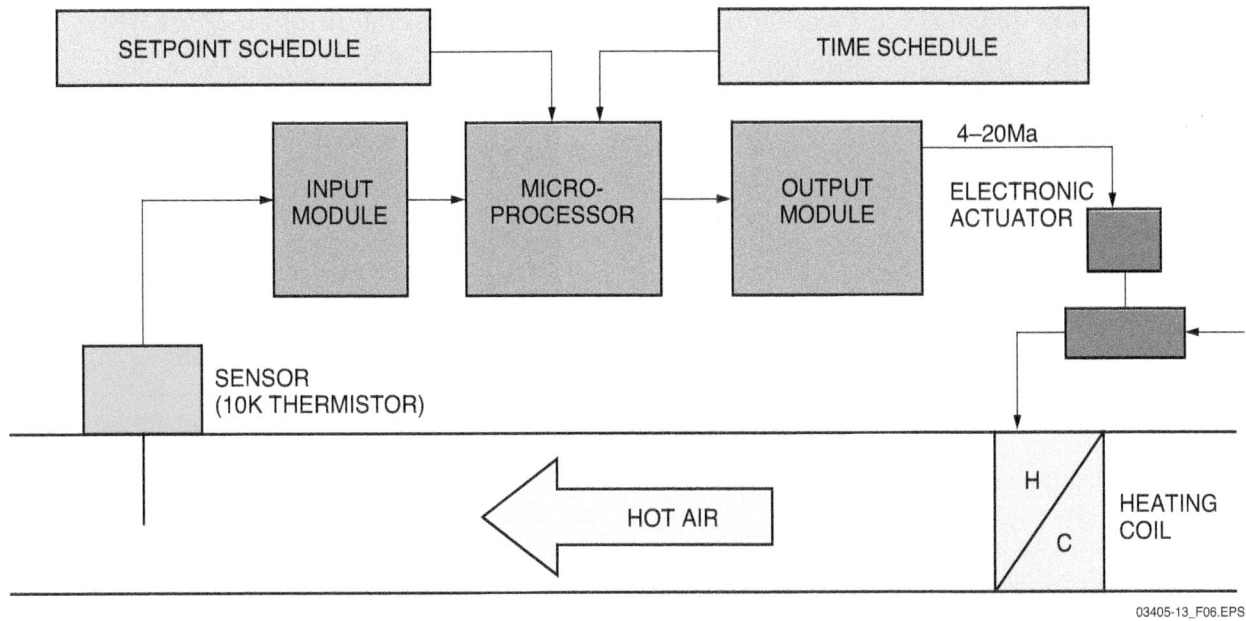

*Figure 6* Closed control loop.

Closed control loops are executed by running multiple algorithms at the same time within the microprocessor. Each algorithm controls one closed control loop. The algorithms interpret input conditions and decide what output signals to send for each control loop. Algorithms consist of a series of logic statements, continuously repeated at high speed. For example, part of the heating coil algorithm for *Figure 6* might consist of something like the following:

*Step 1*  Read the duct temperature.

*Step 2*  Compare it to the setpoint. If it's too low, calculate what output signal to send to the hot-water valve actuator.

*Step 3*  Send X milliamps to the hot-water valve actuator.

*Step 4*  Wait for Y seconds and then go back to Step 1.

Many algorithms also require time schedules to determine when specific equipment should be turned on or off. Setpoint schedule data are also needed to determine the temperatures at which valves and dampers should be opened or closed. Using setpoints, time schedules, and point data, a microprocessor can use algorithms to perform a wide range of control functions. Some of the most typical HVAC algorithms are as follows:

- Heating/cooling coil control
- Humidification/dehumidification
- Mixed air damper optimization
- VAV fan supply control
- Indoor air quality (IAQ)

- Time of day scheduling
- Electric heat staging
- Primary/secondary pump control

The following three strategies are used in operating DDC control loops:

- Proportional (P)
- Proportional-integral (PI)
- Proportional-integral-derivative (PID)

### 1.3.1 Proportional Control Algorithms

Proportional control is much like the early cruise control used on automobiles. Setting the cruise at 55 miles per hour (mph) meant the car stayed at 55 on a level road. However, when the car started up a hill, it would slow to 50. Missing the setpoint on the low side is called undershoot. When the car traveled down a hill, it would increase to 60 mph. Missing the setpoint on the high side is called overshoot. Due to control inefficiencies as well as overshoot and undershoot, the speed fluctuated within a wide range using proportional control.

Even though the setpoint was 55 mph, in proportional control the speed of the car did not stay constant. The difference between the actual speed of the car when it settled at a fixed speed (the control point) and the desired speed of the car (setpoint) is called offset. Offset is a measure of the inefficiency of proportional control.

In order to respond, proportional control must have a change in the controlled variable. After responding, proportional control will always arrive at a control point that is different from the original point.

### 1.3.2 Proportional-Integral Control Algorithms

An integral control function is added to the control loop of a modern cruise control to eliminate offset over time. When the speed falls below 55 mph while climbing a hill, the integral function moves the accelerator pedal down and causes the car to increase speed until it returns to 55 mph again. The reverse is true when the automobile travels down a hill. The accelerator id released, slowing the engine speed. Thus, the integral term eliminates offset over time, and keeps the car at 55 mph regardless of the terrain being traveled.

One problem remains, however. Going up a very steep hill will cause the car to decelerate at a faster rate than going up a mild grade hill. The reverse is true for a steep downhill. Utilizing the PI function, the car's cruise control will take longer to bring the car back to 55 mph on a steep hill than it will for a moderately inclined hill. PI control does not respond to the rate at which the speed change occurs (the acceleration of the error); it simply eliminates the error over time.

PI control is used in most HVAC applications where the controlled variable changes slowly over time. The control of room temperature is one such example because room temperature changes gradually with time.

### 1.3.3 Proportional-Integral-Derivative Algorithms

Adding a derivative term to the cruise control function allows the control to respond to the car's acceleration or deceleration when it encounters hills. The derivative function measures the rate of change of the variable (speed) and causes the gas pedal to overcompensate beyond that required by the PI function. The net result is that the cruise control now rapidly brings the car back to 55 mph, regardless of the incline of the hill.

PID control is used in HVAC systems when the controlled variable is changing rapidly over time. The control of duct pressure in a VAV system is one such example. *Figure 7* summarizes the three functions involved in PID control and shows the benefits of each function.

Most control applications will use the PI function and not the full PID function because the controlled variable changes slowly over time. In fact, using the D function on a slowly changing process will actually make the control response worse. In most HVAC applications, the D function is simply not necessary to maintain the needed conditions.

| ELEMENT | DEALS WITH | CONTROL CONCEPT | ADDED BENEFIT | RESPONSE TO PERTURBANCE |
|---------|-----------|-----------------|---------------|-------------------------|
| Proportional (P) | Error | Signal proportional to error | Control signal proportional to error (demand-based) | 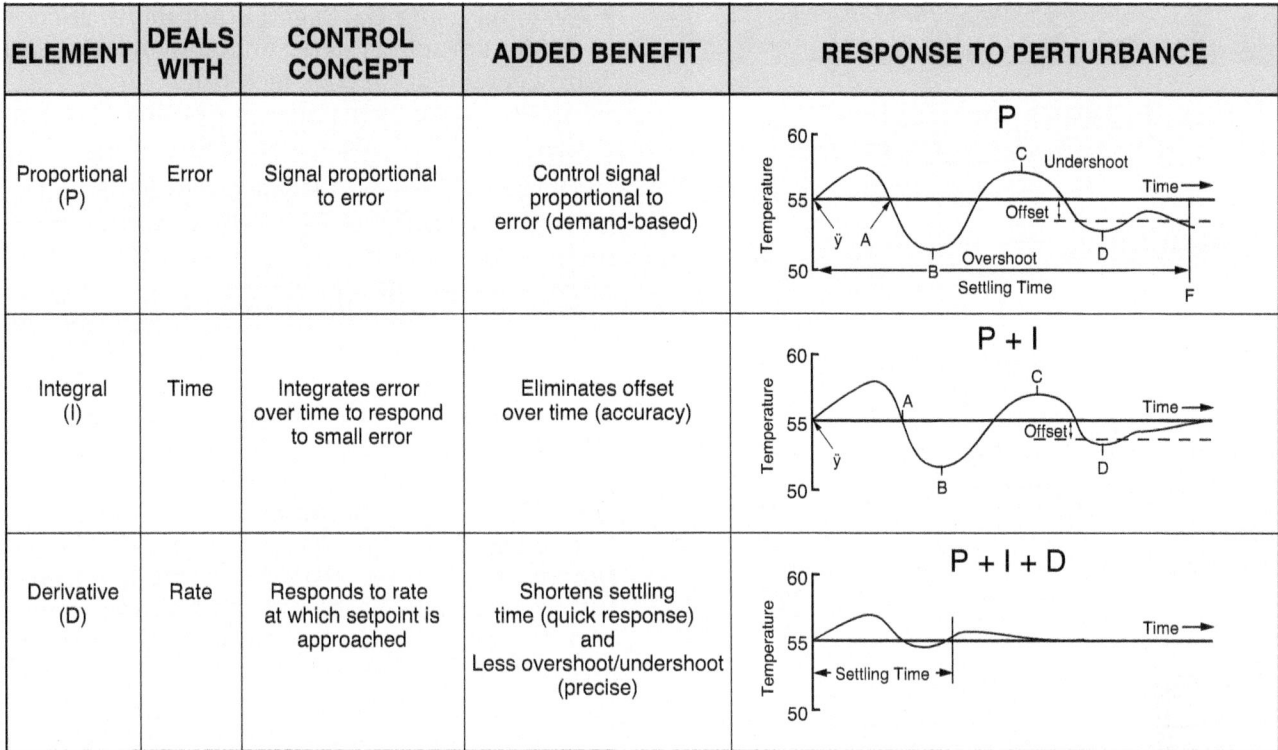 |
| Integral (I) | Time | Integrates error over time to respond to small error | Eliminates offset over time (accuracy) | |
| Derivative (D) | Rate | Responds to rate at which setpoint is approached | Shortens settling time (quick response) and Less overshoot/undershoot (precise) | |

03405-13_F07.EPS

*Figure 7* PID control summary.

## 1.0.0 Section Review

1. Binary control points are also called _____.

   a. discrete points
   b. analog points
   c. single points
   d. bi-points

2. The most common analog input devices sense _____.

   a. pressure
   b. voltage
   c. temperature
   d. resistance

3. A closed control loop is characterized by automatic feedback.

   a. True
   b. False

## 2.0.0 BUILDING MANAGEMENT SYSTEM ARCHITECTURE

### Objectives

Describe the architecture of a building management system.

   a. Describe a DDC peer-to-peer network.
   b. Describe the functions of a packaged unit digital controller.
   c. Describe BMS control of an applied VAV system.

### Trade Terms

**Application-specific controller**: A digital controller installed by a manufacturer on a specific product at the factory.

**Baud rate**: The rate at which information is transmitted across communication lines.

**Bit**: Short for binary digit. The smallest element of data that a computer can handle. It represents an off or on state (zero or one) in a binary system.

**Bus**: A multi-wire communication cable that links all the components in a hard-wired computer network.

**Network**: A means of linking devices in a computer-controlled system and controlling the flow of information among these devices.

**Protocol**: A convention that governs the format and timing of communication between devices in a computer network.

Building management systems usually start with the computer-controlled automation of a building's HVAC system, and then expand as a building grows in size to include the building's lighting, security access, computer room air conditioning, and smoke control systems. The objective of the BMS is to use the power of the computer to provide energy management routines to save energy; provide monitoring and reporting capability for planned maintenance; and permit human access and intervention from a central point in the building or from a remote location. *Figure 8* shows an example of building management system architecture.

The automation of the building's HVAC system is divided into two broad categories:

- Systems using packaged equipment, such as residential split systems or commercial rooftop units and vertical packaged units
- Applied systems that use air terminals, air handlers, chillers, boilers, and cooling towers

The lower portion of *Figure 8* shows the controllers the BMS will interface with to control the individual pieces of HVAC equipment. The portion above shows the next level of building control, the interface of the building's other system management functions. Finally, the very top of *Figure 8* shows the final level of building control—human access capability through the use of local or remote computers, and/or the use of the internet. All of the building's digital controllers are connected together by the network communication bus.

The key to all building management system functions is the use of digital controllers to control equipment in all the various building systems and the use of a communication network to link all controllers in the building to a centralized computer for human access and integrated building system control strategies. BMS functions can be applied to installations ranging in size from a residence to a multi-building campus.

### 2.1.0 Peer-to-Peer Networks

The simplest DDC control system is a single, standalone digital controller wired to the input and output devices of a single HVAC system. By itself, however, a single digital controller can only control a finite number of input and output points. The number of points is a function of the microprocessor design and varies with each manufacturer. Thus, a standalone general-purpose digital controller can only provide comfort and energy management strategies for individual HVAC units within the point limitation of the controller. In many small buildings, this is sufficient.

Multiple digital controllers are required for applications involving multiple pieces of equipment and a large number of points. The ability to share information between controllers becomes important, so a DDC network is required.

A DDC network is a system that allows its individual controllers to communicate with each other and share information. Information sharing can take place between two controllers, or it may be broadcast by one controller to all other controllers. For example, multiple controllers in a building may need to know the outdoor air temperature in order to run their control algorithms. Only one outside air temperature sensor needs to be wired to one controller. This controller then broadcasts

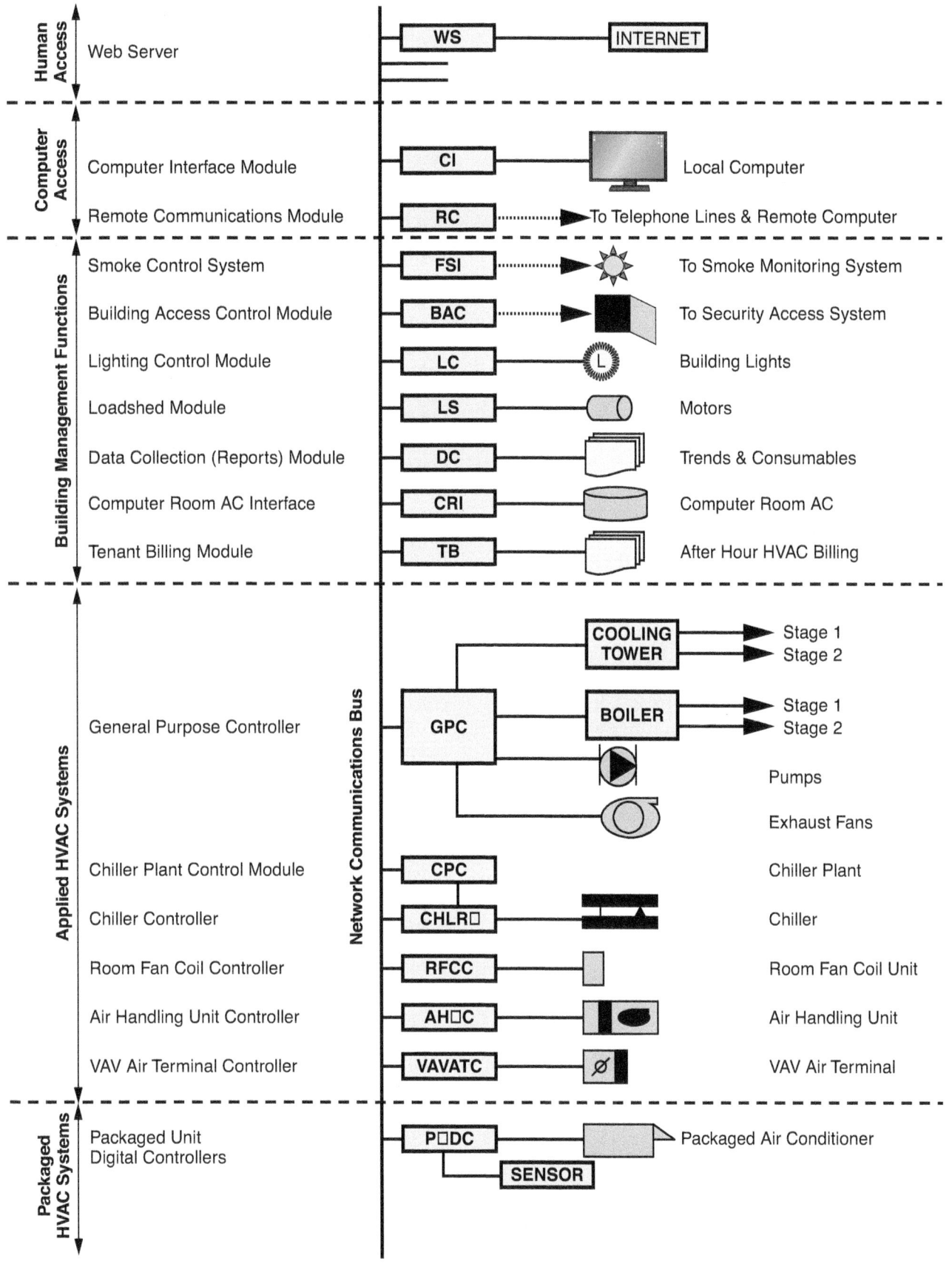

*Figure 8* BMS architecture.

the sensor value along the communication bus to all other controllers in the network. In this manner, all controllers on the network share the same outside air temperature value, saving sensor and wiring expense. When necessary, sensors can be networked and their values averaged for system use instead. This may be used for a very large and open area that requires temperature control.

Information sharing can also be used to achieve HVAC system coordination between water-side and air-side equipment controllers. Air terminals can share information with air handlers that in turn can share information with chillers. In such a system, the air handlers continually vary the air quantity and temperature to meet changing air terminal needs. Likewise, chillers continually vary the water volume and temperature to meet varying air handler needs. Thus, sharing information allows the HVAC system to continually provide optimal comfort while minimizing system operating costs.

Peer-to-peer networks (*Figure 9*) get their name from the way in which communication occurs between controllers. Each controller can communicate with any other controller connected on the network and all controllers are peers, having equal priority to use the communication bus at any time. Each component in the network is both a provider and a user of shared information. In such a network, any controller can communicate at any time by going through a four-step process:

*Step 1*  Wait (in milliseconds) to see that the communication bus is quiet.

*Step 2*  When sufficient quiet time has occurred, send a packet of information to another controller(s).

*Step 3*  Wait to receive a confirmation that the other controller(s) received the message.

*Step 4*  Repeat the transmission if confirmation has not been received in a timely manner.

All controller information may also be viewed or modified from a single user interface (computer). In this way, all controllers on all floors of a building can be interfaced from a single computer located anywhere within the building. The computer is not the central point of the network; instead, it simply functions as another peer.

The bus cables and cable connectors are similar to those used in hooking up telephones and video equipment. The bus is a simple three-wire cable with a shield that is physically wired to each controller in the network. Fiber-optic cable is also used.

The controllers connected in a network can share information because they use a common language or set of rules known as a protocol. In a protocol, the information traveling along the bus must be configured in a specific way in order to be recognized. The protocol is designed to check the data for errors and make corrections, provide access for all devices on the network, and make sure that only one device transmits at a time. Each DDC control system manufacturer has a unique protocol that governs communication along the communication bus.

The rate at which information moves on the network is known as the baud rate. The baud rate is stated in bits per second (BPS). The higher the baud rate, the faster the data transfers. In DDC networks, data can be transferred at baud rates

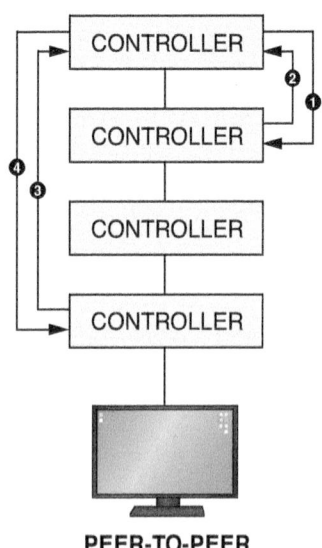

**PEER-TO-PEER**

03405-13_F09.EPS

*Figure 9* Peer-to-peer network.

## Polling Networks

Polling networks work differently than peer-to-peer systems. A controller looks for votes from the members of the network and makes a decision as to what to provide in terms of heating or cooling. Polling networks are sometimes used in commercial applications as a means of zoning.

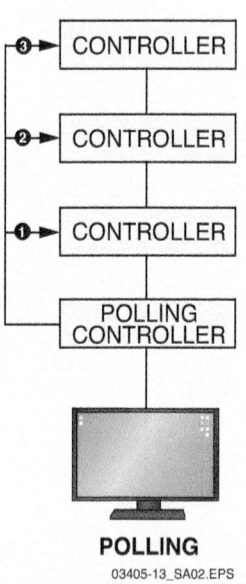

**POLLING**

03405-13_SA02.EPS

upwards of 100 million bits (megabits, or Mbps) per second. Even higher transfer rates are available with fiber-optic cable.

## 2.2.0 Packaged Unit Digital Controllers

A further level of BMS sophistication exists when the manufacturer of packaged equipment installs a communicating digital controller on the unit at the factory (*Figure 10*). This type of controller is also referred to as a product-integrated controller (PIC), or as an application-specific controller. With this controller, all information regarding the packaged unit operating and safety systems is available in the controller. Such a controller is typically provided with a sophisticated 365-day clock with up to eight occupancy periods and setpoints for occupied and unoccupied periods. Holiday and daylight savings time programming is also provided. In addition, a sensor, not a thermostat, is located in the zone.

The sensor is provided with a setpoint adjustment slide bar and an override button, so the occupant can adjust the comfort level and manually start the packaged unit during unoccupied hours. The use of communicating digital controllers can be found in packaged units with cooling capacities as low as 5 tons. However, the larger the capacity of the packaged unit, the more likely it will be equipped from the factory with this type of controller.

Thus, connecting the packaged unit controller to the building's peer-to-peer network bus makes all this internal unit information and scheduling available from a remote central computer. This is very attractive to owners who are interested in energy management, servicing, and troubleshooting.

## 2.3.0 Applied HVAC Systems

General-purpose controllers are typically used to integrate the boiler, cooling tower, pumps, and exhaust fans into the HVAC system controls.

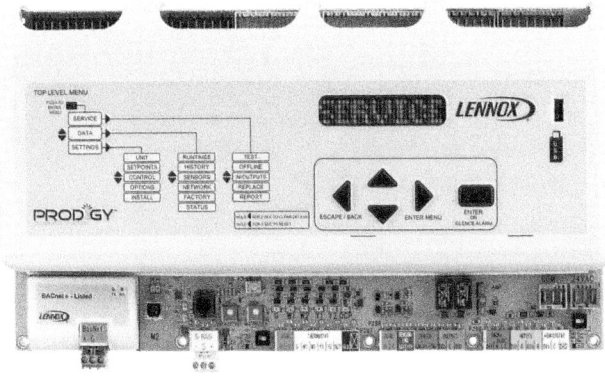

03405-13_F10.EPS

*Figure 10* Product-integrated controller (PIC).

Sometimes general-purpose controllers are also installed on air handlers in the field by the automatic temperature controls contractor. They are more flexible in design than PICs, since they must be capable of interfacing with and controlling a diverse selection of equipment.

The following sections describe the use of digital controllers in a variable air volume (VAV) system served by chillers.

### 2.3.1 Chilled-Water Systems

*Figure 11* shows a typical chilled-water system. Two parallel water-cooled chillers supply chilled water to three air handling units (loads) equipped with two-way control valves. Each chiller has its own chilled-water pump and leaving chilled-water sensor (LCHW). Sensors are also installed in the chilled-water plant leaving pipe (CHWST) and in the chilled-water return pipe (CHWRT). Both chillers are typically sized for 50 percent of the building load.

---

# Communication Bus

The communication bus is a simple 20-gauge three-wire (stranded) cable. The three wires are insulated with a PVC, nylon, or Teflon® coating. The three wires are wrapped in an aluminum/polyester foil shield and covered with an outer jacket made of PVC, nylon, chrome, vinyl, or Teflon®. The shield helps protect the bus from external noise and interference.

---

# Commercial HVAC Systems

There is a tremendous variety of applied commercial HVAC zoning systems in the market. However, over the past 25 years, both VVT and VAV systems have dominated the commercial market due to the comfort they provide and their low operating costs. VVT has dominated in buildings less than 25 tons, while VAV systems have dominated in buildings greater than 25 tons. VVT systems are used with constant-volume packaged units. The VAV system, in buildings between 25 and 100 tons, is typically used with a VAV-type packaged rooftop unit. In larger high-rise buildings, a VAV is often used with a chilled-water system.

---

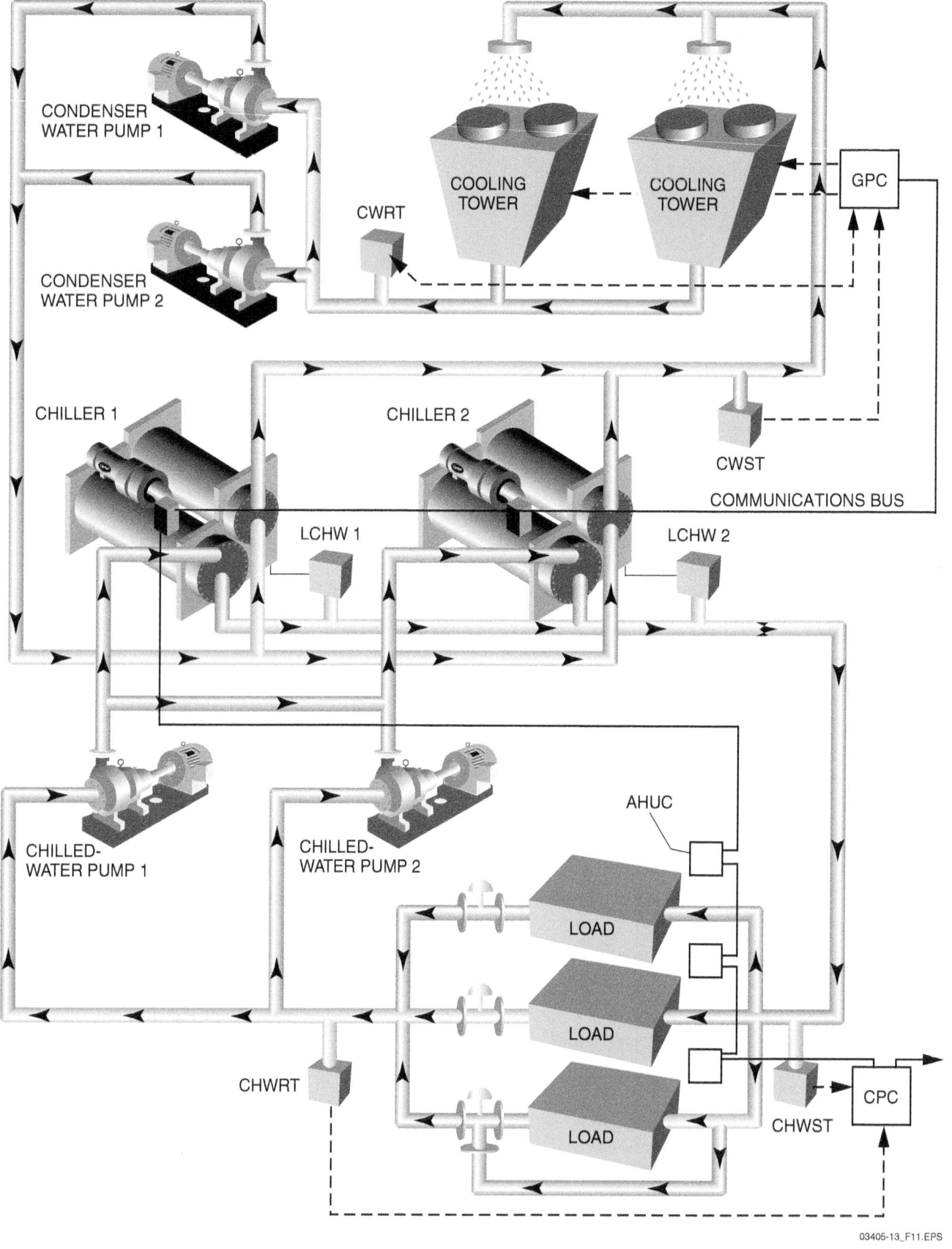

*Figure 11* Chilled-water system.

03405-13_F11.EPS

Chiller condensers are piped in parallel, with each chiller having its own condenser pump. Each chiller will also have its own cooling tower or tower cell. Sensors will be installed in the piping entering (CWST) and leaving (CWRT) the tower.

Digital controllers are mounted on each chiller and on each air handler. In addition, a general-purpose controller (GPC) is used to control the cooling tower fan(s). Finally, a chiller plant controller (CPC) is provided on the job to manage the two chillers into a single chilled-water plant. All digital controllers are connected together with a communication bus.

Each chiller controller contains its own operating system, safety circuits, time clock, occupancy schedule, and chilled-water setpoint. When requested to start by the chiller plant control module (CPC), each chiller starts its own chilled-water and condenser water pumps. Each chiller also reads its own chilled-water temperature sensor (LCHW) and modulates its refrigeration capacity to maintain its leaving chilled water at a constant level, typically 45°F.

Digital controllers mounted on each air handling unit start the air handler based on zone needs and time, and they modulate their chilled-water valves to maintain a supply air temperature of about 55°F. This air is supplied to the VAV air distribution system.

### 2.3.2 Chiller Plant Control Module

Because most buildings will have more than one chiller, a chiller plant control module (*Figure 12*) is required on the job to coordinate their operation. This controller monitors the needs of the building and starts the chillers based on various inputs. The controller then uses both the plant's leaving water temperature (CHWST) sensor and the return water temperature (CHWRT) sensor to determine how many chillers need to be operating and what leaving chilled-water temperature each need to produce. Rather than operate several chillers at partial loading using capacity control, thus increasing energy consumption, the system will try to operate one chiller fully loaded and cycle another on when necessary.

Besides starting and stopping chillers, the chiller control module also equalizes their run hours, keeps them equally loaded when both are operating, starts another chiller when one fails, and performs other such chiller plant management functions.

### 2.3.3 General-Purpose Controller Functions

Other components associated with an applied HVAC system are the boiler, cooling tower, and exhaust fans. A general-purpose digital controller (*Figure 13*) is installed to integrate these pieces of equipment into the HVAC system. Algorithms in the general-purpose controller (GPC) cycle the cooling tower fans to maintain the water temperature leaving the tower (CWRT).

## Chillers

Most commercial buildings equipped with chilled-water systems have one or two chillers in their plant room. Large facilities may have several in the mechanical room. The chiller plant module provided by manufacturers can typically control up to eight chillers, and the chillers can be of a mixed variety—some screw machines, some reciprocating, and some centrifugal. A screw chiller using R-134a refrigerant is shown.

03405-13_SA03.EPS

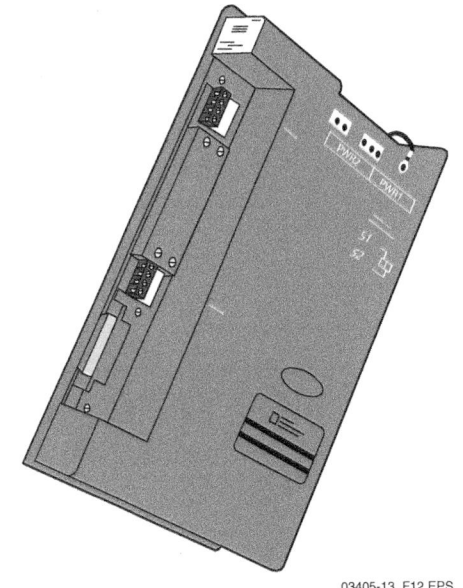

03405-13_F12.EPS

*Figure 12* Chiller plant control module.

 03405 **Building Management Systems**

03405-13_F13.EPS

*Figure 13* General-purpose controller.

Typically, the boiler for such a building comes with its own control panel mounted at the factory. The boiler controller contains the operating system and safety circuits for the boiler. The GPC is used to integrate the cycling of the boiler with the HVAC system heating needs and to send the boiler's controller a hot-water temperature reset signal as necessary.

Similarly, the GPC is used to cycle building exhaust and pressurization fans as required by the building systems. Finally, the GPC is used to control chilled-water pumps when they are ganged and not started by the individual chiller controls.

### 2.3.4 VAV Systems

VAV systems (*Figure 14*) are used in buildings with loads in excess of 25 tons that require simultaneous cooling and heating at the zone level. A rooftop unit or air handling unit provides a variable volume of air to individual VAV boxes, one per zone. If an air handler is used, a chilled-water coil is typically provided in the air handler, thus requiring a chiller in the building. The air source is equipped with a means of fan volume control, either variable frequency drive or inlet guide vanes. A digital controller mounted on the air source controls all air source functions. Each VAV box is installed with a zone heater, a digital controller, and a zone sensor. Finally, a communication bus connects all VAV box controllers with the air source controller.

During occupied periods, the primary air handling unit supplies cool air in the supply duct.

VAV systems typically remain in the cooling mode of operation during all occupied periods year-round. The inlet air temperature is typically maintained between 50°F and 60°F year-round, depending on a variety of conditions. Individual zone VAV boxes vary the amount of air necessary to satisfy zone needs, throttling down to a minimum quantity for ventilation when cooling needs are satisfied. If any zone requires heat, the VAV box controller activates the heater located on the discharge of the VAV box. In addition, the VAV box fan starts, drawing air from the ceiling plenum and mixing it with the small amount of cool air still entering from the main system. Since the cooled primary air is partially comprised of fresh air for ventilation, it must continue to move through the box at all occupied times. *Figure 15* shows the heating and cooling operating modes of a parallel VAV box.

Individual zone setpoints and occupancy schedules are programmed at the computer. The individual box controllers send operating status information to it for reporting purposes.

The primary air handling generally operates independently of the VAV boxes it serves. It actually has little need of sharing or gathering information with them directly. As VAV boxes open and close to satisfy the demands of the space, the air pressure in the primary air duct changes as well. The air handling unit simply responds to an input from a static pressure sensor located in the supply duct. As the pressure falls, indicating more boxes need air, the blower motor speeds up or inlet dampers open wider. As the pressure rises, indicating that VAV boxes are closing their dampers, the central blower motor slows down or inlet dampers close a bit. Very precise air pressures in the system can be maintained this way. The mechanical cooling portion of the system, such as a chilled-water valve or compressors connected to evaporator coils, respond to maintain the temperature of the supply airstream at the setpoint.

Each room sensor is provided with a setpoint adjustment slide bar and an override button, so the occupant can adjust the comfort level and manually start the air source during unoccupied hours. Such a system provides optimal zoning comfort while minimizing the use of fan energy.

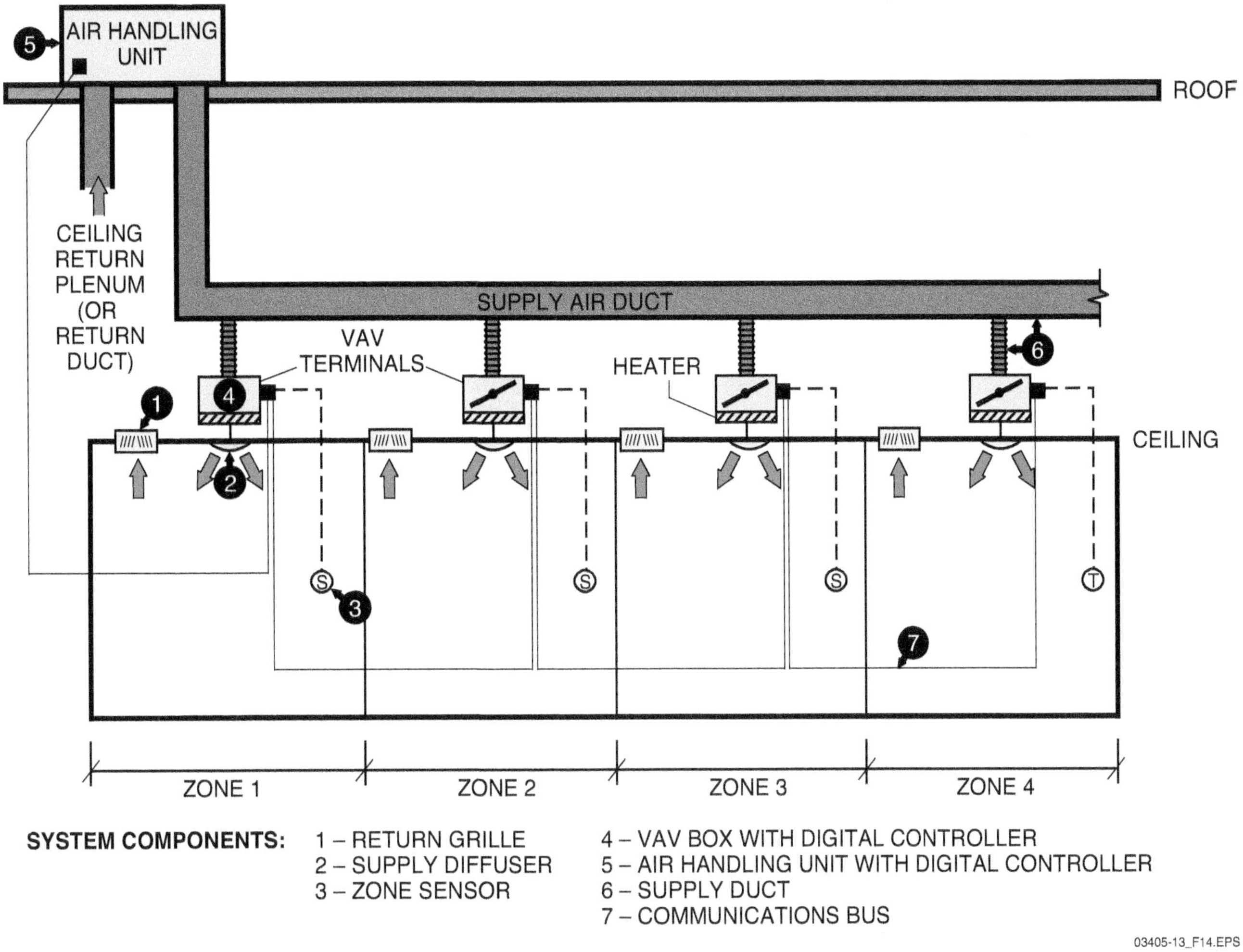

The figure shows a VAV system diagram with the following labels:

AIR HANDLING UNIT ⑤

ROOF

CEILING RETURN PLENUM (OR RETURN DUCT)

VAV TERMINALS

SUPPLY AIR DUCT

HEATER

⑥

① ④ ②

CEILING

Ⓢ ③   Ⓢ   Ⓢ   Ⓣ

⑦

ZONE 1    ZONE 2    ZONE 3    ZONE 4

**SYSTEM COMPONENTS:**
1 – RETURN GRILLE
2 – SUPPLY DIFFUSER
3 – ZONE SENSOR
4 – VAV BOX WITH DIGITAL CONTROLLER
5 – AIR HANDLING UNIT WITH DIGITAL CONTROLLER
6 – SUPPLY DUCT
7 – COMMUNICATIONS BUS

03405-13_F14.EPS

*Figure 14* VAV system.

# Parallel VAV Terminal

A fan-powered, parallel-flow VAV terminal typical of those commonly used to control the conditioned air in each exterior zone of a commercial building VAV system is shown. The tubes crossing the round air inlet are monitoring air velocity, which is then used by the VAV controller to calculate the primary air volume entering the box. The information is also reported to the BMS system for viewing by the user. The motorized damper controlling primary airflow can also be seen inside the inlet opening. Since the primary air duct is at a significantly higher pressure, 2.5" wc, for example, than the outlet duct, the inlet duct connection can be much smaller than the air outlet of the box (on the opposite side).

03405-13_SA04.EPS

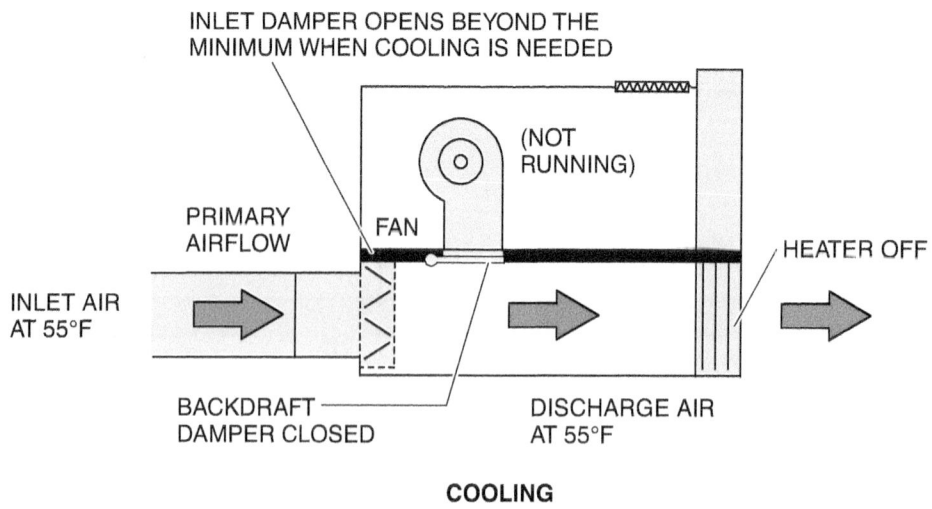

INLET DAMPER OPENS BEYOND THE
MINIMUM WHEN COOLING IS NEEDED

(NOT RUNNING)

PRIMARY AIRFLOW

FAN

HEATER OFF

INLET AIR AT 55°F

BACKDRAFT DAMPER CLOSED

DISCHARGE AIR AT 55°F

**COOLING**

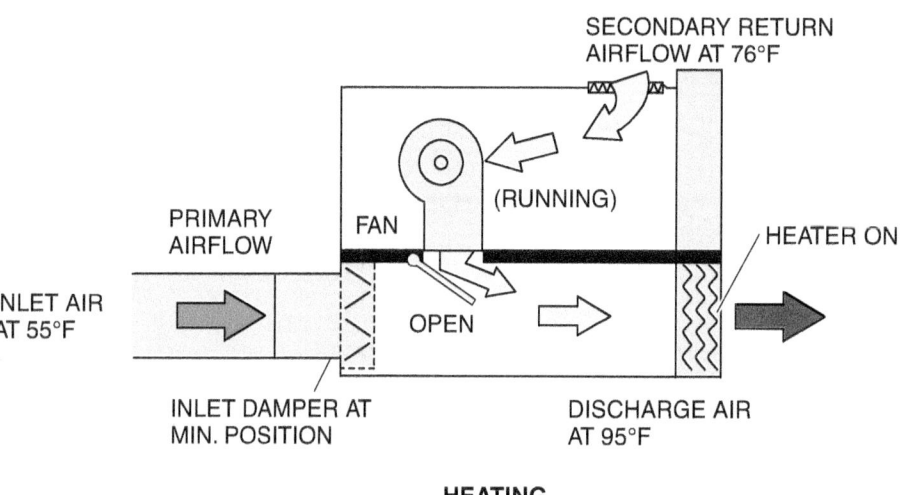

SECONDARY RETURN AIRFLOW AT 76°F

(RUNNING)

PRIMARY AIRFLOW

FAN

HEATER ON

INLET AIR AT 55°F

OPEN

INLET DAMPER AT MIN. POSITION

DISCHARGE AIR AT 95°F

**HEATING**

03405-13_F15.EPS

*Figure 15* Parallel VAV box heating and cooling operating modes.

## 2.0.0 Section Review

1. In a peer-to-peer network, before a controller can communicate, it must _____.

   a. send out a query signal
   b. confirm the communication bus is quiet
   c. wait for a confirmation signal
   d. confirm the communication bus is active

2. A product-integrated controller is likely to be found in a _____.

   a. packaged rooftop unit
   b. residential split-system
   c. cooling tower
   d. hot-water boiler

3. Chilled-water valves modulate to maintain supply air temperature in a VAV system at about _____.

   a. 36°F
   b. 44°F
   c. 55°F
   d. 60°F

## 3.0.0 BUILDING MANAGEMENT SYSTEM USER FUNCTIONS

### Objectives

Describe various user-related tasks that can be achieved through a building management system.

a. Describe the ways in which users interface with and access the system.
b. Identify various tasks that are not related to temperature control that can be accomplished through the system.

### Performance Task

1. Interpret operating data received through building management system software.

### Trade Terms

**Data collection**: The collection of trend, runtime, and consumable data from the digital controllers in a building.

**Ethernet**: A family of frame-based computer networking technologies for local area networks (LANs). It defines a number of wiring and signaling standards.

**Hypertext transfer protocol (http)**: The base protocol used by the worldwide web.

**Internet protocol (IP)**: A data-oriented protocol used for communicating data across a packet-switched network.

**Internet protocol address (IP address)**: A unique address (computer address) that certain electronic devices use in order to identify and communicate with each other on a computer network utilizing the internet protocol (IP) standard.

**Load-shedding**: Systematically switching loads out of a system to reduce energy consumption.

**Local area network (LAN)**: A server/client computer network connecting multiple computers within a building or building complex.

**Local interface device**: A keypad with an alphanumeric data display and keys for data entry. It is connected to the digital controller with a phone-type cable that provides power and communication.

**Tenant billing**: The ability to charge a building tenant for after-hours use of the building's HVAC system.

**Wide area network (WAN)**: A server/client computer network spread over a large geographical area.

Digital control modules or software programs running on a central computer are used to achieve a variety of building subsystem management functions. The following sections describe how users interface with the systems and some of the tasks that can be accomplished.

### 3.1.0 User Interfaces

Users can interface with building management system controllers in the following ways:

- A handheld local interface device, connected directly to a controller
- A locally connected (within the facility) personal computer
- A local area network (LAN) or a wide area network (WAN)
- A remote computer and the telephone lines
- The internet with web browser or smart phone technology

A local interface device (*Figure 16*) is a keypad with an alphanumeric data display and keys for data entry. It is connected to the digital controller with a cable that provides power and communication. Handheld local interface devices may be moved from one controller to another as needed for viewing, diagnosing, and changing controller information. Many controller manufacturers provide such a device that is designed for use with their products. Newer interface strategies include the use of common tablet computers.

More controllers are also being developed with a built-in interface (*Figure 17*). These controllers do not require an interface device. The program and information can be viewed directly on an integral graphics display.

Another popular method for interfacing with the network controllers is to use a single computer as shown in *Figure 18*. Through the single computer, loaded with appropriate control system software, the operator can interact with all controllers on the connected network. Most provide a graphical view of the entire system, or portions thereof, as desired.

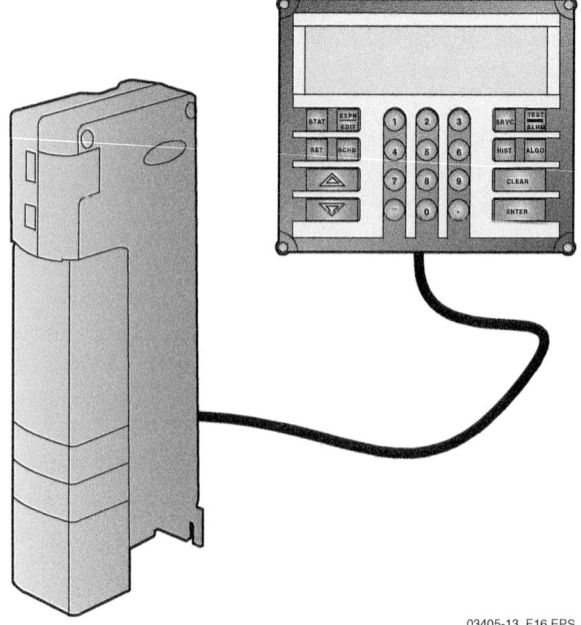

*Figure 16* Local interface device.

In larger applications, where multiple computer users need simultaneous access to the DDC network, a computer network (*Figure 19*) may be used as the user interface. The use of a local area network (LAN) links computers into a server/client format, where the server computer provides central data storage for all network controller data. Client computers access the network through the server and change the information stored on the server. Thus, all changes made to the network are current and located in one place, making multiple-PC user database management easy.

A LAN is typically used with a single building or building complex, such as a college campus. All computers in the LAN can access any building BMS through the server database.

Computers may also be arranged in a wide area network (WAN). A WAN is a server/client computer network spread over a larger geographical area. A WAN is unlimited in geographical size and may connect several LANs together through telephone lines or satellite links.

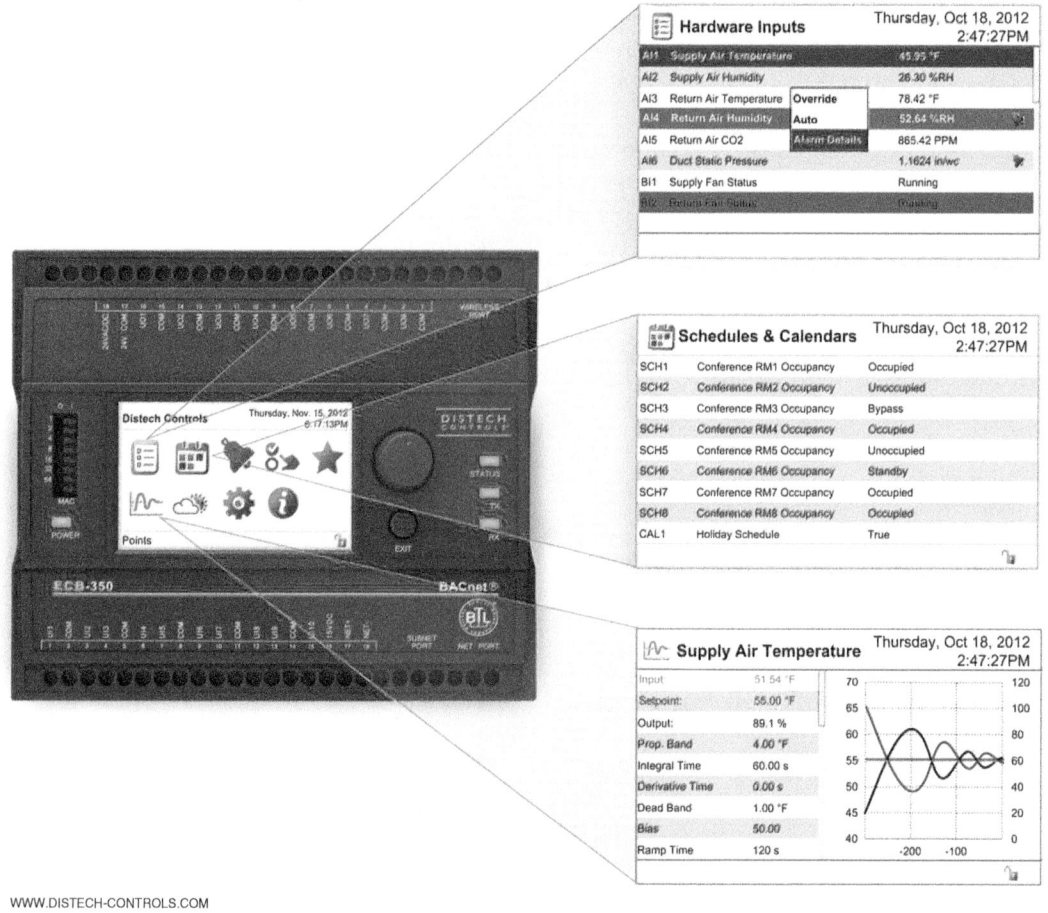

WWW.DISTECH-CONTROLS.COM

*Figure 17* Digital controller with integrated interface.

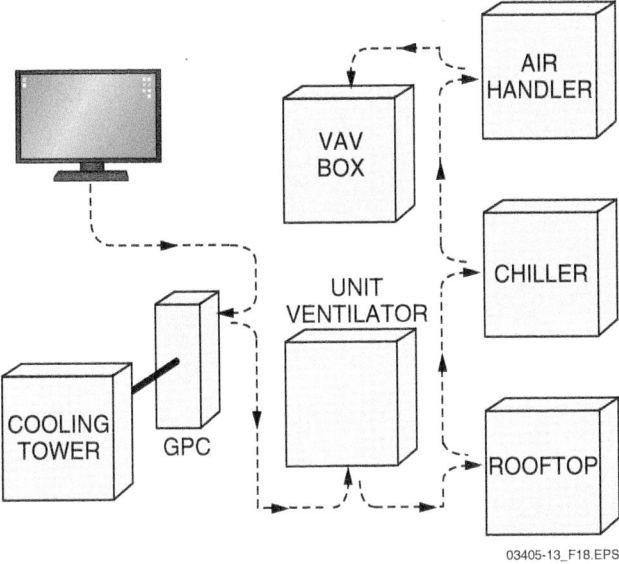

*Figure 18* Local computer connection.

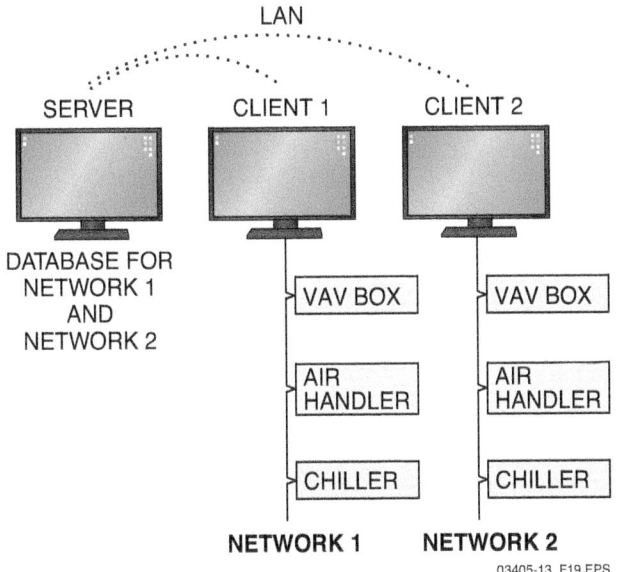

*Figure 19* Local area network.

In today's market, the internet is the most popular way to access the BMS network from any site outside of the building's Ethernet LAN (*Figure 20*). A server is wired to any unique protocol bus as well as the building's Ethernet LAN. The web page server is programmed to scan the unique protocol bus and create web pages for each controller on the bus. These web pages may be tabular or graphical in nature, and are dynamically updated regularly. In addition, the open hypertext transfer protocol (http)/internet protocol (IP) may be used to interface with the web pages. Another advantage to this approach is that the computer used to access the system this way does not usually require the BMS software to be

installed; any computer with internet access will work.

New technology is making it much easier to remotely connect to a building management system through the internet. Smart phones and tablet-style devices (*Figure 21*) use wireless (cell phone) technology to access the internet and act as the user interface. These devices use touch-screen technology so they do not require a mouse or keyboard. There are many applications (called apps) now available for smart phones that enable a technician to access the BMS from any location where cell phone service is available. Once the system is accessed, it can be controlled in much the same way as it would if connected by way of a PC. Any user equipped with such a device who knows the proper access codes and internet protocol addresses (IP addresses), can communicate with the building web page server computer from a remote location. This system is especially attractive to companies that have a number of technicians on the road. Using wireless technology, a technician can have access to any number of buildings from home, or wherever the technician is located.

### 3.1.1 Sign On/Off and Operator Management

In the sign on/off portion of the software, security controls, including passwords, are used to limit access. In the operator management portion, different operators are given different levels of authority. Some can only look at data, some can run diagnostic tests, and others can change the operating parameters or programming. The higher the level of authority, the more control the individual will have over the system.

### 3.1.2 Graphical Interface

In the graphical interface portion of the software, graphic images or photographs are presented for each piece of equipment in the building. Dynamic, live information about the operation of the unit is superimposed on these images.

*Figure 22* shows a graphical image of the control system for a typical air handling unit. The diagram shows all the controllable devices in the air handler, including the cooling and heating coil, supply and return fan, mixed air dampers, and spray humidifier. In addition, all the input and output points are represented by symbols and abbreviated descriptions. Next to each point symbol is dynamic data that the software uploads from the associated controller in the network about every five seconds.

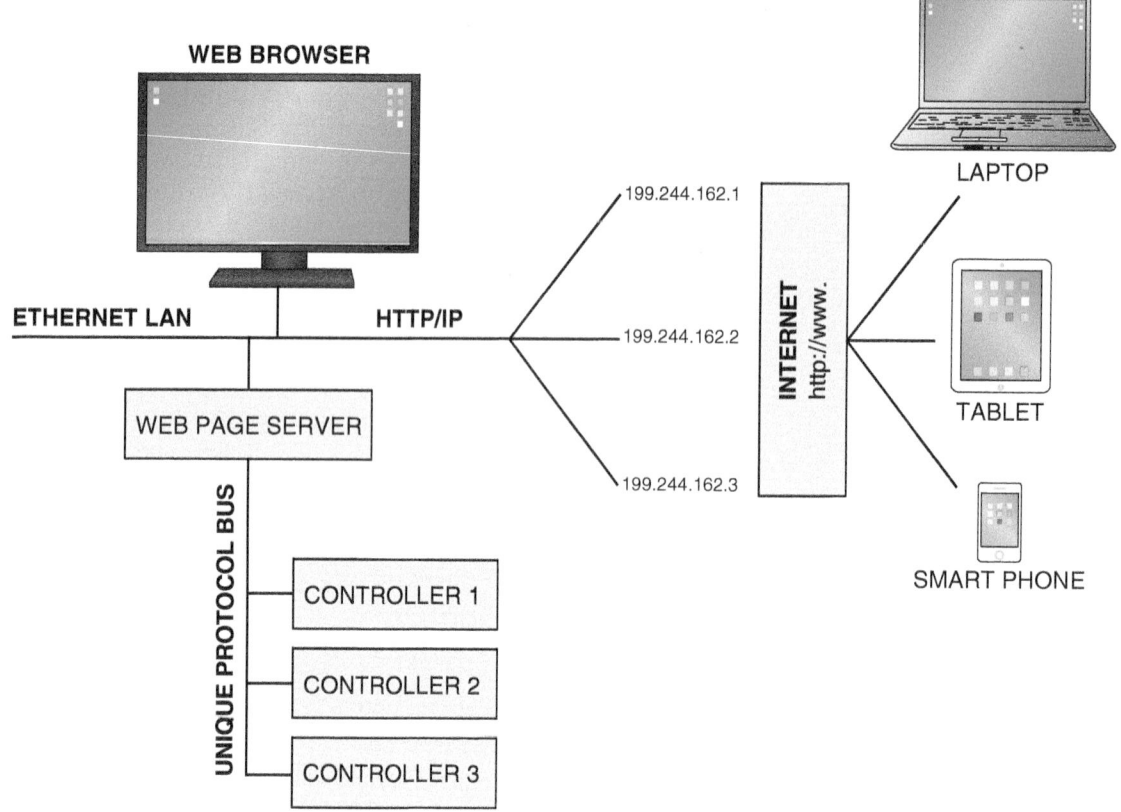

WEB BROWSER

ETHERNET LAN          HTTP/IP

199.244.162.1

199.244.162.2

199.244.162.3

WEB PAGE SERVER

UNIQUE PROTOCOL BUS

CONTROLLER 1

CONTROLLER 2

CONTROLLER 3

INTERNET
http://www.

LAPTOP

TABLET

SMART PHONE

03405-13_F20.EPS

*Figure 20* Internet access to a BMS.

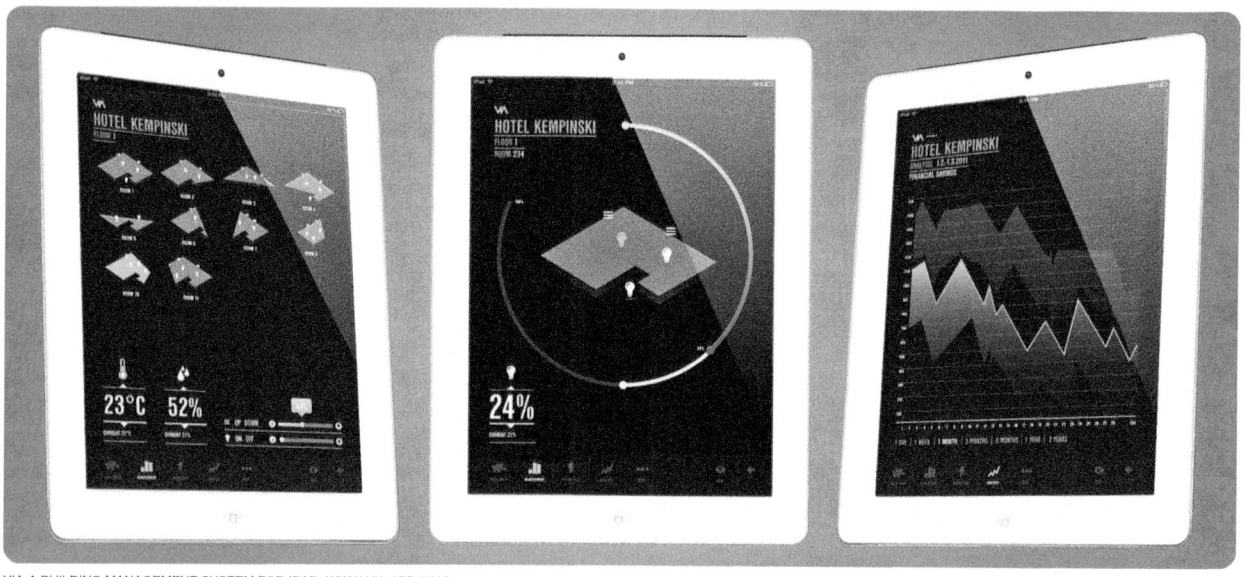

VIA-A BUILDING MANAGEMENT SYSTEM FOR IPAD, WWW.VIA-APP.COM.

03405-13_F21.EPS

*Figure 21* Tablet-style device with BMS application.

This information is very useful for monitoring and troubleshooting. Also, the setpoint and time schedules associated with the air handler can be displayed and modified by using the mouse or touchscreen to access the value of any point on the screen.

Screens similar in format are available for all the building systems. In fact, the trend is to provide simple graphics that provide key information from multiple building systems on a single graphic. A click of the mouse or the swipe of a fin-

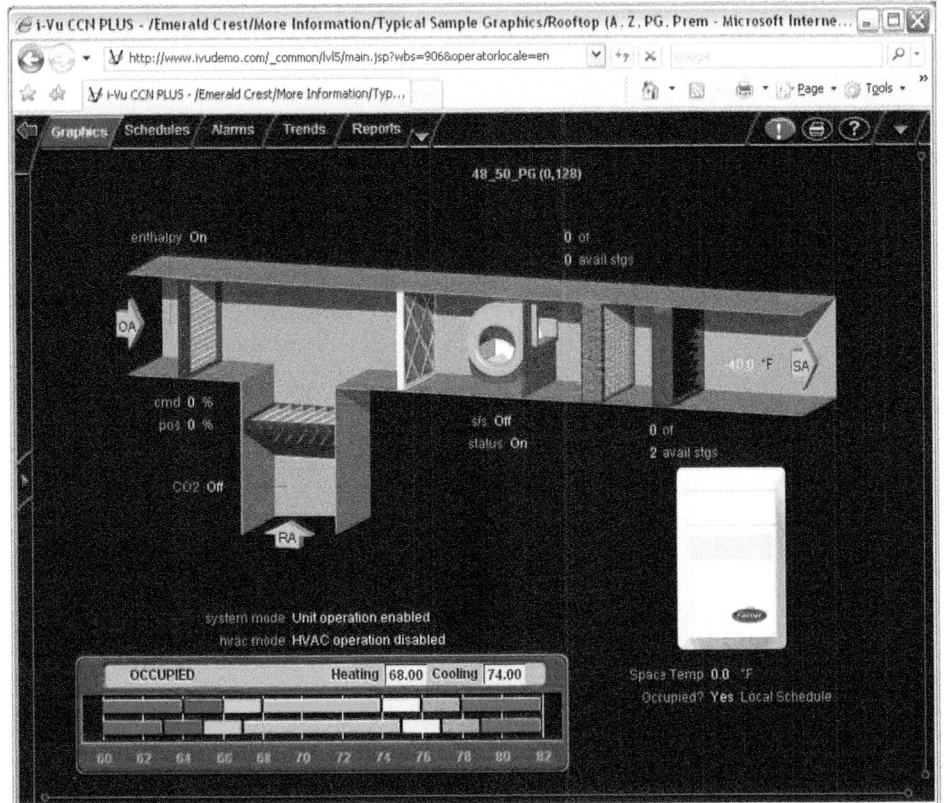

*Figure 22* AHU control system.

ger on a touchscreen shows the operator greater detail of any of the building systems.

*Figure 23* shows a typical time schedule. The schedule has eight periods, with each period defining when the building is considered occupied and when it is unoccupied. Different periods can apply for different days of the week. The right side of the image is a graphical display of the occupied hours for each period. Once the operator makes a change, the change is sent down to the actual controller containing the schedule and then saved in the controller's database on the central computer.

These time schedules can be specific to one controller, grouped to a number of controllers, or it can be global in nature and apply to all controllers in a network or networks. Thus, the operator may impact controller operation across building systems through a simple graphical interaction.

## 3.2.0 Other Building Management System Functions

Building management systems have features and functions that go far beyond simply maintaining the desired conditions within a space. They include the following:

- Database management
- Alarm management
- Data collection
- Tenant billing
- Load shedding
- Lighting control
- Building access
- Smoke control

### 3.2.1 Database Management

In the database management portion of front-end software systems, files from all the controllers in the local building are uploaded and stored according to the networks in which they exist. If the computer is used to dial other remote buildings, then the files from those networks are also stored in the database. These files are updated whenever an operator changes any parameter in any controller. In web-based systems, each controller database file exists at the controller level and at the web page server level.

The operator can view and modify the algorithms and points for any controller (*Figure 24*). This information is displayed on dynamic screens that interrogate the highlighted controller through the network bus. Any changes made are automatically stored at the controller level and at the computer that stores a mirror image of the controller file.

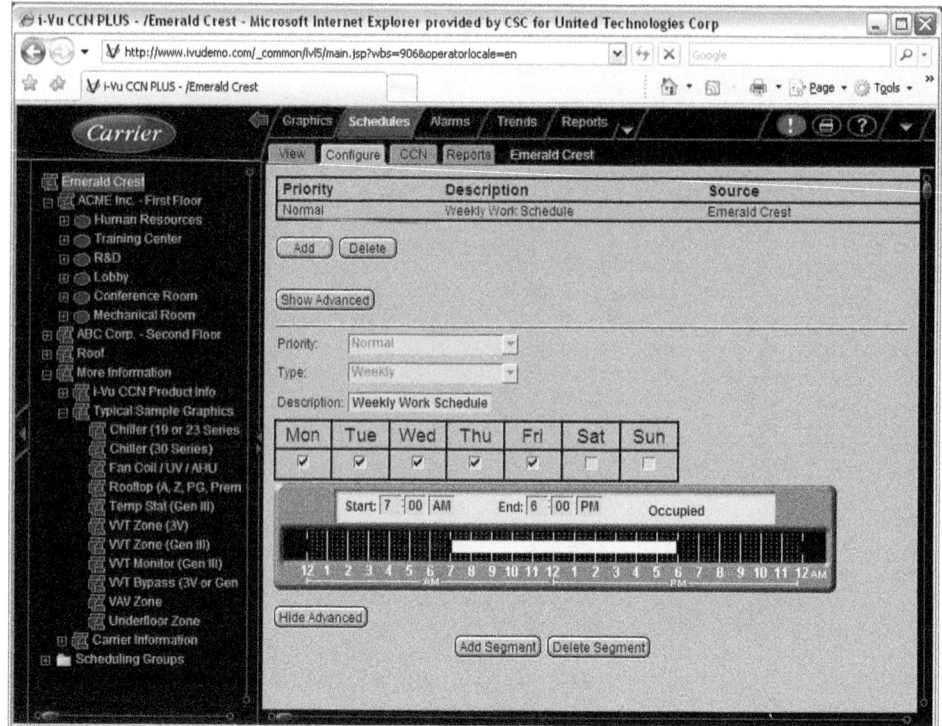

*Figure 23* Time schedule.

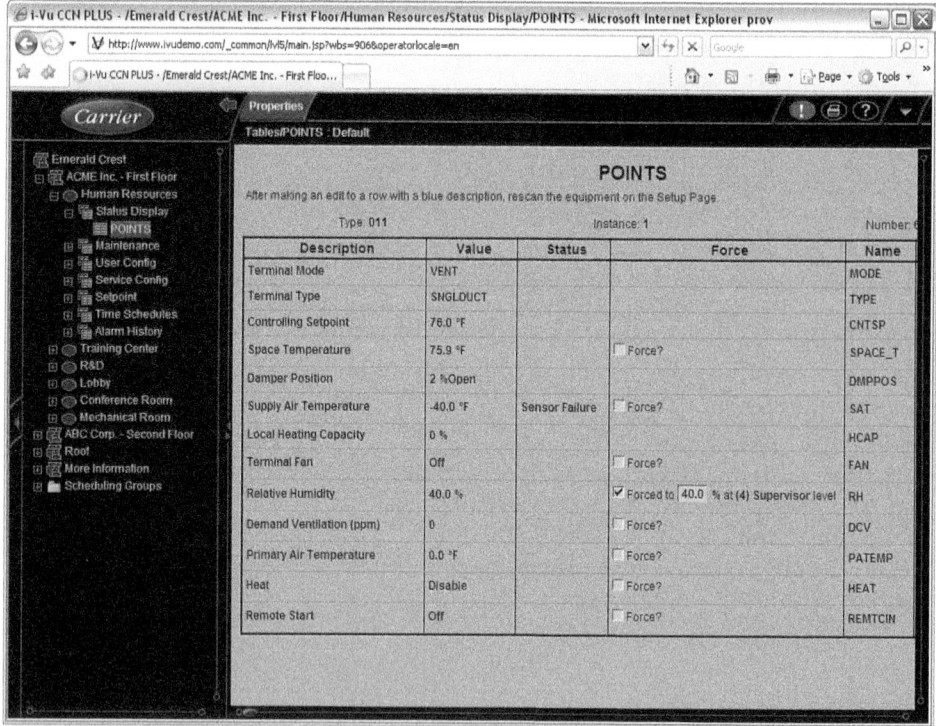

*Figure 24* Controller and point list screens.

---

NCCER – *HVAC Level Four*   03405

### 3.2.2 Alarm Management

Whenever an alarm condition occurs in a controller, the alarm is transmitted to the central computer and stored in an alarm file (*Figure 25*). The operator is notified of the incoming alarm visually and audibly and can then deal with the alarm condition by reviewing the alarm file and acting accordingly. The alarms remained stored so that operators or technicians can spot trends. For example, a particular alarm may repeatedly occur at certain times of day or under specific environmental conditions. This information can often help track down the problem.

### 3.2.3 Data Collection

Access to historical data is very important to building managers. In nationwide retail chains, hotels, and industries, energy managers constantly need to look at operating data from all of their locations to determine trends, identify problems, and develop energy management strategies. Installing a data collection module on the building network and programming it to collect desired information satisfies these needs. The data collection module is capable of collecting trend information on temperatures and pressures, and recording the use of consumables, such as gallons of water and kW of power usage. This information is collected on the network and does not require the central computer to be operational.

The central computer is programmed to use the building's communication network to automatically upload stored data from the data collection module on a daily basis. This information is then archived on the central computer into weekly, monthly, and yearly historical files for generating reports needed to manage the building.

Many types of reports are available to the operator of the system. They include operator activity, how alarms were handled, trends of temperatures and pressures, equipment runtimes, tenant billing, and consumable information. *Figure 26* shows a typical daily space temperature report.

Data from the network can be stored on the central computer in daily, weekly, monthly, and yearly files. Reports can then be set up and run by the operator to cover any particular period of time.

With information like that shown in the previous examples, the building operator is well equipped to monitor the building performance, adjust operating parameters, and troubleshoot problems when they arise. In addition, building management is well equipped with historical information to invest in building management strategies that, when implemented across building systems, reduce overall building energy use and operating costs.

### 3.2.4 Tenant Billing

In large commercial buildings, central HVAC systems like VAV or chilled-water systems centralize the production of chilled water and hot water to improve efficiency and reduce operating costs. In

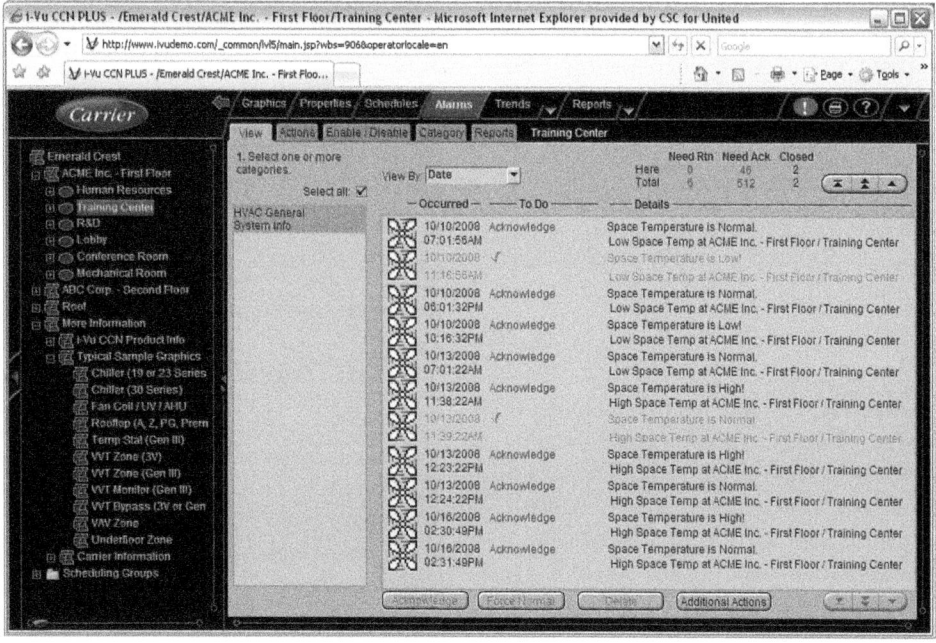

03405-13_F25.EPS

*Figure 25* Alarm file.

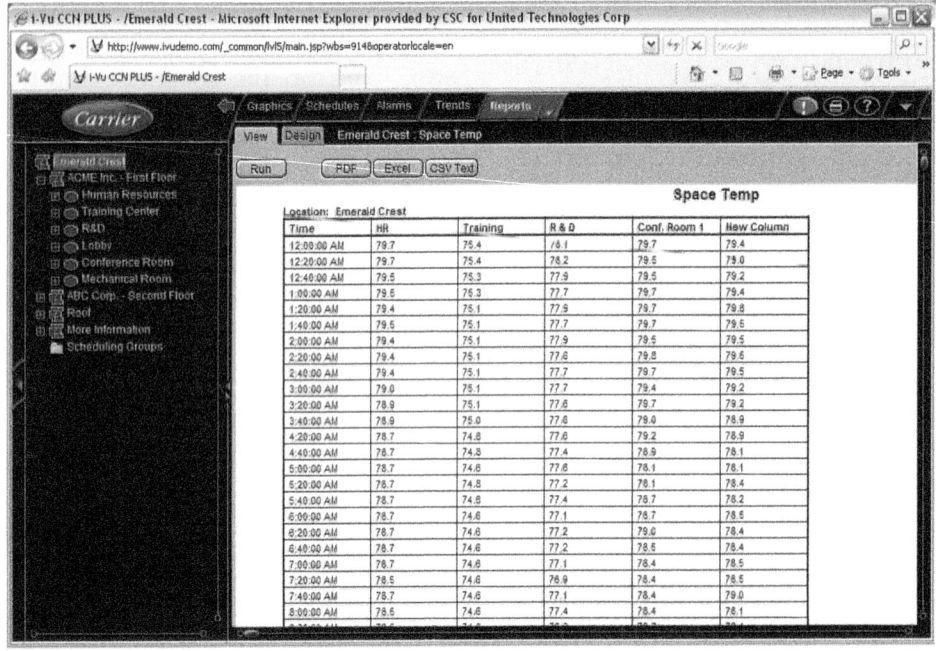

*Figure 26* Space temperature report.

03405-13_F26.EPS

addition, each building has an occupancy schedule, such as becoming occupied at 6:00 AM and unoccupied at 6:00 PM. Tenants of the building are charged for their occupied heating and cooling usage by the square footage of the building they occupy. However, when tenants activate the HVAC system during unoccupied periods, building owners want to charge the individual tenants for that usage.

Room sensors in each tenant zone have an override button. When pushed during unoccupied hours, the override button causes the associated air handler and chiller or boiler to activate and provide cooling or heating as needed. The tenant billing module, located on the communication bus, registers this after-hours demand and stores the zone that requested the extra HVAC usage and how many hours were used. At the end of the month, the tenant billing module works with the central computer software to produce a bill for each tenant's use.

### 3.2.5 Load-Shed

One of the most important BMS functions is load-shedding, which is a key element of energy cost control. Most utilities charge their commercial and industrial customers higher rates during peak use periods and for consumption above a pre-established demand limit. Therefore, it is in the building owner's best interest to minimize energy consumption during these periods and to keep track of the energy consumption in relation to the demand limit. A load-shed control module uses one or more power meters to track energy consumption. As the energy consumption approaches the demand limit, the system is programmed to exercise one or more strategies to shed loads. The following are examples of these strategies:

- Override the zone thermostat setpoints so that the equipment is off for longer periods.
- Deactivate auxiliary heating devices, which use a lot of energy.
- Shut off electrical devices in a prearranged order of priority.

If energy consumption exceeds the demand limit, an alarm may sound. At that time, building managers may implement even more severe measures to curtail the use of energy in order to avoid cost penalties.

## Management Modules

In buildings less than 50 tons, you are not likely to see the use of tenant billing, load-shed, or data collection software. These functions are typically encountered in larger buildings where the investment of time and resources is often necessary.

### 3.2.6 Lighting Control

Among the major energy users in a commercial building are the lights, both internal and external. Algorithms are used to schedule their use. For example, programmed daily scheduling can minimize the use of parking lot lights. These algorithms, depending on the control system manufacturer, may be run on the central computer, in a separate control module, or in a general-purpose controller.

Commercial building lights are controlled by their own digital controller system and use one of several unique protocols like digital addressable lighting interface (DALI), ZigBee Wireless, or building automation and control network (BACnet). A protocol is a convention that governs the format and timing of communication between devices in a computer network. In order for the BMS to access information from this system, a lighting interface module is used. The interface module translates between the protocol of the BMS and the protocol of the lighting system.

### 3.2.7 Building Access

For security reasons, many buildings have limited access. Only individuals with access cards or keys may gain entry to the building, and their access can often be tracked. The building access system software has its own protocol. In order for the BMS to gather information from the building access system, an interface module is used. The interface module translates between the protocol of the BMS and the protocol of the building access system.

### 3.2.8 Smoke Control System

In larger commercial buildings, the ability to sense smoke and control its spread (and therefore the

spread of a fire) throughout the building is significant. Typically, this function is controlled by a specialized software system separate from the building's HVAC control system. *Figure 27* shows a typical smoke control system with smoke detectors, detection and alarm panels, and remote control panels located appropriately in the building. The smoke control system has its own unique communication protocol, typically Modbus or BACnet.

The remote control panels position ventilation and exhaust dampers and operate fans to exhaust smoke and pressurize floors on either side of the floors following a fire. Operating ventilation systems in the presence of a fire simply feeds it with fresh oxygen. Ventilation and the supply of fresh oxygen must be stopped during a fire event. An operator's control panel allows building personnel to interface with the system, and a firefighter's control panel allows the fire marshal to view equipment status and override the smoke control system.

In order for the BMS to access information from the fire alarm system, an interface module is used. The interface module translates between the protocol of the BMS and the protocol of the fire alarm system.

## Load-Shedding

Historically, load-shedding was used to conserve energy by turning off selected heating and cooling equipment and other loads when the building power consumption exceeded the pre-established demand limit. However, because of current increased concerns about indoor air quality (IAQ) and occupant comfort, this practice is falling out of favor. Before programming and initiating load-shedding sequences of operation into a building system, the effect they will have on the building IAQ and tenant comfort must be taken into consideration.

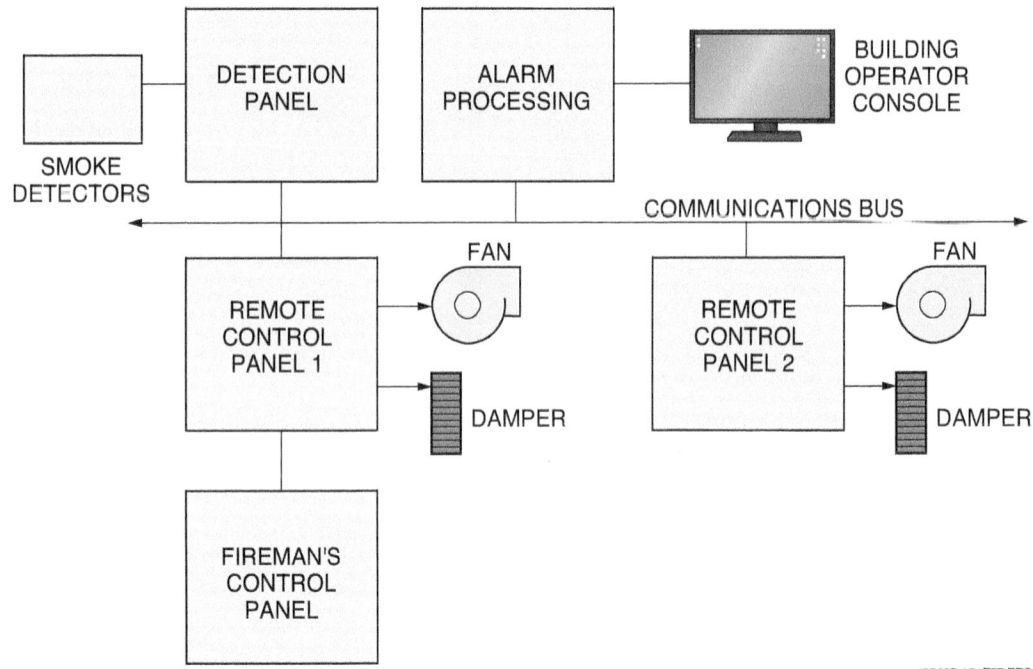

Figure 27 Smoke control system.

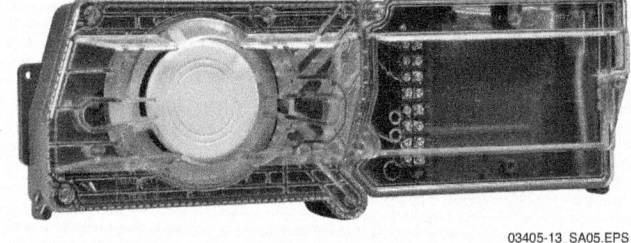

## Smoke Control

The control of smoke and fire is a rather sophisticated function in a building. A limited number of manufacturers specialize in this area. This system is typically a separate network from that used for HVAC, and the two systems cannot be controlled through a single software package. This is a requirement of *NFPA 101, Life Safety Code*. The smoke control system interacts with the HVAC system through smoke detector safeties or through relay contacts wired into the HVAC equipment control systems.

A typical smoke detector is shown here. It is designed to detect smoke in HVAC air distribution duct systems. A number of these units can be networked together, feeding information to a central panel. When one detector goes into alarm, all of the other detectors on the network activate in order to control the connected fans, blowers, and dampers. However, only the duct detector that initiated the alarm shows an alarm indication to identify it as the source of the alarm.

## 3.0.0 Section Review

1. An advantage that a smart phone has over a conventional laptop computer for remote BMS access is that _____.

   a. no internet connection is needed
   b. no keyboard is required
   c. connection speed is faster
   d. graphic displays are better

2. Load-shedding is activated when _____.

   a. zone thermostats are set too low
   b. all zones are fully occupied
   c. zone thermostats are set too high
   d. electrical energy is expensive

## 4.0.0 SYSTEM CONTROL STRATEGIES

### Objectives

Describe various building management system control strategies.

a. Describe occupied building temperature control strategies.
b. Describe unoccupied building temperature control strategies.
c. Describe other building control strategies that are not related to temperature control.

This section builds on discussions in previous modules related to VAV systems that are likely to be installed in buildings managed with BMS software. Understanding these control sequences will give you insight into how an HVAC system can be integrated with other building systems using BMS software. The strategies of controlling other, non-temperature-related systems are also discussed.

### 4.1.0 Occupied Building Temperature Control

This section covers the sequence of operation of a single duct box VAV system (*Figure 28*). Slight variations in this sequence do occur when fan-powered boxes are used. A central air source unit provides air through ductwork to single-duct VAV boxes equipped with zone reheaters. System control modes are established through communication between digital controllers that are mounted on all pieces of equipment and connected by a communication bus. Times when a building is occupied and the demand for cooling or heating are factors that help determine occupied building BMS control strategies.

### 4.1.1 Occupancy Period Scheduling

For VAV systems that do not use a DDC control system, the central air source unit determines when the building's occupied period begins and ends in one of two ways: sensing a set of contacts wired to a remote time clock or energy management system; or responding to its own internal programmable time clock. The user programs the clock's occupied and unoccupied periods.

When a DDC control system is used, the occupancy period control typically resides in the zone controllers at the zone level. Any zone's controller can go to an occupied mode and override the central unit's occupancy schedule, thus asking the central unit to start. Likewise, the last zone to go unoccupied determines when the central unit stops.

Each zone sensor can be provided with a manual override button. Pushing the button allows an occupant to manually start the system when the zone is unoccupied. The system will start and remain occupied for a preconfigured period of time, and then return to its unoccupied mode.

### 4.1.2 Occupied Cooling

When an occupied period begins, the central air handling unit fan starts. Inlet guide vanes or a variable-speed drive controls the amount of air delivered to the system by maintaining a supply duct static pressure setpoint. When the VAV terminals throttle to match a falling building load, the supply duct pressure will increase. The air source control module modulates the supply air volume to maintain the duct pressure setpoint. Along with the start of the fan, the central unit's outdoor air dampers open to a preconfigured minimum ventilation position. If the central unit is an air handler, the associated chiller or condensing unit is also started at this time.

The central unit will then add cooling capacity stages as necessary to maintain an adjustable supply air temperature setpoint, typically 55°F. Many VAV systems feature a supply air reset function, which normally resets the supply air temperature setpoint based on outdoor conditions and other factors. In a large building, moving the supply air temperature setpoint just several degrees higher can save a significant amount of energy when 55°F air is cooler than necessary to achieve comfort. If the unit is equipped with an economizer, and the outdoor air is suitable, the first stage of cooling will be using up to 100 percent outdoor air to maintain this setpoint. If more cooling is desired, stages of compression will be added to maintain the supply air temperature setpoint.

*Figure 29* shows how a VAV terminal controls its damper position during cooling and heating. The vertical axis shows zone cfm settings (percent airflow), while the horizontal axis shows zone temperature setpoints. In response to a zone sensor, the zone controller modulates the primary air cfm to meet the control zone occupied cooling setpoint, typically 70°F to 76°F. The controller is set to limit the primary air to a maximum cooling cfm to match the zone's design cooling load. The

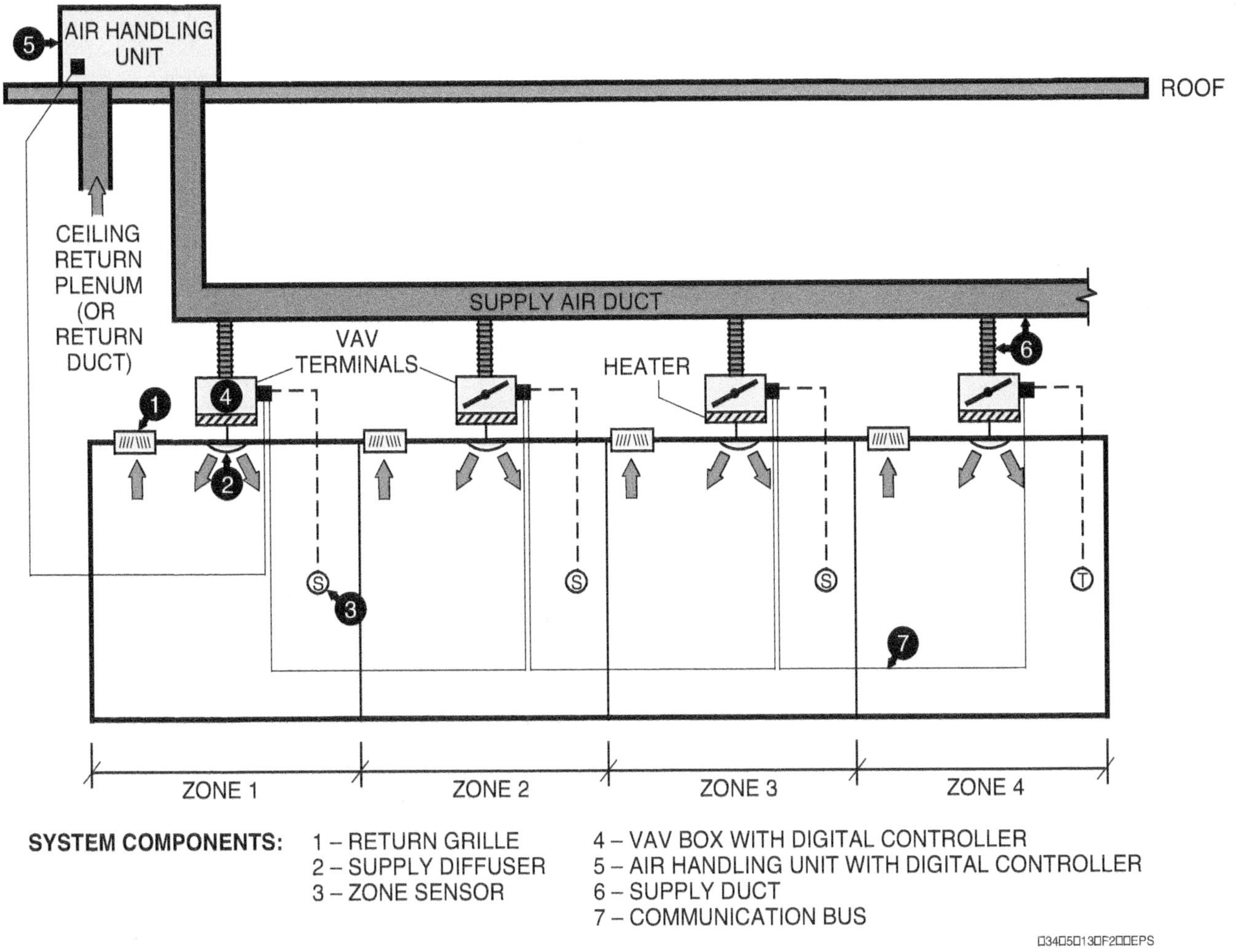

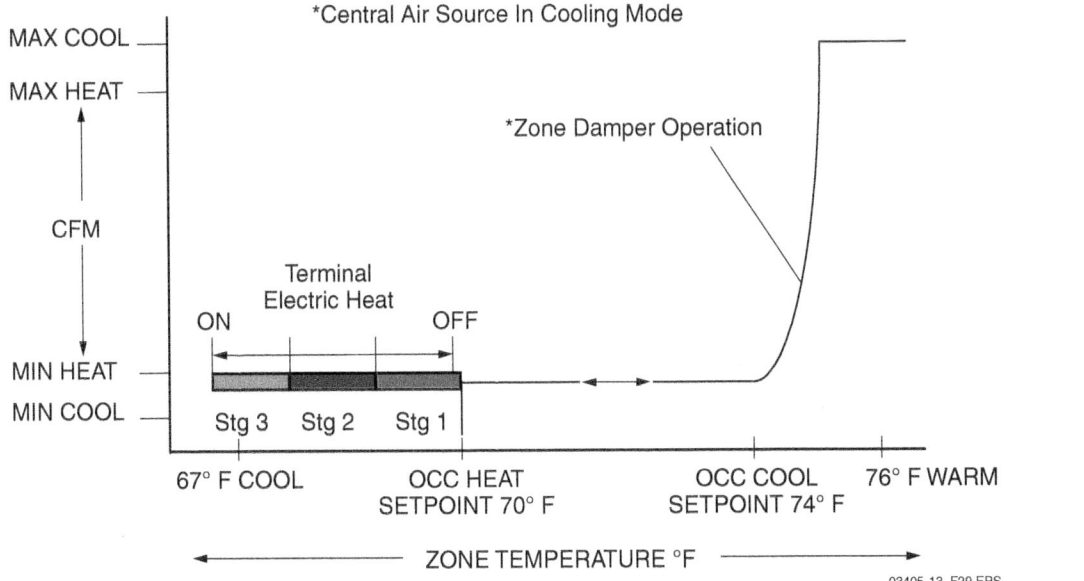

Figure 28 VAV system.

SYSTEM COMPONENTS:
- 1 – RETURN GRILLE
- 2 – SUPPLY DIFFUSER
- 3 – ZONE SENSOR
- 4 – VAV BOX WITH DIGITAL CONTROLLER
- 5 – AIR HANDLING UNIT WITH DIGITAL CONTROLLER
- 6 – SUPPLY DUCT
- 7 – COMMUNICATION BUS

03405013F200EPS

*Central Air Source In Cooling Mode

*Zone Damper Operation

MAX COOL
MAX HEAT

CFM

Terminal
Electric Heat

ON          OFF

MIN HEAT
MIN COOL

Stg 3   Stg 2   Stg 1

67° F COOL          OCC HEAT              OCC COOL          76° F WARM
                    SETPOINT 70° F        SETPOINT 74° F

ZONE TEMPERATURE °F

03405-13_F29.EPS

Figure 29 VAV damper operation.

controller is also set to limit the primary air to a minimum cfm to maintain the design ventilation rate for the zone. When the central unit is in the economizer mode, the percentage of ventilation air to the zone may vary from the minimum setting up to 100 percent outdoor air.

### 4.1.3 Occupied Heating

When heating is required, most VAV systems rely on electric or hydronic heat from either fan-powered VAV boxes and/or heat circulated from the ceiling plenum to the space below.

A few DDC control systems have the optional ability to use the central unit as a heat source for the building. If the average zone temperature or night low-limit controller indicates heating is needed, the air source control module can be configured to switch the central unit into the heating mode. Zones requiring heat simply modulate their air terminal dampers accordingly, and all zone heaters are deactivated. When the zones have been heated sufficiently, the central unit returns to its normal mode. However, this strategy is quite rare for VAV systems. It is the common approach for VVT systems in smaller applications.

### 4.2.0 Unoccupied Building Temperature Control

Times when a building is unoccupied and methods that can satisfy the heating/cooling demand during unoccupied periods are factors that help determine unoccupied building BMS control strategies.

### 4.2.1 Unoccupied Period Initialization

When the system determines that all zones have become unoccupied, it stops the central air source. The central unit supply fan is stopped and the outside air dampers close. If the system is programmed with setback temperatures, the central unit may operate just like the occupied period except that the outside air dampers remain closed; temperature setpoints for all zones are increased. A more common approach though, is to shut the system down and have the air handling unit respond to the temperature of a single, representative zone. When that zone requests cooling, the air handling unit starts, and most or all VAV boxes open to provide some cool air to every zone. The unoccupied setpoint may be as high as 85°F. Central chillers operate only as necessary to respond to this demand.

When heating is desired during an unoccupied period, the air handling unit starts and allows the VAV boxes to provide whatever form of heat they have available. The outside air dampers remain closed. If heat is available from the central unit, that heat will be used instead to keep the building reasonably warm.

### 4.2.2 Morning Warm-Up

Before the building is occupied in the winter, the VAV central unit enters into a morning warm-up cycle to bring the temperature back up, at or near the occupied setpoint. During morning warm-up, the central air handler's digital controller modulates the heating coil to match the building heating needs. This is accomplished by maintaining the return air temperature from the zones at a predetermined setpoint. Heat from the VAV boxes is used if the central system has no heating capability. During morning warm-up, the outside air dampers remain closed to minimize the heating load.

When the zone controller is informed that the central unit is in morning warm-up, it cycles the VAV terminal damper as necessary to bring the space up to the occupied heating setpoint.

Morning warm-up often ends before the building reaches the occupied setpoint of each zone. Temperature-control of individual zones is suspended, and a single zone thermostat is used to monitor morning warm-up progress. At that time, the programming for an occupied building mode takes effect.

### 4.2.3 Night-Time Free Cooling

VAV systems can use an energy-saving routine called night-time free cooling (NTFC). During the unoccupied cycle (typically between 3:00 AM and 7:00 AM), the central air handler's digital controller can initiate the NTFC routine if there is sufficient difference between the outside air and the inside building temperatures. When activated, the central unit opens the outdoor and exhaust dampers 100 percent, starts the supply fan, and prevents mechanical cooling from being activated.

When activated by the central unit's NTFC mode, the zone controllers position their primary air dampers to the maximum cooling cfm limit until the NTFC setpoint is reached. When the NTFC setpoint is reached, each zone controller repositions its primary air damper to the minimum cooling cfm limit. Each zone's primary damper is thus cycled between these two cfm settings until the building becomes occupied or all zone NTFC setpoints have been reached.

## 4.3.0 Other Control Strategies

In addition to controlling temperatures in a building, a good building management system is capable of controlling $CO_2$ and humidity levels, which affect indoor air quality.

### 4.3.1 Demand-Controlled Ventilation

VAV systems are capable of controlling indoor air quality (IAQ) at the zone level by employing a control strategy called demand-controlled ventilation (DCV). With DDC controls, a carbon dioxide ($CO_2$) sensor is mounted in each zone (*Figure 30*). The sensor is typically wall-mounted and wired back to a zone controller. The sensor is also available as a combination temperature, humidity, and $CO_2$ sensor to minimize installation costs.

When occupied zone $CO_2$ levels exceed an adjustable setpoint (measured in parts per million), the zone controller modulates the primary air damper between the maximum and minimum cool cfm limits to satisfy the $CO_2$ setpoint. More primary air, which is partially comprised of fresh outdoor air, is pushed into the space to dilute the $CO_2$ concentration.

If overcooling of the zone results from ventilation, the primary air damper is temporarily positioned to a minimum IAQ reheat cfm limit and terminal reheat is added to return the room temperature to setpoint. A fan-powered box will provide heat along with the primary air.

The central air handling unit controller continually monitors the zone $CO_2$ levels and adjusts the outdoor air dampers accordingly. If the zone $CO_2$ levels are satisfactory, the central unit maintains the base ventilation cfm. If the discharge temperature falls below the supply air setpoint (typically 55°F) in some central packaged units during DCV operation, the central unit heater is activated to provide reheat and prevent operational problems. If the central unit is equipped with economizer control, the economizer routine overrides DCV and modulates the outdoor air dampers to provide cooling. DCV operation is disabled whenever the central unit is in the unoccupied mode.

### 4.3.2 Humidity Control

VAV systems are sometimes capable of controlling relative humidity at the zone level instead of IAQ control.

When the zone is occupied, the operation of the central unit normally prevents humidity buildup in the zone by dehumidifying the supply air. However, this method may be insufficient at part load. Humidity control may be improved by sensing humidity at the zone level. An optional relative humidity sensor may be wired to the zone controller. With DDC controls, zone controllers equipped with humidity sensors send humidity information back to the central unit controls. In response, the central unit adds cooling capacity to lower the supply air temperature, creating drier supply air.

At the zone level, each zone controller modulates its primary air damper between minimum and maximum cool cfm limits to maintain a zone humidity setpoint. If the maximum zone temperature setpoint is exceeded while controlling zone relative humidity, the zone controller positions the primary damper to a reheat cfm limit and adds terminal reheat to return the zone to normal temperature.

### 4.3.3 Building Pressurization

Many buildings require some form of pressurization control, which can be done through the VAV system. In fact, in most cases, the pressure must be controlled because of the VAV system's introduction of outside air into the building. As a higher volume of outside air is drawn into the building, the building pressure increases.

The central air handling unit can be provided with a powered exhaust system to accomplish this. Responding to a space pressure sensor located in a key building space, the central unit controls activate a powered exhaust fan or variable-speed fan to modulate building exhaust air accordingly. Alternatively, dampers can be provided at the relief air outlet. The dampers are positioned based on building pressure. As the pressure increases, the dampers open wider to allow more air to escape the building.

03405-13_F30.EPS

*Figure 30* Wall-mounted $CO_2$ sensor with display.

VAV central units are sometimes equipped with a return fan. When this is the case, some form of fan tracking control is required to keep the cfm leaving the conditioned space in a controlled relationship to the cfm being supplied. In one method, the central unit controls maintain a constant cfm differential between the supply and return airflow. This maintains a positive pressure in the conditioned space. In another method, the central controls vary the return fan cfm to maintain a predetermined space pressure. The designer must choose the method based on the application needs. Air terminal controls do not participate in this function.

### 4.3.4 Smoke Control

It is common practice to use the central unit supply and exhaust fan to provide zone smoke control following a fire or smoke event. Under the command of a fire marshal's panel (part of the building's fire/life safety system), the central unit may be manually placed in one of three modes:

- *Pressurization* – The DDC control system provides excess air into the zones served by the central unit by opening outdoor air dampers 100 percent and running the supply fan. During this mode the return and exhaust dampers are closed to prevent or delay smoke entry from adjacent zones. Each zone damper is positioned to its maximum cool cfm limit. If an air terminal series fan is installed, it is started.

- *Evacuation* – This mode removes smoke from the control zones by running the powered exhaust fan only. The return and outdoor air dampers are held closed. Each zone damper is moved to its closed position. This position will be maintained until the central unit changes operating mode. If an air terminal series fan is installed, it is stopped.

- *Smoke purge* – This mode flushes smoke and/or contaminated air from the controlled zones by running both the supply and powered exhaust fans, utilizing 100 percent outdoor air. Return dampers are closed. Each VAV damper is positioned to its maximum cool cfm limit. This cfm will be maintained until the central unit changes operating mode. If an air terminal series fan is installed, it is started.

### 4.0.0 Section Review

1. Zone sensors are often equipped with a _____.

    a. humidistat
    b. manual override
    c. manual reset
    d. timer

2. When is night-time free cooling activated?

    a. When enough temperature difference exists between indoor and outdoor air.
    b. When outdoor air temperature is above 40°F but below 70°F.
    c. Only when utility power rates are lower during the evening.
    d. Only when the air handler is equipped with an economizer.

3. What happens during the smoke control evacuation mode?

    a. Outdoor air dampers close
    b. Supply fan is turned off
    c. Return dampers close
    d. Exhaust dampers close

# SECTION FIVE

## 5.0.0 INTEROPERABILITY

### Objectives

Define the concept of interoperability and describe the various related protocols.

   a. Describe the four primary protocols in use.
   b. Define and describe web browser system integration.
   c. Project the course of interoperability in building management.

### Trade Terms

**Firmware**: Computer programs that are permanently stored on the computer's memory during a manufacturing process.

**Gateway**: A link between two computer programs allowing them to share information by translating between protocols.

**Interoperability**: The ability of digital controllers with different protocols to function together accurately.

Facility managers would like to have a complete real-time model of all the buildings under their care, providing every important detail at their fingertips. They want to know what building space each tenant occupies, when it is being used, what everyone's phone number is, what phones are or are not operational, what furniture (including color and fabric) is assigned to each building space, and what kind of special assistance might be required in case of an emergency. They want easy access to the actual temperature in each building space, the ability to adjust each space's control temperature at will, and the ability to match the operational schedule of each HVAC system in their facilities to its real-time occupant use. They also want to be able to optimize their building's utility costs to match real-time utility rate structures.

Building managers like to have status information about video teleconferencing sites throughout the facilities, and have security video feeds at their fingertips from cameras located throughout the sites. Additionally, they would like to have data exchanged between all computers and voice over IP (VOIP) phone connections.

They want to know who manufactured each piece of equipment, who installed it, when its warranty period ends, what preventive maintenance has been performed, and when its next maintenance is scheduled. They also want to know where all the equipment is located, where all conduits and piping are, and where all communication cables are located.

The accurate transfer of information and commands required to match the facility manager's desires across building systems with different protocols is called interoperability. These desires are becoming more and more possible today with proper planning by a qualified system integration company and utilization of manufacturer's equipment that is becoming more and more interoperable.

With the evolution of open communication protocols in the market place, it is now possible to integrate an HVAC control system with other building systems such as security, lighting, and fire/life safety (*Figure 31*). Open protocols are defined as communication standards that allow different manufacturer's control systems to share information. Each protocol has a unique set of rules to ensure that good communication take place between controllers.

Building owners want flexibility when it comes to selecting building systems equipment. Thus, through various industry organizations and groups, they have moved toward the use of open protocols. Unfortunately, none of the leading protocols are perfect and none are able to satisfy the needs of all building systems. Thus, a unique protocol does not exist in today's market that is the best. Each protocol has advantages and disadvantages. With these limitations in mind, however, there are several open protocols that are tending to become most used by commercial building owners. Real-time HVAC usage and demand response are two examples of interoperability.

Real-time HVAC usage allows a building owner to operate the lights and HVAC system only when the building is occupied by tenants. Thus, when a tenant's access card is scanned upon entering the building, an open communication protocol is used to monitor the card access, and then command the lights and HVAC system on in that portion of the building occupied by that tenant. At the end of the day, when all employees associated with a particular portion of the building leave, the associated lighting and HVAC system shuts down.

The critical shortage of electricity in the United States is driving utilities to offer incentives to curtail electrical loads at peak times. In addition, utilities have rate structures that vary by time of day, day of week, etc. Using web-based communications, they have the ability to provide build-

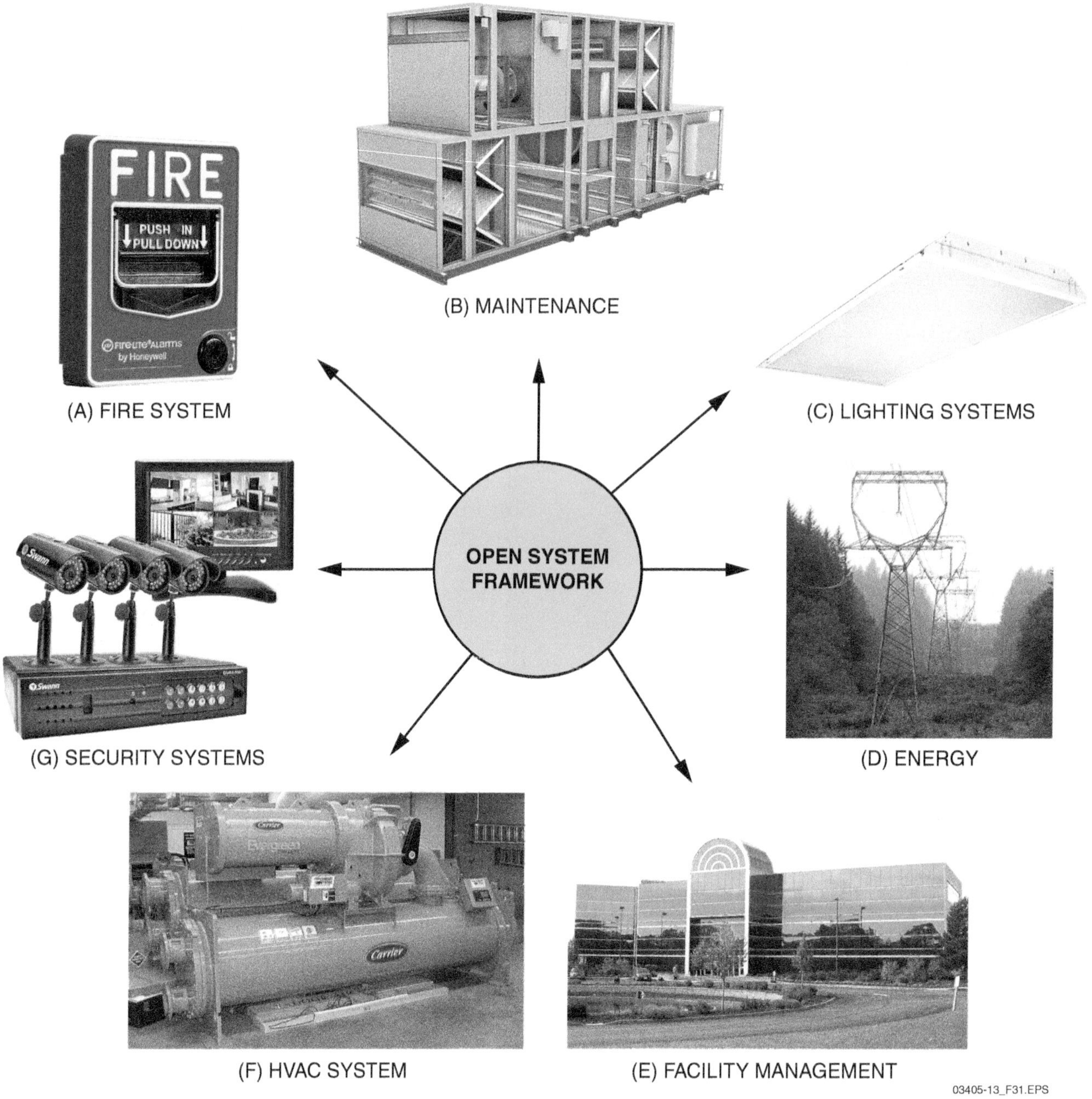

(B) MAINTENANCE

(A) FIRE SYSTEM

(C) LIGHTING SYSTEMS

**OPEN SYSTEM FRAMEWORK**

(G) SECURITY SYSTEMS

(D) ENERGY

(F) HVAC SYSTEM

(E) FACILITY MANAGEMENT

03405-13_F31.EPS

*Figure 31* Interoperability.

ing owners with real-time utility rates. Demand response is the ability of a BMS system to read the incoming real-time utility rate structure and to modify the operation of HVAC and other building systems to optimize the building's energy usage and operational cost. Thus, by integrating demand response as an interoperable control strategy, a building can be optimized for both utility cost and comfort while resulting in the lowest possible energy consumption.

## 5.1.0 Building System Protocols

The following sections describe the most popular protocols being used in building systems today. It should be understood that the list is not exhaustive, and does not cover other industries such as factories and automotive industries.

### 5.1.1 BACnet

BACnet was developed by ASHRAE and approved as a standard in 1995. The latest version of the standard, dated 2012, was published in 2013. This building automation and control network (BACnet) protocol provides a comprehensive set of message rules that, if implemented, provide interoperability between the various building systems. As *Figure 32* shows, each manufacturer would design and make available a BACnet in-

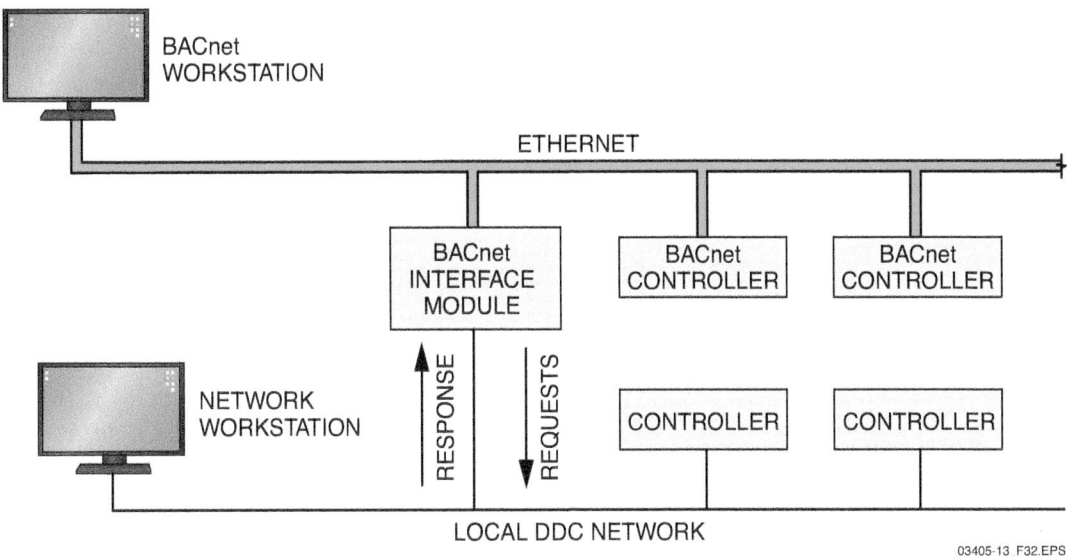

*Figure 32* BACnet overview.

terfacing module that would serve as a gateway to the manufacturer's proprietary protocol bus. This module would have the capability of two-way communication between the manufacturer's network and the BACnet Ethernet. Each BACnet module would provide a limited amount of information transfer as determined by the module's manufacturer. Depending on the amount of information to be exchanged, multiple BACnet modules would be required.

### 5.1.2 Lon Technologies

Another contender for standard protocol is the Echelon Company, with its local operating network (LON) technologies, including LonTalk™, LonWorks™, and LonMark™.

Echelon has developed LON technology based on a neuron chip microprocessor. The neuron chip contains hardware and firmware (instructions embedded in the chip) that operate on a seven-layer communication protocol. LonTalk™ is the seven-layer protocol that allows controllers (nodes) to operate using an efficient, reliable communication structure. Each node in the network contains a chip with embedded intelligence that implements the protocol and performs control functions.

LonWorks™ is Echelon's DDC network that uses LonTalk™ protocols, and is part of the ANSI-approved ASHRAE BACnet standard for building automation.

Echelon's LON technology products were designed to provide solutions for a general manufacturing environment; however, the technology has been equally applied to the HVAC industry. Because each node sold by Echelon has a LON

chip in it, any product with this chip has the built-in protocols to communicate with any another product. Likewise, any manufacturer who uses LON chips in their products ensures interoperability with any other manufacturer's devices that include LON chips.

### 5.1.3 Modbus

The Modbus standard was developed by the Modicon Company in 1979 and later became the trademark of Schneider Electric. Schneider made Modbus available for free and gave the Modbus TCP/IP protocol specifications to the Internet Engineering Task Force. Thus Modbus TCP/IP has become widely used and is popular in the industrial, metering, power generation, and electrical equipment industries. Modbus is typically used as a translation protocol between two other proprietary protocol networks.

The Modbus standard defines both hardware and software standards. Modbus Remote Terminal Unit (Modbus RTU) communicates serially over Electrical Industry of America (EIA) RS232 and RS485 hardware connections. Modbus Transmission Control Protocol (Modbus TCP) communicates over Ethernet LAN type hardware connections.

### 5.1.4 Hypertext Transfer Protocol (http)

Hypertext Transfer Protocol (http) is the base protocol used by the world wide web. This standard is published and updated by the World Wide Web Consortium (W3C), and uses TCP/IP as the means of transporting information over the internet. The http protocol allows a computer operator to use a web browser program to view transmitted data in the form of web pages. The http protocol defines how messages are formatted and transmitted. The http protocol also defines what actions web server computers, browsers, and network controllers take in response to various commands.

Thus, control systems can take advantage of this well-defined protocol to share data with end users as well as other controllers. Since the use of this standard extends beyond the HVAC industry, BMS control systems can use http to easily share information, such as corporate financial data, energy data, maintenance management data, and HVAC system operational data across building systems.

### 5.2.0 Web Browser System Integration

Approximately 80 percent of all commercial building projects will be satisfied with an integration system like that shown in *Figure 20*, but the remaining 20 percent require more complex interoperability to achieve their operational needs. To meet this greater complexity, many manufacturers now provide a low-cost web-based connection to their control systems. Thus, building owners and operators can now use the common http protocol and a standard web browser to interface with their multiple building systems.

*Figure 33* shows the network architecture of such a system. In this system, a unique protocol can be used where its qualities best suit the control needs involved. Thus, separate control buses (and protocols) for lighting, HVAC, building security, and other systems, may still exist, but they are seamless to the building operator. In addition, the integrators are capable of implementing control strategies across building systems.

System integration is achieved by the use of multiple industrial integrator computers, known as gateways, located on the building's Ethernet LAN. These integrators do not have keyboards or monitors. They are powerful central processing units with multiple cable connections that match the transport medium requirements of the common open protocols (BACnet, LON, and Modbus). In addition, the integrators have a bus connection that matches the manufacturer's own unique proprietary protocol. The integrators are then programmed to regularly extract data from controllers located in the building's lighting, HVAC, security, and other systems. This extracted data is then complied and temporarily stored in two-way information tables recognized by a common protocol normally used on the Ethernet LAN. BACnet/IP is becoming the common protocol used on the Ethernet LAN.

Also located on the LAN is a web page server computer. The web page server computer uses the common protocol (typically BACnet/IP) to regularly extract table information from the integrator computers and to create web pages of information.

These web pages are then viewable in tabular or graphical format by any operator through a web browser located on any computer on the LAN. The communication is also two-way, so that any operator may manually intervene at the controller level in any building system. Additionally, the operator may communicate with the web page server over the internet via a computer, or using a wireless device such as a smart phone or tablet.

Finally, integrated control strategies across building systems may be achieved through special routines written in the web page server computer. For example, data from a card reader can be sent to the lighting and HVAC system controllers so that they are activated when an employee enters the building, and shut down when the employee leaves.

### 5.3.0 The Future of Interoperability

In the future, most manufacturers will build their equipment with controllers that have bus connections to support four typical protocols: their own, BACnet, LON, and Modbus. A simple switch setting on the controller will make it ready to communicate with the protocol of choice. This will make it easy to integrate equipment of multiple vendors into a single, seamless system.

In addition, within each building system industry, equipment manufacturers are all moving toward the elimination of their own unique communication bus protocol and tending to standardize around a common open protocol that meets the needs of that industry.

Using the HVAC industry as an example, building owners can buy a chiller from Manufacturer

---

## The Growth of Wireless Technology

The late Steve Jobs of Apple fame had a vision that has changed the way people communicate. Using wireless technology, the simple cell phone has become a device for communicating all sorts of information. Several designers have developed applications (apps) for smart phones and tablets that enable them to, among many other things, act as a building management system user interface.

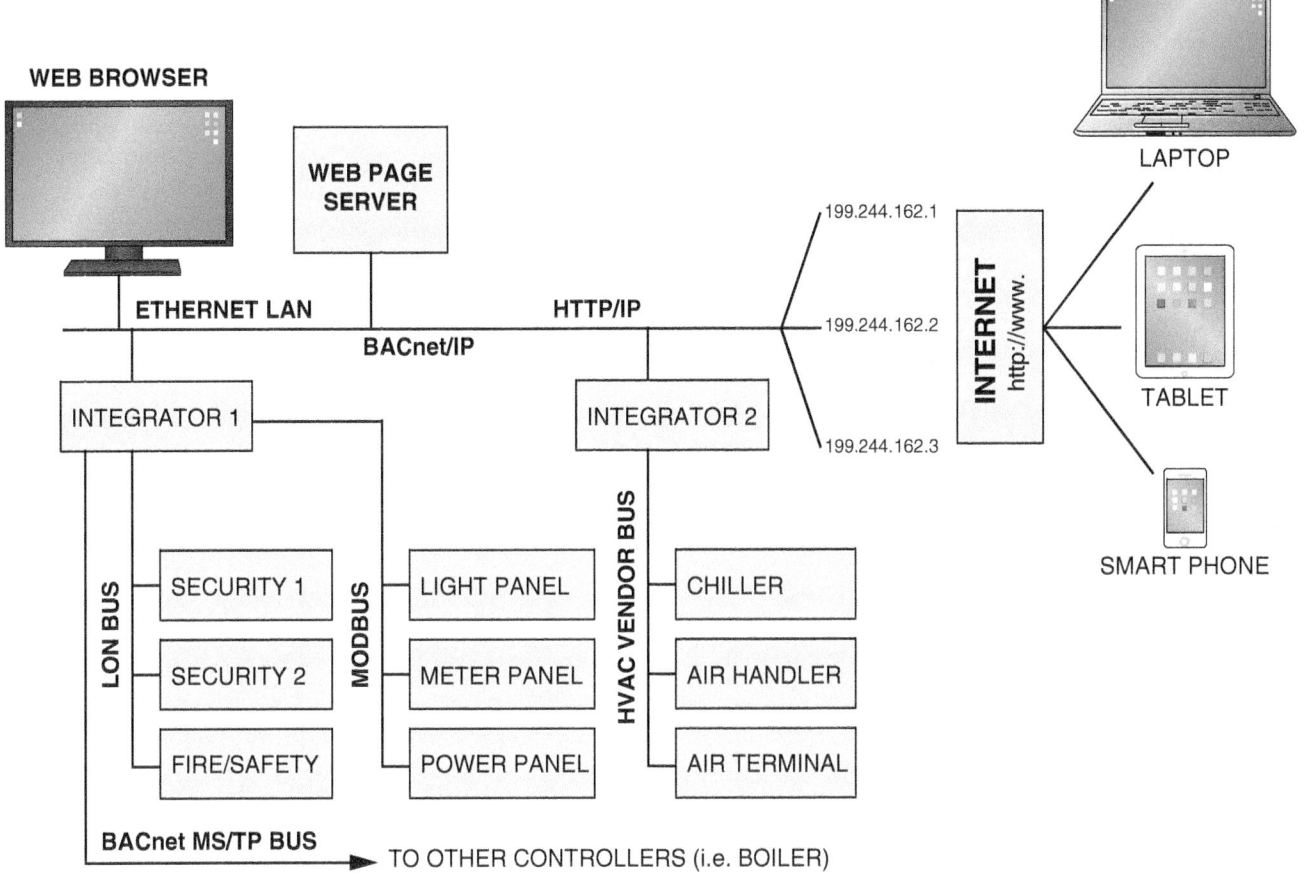

**WEB BROWSER**

**WEB PAGE SERVER**

**LAPTOP**

**ETHERNET LAN**

**HTTP/IP**

**BACnet/IP**

199.244.162.1

199.244.162.2

199.244.162.3

**INTERNET** http://www.

**TABLET**

**SMART PHONE**

**INTEGRATOR 1**

**INTEGRATOR 2**

**LON BUS**

SECURITY 1

SECURITY 2

FIRE/SAFETY

**MODBUS**

LIGHT PANEL

METER PANEL

POWER PANEL

**HVAC VENDOR BUS**

CHILLER

AIR HANDLER

AIR TERMINAL

**BACnet MS/TP BUS**

→ TO OTHER CONTROLLERS (i.e. BOILER)

03405-13_F33.EPS

*Figure 33* Web browser integration technology.

A, an air handler from Manufacturer B, and a set of air terminals from Manufacturer C that can talk together on a common HVAC communication bus. The emerging use of the common open protocol (BACnet) in the HVAC industry will make it easy to integrate the chiller, air handler, and air terminals into a cohesive HVAC system.

Because of these trends, the building owner will no longer be locked into a single service vendor because of the control system. In addition, during later building expansions or renovations, the owner can choose freely between equipment manufacturers within a type of building system without having to worry about interoperability issues. In the past, this was a significant problem.

Wireless technology has also emerged in the BMS industry. As the air waves are being managed differently and frequencies are opened up to support more wireless communication, it has become easier to apply wireless components in building control. However, wireless devices do have their technical problems, the most notable

being the problem of broadcast range. Interference from other wireless devices in a building can also cause trouble. But new wireless communication standards are being adopted at this writing, and the dependability and range of wireless devices continues to improve. The impact can be very significant, as routing wire and cables to support BMS throughout a large building is very expensive and labor-intensive. With a wireless device, placement is as easy as taking it out of the box and mounting it in the chosen location.

When it comes to interoperability across building systems, however, it is highly unlikely that any one protocol will be developed to meet the needs of all systems. However, as shown above in *Figure 33*, the building owner will be able to use the power of http internet protocol and web browsers to more easily manage the building through a single seamless interface. The use of web-based technology is already growing rapidly today, and will continue even more rapidly in the future.

# Forewarned is Forearmed

Many problems can be solved without dispatching a technician to the site. The ability to troubleshoot a building from a central location within the building or from across town through a wireless internet connection can save a service technician time and save the building owner/manager money. With this capability, the service technician can arrive at a service call with advance knowledge of the problem and the tools, equipment, and spare parts needed to solve it.

## 5.0.0 Section Review

1. The BACnet protocol provides _____.
   a. a comprehensive set of message rules
   b. priority to HVAC control signals
   c. BMS operator access over the internet
   d. BMS operator access over the Ethernet

2. The http protocol allows building owners to access their BMS control systems over the internet.
   a. True
   b. False

3. In the future, which communication protocol will enable equipment from different manufacturers to be easily integrated into a single building management system?
   a. LANworks
   b. http
   c. LonMark
   d. BACnet

## SUMMARY

In today's environment, it is absolutely essential for the HVAC technician to become familiar with computer-controlled systems. At some point in the future, all HVAC systems will be controlled by digital controllers.

Not only will heating, cooling, and ventilation be controlled, but all building systems will be tied together with control strategies that cross over between each system's controls. As you have seen, integrated building management systems are commonplace in commercial applications. Simpler building management systems are now available for residential use. Today, lighting, entertainment, security, cooking, and comfort systems can be controlled from a computer or smart phone accessed by the homeowner from anywhere in the world.

1. A digital controller consists of three major components: input module, microprocessor, and _____.
   a. output module
   b. actuator
   c. valve
   d. sensor

2. A general-purpose controller is _____.
   a. also known as a product-integrated controller
   b. often installed by the equipment manufacturer
   c. installed where simple operating schemes are used
   d. often used for chiller or boiler pressure control

3. A discrete input or output is one that _____.
   a. is proportional
   b. has two states (conditions)
   c. varies within a wide range
   d. has several states (conditions)

4. The type of sensor that uses a long resistance-sensing element to sense average duct temperature is a(n) _____.
   a. averaging thermistor
   b. PID
   c. analog thermistor
   d. RTD

5. A standard analog output signal used in computer-controlled systems is _____.
   a. 4–20mA
   b. 24VAC
   c. 120VAC
   d. 20VDC

6. The most accurate algorithm used for processing information when the controlled variable is changing rapidly over time is _____.
   a. proportional
   b. proportional-integral
   c. proportional-integral-derivative
   d. proportional-derivative

7. In *Review Question Figure 1*, the device that connects all the building digital controllers together is the _____.
   a. local area network
   b. network communication bus
   c. remote communication module
   d. general-purpose controller

8. The rate at which information travels on a network is known as the _____.
   a. bit rate
   b. baud rate
   c. bus rate
   d. LAN rate

9. In a VAV system, what mode does the system normally remain in as long as the building is occupied?
   a. Heating
   b. Standby
   c. Cooling
   d. Economizer

10. In order for remote computers to access a local BMS network today, all that is needed is a _____.
   a. bridge
   b. web browser
   c. gateway
   d. router

11. A locally connected PC is a valid user interface device to a building automated system.
   a. True
   b. False

12. Which protocol allows the user to use a web browser to view a BMS system's web pages?
   a. BACnet
   b. LON
   c. http
   d. Modbus

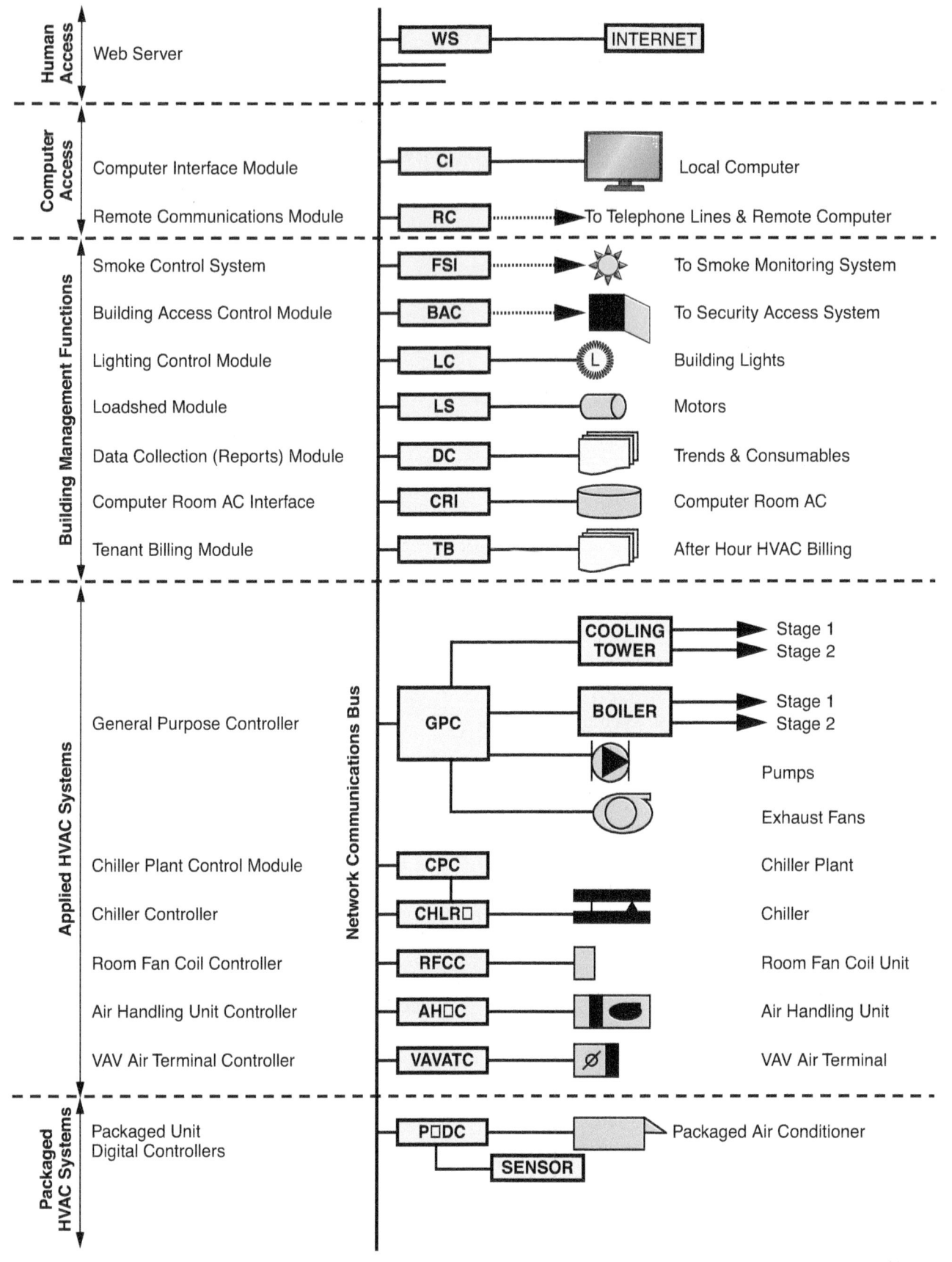

**Human Access**

Web Server

**Computer Access**

Computer Interface Module

Remote Communications Module — To Telephone Lines & Remote Computer

**Building Management Functions**

Smoke Control System — To Smoke Monitoring System

Building Access Control Module — To Security Access System

Lighting Control Module — Building Lights

Loadshed Module — Motors

Data Collection (Reports) Module — Trends & Consumables

Computer Room AC Interface — Computer Room AC

Tenant Billing Module — After Hour HVAC Billing

**Applied HVAC Systems**

General Purpose Controller

Chiller Plant Control Module — Chiller Plant

Chiller Controller — Chiller

Room Fan Coil Controller — Room Fan Coil Unit

Air Handling Unit Controller — Air Handling Unit

VAV Air Terminal Controller — VAV Air Terminal

**Packaged HVAC Systems**

Packaged Unit Digital Controllers — Packaged Air Conditioner

Network Communications Bus

WS — INTERNET

CI — Local Computer

RC

FSI

BAC

LC

LS

DC

CRI

TB

GPC

COOLING TOWER — Stage 1, Stage 2

BOILER — Stage 1, Stage 2

Pumps

Exhaust Fans

CPC

CHLR☐

RFCC

AH☐C

VAVATC

P☐DC

SENSOR

Figure 1

03405-13_RQ01.EPS

13. In *Review Question Figure 2*, which image could the operator click on to change a time schedule or setpoint schedule for the air handler?

   a. Supply fan
   b. Room sensor
   c. Cooling coil
   d. Room temperature point value

14. Which portion of the front-end software deals with uploading all the files from the network controllers and storing them according to the networks in which they exist?

   a. Report management
   b. Alarm management
   c. Operator management
   d. Database management

15. During the occupied cooling period with suitable outdoor air, what is used for first-stage cooling?

   a. 100 percent outdoor air
   b. First-stage compressor
   c. 50 percent outdoor air
   d. 100 percent indoor air

16. The building morning warm-up begins _____.

   a. at 3:00 AM
   b. at 7:00 AM
   c. before people arrive
   d. between 3:00 AM and 7:00 AM

17. What occurs during night-time free cooling?

   a. Exhaust dampers close.
   b. There is no compressor operation.
   c. Outdoor dampers open 50 percent.
   d. Supply fan is turned off.

18. Control of individual zone temperature is suspended during _____.

   a. morning warm-up
   b. evening hours
   c. demand-controlled ventilation
   d. humidity control periods

19. Real-time HVAC usage only allows the equipment to operate when the conditioned space is occupied.

   a. True
   b. False

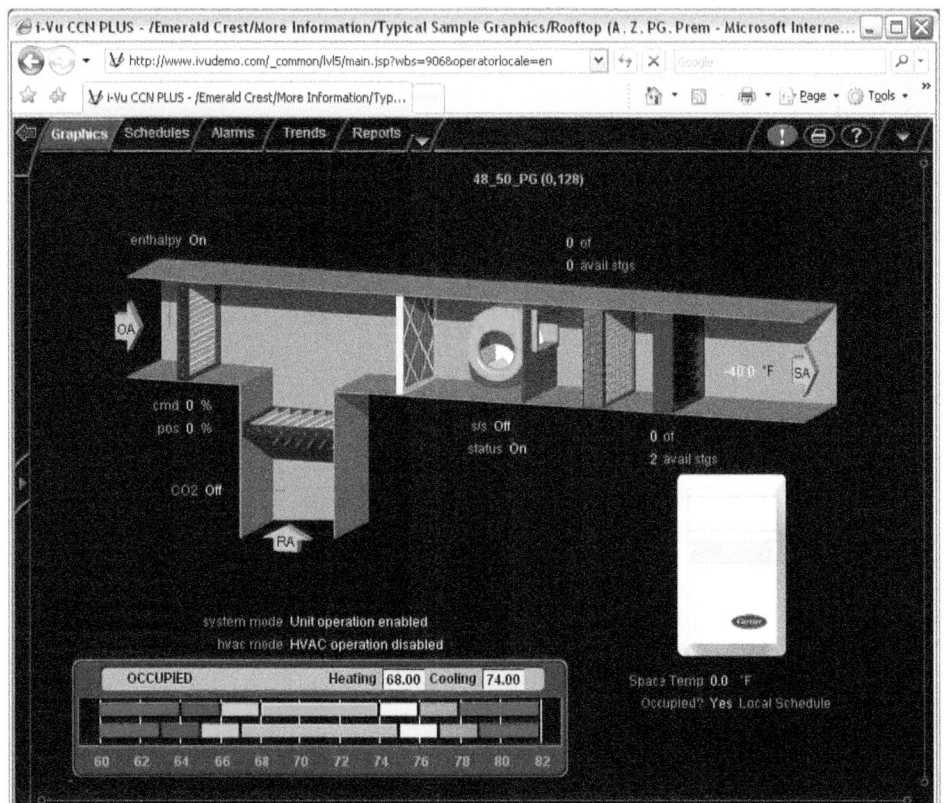

**Figure 2**

03405-13_RQ02.EPS

20. The ability of a BMS system to read the incoming real-time utility rate structure and modify the operation of HVAC and/or other building systems to optimize the building's energy consumption and operational costs is called _____.

    a. load-shed
    b. demand response
    c. night-time free cooling
    d. time guard

21. LON is a valid candidate for use as a standard open protocol.

    a. True
    b. False

22. What is the Modbus standard typically used for?

    a. Translation protocol
    b. Web-based protocol
    c. Protocol standard
    d. BACnet programming

23. What is the Hypertext Transfer Protocol (http) used for?

    a. BACnet translation
    b. Modbus to LAN translation
    c. Worldwide web protocol
    d. LON transfer protocol

24. Which of the following is *not* considered an open communication protocol?

    a. BACnet
    b. LON
    c. LAN
    d. Modbus

25. What percentage of commercial building management systems require complex interoperability to achieve operational needs?

    a. 80
    b. 60
    c. 40
    d. 20

# Trade Terms Introduced in This Module

**Algorithm**: A mathematical equation consisting of a series of logic statements used in a computer or microprocessor to solve a specific kind of problem. In HVAC applications, algorithms are typically used in microprocessor-controlled equipment to control a wide range of control function operations based on the status of various system sensor input signals.

**Application-specific controller**: A digital controller installed by a manufacturer on a specific product at the factory.

**Baud rate**: The rate at which information is transmitted across communication lines.

**Bit**: Short for binary digit. The smallest element of data that a computer can handle. It represents an off or on state (zero or one) in a binary system.

**Building management system (BMS)**: A centralized, computer-controlled system for managing the various systems in a building. Also known as a building automation systems (BAS).

**Bus**: A multi-wire communication cable that links all the components in a hard-wired computer network.

**Control point**: The name for each input and output device wired to a digital controller.

**Data collection**: The collection of trend, run-time, and consumable data from the digital controllers in a building.

**Digital controller**: A digital device that uses an input module, a microprocessor, and an output module to perform control functions.

**Direct digital control (DDC)**: The use of a digital controller is usually referred to as direct digital control or DDC system.

**Ethernet**: A family of frame-based computer networking technologies for local area networks (LANs). It defines a number of wiring and signaling standards.

**Firmware**: Computer programs that are permanently stored on the computer's memory during a manufacturing process.

**Gateway**: A link between two computer programs allowing them to share information by translating between protocols.

**Hypertext transfer protocol (http)**: The base protocol used by the worldwide web.

**Internet protocol (IP)**: A data-oriented protocol used for communicating data across a packet-switched internetwork.

**Internet protocol address (IP address)**: A unique address (computer address) that certain electronic devices use in order to identify and communicate with each other on a computer network utilizing the internet protocol (IP) standard.

**Interoperability**: The ability of digital controllers with different protocols to function together accurately.

**Load-shedding**: Systematically switching loads out of a system to reduce energy consumption.

**Local area network (LAN)**: A server/client computer network connecting multiple computers within a building or building complex.

**Local interface device**: A keypad with an alphanumeric data display and keys for data entry. It is connected to the digital controller with a phone-type cable that provides power and communication.

**Network**: A means of linking devices in a computer-controlled system and controlling the flow of information among these devices.

**Product-integrated controller (PIC)**: A digital controller installed by a manufacturer on a product at the factory.

**Product-specific controller**: A digital controller designed to control specific equipment that may be installed by the equipment manufacturer. The controllers for chillers or boilers are examples of product-specific controllers. Also known as a product-integrated controller.

**Protocol**: A convention that governs the format and timing of communication between devices in a computer network.

**Software**: Computer programs transferred to the computer from various media and stored in an erasable memory.

**Sustainable**: Designed to reduce impact on the environment.

**Tenant billing**: The ability to charge a building tenant for after-hours use of the building's HVAC system.

**Wide area network (WAN)**: A server/client computer network spread over a large geographical area.

# Figure Credits

| Answer | Section Reference | Objective |
|---|---|---|
| **Section One** | | |
| 1. a | 1.1.0 | 1a |
| 2. c | 1.2.1 | 1b |
| 3. a | 1.3.0 | 1c |
| **Section Two** | | |
| 1. b | 2.1.0 | 2a |
| 2. a | 2.2.0 | 2b |
| 3. c | 2.3.1 | 2c |
| **Section Three** | | |
| 1. b | 3.1.0 | 3a |
| 2. d | 3.2.5 | 3b |
| **Section Four** | | |
| 1. b | 4.1.1 | 4a |
| 2. a | 4.2.3 | 4b |
| 3. d | 4.3.4 | 4c |
| **Section Five** | | |
| 1. a | 5.1.1 | 5a |
| 2. a | 5.1.4 | 5b |
| 3. d | 5.3.0 | 5c |

# NCCER CURRICULA — USER UPDATE

NCCER makes every effort to keep its textbooks up-to-date and free of technical errors. We appreciate your help in this process. If you find an error, a typographical mistake, or an inaccuracy in NCCER's curricula, please fill out this form (or a photocopy), or complete the online form at **www.nccer.org/olf**. Be sure to include the exact module ID number, page number, a detailed description, and your recommended correction. Your input will be brought to the attention of the Authoring Team. Thank you for your assistance.

*Instructors* – If you have an idea for improving this textbook, or have found that additional materials were necessary to teach this module effectively, please let us know so that we may present your suggestions to the Authoring Team.

**NCCER Product Development and Revision**
13614 Progress Blvd., Alachua, FL 32615

**Email:**  curriculum@nccer.org
**Online:**  www.nccer.org/olf

❏ Trainee Guide   ❏ Lesson Plans   ❏ Exam   ❏ PowerPoints   Other _____

Craft / Level: _____   Copyright Date: _____

Module ID Number / Title: _____

Section Number(s): _____

Description: _____

_____

_____

_____

Recommended Correction: _____

_____

_____

_____

Your Name: _____

Address: _____

_____

Email: _____   Phone: _____

# System Air Balancing

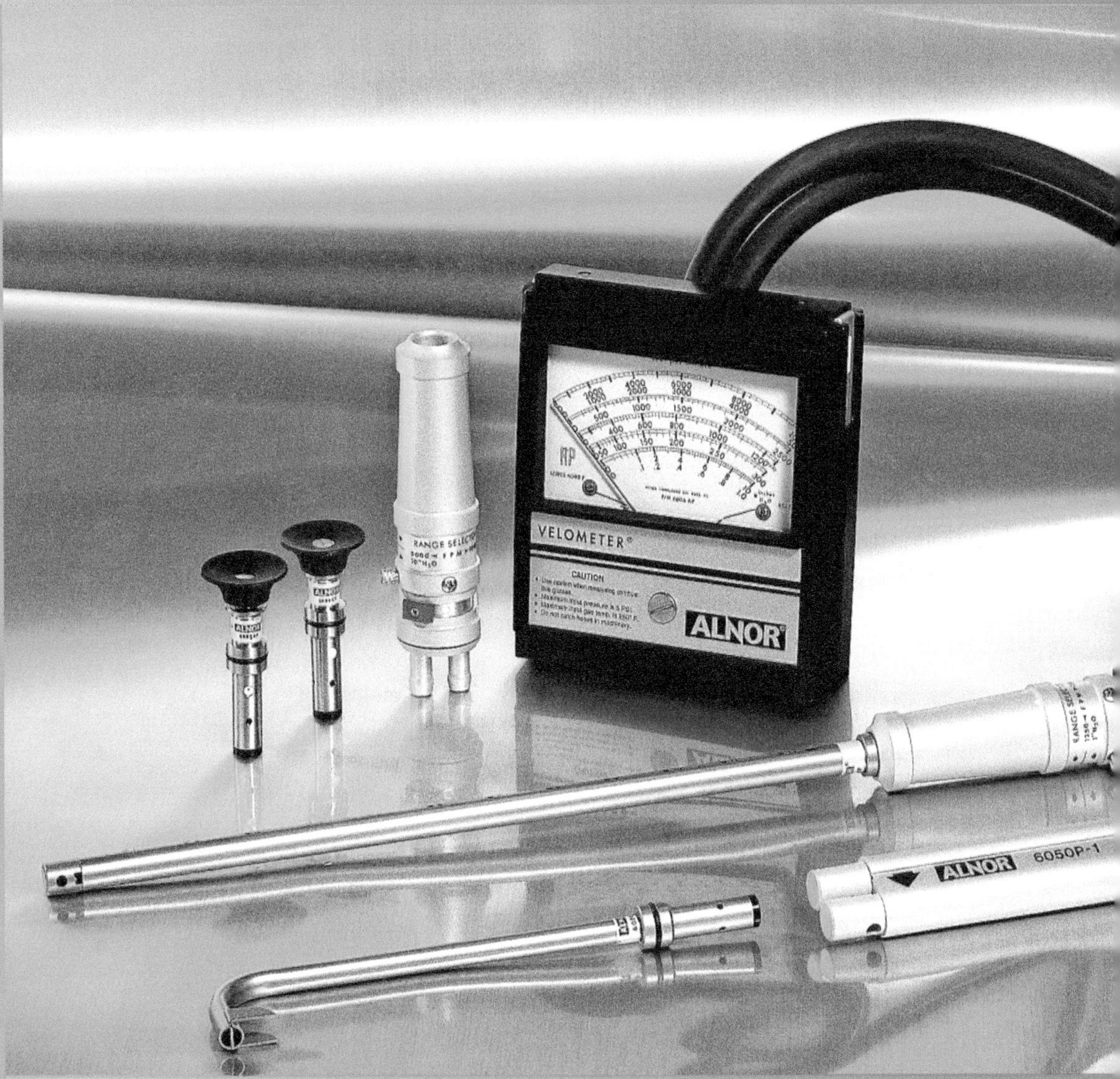

## OVERVIEW

Even when a forced-air comfort system is properly installed and operating, the air must be properly distributed to the spaces. Therefore, after the system is installed and running, the air must be properly balanced. This process requires specialized knowledge and the ability to use specialized test instruments to adjust air flow for optimum performance. Air balancing also requires knowledge of the physical properties of air and the ability to interpret psychrometric charts.

# Module 03402

*03402 V5*

## Objectives

When you have completed this module, you will be able to do the following:

1. Describe the properties of air and the laws related to its temperature, pressure, and volume.
   a. Describe the basic properties of air that are related to airflow and balancing.
   b. Explain Dalton's, Boyle's, and Charles' laws.
2. Describe the study of psychrometrics and how to use the psychrometric chart.
   a. Describe psychrometrics and the related properties of air.
   b. Describe the structure of the psychrometric chart.
   c. Explain how to use the psychrometric chart to determine specific air properties.
3. Describe the air balancing process and identify the required tools and instruments.
   a. Describe air balancing and define common terminology.
   b. Identify the tools and instruments used in air balancing.
   c. Describe the fan laws and explain how to make changes to the supply air volume.
4. Explain how to balance an air distribution system.
   a. Describe the steps to take prior to beginning an air balancing task.
   b. Explain how to measure temperature rise and drop and then use the acquired information.
   c. Explain how to measure system and terminal airflow and adjust as required.
   d. Explain how to balance using the thermometer methods.

## Performance Tasks

Under the supervision of your instructor, you should be able to do the following:

1. Select and properly use test instruments for balancing air distribution systems.
2. Measure the temperature rise and drop across ducted heating and cooling equipment.
3. Adjust supply fan speed to provide higher or lower air quantities.
4. Measure airflow at air supply outlets.
5. Adjust dampers in branch supply ducts and at air terminals and diffusers.

## Trade Terms

| | | |
|---|---|---|
| $A_k$ factor | Entrained air | Throw |
| Atmospheric pressure | Induction unit system | Total heat |
| Boyle's law | Primary air | Total pressure |
| Charles' law | Psychrometrics | Traverse readings |
| Dalton's law | Specific density | Velocity pressure |
| Dew point | Specific volume | |
| Drop | Spread | |

## Industry-Recognized Credentials

If you are training through an NCCER-accredited sponsor, you may be eligible for credentials from NCCER's Registry. The ID number for this module is 03402. Note that this module may have been used in other NCCER curricula and may apply to other level completions. Contact NCCER's Registry at 888.622.3720 or go to **www.nccer.org** for more information.

# Contents

## Figures and Tables

## Figures and Tables (continued)

## 1.0.0 AIR PROPERTIES AND LAWS

### Objectives

Describe the properties of air and the laws related to its temperature, pressure, and volume.

 a. Describe the basic properties of air that are related to airflow and balancing.
 b. Explain Dalton's, Boyle's, and Charles' laws.

### Trade Terms

$A_k$ factor: The actual free area of an air distribution outlet or an inlet stated in square feet. The terminal device manufacturer specifies this value.

Atmospheric pressure: The pressure exerted on all things on the surface of the earth as the result of the weight of the atmosphere.

Boyle's law: With a constant temperature, the pressure on a given quantity of confined gas varies inversely with the volume of the gas. Similarly, at a constant temperature, the volume of a given quantity of confined gas varies inversely with the applied pressure.

Charles' law: With a constant pressure, the volume for a given quantity of confined gas varies directly with its absolute temperature. Similarly, with a constant volume of gas, the pressure varies directly with its temperature.

Dalton's law: The total pressure of a mixture of confined gases is equal to the sum of the partial pressures of the individual component gases. The partial pressure is the pressure that each gas would exert if it alone occupied the volume of the mixture at the same temperature.

Total pressure: The sum of the static pressure and the velocity pressure in an air duct. It is the pressure produced by the fan or blower.

Traverse readings: A series of velocity readings taken at several points over the cross-sectional area of a duct or grille.

Velocity pressure: The pressure in a duct that results from the movement of the air. It is the difference between the total pressure and the static pressure.

The efficient and proper operation of a building air conditioning system requires more than just properly operating heating, cooling, and electrical systems. Of equal importance is the delivery of the correct quantity of conditioned air to the occupied space. This requires that the related air distribution system be properly installed and balanced. Having an understanding of air and its properties will enable technicians to better analyze and/or predict the operating conditions for an air conditioning system.

Testing and balancing is the total process of measuring, adjusting, and documenting the performance of a building's air conditioning system to make sure that it meets its design specifications. System balancing is necessary because human comfort depends on having the right volume of heated or cooled air at the right temperature delivered to each space in the building. A building that is not properly air balanced is not operating efficiently, which usually results in higher operating costs.

Before you can study how to balance an air system, it is necessary to have a basic understanding of the properties of air and terms used when describing air. Also, technicians must become familiar with basic laws of physics that apply to air systems. This section covers these subjects at an introductory level. For more advanced study, refer to the references and acknowledgments section of this module.

### 1.1.0 Air Properties

The earth is surrounded by a mixture of gases composed of 78 percent nitrogen, 21 percent oxygen, and a 1 percent mix of other gases. Together, they form our atmosphere, which extends about 600 miles above the earth and is held to the planet by the force of gravity.

The atmosphere is a gas that has weight. That weight is measured in pounds per square inch (psi). At sea level, a column of air with a base of one square inch, extending 600 miles above the earth, exerts a weight and pressure of 14.7 psi. The pressure exerted on all things on the earth's surface as a result of the weight of our atmosphere is called atmospheric pressure (*Figure 1*).

Atmospheric pressure can be measured with a barometer. *Figure 2* shows a basic mercury-tube barometer. It consists of an open dish of mercury, with a mercury-filled tube sealed at one end and inverted vertically into the dish.

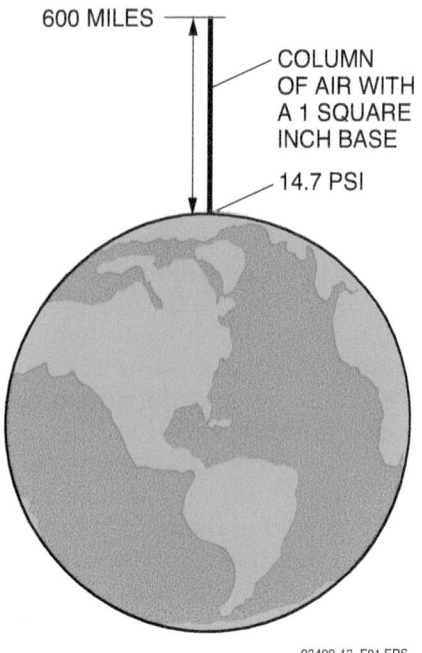

**Figure 1** Atmospheric pressure.

If the 14.7-psi column of air is applied to the mercury-tube barometer at sea level, it will cause the mercury in the tube to rise to a height of 29.92". This means that the standard atmospheric pressure at sea level is also equal to 29.92 inches of mercury, or 29.92" Hg.

These values of 14.7 psi and 29.92" Hg for dry air at sea level, at 70°F, are standards that are used frequently in HVAC work. One pressure scale, called the absolute pressure scale, is based on the barometer measurements just described.

On this scale, pressures are expressed in pounds per square inch absolute (psia), starting from zero (0) psia. At 0 psia, atmospheric pressure is completely absent. Any area, such as a refrigerant circuit, where the pressure measurement has been reduced below 14.7 psia is said to be in a vacuum. Of course, the level of vacuum can vary.

Another scale, called gauge pressure, is frequently used to define air pressure levels. Gauge pressure scales use atmospheric pressure as their zero starting point. Positive gauge pressures, those above zero (14.7 psi), are expressed in pounds per square inch gauge (psig). Negative pressures, those below 0 psig, are expressed in inches of mercury vacuum (in Hg vacuum). 0 psig and 14.7 psia represent the same pressure value.

Gauge pressures can easily be converted to absolute pressures by adding 14.7 to the gauge pressure value. For example, a gauge pressure of 10 psig equals an absolute pressure of 24.7 psia (10 + 14.7). A comparison of the gauge and absolute pressure scales is shown in *Figure 2*. Conversion between the absolute and gauge pressure scales is often necessary when making calculations concerned with air pressure relationships.

In HVAC work, inches of water column (in wc), instead of inches of mercury, are frequently used to measure pressure. One example is when measuring the air pressure in an air distribution duct system. Inches of water column express the height, in inches, that an applied pressure will lift a column of water. The atmospheric pressure of 14.7 psia at sea level, with 70°F dry air, will support a column of water 406.9 inches (33.9 feet)

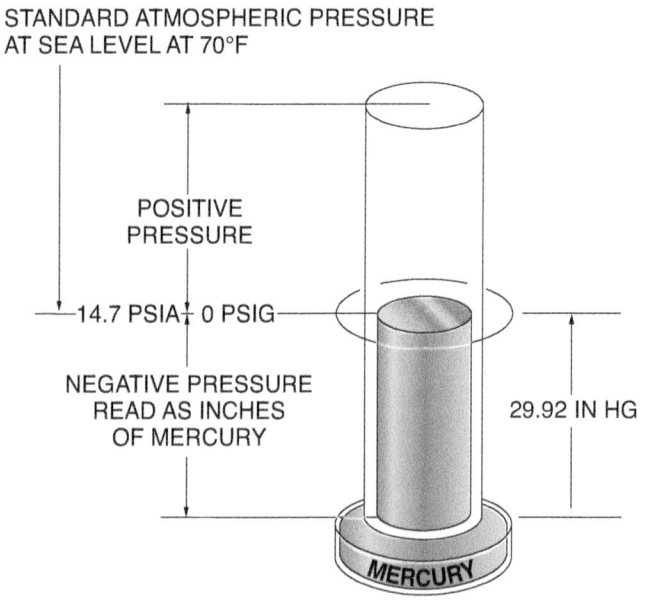

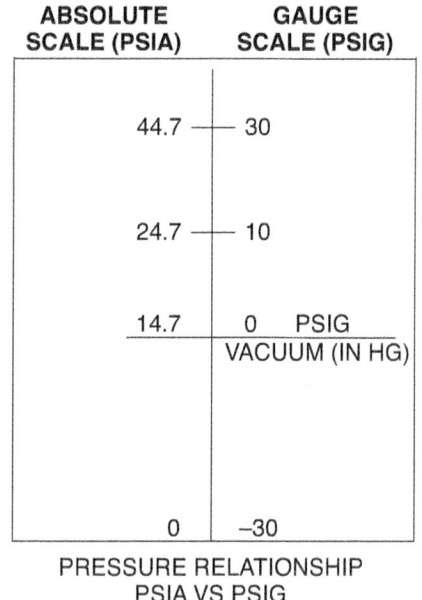

**Figure 2** Atmospheric, absolute, and gauge pressure relationships.

NCCER – *HVAC Level Four*  03402

high. Therefore, for every one pound per square inch of pressure, a column of water will rise to a height of 27.68 inches (406.9 ÷ 14.7), or about 2.3 feet.

Pressures expressed in psi can therefore be converted to inches of water column by multiplying the psi value by 27.68. Conversely, inches of water column can be converted to psi by dividing the in wc value by 27.68. For example, 406.896 in wc ÷ 27.68 = 14.7 psi.

The metric Pascal (Pa) is also frequently used as a unit of measure for air pressure. Normal pressure of the atmosphere at sea level is 101.4 kilopascals (kPa), often rounded to 100 kPa. Since atmospheric pressure is 14.7 psia, dividing 101.4 by 14.7 provides a value of 6.897kPa for every 1 psia. Pressures expressed in psi can be therefore be converted to kPa by multiplying the psi value by 6.897. Pressures expressed in kPa can be converted to psi by dividing the kPa value by 6.897. For example, 101.39kPa ÷ 6.897 = 14.7 psi.

A pitot tube and an inclined tube manometer (*Figure 3*) are tools that are used to measure duct pressures. The manometer shown is calibrated in in wc. Manometers that read in kPa are also available, but are not widely used in the US. The velocity pressure readings in in wc measured with this manometer can be converted to airflow velocity readings in feet per minute (fpm) by using the following formula:

$$\text{Velocity (fpm)} = 4.500 \times \sqrt{\text{pressure (in wc)}}$$

For example, a pressure reading of 0.04 in wc is equal to 801 fpm, calculated as follows:

$$4{,}005 \times \sqrt{0.04} = 4{,}005 \times 0.2 = 801$$

### 1.1.1 Air Weight

Air is compressible and has weight. Its density and weight vary with the elevation above the earth's surface. At higher elevations, it weighs less than at lower elevations. Standard air is considered to be air with a density of 0.075 lb/cu ft at 70°F and a barometric pressure of 29.92 in Hg. Thus, the layer of air is not as deep or as heavy at Denver, Colorado (elevation 5,500 feet), as it is at sea level. The atmospheric pressure at higher altitudes is also less. Using standard air in many calculations provides a consistent baseline, but the value must be changed to maintain accuracy at different conditions.

Heating and cooling loads and equipment capacities are based on British thermal units per hour (Btuh). Some calculations used in HVAC work are based on the pounds of air to be handled in an hour, rather than the volume. Therefore, it is important to know how to convert the volume of

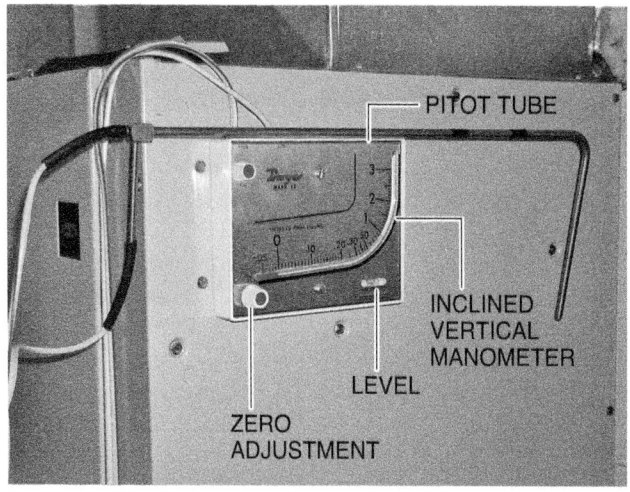

03402-13_F03.EPS

*Figure 3* Pitot tube and inclined tube manometer.

air flowing in a system from cubic feet per minute (cfm) to pounds per hour.

At sea level, each cubic foot per minute represents 4.5 pounds of air per hour. For example, if a system is handling 450 cfm/ton of air conditioning, and the cfm/ton is multiplied by 4.5 pounds of air per hour, 2,025 pounds of air per hour is being moved for each ton of cooling capacity. Although it may seem hard to believe, it is true.

Since air is lighter at higher elevations, the weight per hour of air for each cubic foot of air moved per minute is less. The air's ability to carry heat is less as well, due to the lower density. At sea level, it takes 1.08 Btuh to raise the temperature of one cfm (4.5 pounds per hour) 1°F; at 5,000 feet of elevation, it takes only 0.892 Btuh to raise the temperature of this lighter air 1°F for each cfm. 3.717 pounds per hour is the weight of air at this altitude. The result is that, at higher elevations, more air must be moved to do the same amount of work as that done at sea level.

This greater air quantity can be determined mathematically by using appropriate altitude correction factors. The cfm required to satisfy a heating or cooling load increases as the elevation increases. One manufacturer uses the following rule for air conditioning equipment:

- Sea level – 400 cfm per ton of cooling
- 2,500 feet elevation – 440 cfm/ton
- 3,500 feet elevation – 460 cfm/ton
- 4,500 feet elevation – 475 cfm/ton
- 5,500 feet elevation – 500 cfm/ton

These differences are most significant in the system design phase. However, technicians conducting an air balance at higher altitudes should be aware of this as well.

### 1.1.2 Airflow

In a duct system, air flows from a high-pressure region to a low-pressure region. The blower in the air handler creates the pressure differential necessary to cause air to move in the duct system. In a typical system (*Figure 4*), the highest air pressure in the system is found at blower outlet C. This pressure gradually and constantly decreases throughout the supply duct until it becomes equal to the atmospheric pressure in the space to be conditioned after it leaves supply grille A.

The lowest pressure in the system is at the blower inlet B. The pressure of the air in the return duct constantly decreases from atmospheric pressure from return grille D in the room to a pressure lower than atmospheric at the blower. The total system pressure can be measured across the blower inlet B and outlet C.

Pressure losses in the duct run result from several factors that create resistance to airflow in the system. The blower fan must overcome these pressure losses as air moves through the system. Most of the pressure losses in a duct result from friction caused by the air molecules rubbing against the walls of the duct. At the walls, they travel at a slower speed than the air molecules moving in the center of the duct. This reduced speed along the walls may be referred to as laminar flow.

Losses also occur when air molecules bump into each other as they move along the duct. This happens because the air molecules tend to tumble and mix rather than flow in a straight pattern. Friction losses will also increase with any increase in air speed or change in air direction. The velocity of air moving through a duct is measured in feet per minute. Doubling air velocity increases the friction losses by almost four times.

To overcome additional friction losses within a duct system, the blower must exert greater pressure to keep the air moving through the duct. This greater pressure exerted by the blower results in a higher air pressure within the duct. This pressure, called static pressure, is exerted equally in all directions throughout the duct to which it is applied.

The volume of air moving in a duct system is measured in cubic feet per minute. It can be determined if the velocity (speed of air moving in a duct) and the cross-sectional area of the duct are known. The velocity is measured at several points across the duct (traverse readings) and averaged (*Figure 5*). Since the velocity is slower at the walls and higher towards the center, it is best to work with the average velocity. To measure the velocity in a duct, a velometer is inserted through the wall and stopped at various points, such as those shown in *Figure 5*. The readings at each point are then averaged.

Velocity measurements can also be taken at the face of a return or supply air grille or diffuser. The cross-sectional area of a grille is not equal to its actual measured area because the register or grille vanes restrict the opening. Grille manufacturers always provide data pertaining to the equivalent open area rather than the actual grille measurement. This factor is known as an $A_k$ factor. The A stands for area.

Total system airflow can be determined by using the manufacturer's performance data per-

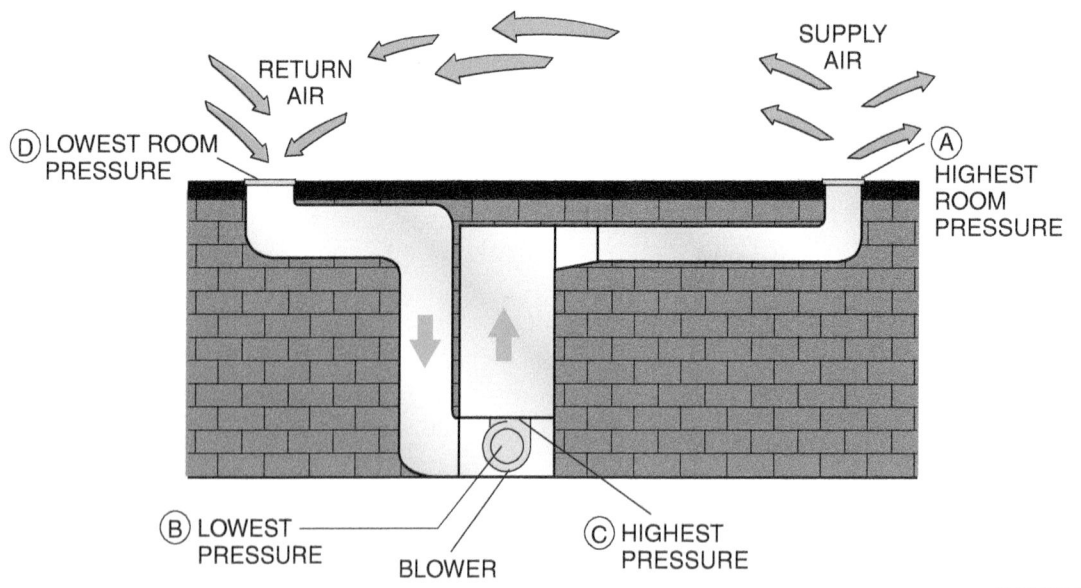

AIR FLOWS FROM HIGHER TO LOWER PRESSURE AREA

*Figure 4* Airflow principles.

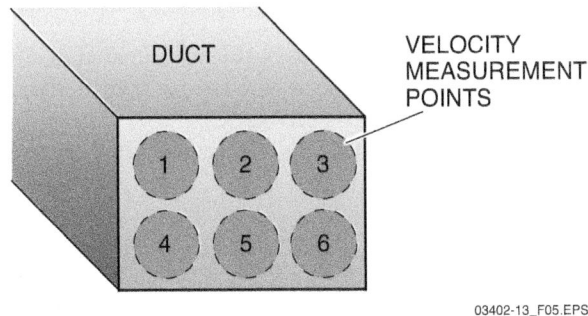

*Figure 5* Velocity measurement points.

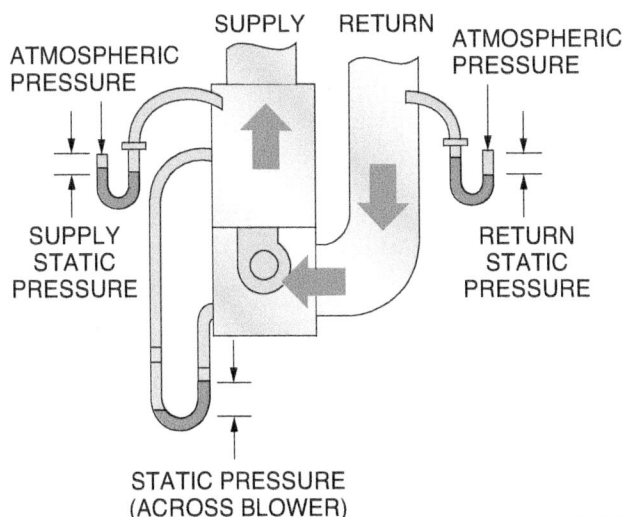

*Figure 6* Determining static pressure.

taining to a particular size or model blower. This method of determining total system airflow requires measurement of total system static pressure (*Figure 6*). A manometer is usually used for this purpose. The manufacturer's service manuals give data indicating the cfm air handling capabilities of their units. This data may indicate the capabilities over the normal operating range of static pressure for each blower listed for use in their equipment. It may be presented on a chart or a graph. By comparing the static pressure to the chart, the amount of air being moved can be determined.

In systems that require the addition of outside air and exhaust ventilation, proper air balancing helps to ensure that the correct amount of outside air is entering the facility. It also ensures that building pressures are appropriate. In many cases, poor air balance can cause one or more areas to operate at negative pressures, causing uncontrolled infiltration from process areas or the outdoors to occur. This further increases energy use while impacting comfort and human health, potentially allowing harmful fumes to enter an area. Although indoor air quality and energy-efficient operation of the HVAC system are priorities that are at odds with each other, effective air balancing can substantially reduce energy use and help to cover the operating costs of providing adequate indoor air quality.

### 1.2.0 Relevant Air Laws

Air is a gas. As such, air follows the gas laws of Dalton, Boyle, and Charles. These laws pertain to air temperature, pressure, and volume relationships, and are discussed in the following sections.

### 1.2.1 Dalton's Law

Dalton's law states that the total pressure of a mixture of confined gases is equal to the sum of the partial pressures of the individual component gases (*Figure 7*). The partial pressure is the pressure that each gas would exert if it alone occupied the volume of the confined area at the same temperature.

### 1.2.2 Boyle's Law

Boyle's law states that with a constant temperature, the pressure on a given quantity of confined gas varies inversely with the volume of gas. Similarly, at a constant temperature, the volume of a given quantity of gas varies inversely with the applied pressure. Formulas 1 through 3 describe this law mathematically.

Formula 1:     $P_o \times V_o = P_n \times V_n$

Formula 2:     $V_n = V_o \times P_o \div P_n$

Formula 3:     $P_n = P_o \times V_o \div V_n$

---

## Air Balance and Energy Efficiency

Air balance, especially in commercial and industrial applications, plays a significant role in the energy efficiency of the HVAC system. Delivering excess air to areas that do not require it adds significantly to operating costs. Proper air balance will reduce not only the energy expended in blower operation and air movement, but typically the energy used to condition the air as well. During the process, other problems such as duct leakage or unusual restrictions may be discovered. These are also significant energy-consuming factors.

---

## DALTON'S LAW

- EACH GAS IN A MIXTURE OF GASES ACTS INDEPENDENTLY.

- TOTAL PRESSURE = SUM OF THE PRESSURES
  CREATED IN A     OF EACH GAS IN THE
  CYLINDER         MIXTURE

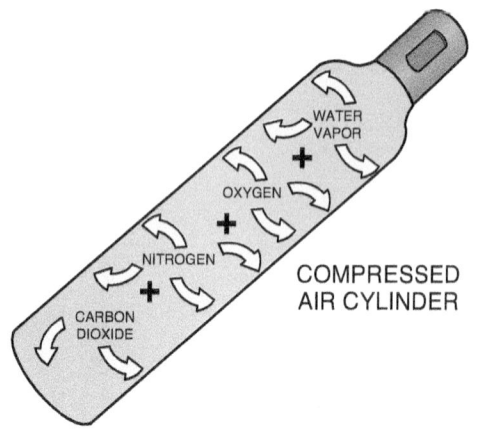

03402-13_F07.EPS

*Figure 7* Dalton's law.

*Where:*

$P_o$ = original absolute pressure (psia)

$P_n$ = new pressure (psia)

$V_o$ = original volume (cubic feet)

$V_n$ = new volume (cubic feet)

Psia = psig + 14.7

**Study Example:**

What is the new volume of 2 cubic feet of gas at 30 psig if it is compressed to 60 psig, providing the temperature remains constant?

Using Formula 2:

$$V_n = V_o \times P_o \div P_n =$$
$$\text{2 cu ft} \times \text{(30 psig + 14.7 psia)}$$
$$\div \text{(60 psig + 14.7 psia)} =$$
$$2 \times 44.7 \div 74.7$$
$$= \text{1.2 cubic feet (rounded off)}$$

### 1.2.3 Charles' Law

Charles' law states that with a constant pressure, the volume for a given quantity of confined gas varies directly with its absolute temperature. Simi-larly, with a constant volume of gas, the pressure varies directly with its absolute temperature. Formulas 4 through 6 describe this law when working with a constant pressure. Formulas 7 through 9 describe it when working with a constant volume.

With constant pressure:

Formula 4:    $V_o \times T_n = V_n \times T_o$

Formula 5:    $V_n = V_o \times T_n \div T_o$

Formula 6:    $T_n = V_n \times T_o \div V_o$

*Where:*

$T_o$ = original absolute temperature

$T_n$ = new absolute temperature

$V_o$ = original volume (cubic feet)

$V_n$ = new volume (cubic feet)

Absolute temperature = °F + 460

With constant volume:

Formula 7:    $P_o \times T_n = P_n \times T_o$

Formula 8:    $T_n = P_n \times T_o \div P_o$

Formula 9:    $P_n = P_o \times T_n \div T_o$

*Where:*

$T_o$ = original absolute temperature

$T_n$ = new absolute temperature

$P_o$ = original absolute pressure (psia)

$P_n$ = new absolute pressure (psia)

Absolute temperature = °F + 460

psia = psig + 14.7

---

**Think About It**

1. What is the new volume of 8 cubic feet of gas at 40°F if the temperature is raised to 80°F, at a constant pressure?

2. What is the new pressure (in psig) of a quantity of gas at 40°F and 35 psig if its temperature is raised to 60°F, and if it is at a constant volume?

---

## Additional Resources

*ASHRAE HVAC Applications Handbook*, 2011. Atlanta, GA: American Society of Heating and Air Conditioning Engineers, Inc.

## 1.0.0 Section Review

1. A zero value on an HVAC pressure gauge is equal to _____.

    a. 14.7 psia
    b. 0 in Hg
    c. 0 psia
    d. 29.92 in wc

2. Boyle's law states that with a constant temperature, the pressure on a given quantity of gas confined in a cylinder varies directly with the volume of gas.

    a. True
    b. False

### 2.0.0 PSYCHROMETRICS

#### Objectives

Describe the study of psychrometrics and how to use the psychrometric chart.
a. Describe psychrometrics and the related properties of air.
b. Describe the structure of the psychrometric chart.
c. Explain how to use the psychrometric chart to determine specific air properties.

#### Trade Terms

**Dew point**: The temperature at which water vapor in the air becomes saturated and starts to condense into water droplets.

**Psychrometrics**: The study of air and its properties.

**Specific density**: The weight of one pound of air. At 70°F at sea level, one pound of dry air weighs 0.075 pound per cubic foot.

**Specific volume**: The space one pound of dry air occupies. At 70°F at sea level, one pound of dry air occupies a volume of 13.33 cubic feet.

**Total heat**: Sensible heat plus latent heat.

Psychrometrics is the study of dry air and water vapor mixtures. Proper measurement and control of this mixture is necessary so that the human body can feel comfortable in a room environment. It is also critical to many commercial and manufacturing processes, and to the development of various molds and fungi. As an HVAC technician, you need to understand air and the relationships that exist between its various properties. With an understanding of psychrometrics, system problems that might elude detection using conventional troubleshooting techniques can be diagnosed.

### 2.1.0 Psychrometric-Related Air Properties

Air has weight, density, temperature, specific heat, and heat conductivity. In motion, it has momentum and inertia. It holds substances in suspension and in solution. Dry air is a mixture of gases composed of about 78 percent nitrogen, 21 percent oxygen, and 1 percent other gases (*Figure 8*).

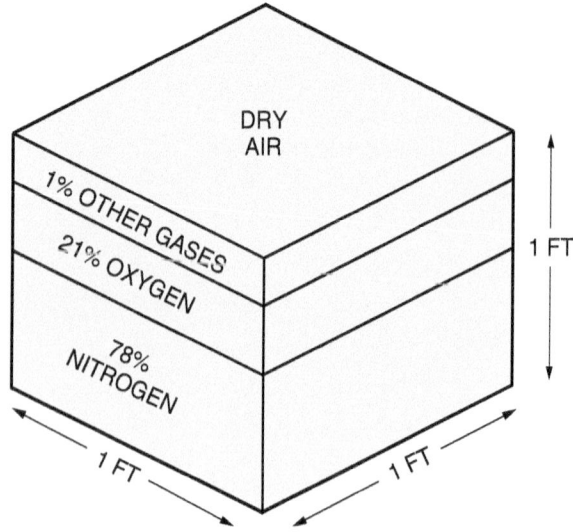

SPECIFIC VOLUME = 13.33 CUBIC FT PER LB

SPECIFIC DENSITY = 0.075 LB PER CUBIC FT

SPECIFIC HEAT = 0.24 BTUS PER LB PER °F

03402-13_F08.EPS

*Figure 8* Standard air – dry air at 70°F at sea level.

Specific volume describes how much space one pound of dry air occupies. At 70°F at sea level (29.92 in Hg), one pound of dry air occupies a volume of 13.33 cubic feet. If air is heated and maintained at its preheated pressure, it will expand and weigh less per cubic foot of volume. This property of air is called specific density. At 70°F at sea level, one pound of dry air weighs .075 pounds per cubic foot. The specific density per pound of dry air is found by dividing the volume of air into one pound.

Specific density of air =
$$1\ lb \div 13.33\ ft^3 = 0.075\ lb/ft^3$$

The specific heat of air determines its ability to warm. Air at sea level has a specific heat of 0.24 Btu/lb/°F. The sensible heat formula can be used to calculate how many Btus are needed to raise the temperature of dry air. It is called the sensible heat formula because sensible heat is the amount of heat which, when added to the air, causes a change in temperature with no change in the amount of moisture present. Sensible heat is measured with a thermometer, and the reading is known as a dry-bulb temperature.

Sensible heat (Btuh) =
specific heat × specific density ×
60 min/hr × cfm × ΔT

Btuh = (0.24 × 0.075 × 60) × cfm × ΔT

Btuh = 1.08 × cfm × ΔT

*Where:*

cfm = velocity of airflow in cubic feet per minute

ΔT = change in temperature (°F)

### 2.1.1 *Water In Air*

Water vapor is almost always suspended in the air around us. This water vapor has many sources. Outdoors, it comes from the evaporation of the oceans and other bodies of water. Inside buildings, it comes from cooking, showers, and similar activities. The amount of moisture that the air will hold depends on the temperature of the air. Warm air will hold more moisture than cold air.

The amount of moisture contained in the air can be expressed as pounds of moisture per pound of dry air, or as grains of moisture per pound of dry air. Grains of moisture per pound of dry air is usually used, because the numbers are larger and easier to work with. They are also used because the amount of water vapor that is present in each pound of air is normally very small. At 70°F at sea level, one pound of water contains 7,000 grains of moisture (*Figure 9*).

A maximum of 110.5 grains of moisture are contained in a pound of saturated air at 70°F at sea level. The air is not able to hold any more moisture than this. If this same amount of mois-

ture were expressed in pounds of moisture per pound of dry air, it would weigh 0.01579 pound.

### 2.1.2 *Relative Humidity*

Relative humidity (RH) is the ratio of the amount of moisture present in a given sample of air to the amount it can hold at saturation. If a volume of air is totally saturated with moisture, its relative humidity is 100 percent. If it contains only half of the moisture it can hold, its relative humidity is 50 percent (*Figure 10*). Relative humidity is expressed as a percentage.

The amount of humidity in the air affects the rate at which perspiration evaporates from the skin. Dry air causes rapid evaporation, which makes the skin feel cooler than the actual temperature. Moist air prevents rapid evaporation, making the body feel warmer than the actual temperature.

In HVAC work, relative humidity measurements are made to determine the level of indoor environmental comfort that exists in the conditioned space. Proper control of relative humidity is also critical to many commercial and manufacturing processes. Most people feel comfortable when the indoor temperature and humidity conditions fall within certain ranges, called comfort zones.

The comfort zone is generally considered to be between 68°F and 78°F, with a relative humidity of 30 to 60 percent. Of course, not every person will agree; there are exceptions to most rules regarding comfort. Properly controlled temperature and

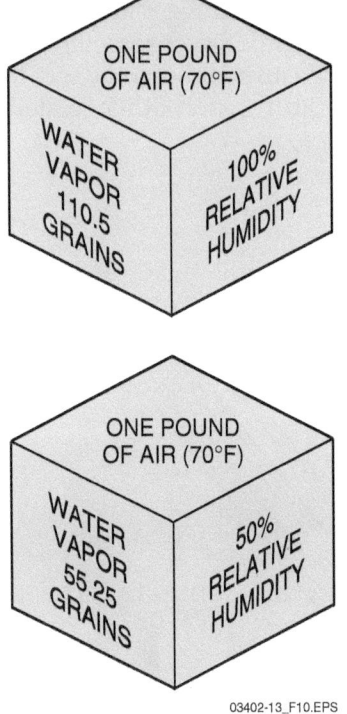

*Figure 9* Water at 70°F at sea level.

*Figure 10* Relative humidity.

---

 03402 **System Air Balancing**

Module Five  9

humidity conditions are important because they improve personal comfort and health conditions in all seasons and reduce the cost of operating heating and cooling equipment. Higher humidity levels make cooler air feel warmer. Conversely, lower humidity levels make warmer air feel cooler. In states like Nevada for example, where the air is much drier than that of Florida, the high temperatures do not feel so uncomfortable. This is because perspiration is rapidly removed from the skin, cooling the body.

Traditional sling psychrometers (*Figure 11*) contain two identical thermometers, one to measure dry-bulb temperatures and the other to measure wet-bulb temperatures. Dry-bulb temperatures measure the amount of sensible heat in the air. Wet-bulb temperatures are taken with a thermometer that has a wick saturated with distilled water wrapped around the sensing bulb. The reading from a wet-bulb thermometer takes into account the moisture content of the air, thus reflecting the total heat content. Evaporation occurs at the wick of the wet-bulb thermometer, giving it a lower temperature reading. The lower the humidity, the lower the wet-bulb temperature will be. This is due to the faster evaporation rate from the wet bulb, resulting in a more significant cooling effect.

The sling psychrometer is fitted with a handle by which it can be whirled with a steady motion through the surrounding air. The whirling motion is periodically stopped to take readings of the wet-bulb and dry-bulb thermometers (in that order) until consecutive readings become steady. Modern electronic psychrometers, like the one in *Figure 12,* measure and display wet-bulb and dry-bulb temperatures, and other related values such

as dew point and relative humidity. The benefit of these modern instruments over their traditional counterparts is accuracy and much greater flexibility of use. However, they do require rather frequent calibration to be considered accurate.

Wet-bulb and dry-bulb temperatures measured with a sling psychrometer are used with a psychrometric chart to find the percent of relative humidity in the measured air. An aid in determining the relative humidity is often integrated into the sling psychrometer itself (*Figure 13*). In this photo, the lines of 95°F dry-bulb and 89°F wet-bulb are aligned, resulting in a relative humidity reading of 82 percent. Note that the 82 percent RH value is true for a number of other dry-bulb/wet-bulb conditions. Air at 83°F dry-bulb and 78°F wet-bulb are also aligned. Electronic thermometers and hygrometers usually give a direct reading of both the measured temperature and relative humidity.

03402-13_F12.EPS

*Figure 12* Electronic psychrometer.

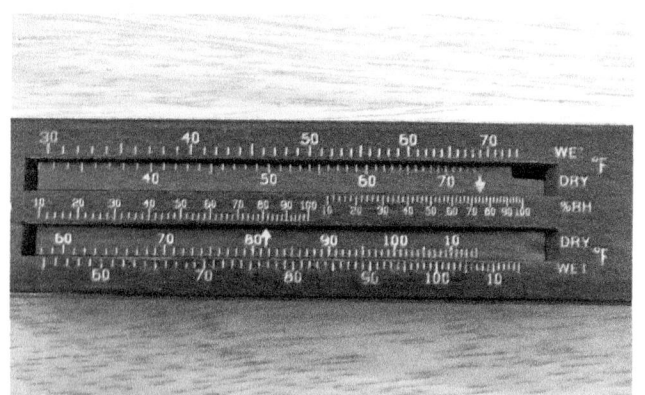

03402-13_F13.EPS

*Figure 13* RH calculator on a sling psychrometer.

WET-BULB TEMPERATURE (°F, WB)

WICK

DRY-BULB TEMPERATURE (°F, DB)

03402-13_F11.EPS

*Figure 11* Traditional sling psychrometer.

## Energy-Saving Humidifiers

The benefits of using a humidifier to add moisture to indoor air during the winter are well known. The humidifier can also act as an energy-saving device. Indoor air that is low in humidity causes moisture on the body to evaporate. This evaporation causes people to feel cold, even if the room thermostat temperature setting seems to be at a comfortable level. Adding moisture to the air increases the relative humidity and the evaporation of moisture from the body is slowed down. This makes people feel more comfortable. In fact, comfort levels can improve to the point that the room thermostat temperature setting can be lowered a few degrees. The end result is that people are comfortable at a lower thermostat setting, saving energy.

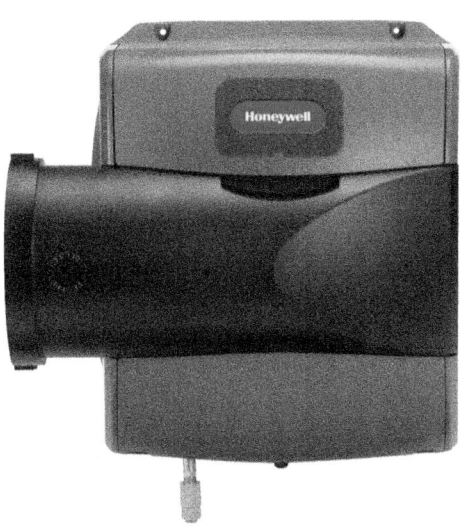

BYPASS

FAN POWERED

03402-13_SA01.EPS

### 2.1.3 Dew Point

The dew point is the temperature at which air becomes saturated with water vapor and the water begins to form water droplets, literally falling out of the air. If the relative humidity of air is 100 percent (saturated), the dew point temperature, wet-bulb temperature, and the dry-bulb temperature are all the same. This is true because the air cannot hold any more moisture, so no water can be evaporated from the wet-bulb thermometer. At night, if and when the outdoor air temperature falls to the dew point, the excess moisture falls to the ground as dew.

### 2.1.4 Enthalpy

Moisture vapor in the air has its own heat content. This latent heat added to the sensible heat of a quantity of dry air yields the total heat of the air. The latent heat referred to here is the heat the water vapor absorbed as it changed from liquid water to water vapor (evaporated). The total heat content of the air and water vapor mixture, as measured from a predetermined base or point, is called enthalpy. Enthalpy is measured using a wet-bulb thermometer and is expressed in Btus per pound (Btu/lb).

When dealing with changes in both sensible and latent heat, such as in heating and humidification or cooling, the difference in enthalpy can be used to calculate how many Btus per hour (Btuh) the temperature of the air has been raised or lowered. This is useful when you need to determine the capacity of a typical air conditioning system.

Using the measured dry-bulb and wet-bulb temperatures for the air stream entering and leaving the equipment, the values of enthalpy, found with a psychrometric chart, can be used to find the total difference in heat content (change in enthalpy) between the two readings in Btu/lb of air. This value of change in enthalpy, along with the value for the air stream velocity in cubic feet per minute (cfm), is used to calculate the equipment capacity in Btuh using the total heat formula shown below.

$$\text{Total heat (Btuh)} =$$
$$\text{specific density of air} \times 60 \text{ min/hr} \times \text{cfm} \times \Delta H$$

$$\text{Btuh} = (0.075 \times 60) \times \text{cfm} \times \Delta H$$

$$\text{Btuh} = 4.5 \times \text{cfm} \times \Delta H$$

*Where:*

cfm = velocity of airflow in cubic feet per minute

$\Delta H$ = change in enthalpy (Btu/lb)

## 2.2.0 Psychrometric Chart Structure

The psychrometric chart (*Figure 14*) is a graph of air properties. It is used to determine how air properties vary as the amount of moisture in the air changes.

The chart shown in *Figure 14* is based on normal temperatures at sea level (29.92 in Hg). Charts are also available that graph the air properties in low temperatures from –20°F to 50°F, high temperatures from 60°F to 250°F, and at various elevations other than sea level, to correct for changes in barometric pressure. They are also available in metric form.

With easily measured values of dry-bulb and wet-bulb temperatures, the psychrometric chart can be used to find the value for one or more of the remaining properties related to a given sample of conditioned air. For HVAC work, the dry-bulb and wet-bulb temperatures are frequently used to find the relative humidity of a conditioned space,

or the dew point for an air sample. One advantage in using a psychrometric chart instead of a direct-reading instrument is that the chart can be used to estimate the changes in value for various properties as a result of making adjustments to an HVAC system.

### 2.2.1 Psychrometric Chart Layout

The scales shown on the psychrometric chart (*Figure 15*) are as follows:

- Dry-bulb temperature °F
- Grains of moisture per pound of dry air
- Relative humidity
- Dew point temperature °F
- Wet-bulb temperature °F
- Enthalpy
- Specific volume
- Sensible heat factor

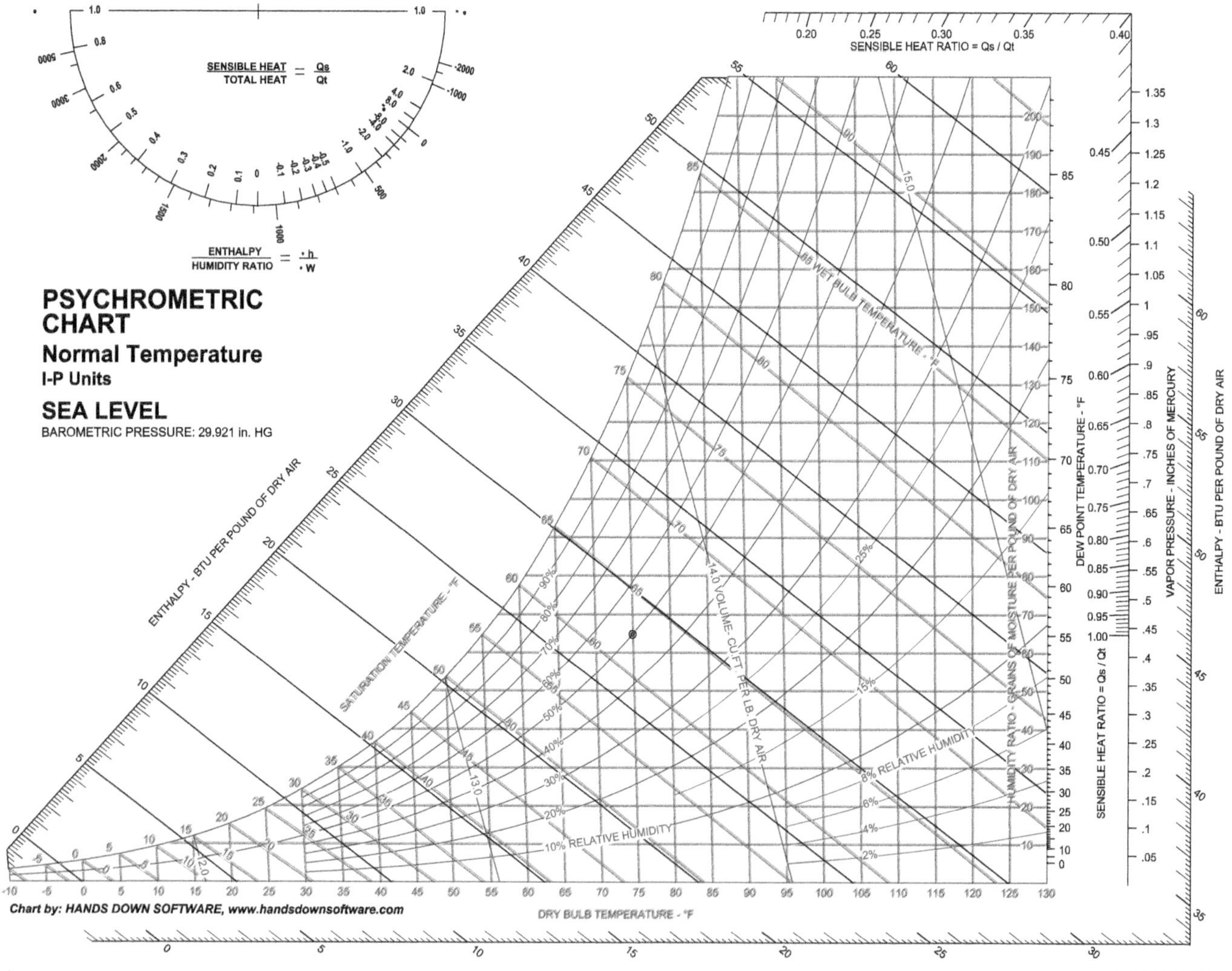

*Figure 14* Psychrometric chart.

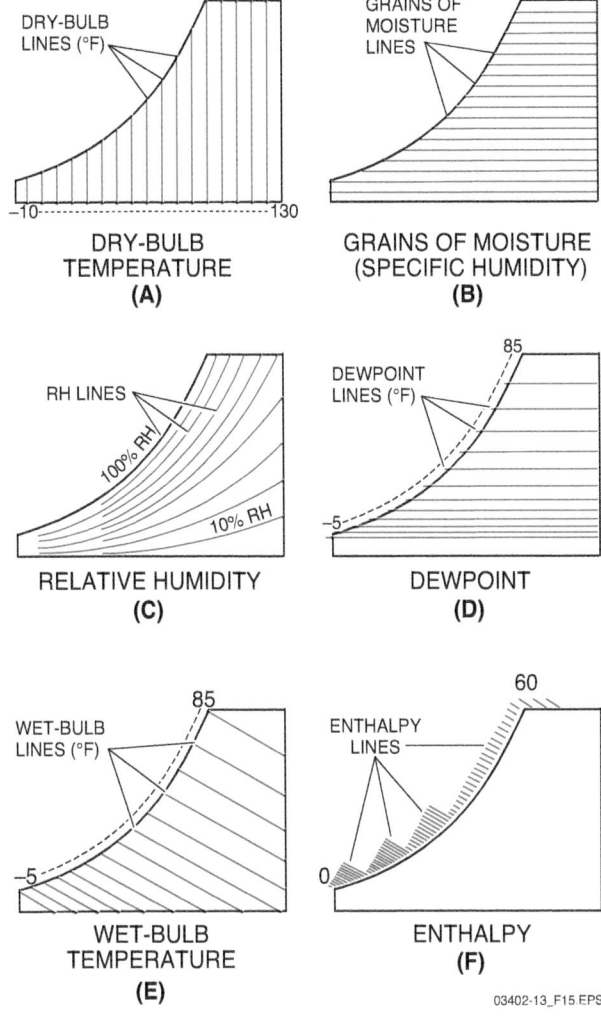

*Figure 15* Psychrometric chart scales.

## 2.2.2 Dry-Bulb Temperature Scale

The sensible heat temperature scale, called the dry-bulb temperature scale, is laid out horizontally across the bottom, with increment lines extended vertically, as shown in *Figure 15A*.

## 2.2.3 Grains of Moisture (Specific Humidity) Scale

The grains-of-moisture scale depicts the amount of water vapor mixed with each pound of air. In

*Figure 15B*, this scale runs vertically with increment lines extended horizontally. Another scale on a full chart, located to the right of the grains of moisture scale, can be used to determine the humidity ratio, in pounds of moisture per pound of dry air.

### 2.2.4 Relative Humidity Lines

Relative humidity is shown as lines from 0 to 100 percent RH in increments of 10 percent (*Figure 15C*). At the 100 percent relative humidity line, the air is fully saturated and can hold no more moisture. Any further addition of moisture would cause condensation.

### 2.2.5 Dew Point Temperature Lines

The temperature at which the moisture content or relative humidity has reached 100 percent is called the dew point. If the temperature drops below the dew point, the moisture vapor begins to condense. The dew point temperature lines run horizontally like the grains of moisture lines (*Figure 15D*). The dew point scale is actually the same scale that is used for the wet-bulb and saturation (100 percent RH) temperatures.

### 2.2.6 Wet-Bulb Temperature Lines

The wet-bulb temperature above 32°F is the temperature indicated by a thermometer whose bulb is covered by a wet wick and exposed to a stream of air moving at a velocity between 1,000 and 1,500 fpm. The wet-bulb temperature lines slant to the left at about a 45-degree angle (*Figure 15E*). The scale is the same one that is used with the dew point and saturation (100 percent RH) temperatures.

### 2.2.7 Enthalpy Scale

Enthalpy is very useful in determining the amount of heat that is added to or removed from air in a given process. Enthalpy is the measurement of the total heat in one pound of air. An approximate

# Psychrometric Chart

In 1911, Dr. Willis Carrier presented his work called the "Rational Psychrometric Formulas" to the American Society of Mechanical Engineers (ASME). This led to the development of the psychrometric chart. A normal temperature psychrometric chart is based on a barometric pressure of 29.92 in Hg. Though based on atmospheric pressure at sea level, the normal temperature chart can be used for field work up to 2,000 feet in elevation. If working at higher elevations, technicians should use a chart for that particular elevation. Dr. Carrier is widely considered to be the father of modern air conditioning.

value for enthalpy is found by extending the wet-bulb temperature line past the 100 percent saturation line on the chart (*Figure 15F*). Enthalpy is the total heat in the air at 100 percent saturation. If the air is not completely saturated, a slight error is present in the enthalpy reading.

## 2.2.8 Specific Volume Scales

Specific volume is used primarily when checking fan performance and determining fan motor sizes. Specific volume is the number of cubic feet occupied by one pound of dry air at any given temperature and pressure. Specific volume lines ranging from 12.5 to 14.5 cubic feet per pound are shown on the chart as almost vertical lines that slant to the left (*Figure 16A*). These lines represent the space occupied in cubic feet per pound of dry air.

Referring again to *Figure 14*, notice that the specific volume of one pound of dry air at 75°F dry bulb displaces a volume of 13.5 cubic feet at sea level. The point represented by 75°F on the bottom horizontal line is halfway between the 13.0 and 14.0 specific volume line intersections with the bottom. If the air is heated to 95°F, it expands and takes up nearly 14 cubic feet. If the air is cooled to 35°F, it occupies roughly 12.5 cubic feet. As described earlier in this module, the volume of a gas (air) varies directly with its absolute temperature. The higher the temperature, the greater its volume; the lower the temperature, the smaller its volume.

## 2.2.9 Sensible Heat Factor

A useful method for finding sensible heat involves the sensible heat factor scale. This scale, located on the right side of the chart, provides the sensible heat factor (*Figure 16B*). The scale range is from 0.35 to 1.00 or 35 to 100 percent, which represents the percentage of sensible heat in the air. This scale is used when plotting processes such as cooling or dehumidification. When the process is plotted, the ratio of sensible heat to total heat (sensible + latent heat = total heat) of the air can be found. The remaining percentage of heat from 100 percent is the percent of latent heat in the air. The dot located at the intersection of the 80°F dry-bulb temperature and the 50 percent humidity lines on the chart is provided for use in conjunction with the sensible heat factor scale. It is used when finding the dew point of a device when the device's sensible heat factor is known.

## 2.3.0 Use of the Psychrometric Chart

The psychrometric chart can be used to show what is happening during a specific heating, ventilating, or air conditioning process. It can be used both when designing systems and when servicing them. The remainder of this section focuses on the use of a psychrometric chart for servicing HVAC equipment. For more advanced study, see the references and acknowledgments section at the back of this module.

### 2.3.1 Using Known Properties of Air to Find Unknown Properties

*Figure 17* shows a psychrometric chart with lines plotted for dry-bulb temperature, wet-bulb temperature, dew point temperature, specific humid-

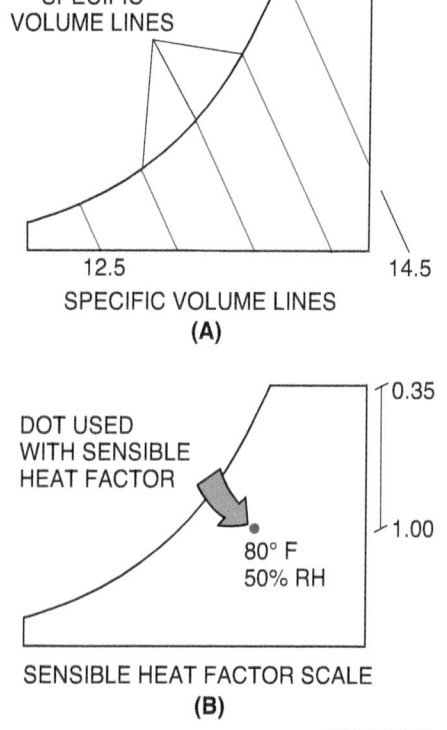

*Figure 16* Psychrometric chart specific volume and sensible heat factor scales.

03402-13_F16.EPS

## Wet-Bulb Temperatures

When using a traditional sling psychrometer to measure wet-bulb temperatures, accurate readings require an air velocity of between 1,000 and 1,500 fpm across the wick. Also, significant errors will result if the wick becomes dirty or dry. This is not a problem with modern electronic psychrometers. However, the handheld sling psychrometer is specifically designed to provide figures based directly on the true thermodynamic principles behind the readings.

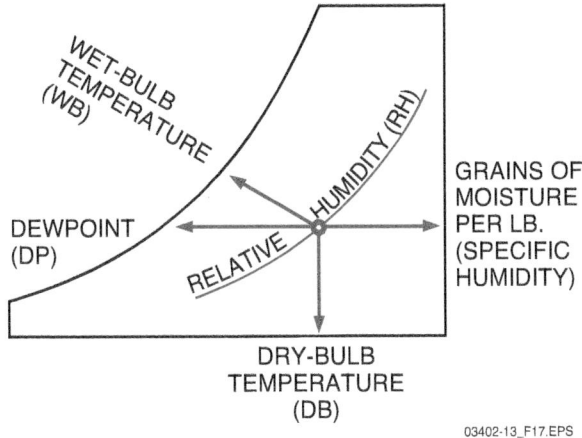

*Figure 17* Example of air properties plotted on a psychrometric chart.

ity, and relative humidity. When any two of these values are known, the exact condition of the air can be located on the chart and the values for all of the other properties can be found from this one point.

For example, assume that the dry-bulb temperature is 75°F and the wet-bulb temperature is 65°F for a given sample of air. Plotting the dry-bulb and wet-bulb temperatures on the psychrometric chart to determine their intersection, it is found that the relative humidity is about 58 percent, the dew point is 59.5°F, and the specific humidity is 76 grains of moisture per pound of dry air. A simplified plot of this example is shown in *Figure 18*.

### 2.3.2 *Working with Changes in Sensible Heat*

The psychrometric chart can be used to evaluate air conditioning processes, such as changes in the sensible heat of air. Sensible heat results in a change in temperature and is indicated by a horizontal line on the chart.

An example of a change in sensible heat caused by heating is plotted in *Figure 19*. Assume that air at 30°F dry bulb and 80 percent RH is heated to 75°F by passing it over a heating coil. If the air is heated with no moisture added or removed, the process results in a horizontal line from point 1 to point 2 as marked on the chart. With the information given, you can determine the values for the

new wet-bulb temperatures, relative humidity, and dew points related to each point.

As shown, the dew point temperature remains the same, because no water vapor has been added or removed. Also, heating air with constant moisture content reduces its relative humidity level. This is because the warmer air can hold more moisture; therefore, the ratio of the moisture in the air to the total amount it is capable of holding decreases. Remember that relative humidity is so-called because it is the amount of moisture in the air relative to its temperature.

An example of a change in sensible heat caused by cooling is plotted in *Figure 20*. Assume that air at 95°F dry bulb and 75°F wet bulb is cooled to 80°F dry bulb by passing it over a cooling coil. If

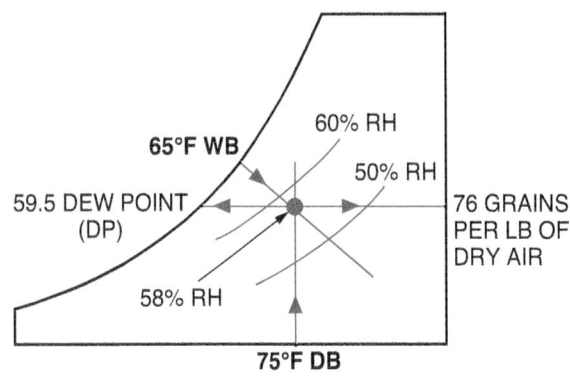

(GIVEN VALUES ARE SHOWN IN BOLD)

*Figure 18* Finding unknown air properties.

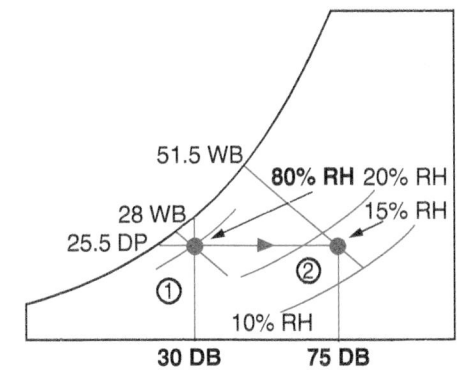

| | DRY-BULB (DB) TEMP °F | WET-BULB (WB) TEMP °F | % RELATIVE HUMIDITY (RH) | DEW POINT (DP) °F |
|---|---|---|---|---|
| AIR AT | **30** | 28 | **80** | 25.5 |
| IS HEATED TO | **75** | 51.5 | 15 | 25.5 |

(GIVEN VALUES ARE SHOWN IN BOLD)

*Figure 19* Psychrometric chart plot of sensible heat increase.

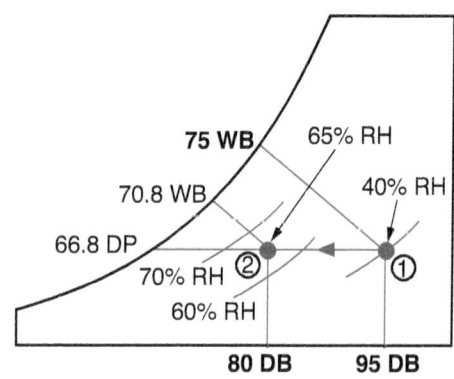

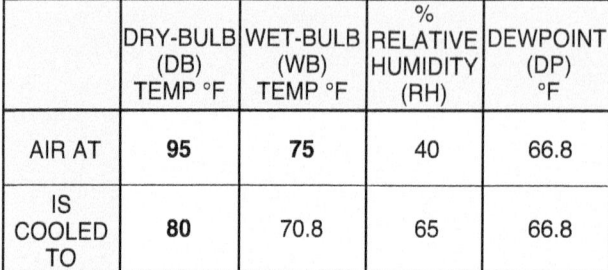

| | DRY-BULB (DB) TEMP °F | WET-BULB (WB) TEMP °F | % RELATIVE HUMIDITY (RH) | DEWPOINT (DP) °F |
|---|---|---|---|---|
| AIR AT | 95 | 75 | 40 | 66.8 |
| IS COOLED TO | 80 | 70.8 | 65 | 66.8 |

(GIVEN VALUES ARE SHOWN IN BOLD)

03402-13_F20.EPS

*Figure 20* Psychrometric chart plot of sensible heat decrease.

the air is cooled without removing any moisture, the process results in a horizontal line from point 1 to point 2 as marked on the chart. With the information given, you can determine the values for the new wet-bulb temperatures, relative humidity, and the dew points related to each point (*Figure 20*). Again, the dew point temperature remains the same because no water vapor has been added or condensed. Also, cooling air with constant moisture content increases its relative humidity level.

### 2.3.3 Working With Enthalpy

Enthalpy is the total heat content of the air and water vapor mixture. It is found on the chart by following along a wet-bulb temperature line, past the saturation line, and out to the enthalpy scale. Enthalpy can be used to determine the total heat that is added to or removed from a volume of air. This is done by reading the scale between the two wet-bulb lines.

For example, *Figure 21* shows a plot of the dry-bulb and wet-bulb temperatures measured for the air entering and leaving an evaporator coil. The measured wet-bulb temperatures can be used to check the capacity of the evaporator.

As shown, the wet-bulb temperatures for the air entering and leaving the evaporator are extended past the saturation line, and out to the enthalpy scale. In this example, the measured

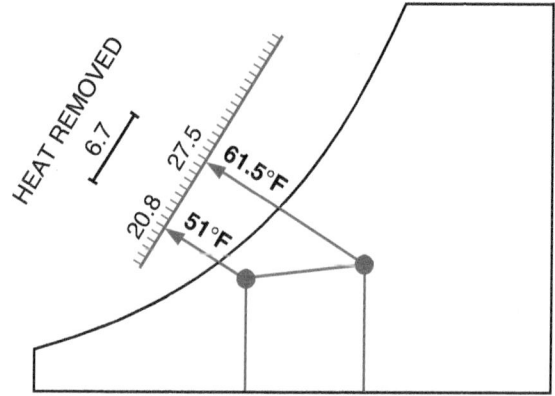

| | WET-BULB (WB) TEMP °F | ENTHALPY (BTU/LB) | AIR-FLOW VOLUME (CFM) |
|---|---|---|---|
| AIR ENTERING EVAPORATOR | 61.5 | 27.5 | 1,000 |
| AIR LEAVING EVAPORATOR | 51 | 20.8 | 1,000 |

(GIVEN VALUES ARE SHOWN IN BOLD)

03402-13_F21.EPS

*Figure 21* Using enthalpy values to find evaporator capacity.

wet-bulb temperatures of 51°F and 61.5°F correspond to enthalpy readings of 20.8 Btu/lb and 27.5 Btu/lb, respectively.

By subtraction, the difference in enthalpy, or the heat removed, is found to be 6.7 Btu/lb. Using the difference value of enthalpy found with the chart, and the value for the measured volume of airflow through the evaporator in cfm, the capacity of the evaporator can be calculated using the following formula:

Capacity (total heat) Btuh = $4.5 \times$ cfm $\times \Delta H$

*Where:*

4.5 = constant (pounds of air per cu ft = 0.075 × 60 minutes/hour)

cfm = volume of airflow in cubic feet per minute

$\Delta H$ = change in enthalpy in Btu/lb (difference in enthalpy between air in and air out)

If 1,000 cfm of air is circulated over the evaporator in this example, then 30,150 Btuh is removed, as follows:

Capacity (total heat) Btuh = $4.5 \times 1,000 \times 6.7 = 30,150$ Btuh

## Additional Resources

*ASHRAE HVAC Applications Handbook*, 2011. Atlanta, GA: American Society of Heating and Air Conditioning Engineers.

*Understanding Psychrometrics*, Second Edition. 2005. American Society of Heating, Refrigeration, and Air Conditioning Engineers.

*ACCA Manual P – Psychrometrics Theory and Applications.* 1983. Air Conditioning Contractors of America.

## 2.0.0 Section Review

1. The amount of moisture contained in air is commonly expressed as _____.

   a. ounces of moisture per pound of dry air
   b. grains of moisture per pound of dry air
   c. grains of moisture per cubic foot of dry air
   d. pounds of moisture per cubic foot of dry air

2. What is the purpose of the specific volume scales on a psychrometric chart?

   a. Determining fan motor sizes.
   b. Calculating total enthalpy.
   c. Finding the volume of a duct run.
   d. Calculating heat loss/heat gain.

3. If the volume of airflow across an evaporator is known, what other value would be needed to calculate the cooling capacity of the evaporator?

   a. Type of refrigerant in the system.
   b. Return airstream dry-bulb temperature.
   c. Change in enthalpy as the air crosses the evaporator.
   d. Grains of moisture in the supply airstream.

### 3.0.0 THE AIR BALANCING PROCESS

#### Objectives

Describe the air balancing process and identify the required tools and instruments.

  a. Describe air balancing and define common terminology.
  b. Identify the tools and instruments used in air balancing.
  c. Describe the fan laws and explain how to make changes to the supply air volume.

#### Performance Tasks

1. Select and properly use test instruments for balancing air distribution systems.
3. Adjust supply fan speed to provide higher or lower air quantities.

#### Trade Terms

**Drop:** The vertical distance that the lower edge of a horizontally projected airstream falls or rises.

**Entrained air:** Also known as secondary air; the induced flow of room air by the primary air from an outlet.

**Primary air:** In the context of air distribution and balancing, it is the air delivered to the room or conditioned space from the supply duct.

**Spread:** The horizontal divergence of an airstream after it leaves the outlet.

**Throw:** The throw of an outlet measured in feet and is the distance from the center of the outlet to a point in the mixed airstream where the highest sustained velocity (fpm) has been reduced to a specified level, usually 50 fpm.

Air balancing means the proper delivery of conditioned air in the correct amounts to each of the areas in the structure being conditioned. The satisfactory distribution of conditioned air depends upon a well-designed duct system and a properly chosen fan.

Air balancing also means that the correct amount of air must be returned to the heating or cooling unit. If the air distribution system is not balanced on both sides of the fan, several problems could arise:

- The spaces or rooms within a structure will have different temperatures.
- Some ducts may be noisy.

- Some spaces may have incorrect humidity.
- Some areas may contain stale air (low airflow).
- Air could be entering the return side of the system from an unconditioned or uncontrolled source.

After the conditioning system is installed, it must be adjusted to make sure the right amount of conditioned air is circulated in the required spaces to meet design conditions. Balancing the air distribution system will often require the adjustment of the fan and the volume dampers so they will deliver enough air at the proper velocity to provide satisfactory heating and cooling. Balancing is simply a method of adjusting air volume to the designed value.

### 3.1.0 Air Balance Terminology

The following information should assist technicians in relating to the terms used with the terminal outlets for space conditioning. These terms are generally broad-based and are not specific to any unit or system. While some of them have been introduced in previous modules, others are likely new to some trainees.

- *Aspect ratio* – The ratio of width to height of the grille core.
- *Diffuser* – A device that discharges air in a flat pattern parallel to the face of the diffuser. They may be square or round and are usually located in the ceiling. Ceiling diffusers are usually placed in the exact center of the room or conditioned space. If more than one diffuser is required, the ceiling area is usually divided into equal parts, and one diffuser is placed in each part to supply the necessary conditioning. They are arranged to promote a mix of primary air with secondary room air (entrainment or induction).
- *Drop* – The vertical distance that the lower edge of a horizontally projected airstream falls or rises (*Figure 22*). It is the result of the combined effect of the density of cool or warm air and the vertical expansion of the air path that results from entrainment.
- *Grille* – A covering for any opening through which conditioned air passes. When grilles have a single flap or several adjustable blades affixed to a collar fastened behind the front face, they are called registers. When grilles and registers deliver supply air, the outlets discharge the air in an outward pattern.
- *Occupied zone* – The area of a conditioned space which extends to within 6 inches of all room surfaces and up to a height of 6 feet.

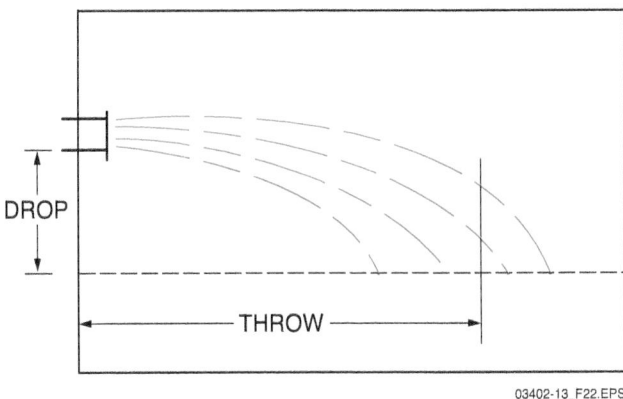

Figure 22 Air drop.

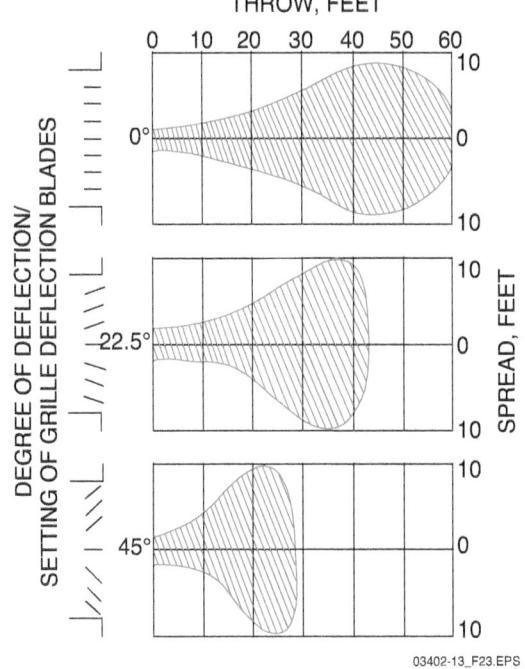

Figure 23 Air patterns.

- Primary air – In the context of air distribution and balancing, it is the air delivered to the room or conditioned space from the supply duct.
- Register – A grille fitted with a damper that is used to control the quantity of air passing through it. Registers are grilles with dampers.
- Return inlet – An opening in a wall, ceiling, or floor through which the space air is returned to the conditioning equipment. The opening is usually fitted with a grille.
- Entrained air – The induced flow of room air by the primary air from an outlet. This creates a mixed air path called secondary air motion. It is also sometimes referred to as induced air. As the primary air leaving the terminal blows into the room, it encourages the movement of air already in the room. This room air entering the mixture is the entrained air. Also known as secondary air.
- Spread – The horizontal divergence (*Figure 23*) of an airstream after it leaves the outlet.
- Throw – The throw of an outlet is measured in feet and is the distance from the center of the outlet to a point in the mixed airstream where the highest sustained velocity (fpm) has been reduced to a specified level, usually 50 fpm (see *Figures 22* and *23*).

To help explain some of the terms associated with air distribution in a room from a given grille or register, refer to *Figure 24*. For example, the performance table for throw (T) is based on wall mounting heights of 8 feet to 10 feet, and a 0-degree setting of both single and double deflection blades. The rated throw is based on the full distance the mixed airstream will travel to a terminal velocity (Vt) of 50 fpm. Note that tables for single- and double-deflection grilles and registers will often show values for the blades being set at other angles besides 0 degrees to show their performance under a variety of conditions.

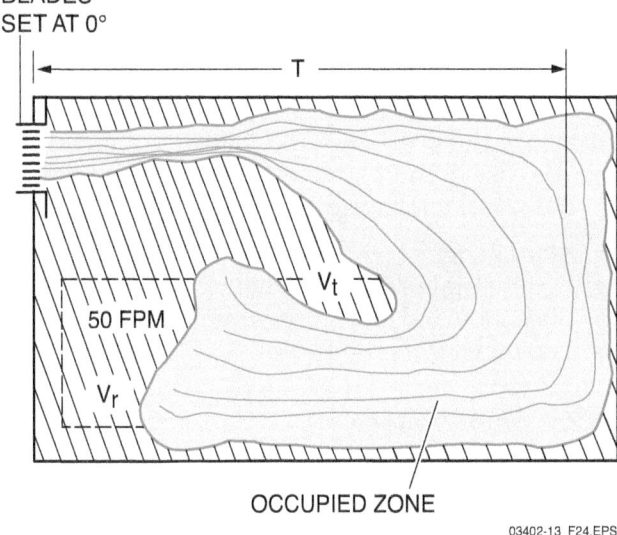

Figure 24 Air pattern terms – elevation.

*Figure 24* graphically shows the T (throw), $V_r$ (room velocity), and $V_t$ (terminal velocity). In multiple unit applications, grilles of equal capacities should be sized for throw equal to one-half the distance between them. Throw from single outlets should be sized within the range of 75 to 100 percent of the distance to the opposite wall. The room velocity ($V_r$) is affected by changes in the mounting height of the grille on the wall. The throw listed for a particular grille should be decreased by one foot for each foot of mounting height above 10 feet. This is necessary to maintain the $V_r$ standard of 50 fpm.

To aid in understanding the use of the area factor ($A_k$) in determining the actual cfm being discharged from a given outlet, use the following formula:

$$cfm = V_k \times A_k$$

*Where:*

cfm = cubic feet per minute

$V_k$ = average outlet velocity in fpm

$A_k$ = area factor from manufacturer

In HVAC work, air velocity is usually expressed in feet per minute (fpm). You can calculate the volume of air flowing past a point in a duct system in cubic feet per minute (cfm) by multiplying the area of the duct in square feet by the air velocity in fpm as follows:

$$cfm = area\ (ft^2) \times velocity\ (fpm)$$

For example, a rectangular duct with an area of 1.5 square feet and a velocity of airflow of 935 fpm has a volume of airflow equal to 1,403 cfm (1.5 × 935).

### 3.2.0 Air System Measuring Instruments

Several instruments and accessories are used to measure the pressures and velocity of airflow in an air distribution system. Common instruments and devices used for this purpose are as follows:

- Manometer
- Differential pressure gauge
- Pitot tube and static pressure tips
- Velometers/anemometers

### 3.2.1 Manometers

Manometers (*Figure 25*) are used to measure the static, velocity, and total air pressures found in air distribution duct systems. Manometers used for air distribution servicing are calibrated in inches of water column (in wc) for use in the United States. Manometers can use water or oil as the measuring fluid. Popular models use a type of oil that has a specific gravity of 0.826 as the measuring fluid. The manufacturer of the gauge specifies the type of oil to be used; therefore, substitution for their specified oil is not recommended.

Manometers come in many types, including U-tube, inclined, and combined U-inclined. Electronic manometers are also widely used. Pitot tubes or static pressure tips, described later in this section, are almost always used with manometers when measuring pressures in duct systems. Manometers work on the principle that air pressure

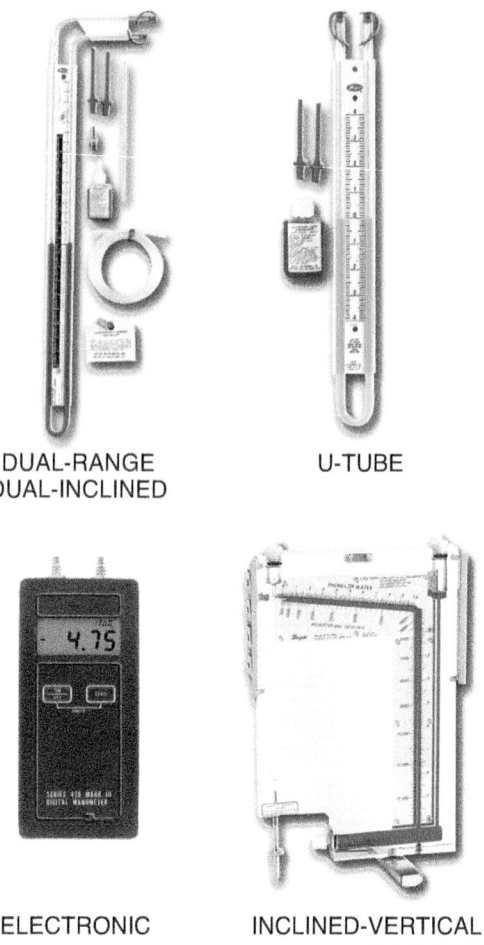

DUAL-RANGE
DUAL-INCLINED

U-TUBE

ELECTRONIC

INCLINED-VERTICAL

03402-13_F25.EPS

*Figure 25* Manometers.

is indicated by the difference in the level of a column of liquid in the two sides of the instrument.

If there is a pressure difference, the column of liquid will move until the liquid level in the low-pressure side is high enough so that its weight and the air pressure being measured will balance with the higher pressure in the other tube.

Individual U-tube and inclined manometers are available in many pressure ranges. Inclined manometers are usually calibrated in the lower pressure ranges and are more sensitive than U-tube manometers. U-inclined manometers combine both the sensitivity of the inclined manometer with the high-range capability of the U-tube manometer in one instrument. Inclined-vertical manometers combine an inclined section for high accuracy and a vertical manometer section for extended range. They also have an additional scale that indicates air velocity in feet per minute.

To get accurate readings with inclined and vertical-inclined manometers, the inclined portion of the scale must be at the exact angle for which it is designed. A built-in spirit level is used for this purpose. Some also have a screw-type leveling adjustment. Before connecting inclined (sloped) manometers to a pressure source, be sure that the gauge is level and the liquid level in the sloping leg is adjusted to zero.

Electronic manometers can typically measure differential pressures of –1 to 10 in wc. Many can also give direct air velocity readings in the range of 300 fpm to 9,990 fpm.

### 3.2.2 Differential Pressure Gauge

The differential pressure gauge (*Figure 26*) provides a direct reading of pressure. These gauges are typically used to measure fan and blower pressures, filter resistance, air velocity, and furnace draft. Some are capable of measuring just pressure or both pressure and air velocity. Single-scale pressure models are calibrated in either in wc or psi. Dual-scale gauges are normally calibrated for pressure in in wc and for air velocity in fpm.

Several models are available covering pressures from 0.0 to 10 in wc, and air velocity ranges from 300 fpm to 12,500 fpm. These gauges are sometimes permanently installed in the equipment. Pitot tubes and/or static pressure tips are normally used with portable models to make air pressure and velocity measurements in air distribution system ductwork.

### 3.2.3 Pitot Tubes and Static Pressure Tips

Pitot tubes (*Figure 27*) and static pressure tips are probes used with manometers and pressure gauges when making measurements inside the ductwork of an air distribution system. The standard pitot tube used for measurements in ducts 8 inches and larger has a 5/16-inch outer tube with eight equally spaced 0.04-inch diameter holes used to sense static pressure.

For measurements in ducts smaller than 8 inches, pocket-size pitot tubes with a 1/8-inch outer tube and four equally spaced 0.04-inch diameter holes are recommended. The pitot tube consists of

an impact tube which receives the total pressure input, fastened concentrically inside a larger tube which receives static pressure input from the radial sensing holes around the tip.

The air space between the inner and outer tubes permits transfer of pressure from the sensing holes to the static pressure connection at the opposite end of the pitot, and then through the connecting tubing to the low- or negative-pressure side of the manometer.

When the total pressure tube is connected to the high-pressure side of the manometer and the static pressure tube to the low-pressure side of the manometer, velocity pressure is indicated directly. To be sure of accurate velocity pressure readings, the pitot tube tip must be pointed directly into the duct airstream. Pitot tubes come in various lengths ranging from 6 inches to 60 inches, with graduation marks at every inch to show the depth of insertion in the duct.

Static pressure tips, like pitot tubes, are used with manometers and differential pressure gauges to measure static pressure in a duct system. They are typically L-shaped with four radially drilled 0.04-inch sensing holes.

### 3.2.4 Velometers/Anemometers

Velometers (*Figure 28*) are used to measure the velocity of airflow. Measurement of air velocity is done to check the operation of an air distribution system. It is also done when balancing system airflow. Most velometers give direct readings of air velocity in fpm. Some can provide direct readings in cfm. Velometers with analog scales and digital readouts are in common use.

Some velometers use a rotating vane (propeller) or balanced swinging vane to sense air movement. When the rotating vane velometer is positioned to make a measurement, the vane rotates at a rate determined by the velocity of the airstream. This rotation is converted into an equivalent velocity reading for display. A rotating vane anemometer (*Figure 29*) can be used to measure air velocity in fpm at large grilles and ducts. The one shown here can measure both air velocity and air volume. It has an internal microprocessor that can be easily programmed to calculate and display air volume in cfm by entering an area factor. Direct readings for instant or averaged fpm or cfm measurements are indicated on a digital display, eliminating the need for using charts or calculations.

In the swinging vane velometer, the airstream causes the balanced vane to tilt at different angles in response to the measured air velocity. The po-

03402-13_F26.EPS

*Figure 26* Portable differential pressure gauge.

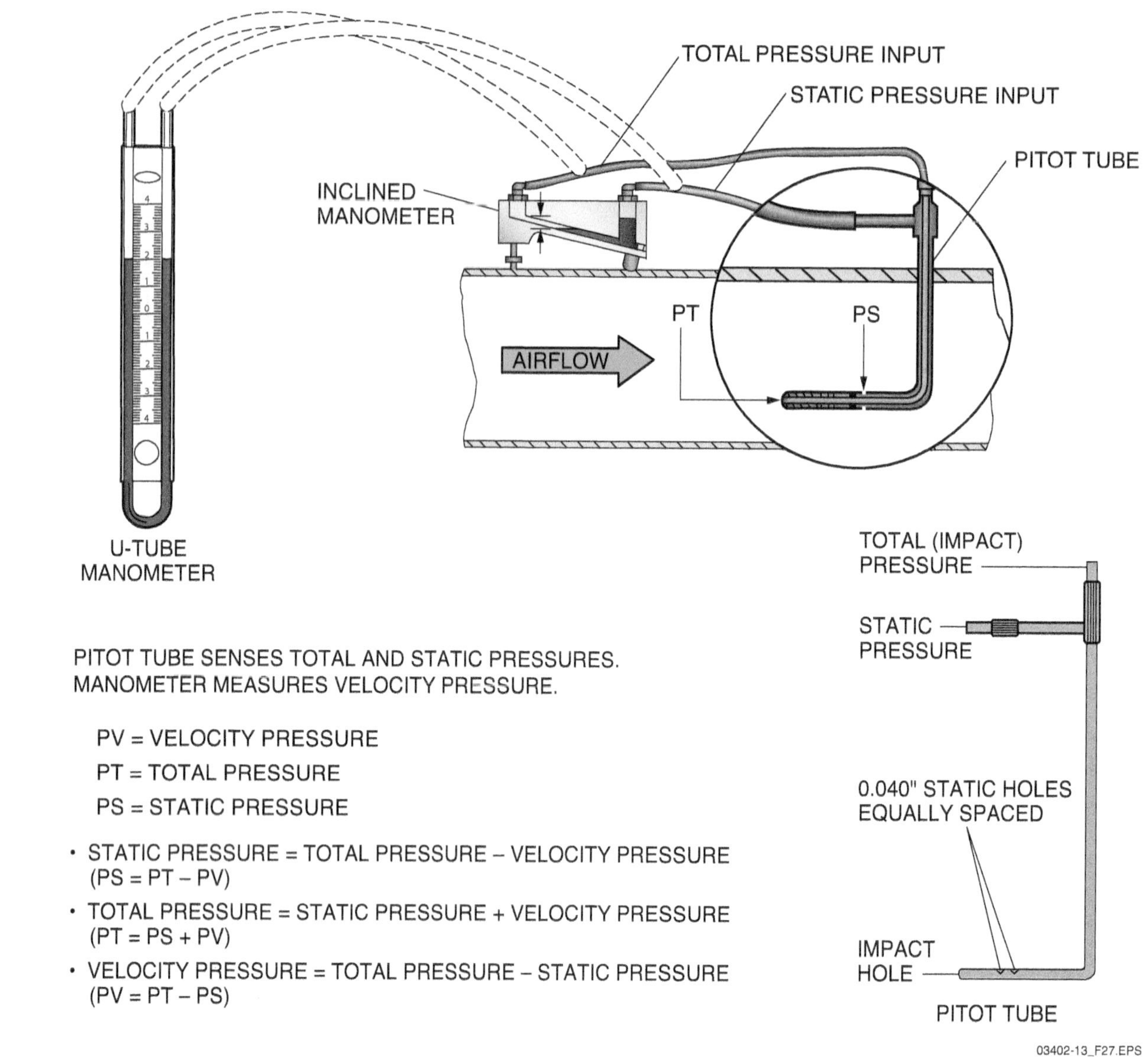

TOTAL PRESSURE INPUT

STATIC PRESSURE INPUT

PITOT TUBE

INCLINED MANOMETER

PT     PS

AIRFLOW

U-TUBE MANOMETER

PITOT TUBE SENSES TOTAL AND STATIC PRESSURES.
MANOMETER MEASURES VELOCITY PRESSURE.

PV = VELOCITY PRESSURE

PT = TOTAL PRESSURE

PS = STATIC PRESSURE

- STATIC PRESSURE = TOTAL PRESSURE − VELOCITY PRESSURE
  (PS = PT − PV)
- TOTAL PRESSURE = STATIC PRESSURE + VELOCITY PRESSURE
  (PT = PS + PV)
- VELOCITY PRESSURE = TOTAL PRESSURE − STATIC PRESSURE
  (PV = PT − PS)

TOTAL (IMPACT) PRESSURE

STATIC PRESSURE

0.040" STATIC HOLES EQUALLY SPACED

IMPACT HOLE

PITOT TUBE

03402-13_F27.EPS

*Figure 27* Pitot tube and manometer used for duct pressure measurement.

sition of the vane is converted into an equivalent velocity reading for display.

Another type of velometer, known as a hot wire anemometer, gives direct readings of air velocity in fpm. This instrument (*Figure 30*) uses a sensing probe containing a small resistance heater element. When the probe is held perpendicular to the airstream being measured, the temperature of the heater element changes because of changes in the airflow. This causes its resistance to change, which alters the amount of current flow being applied to the meter circuitry. There, it automatically calculates the air velocity for display on the meter. These are best used when very low flow rates are being measured.

Some velometers use probes that have a sensitively balanced vane or a small resistance heater element which, when placed in the airstream,

produces a measurement of airflow for display on the velometer meter scale.

Depending on the sensing probe or attachment used, velometers can measure air velocities in several ranges within the overall range of 0 to 10,000 fpm. Some electronic velometers that use a microprocessor can automatically average up to 250 individual readings taken across a duct area to provide an average air velocity. Certain velometers also include an optional microprinter to record the readings.

Special velometers, known as air volume balancers, can be used when balancing air distribution systems. This type of velometer is held directly against the diffuser or register to get a direct reading of air velocity. Still another type of flow measurement instrument, called a flow hood, is frequently used to get direct volume

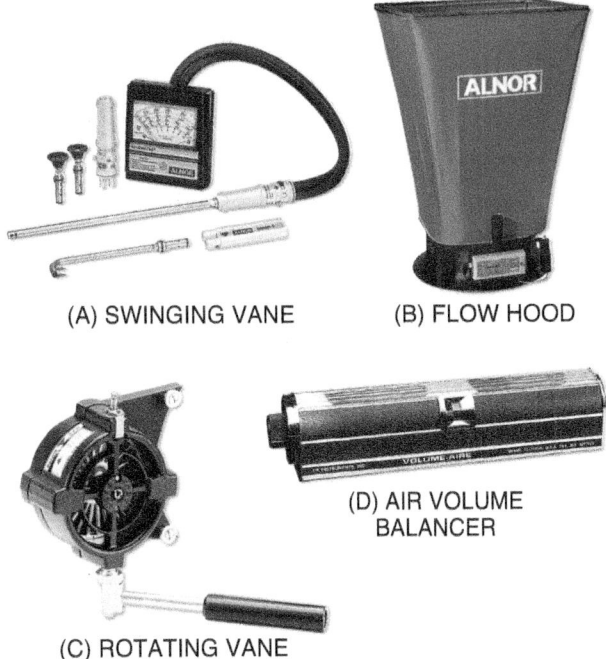

(A) SWINGING VANE    (B) FLOW HOOD

(C) ROTATING VANE    (D) AIR VOLUME
                        BALANCER

*Figure 28* Velometers.

readings in cfm when measuring the output of large air diffusers in commercial systems. Both are pictured in *Figure 28*.

### 3.3.0 Blowers and Fan Laws

The blower or supply fan provides the pressure difference necessary to force the air into the supply duct system, through the grilles and registers, and into the conditioned space. It must overcome the pressure loss involved in the return of the air as it flows into the return air grilles and through the return ductwork system back to the air handler. In addition, the supply fan must also overcome any resistance of other components in the system through which the air passes.

Two types of blowers are commonly used in air distribution systems: belt-drive and direct-drive (*Figure 31*). In belt-drive blowers, the blower motor is connected to the blower by a belt and pulley. The blower speed is changed mechanically by adjusting the diameter of the motor pulley. In direct-drive blowers, the blower wheel is mounted directly on the motor shaft. The blower speed is changed electrically by selecting motor winding speed taps, or changing the settings of motor speed selection switches on a related motor control board.

Today's residential equipment uses multi-speed direct-drive motors. This enables the speed of the motor to be adjusted to match the requirements of the individual heating or cooling air distribution

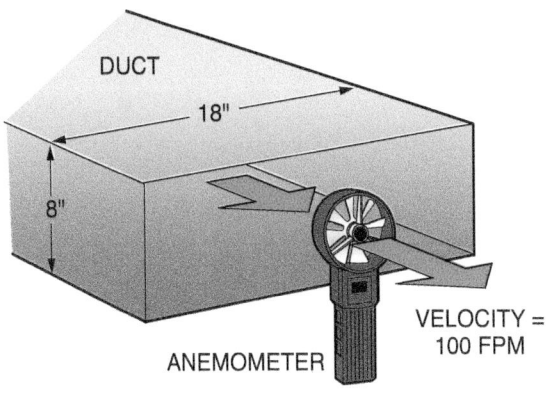

DUCT

18"

8"

ANEMOMETER

VELOCITY =
100 FPM

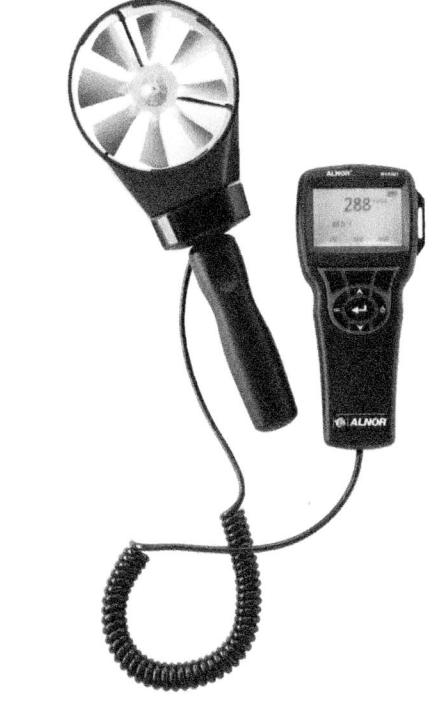

03402-13_F29.EPS

*Figure 29* Rotating vane anemometer.

system. It also allows the speed to be changed automatically between heating and cooling modes, since each mode will have different airflow requirements.

When balancing air distribution systems, adjustment of the fan speed may be required. It is important to remember that any change made in fan speed will affect other system parameters, such as the volume of airflow, static pressure, and motor horsepower. Because of these relationships, it is important to review the three fan laws. Also covered is other information needed for making adjustments to fan speeds.

### 3.3.1 Fan Laws

The performance of all fans and blowers is governed by three rules commonly known as the Fan Laws. Cubic feet per minute (cfm), revolu-

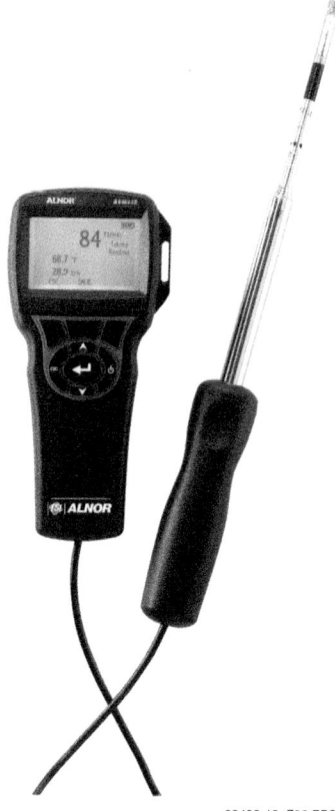

*Figure 30* Hot wire anemometer.

03402-13_F30.EPS

BELT-DRIVE

DIRECT-DRIVE

03402-13_F31.EPS

*Figure 31* Belt-drive and direct-drive motors.

tions per minute (rpm), static pressure (s.p.), and horsepower (hp) are all related. For example, when the cfm changes, the rpm, s.p., and hp will also change. Rpm is the speed at which the shaft of an air-moving device is rotating, measured in revolutions per minute. The easiest way to determine the fan rpm is to measure it directly with a tachometer.

- Fan Law 1 states that the amount of air delivered by a fan varies directly with the speed of the fan. Stated mathematically:

$$\text{New cfm} = (\text{new rpm} \times \text{existing cfm}) \div \text{existing rpm}$$

*or*

$$\text{New rpm} = (\text{new cfm} \times \text{existing rpm}) \div \text{existing cfm}$$

- Fan Law 2 states that the static pressure (resistance) of a system varies directly with the square of the ratio of the fan speeds. Stated mathematically:

$$\text{New s.p.} = \text{existing s.p.} \times (\text{new rpm} \div \text{existing rpm})^2$$

- Fan Law 3 states that the horsepower varies directly with the cube of the ratio of the fan speeds. Stated mathematically:

$$\text{New hp} = \text{existing hp} \times (\text{new rpm} \div \text{existing rpm})^3$$

***Study Example***

Assume that the existing system conditions are 5,000 cfm, 1,000 rpm, and 0.5 in wc static pressure, with a ½ hp fan. With an increase in airflow to 6,000 cfm, find the new rpm, s.p., and hp.

Use Fan Law 1 to calculate the new rpm:

$$\text{New rpm} = (\text{new cfm} \times \text{existing rpm}) \div \text{existing cfm} = (6,000 \times 1,000) \div 5,000 = 1,200 \text{ rpm}$$

Use Fan Law 2 to calculate the new s.p.:

$$\text{New s.p.} = \text{existing s.p.} \times (\text{new rpm} \div \text{existing rpm})^2 = 0.5 \times (1,200 \div 1,000)^2 = 0.72 \text{ in wc}$$

Use Fan Law 3 to calculate the new hp:

$$\text{New hp} =$$
$$\text{existing hp} \times (\text{new rpm} \div \text{existing rpm})^3 =$$
$$0.5 \times (1{,}200 \div 1{,}000)^3 = 0.864 \text{ hp}$$

In this example, the system would require a larger fan motor, and an adjustment of the motor pulley if a belt-driven blower is used.

### 3.3.2 Fan Curve Charts

Manufacturers' fan curve charts can also be used to find the relationships that exist for a set of system conditions involving static pressure, blower/fan rpm, and cfm. *Figure 32* shows an example of a typical fan curve chart. If you know the values for any two of the three characteristics (s.p., rpm, and cfm) shown on the chart, you can easily find the value for the other characteristic.

# Tachometers

Tachometers are used to measure motor or fan rpm. There are two types: contact and non-contact. Use of the non-contact type is safer and it is more convenient when the motor is located in a hard-to-reach place. Some manufacturers make a combination contact/non-contact model tachometer (shown here) that can be used to make rpm measurements either by the contact or no-contact method.

To measure motor rpm with the contact-type tachometer:

- Turn the motor on.
- Contact the end of the motor shaft with the tachometer sensor tip.
- Allow the reading to stabilize, then read the rpm.

To measure motor rpm with the non-contact tachometer:

- Turn the motor off.
- Place a reflecting mark on the motor shaft or object to be measured.
- Turn the motor on.
- Point the tachometer light beam at the shaft or object, then read the rpm.

03402-13_SA03.EPS

For example, assume that the static pressure is 1.4 in wc and the blower is running at 900 rpm. To find the cfm, locate the intersection point of the 1.4 in wc static pressure line and the 900 rpm curve. From this point, drop down vertically to the cfm scale, then read the value of 7,500 cfm.

### 3.3.3 Changing the Speed of the Fan

When balancing a system, the fan speed may need to be changed for one or more reasons, including the following:

- When there is more air being circulated than required and the supply static pressure is higher than specified. The static pressure needs to be reduced while keeping the cfm within specifications. This is done by reducing the supply fan rpm.
- To achieve the airflow needed to balance the system and meet building design requirements.
- When a supply fan motor is not running at the specified rpm.
- When a supply fan motor is overloaded. This is indicated by a motor current draw greater than the full load amps specified on the motor nameplate. Assuming that the high current draw is not the result of some other electrical or mechanical problem, the fan speed must be lowered to reduce the motor load.

If the system supply fan is belt-driven, and the fan motor has a variable-pitch pulley, the speed can be increased by narrowing the width of the pulley V-groove, or decreased by widening the width of the pulley V-groove. When changing blower speed with a variable-pitch pulley, it is important to monitor blower motor current to prevent the motor from overloading.

When checking current draw, always ensure that blower compartment doors, duct access doors, and any other abnormal openings in the air distribution system are closed. To get an accurate picture of the actual load on the motor, the system must be in its normal configuration. If an access door on the supply or return air side is open, for example, the readings will not reflect an accurate picture of performance.

Typically, variable-pitch pulleys allow the speed of the driven fan to be varied by as much as 30 percent.

If the supply fan motor has a fixed-drive pulley, the size of the pulley must be changed to vary the speed. Use of a larger pulley will cause the fan speed to increase; use of a smaller pulley will cause the fan speed to decrease.

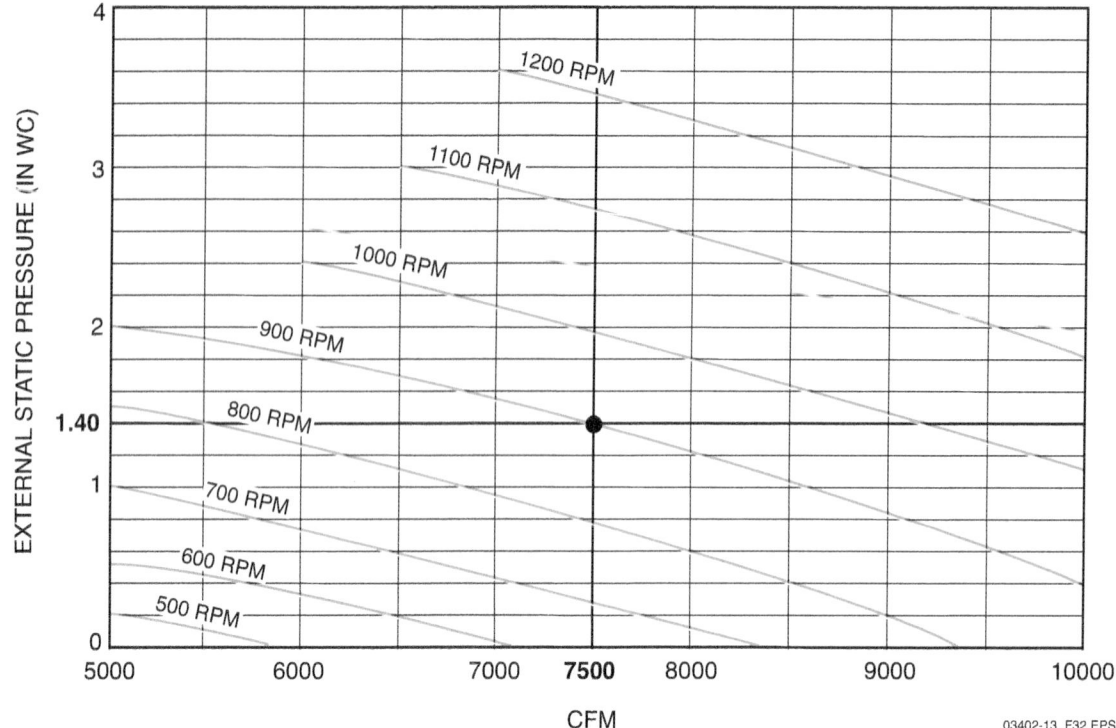

Figure 32 Typical fan curve chart.

Before discussing the sizing of pulleys in depth, it is important to understand what the pitch of a pulley represents. The pitch diameter of a pulley is not the outside diameter, nor is it a measurement of its inside diameter. The pitch diameter is basically considered to be the point in the groove where the most power is transmitted by the belt. Looking at a cross-section of a typical belt, notice there is a row of fibrous material just under the upper surface. This is where most of the power is transmitted—through these strong fibers. The ,remainder of the belt basically provides a contact surface. Therefore, the pitch diameter of a pulley is measured just under the top markings where the belt rides in the groove. Examine a pulley that has had a belt riding in the groove(s). Just below the mark where the top of the belt meets the pulley groove represents the pitch. The pitch diameter then, is the distance from this mark to the same mark on the opposite side. Note that some pulleys can accommodate both 3L and 4L belts, or possibly both A and B belts. Since these belts have

## Changing the Diameter of an Adjustable Motor Pulley

The diameter of the adjustable pulley on a belt-driven blower motor is varied to change the speed of the motor. The pulley consists of a fixed flange and a movable flange with the movable flange being locked in place by a setscrew. To change the pulley diameter, loosen the setscrew, then turn the adjustable pulley flange toward the fixed flange to increase fan speed or away from the fixed flange to decrease the fan speed. After any fan speed adjustment, the fan motor current should always be measured with a clamp-on ammeter to make sure the motor is not overloaded.

03402-13_SA02.EPS

different widths, they will ride in the groove at different heights. This changes the pitch diameter, and changes the speed of what is being driven.

An example of pulley specifications may look like this:

| Pulley Part # | Diameters | | |
|---|---|---|---|
| | Outer | Pitch A Belt | Pitch B Belt |
| PT24 | 24" | 22.85" | 23.25" |

The word *datum* may sometimes by substituted for the word *pitch*. These measurements represent the different pitches when the pulley is used with an A or a B belt. Note that both dimensions are smaller than the actual outside diameter of the pulley.

The size of the pulley pitch diameter needed to achieve the desired speed can be calculated using the following formula:

New pulley pitch diameter =
existing pulley pitch diameter ×
new speed ÷ existing speed

**NOTE**

Do not forget the acronym for the order of operations learned in previous modules: MDAS, which stands for multiplication, division, addition, and subtraction.

### Study Example

Assume that an existing supply fan is running at a speed of 900 rpm with a motor drive pulley that has a 2-inch pitch diameter. What is the pitch of the new drive pulley needed to change the fan speed to 1,000 rpm?

New pulley pitch diameter =
existing pulley pitch diameter ×
new speed ÷ existing speed

New pulley pitch diameter =
2" × 1,000 rpm ÷ 900 rpm

New pulley pitch diameter =
2,000 ÷ 900 rpm

New pulley pitch diameter = 2.22"

The size of the V-belt may need to be changed as a result of changing fan motor pulley sizes. When determining the belt length for most drives, it is not necessary to be exact. This is because of the adjustment built into most drives, and also because the belt selection is limited to the standard lengths available.

To determine the belt length for most drives, use the following calculation as a rule of thumb:

Belt length = A + B + C

*Where:*

A = pitch diameter of first pulley × 1.57

B = pitch diameter of second pulley × 1.57

C = distance between shaft centers (in inches) × 2

**Think About It**

# Belt Sizing

What size V-belt would you need for a belt drive system where the pitch diameter of the first pulley is 2.2", the pitch diameter of the second pulley is 5", and the distance between shaft centers is 12"?

## Additional Resources

*ASHRAE HVAC Applications Handbook,* 2011. Atlanta, GA: American Society of Heating and Air Conditioning Engineers.

*ASHRAE Standard 111-2008 Testing, Adjusting, and Balancing of Building HVAC Systems.* 2008. Atlanta, GA: American Society of Heating and Air Conditioning Engineers.

*Testing and Balancing HVAC Air and Water Systems,* 2001. Samuel C. Monger. Lilburn, GA: The Fairmont Press, Inc.

## 3.0.0 Section Review

1. Aspect ratio is the ratio of area to volume of a grille core.

    a. True
    b. False

2. Standard pitot tubes are used for measurements in _____.

    a. ducts of all sizes
    b. 4-inch and larger ducts
    c. 8-inch and larger ducts
    d. rectangular ducts

3. Variable-pitch pulleys allow blower speeds to be varied by as much as _____.

    a. 10 percent
    b. 20 percent
    c. 30 percent
    d. 40 percent

# SECTION FOUR

## 4.0.0 AIR BALANCING

### Objectives

Explain how to balance an air distribution system.

    a. Describe the steps to take prior to beginning an air balancing task.

    b. Explain how to measure temperature rise and drop and then use the acquired information.

    c. Explain how to measure system and terminal airflow and adjust as required.

    d. Explain how to balance using the thermometer methods.

### Performance Tasks

1. Select and properly use test instruments for balancing air distribution systems.
2. Measure the temperature rise and drop across ducted heating and cooling equipment.
4. Measure airflow at air supply outlets.
5. Adjust dampers in branch supply ducts and at air terminals and diffusers.

### Trade Term

Induction unit system: An air conditioning system that uses heating/cooling terminals with circulation provided by a central primary air system that handles part of the load, instead of a blower in each cabinet. High-pressure air (primary air) from the central system flows through nozzles arranged to induce the flow of room air (secondary air) through the unit's coil. The room air is either cooled or heated at the coil, depending on the season. Mixed primary and room air is then discharged from the unit.

Almost every air delivery duct system requires balancing before it can be considered complete and ready for delivering conditioned air. Balancing involves setting the dampers to deliver the necessary amounts of air and making final adjustments to the registers and grilles to obtain the proper air distribution pattern. The person responsible for balancing can do very little about existing duct design, diffuser locations, and other variables such as grossly oversized equipment that can make it difficult to properly balance the system.

When adjustment is needed to obtain a satisfactory compromise between temperature rise in heating or temperature drop in cooling, total airflow, and/or grille velocities, the only choice is to alter blower speed. Most blower systems are equipped with multi-speed, direct-drive motors, or pulleys on belt-drive systems that can be adjusted to alter their speed.

## 4.1.0 Air Balancing Preparations

A few factors should be considered before the airflow in a duct system is altered, including the following:

- Higher velocities at return and supply grilles can cause objectionable noise.
- Complaints about drafts can be caused by low discharge temperatures leaving terminals at a high velocity.
- A decrease in airflow can cause excessive temperature rise at the furnace, stressing the heat exchanger and possibly opening the limit switch.
- Stratification can be caused by low airflow rates. Stratification results in hot air remaining at the ceiling, while cold air collects at the floor.

A satisfactory compromise must be made between temperature rise or temperature drop through equipment, total airflow, diffuser velocities, and final comfort. The ideal compromise can be achieved with a properly designed duct system and with properly sized heating and cooling equipment. However, these conditions do not always exist, and the person balancing the system must do so within the limitations of the system at hand. For example, many duct systems are equipped with balancing dampers in the branch ducts, but some economy systems have dampers only in branches serving areas closest to the blower. This is not ideal, but it may be adequate in many instances.

Balancing comfort conditioning systems usually requires time and patience as much as sophisticated equipment. If the individual balancing the system follows a systematic procedure, little difficulty should be encountered in obtaining a properly balanced system.

### 4.1.1 Pre-Balance Checks

A systematic balancing procedure begins with preparing a schematic layout of the system. The layout should address the following items:

- Indicate all dampers, regulating devices, terminal units, outlets, and inlets.

- Indicate the sizes and cfm for main ducts, outlets, and inlets. The quantities of both fresh air and return air also need to be indicated.
- Check the shop drawings against design drawings so that construction changes can be shown on the schematic layout.
- Number the outlets for reporting purposes.
- Convert all volume values, in cfm, to velocities, in fpm, because velocities are easier to work with.
- As needed, prepare reporting forms used to record system measurements, such as pitot tube traverse readings (*Figure 33*). Forms available from the National Environmental Balancing Bureau (NEBB) can be used for this purpose. Examples of completed forms are provided in *Appendix A*.

When taking traverse readings, remember that the velocity profile of an airstream is not uniform across the cross section of a duct. Friction slows the air moving close to the walls, resulting in a faster velocity in the center of the duct (*Figure 34*).

For that reason, a traverse pattern such as shown in *Figure 33* allows readings taken at different points in the duct to be averaged. In this figure, the longest duct dimension (18 inches) is divided into 6-inch increments and a hole is drilled into the center of each increment. The purpose is to take a velocity pressure measurement in the center of each 6-inch square (marked X). For the duct shown, six measurements are needed. Regardless

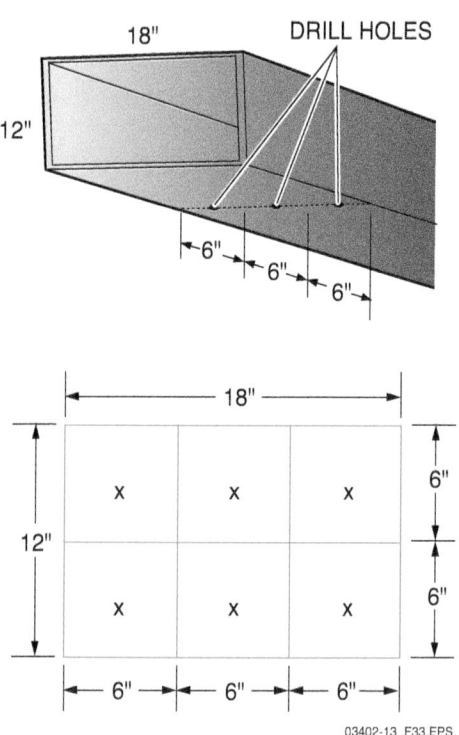

Figure 33 Traverse readings form.

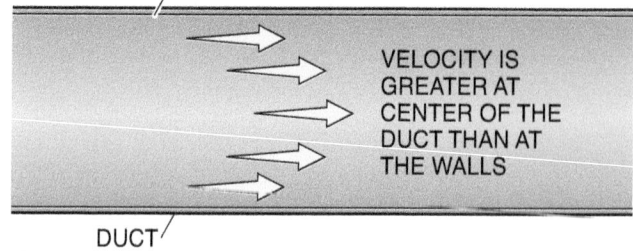

03402-13_F34.EPS

*Figure 34* Velocity profile of an airstream within a duct.

of the duct size, 6 inches should be the maximum distance between measurement points. Mark the pitot tube with insertion depth graduations to aid in placing the pitot tube in the duct at the proper location.

Balancing air distribution systems normally involves the use of ladders to gain access to measurement points and adjustment dampers. Make sure to use ladders in accordance with good safety practices.

Prior to beginning the balancing procedure, make sure that the equipment and duct system components are clean and fully operational. Check with other workers and crafts to ensure that the procedure can begin. Follow the suggested steps below to complete the preparations before flow measurements begin:

*Step 1* Check the blower and its components. Make sure that the blower wheel is free and the housing is clear of obstructions.

*Step 2* Be sure that the bearings are properly lubricated.

*Step 3* Examine the drives for proper belt tension and pulley alignment.

*Step 4* Check any blower assembly vibration eliminators for placement and proper adjustment.

*Step 5* Examine the duct system to make sure that outlet, inlet, and terminal dampers are in their engineered, marked positions. In many cases, they will be fully open, awaiting adjustment by the air balance team.

*Step 6* Make sure that all fire dampers are open.

*Step 7* Check the filters and coils for cleanliness.

*Step 8* Turn the power on and check the motor and fan for the correct direction of rotation.

*Step 9* Energize all exhaust devices that affect each other, such as kitchen, bathroom, and general exhaust.

*Step 10* Close all windows and outside doors, and open any doors within the building that may affect the procedure. However, this situation would be rare, as system designers can rarely depend on doors being in any given position.

*Step 11* Take amperage and voltage readings at blower motors to ensure they are within their normal operating parameters at the beginning of the process. Record this information.

## 4.2.0 Measuring Temperature Rise and Drop

Temperature rise is the difference in air temperature entering and leaving heating or cooling equipment. The rate of airflow has an appreciable effect on the temperature rise. Low airflow can cause excessive furnace temperature rise and result in insufficient heat output from the cycling of the limit switch. High rates of airflow can cause low furnace temperature rise and complaints of drafts.

In cooling systems, the temperature drop across the evaporator coil must be considered. Too little airflow will result in a loss of capacity and possibly a frozen coil. High airflows generally do not have a negative effect on cooling equipment, except that the air temperature delivered to the conditioned space may not be as cool as desired. High airflow rates also increase static loss and possibly increase energy costs.

### 4.2.1 Temperature Rise Measurement

To measure temperature rise, proceed as follows:

*Step 1* Insert a calibrated thermometer in the main supply duct leaving the heating unit (*Figure 35*).

*Step 2* Insert another calibrated thermometer in the return duct entering the heating unit. The same thermometer can also be used in both positions.

*Step 3* The supply thermometer should be out of the line of sight of any heating element or heat exchanger where it will not be affected by radiant heat.

*Step 4* The difference in readings between the two thermometers is the temperature rise.

*Step 5* Measure within 6 feet of an air handler or electric furnace. Do not measure as far away as the return and supply grilles; this procedure is inaccurate.

*Step 6* Use the average temperature when more than one duct is connected to the plenum. Be sure that the air temperature has stabilized prior to measurement.

*Step 7* Measure downstream from any mixed air source.

*Step 8* Record the temperature differential in the return air and supply air runs.

When the temperature rise through an electric furnace or electric heater is known, system airflow can be quickly determined by using the following formula:

$$cfm = volts \times amps \times 3.414 \div (\Delta T)(1.08)$$

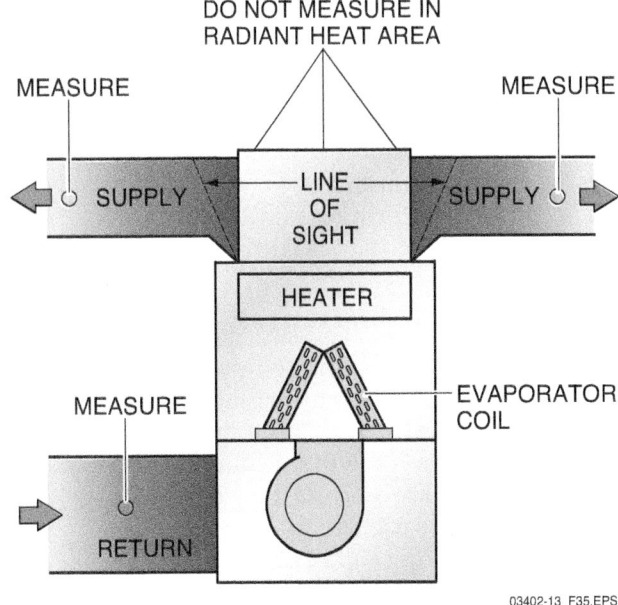

*Figure 35* Measuring temperature rise.

To use the formula for determining airflow, the applied voltage, current draw of the heater, and the temperature rise must be known.

The output in Btuh of an electric resistance heating system may be determined using the following formula, if the applied voltage and current draw of the heater are known:

$$Btuh = volts \times amps \times 3.414$$

### 4.2.2 Temperature Drop Measurement

Temperature drop is associated with cooling equipment and can be used for determining total cooling capacity.

To measure temperature drop, proceed as follows:

*Step 1* Take temperature measurements on cooling equipment in the ducts near the air handling unit. Measurements taken at the grilles are not accurate.

*Step 2* Measure the discharge air from the cooling coils at least 6 feet downstream from the coils.

*Step 3* Take the measurements as near to the center of the duct as possible.

*Step 4* Determine the total cooling capacity:

a. Use a sling psychrometer to find the entering air and leaving air wet-bulb temperatures.

b. Use a psychrometric chart to convert the wet-bulb temperatures to enthalpy (heat content in Btuh/lb).

c. Find the total heating or cooling capacity in Btuh:

$$Btuh = (cfm)(4.5)(H1 - H2)$$

*Where:*
- 4.5 = constant
- cfm = evaporator air volume
- H1 = entering air enthalpy
- H2 = leaving air enthalpy

### 4.2.3 Equipment Adjustment

With temperature rise and temperature drop values now available, that information needs to be analyzed to determine if any adjustments or corrective measures have to be made to the equipment. If values are within the normal range, no corrective actions are needed.

*Temperature rise too high:*

- Check air filters for cleanliness. Clean or replace if dirty.

- Check for blockage or restrictions in the return air duct system.
- Check for excessively high return air temperatures.
- Check burner input rate.
- Increase blower speed.

*Temperature rise too low:*

- Check for excessively low return air temperatures.
- Check burner input rate.
- Decrease blower speed.

*Temperature drop excessive:*

- Check air filters for cleanliness. Clean or replace if dirty.
- Check for blockage or restrictions in the return air duct system.
- Check for excessively low return air temperatures.
- Increase blower speed.

# Electronic Thermometers

When choosing an electronic thermometer, select one that has the capability of accepting multiple temperature probes. The different probes can be placed at different points in the system. To read a temperature at a location, just select the temperature probe that is installed at that location. The use of multiple probes eliminates the need to move a single probe around to different locations.

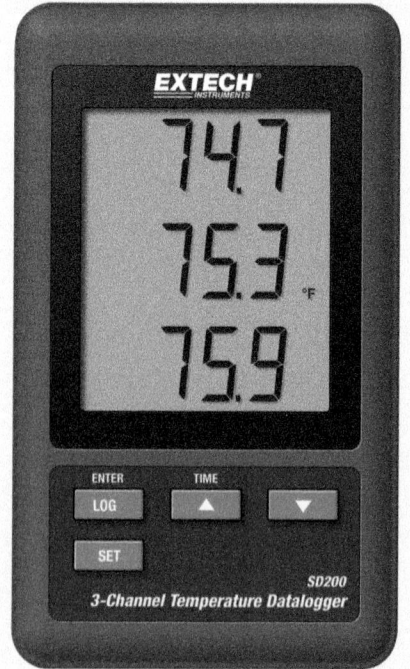

03402-13_SA04.EPS

*Temperature drop inadequate:*

• Check for excessively high return air temperatures.
• Decrease blower speed.
• Check refrigerant charge.

Note that if significant changes are made during the air balancing process that alter the airflow, the temperature rise and/or drop must be checked again to avoid problems.

## 4.3.0 Measuring and Adjusting Airflow

The various portions of the air distribution system have similar and unique balancing requirements. Different areas covered here include:

• Duct system balancing
• Terminal balancing
• Mixed air systems adjustment
• Induction system balancing

Select the instruments that will do the best job and check them for calibration. Whenever possible, use the same instrument for the entire job to minimize possible calibration errors. If more than one instrument of a similar type is used, make a comparison check. The variation between instrument readings should be limited to ±5 percent.

The air balancing process can be a very repetitive one. Since adjustments in one place affect the airflow in other places, it may require that readings for duct and terminals be taken a number of times as adjustments are made.

### 4.3.1 Duct Readings and Balancing

The damper is the primary element in the duct system used for controlling airflow by introducing resistance in the system. Partial closing of the damper increases resistance to airflow, whereas partial opening of the damper decreases resistance to airflow. However, the reduction in airflow with closure of the damper may or may not be proportional to the amount of adjustment of the damper.

In other words, closing the damper halfway does not necessarily mean that the air volume will be reduced to 50 percent of the volume that flows through the damper when it is in the wide-open position. This relationship between the position of the damper and the percentage of air that flows through the damper is called the flow characteristic or mean effective airflow of the damper.

For example, *Figure 36* provides an example of a damper ratio. Note that balancing dampers can be opposed-blade or parallel blade. However, most parallel-blade dampers that are installed for balancing purposes have only one blade, such as those found in round branch lines.

The curve O shows that to obtain 25 percent of the airflow, the damper should be 23 percent open; for 50 percent airflow, the damper is only 38 percent open. The curve P demonstrates the characteristics of parallel-blade dampers. Notice that these damper blades need to open only 20 percent to achieve 50 percent airflow. It is important to remember that these percentages are approximate, and vary according to the system characteristics.

Although it is important for technicians to understand that damper position and the volume of airflow are rarely, if ever, proportional, charts such as the one in *Figure 36* are not needed during the balancing process. Finding the precise damper position for the desired airflow is largely a trial-and-error procedure.

To begin balancing a system, proceed as follows:

*Step 1*  Check the speed of the fan.

*Step 2*  Check the manufacturer's performance tables or curves to determine the cfm being delivered by the fan.

*Step 3*  Adjust the speed to meet the desired conditions if the fan is functioning above or below the design cfm. Recheck the fan after each damper adjustment, if needed.

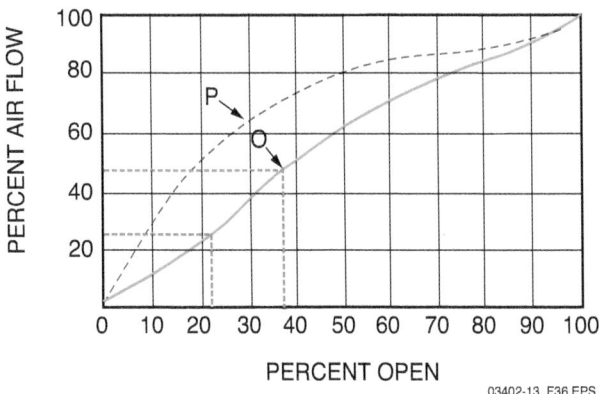

*Figure 36* Damper ratios.

*Step 4* Take traverse velocity readings of the main and branch ducts. Traverse the square and rectangular ducts and record the readings at each point as shown in *Figure 33* and average the readings. Record every reading, even if the value is 0.

The location of the pitot tube in the traverse is not as easy to determine on round duct runs. To achieve accuracy, proceed as follows:

a. Drill two holes at right angles to each other, as shown in *Figure 37*.

b The number of pitot tube readings necessary for accuracy depends on the duct diameter. The larger the duct diameter, the more readings are needed to make sure a good average is obtained. As shown in *Figure 38,* a 4-inch duct diameter requires six readings, whereas a 32-inch duct diameter requires ten readings on each diameter. For each reading, insert the pitot tube and use the graduated scale to determine how far the pitot tube is inserted.

c. The length of straight ductwork before the pitot tube location should be at least 7.5 times the diameter of the duct downstream from any turns or similar obstructions (*Figure 39*). Also, the probe should be at least 3 duct diameters upstream from turns or similar obstructions. Although readings can be made as little as 2 duct diameters downstream and 1 duct diameter upstream, the readings will likely be affected by turbulence and laminar flow against the duct walls. For example, with 6-inch diameter duct, the pitot tube should be at least 45 inches downstream from obstructions or turns to obtain an accurate reading (6 inches × 7.5 diameters = 45 inches).

d. For square and rectangular ducts, the duct size needs to be converted to a round equivalent to determine where the probe should be inserted. A formula can be used to determine this as follows:

$$\text{Equivalent round duct diameter} = \sqrt{4HV/\pi}$$

*Where:*
- H = horizontal duct size (inches)
- V = vertical duct dimension (inches)
- $\pi$ = 3.14

For example, the equivalent round duct dimension of a 15 × 12 duct is calculated as follows:

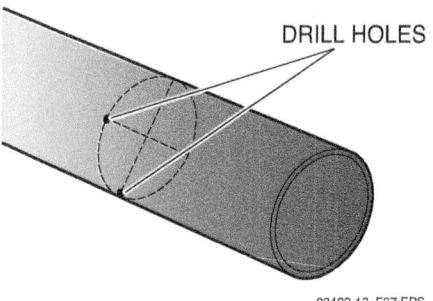

Figure 37 Round duct readings.

$$\text{Equivalent round duct diameter} = \sqrt{[(4 \times 15 \times 12)/3.14]}$$

$$\text{Equivalent round duct diameter} = \sqrt{720/3.14}$$

$$\text{Equivalent round duct diameter} = \sqrt{229.3}$$

$$\text{Equivalent round duct diameter} = 15.1 \text{ (round to 15)}$$

Therefore, for determining how far away from turns the probe should be in this duct, it would be treated as if it was a round duct 16" in diameter.

*Step 5* Once the readings have been taken in the main and branch ducts, adjust the volume and/or splitter dampers to deliver the design cfm. There will likely be some branches that are receiving too much airflow, while others are receiving too little. Reduce the airflow in areas that are receiving too much air flow first, and then take new readings on those that were short of air. Continue the procedure throughout the system. Reconcile the readings taken in the supply and return air ducts with each other, and against the fan curves.

### 4.3.2 Terminal Readings and Balancing

To calculate the volume of air being delivered at a given grille or register, proceed as follows:

*Step 1* Use the recommended velometer, in this case an Alnor velometer equipped with a 2220-A probe. This is a common device for measuring velocity at grilles and registers.

*Step 2* Position the probe as recommended by the manufacturer (*Figure 40*), in this case between the blades with the shank of the probe parallel to the face and across the blades of the grille. As indicated, the probe position remains the same for angled blades and for straight blades. Do not block the opening of the probe with a grille blade.

---

| DUCT DIAMETER (inches) | POINTS IN ONE DIAMETER | DISTANCE FROM INSIDE OF DUCT (POINT A AND B) TO PITOT TUBE TIP | | | | | | | | | |
|---|---|---|---|---|---|---|---|---|---|---|---|
| | | 1 | 2 | 3 | 4 | 5 | 6 | 7 | 8 | 9 | 10 |
| 4 | 6 | – | – | ¼ | ⅝ | 1¼ | 2⅞ | 3½ | 3⅞ | – | – |
| 6 | 6 | – | – | ¼ | ⅞ | 1¾ | 4¼ | 5⅛ | 5¾ | – | – |
| 8 | 6 | – | – | ⅜ | 1¼ | 2⅜ | 5⅝ | 6⅞ | 7¾ | – | – |
| 9 | 6 | – | – | ⅜ | 1⅜ | 2¾ | 6⅜ | 7¾ | 8⅝ | – | – |
| 10 | 8 | – | ⅜ | 1⅛ | 2 | 3¼ | 6¾ | 8⅛ | 9 | 9¾ | – |
| 12 | 8 | – | ⅜ | 1¼ | 2 | 3⅞ | 8⅛ | 9 | 10¾ | 11⅝ | – |
| 14 | 10 | ⅜ | 1⅛ | 2⅛ | 3¼ | 4¾ | 9¼ | 10⅞ | 12 | 12⅞ | 13⅝ |
| 16 | 10 | ½ | 1⅜ | 2⅜ | 3¾ | 5½ | 10½ | 12⅜ | 13⅝ | 14¾ | 15⅝ |
| 18 | 10 | ½ | 1½ | 2⅝ | 4⅛ | 6⅛ | 11⅞ | 14 | 15⅜ | 16½ | 17⅝ |
| 20 | 10 | ½ | 1⅝ | 3 | 4½ | 6⅞ | 13¼ | 15½ | 17⅛ | 18⅜ | 19½ |
| 24 | 10 | ⅝ | 2 | 3½ | 5½ | 8¼ | 15⅞ | 18½ | 20½ | 22⅛ | 22⅜ |
| 28 | 10 | ¾ | 2¼ | 4⅛ | 6⅜ | 9⅝ | 18½ | 21¾ | 24 | 25¾ | 27¼ |
| 32 | 10 | ⅞ | 2⅝ | 4¾ | 7¼ | 11 | 21⅛ | 24¾ | 27¼ | 29⅜ | 31⅜ |

FOR OTHER DUCT DIAMETERS, USE THE FOLLOWING TABLE:

| POINTS IN ONE DIAMETER | CONSTANTS TO BE MULTIPLIED BY DUCT DIAMETER FOR DISTANCES OF PITOT TUBE TIP FROM INSIDE OF DUCT (POINT A AND B) | | | | | | | | | |
|---|---|---|---|---|---|---|---|---|---|---|
| | 1 | 2 | 3 | 4 | 5 | 6 | 7 | 8 | 9 | 10 |
| 6 | – | – | 0.0435 | 0.1465 | 0.2959 | 0.7041 | 0.8535 | 0.9564 | – | – |
| 8 | – | 0.0323 | 0.1047 | 0.1938 | 0.3232 | 0.6768 | 0.8052 | 0.8953 | 0.9677 | – |
| 10 | 0.0257 | 0.0817 | 0.1465 | 0.2262 | 0.3419 | 0.6581 | 0.7738 | 0.8535 | 0.9133 | 0.9743 |

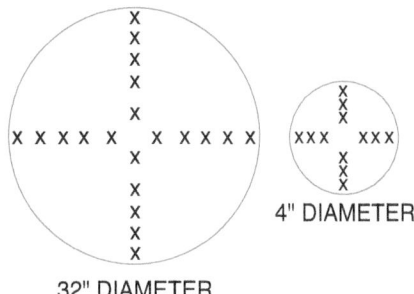

*Figure 38* Round duct velocity reading guide.

03402-13_F38.EPS

*Step 3* Take several readings across the face of the grille (*Figure 41*). Manufacturers will usually provide illustrations of where readings should be taken for each specific series of products. *Figure 41* provides a common example.

*Step 4* Find the average outlet velocity; in this case, 800 fpm.

*Step 5* Use the manufacturer's table for the grille chosen; for example, a sidewall grille, 24-inch by 12-inch, with a blade deflection of 0 degrees. The $A_k$ factor for this grille is 1.32 as determined from the example shown in *Table 1*.

**Step 6** Complete the calculation:

Air volume (cfm) =
800 fpm average velocity × 1.32 $A_k$ factor

*Therefore:*

Air volume (cfm) = 1,056 cfm

All air-measuring devices must be used according to the manufacturer's instructions, be periodically calibrated, and be protected from damage. Correct airflow measurements depend on proper positioning of the probe and using the recommended instrument.

Each manufacturer will recommend the type of equipment and the placement of the probe necessary for determining the actual air volume of their device. *Figure 42* illustrates the use of an Alnor velometer for determining the actual air volume delivered through a stamped steel louvered ceiling diffuser. The field balancing information identified in the text surrounding the figure describes the proper placement of the probe and the number of readings to be taken. This type of velometer can also be used to measure the volume of air flowing into a return air grille. The hose connections between the probe and the meter are simply reversed.

A flow hood (*Figure 43*) can also be used to measure and record the air volume from a terminal. Flow hoods typically read the actual volume of air, in cfm, that is moving through the hood. This eliminates the math of using velocity to determine volume. However, it is not well-suited for all situations. For example, if the grille or register is located on a narrow wall, the hood cannot be sealed firmly against the wall, leaving the sides open. The entire perimeter of the hood must seal against the surface around the grille, so that air cannot escape. They can also be clumsy to handle in sidewall applications. They are well-suited for almost all ceiling diffusers, however.

Now, with the understanding of how to collect readings and determine the air volume from individual terminals, the actual balancing process can begin. Note that this process assumes that the airflow in each branch as a whole has been tested

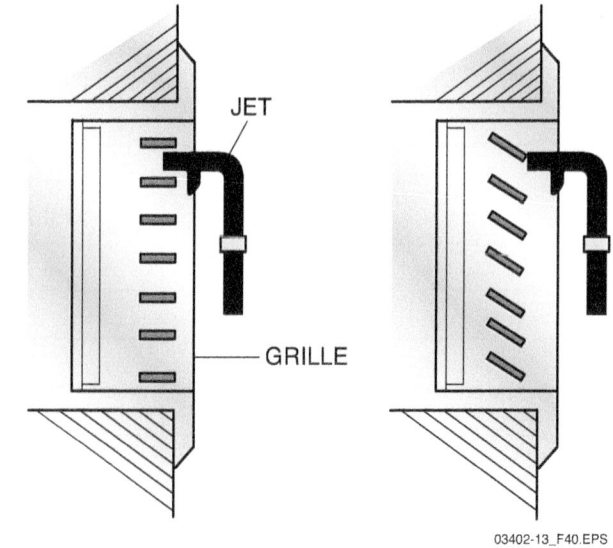

03402-13_F40.EPS

**Figure 40** Probe placement.

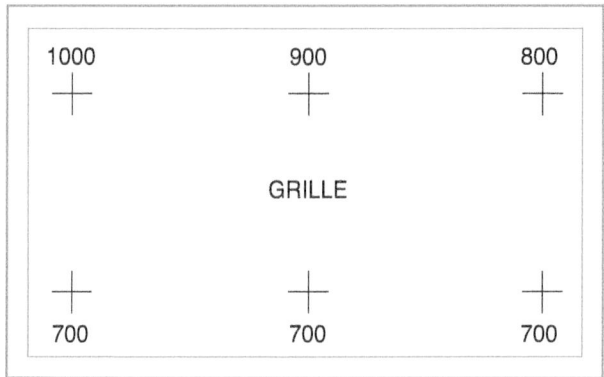

VELOCITY READINGS (fpm)

700 + 700 + 700 + 800 + 900 + 1000 = 4,800

$$\frac{\text{AVERAGE}}{\text{VELOCITY}} = \frac{4,800}{6} = 800 \text{ fpm}$$

03402-13_F41.EPS

**Figure 41** Averaging the outlet velocity readings taken.

and balanced. To balance the terminal and inlet devices, proceed as follows:

**Step 1** Check the farthest outlet in each branch line first. Some branch lines may only have a single terminal attached.

**Step 2** If the farthest outlet is below design fpm, leave the damper fully open and go on to read the next outlet upstream. If the outlet velocity at the farthest device is above design specifications, throttle it down and then proceed to the next outlet.

**Step 3** Repeat the procedure one or more times (some experts recommend at least three), making finer adjustments each time because the adjustment of one outlet affects the others.

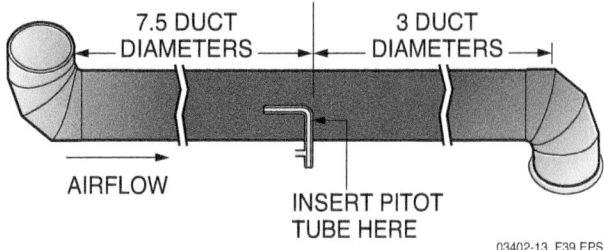

03402-13_F39.EPS

**Figure 39** Pitot tube location.

**Table 1** Example $A_k$ Factors for a Sidewall Grille

| $A_k$ Factors | | Grille Width | | | | | | | | |
|---|---|---|---|---|---|---|---|---|---|---|
| | | 12 | 14 | 16 | 18 | 20 | 22 | 24 | 26 | 28 |
| H | 12 | 0.64 | 0.76 | 0.85 | 0.99 | 1.04 | 1.21 | 1.32 | 1.45 | 1.55 |
| E | 14 | | 0.90 | 1.02 | 1.15 | 1.29 | 1.41 | 1.54 | 1.68 | 1.81 |
| I | 16 | | | 1.17 | 1.33 | 1.46 | 1.64 | 1.78 | 1.96 | 2.15 |
| G | 18 | | | | 1.50 | 1.67 | 1.82 | 2.00 | 2.17 | 2.33 |
| H | 20 | | | | | 1.86 | 2.05 | 2.23 | 2.42 | 2.60 |
| T | 24 | | | | | | 2.24 | 2.42 | 2.63 | 2.86 |

# Changing Rectangular Duct Size To A Round Duct Equivalent

There is a quick way to determine the equivalent round duct size of a rectangular duct without doing any math at all. Using a Ductulator® or similar tool that shows relationship of velocity, static pressure, and volume in various duct sizes, set the rectangular duct size in the window as shown here. Then simply look at the equivalent round duct diameter in the upper window.

READ THE ROUND DUCT DIAMETER
EQUIVALENT TO 15 × 12 DUCT HERE

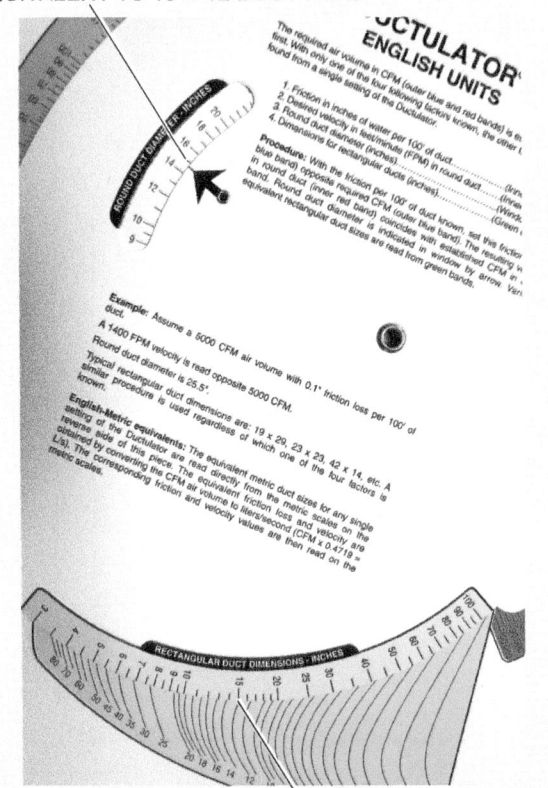

LINE UP THE 15 AND 20 MARKS HERE

**Step 4** Take the necessary readings to ensure the airflow in the branch duct is still appropriate. Assuming the airflow at each terminal is correct, there should be no problem in the branch line volume.

### 4.3.3 Mixed Air System Adjustment

Where outdoor air is introduced into the return air duct system, it is necessary to determine the percentage of outdoor air needed to meet the installation requirements. Dampers are provided for this purpose.

To set the fresh-air damper linkage, proceed as follows:

**Step 1** Remove the damper drive rod from the damper and motor.

**Step 2** Move the damper drive motor to the fully closed position.

**Step 3** Move the damper to the fully closed position.

**Step 4** Install and tighten the drive rod to the fully closed motor and damper.

**Step 5** Find the percent minimum fresh air requirement from the specifications. If the specifications are given in cfm instead of percent, divide the fresh air cfm requirements by the total cfm of the system to obtain percent.

**Step 6** Set the fresh-air control potentiometer to the proper specification (for example, 30 percent).

**Step 7** With the system operating and stabilized in the correct mode, measure the return air temperature in the return air duct at a location upstream from the fresh air inlet (*Figure 44*).

**Step 8** Measure the fresh-air temperature at the fresh air entry to the ductwork.

# STAMPED STEEL LOUVERED DIFFUSERS

$$CFM = V_k \times A_k$$

### AREA FACTOR ($A_k$) TABLE – MODEL SFA

| NECK SIZE | NOMINAL LOUVERED AREA | | |
|---|---|---|---|
| | 12 × 12 | 18 × 18 | 24 × 24 |
| | Horizontal Throw | | |
| 5 | 12 | – | – |
| 6 | 14 | 0.22 | 0.29 |
| 7 | 17 | 0.25 | 0.28 |
| 8 | | 0.27 | 0.28 |
| 10 | | 0.38 | 0.42 |
| 12 | | 0.48 | 0.50 |
| 14 | | – | 0.62 |

### AREA FACTOR ($A_k$) TABLE – MODEL SAA

| NECK SIZE | NOMINAL LOUVERED AREA | | | | | |
|---|---|---|---|---|---|---|
| | Horizontal Throw | | | Vertical Throw | | |
| | 12 × 12 | 18 × 18 | 24 × 24 | 12 × 12 | 18 × 18 | 24 × 24 |
| | 0.09 | – | – | 0.10 | – | – |
| | 0.12 | 0.20 | 0.35 | 0.11 | 0.21 | 0.30 |
| | 0.15 | 0.23 | 0.31 | 0.12 | 0.23 | 0.27 |
| | – | 0.26 | 0.33 | – | 0.24 | 0.36 |
| | – | 0.29 | 0.32 | – | 0.23 | 0.30 |
| | – | 0.38 | 0.39 | – | 0.30 | 0.33 |
| | – | – | 0.60 | – | – | 0.48 |

## FIELD BALANCING

The actual volume of air being discharged from an outlet can be determined by measuring the outlet velocity in feet per minute (fpm) and multiplying by an area factor ($A_k$).

$$CFM = V_k \times A_k$$

The Alnor velometer, with the 2220-A jet is the recommended equipment for balancing Carnes stamped diffusers.

The Alnor Model 6000P with 6070P probe can be used with the same $A_k$ factors.

Place the Alnor jet into the correct louvered space as shown in the sketches below. Point the jet as directly as possible into the air stream and move jet slowly along the lip of the cone to obtain the highest reading. Average the readings from all four sides to obtain $V_k$. Select the correct $A_k$ from the tables and apply the formula to obtain the CFM.

### ALNOR JET POSITION – MODEL SFA

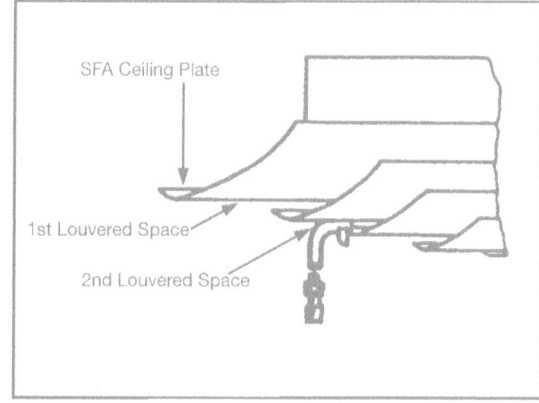

### ALNOR JET POSITION – MODEL SAA

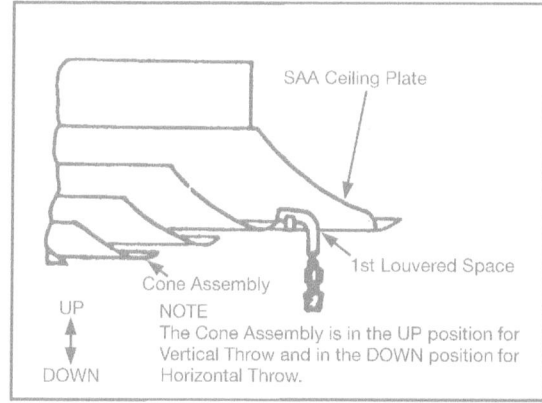

03402-13_F42.EPS

*Figure 42* Field balancing data for a ceiling diffuser.

Figure 43 Flow hood.

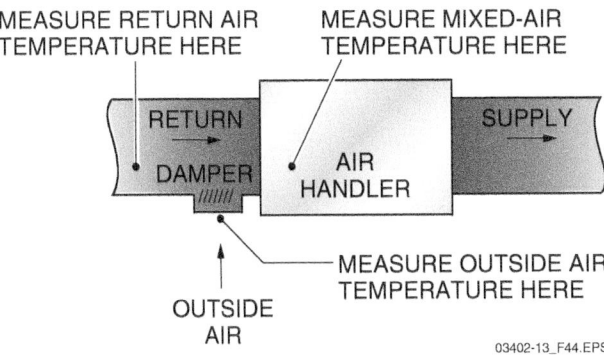

MEASURE RETURN AIR TEMPERATURE HERE

MEASURE MIXED-AIR TEMPERATURE HERE

RETURN

DAMPER

AIR HANDLER

SUPPLY

MEASURE OUTSIDE AIR TEMPERATURE HERE

OUTSIDE AIR

03402-13_F44.EPS

Figure 44 Mixed air system adjustment.

### 4.3.4 Balancing Induction Systems

Induction unit systems have been used extensively for the exterior zones of large, high-rise offices. The induction unit system uses a combination of air and water. The unit shown in *Figure 45* does not have a fan. The fan is replaced by primary air fed to the unit by a high-pressure fan and duct system. Primary air is discharged through a series of nozzles. The streams of primary air induce air from the space to flow over the coil to be either heated or cooled and then discharged into the room.

To balance an induction system, proceed as follows:

*Step 9* Measure the mixed-air temperature downstream from the fresh-air inlet.

*Step 10* Determine the actual fresh air needed by inserting temperature measurements into the following formula:

$$MB = IB + (\%FA)(OB - IB)$$

*Where:*
- FA = fresh air
- OB = outdoor air dry-bulb temperature
- IB = indoor air dry-bulb temperature
- MB = mixed air dry-bulb temperature

Prove this formula by the following example:
FA = 30 percent
OB = 95°F
IB = 78°F
MB = 78 + (0.30)(95 – 78) = 83°F (approx.)

*Step 11* Adjust the damper drive rod until the calculated mixed-air temperature is reached.

*Step 12* Recheck the outdoor and return air temperatures for any change during adjustment.

*Step 1* Check the fan and main trunk capacities for cfm.

*Step 2* Determine the primary airflow from each unit by reading the unit plenum pressure with a portable draft gauge. Refer to the manufacturer's charts to determine cfm from static pressure.

*Step 3* Make a spot check of the air distribution by reading the first and last unit on each riser. If necessary, the riser with the high readings will be cut back to improve airflow in the rest of the system. Do not reset the units; rather, study the results of the checks and then adjust the riser dampers to regulate the proper volume to each riser.

*Step 4* On the first pass around the system, read and adjust along the way. Start on the floor nearest the supply duct. If all unit dampers are open, set the units on the floor next to the supply header about 10 percent under the design specifications.

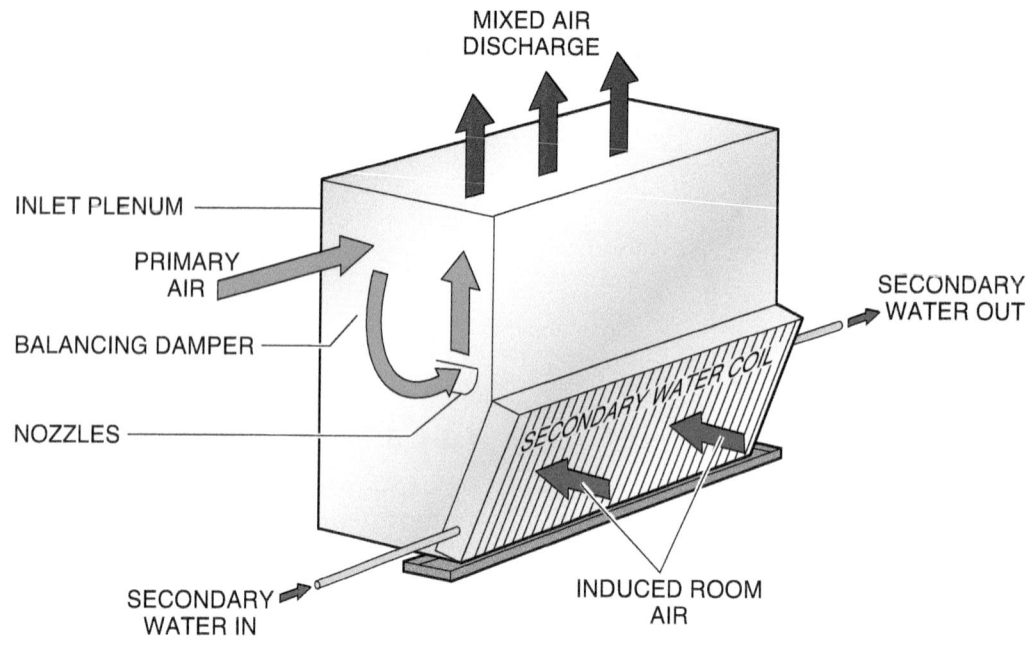

MIXED AIR
DISCHARGE

INLET PLENUM

PRIMARY
AIR

BALANCING DAMPER

NOZZLES

SECONDARY
WATER IN

SECONDARY WATER COIL

SECONDARY
WATER OUT

INDUCED ROOM
AIR

03402-13_F45.EPS

*Figure 45* Induction unit system.

*Step 5* Three complete passes around the entire system are normally necessary for proper adjustment.

*Step 6* If unit dampers are throttled back excessively, objectionable noise may occur. Use branch and riser dampers to prevent this.

*Step 7* The flow of water is automatically controlled to adjust room temperature. On those systems that use the primary air source to power the controls and move the secondary air dampers for adjusting room temperature, it is extremely important to maintain the manufacturer's recommended minimum static pressure.

## 4.4.0 Balancing by Thermometer

The thermometer method and the thermometer and velometer method are two commonly used procedures for balancing residential systems. Using a velometer with a thermometer has the added benefit of knowing the quantity of air being delivered to each room. Keep in mind that there are many factors in residential systems that make system balancing difficult, if not impossible. While similar problems can be found in commercial systems, commercial systems are usually designed and installed to a higher standard than that found in residential systems.

Typical residential system factors include the following:

- Grossly oversized equipment. Oversized equipment is often sold as a cure-all for many residential comfort problems. Oversizing often creates a whole new set of problems.
- Inadequate duct systems. Ducts may be oversized, undersized, poorly insulated, grilles/registers may be improperly placed, and the ducts often leak.
- Inadequate temperature control. Room thermostat may be of low quality, out of calibration, or in the wrong location.
- Equipment may be poorly maintained, if at all.

The problems outlined above are more typically found in older homes.

For proper balancing of duct systems using the different temperature methods, the following pre-checks should be made:

*Step 1* Use a good-quality electronic thermometer. Use the same device for the entire process to avoid calibration issues between two or more instruments.

*Step 2* Make temperature measurements at the center of each room, about 4 feet above floor level.

**Step 3** On supply registers equipped with adjustable vanes, set the vanes for optimum distribution.

**Step 4** Check the space thermostat heat anticipator setting. Ensure that the wire entry hole behind the thermostat is plugged to prevent drafts from affecting the thermostat.

**Step 5** Locate the temperature-sensing element where air from the supply registers will not affect the temperature reading.

**Step 6** Check the air filters and the blower wheel for dirt and obstructions.

> **NOTE**
> Accurate wet-bulb measurements cannot be made in high-velocity duct systems. Entering air wet-bulb measurements can be made near the return air grilles with a sling psychrometer if the system does not contain mixed air inlets, humidifiers, or reheat equipment that could influence the readings.

**Step 7** Check all registers and grilles for blockage by furniture, drapes, or appliances.

**Step 8** Check the main ducts for balancing dampers; place them in the fully open position before starting to balance the duct system.

**Step 9** If airflow appears to be low in a section of the duct run, check for blockage by a duct liner.

**Step 10** Check the ducts for leakage of conditioned air into unconditioned space.

**Step 11** Check the return air ducts for unconditioned air leaking into the return air duct run.

**Step 12** If the system is equipped with an outside air mixing system, check the makeup air and mixed-air percentage.

**Step 13** If the airflow in the system appears to be low, check for dirty filters, blower wheels, or coils, or an iced evaporator.

### 4.4.1 Thermometer Balancing for Heating

When balancing for winter operation, proceed as follows:

**Step 1** Fully open the dampers behind the registers.

**Step 2** Set the balancing dampers in the branches farthest from the blower to the fully open position.

**Step 3** Adjust the thermostat about 2°F above the room temperature at the thermostat. Balancing should be done when the temperature within the space is within 5°F of normal operating temperature.

**Step 4** Operate the system for about 15 minutes before balancing.

**Step 5** Read and record the thermometer reading in each area.

**Step 6** Adjust the balancing damper in the branch supplying the warmest area to throttle the airflow to that area.

**Step 7** Do not reduce the airflow too much at once because the airflow and temperature in the remaining areas will increase. Several adjustments to each damper may be required.

**Step 8** Adjust the remaining dampers. Work from the warmer areas to the cooler areas.

**Step 9** Repeat the balancing process until the temperature in all areas is as near comfort conditions as possible.

**Step 10** Check the temperature rise across the heat exchanger. The temperature rise should not exceed the manufacturer's specifications. If a blower speed adjustment is required because the rise is outside the range, the system must be rebalanced.

### 4.4.2 Thermometer Balancing for Cooling

When balancing for summer operation, proceed as follows:

**Step 1** Position all supply volume dampers in the half-open position. Ensure that all return air grilles are open and not obstructed.

**Step 2** Set the thermostat to call for cooling and allow the system to operate until the room temperature stabilizes. Check to make sure that the blower is delivering airflow near that recommended at the existing external static pressure.

*Step 3* Position all supply volume dampers in the half-open position. Ensure that all return air grilles are open and not obstructed.

*Step 4* Set the thermostat to call for cooling and allow the system to operate until the room temperature stabilizes. Check to make sure that the blower is delivering airflow near that recommended at the existing external static pressure.

*Step 5* When the area thermometers have indicated a stabilized temperature, slightly open the volume damper in those spaces having temperatures higher than the thermostat setting.

*Step 6* Slightly close the volume dampers in those spaces having temperatures lower than the design temperature or thermostat setting.

*Step 7* Repeat Steps 5 and 6 as often as necessary to provide even temperatures within each of the conditioned spaces (±2°F).

### 4.4.3 Thermometer and Velometer Balancing

To balance an air distribution system by the thermometer and velometer method, proceed as follows:

*Step 1* Open the room diffusers or grilles.

*Step 2* Calibrate the thermometers and locate them in the center of each room, about 4' above the floor.

*Step 3* Take several readings of the air velocity from each supply outlet in the building while waiting for the thermometers to stabilize.

## Balancing by Thermometer

A good-quality electronic thermometer should be used for balancing an air system by thermometer. Inexpensive pocket-type dial thermometers are fine for making a quick check to see if temperatures are in the general range, but they lack the accuracy for making precise temperature measurements. Infrared thermometers that measure surface temperatures should also be avoided. For residential air system balancing by thermometer, it is convenient to use a thermistor-type thermometer similar to the one shown here.

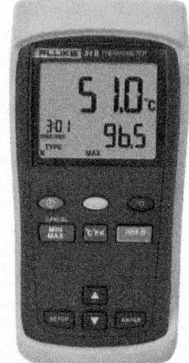
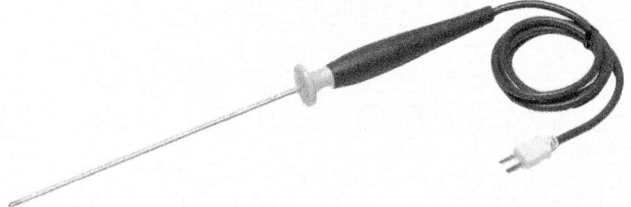

03402-13_SA06.EPS

*Step 4* Make a sketch of the building and locate each supply outlet on the sketch; identify them by a number or symbol.

*Step 5* Record the readings for each outlet as they are taken. Record the velocity readings at each diffuser and identify the section of the diffuser that produced this value.

*Step 6* Take the readings at the same location on the diffuser during succeeding passes.

*Step 7* Take a temperature check after the room temperatures have stabilized within the structure and note which rooms may be overheating and which ones may lack heat.

*Step 8* Reduce the volume controls to each space that is overheating. Recheck the velocity at these diffusers. It is easier to know exactly how much the air volume has been reduced at that diffuser with the use of the velometer.

*Step 9* Make one or more additional passes throughout the structure, rechecking the temperatures and velocities until you are satisfied that the comfort conditioning requirements are being met as closely as possible.

## Additional Resources

*ASHRAE HVAC Applications Handbook*, 2011. Atlanta, GA: American Society of Heating and Air Conditioning Engineers.

*ASHRAE Standard 111-2008 Testing, Adjusting, and Balancing of Building HVAC Systems.* 2008. Atlanta, GA: American Society of Heating and Air Conditioning Engineers

*Testing and Balancing HVAC Air and Water Systems*, 2001. Samuel C. Monger. Lilburn, GA: The Fairmont Press, Inc.

## 4.0.0 Section Review

1. Before altering the airflow in a duct system, consider the factor of air stratification, which can result from _____.
   a. low airflow rates
   b. hot spots near corners
   c. cold spots near corners
   d. high airflow rates

2. Temperature drop measurements are often used to determine _____.
   a. refrigerant charge levels
   b. system airflow volume
   c. system cooling capacity
   d. dehumidification capacity

3. When measuring static pressure in a 8-inch diameter round duct, a 90-degree elbow should be no closer upstream of the pitot tube sampling hole than _____.
   a. 16 inches
   b. 24 inches
   c. 40 inches
   d. 60 inches

4. At what height above the floor should a thermometer be placed when balancing room airflow by thermometer?
   a. 3 feet
   b. 4 feet
   c. 5 feet
   d. 6 feet

# SUMMARY

Psychrometrics is the study of the thermodynamic properties of air and the application of these properties to the environment. Understanding the properties of air enables the HVAC technician to better analyze the operating conditions associated with a selected HVAC system. Gas laws formulated by Dalton, Boyle, and Charles were reviewed in this module. These laws explain basic relationships between the properties of temperature, pressure, and volume as they relate to a confined quantity of air or other gases.

Dry air is a mixture of gases containing mostly nitrogen. However, air in the atmosphere is not dry but contains varying amounts of moisture in the form of water vapor. The percentage of water vapor in the air varies depending on location and other environmental conditions. This air-water mixture is the moist air referred to in the subject of psychrometrics.

The different thermodynamic properties of moist air and how they relate to one another were described in detail in this module. This coverage included how to use and interpret the various scales shown on a psychrometric chart to determine and/or evaluate the different air properties, including:

- Dry-bulb temperature
- Wet-bulb temperature
- Dew point temperature
- Grains of moisture per pound of dry air
- Sensible heat factor
- Enthalpy
- Relative humidity
- Specific volume

Air system balancing involves following a systematic procedure for decreasing and/or increasing the air volume at various points within a duct system when there is too much or too little heating or cooling for a given space. Properly air balancing a system usually requires several passes through the system, during which main and branch duct dampers are adjusted to make sure that all delivered air volumes fall within the design specifications. Air balancing often requires that system blower speed be adjusted to achieve the total needed volume of system airflow, accompanied by acceptable system static pressures.

Balancing of an air distribution system should never be attempted if the system installation is not complete or prior to pre-balance checks having been performed. Air balancing may be accomplished in various ways, depending on the type of equipment or system involved and its application. Regardless of the balancing method used, a system is typically considered balanced when the value of the air quantity of each inlet or outlet is measured and found to be within ±10 percent of the design air quantities.

1. An atmospheric pressure of 0 psia is equal to a pressure of _____.
   a. 0 psig
   b. 30 in Hg vacuum
   c. 30 in Hg
   d. 0 in Hg vacuum

2. In Denver, Colorado, elevation 5,500', the approximate cfm/ton needed to satisfy cooling loads is _____.
   a. 400
   b. 440
   c. 450
   d. 500

3. Within an air distribution system, the highest pressure level is found at the _____.
   a. conditioned space
   b. input to the return duct
   c. output of the blower
   d. input to the blower

4. With a constant temperature, the pressure of gas in a container will _____.
   a. increase if the volume increases
   b. stay the same if the volume increases or decreases
   c. increase if the volume decreases
   d. decrease if the volume decreases

5. With a constant volume, the pressure of gas in a container will _____.
   a. decrease if the temperature is decreased
   b. decrease if the temperature is increased
   c. remain the same if the temperature is increased or decreased
   d. increase if the temperature is decreased

6. The condition in which the dry-bulb, wet-bulb, and dew point temperatures are all the same is when the _____.
   a. enthalpy is at its lowest value
   b. relative humidity is 50 percent
   c. air is considered standard air
   d. relative humidity is 100 percent

7. Plotting a 95°F dry-bulb and a 75°F wet-bulb temperature on a psychrometric chart shows that the relative humidity is _____ .
   a. 30 percent
   b. 40 percent
   c. 50 percent
   d. 60 percent

8. Plotting a 95°F dry-bulb and a 75°F wet-bulb temperature on a psychrometric chart shows that the dew point is _____.
   a. 62°F
   b. 65°F
   c. 67°F
   d. 72°F

9. Plotting a 95°F dry-bulb and a 75°F wet-bulb temperature on a psychrometric chart shows that the enthalpy of the air is _____.
   a. 24.7 Btu/lb
   b. 38.0 Btu/lb
   c. 38.6 Btu/lb
   d. 40.6 Btu/lb

10. Primary air is the _____.
    a. air delivered to a room from the supply duct
    b. same as induced air
    c. air in the heat exchanger
    d. induced flow of room air caused by the airflow from an outlet

11. The term that describes the distance from the center of an outlet to a point in the mixed airstream where the highest sustained velocity has been reduced to a specified level is called the _____.
    a. drop
    b. throw
    c. spread
    d. terminal velocity

12. The static, velocity, and total pressures measured in an air duct system are measured in _____.
    a. psi
    b. inches mercury
    c. inches of water column
    d. inches mercury vacuum

13. Inclined-tube manometers are more precise than U-tube manometers.

    a. True
    b. False

14. When using a manometer and pitot tube to measure velocity pressures in a duct system, the pitot tube tip should be pointed _____.

    a. at right angles to the duct airstream
    b. directly into the duct airstream
    c. away from the duct airstream
    d. in any direction you choose

15. A velometer that gives a direct reading of air velocity when held directly against the surface of a diffuser or register is called a(n) _____.

    a. rotating vane velometer
    b. hot wire anemometer
    c. swinging vane velometer
    d. air volume balancer

16. Fan Law 3 applies to the calculation of _____.

    a. horsepower
    b. static pressure
    c. cfm
    d. rpm

17. If a typical air distribution system fan is running at 900 rpm at a static pressure of 1.2 in wc, what is the airflow in cfm? Hint: Use the fan curve chart shown in *Figure 32*.

    a. 6,250
    b. 7,500
    c. 8,000
    d. 8,500

18. Assume that an existing supply fan is running at a speed of 1,000 rpm with a motor drive pulley that has an 11" pitch diameter. What is the approximate pitch of the drive pulley needed to change the fan speed to 900 rpm?

    a. 9.9"
    b. 11.6"
    c. 12.2"
    d. 13.8"

19. When making traverse readings on a 16" round duct, how many readings should be taken on each diameter?

    a. four
    b. six
    c. eight
    d. ten

20. If the average velocity measured across a grille is 800 fpm, and the area factor ($A_k$) for the grille is 1.5, then the actual airflow (in cfm) is _____.

    a. 533.3
    b. 801.50
    c. 950
    d. 1,200

# Trade Terms Introduced in This Module

$A_k$ factor: The actual free area of an air distribution outlet or an inlet stated in square feet. The terminal device manufacturer specifies this value.

Atmospheric pressure: The pressure exerted on all things on the surface of the earth as the result of the weight of the atmosphere.

Boyle's law: With a constant temperature, the pressure on a given quantity of confined gas varies inversely with the volume of the gas. Similarly, at a constant temperature, the volume of a given quantity of confined gas varies inversely with the applied pressure.

Charles' law: With a constant pressure, the volume for a given quantity of confined gas varies directly with its absolute temperature. Similarly, with a constant volume of gas, the pressure varies directly with its temperature.

Dalton's law: The total pressure of a mixture of confined gases is equal to the sum of the partial pressures of the individual component gases. The partial pressure is the pressure that each gas would exert if it alone occupied the volume of the mixture at the same temperature.

Dew point: The temperature at which water vapor in the air becomes saturated and starts to condense into water droplets.

Drop: The vertical distance that the lower edge of a horizontally projected airstream falls or rises.

Entrained air: Also known as secondary air; the induced flow of room air by the primary air from an outlet.

Induction unit system: An air conditioning system that uses heating/cooling terminals with circulation provided by a central primary air system that handles part of the load, instead of a blower in each cabinet. High-pressure air (primary air) from the central system flows through nozzles arranged to induce the flow of room air (secondary air) through the unit's coil. The room air is either cooled or heated at the coil, depending on the season. Mixed primary and room air is then discharged from the unit.

Primary air: In the context of air distribution and balancing, it is the air delivered to the room or conditioned space from the supply duct.

Psychrometrics: The study of air and its properties.

Specific density: The weight of one pound of air. At 70°F at sea level, one pound of dry air weighs 0.075 pound per cubic foot.

Specific volume: The space one pound of dry air occupies. At 70°F at sea level, one pound of dry air occupies a volume of 13.33 cubic feet.

Spread: The horizontal divergence of an airstream after it leaves the outlet.

Throw: The throw of an outlet is measured in feet and is the distance from the center of the outlet to a point in the mixed airstream where the highest sustained velocity (fpm) has been reduced to a specified level, usually 50 fpm.

Total heat: Sensible heat plus latent heat.

Total pressure: The sum of the static pressure and the velocity pressure in an air duct. It is the pressure produced by the fan or blower.

Traverse readings: A series of velocity readings taken at several points over the cross-sectional area of a duct or grille.

Velocity pressure: The pressure in a duct that results from the movement of the air. It is the difference between the total pressure and the static pressure.

EXAMPLES OF COMPLETED
AIR BALANCE FORMS

Mailing Address
P.O. Box 5782
Greenville, SC 29606-5782

165 South Hammett Road
Greer, SC 29651

Phone (864)877-6832

Fax (864)877-5490

email palmettoab@aol.com

**Palmetto Air&Water**
B A L A N C E
Raising the level of efficiency and comfort

## AIR APPARATUS TEST REPORT

| | |
|---|---|
| PROJECT: | CHARTER COMMUNICATIONS |
| LOCATION: | 1ST FLOOR MECHANICAL ROOM |

SYSTEM/UNIT: SCU-1

### UNIT DATA

| | |
|---|---|
| Make | TRANE |
| Model No. | SCWFN5242DOAo11 |
| Serial No. | T02C18736 |
| Arr./Class | DRAW THRU/1 |
| Discharge | TOP VERITCAL |
| Sheave Diameter | 2B5V124 |
| Sheave Bore | B 2.44 |
| No. Belts/make/size | 2/GOODYEAR/5VX630 |
| No. Filters/type/size | 2/TA/20X20X2  4/TA/16X20X2 |
| | 4/TA/20X25X2  8/TA/16X25X2 |

### MOTOR DATA

| | |
|---|---|
| Make/Frame | BALDOR/254T |
| H.P. | 15 |
| RPM | 1760 |
| Voltage | 460 |
| Phase/Hertz | 3 / 60 |
| F.L. Amps | 20.30 |
| S.F. | 1.25 |
| Sheave Diameter | 2B5V52 |
| Sheave Bore | B 1.63 |
| Sheave Center Distance | 7.25 |

| TEST DATA | DESIGN | ACTUAL |
|---|---|---|
| Total CFM | 17645 | 18235 |
| Fan RPM | | 648 |
| Motor Volts | | 486 |
| Motor Volts | | 486 |
| Motor Volts | | 486 |
| Motor Amps T-1 | | 21.7 |
| Motor Amps T-2 | | 21.7 |
| Motor Amps T-3 | | 21.7 |
| C.F.L.A. | | 19.2 |
| B.H.P. | | 16.9 |

| TEST DATA | DESIGN | ACTUAL |
|---|---|---|
| Make/Frame | | |
| H.P. | | 2.26 |
| RPM | | −0.60 |
| Voltage | | 2.83 |
| Phase/Hertz | | |
| F.L. Amps | | |
| S.F. | | |
| | | 2.2 |
| Sheave Diameter | | |
| Sheave Bore | | |
| Sheave Center Distance | | |

Remarks:
| | |
|---|---|
| Filename: | CCG047 |
| Test Date: | Jan–03 |
| Readings By: | GF |

03402-13_A01A.EPS

Mailing Address
P.O. Box 5782
Greenville, SC 29606-5782

165 South Hammett Road
Greer, SC 29651

Phone (864)877-6832

Fax (864)877-5490

email palmettoab@aol.com

## Palmetto Air&Water
### BALANCE
*Raising the level of efficiency and comfort*

## RECTANGULAR DUCT TRAVERSE TEST REPORT

PROJECT:  CHARTER COMMUNICATIONS

SYSTEM/ UNIT: SCU-1

Zone: DT-1 SUPPLY

Correction Factor:  1.00

| Width: | 30" | Duct Area | | | | | | | | | | Actual | | |
|--------|-----|-----------|---|---|---|---|---|---|---|---|---|--------|---|---|
| Height: | 22" | 4.583 Sq Ft | | | | | | | | | Vel. | 1640 CFM | 7516 | |

| Position | 1 | 2 | 3 | 4 | 5 | 6 | 7 | 8 | 9 | 10 | 11 | 12 | 13 | 14 |
|----------|------|------|------|------|------|------|---|---|---|----|----|----|----|----|
| 1 >>> | 1487 | 1689 | 1893 | 1818 | 1716 | 1933 | | | | | | | | |
| 2 >>> | 1690 | 1656 | 1505 | 1614 | 1730 | 1833 | | | | | | | | |
| 3 >>> | 1619 | 1589 | 1539 | 1416 | 1472 | 1573 | | | | | | | | |
| 4 >>> | 1713 | 1381 | 1467 | 1740 | 1673 | 1610 | | | | | | | | |
| 5 >>> | | | | | | | | | | | | | | |
| 6 >>> | | | | | | | | | | | | | | |
| 7 >>> | | | | | | | | | | | | | | |
| 8 >>> | | | | | | | | | | | | | | |
| 9 >>> | | | | | | | | | | | | | | |
| 10 >>> | | | | | | | | | | | | | | |
| Subtotal | 6509 | 6315 | 6404 | 6588 | 6591 | 6949 | 0 | 0 | 0 | 0 | 0 | 0 | 0 | 0 |

| Total FPM / | Number of Readings = | Average Velocity | X Duct Area = | CFM | Final S.P. |
|-------------|----------------------|------------------|---------------|------|------------|
| 39356 | 24 | 1640 | 4.583 | 7516 | 1.31 |

Remarks:              *SUPPLY CFM TOTAL = DT-1 + DT-2
Filename:             CCG049
Test Date:            Jan–03
Readings By:          GF

03402-13_A01B.EPS

**Palmetto Air&Water**
B A L A N C E
*Raising the level of efficiency and comfort*

## RECTANGULAR DUCT TRAVERSE TEST REPORT

| | | | |
|---|---|---|---|
| PROJECT: | CHARTER COMMUNICATIONS | Zone: DT-2 SUPPLY | |
| SYSTEM/ UNIT: SCU-1 | | Correction Factor: | 1.00 |

| Width: | 34" | Duct Area | | | | Actual | | |
|---|---|---|---|---|---|---|---|---|
| Height: | 24" | 5.667 Sq Ft | | | | Vel. | 1892 CFM | 10719 |

| Position | 1 | 2 | 3 | 4 | 5 | 6 | 7 | 8 | 9 | 10 | 11 | 12 | 13 | 14 |
|---|---|---|---|---|---|---|---|---|---|---|---|---|---|---|
| 1 >>> | 1314 | 2351 | 2436 | 2231 | 1348 | −412 | | | | | | | | |
| 2 >>> | 2745 | 2551 | 2260 | 2308 | 1619 | 335 | | | | | | | | |
| 3 >>> | 2585 | 2337 | 2368 | 2419 | 1572 | 461 | | | | | | | | |
| 4 >>> | 2537 | 2423 | 2509 | 2479 | 1825 | 796 | | | | | | | | |
| 5 >>> | | | | | | | | | | | | | | |
| 6 >>> | | | | | | | | | | | | | | |
| 7 >>> | | | | | | | | | | | | | | |
| 8 >>> | | | | | | | | | | | | | | |
| 9 >>> | | | | | | | | | | | | | | |
| 10 >>> | | | | | | | | | | | | | | |
| Subtotal | 9181 | 9662 | 9573 | 9437 | 6364 | 1180 | 0 | 0 | 0 | 0 | 0 | 0 | 0 | 0 |

| Total FPM | / Number of Readings | = Average Velocity | X Duct Area | = CFM | Final S.P. | |
|---|---|---|---|---|---|---|
| 45397 | | 24 | 1892 | 5.667 | 10719 | 1.62 |

Remarks:
Filename:           CCG050
Test Date:          Jan–03
Readings By:        GF

03402-13_A01C.EPS

Mailing Address
P.O. Box 5782
Greenville, SC 29606-5782

165 South Hammett Road
Greer, SC 29651

Phone (864)877-6832

Fax (864)877-5490

email palmettoab@aol.com

**Palmetto Air&Water**
B A L A N C E
Raising the level of efficiency and comfort

## VAV TEST REPORT

PROJECT: CHARTER COMMUNICATIONS
SYSTEM/ UNIT: SCU-1

| Box Number | Type | Design CFM | All Boxes Maxed | Damper Position | Flow Corr |
|---|---|---|---|---|---|
| FPB-1-1 | SUPPLY | 1015 | 1006 | 67% | 0.729 |
| FPB-1-2 | SUPPLY | 360 | * | 0% | 0.579 |
| FPB-1-3 | SUPPLY | 1980 | 1960 | 66% | 0.647 |
| FPB-1-5 | SUPPLY | 2400 | 2424 | 65% | 0.753 |
| FPB-1-7 | SUPPLY | 2400 | 2156 | 100% | 0.630 |
| FPB-1-8 | SUPPLY | 1465 | 1478 | 70% | 0.587 |
| FPB-1-9 | SUPPLY | 1550 | 1414 | 100% | 0.597 |
| FPB-1-10 | SUPPLY | 1550 | 1630 | 67% | 0.604 |
| VAV-1-2 | SUPPLY | 1005 | 1020 | 55% | 0.737 |
| VAV-1-3 | SUPPLY | 1600 | 1670 | 100% | 0.755 |
| VAV-1-4 | SUPPLY | 360 | 352 | 54% | 0.732 |
| VAV-1-5 | SUPPLY | 1615 | 1540 | 100% | 0.462 |
| VAV-1-6 | SUPPLY | 750 | * | 0% | 0.728 |
| VAV-1-7 | SUPPLY | 730 | 710 | 65% | 0.799 |

TOTAL

Remarks: *CLOSED FOR DIVERSITY
Filename: CCG048
Test Date: Jan–03
Readings By: GF

03402-13_A01D.EPS

Mailing Address
P.O. Box 5782
Greenville, SC 29606-5782

165 South Hammett Road
Greer, SC 29651

Phone (864)877-6832

Fax (864)877-5490

email palmettoab@aol.com

# Palmetto Air&Water

**B A L A N C E**

Raising the level of efficiency and comfort

### AIR OUTLET TEST REPORT

PROJECT:  CHARTER COMMUNICATIONS
SYSTEM/ UNIT:  SCU-1

| Area Served | Number | Type | Design CFM | Preliminary CFM | Final CFM | Percent Design |
|---|---|---|---|---|---|---|
| 128 | 1 | SUPPLY | 800 | 885 | 868 | 109% |
| 131 | 2 | SUPPLY | 800 | 1124 | 852 | 107% |
| 134 | 3 | SUPPLY | 800 | 363 | 720 | 90% |
| FPB-1-7 TOTAL | | | 2400 | 2372 | 2440 | 102% |
| | | | | | | |
| 134 | 1 | SUPPLY | 265 | 156 | 292 | 110% |
| 133 | 2 | SUPPLY | 400 | 416 | 379 | 95% |
| 133 | 3 | SUPPLY | 400 | 411 | 361 | 90% |
| 132 | 4 | SUPPLY | 200 | 266 | 220 | 110% |
| 132 | 5 | SUPPLY | 200 | 309 | 195 | 98% |
| FPB-1-8 TOTAL | | | 1465 | 1558 | 1447 | 99% |
| | | | | | | |
| 135 | 1 | SUPPLY | 280 | 250 | 270 | 96% |
| 135 | 2 | SUPPLY | 280 | 222 | 252 | 90% |
| 135 | 3 | SUPPLY | 280 | 272 | 253 | 90% |
| 135 | 4 | SUPPLY | 355 | 524 | 361 | 102% |
| 135 | 5 | SUPPLY | 355 | 445 | 356 | 110% |
| FPB-1-9 TOTAL | | | 1550 | 1713 | 1492 | 96% |
| | | | | | | |
| 135 | 1 | SUPPLY | 280 | 318 | 279 | 100% |
| 135 | 2 | SUPPLY | 280 | 285 | 272 | 97% |
| 135 | 3 | SUPPLY | 280 | 268 | 255 | 91% |
| 135 | 4 | SUPPLY | 100 | 192 | 105 | 105% |
| 135 | 5 | SUPPLY | 255 | 336 | 237 | 93% |
| 135 | 6 | SUPPLY | 355 | 601 | 363 | 102% |
| FPB-1-10 TOTAL | | | 1550 | 2000 | 1511 | 97% |

| TOTAL |
|---|
| Remarks: |
| Filename:  CCG002 |
| Test Date:  Jan–03 |
| Readings By:  GF |

03402-13_A01E.EPS

# Additional Resources

This module presents thorough resources for task training. The following resource material is suggested for further study.

*ACCA Manual P – Psychrometrics Theory and Applications.* 1983. Air Conditioning Contractors of America.

*ASHRAE HVAC Applications Handbook, 2011.* Atlanta, GA: American Society of Heating and Air Conditioning Engineers.

*ASHRAE Standard 111-2008 Testing, Adjusting, and Balancing of Building HVAC Systems.* 2008. Atlanta, GA: American Society of Heating and Air Conditioning Engineers.

*Testing and Balancing HVAC Air and Water Systems,* 2001. Samuel C. Monger. Lilburn, GA: The Fairmont Press, Inc.

*Understanding Psychrometrics, Second Edition.* 2005. American Society of Heating, Refrigeration, and Air Conditioning Engineers.

# Figure Credits

Courtesy of TSI/Alnor Instruments, Module opener, Figures 28A and B, 29 (photo), 30, 43

Courtesy of Extech Instruments, a FLIR Company, Figures 12, SA03, SA04

Courtesy of Honeywell International, SA01

Courtesy of Hands Down Software, www.hands-downsoftware.com, Figure 14

Carnes Company, Figures 24, 41, 42

Courtesy of Dwyer Instruments, Inc., Figures 25, 26

Fluke Corporation, reproduced with permission, SA06

Palmetto Air & Water Balance, Inc., Appendix

# Section Review Question Answers

| Answer | Section Reference | Objective |
|---|---|---|
| **Section One** | | |
| 1.a | 1.1.0 | 1a |
| 2.b | 1.2.2 | 1b |
| **Section Two** | | |
| 1.b | 2.1.1 | 2a |
| 2.a | 2.2.8 | 2b |
| 3.c | 2.3.3 | 2c |
| **Section Three** | | |
| 1.b | 3.1.0 | 3a |
| 2.c | 3.2.3 | 3b |
| 3.c | 3.3.3 | 3c |
| **Section Four** | | |
| 1.a | 4.1.0 | 4a |
| 2.c | 4.2.2 | 4b |
| 3.d | 4.3.1 | 4c |
| 4.b | 4.4.0 | 4d |

# NCCER CURRICULA — USER UPDATE

NCCER makes every effort to keep its textbooks up-to-date and free of technical errors. We appreciate your help in this process. If you find an error, a typographical mistake, or an inaccuracy in NCCER's curricula, please fill out this form (or a photocopy), or complete the online form at **www.nccer.org/olf**. Be sure to include the exact module ID number, page number, a detailed description, and your recommended correction. Your input will be brought to the attention of the Authoring Team. Thank you for your assistance.

*Instructors* – If you have an idea for improving this textbook, or have found that additional materials were necessary to teach this module effectively, please let us know so that we may present your suggestions to the Authoring Team.

**NCCER Product Development and Revision**
13614 Progress Blvd., Alachua, FL 32615

**Email:**   curriculum@nccer.org
**Online:**   www.nccer.org/olf

❏ Trainee Guide     ❏ Lesson Plans     ❏ Exam     ❏ PowerPoints     Other _____

Craft / Level: _____     Copyright Date: _____

Module ID Number / Title: _____

Section Number(s): _____

Description: _____

_____

_____

_____

Recommended Correction: _____

_____

_____

_____

Your Name: _____

Address: _____

Email: _____     Phone: _____

# System Startup and Shutdown

## OVERVIEW

When boilers, chillers, and other commercial heating or cooling systems are shut down for extended periods and then brought back on line, specific routines must be followed to properly prepare the system for operation. This module provides an overview of the startup and shutdown requirements of these types of equipment; however, the equipment manufacturer's recommended startup and shutdown procedures always take precedence.

## Module 03406

Trainees with successful module completions may be eligible for credentialing through NCCER's National Registry. To learn more, go to **www.nccer.org** or contact us at **1.888.622.3720**. Our website has information on the latest product releases and training, as well as online versions of our *Cornerstone* magazine and Pearson's product catalog.
Your feedback is welcome. You may email your comments to **curriculum@nccer.org**, send general comments and inquiries to **info@nccer.org**, or fill in the User Update form at the back of this module.

This information is general in nature and intended for training purposes only. Actual performance of activities described in this manual requires compliance with all applicable operating, service, maintenance, and safety procedures under the direction of qualified personnel. References in this manual to patented or proprietary devices do not constitute a recommendation of their use.

*03406 V5*

## Objectives

When you have completed this module, you will be able to do the following:

1. Explain how to properly shut down and start up boilers.
   a. Explain how to shut down and prepare boilers for dry storage.
   b. Explain how to shut down and prepare boilers for wet storage.
   c. Explain how to prepare and start up a steam boiler.
   d. Explain how to prepare and start up a hot-water boiler.
2. Explain how to start up and shut down various chillers and water systems.
   a. Explain how to start up and shut down a reciprocating chiller system.
   b. Explain how to start up and shut down a centrifugal or screw chiller system.
   c. Explain how to start up and shut down cooling towers.
   d. Describe the process of inspecting and cleaning various heat exchange components.
3. Explain how to start up and shut down air handling and packaged rooftop systems.
   a. Explain how to start up and shut down air handling units and their associated air distribution system.
   b. Explain how to start up and shut down packaged rooftop units.

## Performance Tasks

Under the supervision of your instructor, you should be able to do the following:

1. Start up and shut down an air handling unit and prepare it for normal operation.
2. Start up and shut down at least one of the following:
   - Steam boiler
   - Hot-water boiler
   - Reciprocating chiller
   - Screw chiller
   - Centrifugal chiller
   - Cooling tower
   - Evaporative condenser

## Trade Terms

Bleed-off
Deadband
Desiccant

Layup
Recycle shutdown mode

## Industry-Recognized Credentials

If you are training through an NCCER-accredited sponsor, you may be eligible for credentials from NCCER's Registry. The ID number for this module is 03406. Note that this module may have been used in other NCCER curricula and may apply to other level completions. Contact NCCER's Registry at 888.622.3720 or go to **www.nccer.org** for more information.

# Contents

## Figures and Tables

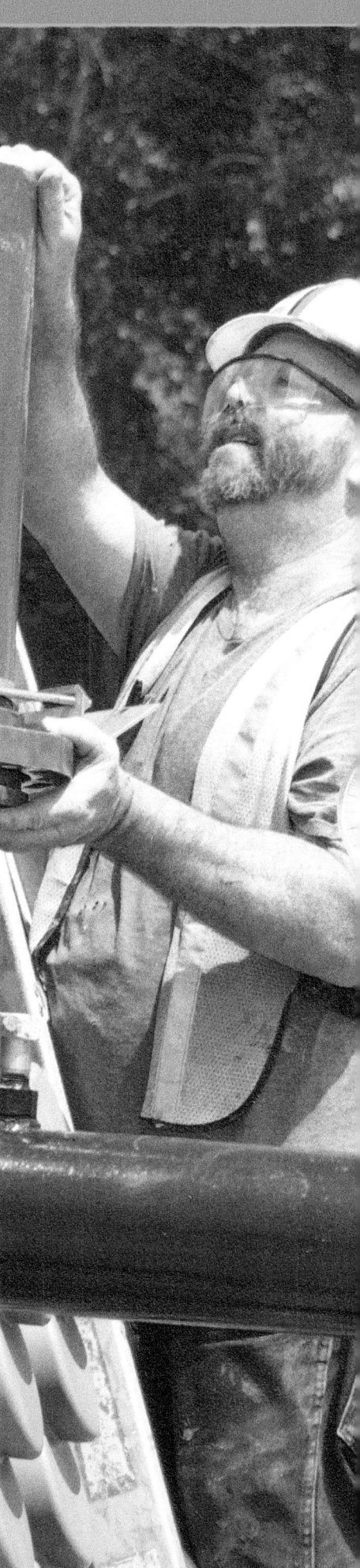

### 1.0.0 BOILER STARTUP AND SHUTDOWN

#### Objectives

Explain how to properly shut down and start up boilers.

   a. Explain how to shut down and prepare boilers for dry storage.
   b. Explain how to shut down and prepare boilers for wet storage.
   c. Explain how to prepare and start up a steam boiler.
   d. Explain how to prepare and start up a hot-water boiler.

#### Performance Task

2. Start up and shut down at least one of the following:
   - Steam boiler
   - Hot-water boiler
   - Reciprocating chiller
   - Screw chiller
   - Centrifugal chiller
   - Cooling tower
   - Evaporative condenser

#### Trade Terms

**Desiccant**: A moisture-absorbing material.

**Layup**: An industry term referring to the period of time a boiler is shut down.

This section describes the tasks needed to prepare larger commercial/industrial steam and hot-water heating boilers for startup and shutdown.

> A boiler should be shut down as directed in the boiler manufacturer's instructions. Do not operate, repair, or dismantle a boiler until the manufacturer's service literature has been read and understood.

### 1.1.0 Shutdown for Dry Storage

Many large boilers used mainly for heating or as standby units may have extended periods of time when they are not used. When idle, these boilers require special attention to make sure that

their waterside and fireside components are not allowed to deteriorate from corrosion and other problems. Unless proper procedures are followed before taking boilers off line, severe corrosion may occur. Oxygen from the atmosphere can enter an idle boiler and combine with condensed moisture to produce extensive pitting of metal surfaces. The proper preparation of a boiler for shutdown is very important. If a boiler is not prepared properly for shutdown, more damage can happen in a month of sitting idle than during an entire season of heating operation.

The maintenance actions described in this module are general in nature. Remember that the startup or shutdown of a specific piece of equipment should always be performed per the manufacturer's service manual. The specific shutdown procedures for different gas-fired and oil-fired boilers vary, but the tasks and their sequences are essentially the same. The shutdown procedure typically used with an oil-fired steam boiler (*Figure 1*) is outlined here.

#### 1.1.1 Boiler Shutdown Concerns

One area of concern during the prolonged shutdown (layup) of an oil-fired boiler is the damage from corrosion caused by the sulfur content of fuel oils in contact with moisture or other residues. During layup, the fireside components are exposed to moisture caused by the condensation of air as it cools below its dew point. This moisture and any sulfur residue form an acid that attacks the boiler metals. With high-humidity conditions, the corrosive effect of the acid can be serious enough to erode through or damage the boiler tubes and other heating surfaces. This acid/moisture condition usually does not exist during normal boiler operation because the high operating temperatures vaporize any condensation.

Corrosion that could occur during the layup period can be greatly reduced by thoroughly cleaning the soot and other products of combustion from the boiler's fireside components immediately after shutdown. Following this cleaning, any remaining moisture should be removed by drying the cleaned areas with a heater. When shutting down a boiler, reduce the load slowly so that the boiler cools at a rate that avoids stresses caused by damaging temperature differentials, also called thermal shock.

#### 1.1.2 Boiler Preparation for Dry Storage

Dry storage of a boiler is typically done when the boiler is subjected to freezing temperatures or a humid environment, or when it will be out of ser-

*Figure 1* Horizontal firetube steam boiler.

03406-13_F01.EPS

vice for more than a month. Dry storage involves completely cleaning the waterside and fireside components and surfaces to remove all deposits, scale, and soot. After cleaning, the fireside surfaces are coated with an anticorrosive compound, and heaters are used to dry any remaining moisture within the boiler. Be careful not to overheat the boiler while drying. Only use enough heat to dry it completely. A moisture-absorbing material known as a desiccant is then placed inside, and the boiler is sealed to maintain a low level of moisture during storage. The specific procedures for the dry storage of the different types of gas-fired and oil-fired boilers vary, but the tasks and their sequences are essentially the same. The following procedure outlines the preparation of a typical oil-fired steam boiler for dry storage.

> **WARNING!**
> Preparing a boiler for storage involves working on energized equipment containing pressurized and/or hot components. Follow all manufacturer's instructions, OSHA regulations, and job-site requirements relating to safety and the use of safety equipment.

*Step 1*  Shut down the boiler per the manufacturer's instructions. General instructions are outlined as follows:

    a. Lock out and tag the boiler power and controls.

    b. Shut off and secure the fire.

    c. After shutting off the fire, cut down on the draft to keep the refractory or brickwork from cooling too fast. Rapid cooling can cause flaking or corrosion.

    d. Watch the water level.

    e. After the boiler has stopped steaming, close the main stop valve.

    f. After the pressure has dropped to roughly 1-5 psig, carefully open the air cock so that no vacuum forms in the boiler as it continues cooling. In extreme cases, vacuum can cause boiler tubes/shells to collapse.

    g. Do not immediately close the main steam valve or the safety valve may still pop open.

> **WARNING!**
> Failure to safely discharge water or steam can result in serious burns, eye injury, or equipment damage. Piping connected to safety, blowdown, and other drain valves must be routed so that the flow of the discharged wastewater or steam goes into the local sewer system or other safe disposal area. Do not drain the boiler until all the pressure is relieved.

## Vapor Phase Corrosion Inhibitors

Controlling metal corrosion in enclosed spaces (a boiler in dry storage) has always been a challenge. The use of volatile corrosion inhibitors, often called vapor phase corrosion inhibitors, offers new ways to control boiler corrosion. These corrosion inhibitors are placed in a confined space and emit a gas that bonds to metal to form a corrosion-resistant coating. One vapor phase corrosion inhibitor product currently available to protect boilers during shutdown (called layup powder) can be applied either wet or dry.

**Step 2** After the boiler is completely cooled, drain the boiler as directed in the manufacturer's service literature. Make sure you follow all federal, state, and local laws, rules, and regulations that govern the discharge of the wastewater into the environment.

**Step 3** After the boiler is drained, flush out the waterside with fresh high-pressure water.

**Step 4** Thoroughly clean the fireside surfaces of all soot and deposits from combustion. Brush out and/or vacuum up the loosened materials. Some oil-fired boilers have a water washing device used to clean convection surfaces. Use this device as instructed by the boiler manufacturer.

> **WARNING!**
>
> Make sure that all steam and other system valves, blow-down valves, and electrical switches are turned off before opening boiler hand-holes and manhole covers. Always adequately vent the boiler before entering and apply lockout/tagout procedures.

**Step 5** Inspect all the fireside metal surfaces for damage or corrosion.

**Step 6** Remove any scale or deposits from the waterside surfaces. Check for internal leakage or corrosion.

**Step 7** Check and clean the following:
- Low-water cutoff piping
- Water level controls and cross piping connections
- Blow-down piping, valves, and drain

**Step 8** Check all water and steam piping, valves, and other components for leaks, wear, corrosion, or other damage. Replace or repair any components, if needed.

**Step 9** Brush the refractories clean and inspect for damage. If cracks over ¼-inch wide exist, clean and fill them with high-temperature bonding mortar.

**Step 10** Wash and coat the refractories using the material recommended by the boiler manufacturer. Often this is a high-temperature air-dry mortar diluted with water to the thickness of light cream.

**Step 11** Coat the fireside surfaces with an anticorrosive material recommended by the boiler manufacturer.

**Step 12** After making sure that the boiler is dry and the fireside is properly coated with anticorrosive, place a desiccant inside as recommended by the boiler manufacturer. Some boiler manufacturers recommend using the following:
- Quick lime at 2 pounds per 3 cubic feet of volume
- Silica gel at 5 pounds per 30 cubic feet of volume
- Calcium chloride at 3 pounds per 100 square feet of surface area

Put the desiccant in half-filled trays to allow room for the water absorbed by the desiccant. Using the boiler access manholes and hand-holes, place half the amount of desiccant inside the firebox and the other half on top of the tubes.

**Step 13** Close and seal all boiler openings, including hand-holes and manholes. Where needed, use new gaskets. Close all feedwater and steam valves. Close the dampers and vents to prevent air from reaching the fireside surfaces of the boiler.

**Step 14** Maintain lockout/tagout so that the unit cannot be started during shutdown.

**Step 15** At six-week intervals, open the boiler hand-holes and manholes and inspect the desiccant in the boiler. Renew any desiccant that is saturated with moisture, or regenerate the desiccant as the manufacturer recommends.

### 1.2.0 Boiler Preparation for Wet Storage

Wet storage of a boiler is typically done when the boiler is held in standby, when storage temperatures will remain above freezing, or when dry storage is impractical for other reasons. The specific procedures for the wet storage of the different kinds of gas-fired and oil-fired boilers vary, but the tasks and their sequences are essentially the same. The following procedure outlines the preparation for wet storage of a typical oil-fired steam boiler:

**Step 1** Drain and clean the boiler according to Steps 1 through 10 of the dry storage procedure.

### Desiccant

Some desiccants used for boiler dry storage change color to indicate that it is time for replacement.

**Step 2** Valve the boiler off from the rest of the system, and then refill it to overflowing with treated water. If deaerated water is not available, fire the boiler for a short time until the water reaches a temperature of about 200°F to drive off most of the dissolved gases. Let the boiler water cool down to room temperature, and then add water until it overflows the top.

**Step 3** Add the water treatment chemicals needed to condition the added amount of water that is now above the boiler's normal water line. Close all boiler openings. The water should be circulated in the boiler to prevent stratification and to make sure the added chemicals are thoroughly mixed in order to be in contact with all surfaces.

**Step 4** Maintain an internal water pressure above atmospheric pressure. Nitrogen gas is sometimes used for this purpose.

**Step 5** Protect the exterior surfaces and components from rust by coating them with mineral oil or boiler paint. Cover all tube and firebox surfaces with a coating of mineral oil or other rust inhibitor.

**Step 6** Keep the control circuit energized to prevent condensation from forming in the control cabinet or on the flame safeguard control device.

**Step 7** Leave the flue and firebox doors wide open during the period of shutdown. This helps to keep them dry.

**Step 8** Keep the boiler room dry and well ventilated.

### 1.3.0 Steam Boiler System Startup for Normal Operation

A steam boiler should be started up as directed in the manufacturer's service instructions. The specific startup procedures for the different kinds of gas-fired and oil-fired boilers vary, but the tasks and their sequences are essentially the same. A typical startup procedure for an oil-fired steam boiler is outlined here.

> **WARNING!**
> Preparing a boiler for startup involves working on energized equipment containing pressurized and/or hot components. Follow all manufacturer's instructions, OSHA regulations, and job-site requirements relating to safety and the use of safety equipment.

**Step 1** If not previously done, inspect the steam piping system to verify the following:
  a. All strainers have been cleaned.
  b. The settings of all safety valves are correct.
  c. All manual and automatic valves are in the required positions.

**Step 2** Check that all tools, equipment, desiccant trays, or other debris are removed from the boiler and firebox. Make sure that all boiler openings, including hand-holes and manholes, are properly secured.

**Step 3** Inspect the boiler and accessories to verify the following:
  a. Fuel and electrical power are supplied to the boiler.
  b. Ensure that the manual reset buttons of all starters and other controls have been reset.
  c. The linkages of all dampers, metering valves, and cams have full stroke and free motion.
  d. The direction of rotation for all motors is correct when momentarily energizing the control switch.
  e. The pressure control settings are set slightly above the highest steam pressure needed. This setting must be at least 10 percent below the setting of the safety valve(s). On a low-pressure boiler, this setting is typically two or three psig above the operating limit.
  f. The float in the low-water cutoff and pump controls can move freely. Also, make sure the controls are level and the piping is plumb.
  g. Discharge piping connected to safety, blowdown, and other drain valves is routed so that the flow of the wastewater or steam goes into the local sewer system or other safe disposal area.

## Boiler Water Treatment Chemicals

Always check with the boiler manufacturer to make sure that any water treatment chemical used is compatible with their product. The wrong chemical could damage the boiler and/or void any warrantees on the boiler.

h. If not a new startup, replace the oil burner nozzle. This is typically done on boilers with nozzles less than 5 gph in size, which are used only in large commercial and industrial applications. These larger nozzles are far less likely to clog or foul. Make sure that the electrode gap and positioning of the electrode relative to the nozzle are correct.

i. All combustion openings and barometric or draft control dampers for the boiler are the proper size for the fuel being used.

j. Any feedwater treatment equipment, such as filters, chemical feeders, demineralizers, softeners, and deaerators, are operational and prepared for use.

> **CAUTION**
> If starting up a newly installed boiler or one that has been in dry storage, do not attempt to start up the boiler until the boiler has first been thoroughly cleaned to remove any accumulated dirt and oil.

*Step 4*  Make sure that the main steam stop valve is closed.

*Step 5*  If not previously done, fill the boiler with properly treated water up to the normal operating level. It is generally recommended, and possibly required, to fill the boiler with water that is at least 70°F; cold water should not be used in a steam boiler. Vent the boiler so that no air is trapped in the steam space.

> **WARNING!**
> Never stand in front of a burner or boiler during startup. Accumulated oil (or gas) vapors present in the combustion chamber or flue passages can explode. If the burners fail to ignite, allow enough time for the boiler's burner and control system to operate and purge accumulated vapors from the boiler system before attempting to relight the boiler's burners.

*Step 6*  Set the boiler controls for a cold startup, and then fire the boiler as directed by the manufacturer's operating instructions. On initial startup, it may be necessary to attempt igniting the burners several times in order to bleed air from the main and/or pilot fuel lines.

*Step 7*  The boiler burner and control system should allow the boiler water to warm up slowly via the use of a burner low-fire flame. In some boilers, this may be done automatically. In others, it is done manually by setting the Manual/Automatic control to the Manual position and the Manual Flame control to the Low-Fire position. During the boiler warm-up, monitor the boiler water level frequently to make sure that it stays at the normal operating level.

*Step 8*  After the boiler water is thoroughly warmed, the boiler burner and control system should function to switch from the low-fire to high-fire mode of burner operation. In non-automatic boilers, this is done manually by setting the Manual Flame control to the High-Fire position.

*Step 9*  When the steam gauge shows pressure in the boiler, blow down the gauge glass, water column, and low-water cutoff as applicable.

*Step 10*  Use a combustion test kit to perform combustion efficiency and analysis tests. Make these tests and any related burner and/or draft regulator adjustments as instructed by the test instrument and boiler manufacturer's service literature.
- Oxygen ($O_2$)
- Carbon dioxide ($CO_2$)
- Excess air
- Stack temperature
- Carbon monoxide ($CO$)
- Nitrous oxide ($NO_X$)
- Other gases

*Step 11*  When the boiler reaches about two or three pounds below the steam header pressure, very slowly open the main boiler stop valve until it is fully open. If opened too quickly, the header can rapidly lose pressure. Also, by opening the valve slowly, carryover can be prevented.

*Step 12*  With the boiler switched to the automatic mode of operation, allow it to operate at normal pressures and temperatures until it is shut down normally by the operation of the boiler burner control system.

*Step 13*  Monitor the boiler through enough cycles of operation to make sure that the boiler, including ignition and control program sequences, is functioning correctly.

**Step 14** Confirm that the safety controls and protective circuits listed below are functioning properly as directed in the manufacturer's service instructions.

- Pressure control
- Electronic and/or mechanical low-water cutoff
- Flame sensor
- Flame failure
- Ignition failure
- Power failure

> **NOTE**
>
> If above-normal amounts of water treatment chemicals were added to a boiler during a period of wet storage, it may be necessary to use more frequent blow-downs when the boiler is returned to service. This is required to reduce the chemicals to normal levels.

**Step 15** Inspect the steam piping system to verify the following:

- The system is free of leaks.
- All pressure-reducing valves/stations are operating properly.
- All steam traps are operating properly on all equipment, at the ends of mains, and at all drip points.

**Step 16** Record boiler and system parameters (*Table 1*).

### 1.3.1 Gas-Fired Steam Boiler Startup Checks

Except for some of the operating and safety controls and their adjustment, the general procedure for the startup of a gas-fired steam boiler is similar to that described for an oil-fired steam boiler. A steam boiler should be started up as directed in the manufacturer's service instructions.

The operation of the boiler controls and safety devices should be checked in accordance with the manufacturer's service literature:

- Automatic gas valve
- Makeup and/or feedwater controls
- Low-water cutoff
- Flow switches
- Pressure controls
- Safety devices

Once the boiler has been turned on, the following combustion parameters should be measured and recorded. If necessary, adjust any controls to meet the required parameter values recommended by the manufacturer.

- Oxygen ($O_2$)
- Carbon dioxide ($CO_2$)
- Carbon monoxide ($CO$)
- Stack temperature
- Stack draft
- Manifold and supply gas pressures

Refer to *Appendix A* for the typical relationships between $O_2$, $CO_2$, and excess air for natural gas and other common fuels. Check the manufacturer's recommendation for excess air. Carbon monoxide ($CO$) should always be minimized. Note that the maximum allowable CO level in flue gas (on a free air basis) is 400 ppm according to the EPA and AGA.

**Table 1** Boiler Data and Operating Parameters

| Boiler Data | |
|---|---|
| Location: _____ | Manufacturer: _____ |
| Model Number: _____ | Serial Number: _____ |
| Type/Size: _____ | Fuel Type: _____ |
| Ignition Type: _____ | Burner Control: _____ |
| Volts/Phase/Hertz: _____ | |
| **Boiler Operating Data** | |
| Pressure/Temperature: _____ | Voltage: _____ |
| No. Safety Valves/Size: _____ | Amperage T1/T2/T3: _____ |
| Safety Valve Setting: _____ | Draft Fan Volts/Amps: _____ |
| Operating Control Setting: _____ | Manifold Pressure: _____ |
| High-Fire Setpoint: _____ | Output-MBH (kW): _____ |
| Low-Fire Setpoint: _____ | Safety Control Check: _____ |

# Combustion Analyzers

Combustion analyzers for commercial and industrial systems usually provide measurements of the stack gas content of oxygen, carbon monoxide, carbon dioxide, oxides of nitrogen, and sulfur dioxide, as well as the stack gas temperature. Some, like the one shown, provide microprocessor sequence of operation while calculating burner efficiency, stack loss, and excess air percentages. Programmed routines lead the operator through the proper procedures, signaling the appropriate action and waiting until the correct steps are taken.

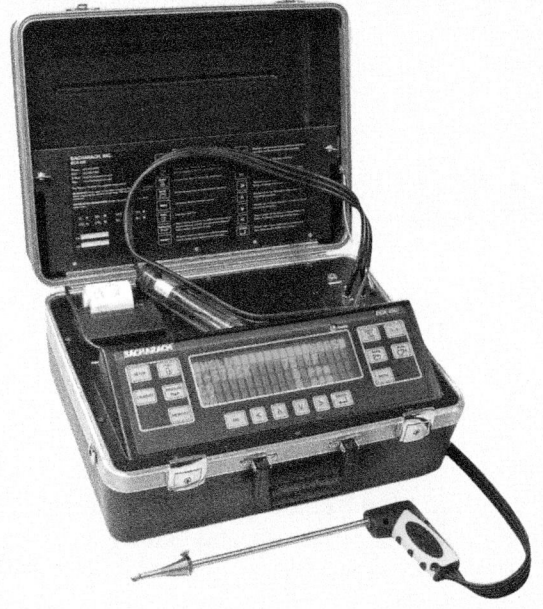

03406-13_SA01.EPS

## 1.4.0 Hot-Water Boiler Startup

Hot-water boilers are constructed and operated similarly to steam boilers, except some of the operating and safety controls used with hot-water boilers are different from those used with steam boilers. Also, hot-water boilers and related system piping are entirely filled with water, while steam boiler systems are not.

A hot-water boiler (*Figure 2*) should be started up as directed in the boiler manufacturer's service instructions.

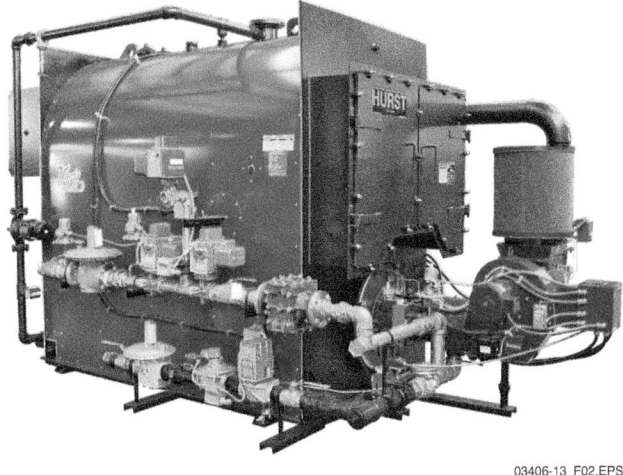

03406-13_F02.EPS

*Figure 2* Hot-water boiler.

> **WARNING!**
>
> Do not attempt to operate a boiler until you have read and understand the manufacturer's operating instructions.

---

## Flue Gas Relationship to Excess Combustion Air

**GOING GREEN**

*Appendix A* shows the typical relationships between $O_2$, $CO_2$, and excess air for fuel oil, natural gas, and other common fuels. As shown in the chart, when the percentage of excess air increases, the percentage of $CO_2$ decreases. If it were possible to have perfect combustion, $CO_2$ would be maximized and $O_2$ would be at (or close to) zero in the flue gas stream. Because perfect combustion is not practically possible, most combustion equipment is set up to have a small percentage of excess $O_2$ present. Check the manufacturer's recommendation for excess air. Carbon monoxide (CO) should always be minimized. Note that the maximum allowable CO level in flue gas (on a free air basis) is 400 ppm according to the EPA and AGA.

---

With the exceptions covered in the following paragraphs, the tasks and sequence for starting gas-fired and oil-fired hot-water boilers are similar to those described for steam boilers.

### 1.4.1 Filling and Venting a Hot-Water System

When filling a hot-water system with treated water, the entire boiler, piping system, and terminals must be filled with water and the manual system air vents opened (if used) to expel any air trapped in the system. If the system uses automatic air vents, they should be checked for leakage. Because a hot-water boiler is completely filled with water, the hot-water outlet usually includes a dip tube that extends 2 or 3 inches into the boiler.

This dip tube traps any air or oxygen that is released from the water during heating at the top of the boiler shell and routes the trapped air into the expansion tank where the air is held.

### 1.4.2 Hot-Water System Operating Temperatures

The minimum boiler water temperature recommended for hot-water boilers is 140°F. At lower temperatures, combustion gases can condense in the fireside of the boiler and cause corrosion. This problem is more severe in systems that are operated intermittently or are greatly oversized for the heating load. The temperature control in a hot-water boiler is typically set between 5°F and 10°F above the boiler's operating limit temperature control setting.

### 1.4.3 Hot-Water Circulation Considerations

The system piping and controls are arranged to prevent the possibility of pumping large volumes of cold water into a hot boiler. If 140°F boiler water is replaced with 80°F water in a short period of time, it causes thermal stress or shock. Thermal shock is the condition in which non-uniform expansion or contraction occurs in the boiler components as a result of sudden water temperature changes. Thermal shock must be prevented after a boiler has been drained, or on the initial firing of a new boiler. This is done by gradually warming the water in the boiler and the water being

## Dual-Rated Pressure Vessels

Boiler manufacturers will often build a single boiler pressure vessel that is pressure-rated for both steam and hot water. When a boiler is built, different operating and safety controls can be added to make that single pressure vessel suitable for either steam or hot water use.

circulated through the piping system. This means that when starting up a hot-water boiler, the entire water content of the boiler and piping system must first be completely warmed at the low-fire level before the fuel input can be switched to the high-fire level. The potential for thermal shock can be reduced by having the circulating pumps interlocked and switched on with the burners, so the burners cannot operate unless the water is being circulated.

During initial startup and subsequent operation, it is important to maintain the pressure/temperature relationships in a hot-water boiler per the boiler manufacturer. *Figure 3* shows the recommended pressure/temperature relationships for one manufacturer's hot-water boilers. The water pressure and temperature gauges on the boiler should indicate values similar to those shown on the chart.

A constant rate of water circulation through a hot-water system helps eliminate water stratification within the boiler and results in more even water temperatures being delivered to the system. In order to establish an accurate water circulation rate for the system, monitor the temperature drop that occurs as the water flows through the system. This can be done by installing a temperature gauge on the water return line, then determining the difference between the supply and return water temperatures.

Boiler manufacturers usually specify a set of recommended minimum and maximum circulating rates based on the system temperature drop and full boiler output. *Figure 4* shows a chart used by one manufacturer to determine the circulation rate in a system based on temperature drop.

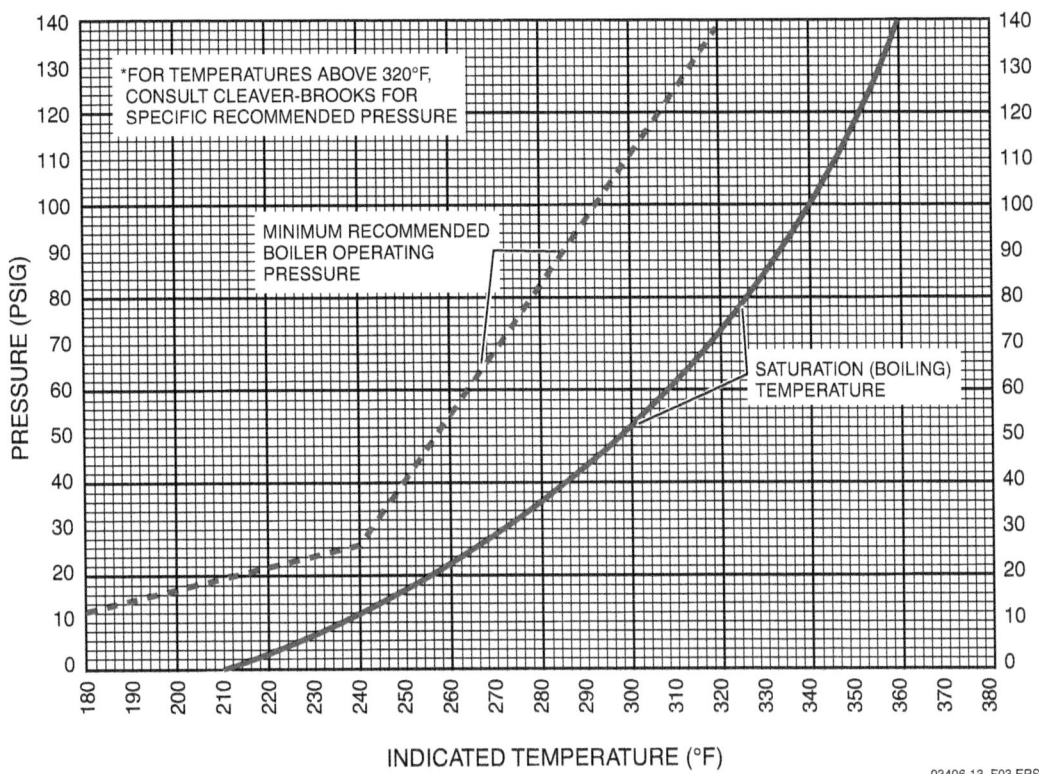

*Figure 3* Typical hot-water boiler pressure/temperature relationships.

# Adjusting for Water Circulation Rate

A common rule of thumb is to adjust the system flow to provide a circulation rate of one-half to one gallon per minute (gpm) per boiler horsepower. One boiler horsepower (BHP or BoHP) is equal to 9.803 kilowatts. Also, it is equal to a heat output of 33,475 Btus per hour.

| Boiler Size (BHP) | Boiler Output (1000) Btu/Hr | System Temperature Drop – °F | | | | | | | | | |
|---|---|---|---|---|---|---|---|---|---|---|---|
| | | 10 | 20 | 30 | 40 | 50 | 60 | 70 | 80 | 90 | 100 |
| | | □a□imum □ir□ulati□□ □ate – □P□ | | | | | | | | | |
| 15 | 500 | 100 | 50 | 33 | 25 | 20 | 17 | 14 | 12 | 11 | 10 |
| 20 | 670 | 134 | 67 | 45 | 33 | 27 | 22 | 19 | 17 | 15 | 13 |
| 30 | 1,005 | 200 | 100 | 67 | 50 | 40 | 33 | 29 | 25 | 22 | 20 |
| 40 | 1,340 | 268 | 134 | 89 | 67 | 54 | 45 | 38 | 33 | 30 | 27 |
| 50 | 1,675 | 335 | 168 | 112 | 84 | 67 | 56 | 48 | 42 | 37 | 33 |
| 60 | 2,010 | 402 | 201 | 134 | 101 | 80 | 67 | 58 | 50 | 45 | 40 |
| 70 | 2,345 | 470 | 235 | 157 | 118 | 94 | 78 | 67 | 59 | 52 | 47 |
| 80 | 2,680 | 536 | 268 | 179 | 134 | 107 | 90 | 77 | 67 | 60 | 54 |
| 100 | 3,350 | 670 | 335 | 223 | 168 | 134 | 12 | 96 | 84 | 75 | 67 |
| 125 | 4,185 | 836 | 418 | 279 | 209 | 168 | 140 | 120 | 105 | 93 | 84 |
| 150 | 5,025 | 1,005 | 503 | 335 | 251 | 1201 | 168 | 144 | 126 | 112 | 100 |
| 200 | 6,695 | 1,340 | 670 | 447 | 335 | 268 | 224 | 192 | 168 | 149 | 134 |
| 250 | 8,370 | 1,675 | 838 | 558 | 419 | 335 | 280 | 240 | 210 | 186 | 167 |
| 300 | 10,045 | 2,010 | 1,005 | 670 | 503 | 402 | 335 | 287 | 251 | 223 | 201 |
| 350 | 11,720 | 2,350 | 1,175 | 784 | 587 | 470 | 392 | 336 | 294 | 261 | 235 |
| 400 | 13,400 | 2,680 | 1,340 | 895 | 670 | 535 | 447 | 383 | 335 | 298 | 268 |
| 500 | 16,740 | 3,350 | 1,675 | 1,120 | 838 | 670 | 558 | 479 | 419 | 372 | 335 |
| 600 | 20,080 | 4,020 | 2,010 | 1,340 | 1,005 | 805 | 670 | 575 | 502 | 448 | 402 |
| 700 | 23,430 | 4,690 | 2,345 | 1,565 | 1,175 | 940 | 785 | 670 | 585 | 520 | 470 |
| 800 | 26,780 | 5,360 | 2,680 | 1,785 | 1,340 | 1,075 | 895 | 765 | 670 | 595 | 535 |

03406-13_F04.EPS

*Figure 4* Typical hot-water boiler circulation chart.

## Additional Resources

*ASHRAE Handbook—HVAC Applications*, Latest Edition. Atlanta, GA: American Society of Heating, Refrigerating, and Air-Conditioning Engineers, Inc.

*ASHRAE Handbook—HVAC Systems and Equipment*, Latest Edition. Atlanta, GA: American Society of Heating, Refrigerating, and Air-Conditioning Engineers, Inc.

*Boilers Simplified*. Troy, MI: Business News Publishing Company.

## 1.0.0 Section Review

1. What substance is responsible for forming acids during an oil-fired boiler's shutdown?

   a. Sulfur
   b. Carbon
   c. Oxygen
   d. Hydrogen

2. When a boiler is put in wet storage, the boiler's water pressure is often maintained above atmospheric pressure using _____.

   a. a pump
   b. nitrogen gas
   c. compressed air
   d. carbon dioxide gas

3. What happens if the main boiler stop valve is opened too quickly as a steam boiler heats up?

   a. Pressure rapidly rises
   b. Steam will condense
   c. Burner will shut off
   d. Pressure rapidly drops

4. If cold water suddenly enters a boiler that has reached operating temperature, the boiler will experience _____.

   a. loss of capacity
   b. air bubbles
   c. thermal shock
   d. pump overload

### 2.0.0 CHILLERS AND WATER SYSTEM STARTUP AND SHUTDOWN

### Objectives

Explain how to start up and shut down various chillers and water systems.

a. Explain how to start up and shut down a reciprocating chiller system.
b. Explain how to start up and shut down a centrifugal or screw chiller system.
c. Explain how to start up and shut down cooling towers.
d. Describe the process of inspecting and cleaning various heat exchange components.

### Performance Task 2

2. Start up and shut down at least one of the following:

- Steam boiler
- Hot-water boiler
- Reciprocating chiller
- Screw chiller
- Centrifugal chiller
- Cooling tower
- Evaporative condenser

### Trade Terms

**Bleed-off**: A method used to help control corrosion and scaling in a water system. It involves the periodic draining and disposal of a small amount of the water circulating in a system. Bleed-off aids in limiting the buildup of impurities caused by the continuous addition of makeup water to a system.

**Deadband**: In a chiller, the tolerance on the chilled-water temperature control point. For example, a 1°F deadband controls the water temperature to within ±0.5°F of the control point temperature (0.5°F + 0.5°F = 1°F deadband).

**Recycle shutdown mode**: A chiller mode of operation in which automatic shutdown occurs when the compressor is operating at minimum capacity and the chilled-water temperature has dropped below the chilled-water temperature setpoint. In this mode, the chilled-water pump remains running so that the chilled-water temperature can be monitored.

Chillers produce cold water that is piped to a conditioned space for comfort cooling or for use in an industrial process. Like boilers, chillers must be started up and shut down in a certain manner to avoid damage. They also have specific requirements for maintenance and cleaning.

### 2.1.0 Reciprocating Chillers and Water Systems

Packaged reciprocating chillers are typically used on jobs requiring chiller capacities ranging from 2 to 200 tons. These chillers can use one or more reciprocating compressors. The chiller's condenser can be water cooled or air cooled. The evaporator (cooler) is usually a direct expansion type in which refrigerant evaporates while it is flowing inside the tubes, and the chilled water is cooled as it flows over the outside of the tubes. *Figure 5* shows a typical installation of a roof-mounted, packaged reciprocating chiller with an air-cooled condenser. The chiller shown is supplying chilled water to the coil of an air handling unit on an all-air system. Depending on the capacity of the chiller, several air handling units are often supplied by a single chiller. The following sections describe the tasks needed to start up and shut down a reciprocating chiller and related chilled-water system.

#### 2.1.1 Startup

Complete the startup of a reciprocating liquid chiller as directed in the manufacturer's service instructions. The specific startup procedures for the different types of reciprocating liquid chillers vary, but the tasks and their sequences are essentially the same. The tasks in the procedure are those that are done when starting up a new chiller after installation or when it has been shut down for an extended period of time. The procedure assumes that the chiller is correctly installed and is properly charged with refrigerant.

> **WARNING!**
> Do not attempt to operate a chiller until you have read and understand the manufacturer's operating instructions.

> **WARNING!**
> Preparing a chiller for startup involves working on energized equipment containing pressurized and/or hot components. Follow all manufacturer's instructions, OSHA regulations, and job-site requirements relating to safety and the use of safety equipment.

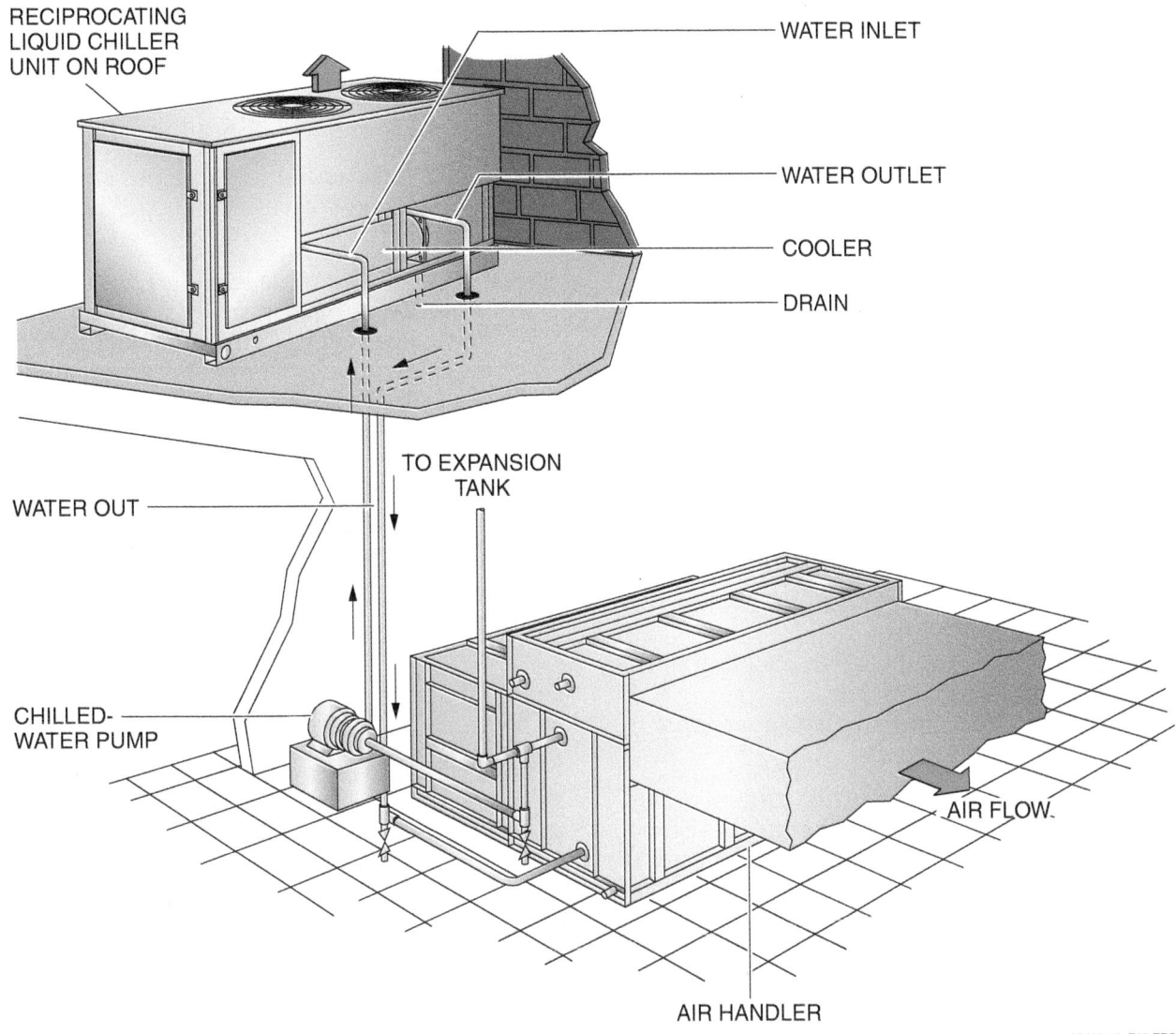

RECIPROCATING
LIQUID CHILLER
UNIT ON ROOF

WATER INLET

WATER OUTLET

COOLER

DRAIN

WATER OUT

TO EXPANSION
TANK

CHILLED-
WATER PUMP

AIR FLOW.

AIR HANDLER

03406-13_F05.EPS

*Figure 5* Chilled-water system with packaged, air-cooled reciprocating chiller.

*Step 1* Check that all tools, equipment, and de-bris are removed from the chiller and the chilled-water system components.

> **CAUTION**
> Apply power to the chiller at least 24 hours prior to chiller startup. This allows the crankcase heaters enough time to warm the oil in the compressor crankcase, driving out any condensed refrigerant.

*Step 2* If not previously done, perform an in-spection of the chilled-water circuit pip-ing to verify the following:

a. All system strainers are installed and clean.

b. All piping connections are connected and tightened.

c. All coil fins and tubing are undam-aged, and the coils are free of debris or other obstructions.

d. All safety and pressure-relief valves are set to the correct setpoint.

e. All balance or isolating valves are set or opened to the required positions.

f. Stop valves in the chiller unit evapora-tor (cooler) circuit are open.

*Step 3* Fill the components and piping of the chilled-water system with clean water that has been treated with inhibitors, as needed. If starting up a system after it has been shut down through the previous heating season, drain any antifreeze solu-tion from the chilled-water components and piping, then flush and refill the sys-tem with clean water.

**Step 4** Turn on the chilled-water circuit circulating pump and other equipment. Verify the following:

   a. No leaks are present in the system coils or piping.

   b. All air is purged from the piping and system terminal units using the vents usually located at the highest point in the circuit.

   c. The water flow, water level, and pressures are correct for the height of the highest terminal unit.

**Step 5** Check the chilled-water system pumps and fans for the following:

   a. Rotation is in the proper direction.

   b. Operation is quiet, and lubrication is adequate.

   c. Any motor-to-pump shaft alignments are correct.

   d. Drive belt type and tightness are correct.

   e. Setscrews on the drive coupling and/or fan blades are tight.

   f. All belt and/or fan blade guards are in place and secured.

**Step 6** Check the pH level of the chilled water. If necessary, use the chemicals recommended by the water treatment specialist to maintain the water pH between 7 and 8.

**Step 7** Perform a preliminary inspection of the air-cooled chiller to verify the following:

   a. The electrical power source voltage meets the unit nameplate requirements.

   b. The compressor crankcase is warm, indicating that the crankcase heaters are energized.

   c. Oil is visible in the compressor sight glass.

   d. The compressor floats freely on its mounting springs or isolators.

   e. The setpoint of the low-water temperature cutout is correct.

   f. The setpoint of the chilled-water temperature controller is correct.

   g. The condenser coils are free of debris from trees and plants.

   h. The condenser fan to venturi (orifice) ring adjustment is correct per the manufacturer's specifications.

**Step 8** Open the chiller compressor suction and discharge valves. Open the liquid line valve.

**Step 9** Use an electronic leak detector to test the chiller components and piping for refrigerant leaks. Oil stains near valves or flanged gasket connections may indicate a leak. Make any necessary repairs, being sure to use the proper refrigerant containment procedures and equipment.

> **NOTE**
>
> Most chillers have a short-cycle protection circuit that provides a delay of about five or six minutes from the time a compressor is stopped before it can be restarted. This delay is activated when the compressor stops at the end of a normal cooling cycle, if a safety device opens, or if there is a power interruption. If applicable, be sure to allow enough time for the short-cycle control circuit to time out before attempting to start the compressor.

**Step 10** Turn on the chiller. The chiller compressor should start after a short delay, during which the condenser fan operates to purge any residual heat from the area of the condenser coil.

**Step 11** If the unit is equipped with a sight glass in the liquid line, check that it is clear with no sign of bubbles; this indicates that the refrigerant is leaving the condenser fully condensed and the unit likely has a sufficient refrigerant charge for the operating conditions. Remember that bubbles may also appear when refrigerant blends are in use.

## pH Definition

The term *pH* means *power of hydrogen* and is a measure of the concentration of hydrogen ions in water. Depending on the amount of hydrogen ions present, the water may be acidic, alkaline, somewhere in between, or neutral. pH test kits are affordable, widely available, and easy to use. They typically include small bottles of testing chemicals along with an indicator scale (often color-based) to determine the pH level of the water under test. Paper test strips are also available that change color when dipped in water. The degree of color change indicates the pH. pH levels below 7 are considered acidic. Those above 7 are considered alkaline. A pH of 7 is considered neutral.

*Step 12* Check to see if the moisture indicator (in the sight glass, if any) shows the presence of moisture in the refrigerant circuit. If it indicates only a slight amount of moisture, shut down the chiller and change the filter-drier(s). If the moisture indicator shows wetness, use an approved moisture test kit to verify its presence. If moisture is present, determine the source of the moisture and correct the problem to prevent a possible chiller failure.

*Step 13* After the compressor is running, check that the temperature of the chilled water leaving the chiller's cooler continues to drop until the chiller automatically shuts off. The temperature at which this occurs should agree with the setting of the temperature controller. If it is not the same as the controller setting, adjust the control point as necessary per the manufacturer's instructions.

*Step 14* With the chiller operating, measure the high-side and low-side operating pressures with a gauge manifold set. Compare these pressures with the unit's superheat charging table to determine if the amount of superheat leaving the cooler is correct. Typically, the superheat level should be between 8°F and 12°F. If necessary, follow the manufacturer's instructions to add or remove refrigerant from the unit in order to obtain the correct level of superheat and/or subcooling. In most cases, charging will be done by subcooling, since very few chilled water units use a fixed restrictor as a metering device. Most use either a thermostatic or electronic expansion valve.

*Step 15* Confirm that the chiller's low-water and temperature cutout safety controls are functioning properly per the manufacturer's instructions.

*Step 16* Check the cooler heater cable. Make sure that the outdoor thermostat causes the heater cable to be energized when the ambient temperature drops below 35°F.

*Step 17* Record the chiller and chilled-water circulating pump data and operating parameters on suitable forms. Records should include the information summarized in *Tables* 2 and 3.

## Refrigerant Gauge Sets

The standard gauge manifold sets shown are available with scales for obsolete CFC and HCFC refrigerants, as well as their replacement refrigerants. Electronic gauge sets can be used with all refrigerants, can measure system temperature and vacuum during evacuation, and can calculate superheat and subcooling.

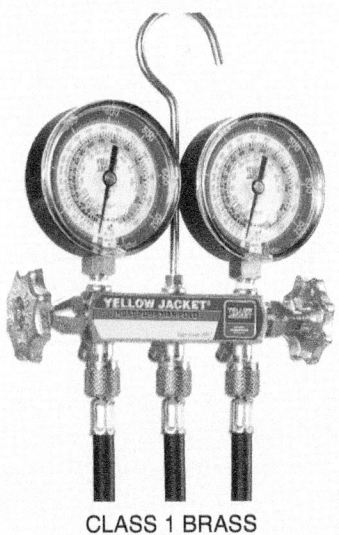

CLASS 1 BRASS
GAUGES

SOLAR-POWERED
ELECTRONIC GAUGES

REFRIGERATION SYSTEM
ANALYZER

03406-13_SA02.EPS

**Table 2** Packaged Chiller Data and Operating Parameters

| Chiller Data | | |
|---|---|---|
| Location: _____ | Manufacturer: _____ | |
| Model Number: _____ | Serial Number: _____ | |
| **Compressor Operating Data** | | |
| Suction Pressure/Temperature: _____ | Crankcase Heater Amperes: _____ | |
| Discharge Pressure/Temperature: _____ | Chilled-Water Control Setting: _____ | |
| Oil Pressure/Temperature: _____ | Condenser Water Control Setting: _____ | |
| Voltage: _____ | Low-Pressure Cutout Setting: _____ | |
| Amperage T1/T2/T3: _____ | High-Pressure Cutout Setting: _____ | |
| kW Input: _____ | Safety Control Check: _____ | |
| **Evaporator (Cooler) Operating Data** | | |
| Entering/Leaving Water Pressure: _____ | Flow in Gallons per Minute (gpm): _____ | |
| Water Pressure Difference: _____ | Refrigerant Temperature: _____ | |
| Entering/Leaving Water Temperature: _____ | Refrigerant Pressure: _____ | |
| Water Temperature Difference: _____ | | |
| **Condenser Operating Data** | | |
| **Water-Cooled Condenser** | **Air-Cooled Condenser** | |
| Entering/Leaving Water Pressure: _____ | Entering/Leaving Dry-Bulb Temperature: _____ | |
| Water Pressure Difference: _____ | Air Temperature Difference: _____ | |
| Entering/Leaving Water Temperature: _____ | Refrigerant Pressure: _____ | |
| Water Temperature Difference: _____ | Refrigerant Temperature: _____ | |
| Flow in Gallons per Minute (gpm): _____ | | |
| Refrigerant Temperature: _____ | | |
| Refrigerant Pressure: _____ | | |

**Table 3** Circulating Pump Data and Operating Parameters

| Chiller Data | | |
|---|---|---|
| Location: _____ | Manufacturer: _____ | |
| Model Number: _____ | Serial Number: _____ | |
| **Compressor Operating Data** | | |
| Pump Off Pressure: _____ | Final Suction Pressure: _____ | |
| Valve Shutoff Differential: _____ | Final Pressure Difference: _____ | |
| Valve Open Differential: _____ | Final gpm: _____ | |
| Valve Open gpm: _____ | Voltage: _____ | |
| Final Discharge Pressure: _____ | Amperage T1/T2/T3: _____ | |

### 2.1.2 Shutdown

Except for turning off the chiller and related chilled-water circuit equipment, only a few tasks are required to prepare a reciprocating chiller with an air-cooled condenser for an extended period of shutdown. Be sure that the chiller unit compressor power is turned off, but do not shut off the control power during the shutdown period. If the chiller is being shut down for the winter season, the chiller's evaporator (cooler) should be isolated from the chilled-water circuit by closing the chilled-water input and output valves. After the cooler is isolated, remove the drain plug from the cooler and drain the cooler of all water.

Put an approved permanent antifreeze solution in the cooler to prevent any residual water from freezing. Use the amount of antifreeze recommended by the manufacturer.

The chiller refrigerant should be tested for the presence of moisture and/or acid contamination. A sealed-tube acid/moisture test kit or an oil acid test kit (*Figure 6*) can be used for this purpose. A sealed-tube acid/moisture test kit is connected to a system service port to obtain a sample of the refrigerant. Oil acid test kits require that the test be performed on a sample of the compressor oil. In either case, follow the test kit manufacturer's instructions to determine the amount of acid/moisture contamination in the system. Should moisture and/or acid be detected, make repairs as necessary during the shutdown period to eliminate the cause of the problem.

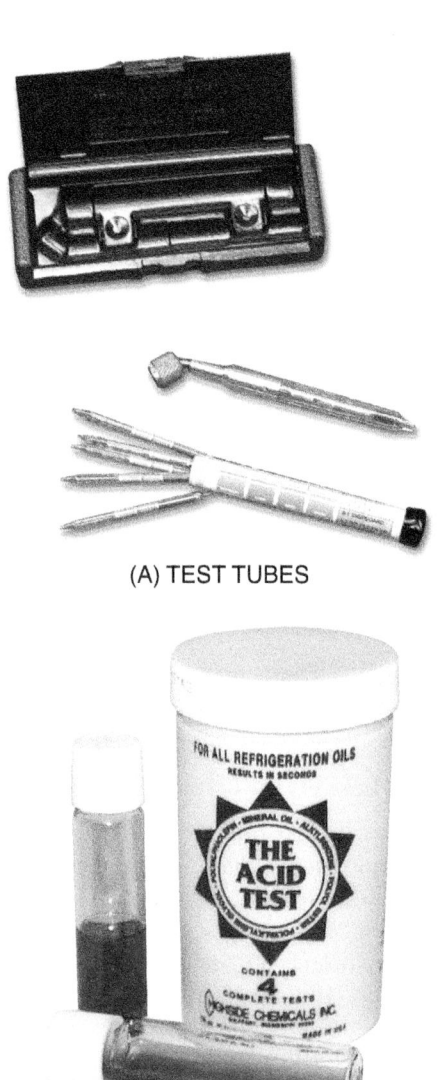

(A) TEST TUBES

(B) ACID TEST KIT

03406-13_F06.EPS

*Figure 6* Acid and moisture test kits.

Some manufacturers recommend that the bulk of the chiller system refrigerant charge be isolated in the condenser or receiver during shutdown. This minimizes the amount of refrigerant that might be lost due to any small leak on the low-pressure side of the system. If the system uses an open compressor, it reduces the amount of any refrigerant that might leak through the crankshaft oil seal. Isolation of the refrigerant in the condenser or receiver usually involves bypassing the low-pressure protective switch and readjusting the chilled-water temperature controller settings. This is necessary in order to operate the system during the procedure. For these reasons, it should always be done only as directed in the chiller manufacturer's instructions.

### 2.2.0 Centrifugal Chillers and Water Systems

Packaged centrifugal chillers are typically used on jobs requiring chiller capacities over 100 tons. They use a centrifugal compressor with one or more stages. The condenser is normally water cooled via a cooling tower with the refrigerant condensing on the outside of the tubes. The evaporator (cooler) can be a direct expansion type in smaller units, but is usually a flooded type in larger units. *Figure 7* shows a packaged centrifugal chiller with a water-cooled condenser. Cooling water for the condenser is supplied by a cooling tower. The following sections describes the tasks needed to start up and shut down a centrifugal chiller and related chilled-water system.

#### 2.2.1 Startup

The startup and operation procedures of a centrifugal chiller are normally much more automated than with reciprocating chillers. Most have a microprocessor-based control center that monitors and controls equipment operation. The microprocessor control system matches the cooling capacity of the chiller to the cooling load. It also monitors system operating conditions to execute capacity overrides or safety shutdowns.

Operator interface with the chiller's control system is usually done at an interface panel (*Figure 8*). There, the operator can input machine setpoints, schedules, setup functions, and options. Typically, these inputs are entered in response to software-driven prompts shown on the panel's displays (screens). The responses are entered using softkeys located on the panel. On some chillers, the responses may be entered using a standard computer keyboard. The operator also can use the in-

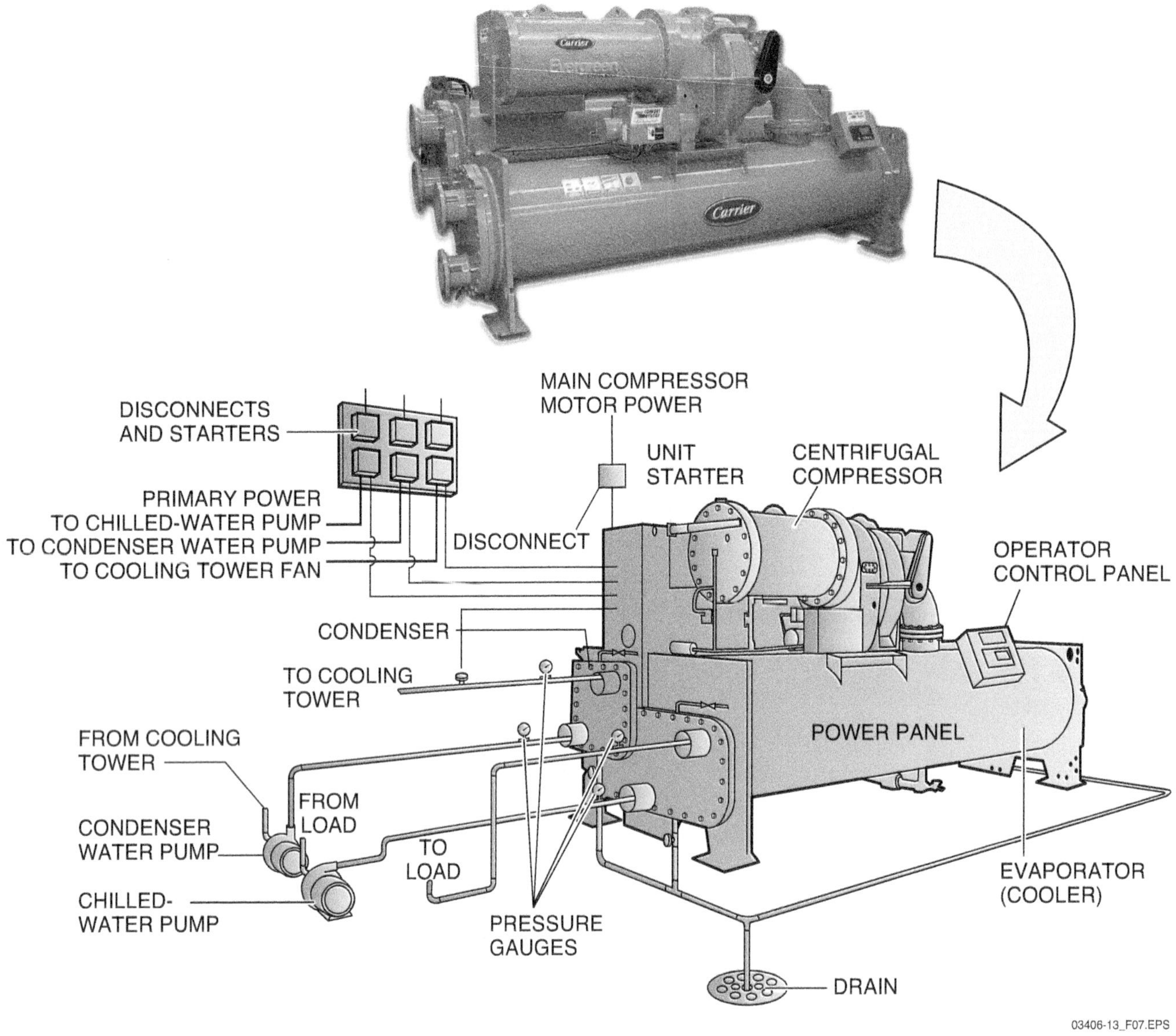

Figure 7 Typical packaged centrifugal chiller.

DISCONNECTS
AND STARTERS

MAIN COMPRESSOR
MOTOR POWER

UNIT
STARTER

CENTRIFUGAL
COMPRESSOR

PRIMARY POWER
TO CHILLED-WATER PUMP
TO CONDENSER WATER PUMP
TO COOLING TOWER FAN

DISCONNECT

OPERATOR
CONTROL PANEL

CONDENSER

TO COOLING
TOWER

FROM COOLING
TOWER

FROM
LOAD

POWER PANEL

CONDENSER
WATER PUMP

TO
LOAD

CHILLED-
WATER PUMP

PRESSURE
GAUGES

EVAPORATOR
(COOLER)

DRAIN

03406-13_F07.EPS

## Impact of LEED on System Commissioning

GOING GREEN

The Leadership in Energy and Environmental Design (LEED) green building rating system initiative was introduced in an earlier module. LEED requirements will have an impact on every phase of building construction, from design through commissioning. Many local energy codes and the guidelines developing through the LEED building certification process are modeling new and varied startup record formats and processes. The building commissioning process, including the component HVAC systems, is becoming far more complex and integrated than HVAC alone. In many cases, a third-party commissioning agent is being employed in addition to the traditional system test and balance contractor. These trends and strategies are designed to become part of the long-term building maintenance process, which will require higher levels of training, documentation, system testing, and data reporting than has been traditionally required.

## TYPICAL STATUS SCREEN

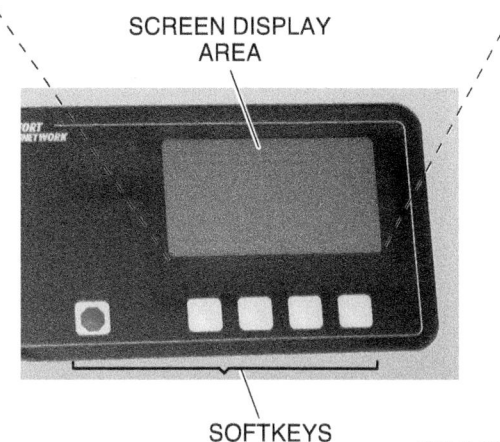

| RUNNING = TEMP CONTROL LEAVING CHILLED WATER | | 05-10-95  9:34 28.2 HOURS |
|---|---|---|
| CHW IN **55.3** | CHW OUT **45.4** | EVAPORATOR REFERENCE **42.1** |
| CDW IN **85.2** | CDW OUT **95.1** | CONDENSER REFERENCE **97.8** |
| OIL PRESSURE **23.1** | OIL TEMPERATURE **135.4** | METER AMPS **98** |
| CCN | LOCAL | RESET   MENU |

SCREEN DISPLAY AREA

SOFTKEYS

03406-13_F08.EPS

*Figure 8* Typical operator interface panel and display screen.

terface panel to monitor system status while the chiller is running or not running.

Because the operational software and the formats of the operator interface display screens vary widely among manufacturers, it is essential to complete the startup of a centrifugal chiller as directed in the manufacturer's instructions.

> **WARNING!**
> Do not attempt to operate a chiller until you have read and understand the manufacturer's operating instructions.

The following startup procedure briefly outlines the tasks used to start up a centrifugal chiller with a water-cooled condenser. The tasks described are those that are commonly performed when starting a new chiller after installation or when it has been shut down for an extended period of time.

> **WARNING!**
> Preparing a chiller for startup involves working on energized equipment containing pressurized and/or hot components. Follow all applicable manufacturer's instructions, OSHA regulations, and job-site requirements relating to safety and the use of safety equipment.

*Step 1* Check that all tools, equipment, or other debris are removed from the chiller, the cooling tower, and the components of the chilled-water system.

*Step 2* Prepare the building chilled-water circuit for operation per Steps 2 through 6 of the reciprocating chiller startup procedure.

*Step 3* Prepare the cooling tower for operation per the cooling tower startup procedure (described in detail later in this module).

*Step 4* If starting a newly installed centrifugal chiller, perform the necessary tasks listed below per the manufacturer's instructions:
  a. Remove the shipping packaging.
  b. Open the oil circuit valves.
  c. Torque all gasketed joints.
  d. Check the machine tightness.
  e. Leak-test the machine.
  f. Perform a standing vacuum test.
  g. Dehydrate the machine.
  h. Inspect the water piping including the pump-out compressor water piping (if so equipped).
  i. Check the relief devices.
  j. Inspect the wiring.
  k. Check the starter.
  l. Check the oil charge.
  m. Energize and check the oil heater operation.
  n. Set up the machine control configuration.
  o. Check the optional pump-out system controls and compressor.
  p. Charge refrigerant into the machine.

*Step 5* If starting a chiller after an extended shutdown, make sure that the chilled-water, condenser water, and any other water drains are closed. If required, flush the water circuits to remove any soft rust that may have formed.

*Step 6* Before attempting machine startup, check the following:

a. Power is turned on to the main starter, tower fan starter, oil pump and heater relays, and the machine control center.

b. Cooling tower water is at the proper level, and its temperature is at or below the machine's design entering temperature.

c. Machine is charged with refrigerant, and all refrigerant and oil valves are in the proper operating position.

d. Oil is at the proper level in the compressor reservoir sight glass.

e. Oil reservoir temperature is within the range recommended by the manufacturer (typically 140°F to 160°F).

f. Valves in the evaporator and condenser water circuits are open. If the circulating pumps are not automatic, make sure they are turned on and water is circulating properly.

g. Solid-state starter checks (if so equipped) are performed.

*Step 7* As applicable, energize the unit and perform the following tasks per the manufacturer's instructions:

a. Dry run to test the startup sequence.

b. Perform a compressor motor rotation check.

c. Conduct oil pressure and compressor stop checks.

d. Calibrate the motor current demand setting.

*Step 8* Start the chiller per the manufacturer's instructions. If the chiller fails to start, check for conditions that prohibit starting, such as activated machine safety devices, no Building Occupied mode schedule selected for the current time period, or short-cycle time delays not timed out from the last shutdown or startup.

Assuming none of these conditions exist, the automatic sequence for the startup of a typical centrifugal chiller is similar to the one shown in *Figure 9* and is described as follows.

NOTE

Note that the times and other values given in the sequence are not specific, but are typical times and values used for descriptive purposes only. Failure to achieve any of the events described in sequence A through E will typically result in the control system automatically aborting the startup sequence and displaying the applicable failure message(s) on the control panel display.

*Event A* – The chilled-water pump starts when startup is initiated if the equipment protective safety limits and control settings are okay. These safety limits and control settings typically include the following:

• Temperature sensors out of range
• Pressure transducers out of range
• Compressor discharge temperature
• Motor winding temperature
• Evaporator refrigerant temperature
• Transducer voltage
• Condenser pressure
• Oil pressure
• Line voltage
• Compressor motor load
• Starter acceleration time

*Event B* – The condenser water pump starts after a short delay of about five seconds.

*Event C* – The system controller begins monitoring the status of the chilled-water and condenser water flow switches. It waits up to five minutes to verify water flow. Note that the waiting period allocated to monitor the water flow usually can be changed at the control panel, if desired. After water flow is verified, the chiller is enabled.

As the sequence continues, the chilled-water temperature is compared to the chiller's leaving (or entering) water temperature control point and deadband. The deadband is the tolerance on the chilled-water temperature control point. For example, a 1°F deadband controls the water temperature to within ±0.5°F.

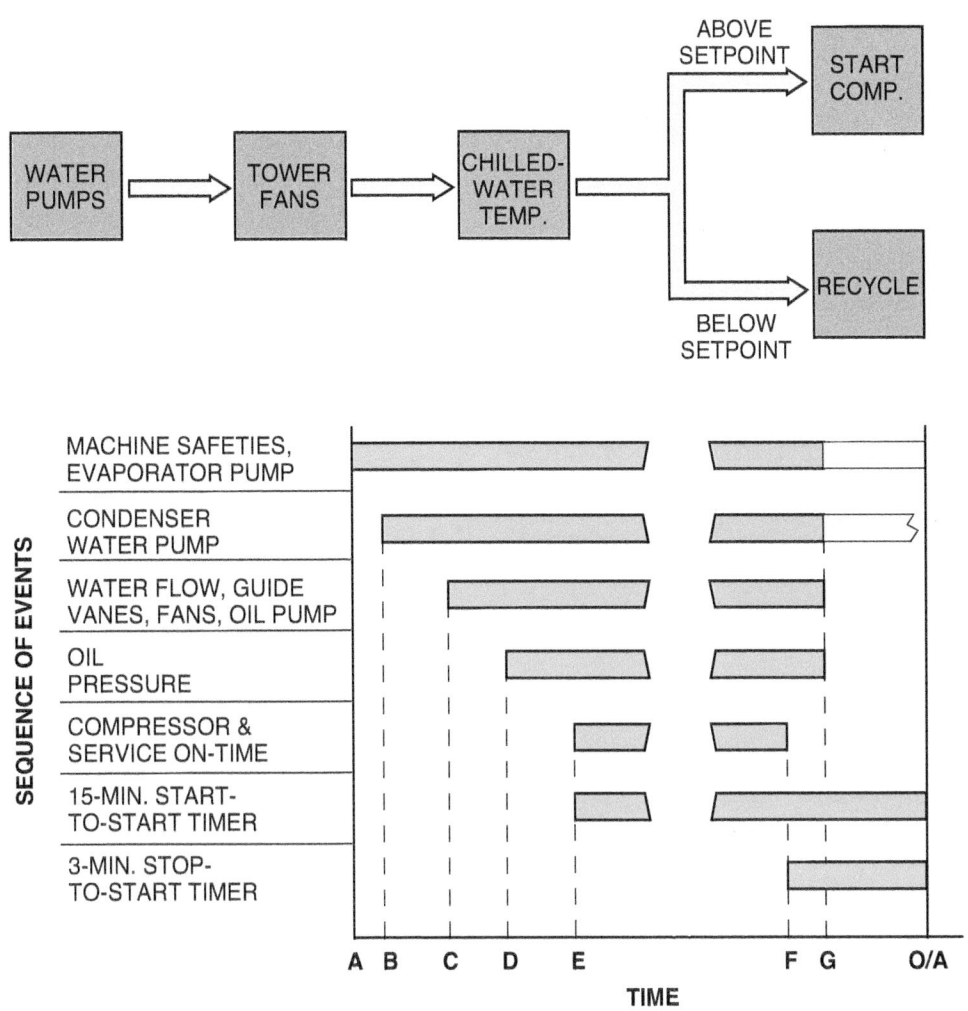

*Figure 9* Typical centrifugal/screw chiller sequence of operation.

03406-13_F09.EPS

Should the chilled-water temperature be 5°F or more below the control point and deadband, the startup sequence stops and the machine control system initiates a normal recycle shutdown mode. This mode is described later in this section.

If the chilled-water temperature is equal to or higher than the control point and deadband, the control system continues the startup sequence and checks the position of the capacity control guide vane and the oil pump pressure. If the vane is closed and the oil pump pressure differential is less than 3 psi, the oil pump relay is energized and the oil pump is turned on.

*Event D* – The system waits until the oil pressure differential reaches a minimum pressure of 18 psi.

*Event E* – 15 seconds after the minimum oil pressure is reached, the compressor relay energizes and the compressor starts.

The 15-minute start-to-start timer is also activated.

*Step 9* After the compressor starts and system conditions have stabilized, monitor the control panel displays and record the chiller data and operating parameters.

### 2.2.2 Normal Shutdown

During normal chiller operation, the chiller can shut down for a variety of reasons, such as when the Stop control is pressed, the time schedule has moved to a Building Unoccupied mode, or the machine is cycling to maintain chilled-water temperature. When a Stop command occurs, the automatic sequence for the shutdown of a typical centrifugal chiller is similar to the one shown for events F and G in *Figure 9* and described below:

*Event F* – The compressor start relay is de-energized, causing the compressor to stop. Also, the inlet guide vanes are brought to the closed position and the 3-minute short-cycle timer is activated.

*Event G* – 60 seconds after the compressor turns off, the oil pump, chilled-water pump, and condenser pumps are disabled. However, if the condenser water temperature is still high, the tower fan(s) may continue to run, as they are commanded by a separate control loop.

During normal operation, the chiller may automatically cycle off and wait until the load increases before it restarts. This mode of operation is called the recycle shutdown mode. It occurs when the compressor is running under a light load and the chilled-water temperature drops 5°F below the chilled-water control setpoint (*Figure 10*). The sequence of shutdown in this mode is the same as described above, except the chilled-water pump remains running so that the temperature of the chilled water continues to be monitored. The chiller control system automatically restarts the chiller when it senses that the chilled-water temperature has increased to 5°F above the chilled-water control setpoint.

### 2.2.3 Extended Shutdown

When the centrifugal chiller is going to be shut down for an extended period, transfer the refrigerant into a storage tank. This helps reduce machine pressure and the possibility of leaks. A holding charge recommended by the manufacturer, usually nitrogen, should be maintained in the machine to prevent air from leaking in. Many chillers are equipped with isolation valves, a pump-out system, and/or an optional storage tank. Following the manufacturer's instructions, use this equipment to transfer the refrigerant charge into the storage tank. For chillers equipped with isolatable condenser and cooler vessels, transfer the refrigerant into the condenser vessel. If the machine may become exposed to freezing temperatures, drain the chilled-water, condenser water, and pump-out condenser unit water circuits to

avoid freeze-up. Also, leave the water box drains open. Leave the oil charge in the machine with the oil heater and controls energized to maintain the minimum oil reservoir temperature.

During extended periods of shutdown, perform the scheduled maintenance procedures recommended by the chiller manufacturer as directed in the service literature. These procedures can include the following:

- Test the refrigerant for acid/moisture contamination.
- Inspect and clean the operator control panel.
- Check the safety and operating controls.
- Change the oil and oil filters.
- Change the refrigerant filter-driers.
- Change the oil reclaim system filters and/or strainers.
- Inspect and clean the refrigerant float system.
- Inspect and clean the relief valves and piping.
- Inspect and clean the cooler and condenser heat exchanger tubes.
- Inspect and clean the starting equipment.
- Check the calibration of the oil, condenser, and cooler pressure transducers.
- Perform pump-out system maintenance.

In closed systems, manufacturers usually allow an appropriate amount of proper antifreeze to be added to the chilled water to prevent freeze-up during operation or shutdown. A benefit of most recommended antifreeze is that it usually contains inhibitors that prevent corrosion, biological growth, and sludge formation. Open systems, such as the condenser water loop, are typically drained. Glycol or other forms of antifreeze are not usually added to open loops.

Most manufacturers recommend using auxiliary electric heating of piping and/or open-system water basins to prevent freeze-up in systems that must operate in the winter. For short-term below-freezing temperatures, some manufacturers allow the fans to be temporarily reversed to prevent ice formation on the tower. If the system is shut down for the winter but not drained, auxiliary heating with provisions for water circulation must be installed.

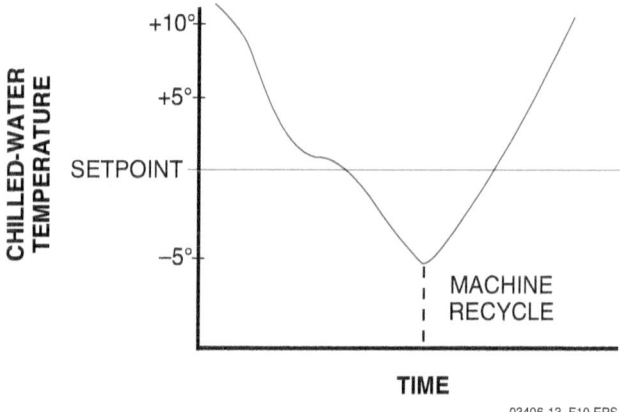

Figure 10 Recycle shutdown mode setpoints.

03406-13_F10.EPS

## Oil/Refrigerant Analysis

On large chillers, it is a common practice to periodically collect oil and/or refrigerant samples for submission to an independent laboratory for analysis.

### 2.2.4 Screw Chillers and Water Systems

Screw chillers are typically used on jobs requiring chiller capacities over 100 tons. They use one or more oil-injected screw compressors. The condenser is normally water-cooled via a cooling tower with the refrigerant condensing on the outside of the tubes. The air- or water-cooled condenser is often installed remotely (outdoors) and piped to the chiller. However, packaged models are available as well. The model shown in *Figure 11* can be provided with or without a water-cooled condenser attached.

The evaporator (cooler) can be either a direct expansion or flooded type. When the direct expansion type is used, such as the one shown in *Figure 11*, the refrigerant usually flows through the tubes. With a flooded type, the refrigerant is normally outside the tubes. Because the screw compressor is oil injected, an oil separator is also part of the package.

Like centrifugal chillers, the startup, operation, and shutdown of most screw chillers is highly automated. Operation is under the control of a microprocessor-based control system that monitors and controls all operations of the machine. The screw chiller operator interface control panel and menu-driven displays are also similar to those used with centrifugal chillers. The sequence and tasks involved in the startup and shutdown of a screw chiller are basically the same as those described earlier for a centrifugal chiller.

### 2.3.0 Cooling Tower Water Systems

Correct startup and operation of a cooling tower (*Figure 12*) water system is important to efficient and trouble free water-cooled chiller and tower operation. It is even more important because of the health and safety issues related to tower water systems. Although rare, poorly prepared and maintained cooling towers are known to support the growth of Legionella bacteria, which cause Legionnaires' disease. The aerosol produced has the potential of infecting not only the people at the equipment site, but also the surrounding community. Note that the sequence and tasks performed to start up, operate, and maintain an evaporative condenser are the same as described for a cooling tower.

> **WARNING!**
> Preparing a cooling tower for startup involves working on energized equipment containing pressurized and/or hot components. Follow all applicable manufacturer's instructions, OSHA regulations, and job-site requirements relating to safety and the use of safety equipment.

### 2.3.1 Startup

*Step 1* Check that all tools, equipment, or other debris are removed from the cooling tower and other components of the tower water system.

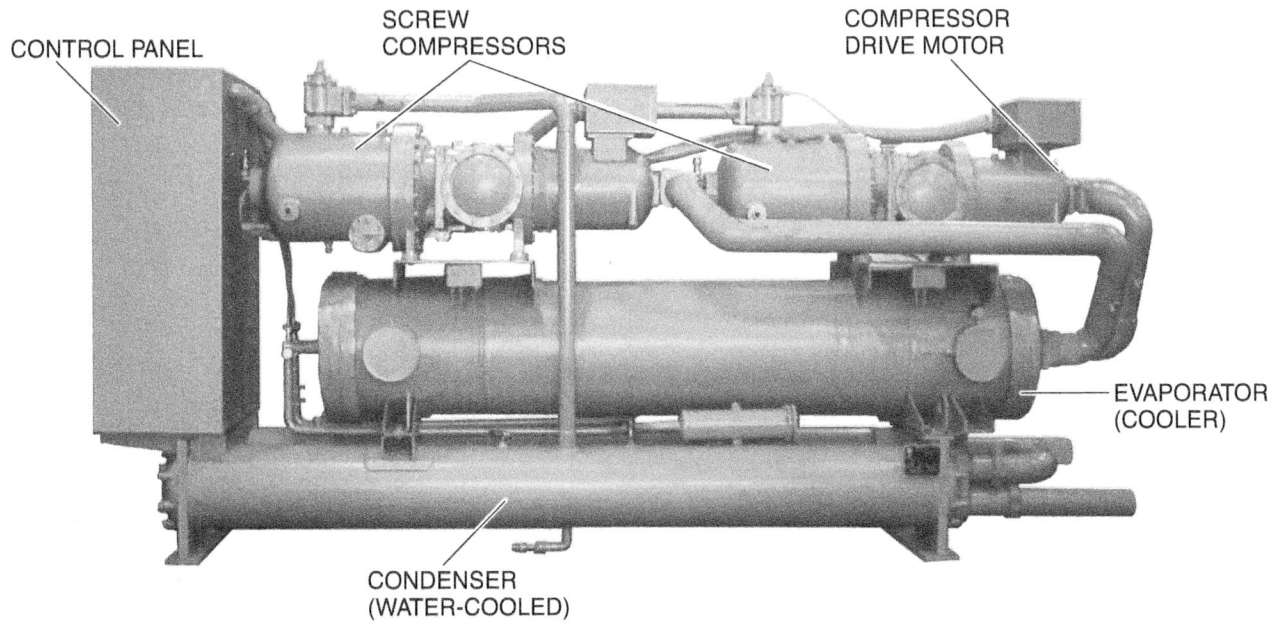

CONTROL PANEL

SCREW COMPRESSORS

COMPRESSOR DRIVE MOTOR

EVAPORATOR (COOLER)

CONDENSER (WATER-COOLED)

03406-13_F11.EPS

*Figure 11* Typical packaged screw chiller.

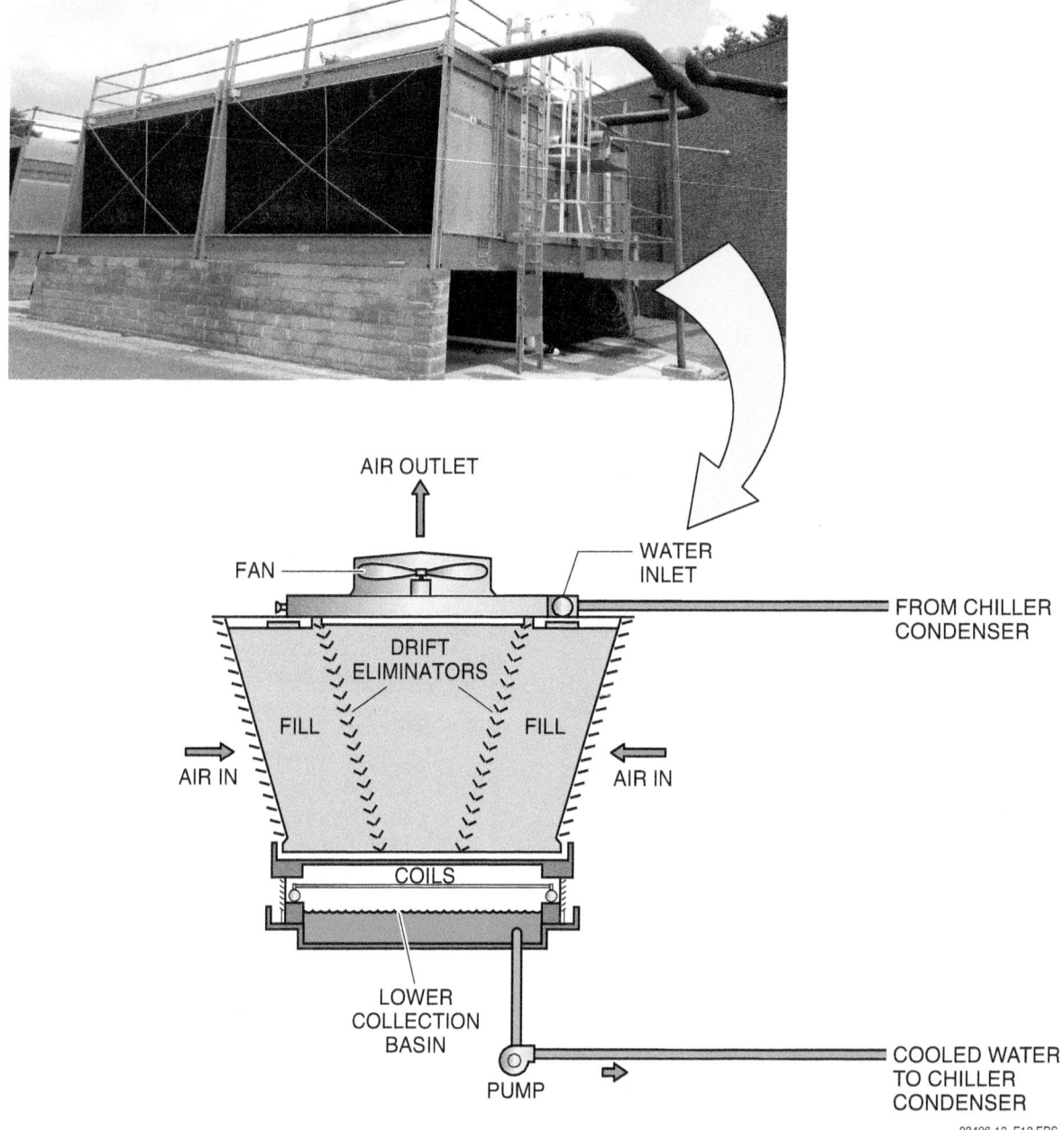

*Figure 12* Typical induced-draft cooling tower.

*Step 2* If not previously done, perform an inspection of the cooling tower and circuit piping to verify the following:

   a. All system strainers are installed and clean.

   b. All piping connections are connected and tightened.

   c. All balance or isolating valves are set or opened to the required positions.

   d. Motor and pump drive shafts are correctly aligned.

   e. Motor and/or fan drive belt tension is correct.

   f. Driveshaft couplings and/or fan blade setscrews are tight.

   g. All belt and/or fan blade guards are in place and secured.

   h. Stop valves in the chiller unit water-cooled condenser circuit are open.

   i. There is no mechanical damage, and the tower components are free of visible algae growths and deposits.

   j. There is no slimy feel to the tower fill pack, tower sump, or sidewalls.

   k. There is no visible corrosion on the outside of the sump heater.

l. The motors and bearings are lubricated.

*Step 3* Fill the components and piping of the tower water system with clean water that has been treated with inhibitors as needed. If starting up a system after it has been shut down during the previous heating season, first drain any antifreeze solution from the system components and piping, then flush and refill the system with clean, treated water.

    If the system remained filled with treated water during shutdown, clean and replace or recharge the water treatment chemical feeding devices. Test the water to determine its condition, and treat as necessary to achieve an acceptable quality.

*Step 4* Turn on the tower make-up water system and adjust the float valve for the correct water level. Make sure that the water level is high enough to prevent the condenser water pump from drawing air into the water.

*Step 5* Check that no leaks are present in the tower components and system piping.

*Step 6* Check the pumps and fans for proper direction of rotation and quiet operation.

## Use an Approved Treatment for Cooling Tower Water

Operating a cooling tower with untreated or improperly treated water can result in rapid corrosion, scaling, erosion, or algae in the tower and the related chiller's water-cooled condenser. Make sure to use a water treatment method approved and/or recommended by a qualified water treatment specialist for the specific cooling tower system being serviced.

*Step 7* Check that the sump immersion heaters turn on when the setpoint on the thermostat is reached.

*Step 8* Check the operation of spray nozzles and drift eliminators. The bypass of aerosol droplets should be minimal when the fans are operating.

*Step 9* Adjust the bleed-off rate to match the rate set by the system water treatment program. Bleed-off is a method used to control scaling in the system by periodically draining and disposing of a small amount of the water circulating in a system. This helps limit the buildup of mineral deposits caused by the continuous adding of makeup water. As water evaporates from the tower, the minerals remain behind. Without bleed-off, the concentration of minerals in the remaining water can be become extremely high.

*Step 10* Record the cooling tower data and operating parameters on a suitable form. The records should include the information summarized in *Table 4*.

The efficient operation of a cooling tower depends not only on mechanical maintenance, but also on cleanliness. Recirculating and spraying water in a cooling tower allows the water to come into contact with air. As a result, the composition of the water is drastically changed by evapora-

### Bleed-Off Water Control

GOING GREEN

    Older cooling tower systems were often fitted with a simple manual valve to allow for bleed-off. In most cases the valve was set by HVAC technicians and system operators to a flow rate far greater than required. In some cases, calculations were never made to determine the correct volume, nor was the volume of water flowing to bleed-off ever measured. As long as the objective of reducing impurities was reached, this method seemed sufficient, even though it wasted a great deal of water.

    Today, automated controls coupled with precise methods of monitoring the levels of impurities in water, allow bleed-off to be accurately controlled. The automated controls open and close a solenoid valve in the bleed-off line, opening it only when impurity levels indicate the need for dilution. The end result is that millions of gallons of precious water are saved each year.

**Table 4** Cooling Tower Data and Operating Parameters

| Cooling Tower Data | |
|---|---|
| Location: _____ | Manufacturer: _____ |
| Model Number: _____ | Serial Number: _____ |
| **Pump Data** | |
| Make/Model: _____ | Motor HP/rpm: _____ |
| Serial Number: _____ | Volts/Phase/Hertz: _____ |
| Motor Make/Frame: _____ | Gallons per minute (gpm): _____ |
| **Fan Data** | |
| Number of Fan Motors: _____ | Motor Sheave Diameter/Bore: _____ |
| Motor Make/Frame: _____ | Fan Sheave Diameter/Bore: _____ |
| Motor HP/rpm: _____ | Sheave C/L Distance: _____ |
| Volts/Phase/Hertz: _____ | Number Belts/Make/Size: _____ |
| **Operating Water Flow Parameters** | |
| Entering/Leaving Water Pressure: _____ | Water Temperature Difference: _____ |
| Water Pressure Difference: _____ | Flow in Gallons per Minute (gpm): _____ |
| Entering/Leaving Water Temperature: _____ | Temperature Bleed-Off gpm: _____ |

tion, aeration, and other chemical and/or physical processes. Also, acidic gases and other contaminants in the air are absorbed into the water. Cooling towers are highly susceptible to the growth of algae, bacteria, and other living organisms, especially if they are located where the water surface is exposed to sunlight.

Because of these harmful conditions, periodic maintenance on cooling towers must be done faithfully to prevent the buildup of corrosion and scale and to eliminate the growth of algae, bacteria, and other living organisms in the system. The tower manufacturer often furnishes service manuals that recommend scheduled maintenance tasks that should be performed on the tower. If manufacturer information is not available, *Table 5* can be used as a guide.

### 2.3.2 Winter Operation

When a cooling tower is used in subfreezing temperatures, the open recirculating water, closed recirculating water, and sump water systems need to be winterized. Equipment used for this winterization should be already installed in the system. It should be operated as directed by the tower manufacturer's service literature. During cold weather operation, more frequent visual inspections should be made and routine maintenance performed.

This helps to make sure all the controls are operating properly and aids in the detection of icing conditions before they become serious.

## Acid Cleaning of a Cooling Tower or Air-Cooled Condenser

If using an acid solution to clean external cooling tower components or air-cooled condenser fins, a commercial acid cleaner formulated for the task and metals involved usually has two benefits. The first is that inhibitors have been added to protect the specific metals being cleaned. The second is that some contain a pH color indicator that indicates when the acid is spent or neutralized. When the acid is spent while cleaning a cooling tower, more acid is added until the correct color is maintained for 20 to 30 minutes of cleaning. Then, soda ash can be added to neutralize the acid, as indicated by the color, before the solution is flushed into drains.

**Table 5** Cooling Tower/Open Recirculating Water System Maintenance

| Maintenance Action/Test | Frequency |
|---|---|
| Test and record bacteriological quality of the system water. | Biweekly |
| Test and record biocide and inhibitor reserves, pH, and conductivity of the system water. | Biweekly |
| Check that dosing equipment containers are full, pumps are operating properly, and supply lines are not blocked. | Weekly |
| Check the bleed-off control equipment to make sure it is operating properly and the controller is in calibration. The solenoid valve should be manually operated to confirm that the flow of water to the drain is clear. | Weekly |
| Check the system for growths and deposits. There should be no algae growth in the towers or slimy feel to the fill pack, tower sump, or side walls. | Weekly |
| Check the operation of sump immersion heaters. There should be no visible corrosion on the outside of the heater, and the unit should activate when the setpoint on the thermostat is reached. | Monthly |
| Check the operation of sprays, fans, and drift eliminators. There should be no mechanical damage, and the components should be free of visible deposits. The bypass of aerosol droplets should be minimal when the fans are operating. The distribution system should have no deposits, with an even flow of water to all areas of the tower. | Monthly |
| Drain, clean, and disinfect cooling towers and associated pipe work in accordance with the method approved for the site. The chlorination period should be a minimum of five hours. Free chlorine residuals should be checked regularly. If possible, the tower pack should be removed for cleaning. Post-cleaning chlorination should be monitored to make sure that free chlorine residuals are maintained. | Semiannually |
| Review maintenance and water treatment program performance, including the quality of results obtained and the cost of system operation. | Annually |

### 2.3.3 Shutdown

After the system chiller and cooling tower equipment have been shut down, the cooling tower and related water piping system can be prepared for an extended period of shutdown, as follows:

> **WARNING!**
> Use the proper face mask, goggles, rubber gloves, and protective clothing as required when draining tower wastewater and working on tower equipment and surfaces that have been in contact with water treatment chemicals.

*Step 1* Turn off power to all cooling tower equipment, including the water treatment equipment.

*Step 2* If exposed to freezing temperatures during shutdown, turn off the supply water and drain the supply water line. Drain all recirculating water lines. Be sure to follow all local, state, and federal laws pertaining to the discharge and disposal of the drained wastewater.

*Step 3* Remove all chemicals from the various water treatment equipment, and store or dispose of them as directed by the equipment or chemical manufacturer.

Prepare the water treatment equipment for extended shutdown as directed by the applicable equipment manufacturer.

*Step 4* Remove and clean the screens, strainers, and strainer baskets.

*Step 5* Clean the makeup water system float valve assembly.

*Step 6* Drain the water circulation pumps. Inspect and perform any scheduled maintenance on the pumps and motors specified in the pump manufacturer's service literature. When completed, add a corrosion inhibitor and lubricant to the pumps so that they will restart easily.

*Step 7* Inspect and perform any scheduled maintenance on the fans and fan motors specified in the fan manufacturer's service literature.

*Step 8* Inspect and clean the distribution basins or spray nozzles, cold water basin, and pipelines.

*Step 9* Inspect for clogging and clean the drift eliminators and fill and/or decking.

*Step 10* As needed, paint metal parts and components exposed to weather or wetting during operation. Do not paint spray nozzles or heat exchanger surfaces. Use a paint that is suitable for contact with water and the water treatment chemicals being used.

*Step 11* To reduce the airborne dirt and debris that can be carried into the tower during shutdown, cover or screen the fan and louver openings.

## 2.4.0 Inspecting and Cleaning Heat Exchangers

Water-cooled shell and tube condensers that use water supplied from open cooling towers are subject to contamination and scaling. Heat exchangers can also become excessively fouled. Condensers and heat exchangers should be periodically inspected and cleaned per the manufacturer's instructions.

### 2.4.1 Inspecting and Cleaning Shell and Tube Condensers and Evaporators

Typically, shell and tube condensers and evaporators should be cleaned at least once a year, and more often if the water is contaminated. Normally, the tubes and heat exchanger surfaces require cleaning using brushes designed to prevent scraping and scratching of the tube walls. Wire brushes should not be used. If cleaning the condenser tubes and heat exchanger surfaces with brushes fails to adequately remove the corrosion or scaling, chemical cleaning may be required. In this situation, the condenser manufacturer and/or a water specialist should be contacted in order to determine what chemicals should be used and how they are used to clean the condenser.

Inspect the entering and leaving condenser water sensors also for signs of corrosion and scaling. If corroded, the sensors should be replaced. If scale is found, they should be cleaned.

The waterside inspection and cleaning process used with shell and tube evaporators (coolers) in

chilled-water systems is similar to the inspection and cleaning process used for water-cooled condensers.

### 2.4.2 Inspecting and Cleaning Heat Exchangers

In general, heat exchangers require maintenance only in the form of periodic inspection of their heating surfaces to determine if excessive fouling has occurred. The amount of fouling usually depends on water temperatures and local feedwater conditions. If either side of the heat exchanger is part of a closed system, it is unlikely that any cleaning of that portion of the heat exchanger will be needed, as long as the water has been properly treated.

If the heat exchanger is used to heat fresh water, fouling can occur. The amount of fouling depends on the water temperature and the amount of dissolved and entrained minerals contained in the water. Some of the entrained minerals break down into insoluble compounds when heated. They precipitate out of the water and stick to the tubing walls. The higher the temperature to which the water is heated, the more rapidly scaling occurs.

Heat exchanger tubing does not form scale in a uniform manner. Most of the scaling takes place on the tubing immediately adjacent to the cold-water inlet. As the scale builds up at the entry of the tubes, it becomes an insulator, and heating will take place further back in the tubing, causing progressive scaling. Inspection of the tubing at the cold-water inlet will indicate when scale has formed. Use a brush-type tube cleaner (*Figure 13*) to clean heavily fouled smoothbore or internally enhanced chiller, heat exchanger, and condenser tubes. The shaft speed can be adjusted to match the internal spiral of an internally enhanced tube, causing the brush to be self-propelled through the tube. High-pressure flushing water and cleaning solution can be applied through the shaft casing to remove deposits as they are loosened. In large condensers or heat exchangers, high-pressure water jetting (10,000 to 20,000 psig) is increasingly being used to remove scale and deposits from inside the tubes.

Any one of the many available chemical coil cleaners may be used for cleaning if the manufacturer's recommendations on concentration and application are followed. When scaling is not severe, standpipes can be installed in the head openings and the tubes filled with cleaning solution (*Figure 14*). The standpipes allow for enough solution to be added to flood the heat exchanger and soak all the internal surfaces with the solution. It may take several applications to remove all the scale.

After cleaning, these tubes should be flushed with clean water to remove any chemical residue. Larger, multiple-pass heads will have a drain opening that can be used to eliminate spent cleaning solution.

Visual inspection determines when cleaning is complete. On severely fouled tubes, it may be necessary to pump the cleaning solution through the tubes continuously. If the tubes become completely plugged or are severely corroded, a new tube bundle may have to be installed.

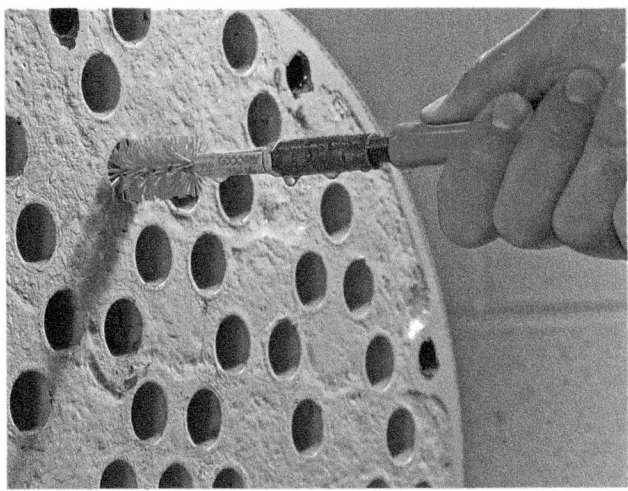

03406-13_F13.EPS

*Figure 13* Brush-type tube cleaner.

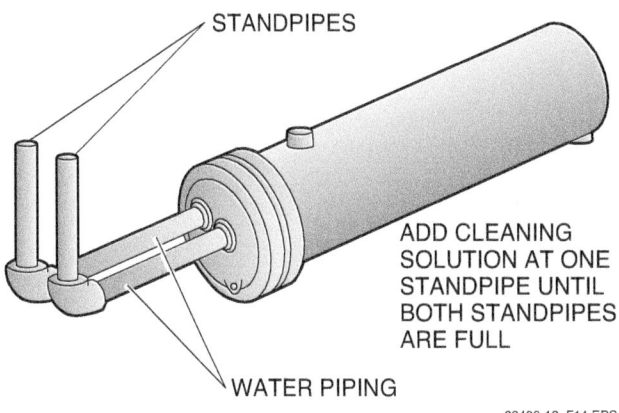

03406-13_F14.EPS

*Figure 14* Typical heat exchanger cleaning setup.

## Additional Resources

*ASHRAE Handbook—HVAC Applications*, Latest Edition. Atlanta, GA: American Society of Heating, Refrigerating, and Air-Conditioning Engineers, Inc.

*ASHRAE Handbook—HVAC Systems and Equipment*, Latest Edition. Atlanta, GA: American Society of Heating, Refrigerating, and Air-Conditioning Engineers, Inc.

## 2.0.0 Section Review

1. The pH level of water in a chilled water system should be _____.

    a. acidic
    b. between 7 and 8
    c. alkaline
    d. less than 7

2. Recycle shutdown mode occurs when the chiller compressor is running _____.

    a. under a heavy load
    b. with water below freezing
    c. under a light load
    d. with water above freezing

3. What will cause the sump pump in a cooling tower to draw air into the water?

    a. Incorrect float valve setting
    b. Incorrect spray nozzle pattern
    c. Excess algae in the sump
    d. Misaligned drift eliminators

4. Wire brushes are commonly used to clean scale from condenser tubes.

    a. True
    b. False

## SECTION THREE

### 3.0.0 AIR HANDLING AND ROOFTOP SYSTEM STARTUP AND SHUTDOWN

### Objectives

Explain how to start up and shut down air handling and packaged rooftop systems.
- a. Explain how to start up and shut down air handling units and their associated air distribution system.
- b. Explain how to start up and shut down packaged rooftop units.

### Performance Task

1. Start up and shut down an air handling unit and prepare it for normal operation.

Large commercial air handlers condition air by passing it across coils that contain refrigerant, chilled water, or hot water. Packaged rooftop units are self-contained. They can provide cooling or both heating and cooling. Like other HVAC products, they must be started up and shut down in a certain manner to avoid damage. They also have specific requirements for maintenance and cleaning.

### 3.1.0 Air Handling and Air Distribution Systems

As a general rule, air handling and air distribution systems are prepared and started before condensing units, chillers, and boilers can be brought on line. Condensing units, for example, cannot be operated without some load on the evaporator coil. That load comes from the movement of air generated by the blower or air handling unit. Although chillers and boilers rely on the movement of water to impose a load, the air handling units and air distribution system still must be in operation before a consistent load is available.

### 3.1.1 Startup

The following procedure describes the startup of a typical constant volume air handling unit (*Figure 15*) connected to a single-duct air distribution system. The sequence and tasks described are ba-

sic to the startup and checkout of almost all types of air distribution systems. The startup of an air handling unit should always be performed per the manufacturer's service instructions. Before starting an air handler, perform the following preliminary checks:

> **WARNING!**
>
> Do not attempt to operate an air handling unit until you have read and understand the manufacturer's operating instructions.

> **WARNING!**
>
> Startup of an air distribution system involves working on energized equipment with rotating and moving parts. It also involves working on the related cooling, heating, or air handling equipment with pressurized and/or hot components. Follow all manufacturer's instructions, OSHA regulations, and job-site requirements relating to safety and the use of safety equipment.

*Step 1* Obtain the building plans and specifications, and then locate all the system components.

*Step 2* Make sure that all electrical power to the equipment is turned off. Open, lock, and tag disconnects.

*Step 3* Check all fans and blowers. Verify the following:
- a. All shipping restraints and protective covers are removed.
- b. All bearings are lubricated.
- c. Fan wheels clear the housings and are correctly positioned. Improper clearance can greatly affect the fan/ blower performance, especially with backward-inclined fans.
- d. Motors and bases are securely fastened.
- e. Motor fan drive shafts are correctly aligned.
- f. All setscrews are tight.
- g. Belt tensions are correct.
- h. Belt guards are in place.

*Step 4* Inspect the complete supply air system from the supply air fan through the main supply ducts and through all branch duct runs and outlet terminals. Also inspect the complete return air duct system. Verify the following:
- a. All tools, equipment, and debris are removed.

 03406 System Startup and Shutdown                    Module Six  31

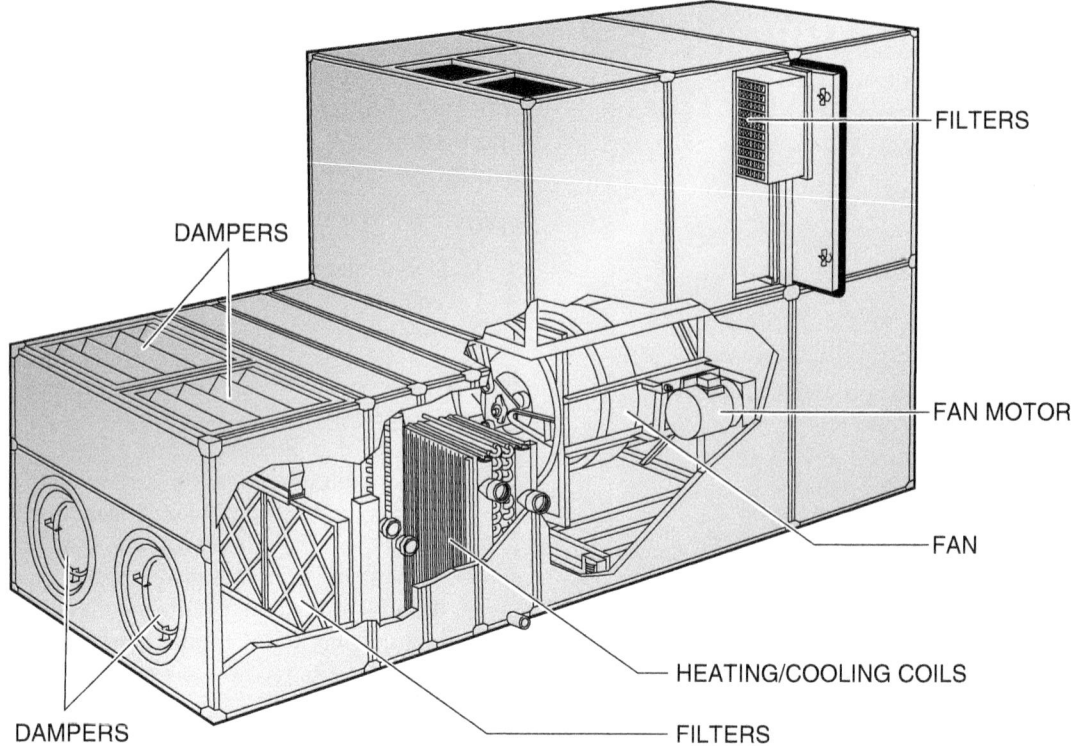

DAMPERS

FILTERS

DAMPERS

FAN MOTOR

FAN

HEATING/COOLING COILS

FILTERS

03406-13_F15.EPS

*Figure 15* Typical air handling unit.

    b. The ductwork is complete with all openings in the duct sealed, all end caps in place, and all access doors closed and secured.

    c. All balancing, volume, and terminal dampers are in their marked positions.

*Step 5* Check the air handling unit filters and all other air filters. Verify that:

    a. The size and types of filters are correct.

    b. The filters are clean.

    c. The filters and filter frames are properly installed and are airtight.

*Step 6* Check that the air handling unit coils are undamaged and clean. Make sure that all piping connections are connected and tightened. Verify that the cooling coil condensate drain pan opening and drain line are clear.

> **NOTE**
>
> Alkaline-based cleaners are always better than acid cleaners for degreasing and cleaning coils. To help coils stay cleaner longer, coat them with a dry coil protectant after cleaning. This will help the coil repel moisture, dirt, grease, and grime.

*Step 7* Check for conditions external to the system that can affect the airflow distribution to the occupied space. Make sure of the following:

    a. Windows or outside doors are closed.

    b. Doors within the building are open.

    c. All ceiling tiles are in place.

    d. All air distribution grilles, registers, and diffusers are unobstructed.

> **CAUTION**
>
> Proper operation of the system dampers during startup is critical. If dampers are closed, restricting airflow, the casings, housings, and ductwork can be damaged. On the initial startup of a new system, a temperature control specialist or system designer should be consulted to make sure all automatic temperature control dampers are properly positioned.

*Step 8* Check that all dampers are open or set so that air travels through the correct components of the system. Normally, dampers should not cause a blocked or restricted condition.

Once the preliminary checks have been successfully completed, proceed with the startup per the following:

*Step 1* Turn on the air handling unit per the manufacturer's service literature. Select the mode of operation that calls for maximum airflow. Normally, this is the cooling mode with wetted coils.

*Step 2* Check that the supply air fan is operating and its direction of rotation is correct.

*Step 3* Check that all automatic dampers are being controlled and are in the proper position.

*Step 4* Measure and record the voltage and current of the supply air fan motor to check for load conditions. If the amperage exceeds the motor nameplate full load amperage, stop the fan to determine the cause or to make necessary adjustments. After any adjustment in supply fan rpm, always recheck the current draw with a clamp-on ammeter to make sure the motor is not overloaded. Make sure the supply air fan access door is in place when making a current measurement.

*Step 5* Measure and record the speed of the supply air fan. If the fan is not running at the specified rpm, or as close as standard pulleys will allow, the problem must be corrected before proceeding further.

### 3.1.2 Shutdown

The shutdown of an air handler does not usually require any special service tasks. Do not leave outdoor air dampers open when the unit is not in use, especially during cold weather when freeze-ups can occur. During extended periods of shutdown, perform the scheduled maintenance procedures recommended by the air handler unit manufacturer as directed in the service literature.

During shutdown, examine and clean the air distribution system ductwork and components. Clean air ducts can help improve indoor air quality. When duct cleaning is done at the same time as the recommended HVAC equipment scheduled maintenance, it can help reduce the threat of indoor air pollution. Clean ductwork using portable and/or truck-mounted HEPA-filtered vacuuming and power brushing equipment to dislodge dirt and debris in the ductwork (*Figure 16*). The method used should follow the guidelines given in the National Air Duct Cleaners Association (NADCA) *Standard ACR 2006*. Before and after cleaning the ductwork, make a visual inspection of the internal ductwork and components using a borescope or video camera. Make a video record of the duct condition and cleanliness. An internal inspection may also detect mechanical damage to the components that is not visible from the outside of the duct. Since duct-cleaning equipment is specialized and expensive, this service is often performed by a specialty contractor.

Any wet fiberglass ductboard and/or insulation internal to the ductwork should be replaced so it does not become a breeding ground for biological contaminants. Detailed information on duct cleaning is provided in other HVAC training modules.

### 3.2.0 Packaged Units

Packaged year-round air conditioning units provide gas or electric heating, electric cooling, and air handling in a single unit. Most are sold as rooftop units similar to the one shown in *Figure 17*. This unit has a 20-ton cooling/235,000-Btuh heating capacity. It is equipped with an optional economizer that allows cool outdoor air to be brought into the building for cooling. When in the economizer mode, one or more compressors do not operate, thus saving energy. In most applications, more than one YAC is installed on a building. Like air handling units, YACs must be properly installed and maintained for efficient operation.

### 3.2.1 Packaged Unit Startup

The startup procedure outlined here describes the startup of a typical packaged unit connected to a supply air and return air duct distribution system. The sequence and tasks described are basic to the startup and checkout of almost all types of heating, cooling, and air distribution systems. Startup in both the cooling and heating modes is described.

> **WARNING!**
>
> Startup of a packaged unit involves working on energized equipment with rotating and moving parts. It also involves working on the related cooling, heating, or air handling equipment with pressurized and/or hot components. Follow all manufacturer's instructions, OSHA regulations, and job-site requirements relating to safety and the use of safety equipment.

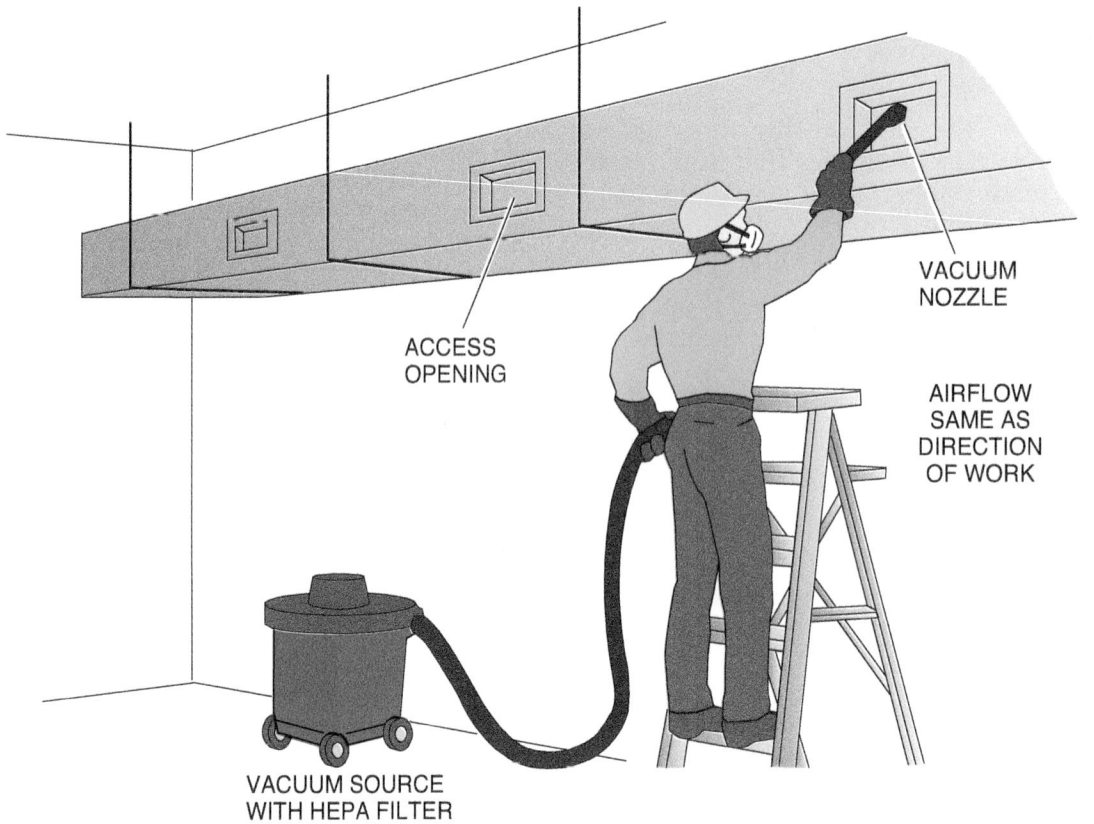

VACUUM
NOZZLE

ACCESS
OPENING

AIRFLOW
SAME AS
DIRECTION
OF WORK

VACUUM SOURCE
WITH HEPA FILTER

03406-13_F16.EPS

*Figure 16* Duct cleaning.

ELECTRIC
DISCONNECT SWITCH

FLUE GAS VENT FROM
COMBUSTION AIR INDUCER

CONDENSATE TRAP AND VENT

GAS SUPPLY
SHUTOFF VALVE

OPTIONAL ECONOMIZER
DAMPERS AND HOOD

OPTIONAL GRAVITY
EXHAUST DAMPERS AND HOOD

CONDENSER COIL

03406-13_F17.EPS

*Figure 17* Typical packaged unit.

Use the following procedure to perform preliminary checks prior to startup in either mode:

*Step 1*  Obtain the building plans and specifications, and then locate all the system components.

*Step 2*  Make sure that all electrical power to the equipment is turned off. Open, lock, and tag disconnects.

> **WARNING!**
>
> Do not attempt to operate a packaged unit until you have read and understand the manufacturer's operating instructions.

*Step 3*  Check all fans and blowers and compressors. Verify the following:
   a. All shipping restraints, lock-down bolts or nuts, and protective covers are removed.
   b. All bearings are lubricated.
   c. All fan wheels clear housings and are correctly positioned. Improper clearance can greatly affect the fan/ blower performance, especially with backward-inclined fans.
   d. Motors and bases are securely fastened.
   e. Condenser fan motors and fan blades are correctly aligned.
   f. All setscrews are tight.
   g. Belt tensions are correct.
   h. Belt guards are in place.

*Step 4*  Inspect the complete supply air system from the supply air fan through the main supply ducts and through all branch duct runs and outlet terminals. Also, inspect the complete return air duct system. Verify the following:
   a. All tools, equipment, and debris are removed.
   b. The ductwork is complete with all openings in the ducts sealed, all end caps in place, and all access doors closed and secured.
   c. All balancing, volume, and terminal dampers are in their marked positions.

*Step 5*  Check the air handling unit filters and any other air filters. Verify the following:
   a. The size and type of filters are correct.
   b. The filters are clean.
   c. The filters and filter frames are properly installed and are airtight.

*Step 6*  Check that the air handling unit coils are undamaged and clean. Check that refrigerant lines do not rub against the cabinet or against other refrigerant lines. Verify that the cooling coil condensate drain pan opening and drain line are clear.

*Step 7*  Check for conditions external to the system that can affect the airflow distribution to the occupied space. Check the following:
   a. Windows or outside doors are closed.
   b. Doors within the building are open.
   c. All ceiling tiles are in place.
   d. All air distribution grilles, registers, and diffusers are unobstructed.

> **CAUTION**
>
> Proper operation of the system dampers during startup is critical. If dampers are closed, restricting airflow, casings, housings, and ductwork can be damaged. On the initial startup of a new system, a temperature control specialist or system designer should be consulted to make sure all automatic temperature control dampers are properly positioned.

*Step 8*  Check that all dampers are open or set so that air travels through the correct components of the system. Normally, dampers should not cause a blocked or restricted condition.

*Step 9*  Turn on internal and external heating fuel valves. If heating is gas-fired, check for and correct any leaks. With the internal gas valve on, check for proper manifold gas pressure at the manifold pressure port on the gas valve.

> **WARNING!**
>
> Gas-fired heating units present an explosion hazard. If the odor of gas is detected, do not try to fire the heating unit. Extinguish any open flame. Do not activate any electrical switch or device. Shut off the gas supply and correct any leaks.

*Step 10*  If equipped with electronic ignition, check for a proper spark gap at the igniter and proper spacing and location of the flame sensor per the manufacturer's instructions.

### 3.2.2 Packaged Unit Startup – Cooling Mode

Startup in cooling mode for a packaged unit is as follows:

*Step 1*   Set external control(s) to activate the cooling mode and turn on the unit power. The unit blower should generally start immediately. In packaged heat pumps especially, the blower will usually operate at the same speed for both the heating and cooling modes. In units equipped with fossil-fuel heating, the blower may operate at a higher speed for cooling. This is less common in commercial applications. It is not uncommon for commercial packaged units to operate at one blower speed for both modes. This is especially true of belt-driven blowers, for obvious reasons.

*Step 2*   If more than one compressor is contained in the unit, check that all three-phase compressors are rotating in the correct direction by monitoring system suction and discharge pressures. Normal suction and discharge pressures usually indicate correct phasing. A phasing meter can also be used to check for correct rotation.

*Step 3*   Check for the proper refrigerant charge per the manufacturer's instructions after the system has stabilized. This is normally accomplished by checking for normal pressures, superheat, and subcooling at each refrigerant circuit.

*Step 4*   If not already done automatically, set external controls to activate cooling mode airflow.

*Step 5*   Check that the supply air blower is operating and its direction of rotation is correct.

*Step 6*   Check that all automatic dampers are being controlled and are in the proper position.

*Step 7*   Ensure that the blower motor current and speed is correct.

*Step 8*   Terminate cooling operation and prepare the unit for heating operation.

### 3.2.3 Packaged Unit Startup – Heating Mode

Many packaged units contain gas heating sections that have similar operating sequences, regardless of the manufacturer. The following is a normal operating sequence for a packaged unit with pilot re-ignition and single-stage heat.

1. Thermostat (or controller) calls for heat.
2. Induced draft motor starts.
3. Pilot gas and spark turn on.
4. Pilot ignites, flame is proved, and pilot spark turns off.
5. Gas is turned on to burners, and burners ignite.
6. Evaporator fan turns on.
7. Thermostat is satisfied.
8. Burners and induced draft motor turn off.
9. Blower turns off.
10. Unit is ready for next heating cycle.

Startup in heating mode for a packaged unit is as follows:

*Step 1*   Set external controls to activate the unit in the heating mode. This may result in a lower (normal) blower speed than used for cooling mode.

*Step 2*   Observe correct burner firing sequence and proper burner flame per the manufacturer's instructions.

*Step 3*   Check that the supply air blower is operating and its direction of rotation is correct.

*Step 4*   Check that all automatic dampers are being controlled and are in the proper position.

*Step 5*   Ensure that the blower motor current and speed is correct for heating operation.

*Step 6*   Terminate heating operation.

### 3.2.4 Packaged Unit Shutdown

If required, periodic shutdown of a packaged unit usually does not require any special service tasks. Do not leave outdoor air dampers open when the unit is not in use, especially during cold weather when freeze-ups can occur. If freezing temperatures will occur during shutdown or during the heating season, the condensate drain traps and condensate pans must be drained.

During extended periods of shutdown, perform scheduled maintenance procedures recommended by the manufacturer per the service literature. These tasks can include the following:

• Motor and/or blower bearing lubrication
• Blower belt replacement
• Filter replacement

- Condenser, evaporator, and blower wheel cleaning
- Inspection of the burner flame and cleaning of burners
- Inspection and cleaning of the combustion air inducer and vent cap

- Inspection and cleaning of the flue box and any flue tube baffles in the heat exchanger tubes

In addition, a shutdown period provides an opportunity to examine and clean the air distribution system ductwork and components. These steps can help improve indoor air quality.

### Additional Resources

*ASHRAE Handbook—HVAC Applications*, Latest Edition. Atlanta, GA: American Society of Heating, Refrigerating, and Air-Conditioning Engineers, Inc.

*ASHRAE Handbook—HVAC Systems and Equipment*, Latest Edition. Atlanta, GA: American Society of Heating, Refrigerating, and Air-Conditioning Engineers, Inc.

### 3.0.0 Section Review

1. When shutting down an air handler for winter, leave outdoor air dampers _____.

    a. fully closed
    b. fully open
    c. open 50 percent
    d. open 10 percent

2. Three-phase compressors used in packaged units can be checked for correct rotation using a _____.

    a. tachometer
    b. clamp-on ammeter
    c. phasing meter
    d. strobe light

## SUMMARY

Proper startup and shutdown of HVAC systems is necessary in order to make sure the systems operate efficiently without premature failure. Although the procedures vary from one type of equipment to another, the goal is basically the same.

Systems that condition water, such as boilers and chillers, generally require more attention than those that are refrigerant-based only. Steam and hot-water boilers require a significant amount of attention, since both the water and combustion byproducts can create problems during the shutdown period.

It is essential that startup and shutdown tasks are completed in accordance with manufacturer guidelines. These tasks must also be completed in the proper sequence to ensure a successful procedure and technician safety.

1. When an oil-fired boiler is shut down, corrosion can be greatly reduced by _____.

   a. cleaning away soot and drying the cleaned areas with a heater
   b. applying a heavy-grade lubricant to all surfaces
   c. drying the boiler with a cloth and covering it with a tarp
   d. applying protective paint to areas of corrosion

2. When preparing a steam boiler for startup, the main steam stop valve should be _____.

   a. fully open
   b. fully closed
   c. 50 percent open
   d. slightly open

3. When initially firing a boiler at startup, the burners should be operated at _____.

   a. low-fire
   b. mid-fire
   c. high-fire
   d. standby

4. When the steam gauge just begins to register pressure during the startup of a steam boiler, _____.

   a. fully open the main stop valve
   b. vent the boiler so that no air is trapped
   c. blow down the gauge glass and water column
   d. slightly open the main stop valve

5. When a hot-water boiler includes a dip tube, its purpose is to _____.

   a. indicate the level of hot water in the boiler
   b. allow a check of the fuel level of the boiler
   c. allow the addition of water treatment chemicals
   d. capture any air or oxygen released from the water

6. The minimum boiler temperature recommended with hot-water boilers is _____.

   a. 140°F
   b. 180°F
   c. 190°F
   d. 200°F

7. How many hours should the crankcase heaters in a reciprocating chiller compressor be energized before attempting chiller startup?

   a. 8
   b. 12
   c. 24
   d. 36

8. With the deadband on a centrifugal chiller's water temperature control at 2°F, how close will the chilled water temperature be to the temperature control setpoint?

   a. –2°
   b. ±1°
   c. ±2°
   d. ±4°

9. When preparing a screw chiller for extended shutdown at temperatures below 32°F, _____.

   a. the oil charge should be drained from the machine
   b. a holding charge of nitrogen should be maintained
   c. the refrigerant should be placed in a storage tank
   d. the water circuits should be completely drained

10. Which of the following statements about centrifugal chillers is correct?

    a. The chiller refrigerant circuit should be held in a vacuum.
    b. The entire refrigerant charge should be transferred into the condenser section.
    c. The chiller refrigerant circuit should be filled with oxygen as a holding charge.
    d. The refrigerant charge should be transferred to a storage tank during a prolonged shutdown.

11. A cooling tower and its piping should be drained, cleaned, and disinfected at least _____.

    a. monthly
    b. twice monthly
    c. yearly
    d. twice a year

12. When operating a cooling tower in subfreezing temperature conditions, it is recommended that _____.

    a. salt be added to the water
    b. more frequent visual inspections be made
    c. the treated water be heated
    d. the tower be operated after sundown

13. The minimum interval at which shell and tube condensers should be cleaned is _____.

    a. weekly
    b. monthly
    c. yearly
    d. twice a year

14. To ensure that the coils in an air handler are properly cleaned, it is best to use a cleaner that is _____.

    a. alkaline-based
    b. acid-based
    c. water-soluble
    d. fast-drying

15. Duct cleaning should be done using _____.

    a. steel brushes and pressure washer
    b. high-pressure blowers
    c. a pressure washer and squeegees
    d. HEPA-filtered vacuuming and power brushing equipment

# Trade Terms Introduced in This Module

**Bleed-off**: A method used to help control corrosion and scaling in a water system. It involves the periodic draining and disposal of a small amount of the water circulating in a system. Bleed-off aids in limiting the buildup of impurities caused by the continuous addition of makeup water to a system.

**Deadband**: In a chiller, the tolerance on the chilled-water temperature control point. For example, a 1°F deadband controls the water temperature to within ±0.5°F of the control point temperature (0.5°F + 0.5°F = 1°F deadband).

**Desiccant**: A moisture-absorbing material.

**Layup**: An industry term referring to the period of time a boiler is shut down.

**Recycle shutdown mode**: A chiller mode of operation in which automatic shutdown occurs when the compressor is operating at minimum capacity and the chilled-water temperature has dropped below the chilled-water temperature setpoint. In this mode, the chilled-water pump remains running so that the chilled-water temperature can be monitored.

TYPICAL RELATIONSHIPS BETWEEN $O_2$, $CO_2$, AND EXCESS AIR

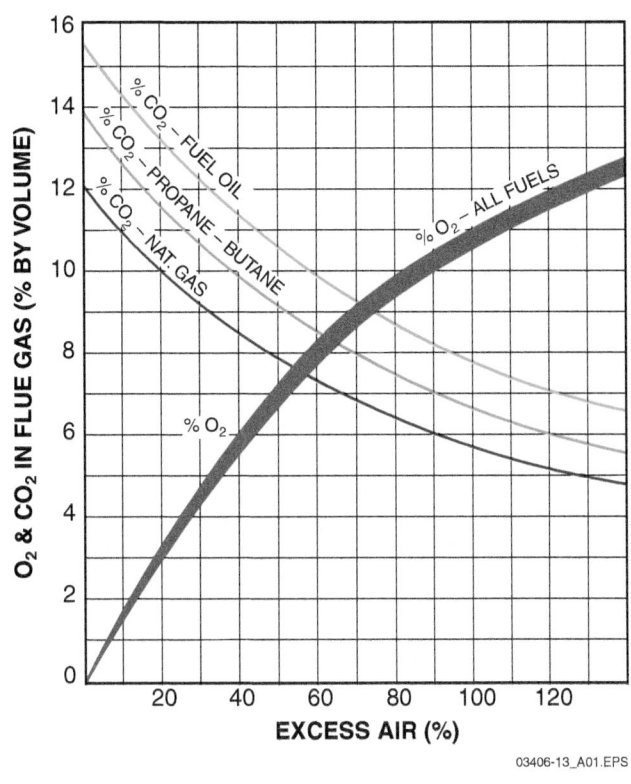

03406-13_A01.EPS

# Additional Resources

This module presents thorough resources for task training. The following resource material is suggested for further study.

*ASHRAE Handbook—HVAC Applications*, Latest Edition. Atlanta, GA: American Society of Heating, Refrigerating, and Air-Conditioning Engineers, Inc.

*ASHRAE Handbook—HVAC Systems and Equipment*, Latest Edition. Atlanta, GA: American Society of Heating, Refrigerating, and Air-Conditioning Engineers, Inc.

# Figure Credits

Courtesy of Fulton Boiler Works, Inc., Figure 1

Courtesy of Hurst Boiler and Welding Company, Inc., Figure 2

Cleaver-Brooks, Inc., Figures 3, 4

Courtesy of Highside Chemicals, Inc., Figure 6B

Courtesy of Bacharach, Inc./www.MyBacharach.com, SA01

Courtesy of Daikin McQuay, Figure 11

Courtesy of Goodway Technologies Corporation, Figure 13

Courtesy of Ritchie Engineering Company, Inc., YELLOW JACKET Products Division, SA02

| Answer | Section Reference | Objective |
|---|---|---|
| **Section One** | | |
| 1.a | 1.1.1 | 1a |
| 2.b | 1.2.0 | 1b |
| 3.d | 1.3.0 | 1c |
| 4.c | 1.4.3 | 1d |
| **Section Two** | | |
| 1.b | 2.1.1 | 2a |
| 2.c | 2.2.2 | 2b |
| 3.a | 2.3.1 | 2c |
| 4.b | 2.4.1 | 3c |
| **Section Three** | | |
| 1.a | 3.1.2 | 3a |
| 2.c | 3.2.2 | 3b |

# NCCER CURRICULA — USER UPDATE

NCCER makes every effort to keep its textbooks up-to-date and free of technical errors. We appreciate your help in this process. If you find an error, a typographical mistake, or an inaccuracy in NCCER's curricula, please fill out this form (or a photocopy), or complete the online form at **www.nccer.org/olf**. Be sure to include the exact module ID number, page number, a detailed description, and your recommended correction. Your input will be brought to the attention of the Authoring Team. Thank you for your assistance.

*Instructors* – If you have an idea for improving this textbook, or have found that additional materials were necessary to teach this module effectively, please let us know so that we may present your suggestions to the Authoring Team.

**NCCER Product Development and Revision**

13614 Progress Blvd., Alachua, FL 32615

**Email:** curriculum@nccer.org
**Online:** www.nccer.org/olf

❏ Trainee Guide ❏ Lesson Plans ❏ Exam ❏ PowerPoints Other _____

Craft / Level: _____ Copyright Date: _____

Module ID Number / Title: _____

Section Number(s): _____

Description: _____

_____

_____

_____

Recommended Correction: _____

_____

_____

_____

_____

Your Name: _____

Address: _____

Email: _____ Phone: _____

# Construction Drawings and Specifications

## OVERVIEW

Anyone involved in the installation of heating and cooling equipment for new construction must be able to interpret the project drawings and specifications. The drawings show the locations of equipment, duct runs, piping runs, and electrical wiring. During the estimating and planning processes, the drawings are used to determine the amount and types of equipment, accessories, and materials needed for the job. Correct interpretation of the drawings is essential in order to determine the correct price for the job, and to have the correct amounts and types of equipment and materials available. A technician or installer who does not learn to interpret them properly is unlikely to advance very far in his or her career.

# Module 03401

Trainees with successful module completions may be eligible for credentialing through NCCER's National Registry. To learn more, go to **www.nccer.org** or contact us at **1.888.622.3720**. Our website has information on the latest product releases and training, as well as online versions of our *Cornerstone* magazine and Pearson's product catalog.

Your feedback is welcome. You may email your comments to **curriculum@nccer.org,** send general comments and inquiries to **info@nccer.org**, or fill in the User Update form at the back of this module.

This information is general in nature and intended for training purposes only. Actual performance of activities described in this manual requires compliance with all applicable operating, service, maintenance, and safety procedures under the direction of qualified personnel. References in this manual to patented or proprietary devices do not constitute a recommendation of their use.

*03401 V5*

## Objectives

When you have completed this module, you will be able to do the following:

1. Describe the types of drawings HVAC technicians work with and how they are used.
   a. Explain the initial approach to viewing a set of drawings.
   b. Describe site plans and their purpose.
   c. Describe plan views, elevations, detail drawings, and section drawings and their purposes.
   d. Describe plumbing, mechanical, and electrical drawings and their purposes.
   e. Describe shop drawings and their purpose.
   f. Describe as-built drawings and their purpose.
   g. Describe schedules and their purpose.
   h. Describe the Request for Information (RFI) and how it is prepared.
   i. Explain the importance of building codes to the design process.
2. Describe the uses of specifications and submittals in construction projects.
   a. Describe specifications and their purpose.
   b. Describe submittals and their purpose.
3. Describe the takeoff process and how it is performed.
   a. Identify and describe the tools and materials used in the takeoff process.
   b. Explain how to conduct a takeoff.

## Performance Tasks

Under the supervision of your instructor, you should be able to do the following:

1. Identify and interpret the following on an architectural drawing:
   - Floor plans and details
   - Elevations
   - Foundation plan
   - Reflected ceiling plan
2. Identify and interpret at least four of the following on a plumbing plan drawing:
   - Sanitary plumbing plans
   - Domestic water plumbing plans
   - Riser diagrams
   - Schedules
   - Specification references
   - Legends
3. Identify and interpret the following on a mechanical plan drawing:
   - Hot- and chilled-water coil piping
   - HVAC piping
   - Chiller piping/installation
   - Refrigeration piping schematics
   - Air handling unit installation/connecting ductwork
   - Hot- and chilled-water flow diagrams
   - Schedules
   - Specification references
   - Legends

4. Identify and interpret the following on an electrical plan drawing:
   - Riser diagrams
   - Schedules
   - Specification references
   - Legends
5. Interpret HVAC-related shop drawings.
6. Perform an HVAC equipment and material takeoff and prepare the takeoff forms.

## Trade Terms

Change order
Coordination drawing
Cut list
Detail drawing
Elevation view
Floor plan
Longitudinal section
Plan view

Riser diagram
Schedules
Section drawing
Shop drawing
Site plan
Takeoff
Transverse section

## Industry-Recognized Credentials

If you are training through an NCCER-accredited sponsor, you may be eligible for credentials from NC-CER's Registry. The ID number for this module is 03401. Note that this module may have been used in other NCCER curricula and may apply to other level completions. Contact NCCER's Registry at 888.622.3720 or go to **www.nccer.org** for more information.

# Contents

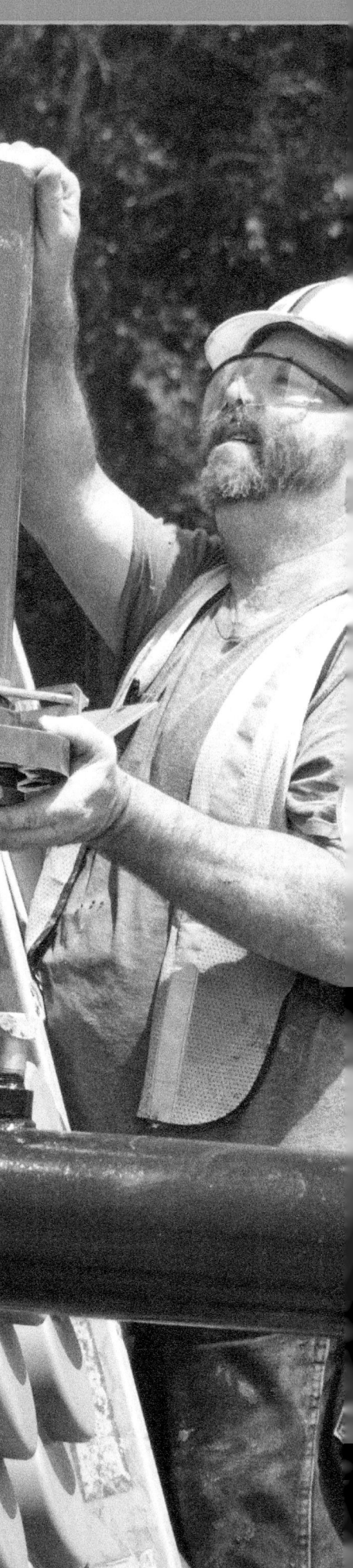

## Figures

## SECTION ONE

### 1.0.0 HVAC DRAWING TYPES

### Objectives

Describe the types of drawings HVAC technicians work with and how they are used.

a. Explain the initial approach to viewing a set of drawings.
b. Describe site plans and their purpose.
c. Describe plan views, elevations, detail drawings, and section drawings and their purposes.
d. Describe plumbing, mechanical, and electrical drawings and their purposes.
e. Describe shop drawings and their purpose.
f. Describe as-built drawings and their purpose.
g. Describe schedules and their purpose.
h. Describe the Request for Information (RFI) and how it is prepared.
i. Explain the importance of building codes to the design process.

### Performance Tasks

1. Identify and interpret the following on an architectural drawing:
   - Floor plans and details
   - Elevations
   - Foundation plan
   - Reflected ceiling plan
2. Identify and interpret at least four of the following on a plumbing plan drawing:
   - Sanitary plumbing plans
   - Domestic water plumbing plans
   - Riser diagrams
   - Schedules
   - Specification references
   - Legends
3. Identify and interpret the following on a mechanical plan drawing:
   - Hot- and chilled-water coil piping
   - HVAC piping
   - Chiller piping/installation
   - Refrigeration piping schematics
   - Air handling unit installation/connecting ductwork
   - Hot- and chilled-water flow diagrams
   - Schedules
   - Specification references
   - Legends

4. Identify and interpret the following on an electrical plan drawing:
   - Riser diagrams
   - Schedules
   - Specification references
   - Legends
5. Interpret HVAC-related shop drawings.

### Trade Terms

**Coordination drawings**: Elevation, location, and other drawings produced for a project by the individual contractors for each trade to prevent a conflict between the trades regarding the installation of their materials and equipment. Development of these drawings evolves through a series of review and co-ordination meetings held by the various contractors.

**Cut list**: An information sheet that is derived from shop drawings. It is the shop guide for fabricating duct runs and fittings.

**Detail drawing**: A drawing of a feature that provides more elaborate information than is available on a plan.

**Elevation view**: A view that depicts a vertical side of a building, usually designated by the direction that side is facing; for example, right, left, east, or west elevation.

**Floor plan**: A building drawing indicating a plan view of a horizontal section at some distance above the floor, usually midway between the ceiling and the floor.

**Longitudinal section**: A section drawing in which the cut is made along the long dimension of the building.

**Plan view**: The overhead view of an object or structure.

**Riser diagram**: A one-line schematic depicting the layout, components, and connections of a piping system or electrical system.

**Schedules**: Tables that describe and specify the types and sizes of items required for the construction of a building.

**Section drawing**: A drawing that depicts a feature of a building as if there were a cut made through the middle of it.

**Shop drawing**: A drawing that indicates how to fabricate and install individual components of a construction project. A shop drawing may be drafted from the construction drawings of a project or provided by the manufacturer.

**Site plan**: A construction drawing that indicates the location of a building on a land site.

**Transverse section**: A section drawing in which the "cut" is made along the short dimension of the building.

This module reviews and builds on the information previously studied in the Core Curriculum module *Introduction to Construction Drawings*. It focuses on techniques for reading various types of HVAC-related construction drawings and project specifications. Construction drawings tell the HVAC technician and installer, as well as other skilled tradespeople, how to build a specific building or structure. A specification is a related contractual document used along with the construction drawings. It contains detailed written instructions that supplement the information shown in the set of drawings. As an HVAC technician, you must be able to interpret drawings and specifications correctly. Failure to do so may result in costly rework and unhappy customers. Depending on the severity of a mistake, it can also expose you and your employer to legal liability.

## 1.1.0 Viewing a Set of Drawings

The following general procedure is suggested as a method of reading a set of drawings. Use this procedure to familiarize yourself with an available set of drawings:

*Step 1* Locate and read the title block. The title block tells you what the drawing is about. It contains critical information about the drawing such as the scale, date of last revision, drawing number, and architect or engineer. If you have to remove a sheet from a set of drawings, be sure to fold the sheet with the title block facing up.

*Step 2* Find the North arrow. Always orient yourself to the structure. Knowing where North is enables you to more accurately describe the locations of walls and other parts of the building.

*Step 3* Check the list of drawings in the set. Note the sequence of the various types of plans. Some drawings have an index on the front cover. Notice that the prints are broken into several categories, as shown in *Figure 1*. However, drawing sets do not have to contain all the categories of drawings shown.

*Step 4* Study the site plan to observe the location of the building. Also notice that the geographic location of the building may be indicated on the site plan.

*Step 5* Check the floor plan for the orientation of the building. Observe the location and other details of entries, corridors, offsets, and any special features.

*Step 6* Study the features that extend for more than one floor, such as plumbing, vents, stairways, elevator shafts, heating and cooling ductwork, and piping. Determine the location of all main electrical runs and fire sprinkler system lines while looking for possible installation conflicts.

*Step 7* Check the floor and wall construction and other details relating to exterior and interior walls.

*Step 8* Check the foundation plan for size and types of footings, reinforcing steel, and loadbearing substructures.

*Step 9* Study the mechanical plans for heating, cooling, and plumbing details.

*Step 10* Observe the electrical entrance and distribution panels, as well as the installation of the lighting and power supplies for special equipment.

*Step 11* Check the notes on the various pages and compare the specifications against the construction details. Look for any variations.

---

**Did You Know?**

# Blueprints

The term *blueprint* is derived from a method of reproduction used in the past. A true blueprint shows the details of the structure as white lines on a blue background.

The method involved coating a paper with a specific chemical. After the coating dried, an original hand drawing was placed on top of the paper. Both papers were then covered with a piece of glass and set in the sunlight for about an hour. The coated paper was developed much like a photograph. After a cold-water wash, the coated paper turned blue and the lines from the drawing remained white.

Today, most drawing reproduction methods produce a black line on a white background. These copies or prints are typically made using a copying machine. However, the term *blueprint* is still widely used, but has been replaced with the terms *drawings* and *prints*.

---

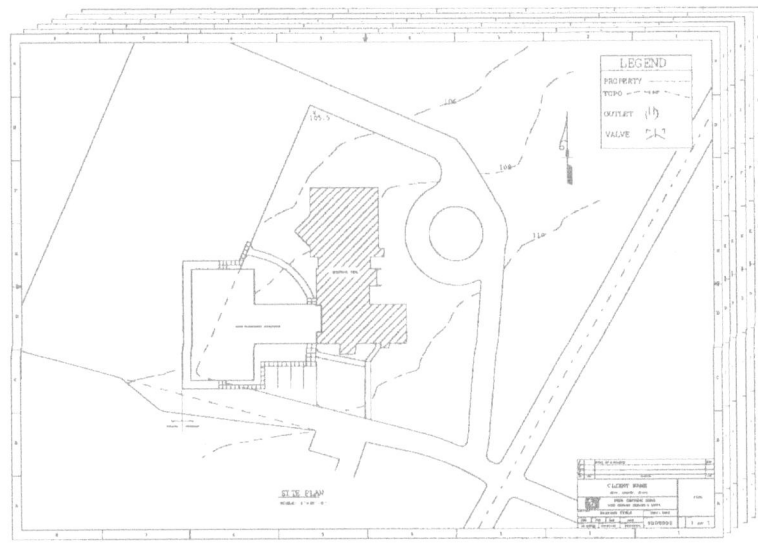

TITLE SHEET(S)
ARCHITECTURAL DRAWINGS
- SITE (PLOT) PLAN
- FOUNDATION PLAN
- FLOOR PLANS
- INTERIOR/EXTERIOR ELEVATIONS
- SECTIONS
- DETAILS
- SCHEDULES

STRUCTURAL DRAWINGS
PLUMBING PLANS
MECHANICAL PLANS
ELECTRICAL PLANS

03401-13_F01.EPS

*Figure 1* Typical categories of drawings in a set of construction drawings.

*Step 12* Thumb through the sheets of drawings until you are familiar with all the plans and structural details.

*Step 13* Recognize applicable symbols and their relative locations in the plans (see *Appendix A* for a listing of common line types, symbols, and abbreviations). Note any special construction details or variations that will affect your trade.

## 1.2.0 Site Plans

The site plan, also called a plot plan, (*Figure 2*) indicates the location of the building on the land site. It may include topographic features such as contour lines, trees, and shrubs. It may also include some construction features such as walks, driveways, curbs, and gutters.

Often the roof plan, if there is one, is also shown on the site plan. General notes pertaining to grading and shrubbery may also be included on the site plan. A separate landscape plan may also exist.

On large commercial jobs, a utility site plan may also be included in the drawing set. It shows the locations for underground facilities such as gas and water pipelines, sanitary sewers, electric power or communication system cables, and other facilities.

## 1.3.0 Plans Views, Elevations, Detail, and Section Drawings

Several types of drawings—including exterior views, cut-aways, enlargements, and various forms of overhead views (plan views)—are used to show different structural aspects and perspectives of a building. Taken together, the drawings show the big picture of how the building is constructed and they provide details for special features such as fixtures and equipment.

### 1.3.1 Plan Views (Floor, Roof, and Ceiling Plans)

The floor plan (*Figure 3*) is one of the most important working drawings. It is the first drawing done by the designer. It shows the length and width of the building and the location of the rooms and other spaces that the building contains. Each floor of the structure has a different floor plan.

A floor plan is a plan view of a horizontal section taken at some distance above the floor, usually midway between the floor and ceiling. However, the cutting plane of the sectional drawing may be offset to a higher or lower level so that it cuts through the desired features such as windows and doors. The cut view crosses all openings for that floor or story. This view gives the dimensions of the window and door openings and indicates which way the doors are to swing.

When supplied, the contractor consults the roof plan for information pertaining to roof slope, roof drain placement, and other pertinent information regarding ornamental sheet metal work and gutter and downspout technical information. *Figure 4* shows an example of a typical roof plan. Where applicable, the roof plan may also contain information on the location of air conditioning units, exhaust fans, and other ventilation equipment.

Some drawing sets include a reflected ceiling plan (*Figure 5*). This view shows the ceiling as if

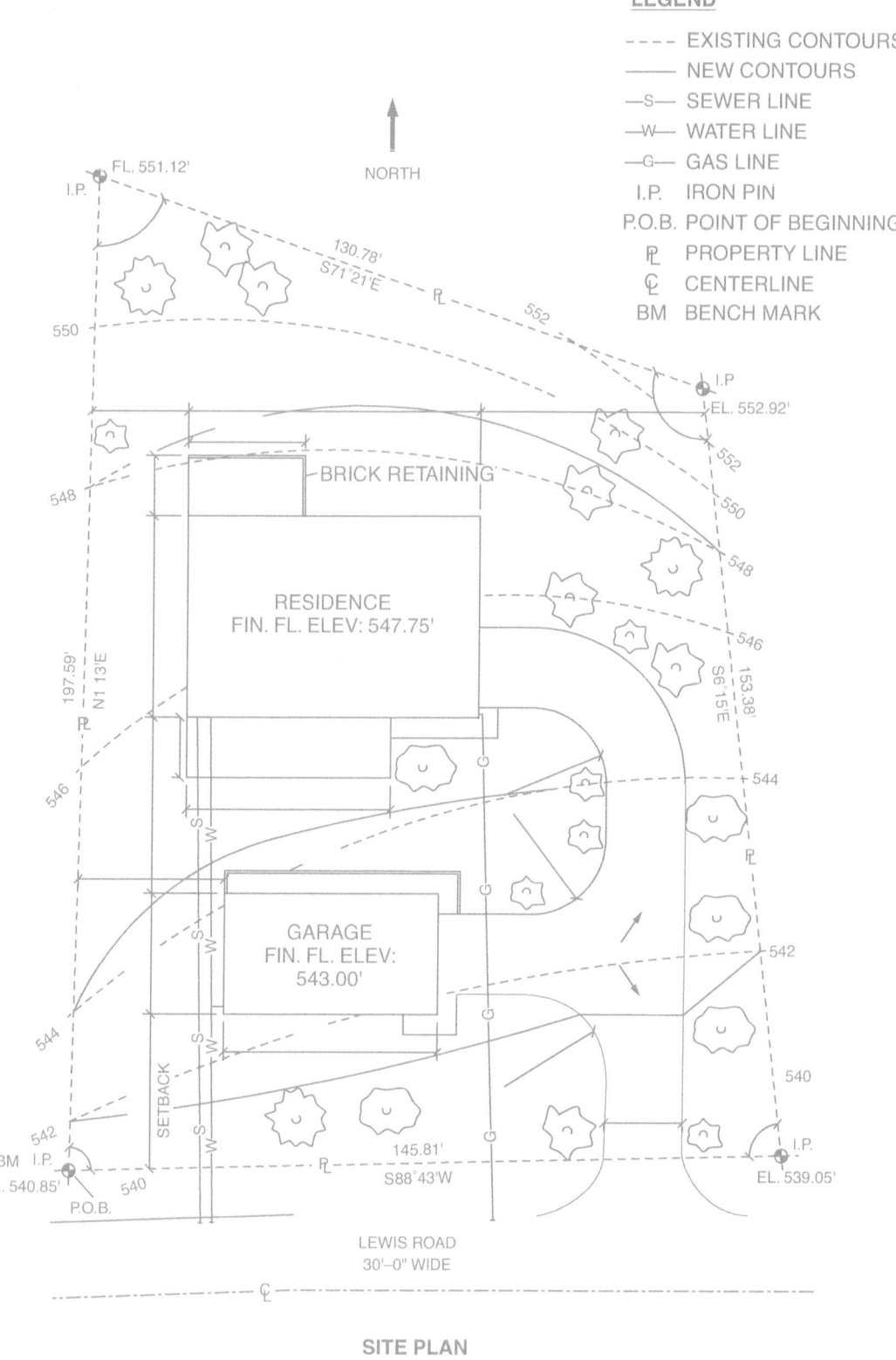

- - - - EXISTING CONTOURS
———— NEW CONTOURS
—S— SEWER LINE
—W— WATER LINE
—G— GAS LINE
I.P. IRON PIN
P.O.B. POINT OF BEGINNING
℗ PROPERTY LINE
℄ CENTERLINE
BM BENCH MARK

NORTH

FL. 551.12'
I.P.

130.78'
S71°21'E

℗

552

550

I.P
EL. 552.92'

BRICK RETAINING

552

548

550

548

RESIDENCE
FIN. FL. ELEV: 547.75'

546

197.59'
N1°13'E

℗

546

153.38'
S6°15'E

544

546

S
W

G

G

℗

542

GARAGE
FIN. FL. ELEV:
543.00'

G

544

540

542

542

540

S
W

SETBACK

BM I.P.
EL. 540.85'
P.O.B.

540

℗

145.81'
S88°43'W

G

I.P.
EL. 539.05'

LEWIS ROAD
30'–0" WIDE

℄

SITE PLAN
SCALE: 1" = 30'–0"

03401-13_F02.EPS

*Figure 2* Site plan.

# Architectural Plans

Architectural plans have been used over the centuries to pictorially describe buildings and structures before they are actually built. In the past, draftspersons would draw these plans by hand. Today, most drawings for buildings and other structures are generated by computer using a process called computer-aided design (CAD). Working with a variety of architectural software, the drafter creates the drawings for the building or structure electronically on the computer. Then, using a computer command, the electronic drawing files are sent to a printer or plotter to be output on paper.

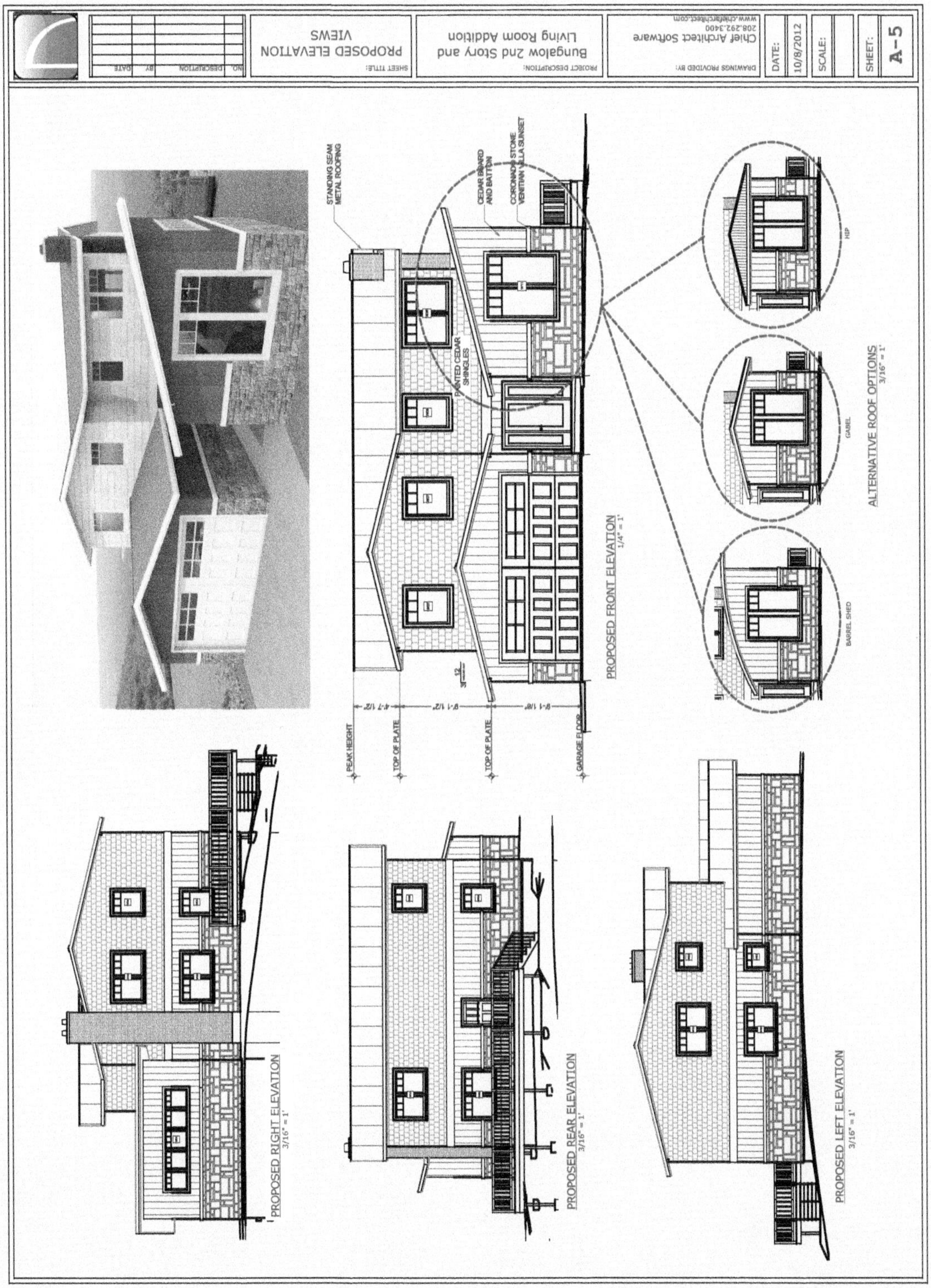

03401-13_SA01.EPS

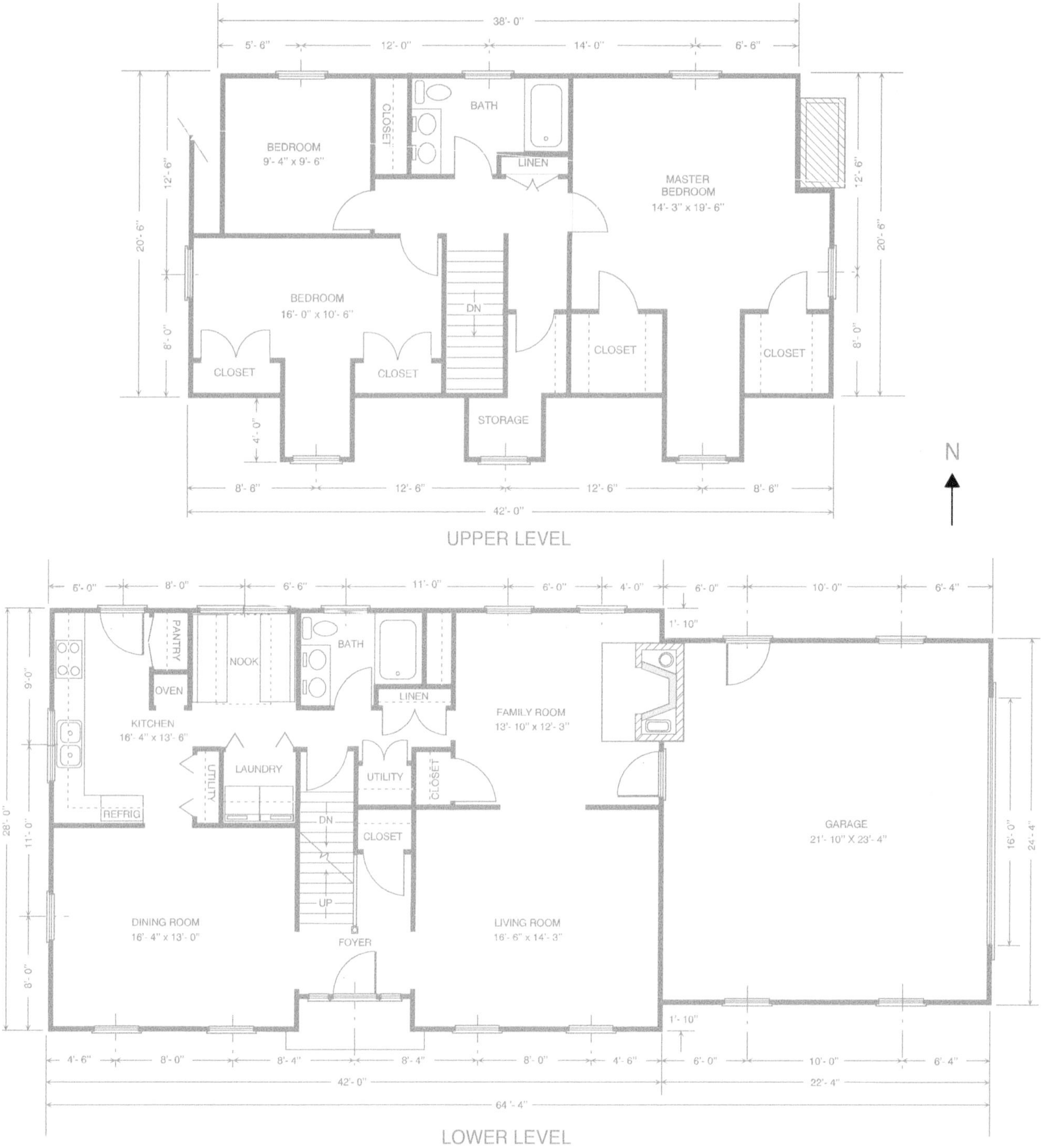

**Figure 3** Floor plans for a building.

it were reflected down into a mirror. A reflected ceiling plan is of particular value to the HVAC contractor, providing information that identifies the location of supply diffusers, exhaust grilles, access panels, and other structural components.

### 1.3.2 Elevation Drawings

The elevation view of a structure (*Figure 6*) shows the exterior features of that structure. Unless one or more views are identical, four views are generally used to show each exposure. With very complex buildings, more than four views may be required. Elevation drawings show the exterior style of the building as well as the placement of doors, windows, chimneys, and decorative trim.

The various views are usually labeled in one of two ways. They may be broken down as front view, right side view, left side view, and rear view, or they may be designated by compass direction.

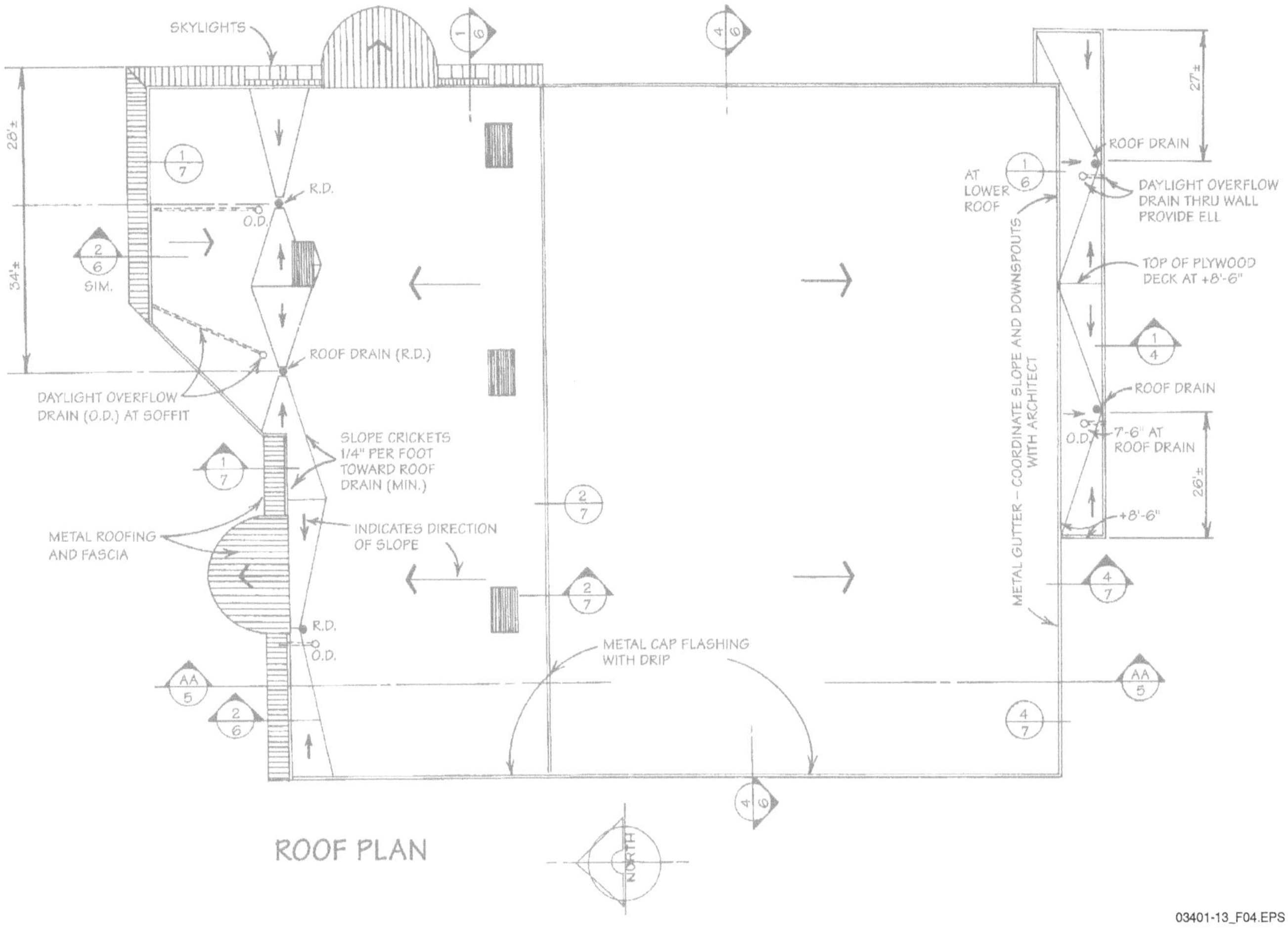

## Figure 4 (Roof Plan)

The following labels appear on the roof plan drawing:

- SKYLIGHTS
- R.D.
- O.D.
- DAYLIGHT OVERFLOW DRAIN (O.D.) AT SOFFIT
- ROOF DRAIN (R.D.)
- SLOPE CRICKETS 1/4" PER FOOT TOWARD ROOF DRAIN (MIN.)
- INDICATES DIRECTION OF SLOPE
- METAL ROOFING AND FASCIA
- R.D.
- O.D.
- METAL CAP FLASHING WITH DRIP
- 28'±
- 34'±
- SIM.
- AT LOWER ROOF
- ROOF DRAIN
- DAYLIGHT OVERFLOW DRAIN THRU WALL PROVIDE ELL
- TOP OF PLYWOOD DECK AT +8'-6"
- ROOF DRAIN
- O.D.
- 7'-6" AT ROOF DRAIN
- +8'-6"
- 27'±
- 26'±
- METAL GUTTER – COORDINATE SLOPE AND DOWNSPOUTS WITH ARCHITECT
- AA / 5
- AA / 5
- NORTH

**ROOF PLAN**

03401-13_F04.EPS

*Figure 4* Roof plan.

If the front of the building faces east, then this becomes the east elevation. The other elevations are then labeled accordingly: west, south, and north.

Materials used for the exterior finish of a building are also indicated on elevation drawings and described in detail in the specifications, either as part of the drawings or as a separate set of written specifications. Foundation walls, footings, and parts of a building hidden from view are shown by broken lines. Elevation drawings often show elements such as the heights of windows, doors, and porches and the pitches of roofs because all of these measurements cannot be shown conveniently on floor plans.

### 1.3.3 Detail Drawings

Detail drawings show enlargements of special features of a building construction, fixtures, or equipment (*Figure 7*). They are drawn to a larger scale in order to make the details clearer. Note that the scale can vary from one detail view to another on the same sheet. For residential and some commercial projects, the detail drawings are often placed on the same sheet where the feature appears in the plan. However, for large or more complex commercial projects, detailed drawings may be drawn on a different sheet than where the feature appears. When this occurs, the detail drawings are referenced to and from the sheets where they apply.

### 1.3.4 Section Drawings

Section drawings (*Figure 8*) are cut-away views that allow the viewer to see the inside of a structure or how something is put together internally. A feature is drawn as if a cut has been made through the middle of it. When a sectional cut is made along the long dimension of a building, it is called a longitudinal section. When it is made through the short dimension, it is called a transverse section. The point on the drawing that the section was taken from, showing where the imaginary cut has been made, is indicated by the section line. A section line is usually a dashed line with arrows on each end to indicate the direction of view. Letters and numbers identifying the section drawing are placed on the arrows or near the line.

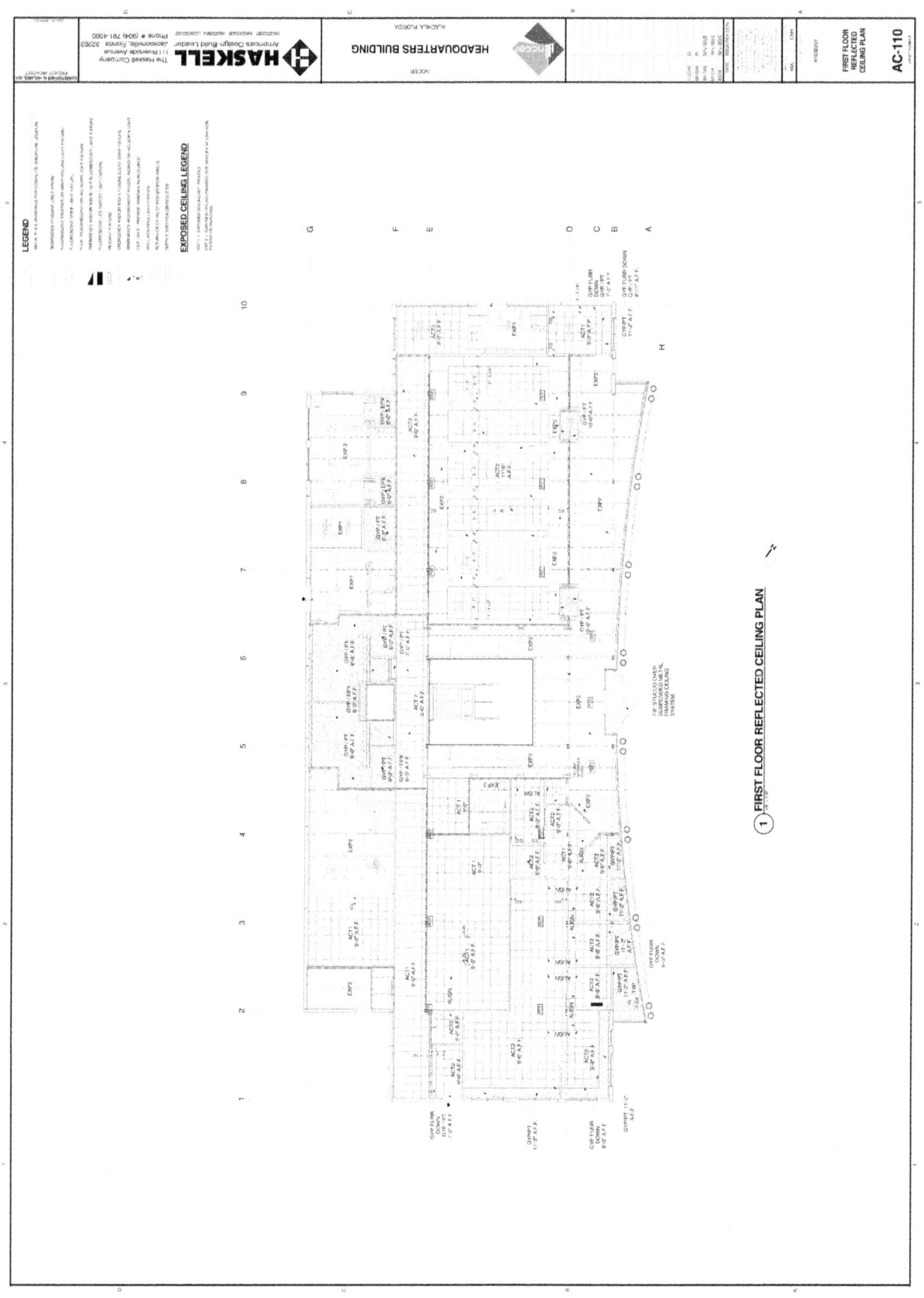

**Figure 5** Reflected ceiling plan.

Like detail drawings, section drawings are drawn to a larger scale than that used in plan views. Section views are commonly given for the construction of walls, stairs, cabinets, and other building features that require more information than is given on the plan views.

Section views on mechanical plans are typically used to show more information about the in-

---

8 NCCER – *HVAC Level Four* 03401

stallation features of a particular fixture or piece of equipment within a building. The detail views in *Figure 9* show locations for the installation of an air handler unit.

### 1.4.0 Plumbing, Mechanical, and Electrical Drawings

Some working drawings, such as floor plans, may also show the plumbing, mechanical, and electrical systems for a building. Often, however, these systems are shown in their own separate plans or drawings. The information that these drawings provide is particularly useful for HVAC technicians, installers, and other tradespeople.

### 1.4.1 Plumbing Plans

Plumbing plans show the layout of fixtures, water supply lines, natural gas piping, and lines to sewage disposal systems. The plans may be included in the floor plan of a regular construction job or on a separate plan for a large commercial structure. When drawn as a separate plan, the plumbing plan details are usually overlaid on tracings of the various building floor plans from which unnecessary details have been omitted to allow the location and layout of the plumbing systems to show clearly. *Figure 10* is a plumbing plan showing the sanitary plumbing in both plan form and as an isometric drawing.

A plumbing legend shows the various symbols pertaining to the plan (*Figure 11*). Some legends provide tabulated plumbing fixture and equipment schedules. Plumbing plans may also show a schedule of plumbing systems and plumbing system specifications. Schedules are discussed in more detail later on in this section.

### 1.4.2 Mechanical Plans

Mechanical plans show the heating, ventilation, and air conditioning systems, as well as other mechanical systems for a building. For some residential jobs, the mechanical plan may be combined with the plumbing plan and show very little detailed information other than the locations of the main HVAC system components. This is because the installation location of duct or piping runs is allowed to be determined on the job by the HVAC contractor. For large commercial jobs, the mechanical plans typically show detailed information about the HVAC system installation and the installation of other equipment. Information

## Floor Plans

The floor plan is a drawing of the structure as if it were cut horizontally about 4' above floor level and viewed from above. This horizontal cutting plane will show most of the windows and doors.

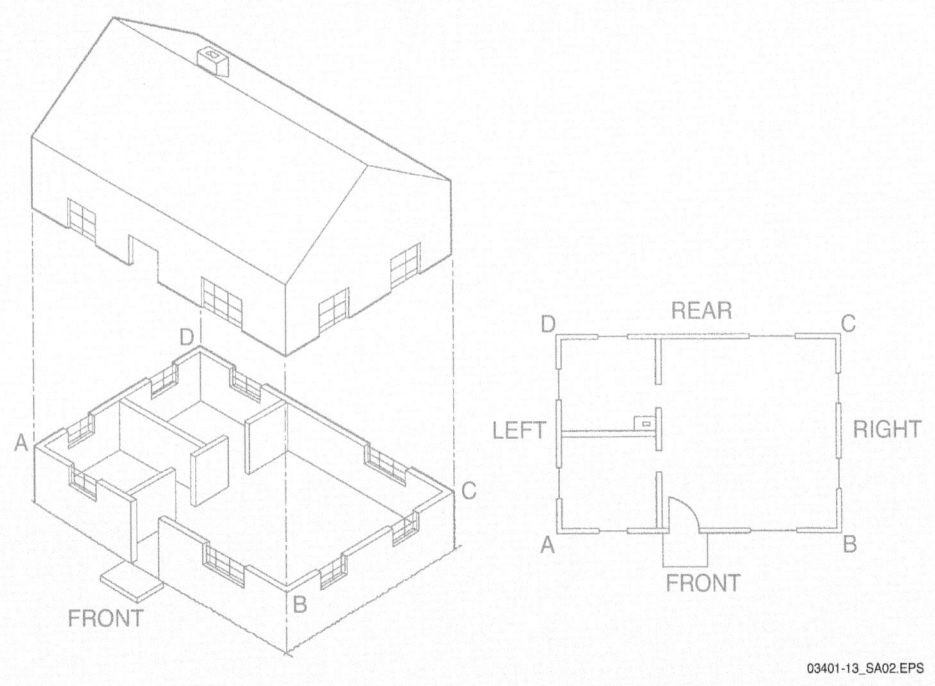

03401-13_SA02.EPS

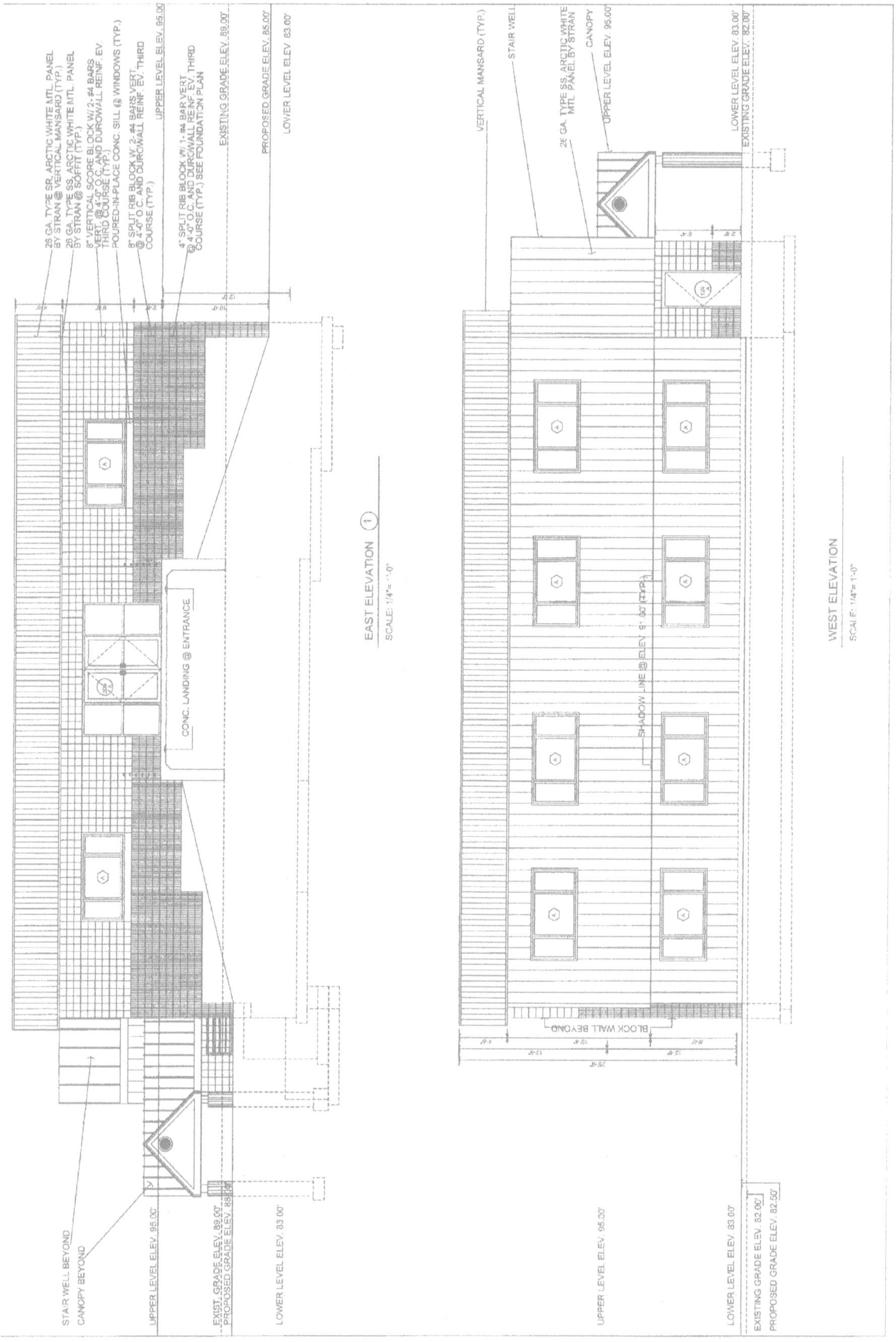


03401-13_F06.EPS

***Figure 6*** Elevation drawing.

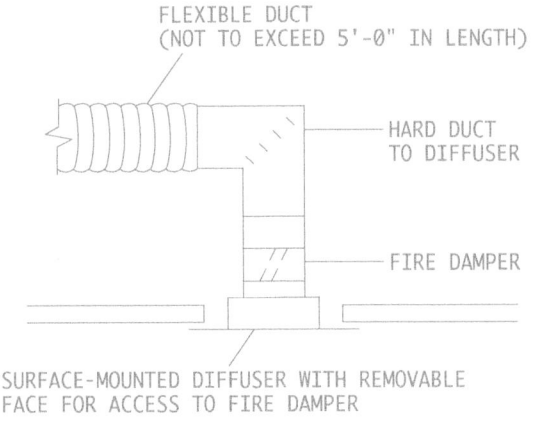

FLEXIBLE DUCT
(NOT TO EXCEED 5'-0" IN LENGTH)

HARD DUCT
TO DIFFUSER

FIRE DAMPER

SURFACE-MOUNTED DIFFUSER WITH REMOVABLE
FACE FOR ACCESS TO FIRE DAMPER

DIFFUSER WITH FIRE DAMPER

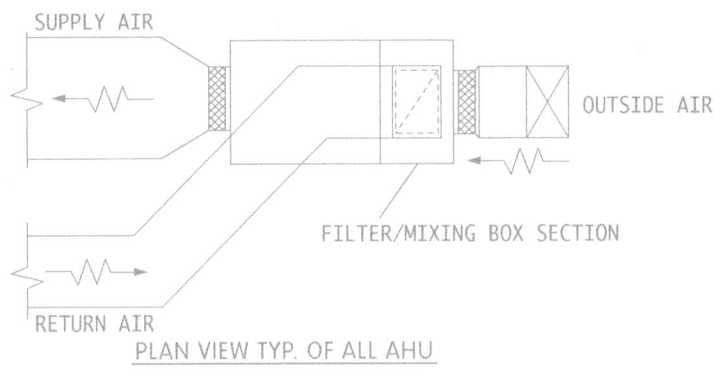

SUPPLY AIR

OUTSIDE AIR

FILTER/MIXING BOX SECTION

RETURN AIR

PLAN VIEW TYP. OF ALL AHU

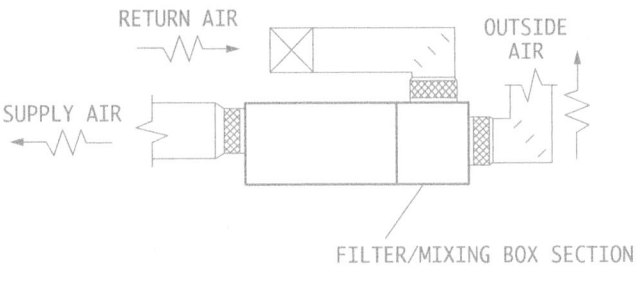

RETURN AIR

OUTSIDE AIR

SUPPLY AIR

FILTER/MIXING BOX SECTION

ELEVATION VIEW TYP. OF ALL AHU

03401-13_F07.EPS

*Figure 7* Equipment-related detail drawings.

# Check the Legend

To avoid making mistakes when reading the drawings, be sure you understand the symbols and abbreviations used in the plumbing, HVAC, and electrical legends on every drawing set. Symbols and abbreviations may vary widely from one drawing set to another.

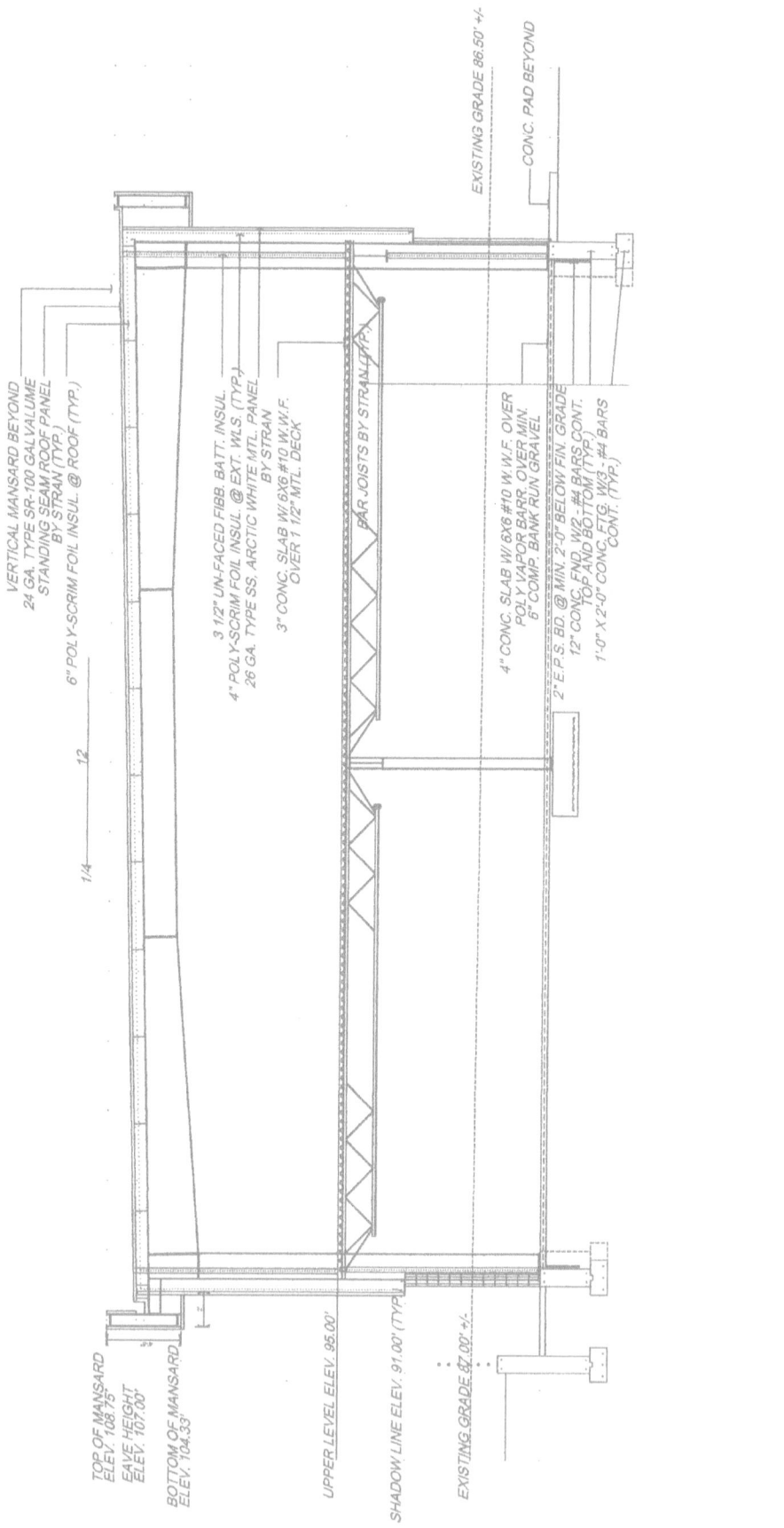

VERTICAL MANSARD BEYOND
24 GA. TYPE SR-100 GALVALUME
STANDING SEAM ROOF PANEL
BY STRAN (TYP.)

6" POLY-SCRIM FOIL INSUL. @ ROOF (TYP.)

3 1/2" UN-FACED FIBB. BATT. INSUL.

4" POLY-SCRIM FOIL INSUL. @ EXT. WLS. (TYP.)

26 GA. TYPE SS. ARCTIC WHITE MTL. PANEL
BY STRAN

3" CONC. SLAB W/ 6x6 #10 W.W.F.
OVER 1 1/2" MTL. DECK

BAR JOISTS BY STRAN (TYP.)

4" CONC. SLAB W/ 6X6 #10 W.W.F. OVER
POLY VAPOR BARR. OVER MIN.
6" COMP. BANK RUN GRAVEL

2" E.P.S. BD. @ MIN. 2'-0" BELOW FIN. GRADE
12" CONC. FND. W/2 - #4 BARS CONT.
TOP AND BOTTOM (TYP.)
1'-0" X 2'-0" CONC. FTG. W/3 - #4 BARS
CONT. (TYP.)

EXISTING GRADE 86.50' +/-

CONC. PAD BEYOND

TOP OF MANSARD
ELEV. 108.75'

EAVE HEIGHT
ELEV. 107.00'

BOTTOM OF MANSARD
ELEV. 104.33'

UPPER LEVEL ELEV. 95.00'

SHADOW LINE ELEV. 91.00' (TYP.)

EXISTING GRADE 87.00' +/-

12
1/4

BUILDING SECTION
SCALE: 1/4" = 1'-0"

03401-13_F08.EPS

*Figure 8* Section drawing showing building construction.

DETAIL SHEET M4-A

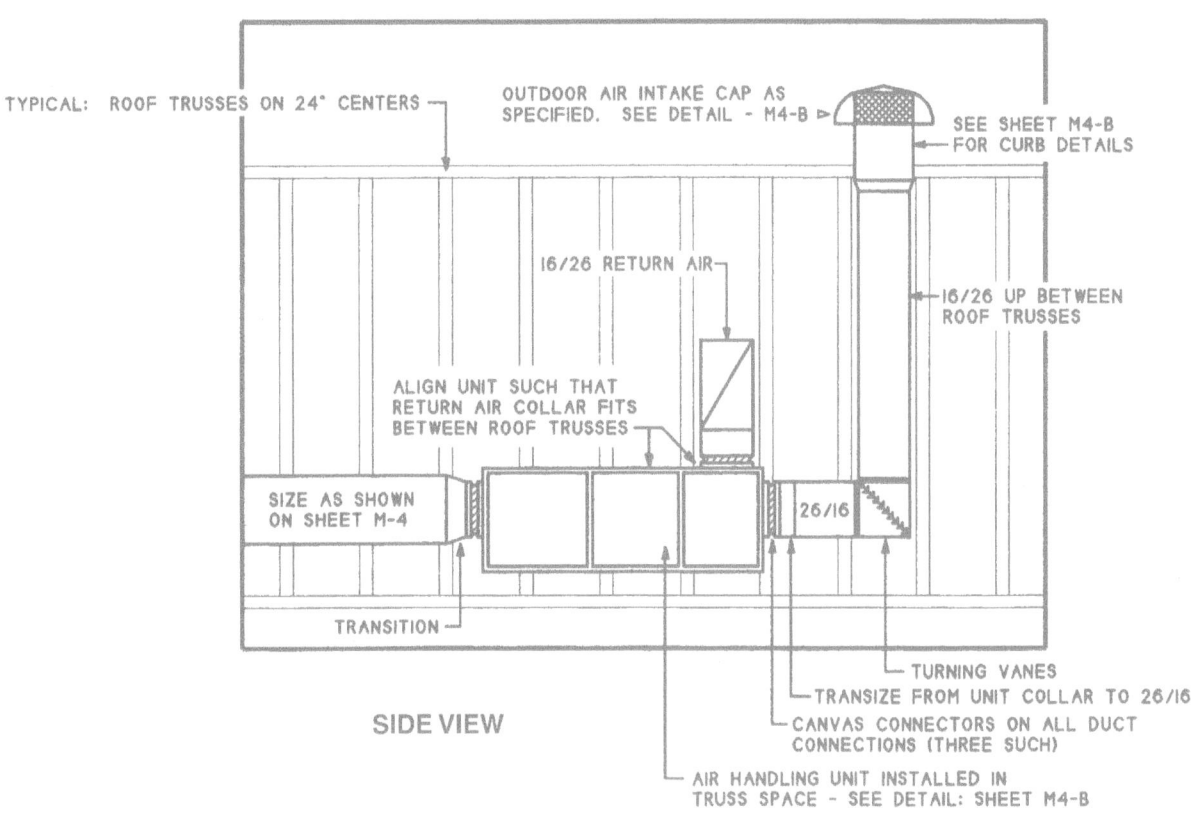

SIDE VIEW

DETAIL SHEET M4-B

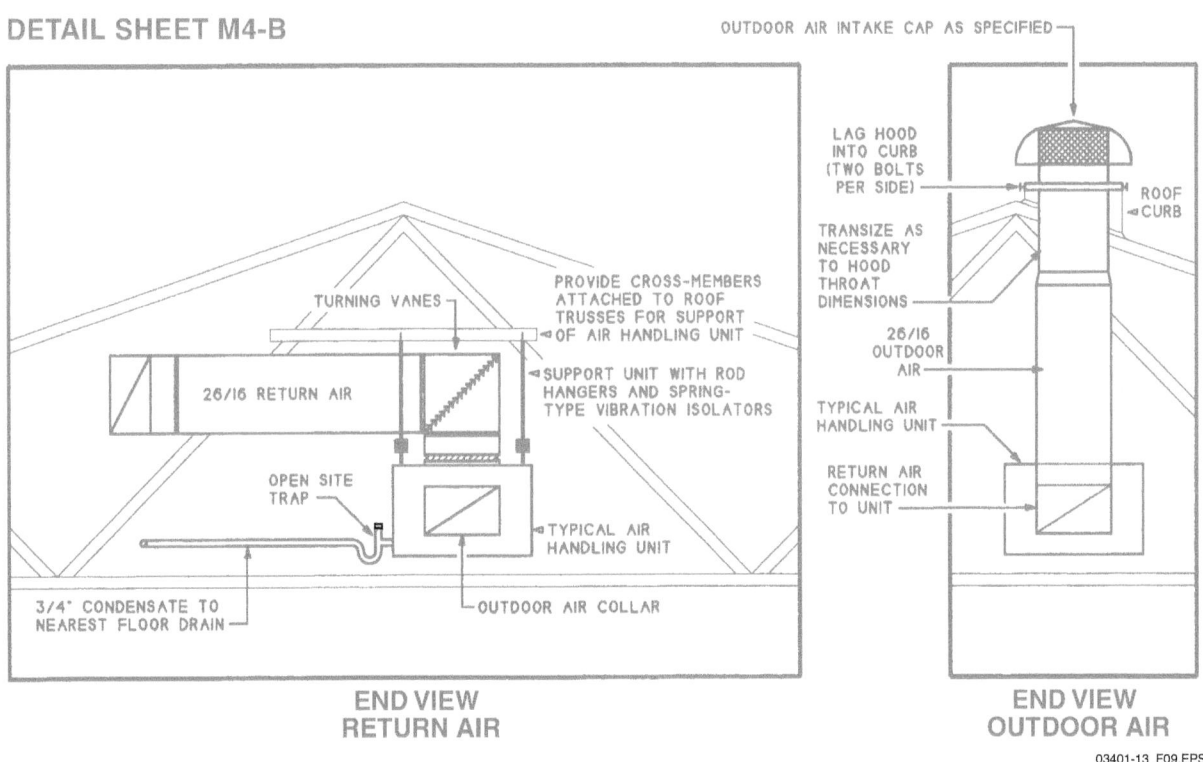

END VIEW
RETURN AIR

END VIEW
OUTDOOR AIR

03401-13_F09.EPS

*Figure 9* Section drawings showing air handling unit installation.

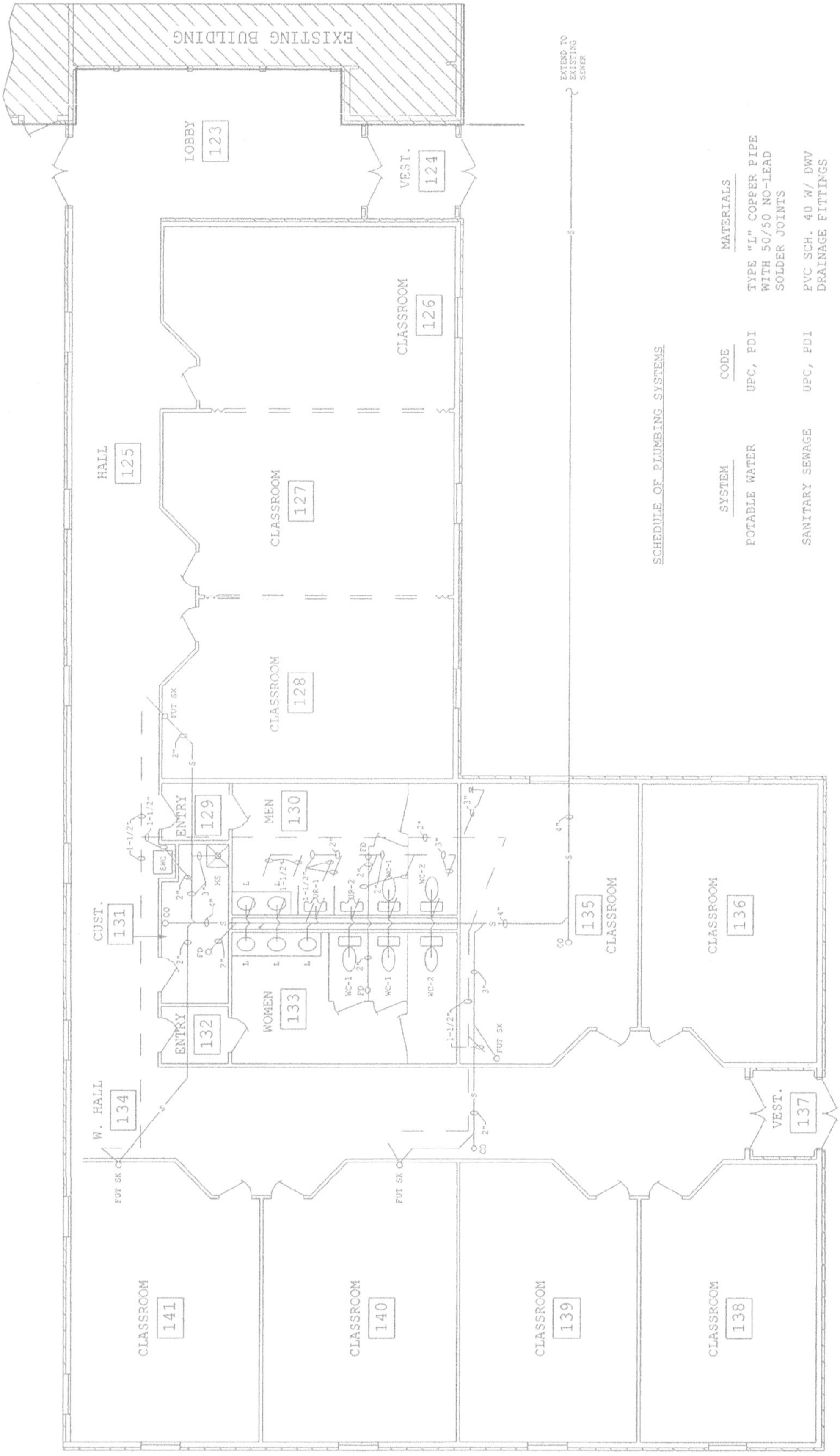

*Figure 10* Sanitary plumbing plan (1 of 2).

NCCER – *HVAC Level Four* 03401

PLUMBING SYSTEM SPECIFICATIONS

COMPLY WITH APPLICABLE STATE AND LOCAL PLUMBING CODES AND STANDARDS PERTAINING TO MATERIALS, PRODUCTS AND INSTALLATION OF POTABLE WATER AND SANITARY SEWAGE SYSTEMS.

TEST EACH PLUMBING SYSTEM IN ACCORDANCE WITH APPLICABLE CODES AND STANDARDS, STERILIZE POTABLE WATER SYSTEMS PER STATE AND LOCAL UTILITY REQUIREMENTS.

SANITARY/VENT ISOMETRIC

03401-13_F10B.EPS

*Figure 10* Sanitary plumbing plan (2 of 2).

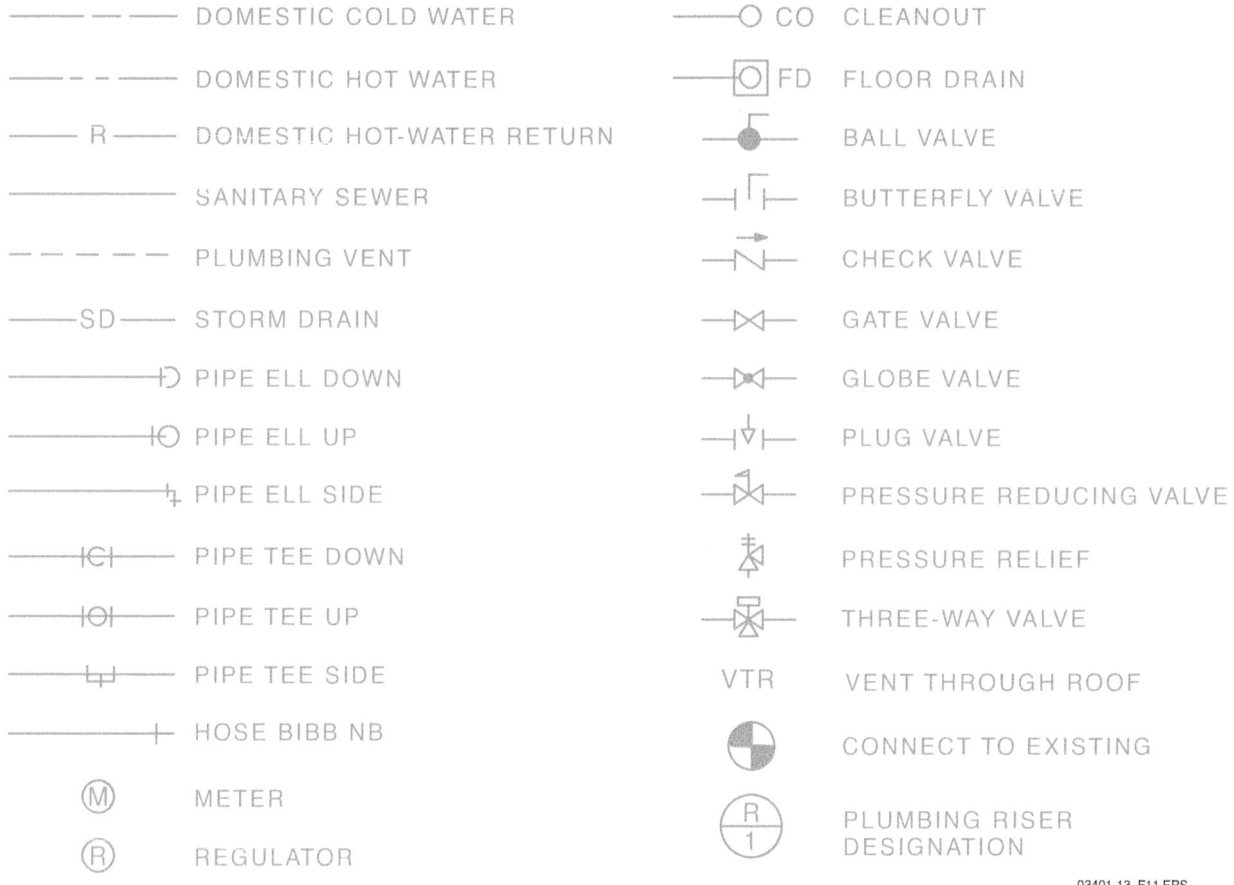

| | |
|---|---|
| ——— — —— DOMESTIC COLD WATER | ——○ CO CLEANOUT |
| ——— – – —— DOMESTIC HOT WATER | —[○] FD FLOOR DRAIN |
| —— R —— DOMESTIC HOT-WATER RETURN | —●— BALL VALVE |
| ———————— SANITARY SEWER | —⌐— BUTTERFLY VALVE |
| — — — — — PLUMBING VENT | —Ň— CHECK VALVE |
| ——SD—— STORM DRAIN | —◁▷— GATE VALVE |
| ———————⊃ PIPE ELL DOWN | —◀▷— GLOBE VALVE |
| ——————⊢○ PIPE ELL UP | —▽— PLUG VALVE |
| —————⌐ PIPE ELL SIDE | —◁▷— PRESSURE REDUCING VALVE |
| ——⟨C⟩—— PIPE TEE DOWN | ⊼ PRESSURE RELIEF |
| ——⟨○⟩—— PIPE TEE UP | —◁▷— THREE-WAY VALVE |
| ——⌐⌐—— PIPE TEE SIDE | VTR VENT THROUGH ROOF |
| ————————+ HOSE BIBB NB | CONNECT TO EXISTING |
| Ⓜ METER | ⓇR₁ PLUMBING RISER DESIGNATION |
| Ⓡ REGULATOR | |

03401-13_F11.EPS

*Figure 11* Plumbing legend.

about the installation of the plumbing and electrical systems is shown on separate drawings.

As with separate plumbing plans, details of the mechanical plan are usually overlaid on tracings of the various building floor plans from which unnecessary details have been omitted to allow the location and layout of the HVAC system equipment to show clearly. *Figure 12* shows an example of a typical HVAC mechanical plan. *Figure 12A* is a plan view, which is the view most often used for mechanical plans.

Mechanical plans typically contain tabulated schedules that identify the different items and types of HVAC equipment. As appropriate, detailed views describing the installation of the HVAC equipment are shown. Depending on the nature of the project, these views can include refrigeration piping schematics (*Figure 13*), chilled-water coil and/or hot-water coil piping schematics, and views detailing piping runs and pipe sizes for major items of HVAC equipment.

Mechanical plans also normally include an HVAC legend listing the various symbols pertaining to the plan. *Figure 14* shows a partial HVAC legend.

Some mechanical plans also contain a schedule of HVAC systems (*Figure 15*) and information about relevant HVAC system specifications (*Figure 16*). It is important to point out that specifications are job-specific. Always be sure to read the specifications for the particular job you are working on.

### 1.4.3 Electrical Plans

For smaller construction jobs, the electrical plans are usually shown on the architectural floor plans. For large commercial jobs, the electrical plans are typically stand-alone drawings that show only information about the electrical system installation. Like the separate plumbing and mechanical plans, electrical plans typically overlay tracings of the various building floor plans from which unnecessary details have been omitted to allow the location and layout of the electrical system equipment to show clearly. *Figure 17* shows an example of a typical electrical plan.

Electrical plans show the locations of the meter, distribution panels, light fixtures, switches, and other electrical equipment. Also shown are equipment and fixtures schedules, an electrical legend listing the various symbols pertaining to the plan, specifications for load capacities, and wire sizes.

---

NCCER – *HVAC Level Four*  03401

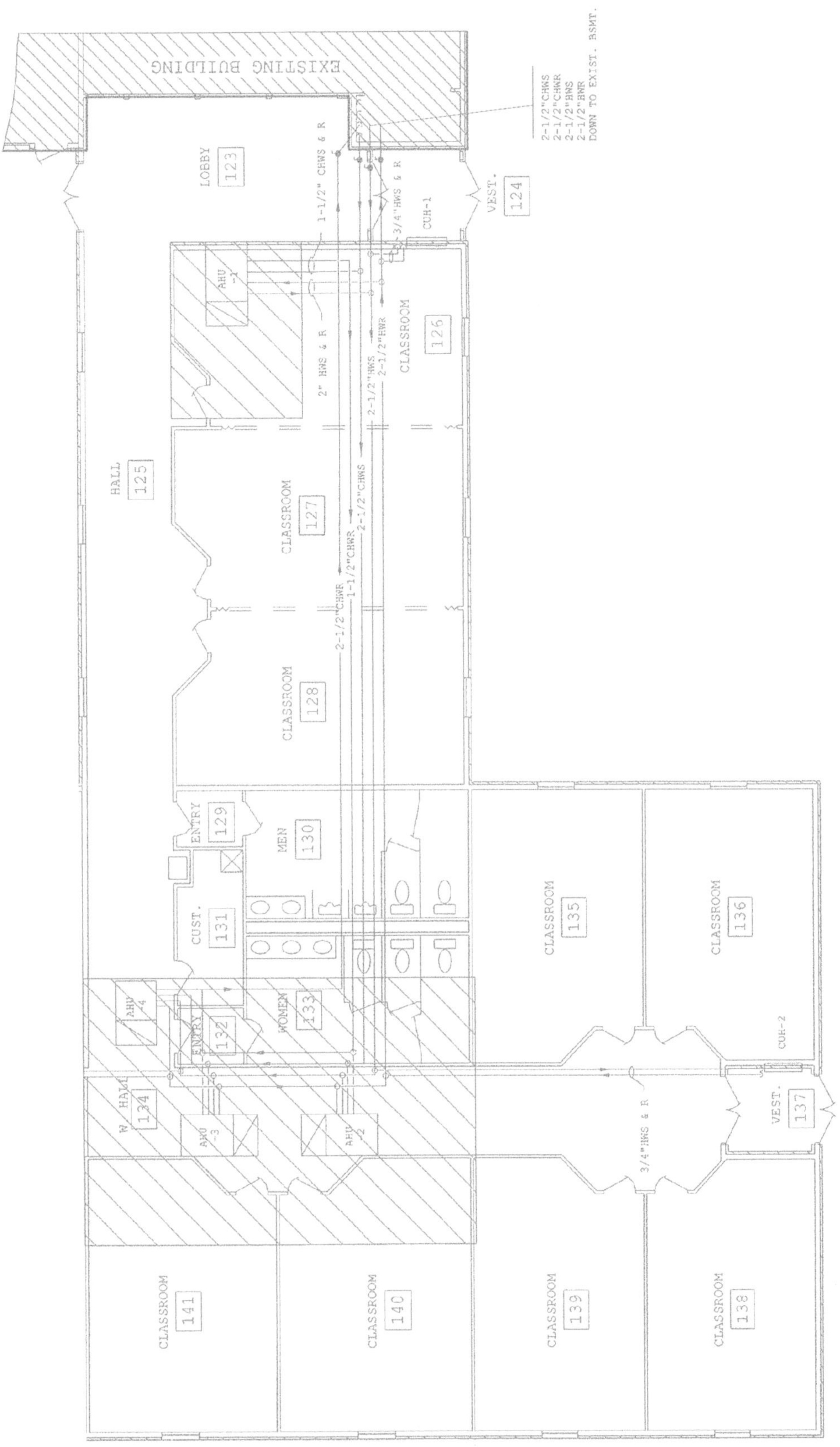

*Figure 12* HVAC mechanical plan (1 of 3).

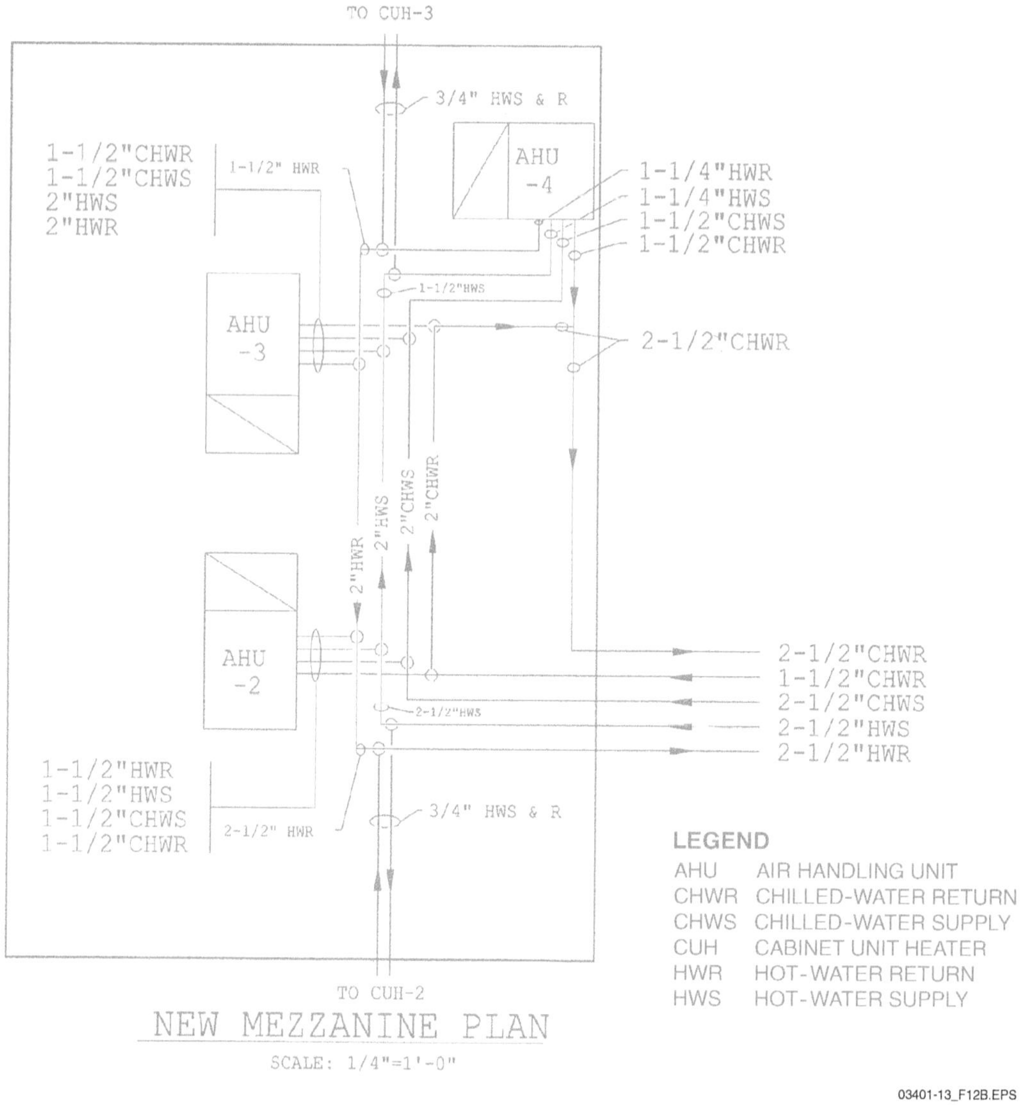

TO CUH-3

3/4" HWS & R

1-1/2"CHWR
1-1/2"CHWS
2"HWS
2"HWR

1-1/2" HWR

AHU
-4

1-1/4"HWR
1-1/4"HWS
1-1/2"CHWS
1-1/2"CHWR

1-1/2"HWS

AHU
-3

2-1/2"CHWR

2"CHWR
2"CHWS
2"HWS
2"HWR

AHU
-2

2-1/2"CHWR
1-1/2"CHWR
2-1/2"CHWS
2-1/2"HWS
2-1/2"HWR

2-1/2"HWS

1-1/2"HWR
1-1/2"HWS
1-1/2"CHWS
1-1/2"CHWR

2-1/2" HWR

3/4" HWS & R

**LEGEND**

AHU   AIR HANDLING UNIT
CHWR   CHILLED-WATER RETURN
CHWS   CHILLED-WATER SUPPLY
CUH   CABINET UNIT HEATER
HWR   HOT-WATER RETURN
HWS   HOT-WATER SUPPLY

TO CUH-2

NEW MEZZANINE PLAN

SCALE: 1/4"=1'-0"

03401-13_F12B.EPS

*Figure 12* HVAC mechanical plan (2 of 3).

Electrical plans usually have a power riser diagram that shows all the major pieces of electrical equipment, including HVAC equipment, as well as the connecting lines used to indicate service-entrance conductors and feeders. Electrical plans may also contain information about the electrical specifications.

### 1.5.0 Shop Drawings

There are three types of shop drawings. One type is a detail drawing that a drafter creates after the engineer designs the structure. It illustrates the connections used, shows the location of all holes and openings, and provides notes specifying how each part is to be made. Assembly instructions are also included and are used principally for structural steel members.

A second type of shop drawing (or submittal) pertains to the purchase of special items of equipment for installation in a building. This type of shop drawing is usually prepared by the equipment manufacturers. This drawing shows overall sizes, details of construction, methods for securing the equipment to the structure, and all other pertinent data that the architect and contractor need to know for the final placement and installation of the equipment. The contractor must check the dimensions of a submitted item of equipment to make sure it will fit into the space provided. If the space is not adequate, a Request for Information (RFI) form is filled out so that the architect or engineer responsible can correct the problem.

A third type of shop drawing very similar in development to the structural steel shop draw-

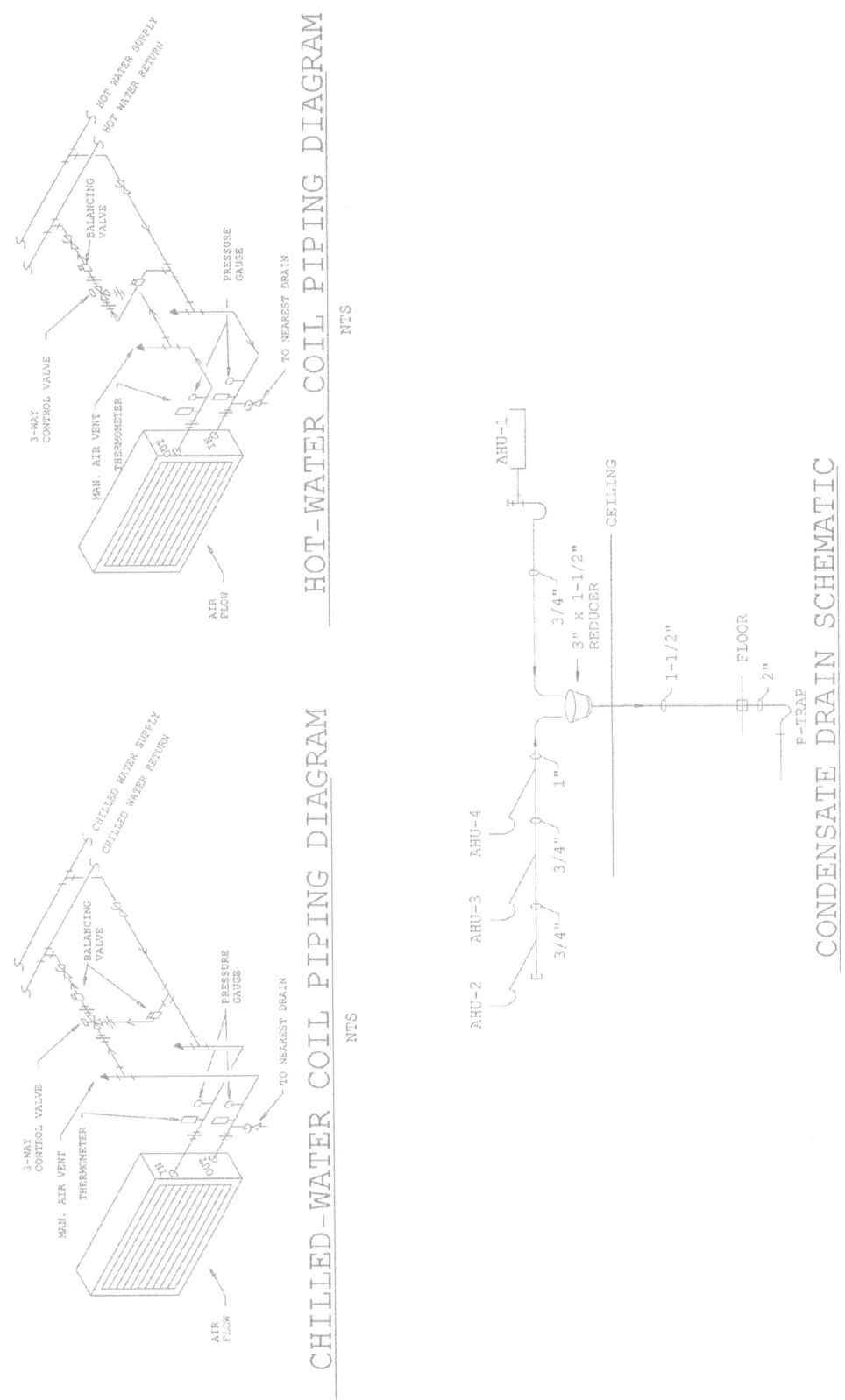

*Figure 12* HVAC mechanical plan (3 of 3).

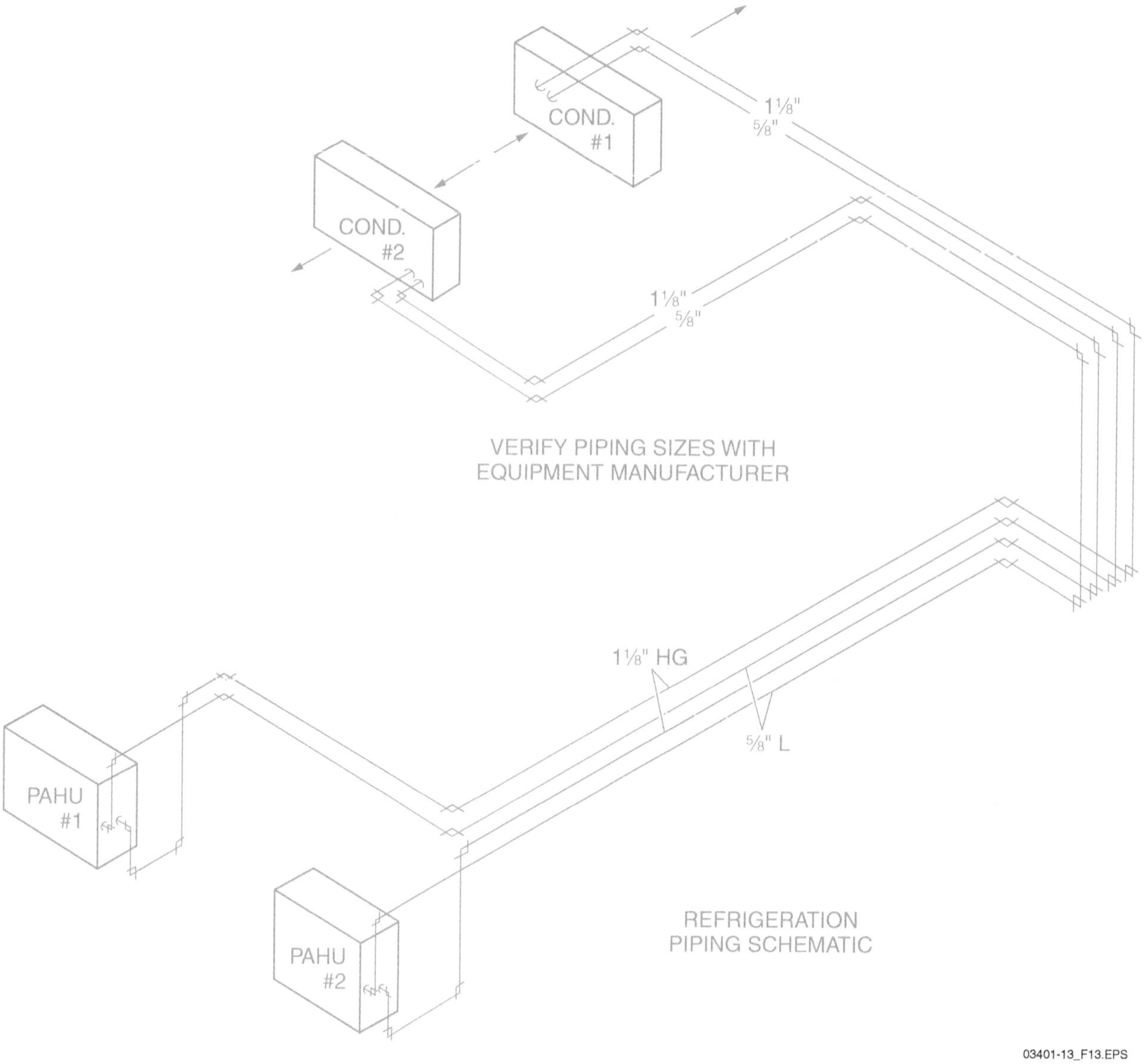

1⅛"
⅝"

COND.
#1

COND.
#2

1⅛"
⅝"

VERIFY PIPING SIZES WITH
EQUIPMENT MANUFACTURER

1⅛" HG

⅝" L

PAHU
#1

PAHU
#2

REFRIGERATION
PIPING SCHEMATIC

03401-13_F13.EPS

*Figure 13* Refrigeration piping schematic.

ing is the one used by sheet metal fabricators and installers. *Figure 18* shows an example of a shop drawing. This shop drawing involves sheet metal drafting techniques and is developed from the design drawings. This shop drawing is usually drawn to a scale that is several times larger than the design drawing. A sheet metal shop drawing shows the exact layout of the ductwork, the sizes and types of fittings, the types of connectors and hangers required for installation, and notes that will assist in fabrication or installation. Although the plan in *Figure 18* may seem difficult to read

## Verifying Dimensions

When installing HVAC equipment and piping according to the mechanical plans, you should always check the related architectural plans to verify the accuracy of rough-in information and drawing dimensions. When given, always use the dimensions shown on a drawing rather than those obtained by scaling the drawing. This is because some reproduction methods used to make copies of drawings can introduce errors in the reproduced image.

Also, when devices are to be located at heights specified above the finished floor (AFF), be sure to find out the actual height of the flooring to be installed. Some materials such as ceramic tile can add significantly to the height of the finished floor.

| | |
|---|---|
| ⊠ | CEILING DIFFUSER (ARROWS INDICATE DIRECTION OF AIR FLOW) |
| ⊠ | RETURN AIR GRILLE |
| ⊠ | SUPPLY DUCT UP |
| ⊠ | SUPPLY DUCT DOWN |
| ◻ | RETURN DUCT UP |
| ◻ | RETURN DUCT DOWN |
| 6"φCD / 200¢ | NECK SIZE/AIR DEVICE CFM |
| | SQUARE TO ROUND TRANSITION |
| | PARALLEL BLADE DAMPER |
| | FIRE DAMPER FD (WALL) (FLOOR) |
| | AIRFOIL BLADE TURNING VANES |
| | AIR EXTRACTOR |
| Ⓣ | THERMOSTAT |
| φ | DIAMETER |
| ¢ | CFM (CUBIC FEET PER MINUTE) |
| RA | RETURN AIR |
| OSA | OUTSIDE AIR |
| CD | CONDENSATE DRAIN |

03401-13_F14.EPS

*Figure 14* HVAC legend.

here, the details are much more visible on a full-size drawing sheet.

The first and second types of shop drawings usually come from the contractor or fabricator and are submitted to the owner or architect for approval and/or revisions or corrections. The design drawing is often put on the same sheet as the shop drawing. Shop drawings are drawn to a large enough scale so that they are clear, but they must not be crowded. They are usually dimensioned to the nearest $1/16$ of an inch.

Contractors commonly use the following approach for the development of shop drawings. When the contract is signed, a full set of drawings is given to the mechanical contractor's drafting department. Upon receipt of the mechanical drawings, the drafting department, depending on the workload, immediately begins developing the shop drawings. As the shop drawings are taken from the mechanical prints, the drafting department also researches the plumbing, electrical, HVAC piping, architectural, and structural prints to see if there will be any conflicts. These drawings are commonly called coordination drawings.

Coordination drawings are drawings produced for a project by the individual contractors for each trade to prevent a conflict between the trades in the installation of their materials and equipment. They are produced prior to finalizing shop drawings, cut lists (shop guides for fabricating duct runs and fittings), and other drawings and before the installation process begins. Development of these drawings evolves through a series of review and coordination meetings held by the various contractors.

Some contracts mandate that coordination drawings be drawn, while others only recommend it. In the case where one contractor elects to make coordination drawings and another does not, the contractor who made the drawings may be given the installation right-of-way by the presiding authority. As a result, the other contractor may have to bear the expense of removing and

## Drawing Revisions

When a set of construction drawings has been revised, always make certain that the most up-to-date set is used for all future layout work. Either destroy the old, obsolete set of drawings or else clearly mark on the affected sheets Obsolete Drawing—Do Not Use. A good practice is to remove the obsolete drawings from the set and file them as history copies for possible future reference. Also, when working with a set of construction drawings and written specifications for the first time, thoroughly check each page to see whether any revisions or modifications have been made to the originals. Doing so can save time and expense for everyone concerned with the project.

SCHEDULE OF HVAC SYSTEMS

| SYSTEM | CODE | MATERIALS | INSULATION |
|---|---|---|---|
| METAL DUCTWORK | SMACNA NFPA 90 A & B | ASTM A527 GAL. SHEET STEEL W/ASTM A525 G90 ZINC COATING | 1" DUCTLINER TIMA AHC-101 |
| ROUND DUCTWORK | SMACNA NFPA 90 A & B | ASTM A527 GAL. SHEET STEEL W/ASTM A525 G90 ZINC COATING | 1$1/2$" DUCTWRAP W/ VAPOR JACKET |
| HOT WATER | ASME 31.9 | COPPER TUBE TYPE L WROT COPPER FITTINGS SOLDERED JOINTS 95/5 | 1" FIBERGLASS |
| CHILLED WATER | ASME 31.9 | COPPER TUBE TYPE L WROT COPPER FITTINGS SOLDERED JOINTS 95/5 | 1" FIBERGLASS |
| CONDENSATE | UPC, PDI | PVC SCH 40, W/ DWV DRAINAGE FITTINGS | $1/2$" FLEXIBLE UNICELL |

03401-13_F15.EPS

*Figure 15* Schedule of HVAC systems.

HEATING, VENTILATING & AIR CONDITIONING SYSTEM SPECIFICATIONS

COMPLY WITH APPLICABLE MECHANICAL CODES AND STANDARDS PERTAINING TO MATERIALS, PRODUCTS, AND INSTALLATION OF AIR HANDLING, METAL DUCTWORK, HOT-WATER SYSTEMS, AND CHILLED-WATER SYSTEMS.

SUBMIT MANUFACTURER'S TECHNICAL PRODUCT DATA TAILORED TO THE PROJECT, ASSEMBLY-TYPE SHOP DRAWINGS, WIRING DRAWINGS, AND MAINTENANCE DATA FOR EACH COMPONENT OF EACH HVAC SYSTEM.

PROVIDE FACTORY-FABRICATED AND FACTORY-TESTED EQUIPMENT AND MATERIALS OF SIZES, RATINGS, AND CHARACTERISTICS INDICATED. REFERENCED EQUIPMENT AND MATERIALS INDICATE STYLE AND QUALITY DESIRED. CONTACT ENGINEER PRIOR TO SUBMITTAL OF ANY OTHER MANUFACTURER FOR PRELIMINARY APPROVAL. PROVIDE PROPER QUANTITY OF MATERIAL AND EQUIPMENT AS REQUIRED FOR COMPLETE INSTALLATION OF EACH HVAC SYSTEM.

IDENTIFY EACH HVAC SYSTEM'S COMPONENTS WITH MATERIALS AND DESIGNATIONS AS DIRECTED.

INSTALL EACH HVAC SYSTEM IN ACCORDANCE WITH APPLICABLE MECHANICAL CODES AND STANDARDS, RECOGNIZED INDUSTRY PRACTICES, AND MANUFACTURER'S RECOMMENDATIONS.

TEST AND BALANCE EACH HVAC SYSTEM IN ACCORDANCE WITH APPLICABLE MECHANICAL CODES AND STANDARDS. BALANCE AIR CONDITIONING SYSTEM TO CFM'S SHOWN ON DRAWINGS. REPORT FINDINGS TO ENGINEER USING APPROVED FORMS.

03401-13_F16.EPS

*Figure 16* Example of HVAC system specifications information.

reinstalling equipment if the equipment was installed in a space designated for use by the contractor who produced coordination drawings.

### 1.5.1 Cut Lists

Another function of the shop drawing is to assist the subcontractor in identifying the number and sizes of the fittings and duct run sections that must be fabricated and subsequently installed on the job.

After the shop drawings are complete, or as they are drawn (depending on the workload), the drafter makes a cut sheet of each individual fitting and assigns a fitting number to it that matches the numbers on the shop drawing.

The straight duct sections are given another number that stays the same until the duct size

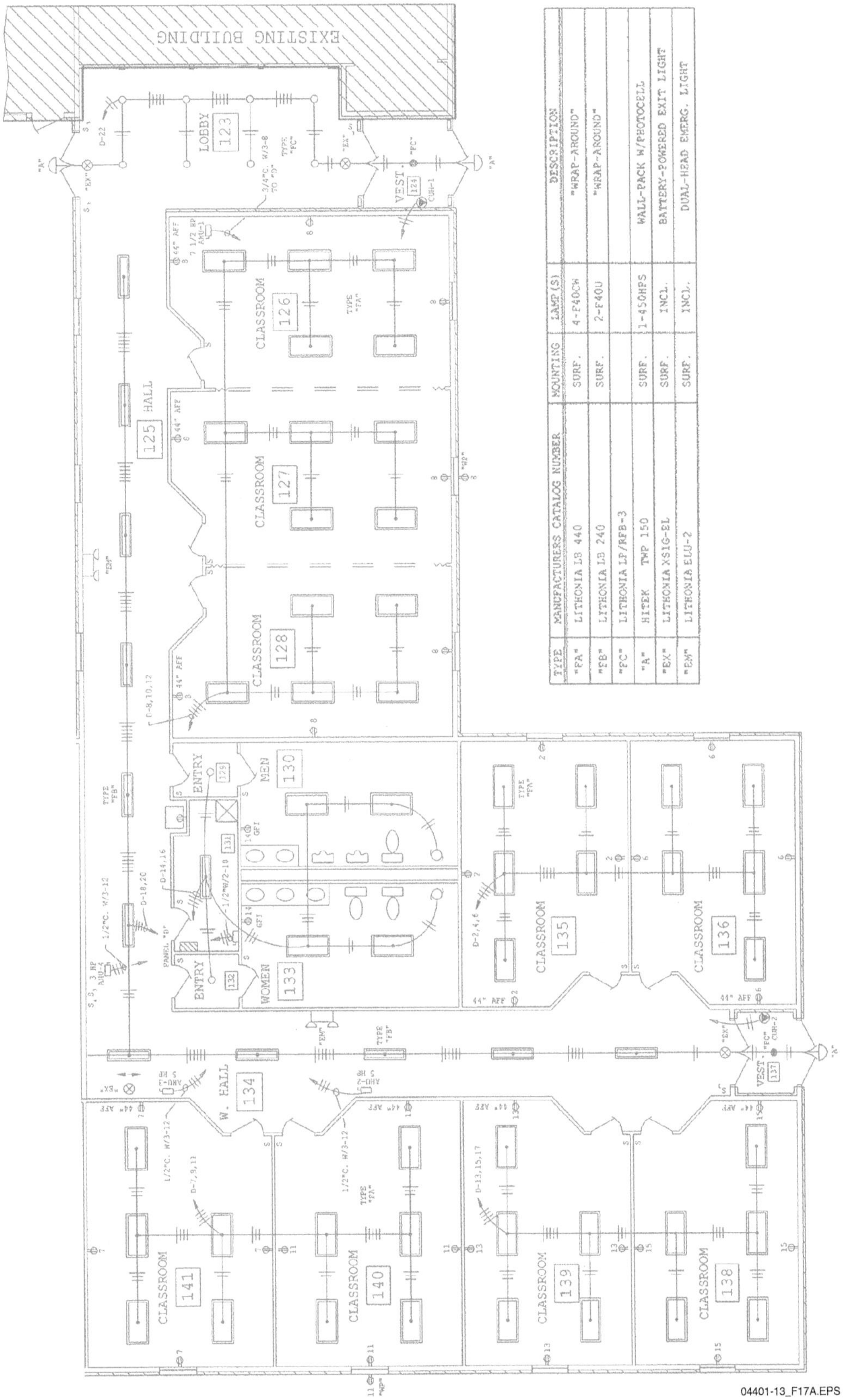


04401-13_F17A.EPS

***Figure 17*** Electrical plan (1 of 2).

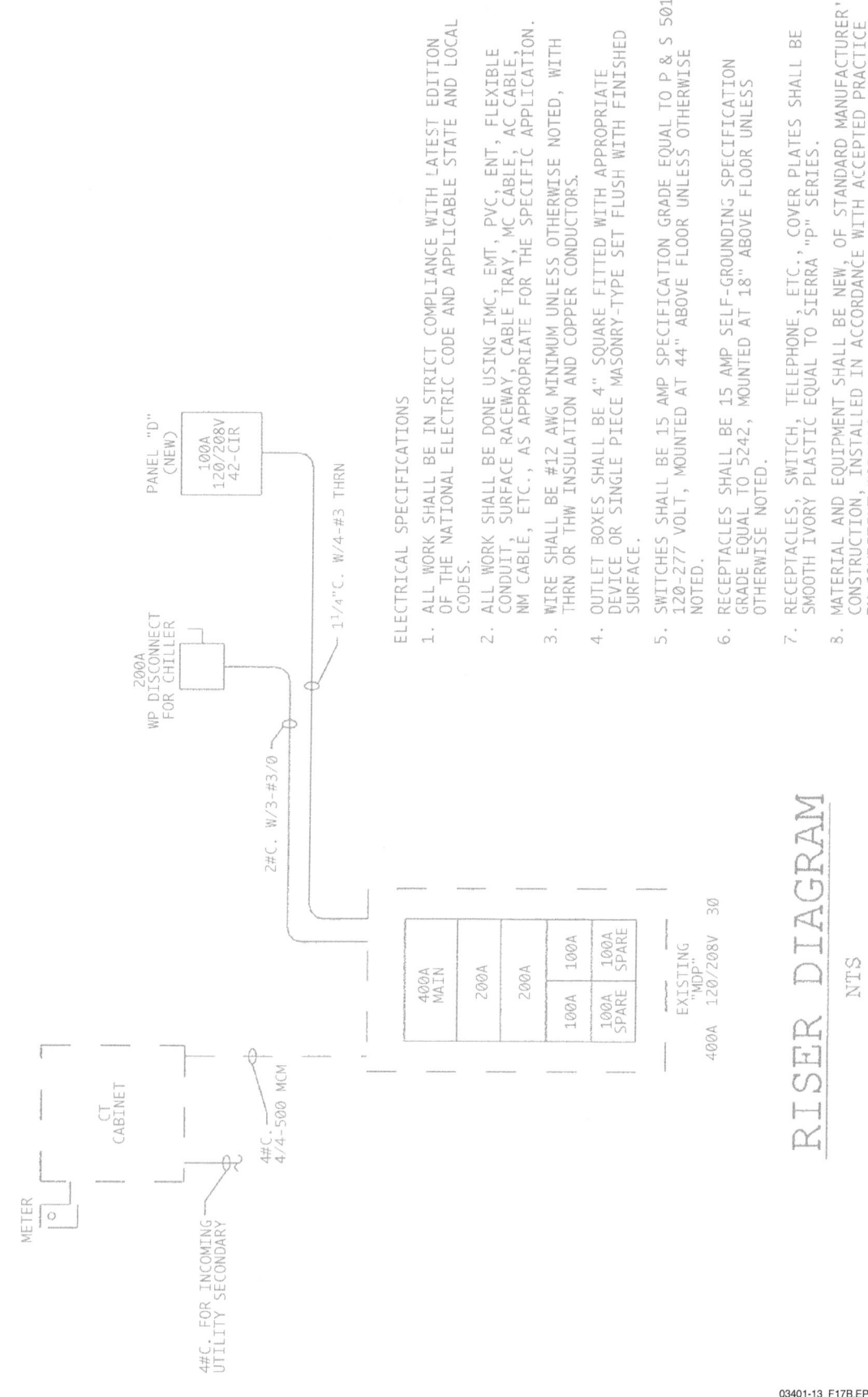

ELECTRICAL SPECIFICATIONS

1. ALL WORK SHALL BE IN STRICT COMPLIANCE WITH LATEST EDITION OF THE NATIONAL ELECTRIC CODE AND APPLICABLE STATE AND LOCAL CODES.

2. ALL WORK SHALL BE DONE USING IMC, EMT, PVC, ENT, FLEXIBLE CONDUIT, SURFACE RACEWAY, CABLE TRAY, MC CABLE, AC CABLE, NM CABLE, ETC., AS APPROPRIATE FOR THE SPECIFIC APPLICATION.

3. WIRE SHALL BE #12 AWG MINIMUM UNLESS OTHERWISE NOTED, WITH THHN OR THW INSULATION AND COPPER CONDUCTORS.

4. OUTLET BOXES SHALL BE 4" SQUARE FITTED WITH APPROPRIATE DEVICE OR SINGLE PIECE MASONRY-TYPE SET FLUSH WITH FINISHED SURFACE.

5. SWITCHES SHALL BE 15 AMP SPECIFICATION GRADE EQUAL TO P & S 501, 120-277 VOLT, MOUNTED AT 44" ABOVE FLOOR UNLESS OTHERWISE NOTED.

6. RECEPTACLES SHALL BE 15 AMP SELF-GROUNDING SPECIFICATION GRADE EQUAL TO 5242, MOUNTED AT 18" ABOVE FLOOR UNLESS OTHERWISE NOTED.

7. RECEPTACLES, SWITCH, TELEPHONE, ETC., COVER PLATES SHALL BE SMOOTH IVORY PLASTIC EQUAL TO SIERRA "P" SERIES.

8. MATERIAL AND EQUIPMENT SHALL BE NEW, OF STANDARD MANUFACTURER'S CONSTRUCTION, INSTALLED IN ACCORDANCE WITH ACCEPTED PRACTICE BY COMPETENT WORKERS.

RISER DIAGRAM
NTS

PANEL "D"
(NEW)
100A
120/208V
42-CIR

200A
WP DISCONNECT
FOR CHILLER

1¼"C. W/4-#3 THRN

2#C. W/3-#3/0

400A MAIN

200A

200A

100A    100A

100A    100A
SPARE   SPARE

EXISTING "MDP"
400A  120/208V  3Ø

METER

CT CABINET

4#C. FOR INCOMING UTILITY SECONDARY

4#C. 4/4-500 MCM

03401-13_F17B.EPS

*Figure 17* Electrical plan (2 of 2).

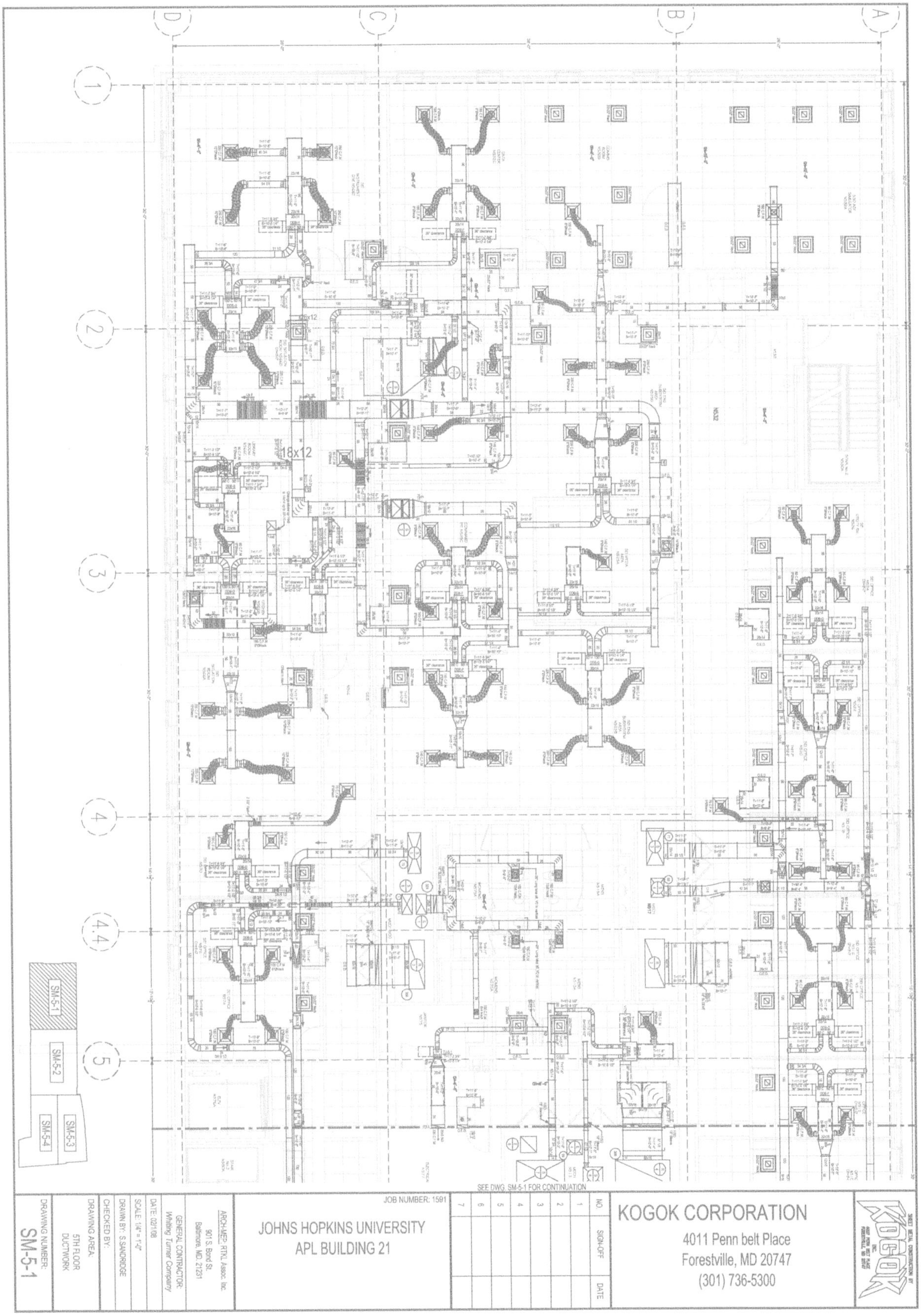

**Figure 18** Shop drawing.

changes. The cut sheets are then given to the fabrication shop. These cut sheets contain the job number, the quantity of ducts required, and the gauge of metal from which they are to be fabricated. Some contractors have their own methods

for tracking fabricated duct pertaining to the duct line, delivery date, and job number.

Cut lists are a function of the production phase in the fabrication shop and may either be generated by the drafter (*Figure 19*) or generated by a computer (*Figure 20*). The cut lists identify the

*Figure 19* Drafter's cut list.

03401-13_F19.EPS

Figure 20 Computer-generated cut list.

fittings and sections by number, type of fitting, amount required, width, depth, length, type of flange or connection to be used, number of parts required for the fitting, cut size, and type and gauge of metal.

### 1.5.2 General Procedures

In large shops, the sheet metal drafters are usually sheet metal mechanics who have been trained in the use of drawing instruments and layout procedures. In smaller shops, the owner or journeyperson may be required to develop the shop drawings. Freehand sketches from field measurements are often passed on to the drafter to develop a shop drawing. Sometimes, written notes and descriptions are provided, and the drafter must translate that information into shop drawings. A general procedure for producing shop drawings is as follows:

Step 1 Select a scale two to four times larger than the scale used for the design drawing. The scale selected should suit the intended purpose of the drawing. If the drawing is to be used as a coordination drawing, then each contractor should use the same scale in order to make comparisons among the drawings easier. Sometimes the engineers or architects will provide a basic template for the drawing that helps achieve accuracy in the layout.

Step 2 Arrange the layout to be evenly spaced on the sheet; it may be desirable to center the layout.

Step 3 Use the standard symbols on the drawing.

Step 4 Add notes when and where necessary.

Step 5 Draw in partitions, exterior walls, beams, columns, hanging ceilings, and any other obstructions that appear on the architectural plan.

Step 6 Use the design drawing to calculate all measurements needed to properly locate the ductwork.

Step 7 When dimensioning, indicate:

- The measurements from the finished floor to the bottom of the duct

- The duct height

- The clearance from the top of the duct to the bottom of the slab, if applicable
- The measurement from the ceiling to the bottom of the slab if a suspended ceiling is part of the construction

*Step 8* Properly locate all access doors, volume dampers, fire and smoke dampers, boots, registers, duct linings, thermostats, and other accessories on the drawing.

*Step 9* Allow sufficient clearance around all ducts so they can pass through walls easily. Check the specifications for sealing around ducts passing through walls and around ducted fire and smoke dampers.

*Step 10* To properly locate ducts, dimension from the centerline of the nearest column.

*Step 11* Refer periodically to the design drawing to check for interference with other trades.

*Step 12* Number all pieces of ductwork on the shop drawing according to the practices in your shop.

*Step 13* Make up a tally sheet or cut sheet that indicates each piece's identification number, the size, description, quantity, type, and gauge of material, and any other pertinent information that will help the fabricator and/or installer.

In addition, consider the following factors when you prepare shop drawings:

- Carefully check the electrical and plumbing mechanical drawings when preparing drawings for ducts.
- Note that the types of connections used for conventional low-pressure or high-pressure duct or for heavy-gauge duct affect the length of the joints and fittings.
- Where necessary, if provided, note on the shop drawing the thickness and type of acoustical lining and external insulation.
- Be aware that duct sizes are usually increased to allow for the thickness of acoustical lining, but plans and specifications should be checked for verification of the designer's intent.
- Refer to layout dimensions regarding approved HVAC equipment submittal cuts.
- Carefully check gauge specifications and types of materials for boiler breechings, exhaust and fume hoods, and kitchen exhaust components.
- Note that watertight duct construction is generally necessary for shower rooms and dishwasher equipment.

- Note that horizontal ducts may be pitched downward slightly to drain connections when run through moist environments.
- Notice that fire and fire/smoke dampers are required to be shown on the HVAC drawings and generally include a note in reference to the applicable installation and material code. Coordinate this information with RFIs and life safety drawings.
- Include a note that requests necessary information from the architect or designer if the location of a fire damper is doubtful.
- Be sure each fire damper is accessible through a properly sized access door for fusible link inspections and maintenance.
- Include additional notes to the attention of the architect or designer if information is necessary to confirm a dimension on the shop drawing in the following cases:
  - When it is not shown on the design or architectural plan
  - When it is necessary to indicate a specific location that is inadequate for duct clearance
  - When work must be done by others
  - When it is necessary to identify locations where coordination of the work with other trades is essential

### 1.6.0 As-Built Drawings

As-built drawings must be made on alteration or addition jobs, on jobs where modifications must be made to make way for other mechanical trades, or to alter the location of a component. In some cases, particularly on additions or alterations, these drawings may be available from the building or plant engineer. These drawings indicate actual installations by the various mechanical trades and must be used for reference by the drafter when called upon to provide a shop drawing for the modified system or components. As-built drawings usually use dashed lines to indicate ducts, piping, and equipment at close proximity to the work. Separate symbols or notes must be used to distinguish ducts that are to be removed and discarded from those that will be relocated.

The as-built drawings are then placed on the architect's plan (sometimes in another color, such as red), but are more often stamped AS-BUILT. The as-built drawings become a permanent part of the building's drawings. In addition, notes should be made as to the types of connections and existing duct locations that will be reconnected. Duct openings through existing concrete or masonry walls should also be located, checked, and indicated.

On larger projects, the various contractors and subcontractors often maintain their own sets of as-builts. This enables electricians, plumbers, HVAC technicians/installers, and other skilled tradespeople to make changes without holding up the job to locate a single set of plans. At the end of the project, however, all of the sets of as-builts are combined into one comprehensive set that is delivered to the project owner.

As-built drawings are often helpful in several additional ways. The information that they provide about the locations of wiring, plumbing, and other hidden components makes it easier to perform repairs and maintenance. As-builts can also be used as a base when creating remodeling plans for future renovations. In some cases, local governments or permitting agencies may require a copy of the as-builts to show locations of sprinkler pipes, fire alarms, and other safety devices.

### 1.7.0 Schedules

Schedules are not drawings. They are tables shown on the various drawings throughout the drawing set. These tables identify the types and sizes of items used by the different trades in the construction of a building. Any or all of the following information about an item may be included in a schedule: vendor's name, product name, model number, quantity, size, rough opening size, and color. Placing these details in simple tables keeps the drawings from becoming cluttered and crowded with textual notes.

Each schedule lists information related to a similar group of items, such as doors, windows,

# Protection of Underground Facilities

To avoid damage or interruptions to underground utilities caused by digging, it is mandatory to contact the various companies for a utility stakeout prior to any digging. Most states have a One Call Notification System center that makes digging notification easy by calling a single phone number or by contacting the center on the Internet. Typically, the call must be made at least two or three working days before the digging is to begin.

A Dig Safely card and other materials are readily available from state Dig Safely notification centers to remind contractors and excavators of this requirement. Shown here is a Dig Safely card available from the Dig Safely New York notification center. In addition to giving the procedure for contacting the center, it shows the American Public Works Association (APWA) universal color codes used for the temporary marking of underground facilities. All other states have cards similar to the one shown for New York.

FRONT

BACK

03401-13_SA03.EPS

and plumbing fixtures. For example, schedules that identify the windows and doors in a building are shown on the floor plans. Items in the schedules are keyed to the drawing with identification letters, numbers, and symbols. So, on a floor plan, doors may be shown as circles that have a letter inside them, and windows may be shown as hexagons that have a number inside them. The letters and numbers are then keyed to the door and window schedules. For instance, a circle with E inside it on the floor plan is described on the door schedule as a hollow-core, 6-panel door, size 2'-8" × 6'-8"; and a hexagon with 4 inside it on the floor plan is described on the window schedule as an aluminum, single-hung window, size 2'-0" × 4'-0".

Of importance to the HVAC technician are the schedules for the mechanical components and equipment shown on the mechanical plans. Schedules shown on the related plumbing plans and electrical plans may also be relevant to the HVAC technician. *Figure 21* shows schedules typical of those found on mechanical plans.

## 1.8.0 Request for Information

The request for information (RFI) is used for clarification. If a discrepancy, conflict, or incomplete information is noted on the plans, an RFI may be issued to the architect or engineer. There is a hierarchy that is usually followed. For example, if you notice a discrepancy on the plans, you should notify your immediate supervisor. That will likely be your foreman on a job site. The foreman generates the RFI, being as specific as possible and making sure to put the time, date, and company RFI number on it. The foreman gives the RFI to the superintendent or project manager, who passes it on to the general contractor. The general contractor then relays the RFI to the architect or engineer. A sample RFI form is shown in *Figure 22*. The person who finds a discrepancy on the plans should clearly describe the conflicting information and, if possible, suggest how to correct the problem.

## 1.9.0 Building Codes

Building codes that are national in scope provide minimum standards to guard the life and safety of the public by regulating and controlling the design, construction, and quality of materials used in modern construction. They have also come to govern use and occupancy, location of a type of building, and the ongoing maintenance of all buildings and facilities. Once adopted by a local jurisdiction, these model building codes then become law. It is common for localities to change or

add new requirements to any model code requirements adopted in order to meet more stringent requirements and/or local needs. The provisions of the model building codes apply to the construction, alteration, movement, demolition, repair, structural maintenance, and use of any building or structure within the local jurisdiction.

The model building codes are the legal instruments that enforce public safety in construction of human habitation and assembly structures. They are used not only in the construction industry but also by the insurance industry for compensation appraisals and claims adjustments, and by the legal industry for court litigation.

Up until 2000, there were three model building codes. The three code writing groups, Building Officials and Code Administrators (BOCA), International Conference of Building Officials (ICBO), and Southern Building Code Congress International (SBCCI), combined into one organization called the International Code Council (ICC) with the purpose of writing one nationally accepted family of building and fire codes. The first edition of the *International Building Code*® was published in 2000, a second edition was published in 2003, and a third in 2006. It is intended to continue on a three-year cycle. At the time of this writing, 2012 is the latest edition.

In 2002, the National Fire Protection Association (NFPA) published its own building code, *NFPA 5000*®. At the time of this writing, 2012 is the latest edition and the next edition is due in 2015.

Thus, there are now two nationally recognized codes competing for adoption by the 50 states—the *International Building Code*® and *NFPA 5000*®. The format and chapter organization of the two codes differ, but the content and subjects covered are generally the same. Both codes cover all types of occupancies from single-family residences to high-rise office buildings, as well as industrial facilities. They also cover structures, building materials, and building systems, including life safety systems.

When states, counties, and cities adopt a model code as the basis for their own code, they often change it to meet local conditions. They might add further restrictions, or they might adopt only part of the model code. An important general rule to remember about codes is that in almost every case the most stringent local code will apply.

The HVAC technician should be aware of the laws, local building codes, and restrictions that affect the specific job being constructed. This should also include a basic understanding of other codes, such as the NFPA gas and electrical codes.

## CABINET UNIT HEATER SCHEDULE

| UNIT HEATER NO. | LOCATION | C.F.M. | FAN MOTOR | | | | MBH | GPM | EWT | EAT | MAX. WATER P.D. | REMARKS |
|---|---|---|---|---|---|---|---|---|---|---|---|---|
| | | | H.P. | VOLTS | PHASE | Hz | | | | | | |
| CUH-1 | 124 | 400 | 1/12 | 115 | 1 | 60 | 23 | 2.3 | 180°F | 60°F | 2.7 | McQUAY #CHF004 SEMI-RECESSED, R.H COIL |
| CUH-2 | 137 | 400 | 1/12 | 115 | 1 | 60 | 23 | 2.3 | 180°F | 60°F | 2.7 | McQUAY #CHF004 SEMI-RECESSED, L.H COIL |
| CUH-3 | 143 | 400 | 1/12 | 115 | 1 | 60 | 23 | 2.3 | 180°F | 60°F | 2.7 | McQUAY #CHF004 SEMI-RECESSED, R.H COIL |

NOTES:
1. 3 Speed Control
2. Front Discharge
3. With Return Air Filters

## PUMP SCHEDULE

| UNIT NO. | LOCATION | SERVICE. | GPM | MBH | MOTOR | | | | | TYPE | REMARKS |
|---|---|---|---|---|---|---|---|---|---|---|---|
| | | | | | RPM | H.P. | VOLTS | PHASE | Hz | | |
| P-1 | MECH. ROOM | NAVE | 45 | 41' | 1750 | 1 1/2 | 208/230 | 3 | 60 | IN-LINE | B/G #60-20T SERVICE 40% GLYCOL SOLUTION |
| P-2 | MECH. ROOM | CHW TO AHU 1-4 | 65 | 37' | 1750 | 1 1/2 | 208/230 | 3 | 60 | IN-LINE | B/G #60-20T SERVICE 40% GLYCOL SOLUTION |
| P-3 | MECH. ROOM | RECIR. TO TANK | 40 | 17' | 1750 | 1/2 | 208/230 | 3 | 60 | IN-LINE | B/G #60-13T SERVICE 40% GLYCOL SOLUTION |
| P-4 | EXIST. MECH. ROOM | HW | 73 | 31' | 1750 | 1 1/2 | 208/230 | 3 | 60 | IN-LINE | B/G #60-20T HOT WATER |

NOTES:
1. Starters And Disconnects By E.C.

## GRILLE, REGISTER AND DIFFUSER SCHEDULE

| ITEM | MANUFACTURER | MODEL NO. | QTY. | LOCATION | CFM EACH | AIR PATTERN | SIZE | | FINISHES | REMARKS |
|---|---|---|---|---|---|---|---|---|---|---|
| | | | | | | | FRAME | NECK | | |
| A | BARBER COLMAN | SFSV | 8 | 126,127,128 144 | 245 | 4-WAY | 12"× 12" | 8"Ø | #7 OFF-WHITE | |
| B | BARBER COLMAN | SFSV | 2 | 142 | 275 | 4-WAY | 18"× 18" | 10"Ø | #7 OFF-WHITE | |
| C | BARBER COLMAN | SFSV | 4 | 140, 141 | 240 | 4-WAY | 12"× 12" | 8"Ø | #7 OFF-WHITE | |
| D | BARBER COLMAN | SFSV | 2 | 139 | 270 | 4-WAY | 18"× 18" | 10"Ø | #7 OFF-WHITE | |
| E | BARBER COLMAN | SFSV | 2 | 138 | 280 | 4-WAY | 18"× 18" | 10"Ø | #7 OFF-WHITE | |
| F | BARBER COLMAN | SFSV | 2 | 136 | 250 | 4-WAY | 12"× 12" | 6"Ø | #7 OFF-WHITE | |
| G | BARBER COLMAN | SFSV | 2 | 135 | 235 | 4-WAY | 12"× 12" | 6"Ø | #7 OFF-WHITE | |
| H | BARBER COLMAN | SFSV | 3 | 134 | 100 | 4-WAY | 12"× 12" | 6"Ø | #7 OFF-WHITE | FIRE DAMPER SEE DETAIL A |
| I | BARBER COLMAN | SFSV | 1 | 134 | 190 | 4-WAY | 12"× 12" | 8"Ø | #7 OFF-WHITE | FIRE DAMPER SEE DETAIL A |
| | MAN | SFSV | 1 | 134 | WAY | | 12" | | | FIRE DAMPER |

03401-13_F21.EPS

*Figure 21* Mechanical equipment schedules.

XYZ, Inc
General Contractors
123 Main Street
Bigtown, USA 10001
(111) 444-5555

**R.F.I.**

Request for Information

XYZ Project # _____
Date: _____
R.F.I. # _____

PROJECT:

TO:

RE:

**Specification
Reference:** _____

**Drawing
Reference:** _____

**SUBJECT:**

**REQUIRED:**

**Date Information
is Required:** _____

**XYZ, Inc
By:** _____

**REPLY:**

Distribution: Superintendent
               Field File

By: _____

Date: _____

03401-13_F22.EPS

*Figure 22* Request for information form.

# Fire and Smoke Dampers

It is important when reading drawings to know the difference between fire dampers and fire/smoke dampers. Fire dampers are simple spring-return devices. When the lead fusible link melts at the set temperature (usually about 165°F to 185°F), the damper spring is released and slams closed. These dampers must be manually reset once they are tripped, and a new fusible link installed to hold the spring. These dampers are much lower in cost and do not require any other coordination for installation outside of the sheet metal trade.

Combination fire/smoke dampers include the same components as a fire damper, but add the complexity of motorized control. This damper is also a spring damper, but includes a motor, and must be powered open by a signal from the fire alarm system. If that signal is lost, the spring automatically closes the smoke damper whether or not the fusible link in the fire damper is affected. Smoke dampers can be installed separately from fire dampers, but they cannot be separated by any more distance than allowed by code, which is usually about 2 feet, or two to four duct diameters. Check local codes. The added cost of the smoke damper to a system is significant, as several trades are needed for a completed installation (electrical, controls, sheet metal, and sometimes others). The cost and size of the actual fire/smoke damper impacts all aspects of its installation. The cost of fire/smoke dampers may be four to six times that of fire dampers.

Access to fire/smoke dampers is critical. They must be tested with visual verification of their operation on an annual or semi-annual basis, depending on local fire marshal requirements and recordkeeping.

# As-Built Drawings

The specifications for a project typically have a section describing the format or method to use when making as-built drawings for delivery to the general contractor, architect, or owner. Construction changes made by the various contractors as the job progresses are typically marked on a master set of working drawings set aside for this purpose. This marked-up set of drawings forms the basis for the final as-built drawings. The drawing change should be made as soon as possible. Fortunately, CAD technology enables the drawings to be updated more easily. On many jobs, however, any changes to the original design can only be made after a formal document called a change order has been generated and approved by the architect or other designated person.

## Additional Resources

*Blueprint Reading for Construction,* 2003. Second Edition. James A. S. Fatzinger, Upper Saddle River, NJ: Pearson/Prentice Hall.

*Reading Architectural Work Drawings,* 2003. Sixth Edition. Edward J. Muller and Phillip A. Grau III. Upper Saddle River, NJ: Prentice Hall.

## 1.0.0 Section Review

1. To find the information that you need, it may be necessary to view several drawings in a set.
   a. True
   b. False

2. Site plans are sometimes called _____.
   a. elevations
   b. sections
   c. plot plans
   d. foundation plans

3. Elements such as the heights of windows, doors, and porches are often shown on which type of drawing?
   a. Floor plan
   b. Riser diagram
   c. Detail drawing
   d. Elevation view

4. The locations where fixtures are installed in a large commercial building are normally shown on what type of plan?
   a. Structural
   b. Floor
   c. Mechanical
   d. Plumbing

5. The shop guides for fabricating duct runs and fittings are known as _____.
   a. cut lists
   b. gaucho documents
   c. detail drawings
   d. blackmore schematics

6. Dashed lines on as-built drawings indicate ducts, piping, and equipment that are _____.
   a. optional items for consideration
   b. at close proximity to the work
   c. no longer functional features
   d. being removed or altered

7. Details about air diffusers, such as size and model number, would generally be found on a _____.
   a. submittal
   b. detail drawing
   c. section drawing
   d. schedule

8. An RFI is used for _____.
   a. authorization
   b. investigation
   c. clarification
   d. modification

9. Model building codes are legal instruments that are not subject to revision.
   a. True
   b. False

## SECTION TWO

### 2.0.0 SPECIFICATIONS AND SUBMITTALS

### Objectives

Describe the uses of specifications and submittals in construction projects.

a. Describe specifications and their purpose.
b. Describe submittals and their purpose.

### Trade Terms

Change order: The documentation of a change management process, where changes in the Scope of Work that were originally agreed to by owner(s), contractor(s), and architects or engineer are recorded. Typically, the order is related to additions or deletions from the Scope of Work, and alters the completion date and/or the contract value.

Specifications and submittals perform interrelated roles in new construction projects. Specifications are formatted to spell out the exact standards for the work to be done and the materials and equipment to be used. If project specifications call for special pieces of equipment or accessories to be furnished and installed by the subcontractor, a submittal document must be provided for such items.

### 2.1.0 Specifications

Specifications are written statements provided by the architectural and engineering firm to the general contractor and, consequently, to the subcontractors. Specifications define the quality of work to be done and describe the materials to be used. They are very important to the architect and owner because they guarantee compliance by the contractors to the standards set for the project. Specifications consist of various elements that may differ somewhat for particular construction projects. An example of the abbreviated specifi-

cations for a hypothetical construction project is provided in *Appendix B*. The example in the appendix follows the specification-writing format developed by the Construction Specifications Institute (CSI).

### 2.1.1 Purpose

Specifications have several important purposes, such as the following:

- They clarify information that cannot be shown on the drawings.
- They identify the work standards, types of materials to be used, and responsibilities of various parties to the contract.
- They provide information on details of construction.
- They serve as a guide for contractors bidding on the construction job.
- They serve as a standard of quality for materials and workmanship.
- They serve as a guide for compliance with building codes and zoning ordinances.
- They serve as the basis of agreement between the owner, architect, and contractors in settling any disputes.

There are two types of information contained in a set of specifications: non-technical (special and general conditions) and technical aspects of construction.

### 2.1.2 Special and General Conditions

Special and general conditions cover the non-technical aspects of the contractual arrangements. Special conditions cover topics such as safety, temporary construction, shop drawing(s) required, and so on. General conditions cover the following points of information:

- Contract terms
- Responsibilities for examining the construction site
- Types and limits of insurance
- Permits and payment of fees
- Use and installation of utilities
- Supervision of construction
- Other pertinent items

## Notes on Drawings

Sometimes notes on drawings contradict written specifications or are inconsistent with other requirements. Even though the specifications usually take precedence, the notes are often closer to the true intent. In any case, initiate an RFI to clarify such a discrepancy when you see it.

The general conditions section is the area of the construction contract where misunderstandings often occur. Therefore, these conditions are usually much more explicit on large, complicated construction projects.

### 2.1.3 Technical Aspects

The technical aspects section of the specifications covers the work to be done by the major divisions and identifies the standards for each part. The divisions are usually in the order in which the work will be performed; for example, site work is listed before carpentry work.

The technical aspects section includes information on materials that are specified by standard numbers and by standard national testing organizations, such as the American Society of Testing Materials (ASTM) International. The specifications are usually written around some standard format published by the American Institute of Architects (AIA).

### 2.1.4 Format

For convenience in writing, speed in estimating work, and ease in reference, the most suitable organization of the specifications is a series of sections dealing successively with the different trades. All the work of each trade should be incorporated into the section devoted to that trade. Those people who use the specifications must be able to find all needed information quickly.

The most commonly used specification-writing format in North America is the MasterFormat™. This standard was developed jointly by the Construction Specifications Institute (CSI) and Construction Specifications Canada (CSC). For many years prior to 2004, the organization of construction specifications and suppliers catalogs was based on a standard with 16 sections, otherwise known as divisions, where the divisions and their subsections were individually identified by a five-digit numbering system. The first two digits represented the division number and the next three individual numbers represented successively lower levels of breakdown. For example, the number 13213 represents division 13, subsection 2, sub-subsection 1 and sub-sub-subsection 3.

In this older version of the standard, all things related to the HVAC systems were lumped together under Division 15 – Mechanical. In construction and on drawings, HVAC systems are often referred to as the mechanical systems; thus the term mechanical contractor. Today, some specifications conforming to the 16-division format may still be in use. The older version of the standard contains the following divisions:

- *Division 1* – General Requirements
- *Division 2* – Site Work
- *Division 3* – Concrete
- *Division 4* – Masonry
- *Division 5* – Metals
- *Division 6* – Wood and Plastics
- *Division 7* – Thermal and Moisture Protection
- *Division 8* – Doors and Windows
- *Division 9* – Finishes
- *Division 10* – Specialties
- *Division 11* – Equipment
- *Division 12* – Furnishings
- *Division 13* – Special Construction
- *Division 14* – Conveying Systems
- *Division 15* – Mechanical
- *Division 16* – Electrical

In 2004, the MasterFormat™ standard underwent a major change. The 16 original divisions were expanded to four major groupings and 49 divisions (50 if Division 00 – Procurement and Contracting Requirements is counted) with some divisions reserved for future expansion. The first 14 divisions are essentially the same as the old format. Subjects under the old Division 15 – Mechanical have been relocated to new divisions 21, 22, and 23. In addition, the numbering system was changed to 6 digits to allow for more subsections in each division for finer definition. Updates to the MasterFormat™ standard were made in 2010 and 2012 with minor effects on the divisions that were introduced in 2004. At the time of this writing, the 2012 update of the standard remains the most recent one.

### 2.2.0 Submittals

Submittals are documents that illustrate special pieces of equipment or accessories that are to be furnished and installed by the subcontractor. The

## Addenda and Change Orders

Addenda and change orders are contractual documents used to correct or make changes to original construction drawings and specifications. The difference between the two documents is a matter of timing. An addendum is written before and a change order is drawn up after the contract is awarded.

submittal document is received from the equipment or accessory supplier by the subcontractor, and then submitted to the general contractor after the bid has been accepted and the contract signed. Subcontractors receive submittals for any major pieces that they are contractually obligated to install. The information from the submittals is used as a resource to complete submittal sheets.

The submittal sheet (*Figure 23*) illustrates the accessory or piece of equipment that has been defined in the specifications and that must conform to the standards as outlined in the specifications manual.

## Permits

Always make sure you have all the required permits before beginning an installation task. Failure to do so can result in increased costs and lost profits for your employer. This is because the laws in some states give owners the right to refuse payment for any work that is done prior to receiving the proper permits to perform that work. Some states also require that changes to a project be formally documented by an approved change order prior to beginning work on the change. A revised permit may also be required.

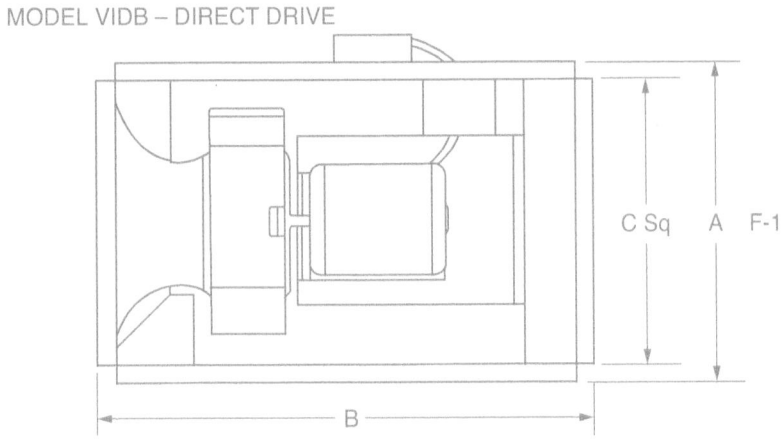

SUBMITTAL SHEET   HV 4-8
IN LINE CENTRIFUGAL DUCT FANS

MODEL VIDB – DIRECT DRIVE

| UNIT SIZE | A | B | C | INLET OR OUTLET AREA | WHEEL DIAMETER | WT (lbs) |
|---|---|---|---|---|---|---|
| 06 | 13⅞" | 20" | 12" | 0.979 sq ft | 10¾" | 30 |
| 08 | 13⅞" | 20" | 12" | 0.979 sq ft | 10¾" | 40 |
| 10 | 13⅞" | 20" | 12" | 0.979 sq ft | 10¾" | 40 |
| 12 | 17⅞" | 27⅜" | 16" | 1.750 sq ft | 11¹³⁄₁₆" | 75 |
| 15 | *21⅞" | 31" | 20" | 2.740 sq ft | 14⅞" | 90 |
| 18 | *21⅞" | 33⅜" | 26" | 4.650 sq ft | 17¹³⁄₁₆" | 140 |

\* A-1 Larger on access door sides

| JOB NAME AND LOCATION | SUBMITTED BY |
|---|---|
| | |
| SECTION 15000          2.4.1 | |

03401-13_F23A.EPS

*Figure 23* Submittal sheet (1 of 2).

## MODEL VIDB 10 – DIRECT DRIVE UNITS

| SIZE 10 | MOTOR HP | RPM | TIP SPEED | STATIC PRESSURE, INCHES W.G. | | | | | | |
|---|---|---|---|---|---|---|---|---|---|---|
| | | | | 1/8 CFM | 1/4 CFM | 3/8 CFM | 1/2 CFM | 5/8 CFM | 3/4 CFM | 1 CFM |
| 10-D3S | 35 MHP | 700 | 1970 | 200 | | | | | | |
| with | Standard | 800 | 2250 | 295 | | | | | | |
| ES60 | Motor | 900 | 2535 | 365 | 160 | | | | | |
| Speed | | 1000 | 2815 | 430 | 290 | | | | | |
| Control | 100W Max. | | | | | | | | | |

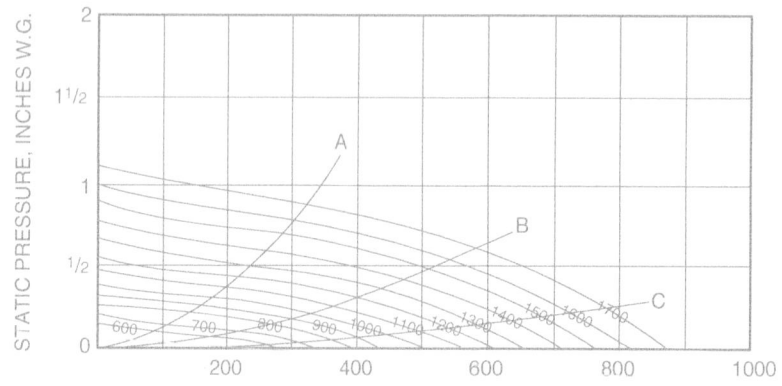

## SOUND LEVEL DATA   VIDB  SIZE 10

| SIZE RPM | * | SOUND POWER LEVEL BY OCTAVE BANDS | | | | | | | | LWA | SONE |
|---|---|---|---|---|---|---|---|---|---|---|---|
| | | 1 | 2 | 3 | 4 | 5 | 6 | 7 | 8 | | |
| 10-10D 1105 | A | 62.5 | 72.5 | 57.5 | 54.5 | 51.0 | 49.5 | 48.0 | 47.0 | 60.5 | 5.6 |
| | B | 62.5 | 69.0 | 58.0 | 56.5 | 53.5 | 50.0 | 48.0 | 47.0 | 60.0 | 5.2 |
| | C | 62.5 | 68.0 | 59.0 | 58.5 | 57.0 | 51.5 | 48.5 | 47.0 | 62.0 | 5.5 |
| 10-10H 1710 | A | 72.0 | 82.0 | 67.0 | 64.0 | 60.5 | 59.0 | 57.5 | 56.5 | 69.5 | 10.7 |
| | B | 72.0 | 78.5 | 67.5 | 66.0 | 63.0 | 59.5 | 57.0 | 56.5 | 69.5 | 9.8 |
| | C | 72.0 | 77.5 | 68.5 | 68.0 | 67.0 | 61.0 | 58.0 | 56.5 | 71.0 | 10.0 |

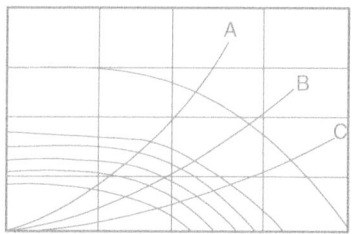

| PROJECT: | AGENT: | |
|---|---|---|
| | SUBMITTAL DATE: | JOB NO: |
| The CARNES Company Manufactured Products | DRAWING NO: | |
| | Date: | Rev: |

03401-13_F23B.EPS

*Figure 23* Submittal sheet (2 of 2).

For example, the specifications for an in-line centrifugal duct fan may have been stated as follows:

"Centrifugal in-line duct fans shall be ACME Company, Model VIDB direct drive or Model VIBA belt drive, as shown on the plans and schedules. Fans shall be constructed of heavy-gauge steel and electro-coated acrylic enamel finish over iron phosphate primer. Wheels 12 inches in diameter and larger shall have median foil blades to assure quiet, efficient operation. The motor drive compartment shall be isolated from the airstream and externally ventilated. Bearings shall be pre-lubricated and sealed for minimum maintenance and designed for 200,000 hours of operation. Internal parts (wheels, shaft, bearings, motor, and drive) shall be accessible for inspection and repair or replacement without disturbing inlet or outlet ductwork. Fans shall be furnished with a mounted safety disconnect. Single-phase motors shall have integral overload protection. V-belt drives shall be adjustable. Horsepower and noise levels shall not exceed the values shown, and oversized motors will not be acceptable. Performance ratings shall be certified for air and sound."

The submittal is commonly used to describe the unit specified by the architect or engineer.

If the specification allows for substitutions, however, the subcontractor may choose an alternate piece of equipment. In that case, a submittal for the new equipment is acquired by the subcontractor, who submits copies of it to the general contractor, who then submits it to the owner, architect, and any code enforcement authorities. These people either accept or reject the submittal. If the submittal is accepted by the general contractor, owner, and architect, the item may then be installed, as agreed upon, by the subcontractor.

The submittal sheet, as shown, includes the pertinent information that meets the specifications for the construction project. It includes such information as the size of the unit and rough opening, the specifications relating to the size of the inlet or outlet, the cubic feet per minute (cfm), and the sound ratings. The type of mounting may also be stated along with any accessories that would be applicable, such as electronic speed control, spark-resistant wheels, and explosion-proof motors.

When agreed upon by all, the submittal contains the genuine specifications for the unit or accessory with no deviations without approval by a change order from all parties.

Submittal drawings, which are considered a type of shop drawing, are typically the responsibility of equipment and accessory manufacturers.

## Specifications

Written specifications supplement the related working drawings in that they contain details not shown on the drawings. Specifications define and clarify the scope of the job. They describe the specific types and characteristics of the components that are to be used on the job and the methods for installing some of them. Many components are identified specifically by the manufacturer's model and part numbers. This type of information is used to purchase the various items of hardware needed to accomplish the installation in accordance with the contractual requirements.

## Making Submittals

When making submittals, make enough copies so that at least one copy of the original is available as an office file copy. It is common for contractors to scan submittal documents and store them in an electronic file. A log book in which each submittal is recorded should be kept by the submitting contractor. As a minimum, the log entries should include the name of the recipient, name of the submittal item(s), the date submitted, and the section of the specification that applies. All submittals should clearly show the company name so that an approved submittal can be returned to that organization. When an approved submittal is received, the date received should be noted in the submittal log, and the approved copy should be filed in a safe location. If the company has subcontracted work to other contractors (such as an air and water balance contractor), it is usually necessary that copies of the applicable approved submittals be provided to the subcontractor for incorporation into the subcontractor's work.

## Additional Resources

*Blueprint Reading for Construction*, 2003. Second Edition. James A. S. Fatzinger, Upper Saddle River, NJ: Pearson/Prentice Hall.

*Construction Specifications Writing: Principles and Procedures*, 2010. Sixth Edition. Mark Kalin, Robert S. Weygant, Harold J. Rosen, and John R. Regener. Hoboken, NJ: John Wiley & Sons, Inc.

*Reading Architectural Work Drawings*, 2003. Sixth Edition. Edward J. Muller and Phillip A. Grau III. Upper Saddle River, NJ: Prentice Hall.

## 2.0.0 Section Review

1. Specifications serve as a basis for settling disputes when the owner, architect, and contractors disagree about the details of a construction project.

   a. True
   b. False

2. To obtain approval for making an alteration to a piece of equipment for which the specifications do not allow for substitutions and a submittal has been agreed upon, you must _____.

   a. petition the appropriate code enforcement authorities
   b. use a change order to get approval from all parties
   c. consult with the owner and architect for the project
   d. request authorization from the general contractor

## 3.0.0 TAKEOFFS

### Objectives

Describe the takeoff process and how it is performed.
  a. Identify and describe the tools and materials used in the takeoff process.
  b. Explain how to conduct a takeoff.

### Performance Task 6

6. Perform an HVAC equipment and material takeoff and prepare the takeoff forms.

### Trade Terms

Takeoff: The process of surveying, measuring, itemizing, and counting all materials and equipment needed for a construction project, as indicated by the drawings.

The takeoff procedure involves surveying, measuring, and counting all materials and equipment indicated on a set of drawings. Although there are many sophisticated software programs available today to help the process, they are best used for large projects and by someone who is skilled at the takeoff process in general. Therefore, it is best to learn and experience the takeoff process using manual methods first.

### 3.1.0 Takeoff Tools and Materials

Open standard takeoff sheets are available for the HVAC technician to use for accurate takeoffs. The sheets are useful because they provide standardization, continuity, and a permanent record. In addition, they reduce the workload as well as the potential for error.

There are several types of takeoff sheets available for use by the HVAC contractor. These sheets include the following:

- *Air devices and equipment takeoff sheets* – These are usually 8½" × 11" standard sheets. They are ruled in four columns, with room to indicate the item number, the classification of the equipment, the quantity, and any necessary remarks.
- *Piping and accessories takeoff sheets* – These often include individual sheets for pipe (*Figure 24*), fittings (*Figure 25*), valves (*Figure 26*), hangers (*Figure 27*), and strainers, traps, and joints.

- *Ductwork takeoff sheets* – The information required includes duct size and length, area per foot of run, total area of duct, maximum duct size, and suggested gauge (*Figure 28*).

The following materials will make the measuring, counting, and calculating tasks easier:

- Colored pencils for checking off items on drawings as they are taken off. The same color can be used for all the material contained within a system to simplify following up on that system.
- An automatic mechanical counter for counting similar diffusers, grilles, and registers
- An electronic wheel scaler or similar device for measuring duct and piping runs (should have scales of ⅛", ¼", and ½")
- Metallic tape with ⅛" scale on one side and a ¼" scale on the other for measuring duct and piping runs
- Two drafting scales (one architect's and one engineer's)
- An adding machine or calculator
- A magnifying glass for examining details on the drawings
- A large collection of manufacturers' catalogs, technical books and manuals, and previous job files for reference

| PROJECT NAME | | | | JOB NO. | | | |
|---|---|---|---|---|---|---|---|
| PIPING SYSTEM | | | | MATERIAL | | | |
| PIPE SIZES (MEASURED IN INCHES) | | | | | | | |
| FEET | NUMBER OF JOINTS | FEET | NUMBER OF JOINTS | FEET | NUMBER OF JOINTS | FEET | NUMBER OF JOINTS |
|  |  |  |  |  |  |  |  |
| TOTAL |  |  |  |  |  |  |  |

03401-13_F24.EPS

*Figure 24* Piping takeoff sheet.

PROJECT NAME__________ JOB NO. __________

MATERIAL__________ CLASS __________

| FITTING TYPE | FITTING SIZES (MEASURED IN INCHES) | | | | | | |
|---|---|---|---|---|---|---|---|
| | | | | | | | |
| | | | | | | | |
| NO. OF JOINTS | | | | | | | |

03401-12_F25.EPS

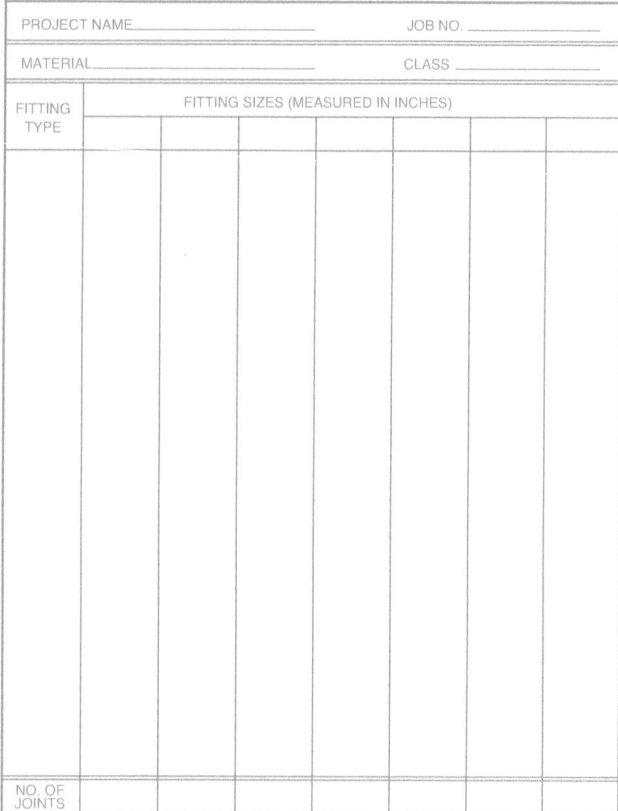

*Figure 25* Piping fitting takeoff sheet.

PROJECT NAME__________ JOB NO. __________

MATERIAL__________ CLASS __________

| TYPE OF VALVE DEVICE | VALVE AND DEVICE SIZES (MEASURED IN INCHES) | | | | | | |
|---|---|---|---|---|---|---|---|
| | | | | | | | |
| | | | | | | | |

03401-13_F26.EPS

*Figure 26* Valve takeoff sheet.

PROJECT NAME__________ JOB NO. __________

| HANGER TYPE | PIPE SIZES (MEASURED IN INCHES) | | | | | | |
|---|---|---|---|---|---|---|---|
| | | | | | | | |
| | | | | | | | |

03401-13_F27.EPS

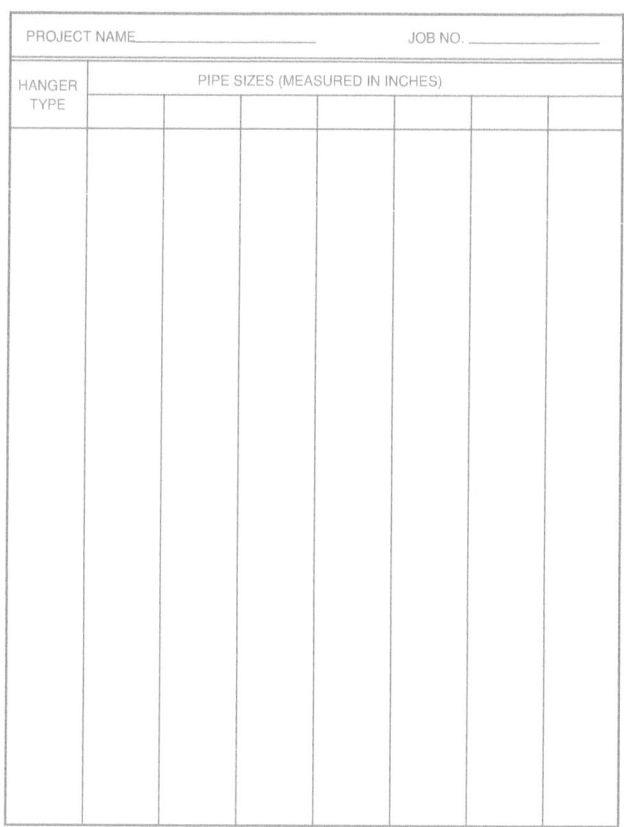

*Figure 27* Hanger takeoff sheet.

PROJECT NAME__________ JOB NO. __________

DUCTWORK SYSTEM__________ MATERIAL __________

| DUCT SIZE (INCHES) | DUCT SIZE (FEET) | SQ. FT. PER FT. OF RUN | TOTAL AREA OF DUCT (SQUARE FEET) | | | | |
|---|---|---|---|---|---|---|---|
| | | | MAXIMUM DUCT SIZE AND SUGGESTED GAUGE | | | | |
| | | | | | | | |
| | | | | | | | |

03401-13_F28.EPS

*Figure 28* Ductwork takeoff sheet.

After the takeoff is complete, record all material and equipment classifications on the job estimating sheets and copy the total quantity of each item as it appears on the takeoff sheets to the job estimating sheets.

The drawings and specifications are used to identify the types and quantities of equipment and materials required for the project. The selection of heating, ventilating, and air conditioning (HVAC) equipment is based on an analysis of heating and cooling loads performed by the architect or engineer. The purpose of the analysis is to select the equipment that is physically suited to the structure and is sized to meet its heating and cooling loads and air flow requirements. Load estimating and equipment selection are explained in a later module.

The selection of materials for HVAC systems is based on a study of the conditions of operation. The factors to consider include the following:

- Code requirements.
- The working fluid in the piping system.
- The pressure and temperature of the fluid.
- The environment around the pipe.
- The budget for the installed system.

Piping, ductwork, pipe/duct hangers and supports, and valves are also selected from the drawings and specifications. An overview of the criteria that go into the selection process for these items is provided in the sections that follow.

### 3.2.0 Takeoff Procedures

In an HVAC takeoff, the materials and equipment should be taken off in the following order:

1. Equipment such as boilers, chillers, pumps, air-handing units, fans, and air cleaners
2. Air devices and air terminals
3. Radiant heating system components
4. Piping and accessory systems
5. Ductwork, dampers, and louvers
6. Insulation for piping, ductwork, and equipment
7. Gauges and thermometers
8. Motor starters, motor control centers, and electrical work to be furnished by the HVAC contractor
9. Temperature control systems
10. Other special systems to be furnished and installed by the HVAC contractor

The takeoff procedure for valves, fittings, piping, hangers, and accessories is as follows:

*Step 1* Prepare the required takeoff sheets.

*Step 2* Read the specifications carefully; then list the materials and joints specified for each system.

## Construction Calculators

Using construction calculators to project the cost and time required to complete a project can help save time and money. These calculators are equipped with built-in solutions for completing plans, layouts, bids, and estimates. They allow you to set and store preferences and format essential calculations (stair clearance, roof pitch, framing, circular calculations, etc.) for dimensions that are frequently required or established by codes. Typically, construction calculators like the one shown here work in and convert between feet-inches-fractions and decimals, including metric. To ensure accurate estimates, you must periodically adjust the calculator based on known changes in the project.

As shown in the example, many construction calculators, are also available as applications that can be installed on mobile phones or tablets.

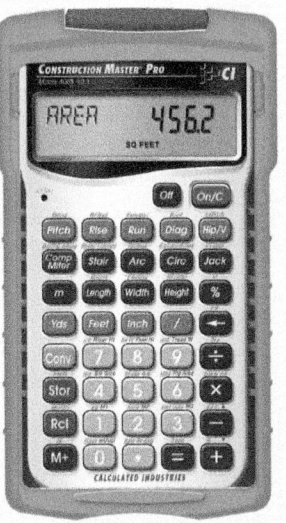
CALCULATOR

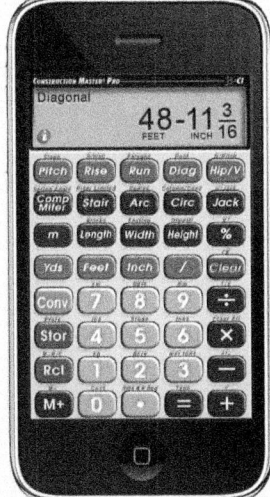
MOBILE PHONE APPLICATION

03401-13_SA04.EPS

*Step 3* Take off each system as indicated on the floor plans, section details, and flow diagrams. The takeoff must include piping, fittings, valves, joints, hangers, and all accessories.

*Step 4* List all quantities of the same size in one column of the takeoff sheet.

*Step 5* Measure pipe lengths in feet; count the fittings, valves, hangers, and so on.

*Step 6* After completing the takeoff, add the quantities of each size, and transfer the sum to a master sheet.

The takeoff procedure for ductwork follows:

*Step 1* Use a standard takeoff sheet or prepare one.

*Step 2* Read the specifications; then list the types of materials, recommended construction, and specified gauges for each system.

*Step 3* Include all the takeoff quantities for all the duct runs that have the same material, construction, acoustic lining, and insulation on one takeoff sheet.

*Step 4* Take off each system as indicated on the drawings.

*Step 5* Place all the findings on a summary sheet.

## Scaling Drawings

Measuring the length of pipe and duct runs on mechanical plans and converting the measurements to actual length using the given scale is called scaling a drawing. Scaling a drawing is commonly done using architect's scales. The task of scaling a drawing when doing a takeoff can be made easier by using an electronic plan wheel scaler like the one shown. Upon input of the scale parameters, this digital plan wheel gives a direct readout of actual length for the scaled distance.

03401-13_SA05.EPS

---

## Additional Resources

*Blueprint Reading for Construction,* 2003. Second Edition. James A. S. Fatzinger, Upper Saddle River, NJ: Pearson/Prentice Hall.

*Reading Architectural Work Drawings,* 2003. Sixth Edition. Edward J. Muller and Phillip A. Grau III. Upper Saddle River, NJ: Prentice Hall.

### 3.0.0 Section Review

1. Types of fittings are usually included in a piping takeoff sheet.

    a. True
    b. False

2. What is the first step in a quantity takeoff procedure?

    a. Read the specifications
    b. List all quantities to be counted
    c. Prepare the required takeoff sheet(s)
    d. Transfer the sum to a master sheet

## SUMMARY

Construction drawings and specifications for the HVAC trade contain the information necessary for the layout, fabrication, and installation of duct runs, HVAC piping, and fittings. Specifications are written descriptions of technical design and performance information that must be used when selecting and installing equipment, systems, and construction components. When there is a conflict between the design specifications and the architectural plans, the specifications that are most stringent usually apply. Often the contract or specification will define which document prevails in case of a conflict. Sometimes which document prevails is determined by all parties at the contract signing.

1. You can find the date of the last revision for a set of drawings by looking at the _____.

   a. index
   b. title block
   c. site plan
   d. submittal

2. Site plans may include some construction features such as _____.

   a. pitches of roofs
   b. water supply lines
   c. curbs and gutters
   d. ductwork layout

3. The distance between the centers of the north window and door in the garage shown on the floor plan in *Review Question Figure 1* is _____.

   a. 6'-0"
   b. 6'-4"
   c. 8'-0"
   d. 10'-0"

4. How many roof drains are called for in the roof plan shown in *Review Question Figure 2*?

   a. Three
   b. Four
   c. Five
   d. Six

5. The vertical mansard (type of roof) shown in the elevation drawing on *Review Question Figure 3* is made of _____.

   a. 26 gauge, Type SR arctic white metal
   b. 26 gauge, Type SS arctic white metal
   c. 8" vertical score block
   d. 8" split rib block

6. In accordance with the plumbing plan shown in *Review Question Figure 4*, how many urinals must be installed?

   a. One
   b. Two
   c. Four
   d. Six

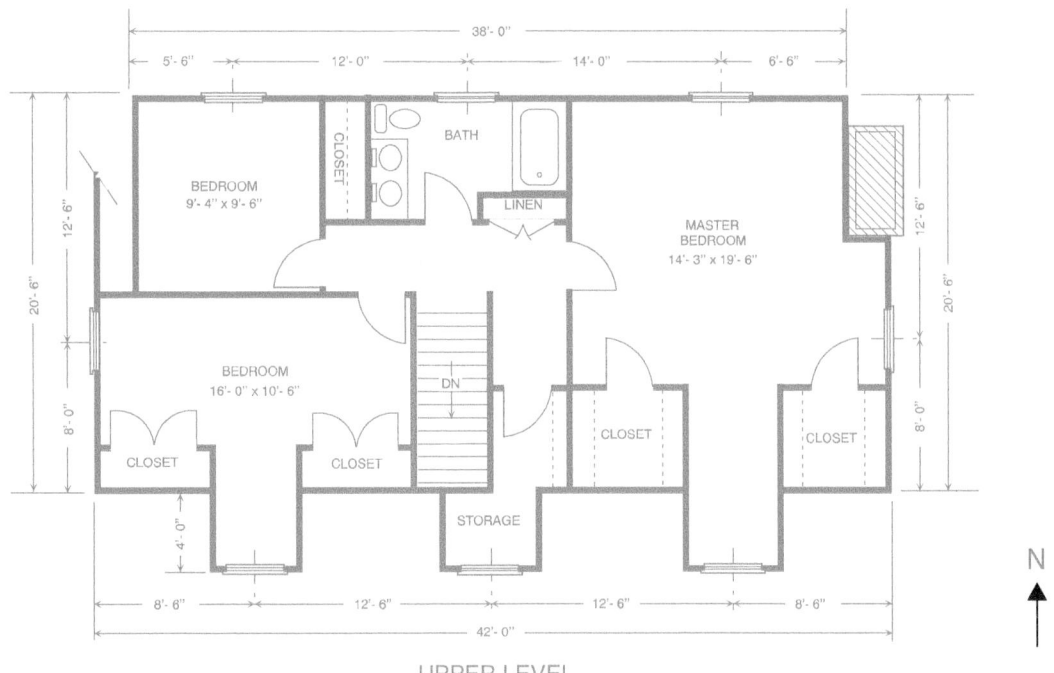

UPPER LEVEL

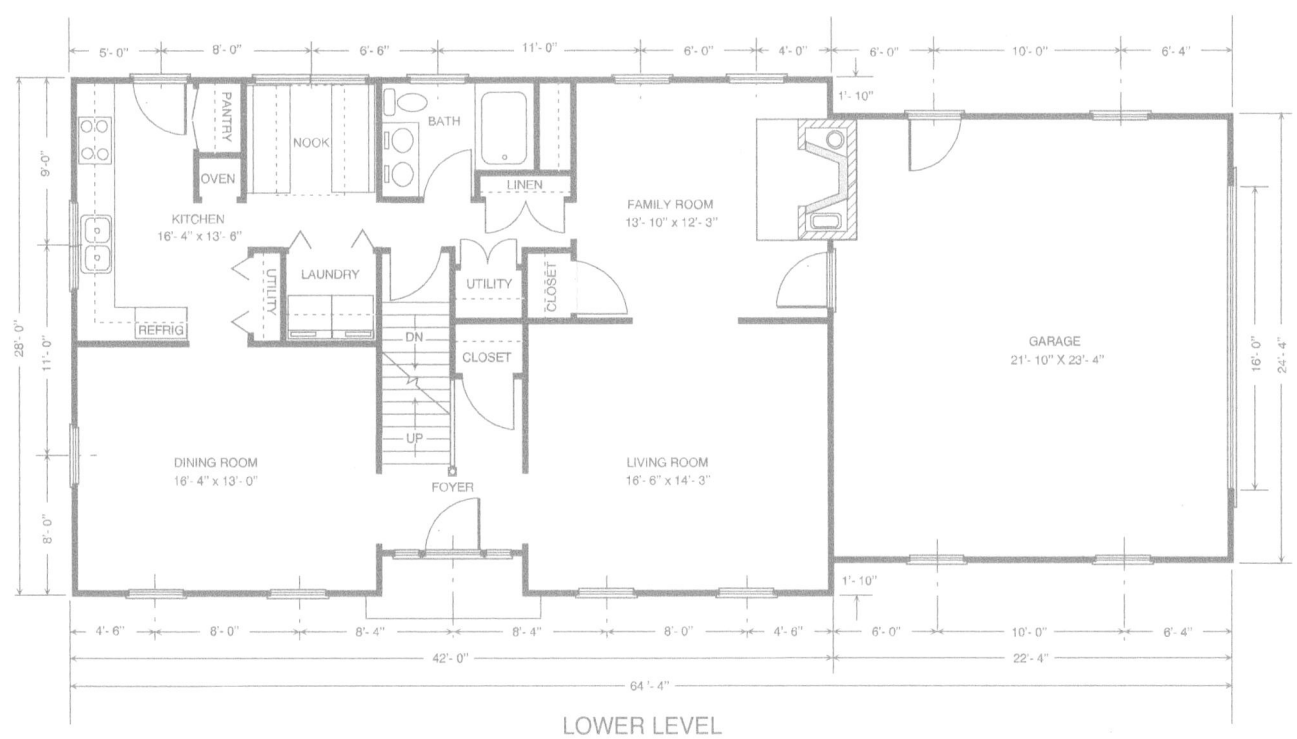

LOWER LEVEL

**Figure 1**

03401-13_RQ01.EPS

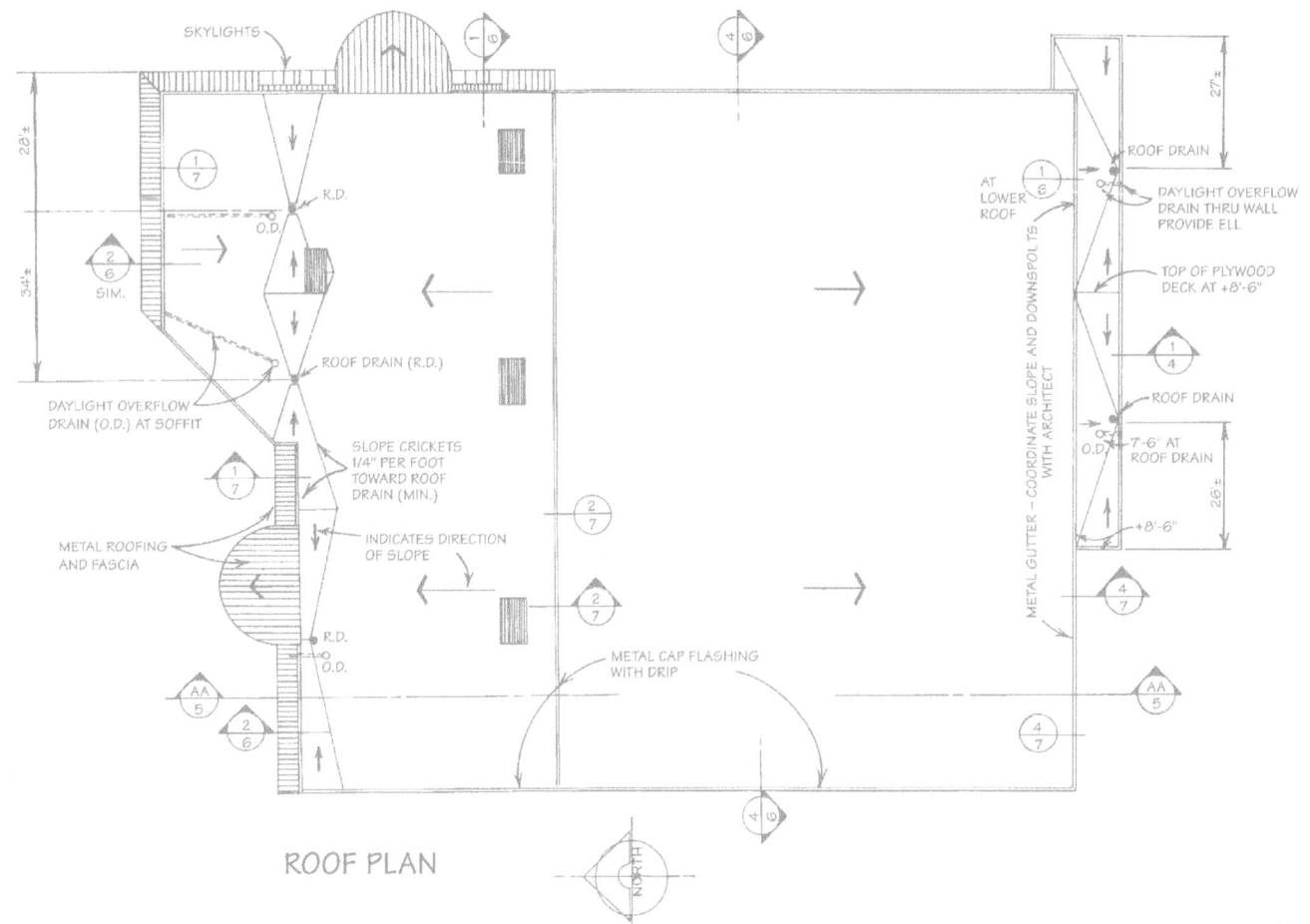

ROOF PLAN

Figure 2

03401-13_RQ02.EPS

Figure 3

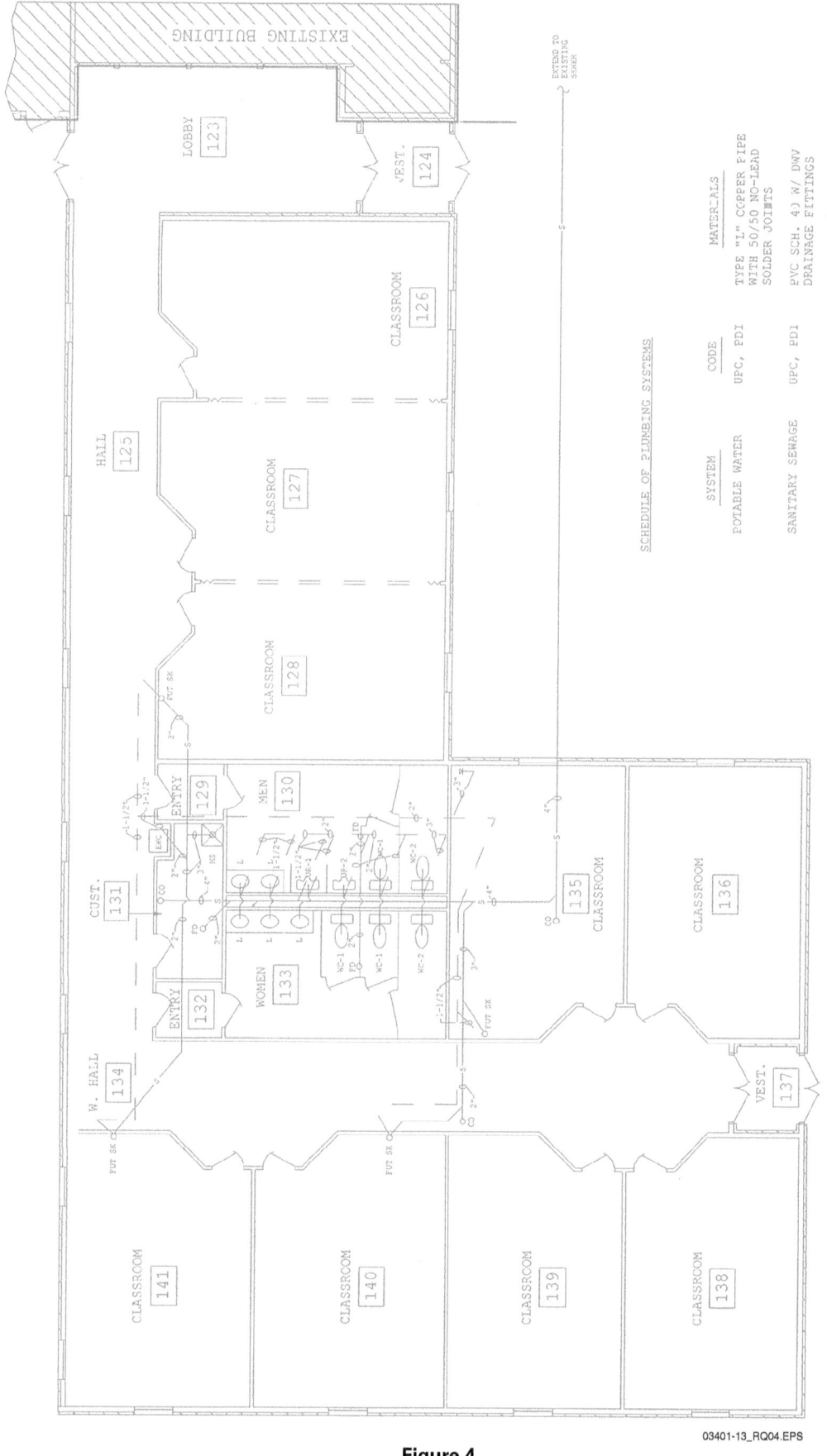

**Figure 4**

03401-13_RQ04.EPS

7. Piping for natural gas is typically shown on the _____.

    a. floor plan
    b. elevation drawings
    c. HVAC drawings
    d. plumbing drawings

8. What size chilled-water supply piping should be connected to Air Handling Unit 1 shown in the mechanical plan on *Review Question Figure 5*?

    a. 1¼"
    b. 1½"
    c. 2"
    d. 2½"

9. The electrical plan in *Review Question Figure 6* shows the new power distribution panel, Panel D, to be installed in _____.

    a. hallway 125
    b. lobby 123
    c. room 126
    d. room 131

10. A shop drawing prepared by an equipment manufacturer typically describes _____.

    a. structural steel members
    b. special equipment
    c. sheet metal components
    d. ductwork layout

11. A type of drawing that shows the exact layout for a run of ductwork, including the sizes and types of duct fittings and the types of connectors and hangers, is a _____.

    a. sheet metal shop drawing
    b. mechanical plan
    c. cut list
    d. detail drawing

12. Cut lists are drawn from _____.

    a. takeoffs
    b. elevations
    c. shop drawings
    d. an RFI

13. A drawing that shows alterations or additions made to an original plan is called a(n) _____.

    a. submittal
    b. cut list
    c. as-built
    d. takeoff

14. What are the size and delivery specifications for Item F listed in the grille, register, and diffuser schedule shown in *Review Question Figure 7*?

    a. 12" × 12"; 240 cfm
    b. 12" × 12"; 250 cfm
    c. 18" × 18"; 235 cfm
    d. 18" × 18"; 280 cfm

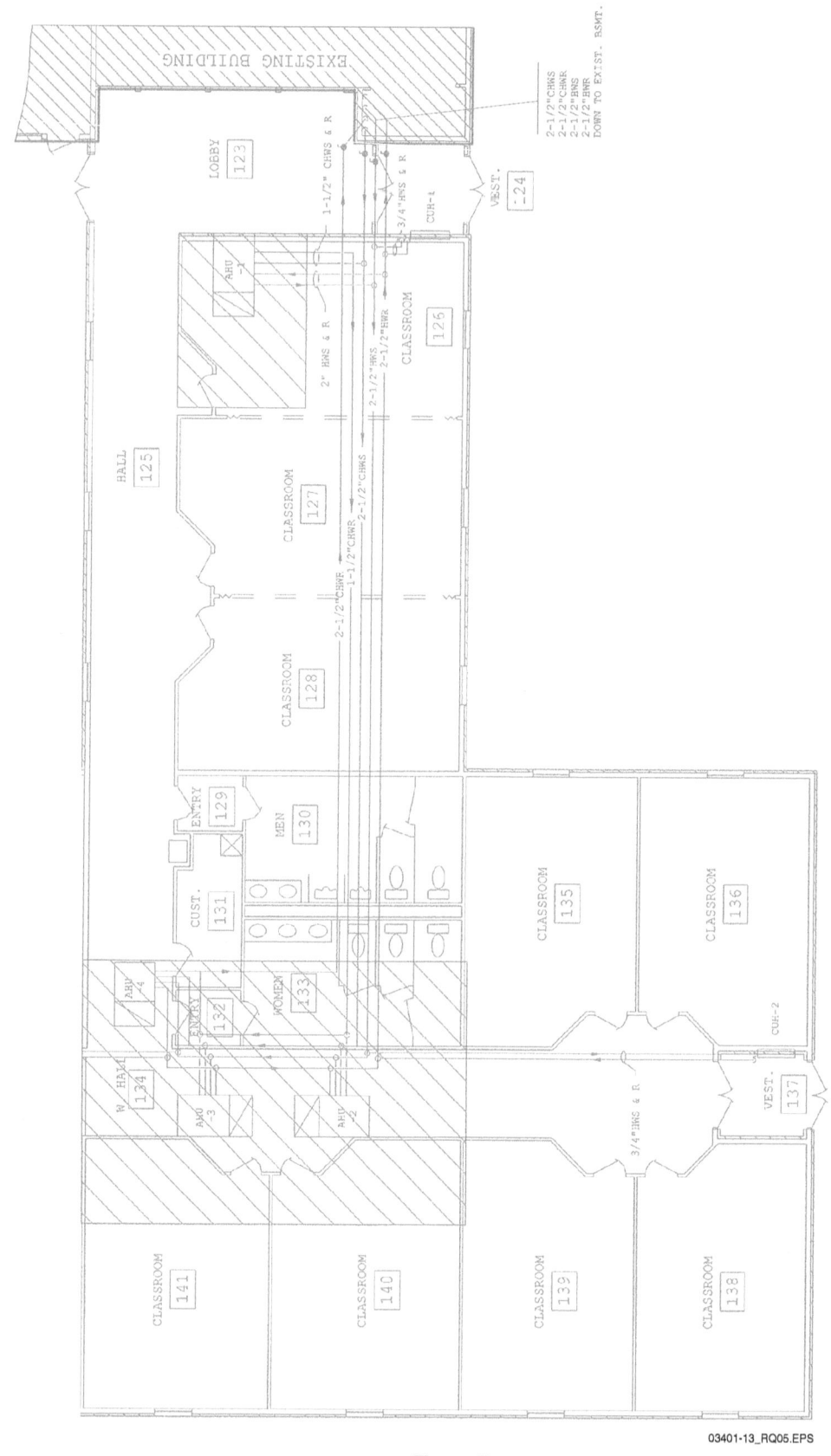

**Figure 5**

03401-13_RQ05.EPS

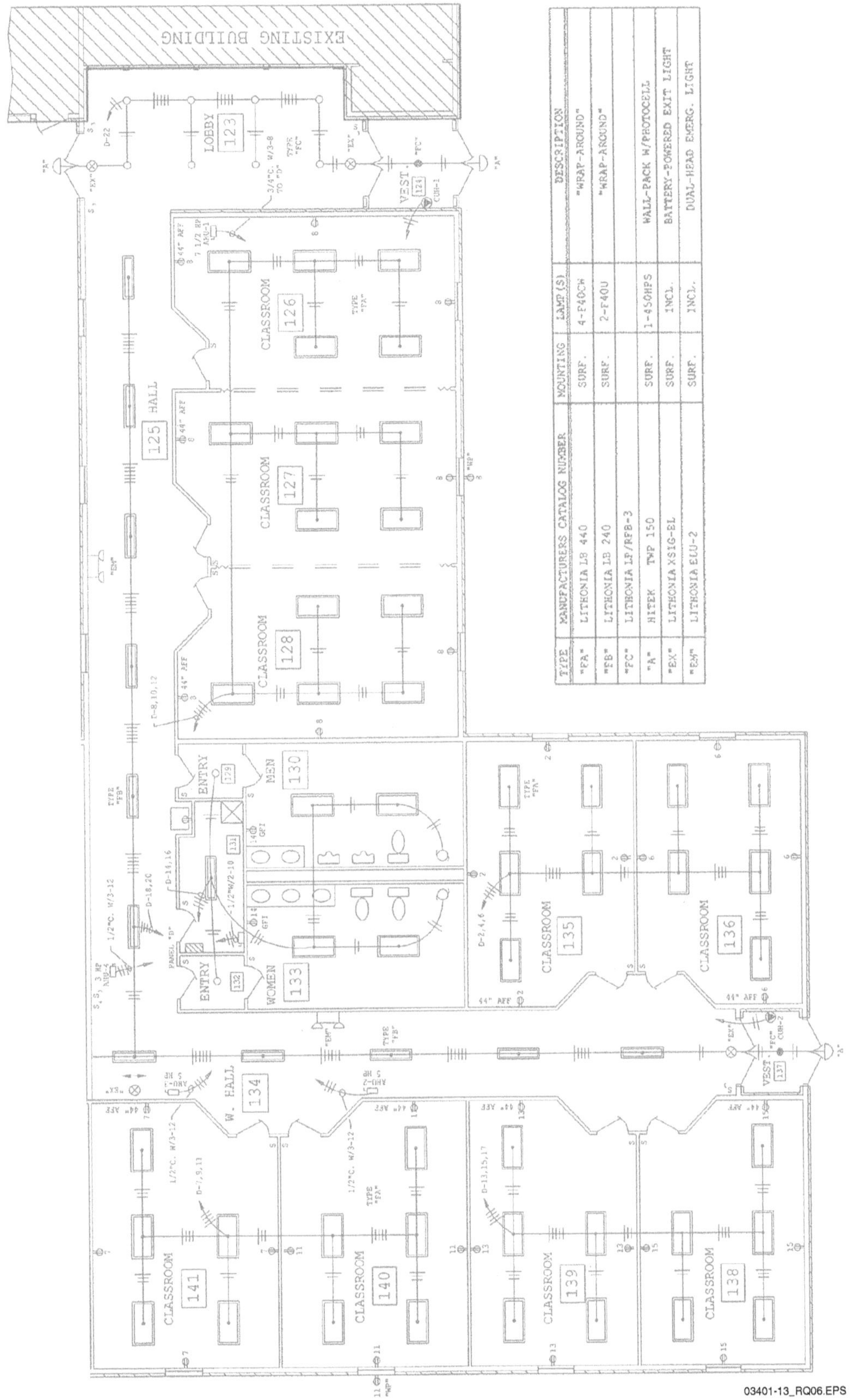

**Figure 6**

03401-13_RQ06.EPS

## CABINET UNIT HEATER SCHEDULE

| UNIT HEATER NO. | LOCATION | C.F.M. | FAN MOTOR | | | | MBH | GPM | EWT | EAT | MAX. WATER P.D. | REMARKS |
|---|---|---|---|---|---|---|---|---|---|---|---|---|
| | | | H.P. | VOLTS | PHASE | Hz | | | | | | |
| CUH-1 | 124 | 400 | 1/12 | 115 | 1 | 60 | 23 | 2.3 | 180°F | 60°F | 2.7 | McQUAY #CHF004 SEMI-RECESSED, R.H COIL |
| CUH-2 | 137 | 400 | 1/12 | 115 | 1 | 60 | 23 | 2.3 | 180°F | 60°F | 2.7 | McQUAY #CHF004 SEMI-RECESSED, L.H COIL |
| CUH-3 | 143 | 400 | 1/12 | 115 | 1 | 60 | 23 | 2.3 | 180°F | 60°F | 2.7 | McQUAY #CHF004 SEMI-RECESSED, R.H COIL |

NOTES:
1. 3 Speed Control
2. Front Discharge
3. With Return Air Filters

## PUMP SCHEDULE

| UNIT NO. | LOCATION | SERVICE. | GPM | MBH | MOTOR | | | | | TYPE | REMARKS |
|---|---|---|---|---|---|---|---|---|---|---|---|
| | | | | | RPM | H.P. | VOLTS | PHASE | Hz | | |
| P-1 | MECH. ROOM | NAVE | 45 | 41' | 1750 | 1½ | 208/230 | 3 | 60 | IN-LINE | B/G #60-20T SERVICE 40% GLYCOL SOLUTION |
| P-2 | MECH. ROOM | CHW TO AHU 1-4 | 65 | 37' | 1750 | 1½ | 208/230 | 3 | 60 | IN-LINE | B/G #60-20T SERVICE 40% GLYCOL SOLUTION |
| P-3 | MECH. ROOM | RECIR. TO TANK | 40 | 17' | 1750 | ½ | 208/230 | 3 | 60 | IN-LINE | B/G #60-13T SERVICE 40% GLYCOL SOLUTION |
| P-4 | EXIST. MECH. ROOM | HW | 73 | 31' | 1750 | 1½ | 208/230 | 3 | 60 | IN-LINE | B/G #60-20T HOT WATER |

NOTES:
1. Starters And Disconnects By E.C.

## GRILLE, REGISTER AND DIFFUSER SCHEDULE

| ITEM | MANUFACTURER | MODEL NO. | QTY. | LOCATION | CFM EACH | AIR PATTERN | SIZE | | FINISHES | REMARKS |
|---|---|---|---|---|---|---|---|---|---|---|
| | | | | | | | FRAME | NECK | | |
| A | BARBER COLMAN | SFSV | 8 | 126,127,128 144 | 245 | 4-WAY | 12"× 12" | 8"Ø | #7 OFF-WHITE | |
| B | BARBER COLMAN | SFSV | 2 | 142 | 275 | 4-WAY | 18"× 18" | 10"Ø | #7 OFF-WHITE | |
| C | BARBER COLMAN | SFSV | 4 | 140, 141 | 240 | 4-WAY | 12"× 12" | 8"Ø | #7 OFF-WHITE | |
| D | BARBER COLMAN | SFSV | 2 | 139 | 270 | 4-WAY | 18"× 18" | 10"Ø | #7 OFF-WHITE | |
| E | BARBER COLMAN | SFSV | 2 | 138 | 280 | 4-WAY | 18"× 18" | 10"Ø | #7 OFF-WHITE | |
| F | BARBER COLMAN | SFSV | 2 | 136 | 250 | 4-WAY | 12"× 12" | 6"Ø | #7 OFF-WHITE | |
| G | BARBER COLMAN | SFSV | 2 | 135 | 235 | 4-WAY | 12"× 12" | 6"Ø | #7 OFF-WHITE | |
| H | BARBER COLMAN | SFSV | 3 | 134 | 100 | 4-WAY | 12"× 12" | 6"Ø | #7 OFF-WHITE | FIRE DAMPER SEE DETAIL A |
| I | BARBER COLMAN | SFSV | 1 | 134 | 190 | 4-WAY | 12"× 12" | 8"Ø | #7 OFF-WHITE | FIRE DAMPER SEE DETAIL A |
| | MAN | SFSV | 1 | 134 | WAY | | 12" | | | FIRE DAMPER |

**Figure 7**

15. The final recipient of a request for information (RFI) form is the _____.

    a. foreman
    b. architect or engineer
    c. superintendent
    d. general contractor

16. The two nationally recognized building codes are the *International Building Code*® and _____.

    a. MasterFormat™ 2004
    b. AIA Standard Format
    c. *NFPA 5000*®
    d. *CSI 13213*

17. In new CSI-formatted specifications, you would expect to find information on HVAC systems in Division _____.

    a. 7
    b. 10
    c. 15
    d. 23

18. In order to secure approval to furnish and install special equipment, a subcontractor must provide the contractor with a(n) _____.

    a. submittal
    b. cut list
    c. as-built
    d. takeoff

19. The purpose of an electronic wheel scaler is to _____.

    a. enlarge details shown on architectural drawings
    b. measure the diameter of pipe and other circular objects
    c. count the number of grilles and diffusers shown on drawings
    d. measure duct and piping runs shown on drawings

20. When performing a ductwork takeoff procedure, you should include the takeoff quantities for all the duct runs that have the same material, construction, acoustic lining, and insulation _____.

    a. in one column
    b. on a master sheet
    c. in an appendix
    d. on one takeoff sheet

# Trade Terms Introduced in This Module

**Change order**: the documentation of a change management process, where changes in the Scope of Work that were originally agreed to by owner(s), contractor(s), and architects or engineer are recorded. Typically, the order is related to additions or deletions from the Scope of Work, and alters the completion date and/or the contract value.

**Coordination drawings**: Elevation, location, and other drawings produced for a project by the individual contractors for each trade to prevent a conflict between the trades regarding the installation of their materials and equipment. Development of these drawings evolves through a series of review and coordination meetings held by the various contractors.

**Cut list**: An information sheet that is derived from shop drawings. It is the shop guide for fabricating duct runs and fittings.

**Detail drawing**: A drawing of a feature that provides more elaborate information than is available on a plan.

**Elevation view**: A view that depicts a vertical side of a building, usually designated by the direction that side is facing; for example, right, left, east, or west elevation.

**Floor plan**: A building drawing indicating a plan view of a horizontal section at some distance above the floor, usually midway between the ceiling and the floor.

**Longitudinal section**: A section drawing in which the cut is made along the long dimension of a building.

**Plan view**: The overhead view of an object or structure.

**Riser diagram**: A one-line schematic depicting the layout, components, and connections of a piping system or electrical system.

**Schedules**: Tables that describe and specify the types and sizes of items required for the construction of a building.

**Section drawing**: A drawing that depicts a feature of a building as if there were a cut made through the middle of it.

**Shop drawing**: A drawing that indicates how to fabricate and install individual components of a construction project. A shop drawing may be drafted from the construction drawings of a project building or provided by the manufacturer.

**Site plan**: A construction drawing that indicates the location of a building on a land site.

**Takeoff**: The process of surveying, measuring, itemizing, and counting all materials and equipment needed for a construction project, as indicated by the drawings.

**Transverse section**: A section drawing in which the cut is made through the short dimension of the building.

## DRAWING SYMBOLS

PROPERTY LINE

BOUNDARY LINE (MATCH LINE)

MAIN OBJECT LINE

HIDDEN LINE

CENTER LINE (Used as finished floor line)

DIMENSION AND EXTENSION LINES

$2\frac{1}{8}$"

LONG BREAK LINE

SHORT BREAK LINE

LEADER LINE

SECTION LINE TYP.

A                                        A'

REF. LINE FOR VARIOUS SECTION TYPES

1
B

1
A

03401-13_A01.EPS

LIGHT FULL LINE

MEDIUM FULL LINE

HEAVY FULL LINE

EXTRA-HEAVY FULL LINE

CENTER LINE

HIDDEN

DIMENSION LINE                3.00"

SHORT BREAK LINE

LONG BREAK LINE

MATCH LINE

SECONDARY LINE

PROPERTY LINE

03401-13_A02.EPS

| MATERIAL | SYMBOL |
|---|---|
| EARTH | |
| CONCRETE | |
| CONCRETE BLOCK | |
| GRAVEL FILL | |
| WOOD | FRAMING FINISH |
| BRICK | FACE COMMON |
| STONE | CUT RUBBLE |

| MATERIAL | SYMBOL |
|---|---|
| STRUCTURAL STEEL BEAM | SPECIFY |
| SHEET METAL FLASHING | SHOW CONTOUR |
| INSULATION | LOOSE FILL or BATT BOARD |
| PLASTER | STUD LATH & PLASTER |
| GLASS | LARGE SCALE SMALL SCALE |
| TILE | |

03401-13_A03.EPS

| DOOR TYPE | SYMBOL |
|---|---|
| SINGLE SWING | |
| SLIDER | |
| BIFOLD | |
| FRENCH | |
| ACCORDION | |

| WINDOW TYPE | SYMBOL |
|---|---|
| AWNING | |
| FIXED SASH | |
| DOUBLE HUNG | |
| CASEMENT | |
| HORIZONTAL SLIDER | |

03401-13_A04.EPS

| | | | | |
|---|---|---|---|---|
| ADD. | addition | N | north |
| AGGR | aggregate | NO. | number |
| L | angle | OC | on center |
| B | bathroom | OPP | opposite |
| BR | bedroom | O.D. | outside diameter |
| BM | bench mark | PNL | panel |
| BRKT | bracket | PSI | pounds per square inch |
| CLK | caulk | PWR | power |
| CHFR | chamfer | REINF | reinforce |
| CND | conduit | RH | right-hand |
| CU FT | cubic foot, feet | RFA | released for approval |
| DIM. | dimension | RFC | released for construction |
| DR | drain | RFD | released for design |
| DWG | drawing | RFI | released for information |
| ELEV | elevation | SHTHG | sheathing |
| ESC | escutcheon | SQ | square |
| FAB | fabricate | STR | structural |
| FLGE | flange | SYM | symbol |
| FLR | floor | THERMO | thermostat |
| GR | grade | TYP | typical |
| GYP | gypsum | UNFIN | unfinished |
| HDW | hardware | VEL | velocity |
| HTR | heater | WV | wall vent |
| " or IN. | inch, inches | WHSE | warehouse |
| I.D. | inside diameter | WH | weep hole |
| LH | left-hand | WDW | window |
| MEZZ | mezzanine | WP | working pressure |
| MO | masonry opening | | |
| MECH | mechanical | | |

03401-13_A05.EPS

## LIGHTING OUTLETS | CEILING | WALL

Surface or pendant incandescent, mercury-vapor, or similar lamp fixture

Recessed incandescent, mercury-vapor, or similar lamp fixture

Surface or pendant individual fluorescent fixture

Recessed individual fluorescent fixture

Surface or pendant continuous-row fluorescent fixture

Recessed continuous-row fluorescent fixture

Bare-lamp fluorescent strip

Surface or pendant exit light

Recessed exit light

03401-13_A06A.EPS

|                                                                                    | CEILING | WALL |
|------------------------------------------------------------------------------------|:-------:|:----:|
| Blanked outlet                                                                     | B       | —B   |
| Fan outlet                                                                         | F       | —F   |
| Drop cord                                                                          | D       |      |
| Junction box                                                                       | J       | —J   |
| Outlet controlled by low-voltage switching when relay is installed in outlet box  | L       | —L   |

RECEPTACLE OUTLETS           GROUNDED

Single receptacle outlet

Duplex receptacle outlet

Waterproof receptacle outlet    WP

Triplex receptacle outlet

Fourplex (Quadruplex) receptacle outlet

Duplex receptacle outlet, split wired

Triplex receptacle outlet, split wired

03401-13_A06B.EPS

## RECEPTACLE OUTLETS

GROUNDED | UNGROUNDED

Floor duplex receptacle outlet

Floor special-purpose outlet

Floor telephone outlet, public

Floor telephone outlet, private

An example of using several floor outlet symbols to identify several gang floor outlets

An example of the use of different symbols to show locations of different types of outlets or connections for underfloor duct or cellular floor systems

Underfloor duct and junction box for triple, double, or single duct system, as indicated by the number of parallel lines

Cellular floor header duct

03401-13_A06C.EPS

## SWITCH OUTLETS

| | |
|---|---|
| Single-pole switch | S |
| Double-pole switch | S2 |
| Three-way switch | S3 |
| Four-way switch | S4 |
| Key-operated switch | SK |
| Switch and pilot lamp | SP |
| Switch for low-voltage switching system | SL |
| Master switch for low-voltage switching system | SLM |
| Switch and single receptacle | ─⊖ S |
| Switch and double receptacle | ═⊖ S |
| Door switch | SD |
| Time switch | ST |
| Circuit breaker switch | SCB |
| Momentary contact switch or pushbutton for other than signaling system | SMC |
| Ceiling pull switch | Ⓢ |

## SIGNALING SYSTEM OUTLETS FOR INDUSTRIAL, COMMERCIAL, AND INSTITUTIONAL OCCUPANCIES

| | |
|---|---|
| Any type of nurse call system device | ─○ |
| Nurses' annunciator | ─① |
| Call station, single cord, pilot | ─② |
| Call station, double cord, microphone speaker | ─③ |
| Corridor dome light, one lamp | ─④ |
| Transformer | ─⑤ |
| Any other item on same system (use numbers as required) | ─⑥ |
| Any type of paging system device | ─◇ |
| Keyboard | ─◇1 |
| Flush annunciator | ─◇2 |
| Two-face annunciator | ─◇3 |
| Any other item on same system (use numbers as required) | ─◇4 |

03401-13_A06D.EPS

Any type of fire alarm system device, including smoke and sprinkler alarm devices

Control panel
1

Station
2

10-inch gong
3

Pre-signal chime
4

Any other item on same system (use numbers as required)
5

Any type of staff register system device

Phone operator's register
1

Entrance register, flush
2

Staff room register
3

Transformer
4

Any other item on same system (use numbers as required)
5

Any type of electric clock system device

Master clock
1

12-inch secondary, flush
2

12-inch double dial, wall mounted
3

Any other item on same system (use numbers as required)
4

Any type of public telephone system device

Switchboard
1

Desk phone
2

Any other item on same system (use numbers as required)
3

Any type of private telephone system device

Switchboard
1

Wall phone
2

Any other item on same system (use numbers as required)
3

03401-13_A06E.EPS

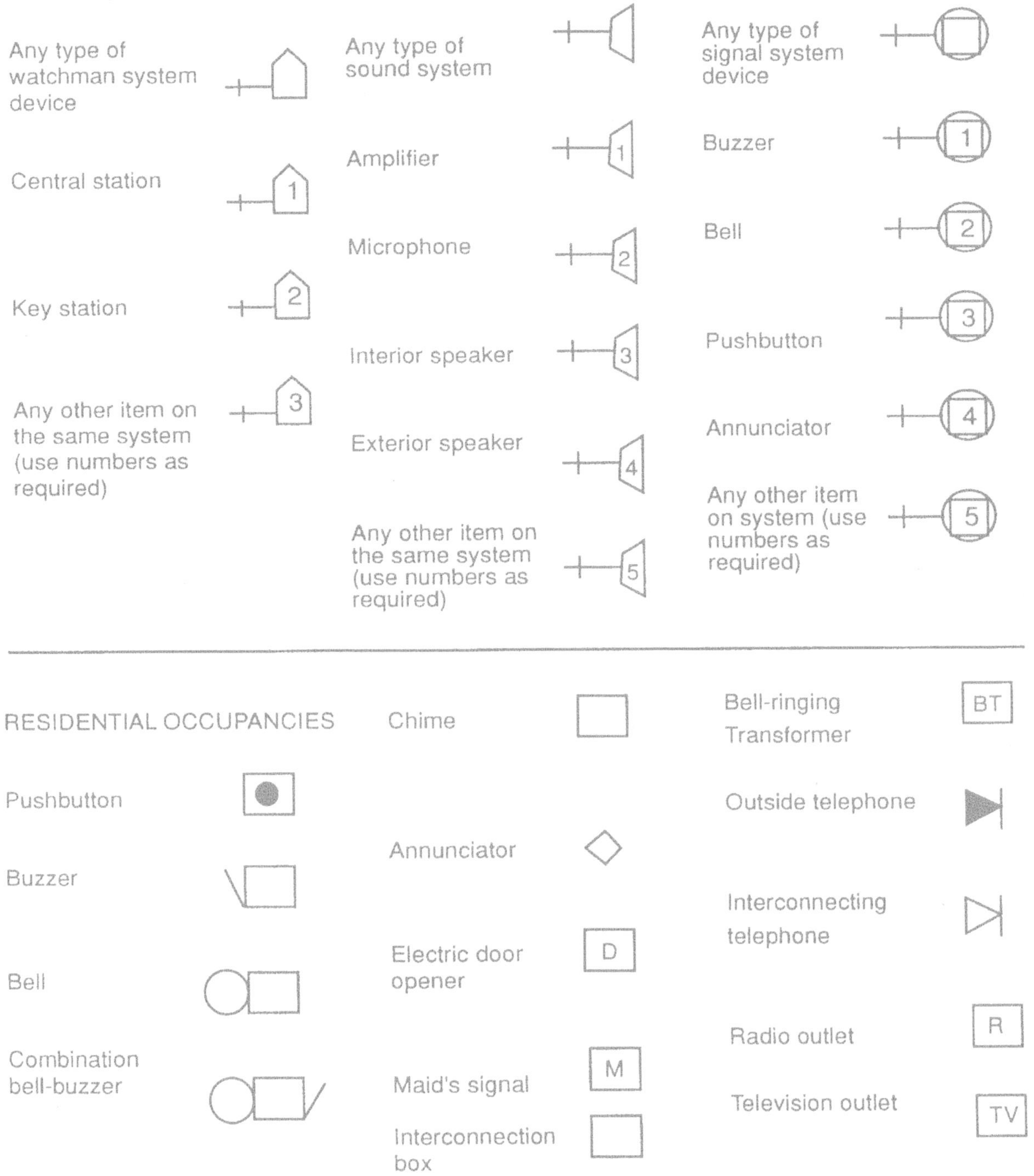

Any type of watchman system device

Any type of sound system

Any type of signal system device

Central station 1

Amplifier 1

Buzzer 1

Microphone 2

Bell 2

Key station 2

Interior speaker 3

Pushbutton 3

Any other item on the same system (use numbers as required) 3

Exterior speaker 4

Annunciator 4

Any other item on the same system (use numbers as required) 5

Any other item on system (use numbers as required) 5

---

RESIDENTIAL OCCUPANCIES

Chime

Bell-ringing Transformer    BT

Pushbutton

Annunciator

Outside telephone

Buzzer

Electric door opener    D

Interconnecting telephone

Bell

Maid's signal    M

Radio outlet    R

Combination bell-buzzer

Interconnection box

Television outlet    TV

03401-13_A06F.EPS

## PANELBOARDS, SWITCHBOARDS, AND RELATED EQUIPMENT

Flush-mounted panel board and cabinet *

Surface-mounted panel board and cabinet*

Switchboard, power control center, unit substations (ANSI recommends drawing to scale)*

Flush-mounted terminal cabinet*

Surface-mounted terminal cabinet*

Pull box—identify in relation to wiring system section and size

Motor or other power controller*

MC

Externally operated disconnection switch*

Combination controller and disconnect means*

*Identify by notation or schedule

## BUS DUCTS AND WIREWAYS

Trolley duct*

| | T | | T | | T | |

Busway (service, feeder, or plug-in)*

| | B | | B | | B | |

Cable trough, ladder, or channel*

| | BP | | BP | | BP | |

Wireway*

| | W | | W | | W | |

* Identify by notation or schedule

## REMOTE CONTROL STATIONS FOR MOTORS OR OTHER

Pushbutton stations in general

Float switch, mechanical — F

Limit switch, mechanical — L

Pneumatic switch, mechanical — P

Electric eye, beam source

Electric eye, relay

Thermostat — T

03401-13_A06G.EPS

# ELECTRICAL DISTRIBUTION
# OR LIGHTING SYSTEMS, AERIAL

Pole*

Pole with street light*

Pole, with down guy and anchor*

Transformer*

Transformer, constant-current*

Switch, manual*

Circuit recloser, automatic*

Line sectionalizer, automatic

Circuit, primary*

Circuit, secondary*

Circuit, series street lighting*

Down guy

Head guy

Sidewalk guy

Service weather head*

# SCHEMATIC CONVENTIONS

Transformer

Switch

Fuse

*Identify by notation or schedule

03401-13_A06H.EPS

EXPOSED WIRING ————— E —————

WIRING CONCEALED IN CEILING OR WALL —————————

WIRING CONCEALED IN FLOOR — — — — —

WIRING TURNED UP —————○

WIRING TURNED DOWN —————●

BRANCH-CIRCUIT HOMERUN TO PANELBOARD* ————▶▶

OR

\*\* 1 2

* Number of arrowheads indicate number of circuits. A number at each arrowhead may be used to identify circuit numbers

** Half arrowheads are sometimes used for homeruns to avoid confusing them with drawing callouts

03401-13_A07.EPS

| JB | Underfloor duct system – junction box and three ducts (one large, two standard) |

Dotted lines indicate black duct

G.E. Type LW223 lighting busway

G.E. Type LW326 lighting busway

G.E. Type LW326 lighting busway

Busway feed-in box

Panel – lighting and/or power

Conduit concealed above ceiling or wall

— — — — — Conduit concealed in floor or in wall

A-1 Homerun to panel; number of arrows indicates number of circuits; letter designates panel; numeral designates circuit number; crossmarks indicate number of conductors if more than two

Motor connection

S_T Switch, toggle with thermal overload protection

- - - - - - - - Conduit exposed

Duplex receptacle, grounded

S_K Switch, key operated

03401-13_A08.EPS

## WASTE WATER

DRAIN OR WASTE – ABOVE GRADE ———————

DRAIN OR WASTE – BELOW GRADE — — —

VENT - - - - - - - ·

COMBINATION WASTE AND VENT —— CWV——

ACID WASTE —— AW——

ACID VENT – – - AV – – -

INDIRECT DRAIN —— D ——

STORM DRAIN —— SD ——

SEWER – CAST IRON _____ S-CI _____

SEWER – CLAY TILE BELL & SPIGOT _____ S-CT _____

DRAIN – CLAY TILE BELL & SPIGOT _____

## OTHER PIPING

GAS – LOW PRESSURE —— G —— G ——

GAS – MEDIUM PRESSURE —— MG——

GAS – HIGH PRESSURE —— HG——

COMPRESSED AIR —— A ——

VACUUM —— V ——

VACUUM CLEANING —— VC——

OXYGEN —— O ——

LIQUID OXYGEN —— LOX——

03401-13_A09.EPS

| | FLANGED | SCREWED | BELL AND SPIGOT | WELDED | SOLDERED |
|---|---|---|---|---|---|
| Bushing | | ⊳ | 6 ← 4 | ✕ | ⊙⊢⊙ |
| Cap | | —] | → | ⊢ | |
| Cross Reducing | | | | | |
| Straight Size | | | | | |
| Crossover | | | | | |
| Elbow — 45-Degree | | | | | |
| 90-Degree | | | | | |
| Turned Down | | | | | |
| Turned Up | | | | | |
| Base | | | | | |
| Double Branch | | | | | |
| Long Radius | | | | | |

| | FLANGED | SCREWED | BELL AND SPIGOT | WELDED | SOLDERED |
|---|---|---|---|---|---|
| Elbow (Cont'd) Reducing | | | | | |
| Side Outlet (Outlet Down) | | | | | |
| Side Outlet (Outlet Up) | | | | | |
| Street | | | | | |
| Joint Connecting Pipe | | | | | |
| Expansion | | | | | |
| Lateral | | | | | |
| Orifice Plate | | | | | |
| Reducing Flange | | | | | |
| Plugs — Bull Plug | | | | | |
| Pipe Plug | | | | | |
| Reducer — Concentric | | | | | |
| Eccentric | | | | | |

03401-13_A10A.EPS

| | FLANGED | SCREWED | BELL AND SPIGOT | WELDED | SOLDERED |
|---|---|---|---|---|---|
| Gate, also Angle Gate (Plan) | | | | | |
| Globe, also Angle Globe (Elevation) | | | | | |
| Globe (Plan) | | | | | |
| Automatic Valve Bypass | | | | | |
| Governor-Operated | | | | | |
| Reducing | | | | | |
| Check Valve (Straight Way) | | | | | |
| Cock | | | | | |
| Diaphragm Valve | | | | | |
| Float Valve | | | | | |
| Gate Valve* | | | | | |

| | FLANGED | SCREWED | BELL AND SPIGOT | WELDED | SOLDERED |
|---|---|---|---|---|---|
| Motor-Operated | | | | | |
| Globe Valve | | | | | |
| Motor-Operated | | | | | |
| Hose Valve, also Hose Globe | | | | | |
| Angle, also Hose Angle | | | | | |
| Gate | | | | | |
| Globe | | | | | |
| Lockshield Valve | | | | | |
| Quick-Opening Valve | | | | | |
| Safety Valve | | | | | |

*Also used for General Stop Valve Symbol when amplified by specification.

03401-13_A10B.EPS

| | FLANGED | SCREWED | BELL AND SPIGOT | WELDED | SOLDERED |
|---|---|---|---|---|---|
| Sleeve | | | | | |
| Tee<br>Straight Size | | | | | |
| (Outlet Up) | | | | | |
| (Outlet Down) | | | | | |
| Double Sweep | | | | | |
| Reducing | | | | | |
| Single Sweep | | | | | |
| Side Outlet<br>(Outlet Down) | | | | | |
| Side Outlet<br>(Outlet Up) | | | | | |
| Union | | | | | |
| Angle Valve<br>Check, also<br>Angle Check | | | | | |
| Gate, also<br>Angle Gate<br>(Elevation) | | | | | |

03401-13_A10C.EPS

| TYPE OF FITTING | | SCREWED OR SOCKET WELD | WELDED | FLANGED |
|---|---|---|---|---|
| | | SINGLE LINE | SINGLE LINE | SINGLE LINE |
| 90°<br>ELBOW | TOP | | | |
| | SIDE | | | |
| | BOTTOM | | | |

03401-13_A11.EPS

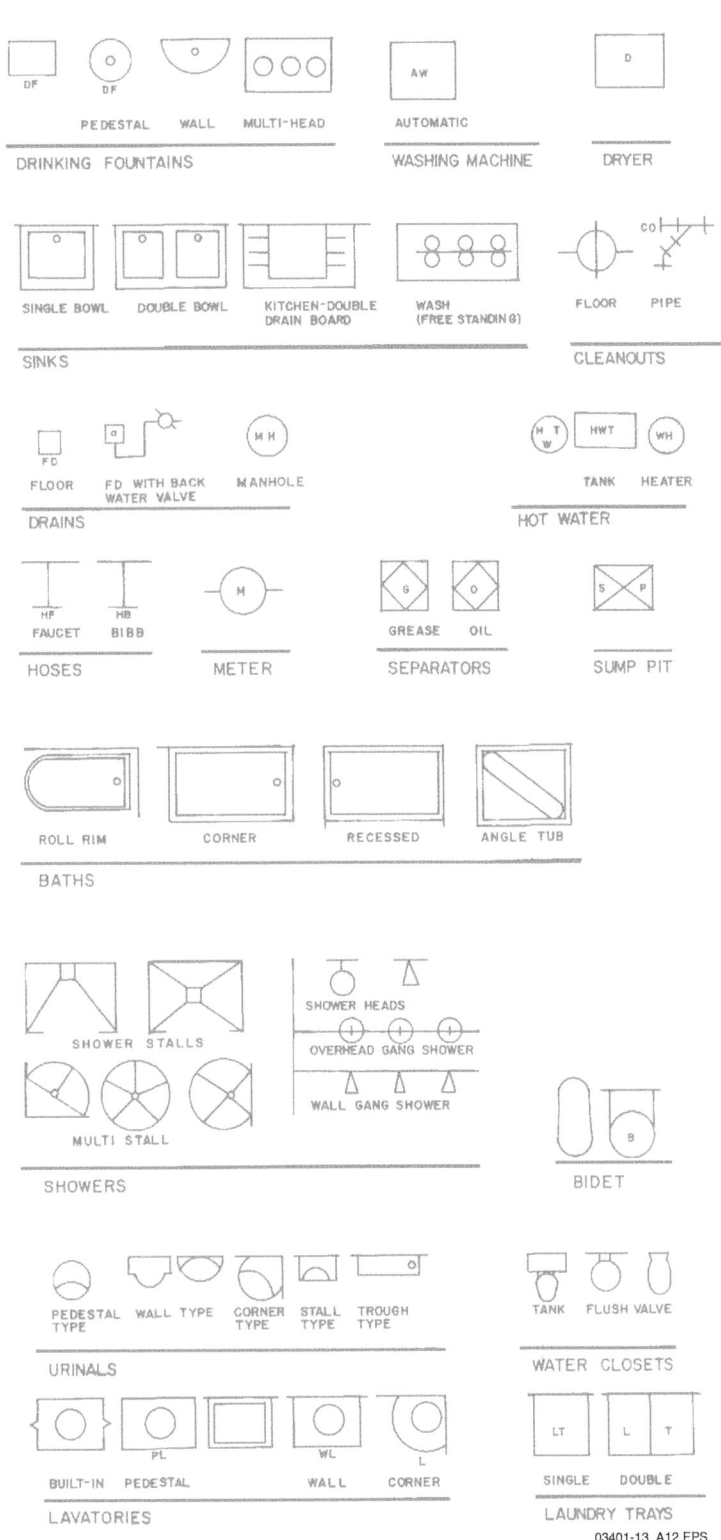

DRINKING FOUNTAINS — PEDESTAL, WALL, MULTI-HEAD

WASHING MACHINE — AUTOMATIC

DRYER

SINKS — SINGLE BOWL, DOUBLE BOWL, KITCHEN-DOUBLE DRAIN BOARD, WASH (FREE STANDING)

CLEANOUTS — FLOOR, PIPE

DRAINS — FLOOR, FD WITH BACK WATER VALVE, MANHOLE

HOT WATER — TANK, HEATER

HOSES — HF FAUCET, HB BIBB

METER

SEPARATORS — GREASE, OIL

SUMP PIT

BATHS — ROLL RIM, CORNER, RECESSED, ANGLE TUB

SHOWERS — SHOWER STALLS, MULTI STALL, SHOWER HEADS, OVERHEAD GANG SHOWER, WALL GANG SHOWER

BIDET

URINALS — PEDESTAL TYPE, WALL TYPE, CORNER TYPE, STALL TYPE, TROUGH TYPE

WATER CLOSETS — TANK, FLUSH VALVE

LAVATORIES — BUILT-IN, PEDESTAL, WALL, CORNER

LAUNDRY TRAYS — SINGLE, DOUBLE

03401-13_A12.EPS

SPECIFICATIONS

**SECTION 23 54 00**
**GAS FIRED FURNACES**

**PART 1 - GENERAL**

**SCOPE**

This section includes specifications for gas fired furnaces. Included are the following topics:

PART 1 - GENERAL
    Scope
    Related Work
    Reference
    Reference Standards
    Quality Assurance
    Energy Efficiency
    Submittals
    Operation and Maintenance Data
    Warranty

PART 2 - PRODUCTS
    Furnaces

PART 3 - EXECUTION
    Installation
    Furnaces
    Construction Verification Items
    Functional Performance Testing
    Agency Training

**RELATED WORK**

Section 01 91 01 or 01 91 02 – Commissioning Process
Section 23 08 00 – Commissioning of HVAC
Section 23 11 00 - Facility Fuel Piping
Section 23 05 13 - Common Motor Requirements for HVAC Equipment
Section 23 51 00 - Breechings, Chimneys, and Stacks

**REFERENCE**

Applicable provisions of Division 1 govern work under this section.

**REFERENCE STANDARDS**

AGA          American Gas Association
ANSI Z21.64   Direct Vent Central Furnaces
GAMA        Gas Appliance Manufacturers Association
NEC          National Electrical Code

**QUALITY ASSURANCE**

Refer to division 1, General Conditions, Equals and Substitutions.

**ENERGY EFFICIENCY**

Provide gas furnaces that bear the ENERGY STAR label and meet the ENERGY STAR specifications for energy efficiency.

**SUBMITTALS**

Refer to division 1, General Conditions, Submittals.

Include specific manufacturer and model numbers, equipment identification corresponding to project drawings and schedules, dimensions, capacities, materials of construction, ratings, weights, power requirements and wiring diagrams, filter information and information for all accessories.

**OPERATION AND MAINTENANCE DATA**

All operations and maintenance data shall comply with the submission and content requirements specified under section GENERAL REQUIREMENTS.

**WARRANTY**
Furnace primary and secondary heat exchangers warranted for 20 years under normal use and maintenance. Remainder of furnace components warranted for 1 year from date of start up.

## PART 2 - PRODUCTS

**FURNACES**
Approved Manufacturers: Bryant, Carrier, Lennox, Trane or York.

Direct-vent, sealed-combustion, condensing-type AGA certified warm-air furnace for use with natural gas. Minimum annual fuel utilization efficiency (A.F.U.E.) of 91 is required. All ratings are to be certified by GAMA. All wiring shall comply with the National Electrical Code.

22 gauge steel casing with baked enamel finish or prepainted galvanized steel. Insulate casing back and side panels with foil faced fiberglass insulation.

Construct primary heat exchanger of aluminized steel. Construct secondary heat exchanger of stainless steel with aluminum fins or of polypropylene laminated steel. Aluminized steel multi-port in-shot burner with hot surface or electronic spark ignition, approved for vertical or sidewall venting.

AGA listed gas controls including manual main shut-off valve, double automatic gas valves for redundancy and gas pressure regulator.

Centrifugal type blower fan statically and dynamically balanced with multiple speed, direct drive or belt drive fan motor. Provide low energy induced draft blower for heat exchanger prepurge and combustion gas venting.

Provide unit(s) with 2" thick 30% efficient disposable type panel air filter and filter holding rack with a maximum filter face velocity of 225 fpm.

Provide solid state integral control unit with all necessary controls and relays including but not limited to:
-Pressure switch for airflow of flue products through furnace and out vent system
-Rollout switch with manual reset to prevent overtemperature in burner area
-Electronic flame sensor
-Blower access safety interlock
-Timed blower start after main burners ignite
-Factory installed 24 v transformer for controls and thermostat
-LED's to indicate status and to aid in troubleshooting

Provide unit with matching cased "A" configuration cooling coil for upflow units, "V" configuration cooling coil for downflow units, and vertical flat face configuration cooling coil for horizontal units.

Minimum 1/2" OD seamless copper tubing mechanically bonded to heavy ripple edged aluminum fins with thermal expansion valve, holding charge and copper tube stubs for field piping.

Non-corrosive stainless steel or polymer drain pan with 3/4" NPT drain connection.

20 gauge steel Coil casing with baked enamel finish and fiberglass insulation.

This Contractor shall provide all temperature control and interlocking necessary to perform the specified control sequence. All wiring is to be in conduit in accordance with Division 26 00 00 - Electrical. All relays, transformers, and controls are to be in enclosures.

Provide a 7-day programmable thermostat with 2 occupied periods per day, automatic changeover, separate heating and cooling set points for both occupied and unoccupied modes. Provide auxiliary controls on sub-base to open minimum outside air damper during occupied mode. Equal to Honeywell model T7300 with Q7300 sub-base.

Provide lockable thermostat guards in public spaces.

During occupied mode run the supply fan continuously, open the outside air damper and cycle the cooling or heating as required to maintain occupied space temperature cooling or heating set point. During

03401-13_A13B.EPS

1    unoccupied mode close the outside air damper and cycle the supply fan and cooling or heating as required
2    to maintain unoccupied cooling or heating space temperature set point.
3
4
5                     **PART 3 - EXECUTION**
6
7 **INSTALLATION**
8
9 Install units as shown on plans, as detailed and according to the manufacturer's installation instructions.
10
11 Pipe vents from gas regulator to outside (where regulators are provided).
12
13 Install remote panels and thermostats where indicated on the drawings. Provide all wiring between remote
14 panels/thermostats and the gas fired item.
15
16 **FURNACES**
17 Install on concrete housekeeping pad, steel stand or suspend unit from structure as indicated on the
18 drawings. Pipe condensate to floor drain.
19
20 Provide schedule 40 PVC, ASTM D1785 combustion air and vent piping and fittings with solvent welded
21 joints as indicated on the drawings. Terminate as recommended by the furnace manufacturer.
22
23 **CONSTRUCTION VERIFICATION**
24 Contractor is responsible for utilizing the construction verification checklists supplied under specification
25 Section 23 08 00 in accordance with the procedures defined for construction verification in Section 01 91
26 01 or 01 91 02.
27
28 **FUNCTIONAL PERFORMANCE TESTING**
29 Contractor is responsible for utilizing the functional performance test forms supplied under specification
30 Section 23 08 00 in accordance with the procedures defined for functional performance testing in Section
31 01 91 01 or 01 91 02.
32
33 **AGENCY TRAINING**
34 All training provided for agency shall comply with the format, general content requirements and
35 submission guidelines specified under Section 01 91 01 or 01 91 02.
36
37                     END OF SECTION

# Additional Resources

This module presents thorough resources for task training. The following resource material is suggested for further study.

*Blueprint Reading for Construction,* 2003. Second Edition. James A. S. Fatzinger, Upper Saddle River, NJ: Pearson/Prentice Hall.

*Construction Specifications Writing: Principles and Procedures*, 2010. Sixth Edition. Mark Kalin, Robert S. Weygant, Harold J. Rosen, and John R. Regener. Hoboken, NJ: John Wiley & Sons, Inc.

*Reading Architectural Work Drawings*, 2003. Sixth Edition. Edward J. Muller and Phillip A. Grau III. Upper Saddle River, NJ: Prentice Hall.

# Figure Credits

# Section Review Question Answers

| Answer | Section Reference | Objective |
|--------|-------------------|-----------|
| **Section One** | | |
| 1.a | 1.1.0 | 1a |
| 2.c | 1.2.0 | 1b |
| 3.d | 1.3.2 | 1c |
| 4.d | 1.4.1 | 1d |
| 5.a | 1.5.1 | 1e |
| 6.b | 1.6.0 | 1f |
| 7.d | 1.7.0 | 1g |
| 8.c | 1.8.0 | 1h |
| 9.b | 1.9.0 | 1i |
| **Section Two** | | |
| 1.a | 2.1.1 | 2a |
| 2.b | 2.2.0 | 2b |
| **Section Three** | | |
| 1.b | 3.1.0 | 3a |
| 2.c | 3.2.0 | 3b |

# NCCER CURRICULA — USER UPDATE

NCCER makes every effort to keep its textbooks up-to-date and free of technical errors. We appreciate your help in this process. If you find an error, a typographical mistake, or an inaccuracy in NCCER's curricula, please fill out this form (or a photocopy), or complete the online form at **www.nccer.org/olf**. Be sure to include the exact module ID number, page number, a detailed description, and your recommended correction. Your input will be brought to the attention of the Authoring Team. Thank you for your assistance.

*Instructors* – If you have an idea for improving this textbook, or have found that additional materials were necessary to teach this module effectively, please let us know so that we may present your suggestions to the Authoring Team.

### NCCER Product Development and Revision
13614 Progress Blvd., Alachua, FL 32615

**Email:**   curriculum@nccer.org
**Online:**   www.nccer.org/olf

❏ Trainee Guide    ❏ Lesson Plans    ❏ Exam    ❏ PowerPoints    Other _____

Craft / Level: _____    Copyright Date: _____

Module ID Number / Title: _____

Section Number(s): _____

Description: _____

_____

_____

_____

_____

Recommended Correction: _____

_____

_____

_____

_____

Your Name: _____

Address: _____

_____

Email: _____    Phone: _____

# Heating and Cooling System Design

## OVERVIEW

The proper design of the air distribution and refrigerant piping systems, along with the selection of the proper heating and cooling equipment, are critically important to the success of an installation. Many variables can influence system design. Contractors who rely on rule-of-thumb methods for equipment sizing and design put themselves at a competitive disadvantage. Successful competitors calculate the heating and cooling losses of the structure before selecting equipment. In many cases, the successful contractor's efforts produce equipment with heating/cooling capacities that are lower and less costly to install than the equipment selected by rule-of-thumb supporters.

# Module 03407

## 03407
# HEATING AND COOLING SYSTEM DESIGN

## Objectives

When you have completed this module, you will be able to do the following:

1. Describe the design process and explain how to evaluate a structure for load estimating.
   a. Describe the design process.
   b. Explain how to evaluate a structure for load estimating.
2. Explain how to complete a heating and cooling load estimate.
   a. Describe how heat is gained or lost through a building structure.
   b. Identify specific cooling and heating load factors.
3. Explain how to select equipment based on the load estimate and describe common support systems to be considered.
   a. Explain how to select cooling equipment.
   b. Explain how to select heating equipment.
   c. Explain how to select heat pumps.
   d. Describe common support systems to be considered.
4. Explain how to design air distribution systems.
   a. Identify basic duct design considerations.
   b. Identify various duct system layouts.
   c. Identify various duct system components.
   d. Describe how to design and size duct systems.
   e. Identify system design factors unique to commercial buildings.

## Performance Tasks

Under the supervision of your instructor, you should be able to do the following:

1. Using plans provided by the instructor, perform a load estimate using a standardized method.
2. Use manufacturer's product data to select the appropriate heating and cooling equipment based on a load estimate and airflow requirements.
3. Determine the number, location, and sizes of supply outlets and return inlets needed in a building.
4. Use standard duct sizing tables, duct design calculator, or software application to size the trunk and branch ducts for a selected low-pressure air distribution system.
5. Calculate the total system friction loss (external static pressure) for a selected air distribution system.

## Trade Terms

Brake horsepower
Equal-friction method
Homogeneous
R-value

Static-regain method
Thermal conductivity
U-factor
Velocity-reduction method

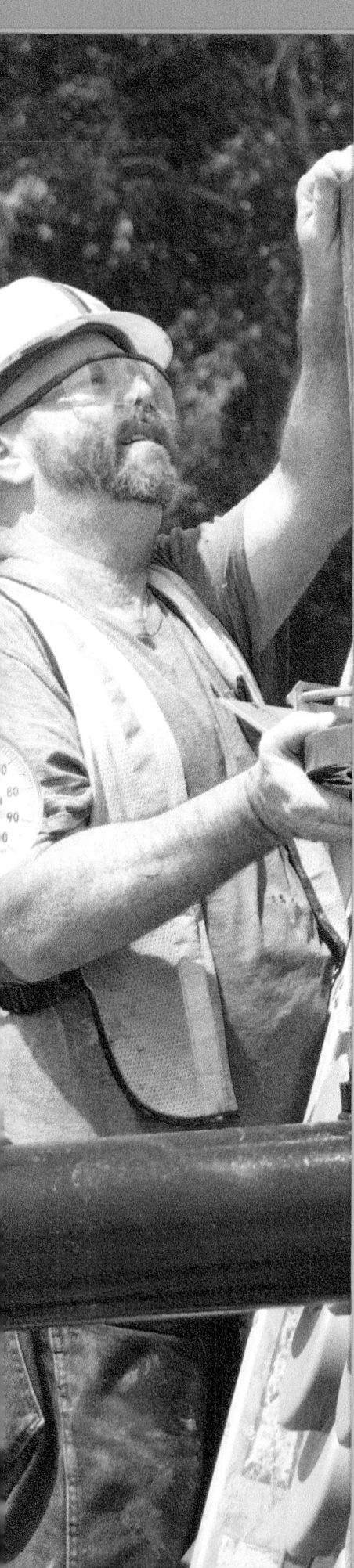

## Industry-Recognized Credentials

If you are training through an NCCER-accredited sponsor, you may be eligible for credentials from NCCER's Registry. The ID number for this module is 03407. Note that this module may have been used in other NCCER curricula and may apply to other level completions. Contact NCCER's Registry at 888.622.3720 or go to **www.nccer.org** for more information.

## Contents

## Figures and Tables

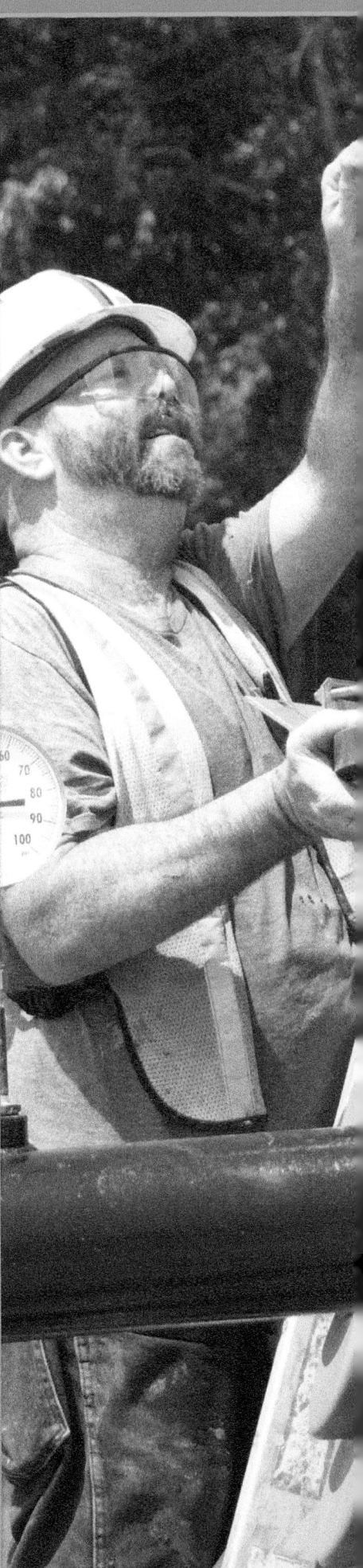

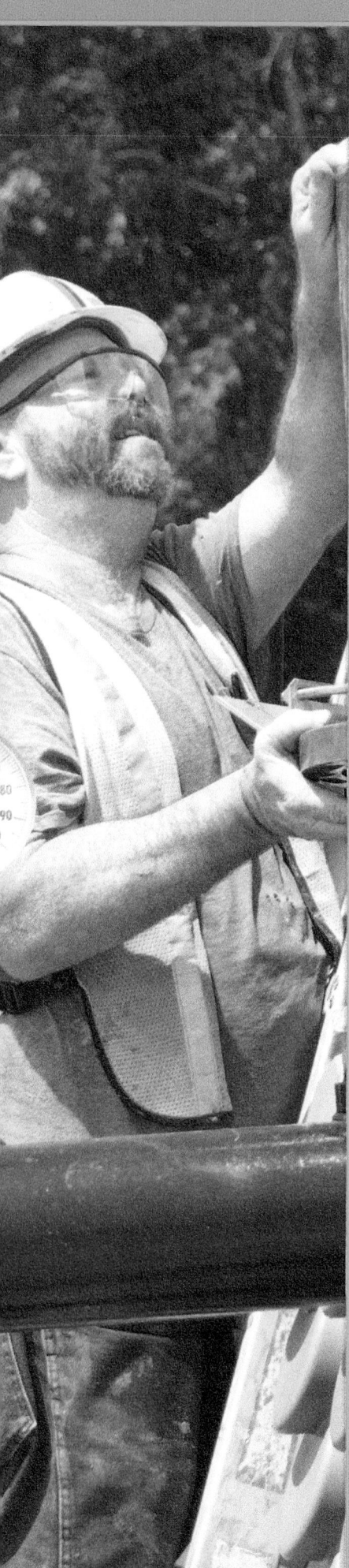

## Figures and Tables (continued)

### 1.0.0 HVAC Design Processes and Structure Evaluation

**Objective**

Describe the design process and explain how to evaluate a structure for load estimating.
  a. Describe the design process.
  b. Explain how to evaluate a structure for load estimating.

**Performance Tasks**

1. Using plans provided by the instructor, perform a load estimate using a standardized method.

This module discusses heating and cooling system design. The source of discomfort, noise, or inefficient operation of an air conditioning system can often be traced to design flaws, poor installation practices, or changes that occurred in the building after the equipment was placed into service. To effectively troubleshoot a comfort air conditioning system, the principles of system design must be understood.

Becoming an effective system designer requires extensive knowledge and experience. This module introduces the factors that affect the selection and design of HVAC systems and ductwork. The focus will be on residential systems, but the coverage will also touch on commercial system design.

Material published by the Air Conditioning Contractors of America (ACCA) is referenced extensively in this module. In addition to this text, it is strongly recommended that *ACCA Manual J, Load Calculation for Residential Winter and Summer Air Conditioning* be reviewed. *ACCA Manual N, Load Calculation for Commercial Winter and Summer Air Conditioning*, is used for commercial system design for trainees who are interested.

The design process begins with gathering information about the construction of the structure and its environment (temperature and humidity). With that information in hand, the heat loss and heat gain of the structure can be calculated. Once these values are known, equipment can be selected, and a duct system can be designed.

### 1.1.0 Overview of the Design Process

The design process has the following two important goals:

1. Selecting the size and type of equipment needed to deliver the correct amount of conditioned air to the building

2. Determining the type and size of air distribution system needed to support the selected equipment.

The first step in this process, as shown in *Figure 1*, is to collect information about the building. If dealing with new construction, the building and site drawings and specifications may be used. If it is a building expansion or a replacement for existing equipment, survey the actual building. A floor plan showing building dimensions, overhangs, number and type of windows and doors, and any other factors, is essential.

The load estimate can be time consuming. In the past, the information gathered at the job site was taken to an office where it was calculated manually. While that may still be true in some cases, computer software can now quickly produce a load calculation in the field or the office, especially on simpler residential jobs. Regardless of how the calculation is performed, it is important that the information gathered during the site survey be as accurate as possible. The primary purpose is to calculate the heating and cooling loads (also referred to as heat loss and heat gain, respectively) and the airflow requirements for both modes of operation. If both heating and cooling are being installed, the load-estimating process will yield five values:

- Sensible cooling load (in Btuh)
- Latent cooling load (in Btuh)
- Cooling airflow
- Sensible heating load (in Btuh)
- Heating airflow

In Step 3, as shown in *Figure 1*, the heating and cooling loads and airflow values calculated are translated into cooling and heating equipment selections using manufacturers' product data. For example, if the cooling load is 52,000 Btuh (about $4\frac{1}{2}$ tons), select equipment of that capacity, or as close to it as possible without under-sizing, from the manufacturer's catalog.

A similar process is used to select heating equipment. The furnace or other heating appliance is selected from the manufacturer's product literature based on the heating load and the required amount of airflow. Factors that affect the type of appliance selected are covered later in the module.

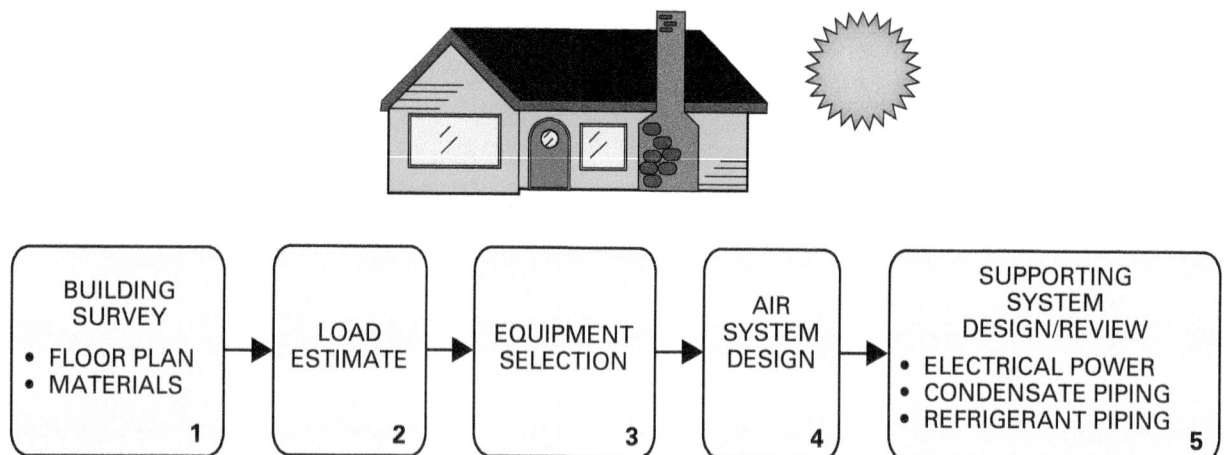

*Figure 1* The design process.

Once the equipment is selected, design the new air distribution system, or verify that an existing one will support the new equipment. This is a very important step. If the duct system is not correctly designed, the system will not perform well and could create noise or comfort problems for the occupants, regardless of capacity.

In the final step, the effect of the new system on other building systems is considered. This step is partly intended to make sure that existing subsystems will support the proposed equipment. In addition, it helps to make sure that all the necessary parts and materials are available when the installation is started. Here are some of the things to consider in this step:

• Will the electrical service support the load?
• How much and what size of wire is needed for power and control wiring?
• How much and what size refrigerant and condensate piping is needed?

Keep in mind that cooling systems both cool and dehumidify. If the designer selects grossly oversized equipment, its excess capacity will bring the space temperature down quickly, but it will not run long enough to remove excess moisture. The occupants will not be comfortable. If the system is undersized, it will run longer, but will not have enough capacity to maintain the thermostat setpoint on hotter days.

The same holds true for heating equipment. An oversized furnace will quickly heat the structure and then shut off for a long cool-down period before the next heating cycle starts. An undersized furnace will maintain indoor comfort in milder weather, but will be unable to heat the structure adequately on the coldest days. The capacity of heating and cooling systems is put to the test at different times of the day. The maximum load a cooling system will usually encounter is late afternoon on a summer day. The sun has been out all day and daytime temperatures tend to peak at that time. The maximum load a heating system will encounter is the early hours of a winter morning, just before dawn. The atmosphere has had all night to cool off, and temperatures tend to be coldest just before dawn.

Oversized equipment costs more to install and operate, so the customer winds up getting less comfort while paying more. Although the entire design process is time-consuming, load-estimating software can save a lot of time. Some

## Equipment Sizes

The majority of HVAC equipment manufacturers make equipment in specific and comparable nominal sizes. However, each model has its own actual performance data. For example, a unit may be considered as a nominal 5-ton unit, but the actual performance at various conditions is something other than the nominal value of 60,000 Btus.

The common nominal sizes of residential and light commercial equipment (excluding ductless systems, which come in much smaller sizes) are as follows:

• 1½ ton
• 2 ton
• 2½ ton
• 3 ton
• 3½ ton
• 4 ton
• 5 ton
• 6¼ ton
• 7½ ton
• 10 ton

programs, especially those provided by equipment manufacturers, can select appropriate equipment or perform other design functions. Many computer software suppliers allow potential customers to download sample versions of their load-estimating program from the internet. The designer gets to manipulate the sample version and explore its various features before deciding to buy.

## 1.2.0 Building Evaluation and Survey

Many variables affect the design of an HVAC system. Some of them have a dramatic effect. If any factor is ignored or miscalculated, the system may not perform properly; it may be undersized or oversized. For that reason, it is very important to obtain as much information as possible about the building. Some of this information is obtained from the blueprints or by observing an existing building.

If possible, it is also important to talk to the building owner or occupants, especially if a replacement or building expansion is being considered.

Occupants can describe problems they have with the current system or about special considerations that must be factored into the design. Unless there are special health considerations, residential designs are straightforward. For a commercial project, it may be more difficult, especially if more than one business will occupy the building.

The types of information needed to prepare the load estimate and select equipment and ductwork are as follows:

- Type of roof material
- Roof color (cooling only)
- Amount and type of insulation in the attic, walls, and basement
- General tightness of the building
- Type of construction material used (frame, masonry, brick)
- Existence, size, and insulation of basements or crawl spaces
- Number of floors
- Direction the building faces (orientation to the sun)
- Number, sizes, and types of windows and doors
- When known, the type(s) of interior window covering, such as draperies, curtains, or blinds
- Exterior dimensions of the building
- Color of the exposed exterior walls
- Length, width, and height of each room
- Shading from trees, adjacent buildings, or large hills
- Size(s) of roof overhang(s) and any window awnings
- Type and size of electrical service
- Walls, floors, and ceilings exposed to the outdoors or to unconditioned spaces such as garages or crawl spaces
- Walls that are below-grade (heating load only)
- Special design considerations such as zoning

These factors, as they affect heat loss and heat gain, are summarized in *Figure 2* and *Figure 3*, respectively.

If the job is an equipment replacement, the type and capacity of the current equipment and whether it is providing adequate comfort must be known.

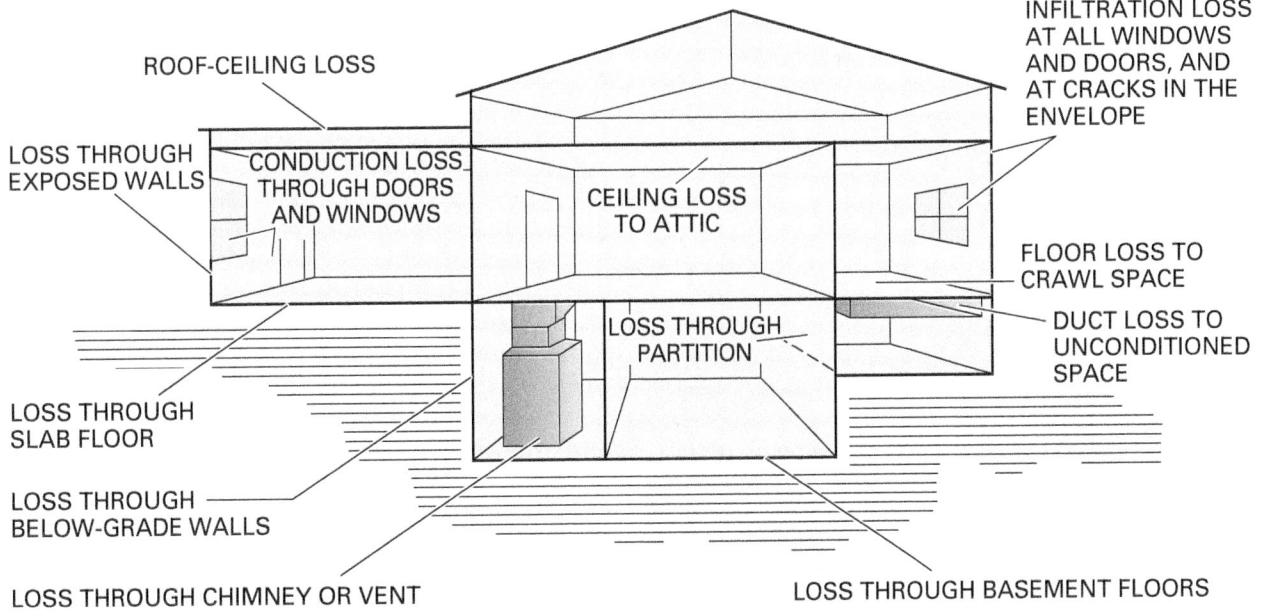

*Figure 2* Winter heating loads (heat losses).

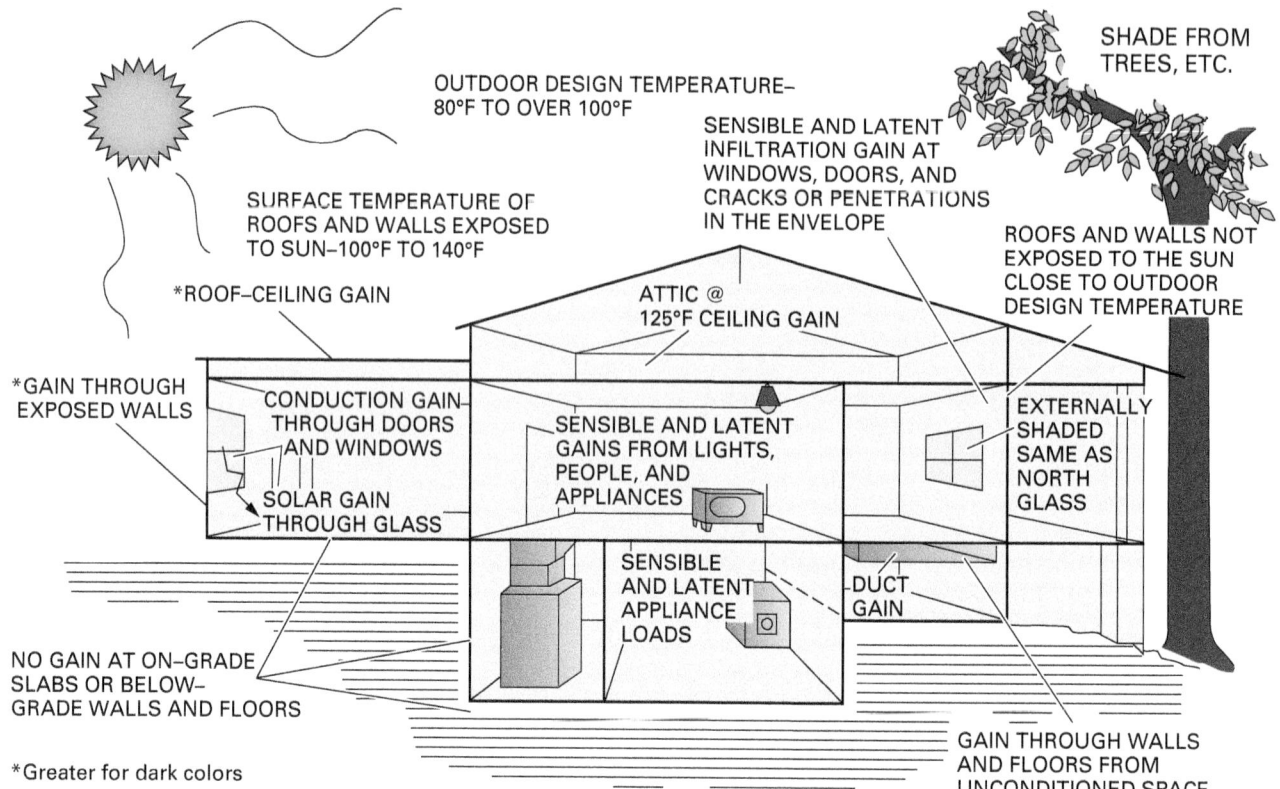

*Figure 3* Summer cooling loads (heat gains).

Never assume that the size (heating/cooling capacity) of existing HVAC equipment is correct. In other words, never size replacement equipment based solely on the size of the existing equipment. To possibly determine if residential equipment is oversized, ask the homeowner how the equipment operates. At a time when the outdoor temperature is at or near design temperature, heating or cooling equipment sized to the design parameters should run almost constantly to maintain the thermostat setpoint. If the equipment cycles on and off often when the outdoor temperature is at or above design, the equipment is likely oversized. The shorter the on cycle and longer the off cycle at the design outdoor temperature, the more oversized the equipment.

Any remodeling done to the structure since the original system was installed must be noted. If rooms have been added or the windows replaced, loads will be affected. It is also important to know the type and size of existing ductwork. Do not assume that the original equipment and ductwork were properly sized.

Oversized equipment and undersized ducts are both common. If the building has a fossil-fuel furnace to be replaced, its vent must be checked for correct size and condition. Keep in mind that old chimneys or vents may not be safe or adequate for use with modern furnaces.

During this step, the designer makes a preliminary decision about the equipment configuration and zoning arrangements, if any. Buildings of all sizes may require more than one system or zone to ensure that all areas of the building are adequately heated and cooled. A building with one large, single-zone system depends on a single room thermostat. Other rooms may be too hot or too cold. Air balancing usually solves most of these issues. The use of separate systems to heat and cool different parts of the building, or the use of a single system with several zones, are other ways to solve that problem.

A checklist designed to record information about the building is an effective way to ensure that all required information needed for calculating a load estimate is collected. *Figure 4* shows an example form that is used with their residential load-estimating process. The system is designed so that the information collected can be directly transferred to the load-estimating forms.

*Figure 5* shows an annotated floor plan. Between the data entered on the survey checklist and this floor plan, enough information should be available to do the load estimate and make an equipment selection for this residence.

The most common load-estimating guidance for residences is *Manual J, Residential Load Calculation,* which is published by the Air

# Survey and Checklist (New Construction, Add-On, Replacement)

## 1. Design Conditions

Location: City ANYTOWN    State U.S.    Latitude 34°N

| Temperature °F | Summer | Winter |
|---|---|---|
| Outside Design DB/WB | 96/77°F | 20°F |
| Daily Range | 22°F | xxxxxx |
| Inside Design DB/%RH | 75/50% | 72/30% |
| Difference | 21°F | 52°F |
| Swing | 3°F | xxxxxx |

Special Internal Loads (Computers, Etc.) _____ NONE _____

Frequent Entertaining ☐    Doors Opened Often ☐

## 2. Orientation and Type

House Faces:   N   NE   E   SE   Ⓢ   SW   W   NW
Single Story ☐   Two Story ☑   Split Level ☐

## 3. Construction

Ceiling Height:   Basement 7½'   1st Floor 8'   2nd Floor 7'

Insulation (Inches)

**Walls:**
                              0  1  2  3½  6  8  12   R Value
Frame ☑ Masonry Above Grade ☑ Bsmt   ☐ ☐ ☐ ☐ ☐ (Frame) R-11
Masonry Below Grade: 0-5'☑ >5'☐      ☑ ☐ ☐ ☐ ☐    R-0

**Roof:**
Ceiling under Ventilated Attic ☑      ☐ ☐ ☐ ☐ ☑ ☐   R-25
Roof-Ceiling Combination ☐            ☐ ☐ ☐ ☐ ☐     R-___
Roof on Exposed Beams or Rafters ☐    ☐ ☐ ☐ ☐ ☐     R-___
Roof Color: Dark ☑   Light ☐

**Floor:**
Slab on Grade ☐   Edge Insulation:   ☐ ☐ ☐    ___
Floor over Garage or Vented Crawl space ☐
   Floor Finish: Hardwood ☐  Carpeted ☐
   Ceiling Below:   Yes ☐  No ☐   ☐ ☐ ☐ ☐ ☐   ___ R-___
Floor over Unheated Basement or Enclosed Crawl space ☐
   Ceiling Below:   Yes ☐  No ☐   ☐ ☐ ☐ ☐ ☐   ___ R-___
Basement Floor >2' Below Grade ☑

## 4. Windows & Doors

**Windows:**
Type:   Movable ☑   Fixed ☐   Jalousie ☐
Glass:  Single ☐  Double (Single + Storm) ☑   Triple ☐
        Clear ☐   Tinted ☐   Reflective ☐   Low "e" ☐
Frame:  Wood ☑   Metal ☐   Metal w/TB ☐

| Number | Size Width | Height | Sq. Ft. | Type |
|---|---|---|---|---|
| 2 | 3' | 3' | 9.0 | 24" O.H. - SOUTH |
| 9 | 3' | 4.5' | 13.5 | SHADED |
| 3 | 3' | 4.5' | 13.5 | 24" O.H. - SOUTH |
| 6 | 3' | 4.5' | 13.5 | 0" O.H. |

Skylight:   Wood ☐   Metal ☐   Metal w/TB ☐

**Doors:**
Type PANEL        Storms METAL

**Shading:**
Internal:   None ☑   Full ☐   Half ☐
Overhangs:  0" ☐   12" ☑   24" ☐   36" ☐   48" ☐   (24" - SOUTH)
Permanent External Shading       NONE

## 5. Infiltration

Weatherstripping:   Windows ☐   Doors ☑
Building Tightness:   Loose ☐   Medium ☑   Tight ☐
Ventilation Fan ☐   Location_____ CFM_____
Fireplaces: No. 1      Loose ☐   Medium ☑   Tight ☐

## 6. Present Equipment Survey

**Indoor System:**
Forced Air Furnace:   Heat Only ☐   Heat/Cool ☑
   Furnace w/Cooling ☐   Heat Pump w/Fan Coil ☐   Hydronic ☐
Fuel:  Natural Gas ☑   LP Gas ☐   Oil ☐   Electricity ☐
Unit:   Upflow ☑   Downflow ☐   Horizontal ☐   Package ☐
Location/Condition: W. END, BSMT.   Good ☐  Avg ☑  Poor ☐  Age 15
Make: BRAND "X"            Model No: XYZ
Capacity:  Input 100,000 Output 76,000 BTU ☑   KW ☐
Blower:  Motor HP ⅓"   Direct Drive ☑   Belt Drive ☐
   Multiple Speeds ✓   Blower Dia. 10½"   Blower Width 10½"

**Vent System:**  Condition:  Good ☐  Avg ☑  Poor ☐  Age___
Metal:   Single Wall ☐   Double Wall Type "B" ☐   Dia._____
Masonry:   Unlined ☐   Lined ☑   Liner Size _____ x _____ "
PVC Plastic: ☐
   Vent Connector:  Dia. 6" Length_____ ft. Corroded: Yes ☐ No ☑
Water Heater: 40,000 BTUH  Common Vent: Yes ☑ No ☐  Replace: Yes ☐ No ☐

**Outdoor:**
Unit:   Split System ☐   Package Unit ☐   Room Air ☐
Condensing Method:  Air Cooled ☐   Water Cooled ☐
   Water Supply:  City ☐   Well ☐   Max Summer Temp_____°F
Location/Condition:_____ Good ☐ Avg ☐ Poor ☐ Age___
Ratings:  Capacity_____ BTUH; EER or SEER_____
Condensate:  Gravity Drain ☐  Pump ☐  Sump ☐  Drywell ☐  Floor ☐
   Emergency Overflow Pan in Attic:  Yes ☐  No ☐
Refrigerant Lines:  Length_____ ft.
   Diameter:  Suction_____"   Liquid_____"

**Other Equipment:**
Central Humidifier:  Yes ☐  No ☑      Electronic Air Cleaner:  Yes ☐  No ☑
Zoning:  Yes ☐  No ☑  Heat Recovery Vent:  Yes ☐  No ☑  Other:_____

## 7. Utilities

Natural Gas Meter Location N.E. CORNER, BASEMENT
LP Gas Tank Location_____
Oil Tank Location_____; Pump Above ☐  Below ☐
   Tank Size_____; Distance from Pump_____
Electrical Service:  Volts 240  Phase 1  Hz 60  Amps 100
   Location of Entrance Panel N.E. CORNER, BASEMENT
Major Elec. Loads: Range/Self Clean ☐ Range/Oven ☑ Range Top ☐ Single Oven ☐
   Double Oven ☐ Dryer ☐ Dish ☑ _____ KW  Other____KW
                              ↳ FUTURE

## 8. Controls

Zones:  Single ☑  Multi ☐  Number 1
Thermostat Type:  Heating ☑  Heat & Cool ☐  Continuous Fan ☑
   Auto Changeover ☐  Clock-type w/Night Setback ☐  Programmable ☐
Location of Master Thermostat: LIVING ROOM - INSIDE WALL

## 9. Air Distribution System

**Supply:**
Location: Basement ☑  In Slab ☐  Crawl space ☐  Ceiling ☐  Attic ☐  Soffit ☐
Exposure: In Unconditioned Space ☐  To Outdoor Temp. ☐
Insulation: 0" ☑  ½" ☐  1" ☑  1½" ☐  2" ☐ (1" @ Wall Stacks)
Plenum: Width 16"  Depth 19"  Height 42  Clearance to Rafters 2"
Main Trunk Duct:  Width 20"  Height 8"
Runout Diameter:  6"
Outlets: Floor Perimeter ☑ Baseboard ☐ Ceiling ☐ High Sidewall ☐ Low Sidewall ☐

**Return:**
Location: Basement ☑  In Slab ☐  Crawl space ☐  Ceiling ☐  Attic ☐  Soffit ☐
Exposure: In Unconditioned Space ☐  To Outdoor Temp. ☐
Insulation: 0" ☑  ½" ☐  1" ☐  1½" ☐  2" ☐
Main Trunk Duct:  Width 16"  Height 8"

CARRIER CORPORATION

*Figure 4* Example of a residential survey form.

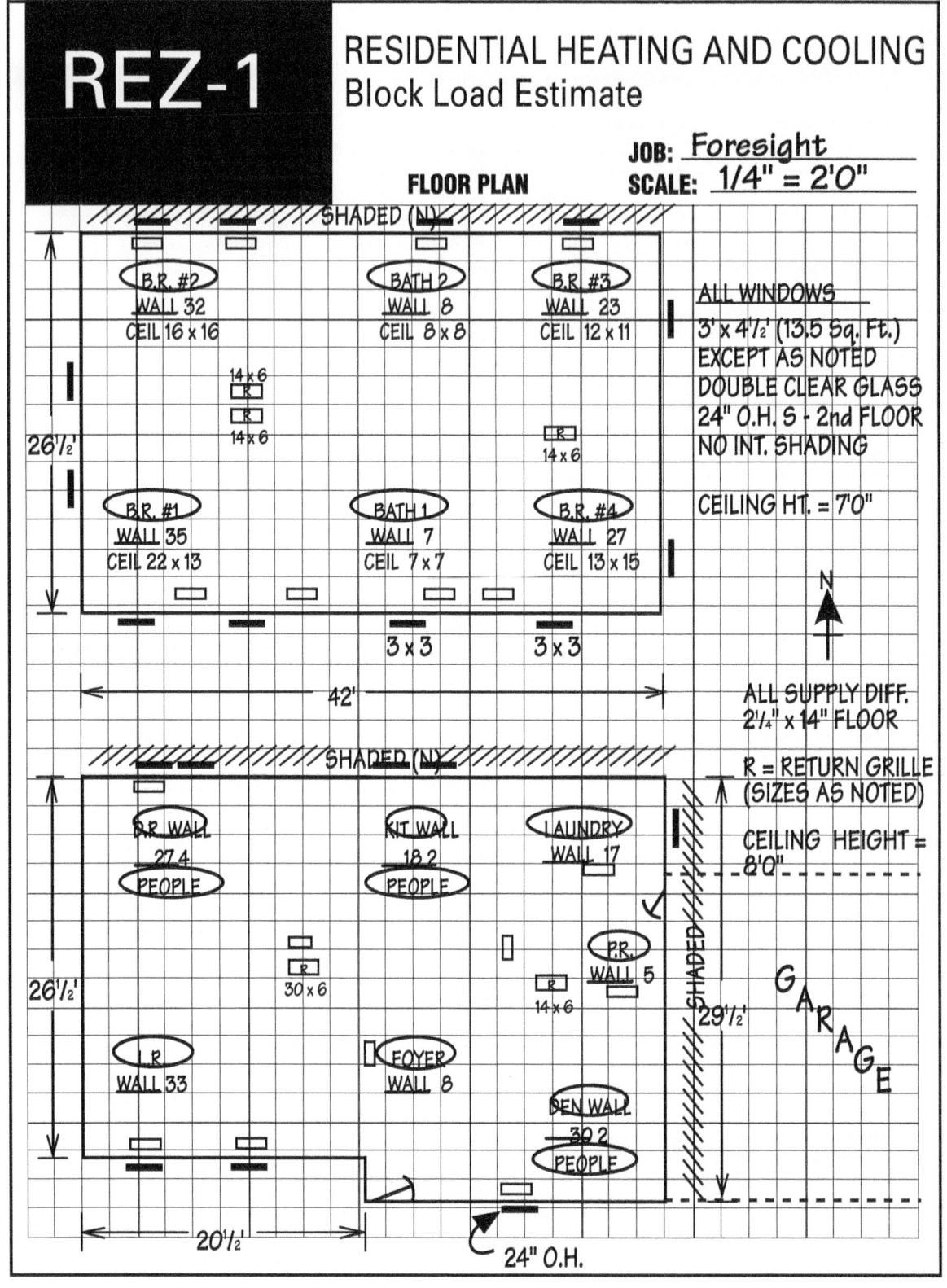

Figure 5 Example of an annotated floor plan.

Conditioning Contractors of America (ACCA). ACCA also publishes *Manual N, Commercial Load Calculation*, which provides load-estimating guidance for commercial structures. Most commercial load estimating, and much of residential load estimating, is now done using software packages. Both *Manual J* and *Manual N* are available in software versions as well as book form.

*Manual J* is the industry standard for residential load-estimating methods and is the method that many utilities and code enforcement officials specify when a residential load estimate is required. It is also the method used in contractor licensing exams in many states. Other load-estimating methods may be as effective as *Manual J*, but because they may not be officially sanctioned, a load estimate developed using one of these other methods may not be acceptable. Always check with the local utility and/or code enforcement office to determine what load-estimating method is acceptable in your area.

Load estimating is a number-crunching process with many variables, making it ideal for software applications. An advantage of software is its ability to quickly compare how different building construction factors might affect the load.

For example, if a building design plan shows standard window glass, it is easy to change the entry to energy-efficient glass and see how it affects the load. This process is a form of "what-if" game. Load-estimating software is often combined with cost-estimating software that can show how a specific change will affect the load as well as the installation cost and return on investment.

## Additional Resources

*Manual J, Residential Load Calculation*. Current Edition. Air Conditioning Contractors of America (ACCA).

*Manual N, Commercial Load Calculation*. Current Edition. Air Conditioning Contractors of America (ACCA).

## 1.0.0 Section Review

1. The goals of the design process are to _____.
   a. determine equipment capacity and an equipment manufacturer
   b. determine the needed equipment capacity and calculate duct sizes
   c. choose an equipment manufacturer and duct material
   d. determine the duct material and duct size

2. When is it acceptable to use the existing equipment's heating/cooling capacity as the sole basis to size replacement equipment?
   a. At the customer's request
   b. When time is critical
   c. When the budget is tight
   d. Never

### 2.0.0  LOAD ESTIMATING

### Objective

Explain how to complete a heating and cooling load estimate.

a. Describe how heat is gained or lost through a building structure.
b. Identify specific cooling and heating load factors.

### Performance Tasks

1. Using plans provided by the instructor, perform a load estimate using a standardized method.

### Trade Terms

**Homogeneous:** A mixture or material that is uniform in its composition throughout.

**R-value:** The thermal resistance of a given thickness of insulating material.

**Thermal conductivity:** The heat flow per hour (Btuh) through one square foot of one-inch thick homogeneous material for every 1°F of temperature difference between the two surfaces.

**U-factor:** The heat-transfer coefficient rating how well a building element conducts heat; the rate of heat transfer (in watts) through one square meter of a structure, divided by the temperature difference. The lower the U-factor, the better the resistance to heat transfer.

It is almost impossible to measure either the actual peak load or the partial load for a space. Thus, these loads must be estimated. The load estimate is based on design conditions inside and outside the building. The outside design conditions, which are the usual extremes of temperature based on National Weather Service data, are readily available. *Figure 6* shows an example of this data as shown in *ACCA Manual J*. The most significant information in the table for the average designer are the *Heating 99% Dry Bulb* and *Cooling 1% Dry Bulb* columns. These are the suggested design temperatures for heating and cooling. Location has a lot to do with the type and intensity of the load. New Mexico is much hotter and drier than northern Maine.

The design high and low temperatures are not often identical to the maximum temperature that could be encountered in an area. If a maximum temperature were to be used instead of the design temperature, the heat loss or heat gain would be exaggerated, resulting in oversizing the equipment. Temperature extremes tend to be isolated and of short duration, so they are not used as the basis for load estimating. The design temperature is closer to the actual temperature that can be expected consistently.

The inside design conditions are the temperature and humidity to be maintained for comfort or material processing. The purpose of a load estimate is to determine the size and balance of the equipment necessary to maintain the inside design conditions during anticipated extremes in outside temperature and humidity.

The initial phase of the load-estimating process yields two kinds of information. One is the size of the building and its parts organized into convenient groups, such as windows, walls, and floors. This information is further divided into subgroups that make it easier to perform the load calculations.

The second type of information is the heat transfer multipliers (HTMs) that represent heat gains and losses that a building is likely to experience based on its construction and location. To obtain the total sensible heat load for the building, the window, wall, door, floor, and ceiling areas for all rooms are added together and then multiplied by their respective HTMs. Examples can be seen in the building load analysis reports provided in the *Appendix*.

Calculating a heat loss or heat gain using a manual, paper-based method can be time consuming and tedious. Various charts and tables must be constantly referenced, and manual calculations must be made, which increases the possibility of error. In today's business environment, many sales professionals do not want, or have the time, to perform a manual load calculation. This often leads to the reliance on old, ineffective rules of thumb that can lead to poorly chosen equipment.

The availability of personal computers revolutionized all aspects of system design, including load estimating. There are dozens of excellent

## Table 1A
### Outdoor Design Conditions for the United States

| Location | Elevation Feet | Latitude Degrees North | Heating 99% Outdoor Dry Bulb | Cooling — Outdoor Air 1% Dry Bulb | Cooling — Outdoor Air Coincident Wet Bulb | Design Grains 55% RH Indoors | Design Grains 50% RH Indoors | Design Grains 45% RH Indoors | Daily Range (DR) | HDD65 / CDD50 Ratio |
|---|---|---|---|---|---|---|---|---|---|---|
| **Alabama** | | | | | | | | | | |
| Alexander City | 686 | 33 | 22 | 93 | 76 | 38 | 45 | 52 | M | 0.55 |
| Anniston Metropolitan AP | 600 | 34 | 25 | 92 | 76 | 36 | 43 | 50 | M | 0.51 |
| Auburn-Opelika AP | 774 | 33 | 28 | 91 | 74 | 28 | 35 | 42 | M | 0.43 |
| Birmingham Municipal AP | 630 | 34 | 24 | 93 | 75 | 32 | 39 | 45 | M | 0.49 |
| Cairns AFB (Ozark) | 299 | 31 | 30 | 93 | 76 | 37 | 44 | 51 | M | 0.28 |
| Centreville WSMO | 453 | 33 | 26 | 92 | 76 | 37 | 43 | 50 | M | 0.45 |
| Decatur (Athens DD) | 592 | 34 | 16 | 93 | 74 | 26 | 32 | 39 | M | 0.79 |
| Dauphin Island | 30 | 30 | 38 | 87 | 77 | 53 | 60 | 66 | L | 0.19 |
| Dothan Municipal AP | 322 | 31 | 31 | 93 | 76 | 35 | 42 | 49 | M | 0.26 |
| Florence AP (Muscle Shoals DD) | 581 | 34 | 21 | 94 | 75 | 30 | 37 | 44 | M | 0.60 |
| Gadsden Municipal AP AWOS | 568 | 34 | 22 | 91 | 75 | 32 | 38 | 45 | M | 0.68 |
| Huntsville IAP, Jones Field | 643 | 35 | 22 | 92 | 75 | 31 | 38 | 45 | M | 0.62 |
| Maxwell AFB (Montgom) | 174 | 32 | 32 | 95 | 77 | 37 | 44 | 51 | M | 0.28 |
| Mobile Regional AP | 220 | 31 | 31 | 92 | 76 | 42 | 48 | 55 | M | 0.25 |
| Mobile CO | 26 | 30 | 29 | 93 | 77 | 43 | 50 | 56 | M | 0.25 |
| Montgomery, Dannelly Field | 203 | 32 | 27 | 94 | 76 | 36 | 43 | 49 | M | 0.35 |
| Muscle Shoals Regional AP | 561 | 35 | 22 | 93 | 75 | 33 | 40 | 46 | M | 0.60 |
| Ozark, Fort Rucker | 356 | 31 | 31 | 94 | 77 | 42 | 49 | 56 | M | 0.28 |
| Selma-Craig AFB | 166 | 32 | 26 | 95 | 77 | 40 | 47 | 53 | M | 0.37 |
| Talladega | 528 | 33 | 22 | 94 | 76 | 36 | 43 | 50 | M | 0.69 |
| Tuscaloosa Municipal AP | 187 | 33 | 25 | 93 | 76 | 38 | 45 | 52 | M | 0.44 |
| **Alaska** | | | | | | | | | | |
| Adak NAS | 20 | 52 | 23 | 57 | 53 | -19 | -13 | -6 | L | 90 |
| Ambler | 289 | 67 | -38 | 73 | 58 | -26 | -19 | -12 | M | 24 |
| ...uk Pass | 2,156 | 68 | -35 | 66 | 53 | -36 | -29 | -22 | M | 77 |
|  | 131 | 61 | -4 | 68 | 57 | -19 | -12 | -6 | L | 14 |
|  | 138 | 61 | -7 | 70 | 58 | -19 | | | L | 11 |
|  | 194 | 61 | -9 | 71 | 58 | | | | L | 13 |
|  | ...42 | 61 | −13 | 71 | 58 | | | | | |
|  | | 55 | 20 | 69 | | | | | | |
|  | | | -22 | 71 | | | | | | |

*Figure 6* Example of design-conditions data.

computer-based, load-estimating programs from which to choose. All of them offer significant time-savings, as well as accurate results. A system designer can take a laptop, smart phone, or tablet to the job site and enter the information as it is collected. All the background information needed to process the input is inherent in the software. Results are available instantly. There is no valid reason today for failing to perform a load calculation.

*Figure 7* shows a typical input screen from load-estimating software. The software shown has features that include the following:

- Calculating heating and cooling loads based on *ACCA Manual J.*
- Calculating duct sizes, system losses, and fan static pressure requirements based on *ACCA Manual D.*
- Calculating the total structure capacity needed and individual room airflow requirements.

In addition to those listed features, the software provides other benefits such as helping to comply with codes such as the *International Residential Code (IRC)* and the *International Energy Conservation Code (IECC)*.

Once all the information is entered, the summary printout (*Figure 8*) provides the completed load estimate for heating and cooling. When printed, the completed load estimate carries great legitimacy when presented to a customer.

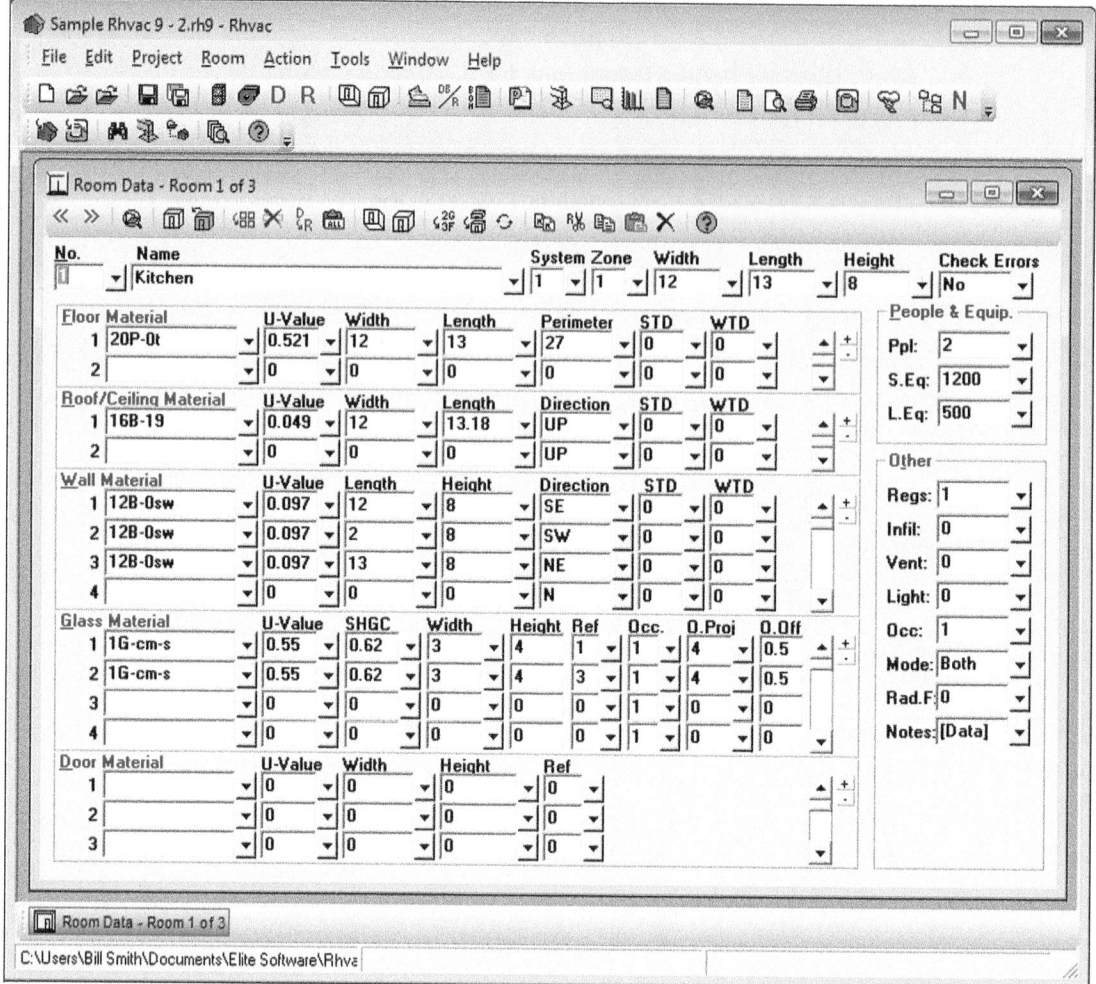

*Figure 7* Example of a data screen from a load-estimating software program.

Load-estimating programs may contain unique features such as equipment selection, operating cost comparisons, and airflow requirements, which enhance the programs and make them more versatile. As with any load-estimating method, whether manual or computer-based, always make sure the method selected complies with local codes and/or utility requirements.

## 2.1.0 Heat Transfer

Heat consists of thermal energy; therefore, it cannot be created or destroyed. It can, however, be moved from one place to another. Heat always flows in one direction—from a higher temperature to a lower temperature. The greater the temperature difference, the greater the quantity of heat that will flow over time. For example, if it is 70°F inside and 10°F outside (21°C and –12°C, respectively), heat will move more rapidly out of the structure because of the large temperature difference. If it is 70°F inside and 65°F outside (21°C and 18°C), the heat loss is nearly insignificant because the temperature difference is very small.

Effective insulation of floors, walls, and ceilings can greatly reduce heat loss and heat gain. The higher the R-value of a material, the lower the conductive heat transfer will be. There are different types of insulation for various applications. Surprisingly, it is possible to over-insulate. Excess insulation does more than just waste money. If a building is over-insulated and lacks the necessary

# Hydronic Heating System Design

Although the load-estimating process for a home served by a hydronic system is exactly the same, the design process differs after that. Instead of calculating airflows and duct sizes, the designer must determine water temperatures, water flow rates, pipe diameters, pump size, and select the terminal devices that will deliver the correct amount of heat to each area of the structure.

# Building Analysis
## *Entire House*
### Bertie Heating & Air Conditioning, Inc.

Job: 6115 NW 117th Place-326...
Date: Jun 28, 2013
By: Amy Kauper
Plan: Creekside Villas Lot 37

1730 NE 23rd Avenue, Gainesville, FL 32609-3904 Phone: 352-331-2005 Fax: 352-371-5770 Email: amy@bertieair.com Web: www.bertieair.com License: CAC-058522

## Project Information

For: Duration Builders, Inc.
11787 NW 61st Terrace, Alachua, FL 32615
Phone: 386-462-0511
Web: www.durationbuilders.com  Email: bjones@durationbuilders.com

## Design Conditions

**Location:**
Gainesville Regional AP, FL, US
Elevation: 164 ft
Latitude: 30°N

| Outdoor: | Heating | Cooling |
|---|---|---|
| Dry bulb (°F) | 33 | 92 |
| Daily range (°F) | - | 18 ( M ) |
| Wet bulb (°F) | - | 76 |
| Wind speed (mph) | 15.0 | 7.5 |

| Indoor: | Heating | Cooling |
|---|---|---|
| Indoor temperature (°F) | 70 | 75 |
| Design TD (°F) | 37 | 17 |
| Relative humidity (%) | 50 | 50 |
| Moisture difference (gr/lb) | 32.5 | 47.4 |

**Infiltration:**
Method: Simplified
Construction quality: Average
Fireplaces: 0

## Heating

| Component | Btuh/ft² | Btuh | % of load |
|---|---|---|---|
| Walls | 3.3 | 4526 | 17.9 |
| Glazing | 37.0 | 4585 | 18.1 |
| Doors | 12.8 | 239 | 0.9 |
| Ceilings | 1.2 | 1501 | 5.9 |
| Floors | 4.8 | 6125 | 24.2 |
| Infiltration | 2.7 | 3548 | 14.0 |
| Ducts | | 4792 | 18.9 |
| Piping | | 0 | 0 |
| Humidification | | 0 | 0 |
| Ventilation | | 0 | 0 |
| Adjustments | | 0 | |
| **Total** | | **25317** | **100.0** |

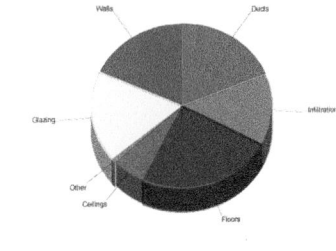

## Cooling

| Component | Btuh/ft² | Btuh | % of load |
|---|---|---|---|
| Walls | 2.1 | 2840 | 13.0 |
| Glazing | 41.9 | 5191 | 23.7 |
| Doors | 10.3 | 192 | 0.9 |
| Ceilings | 1.7 | 2168 | 9.9 |
| Floors | 0 | 0 | 0 |
| Infiltration | 0.6 | 842 | 3.8 |
| Ducts | | 6080 | 27.7 |
| Ventilation | | 0 | 0 |
| Internal gains | | 4600 | 21.0 |
| Blower | | 0 | 0 |
| Adjustments | | 0 | |
| **Total** | | **21912** | **100.0** |

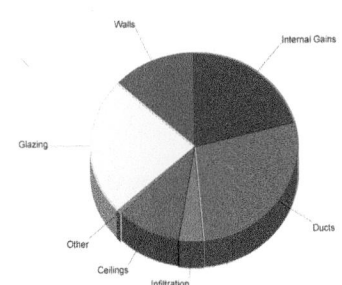

Latent Cooling Load = 6712 Btuh
Overall U-value = 0.114 Btuh/ft²-°F

Data entries checked.

**wrightsoft** Right-Suite® Universal 2013 13.0.03 RSU08133
ACCA ...ders, Inc\Creekside Lot 37\Creekside lot 37.rup  Calc = MJ8  Front Door faces: N

*Figure 8* Example of a printout from a load-estimating program.

# Building Dimensions

Calculating a heat loss or heat gain is fairly easy when up-to-date drawings showing all dimensions and building material specifications are available. But how are calculations for an older structure performed when the drawings are gone?

In those cases, take additional time to make detailed measurements and gain access to the attic, basement, or crawl space to determine the presence and thickness of any insulation. Determining the presence and thickness of insulation in exterior walls can be more difficult. Try removing a cover plate from an electrical outlet or a switch on an exterior wall. The gap between the outlet box and the drywall may allow the insulation to be viewed.

ventilation and vapor barriers, indoor moisture levels rise, promoting the growth of mold and fungi. There are three main ways that heat transfer takes place: conduction, radiation, and convection.

## 2.1.1 Conduction

Conduction is the transfer of heat energy in a substance from particle to particle from the warmer region to the colder region. If, for example, a rod is heated over a flame, heat travels by conduction from the hot end to the cooler end. Heat transfer by conduction occurs not only within an object or substance but also between different substances that may be in contact with one another.

An example can be found in a building constructed of a combination of brick or concrete, insulation, wood, and plaster (*Figure 9*). These materials are often in contact with each other. If it is warmer inside the building than outdoors, heat will pass through these materials to the outdoors by conduction (heat loss). If it is warmer outdoors than inside, heat will be conducted into the building (heat gain). Certain building materials, such as metal, will conduct heat faster than other materials, such as wood.

## 2.1.2 Radiation

Radiant heat does not need a substance to carry it from one object to another. It can travel through a vacuum. Radiant heat exhibits many properties of light. It cannot pass through an opaque object, but it can pass through transparent materials. It can also be reflected from a bright surface, just as light is reflected by a mirror. Radiant heat passing through air does not warm the air through which it passes. For example, a roof with the sun shining on it might be heated to 180°F (82°C) by the rays of the sun, but the air through which the radiant heat travels may only be at 80°F (27°C).

All objects radiate heat. The higher the temperature of the object or substance, the greater the quantity of heat it radiates. The amount of radiant heat energy given off per unit of time depends on both the temperature of the radiating body and on the type and color of the surface. A rough, dark surface, for example, radiates more heat than a smooth, light surface of the same dimensions. The darker a surface, the more solar radiation it will absorb. Thus, dark surfaces always have higher heat gains than light surfaces exposed to the same amount of sunlight.

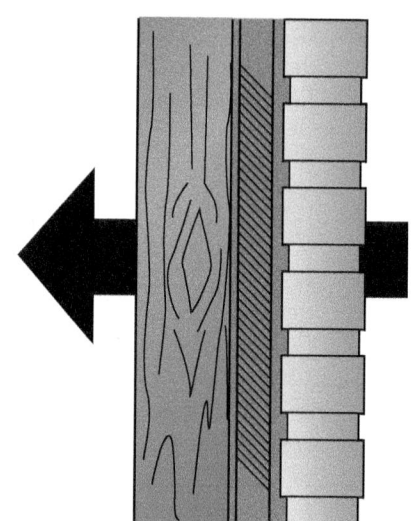

*Figure 9* Conduction.

### 2.1.3 Convection

Convection is the transfer of heat energy due to the movement of fluid. A fluid has been defined as anything that flows; therefore, gases and liquids are both fluids. Air, being a mixture of several gases, can be considered a fluid. As the air moves or circulates, it carries heat from one place to another. In a gas furnace, for example, the energy is released by the combustion process inside the heat exchanger. The walls of the heat exchanger are considerably cooler than the burner flames, so the heat is absorbed by the walls of the heat exchanger. Air from the conditioned space is forced over the heated walls of the heat exchanger by the blower. Because the air is cooler than the walls of the heat exchanger, heat will flow from the warm heat exchanger to the cooler air. The heat exchanger is designed in such a manner that the air passes over its entire surface in a wiping motion (*Figure 10*), causing a large amount of the heat energy to transfer to the moving air.

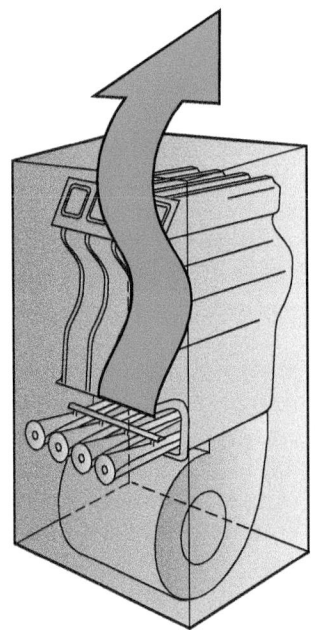

*Figure 10* Convection.

### 2.1.4 Heat Gain and Loss

In the summer, exterior walls transfer heat to the air in a room because they are warmer than the room air. Theoretically, a wall continually losing heat to a room would eventually cool down to the room air temperature. In reality, heat that the wall loses to the room air is continuously replaced with heat from outside the structure.

Heat flow through a wall separating two spaces at different temperatures depends on three factors:

- The area of the wall
- The temperatures of the two spaces (temperature differential)
- The heat-conducting properties of the wall

The larger the area of a wall, the more heat it can conduct. For example, a wall with an area of 200 ft² (18.6 m²) can conduct twice as much heat as a wall with an area of 100 ft² (9.3 m²) in the same conditions. If the difference in temperature between the two spaces is 50°F (28°C), only one-half as much sensible heat will flow through the wall compared to a temperature difference of 100°F (56°C). These heat flow principles also apply to windows, roofs, and other building surfaces. In summary, the flow of heat through any surface is directly proportional to its area and the difference in temperature of the spaces separated by the surface. A third factor affecting heat flow through walls involves the wall material and its thickness.

The terms *conductivity* and *conductance* are used to describe heat flow through building materials. Conductivity (*Figure 11*) is the ability of a material to conduct heat. This ability varies from one material to another. The best conductors of heat are typically metals; the least effective heat conductors are materials such as wood, inert gases, and cork. The ability of a substance to transmit heat by conduction is a physical property of that material. This physical property is called thermal conductivity, which is defined as the heat flow per hour (Btuh) through one square foot of one-inch thick homogeneous material when the temperature difference between the two faces is 1°F.

Conductance is the term used to denote heat flow through materials such as glass blocks, hollow clay tile, concrete blocks, and other composite materials. In such material, each succeeding inch of thickness is not identical with the preceding inch. These materials are not homogeneous. Therefore, conductivity cannot be used to define the heat flow process.

The conductance of a material is defined as the heat flow rate in Btuh through one square foot of a nonhomogeneous material of a certain thickness when there is a 1°F temperature difference between the two surfaces of the material (*Figure 12*).

Conductivity and conductance are not interchangeable terms. Conductivity is the heat flow through one inch of a homogeneous material; conductance is the heat flow through the entire thickness of a nonhomogeneous material. Air-space conductance is another factor that must be considered when calculating the heat gain of a structure via conductance. Air-space conductance is defined as the heat flow in Btuh through one

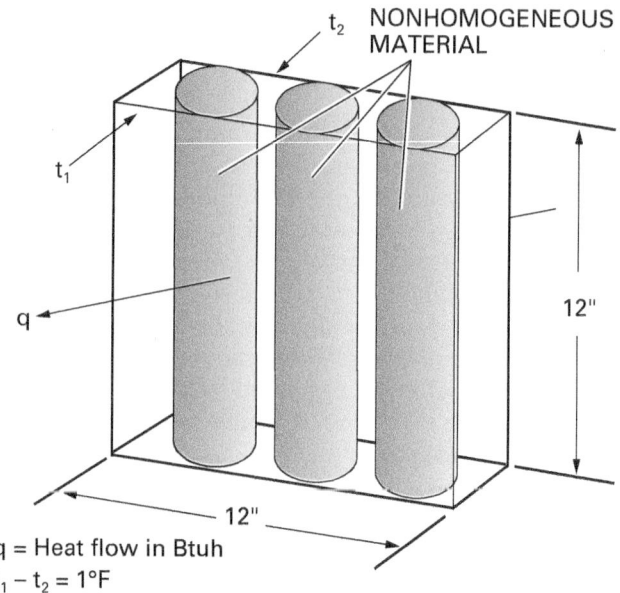

q = Heat flow in Btuh
$t_1 - t_2 = 1°F$

*Figure 12* Conductance.

square foot of air space for a temperature difference of 1°F between the bounding surfaces.

All the preceding factors need to be considered when calculating overall heat transfer.

It is inconvenient to find all the surface temperatures for a wall made up of four or five materials, yet it is easy to find the temperature on both sides of a wall with an ordinary thermometer. Therefore, for nonhomogeneous materials, it is much more convenient to use a heat flow equation written with air temperatures.

The term U-factor is used to simplify this calculation. The U-factor works for homogeneous and nonhomogeneous materials and for a wall or roof made up of several materials. It is a unit of measure for heat flow per hour through one square foot of the material(s) when the temperature difference is 1°F. U-factor tables available from the American Society of Heating, Refrigeration, and Air Conditioning Engineers (ASHRAE) list the U-factors for various types of ceiling, floor, wall, and roofing materials and finishes.

The overall heat transfer formula is:

$$q = A \times U \times (T_2 - T_1)$$
*or*
$$q = A \times U \times \Delta T$$

*Where:*

q = heat flow in Btuh

A = area in square feet

U = overall heat transfer coefficient in Btuh/ft²/°F (U-factor)

$(T_2 - T_1)$ or $\Delta T$ = difference in temperature between the air on each side of the wall or roof

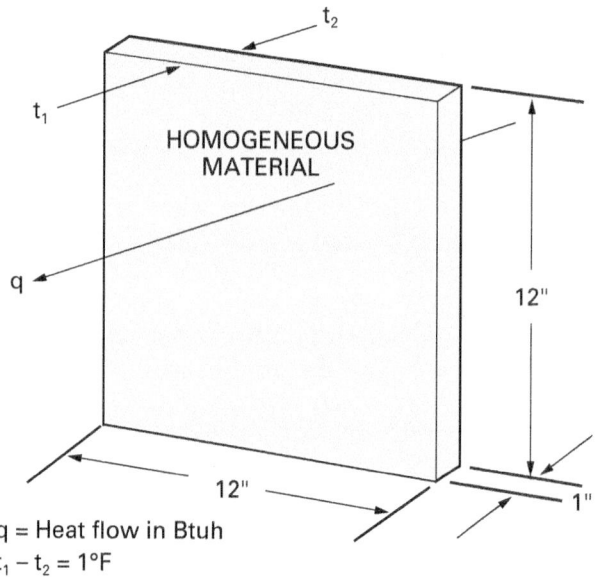

q = Heat flow in Btuh
$t_1 - t_2 = 1°F$

*Figure 11* Conductivity.

## 2.2.0 Cooling and Heating Load Factors

The materials used in constructing a building are not the only factors that affect heat gain and heat loss. The location, design, and positioning of a building also all have an effect, especially on the cooling load.

A residence that is shaded by tall trees or a commercial building that is in the shadow of taller buildings will have less of a cooling load than exposed buildings. This is especially true if the exposed building is oriented so that it has a lot of glass exposed to the sun.

The pie charts in *Figure 13* show the impact of various construction factors on cooling and heating loads. *Table 1* shows how various construction factors affect cooling and heating loads. These factors will be covered in sections that follow. The designer and builder can help provide a more economical installation and better system operation by specifying the use of insulation, close-fitting windows and doors, and double glazing, storm sashes, or reflective glass. Window glass is a significant contributor to both the cooling and heating loads in many buildings. For this reason, load-estimating tools place great emphasis on calculating the glass-related load.

### 2.2.1 Window Glass

Before discussing this subject, it is important to note that the terminology often differs from one load-estimating package to another. In general, the results are comparable, but the factors used to arrive at the answer to each calculation may

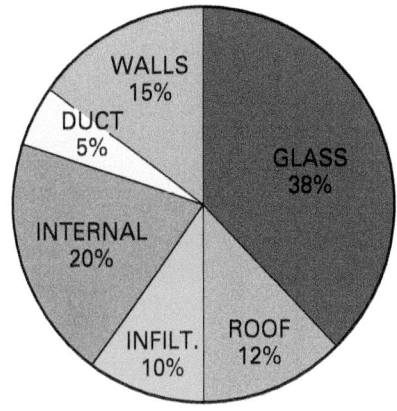

COOLING LOAD COMPONENTS

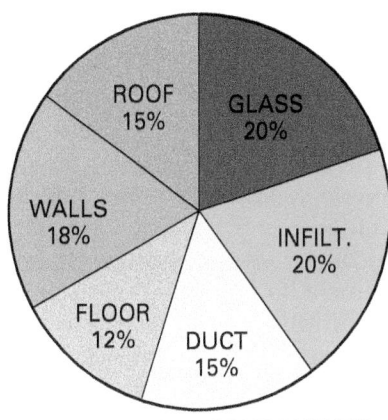

HEATING LOAD COMPONENTS
(UNCONDITIONED BASEMENT
OR ENCLOSED CRAWL SPACE)

**TYPICAL LOAD PERCENTAGES**

*Figure 13* Load factors.

**Table 1** Effects of Construction Factors on Cooling and Heating Loads

| Factor | Cooling Load | Heating Load | Comments |
|---|---|---|---|
| Windows and glass doors | X | X | Major impact on cooling load; significant impact on heating load |
| Shaded glass | X | | |
| Wood and metal doors | X | X | Greater effect on heating load |
| Exterior walls | X | X | Degree of impact depends on insulation thickness |
| Below-grade masonry walls | | X | |
| Floors | | X | |
| Roof | X | | Darker colors create more load |
| Infiltration | X | X | Greater effect on heating load |
| Unconditioned spaces | X | X | |
| Duct gain/loss | X | X | Greater effect on heating load |
| Temperature swing | X | | |
| Internal factors | X | | People, cooking, and bathing |
| Insulation (wall & attic/ceiling) | X | X | Greater effect on heating load |

differ. Most of the examples used in this module were taken from *ACCA Manual J*. For example, the information in *Figure 14* is an example of a table for glass HTMs. However, if the same type of information from ASHRAE is reviewed, the term glass load factor (GLF) will be seen instead. The HTM is based on the projected temperature difference between the indoor and outdoor temperatures. The GLF is based solely on the design outdoor temperature. Be prepared to see different terminology and information presented in various load-estimating packages, and review the definitions for those you do not immediately recognize.

In cooling, window glass is usually the single largest load factor (refer to *Figure 13*). To determine the glass-related load, the area of each type of window glass is multiplied by load factors to determine the amount of load in Btuh. Therefore, the more windows used and the less energy efficient the windows are, the greater the load.

For example, assume a home has ten single-pane, unshaded 3' × 5' windows (15 ft², or 1.4 m², each) with east or west exposure. Refer to *Figure 14* and assume a 25°F (13.9°C) temperature difference. The multiplier is 94 and the related load is 14,100 Btuh, determined as follows:

Glass load = (number of windows × area) × HTM
Glass load = (10 windows × 15 ft²) × 94 Btuh/ft²
Glass load = 150 ft² × 94 Btuh/ft²
Glass load = 14,100 Btuh

That means these ten windows alone create a cooling load exceeding 1 ton. Note that if the same windows have internal shading (lower portions of *Figure 14*), the load would be reduced significantly. Half-open blinds, for example, drop the HTM to 66.

These are some of the tradeoffs that go into both system and building design. If this were a new construction project, the owner could be given the option of installing energy-efficient windows. The initial cost would be higher, but the savings in the size of the cooling system, combined with lower operating costs, would eventually pay back the difference. These tradeoffs are an important aspect of the design and selling process. The effect that window coverings such as draperies or blinds have can also be determined.

If a project is sized for windows with full, insulated draperies and the owner later switches to light, open-style curtains, it can affect the ability of the system to meet the cooling or heating demand on design temperature days, especially if there is a large amount of east- and west-exposure glass. It may not be enough of a difference to cause discomfort throughout the home, but local discomfort

## Plan Ahead

Use caution when considering shading from adjacent buildings or trees for a cooling load estimate. Buildings get torn down and trees age and get cut down. Make sure the buildings or trees you are counting on for shade will be around for a while, and think about the effect on your calculation if the shade is removed. Although you cannot plan and design around every event that could theoretically happen in the future, be sure you are at least aware of all short-term possibilities.

is certainly possible. If windows are fully or partially shaded by roof overhangs, awnings, or shade screens, that fact must be taken into account when calculating the cooling load. A fully shaded east-, west-, or south-exposure window is basically treated like a north-exposure window. If a window is partially shaded, the percent of shading must be factored in. *Manual J* provides a special calculation in which the shaded and unshaded portions are treated as separate entities, and then the results are added to determine the load for the entire window. French doors, sliding glass doors, and other construction features that are primarily glass are treated as windows.

### 2.2.2 Walls, Roofs, Ceilings, and Floors

Any exposed or partially exposed surface must be figured into the load calculation because these surfaces gain and lose heat as described earlier. The amount of load a surface creates depends on the construction material used and is calculated in a similar way as window-glass load. In *Manual J*, tables are provided showing the U-value for each type of construction material. *Figure 15* shows a portion of a *Manual J* table for frame walls and partitions.

A separate table containing similar data is also provided for determining heating load factors. When determining the load represented by exposed walls, the square footage for applicable windows and doors must be subtracted from the wall area. Otherwise, those areas will be counted twice.

As shown in *Figure 15*, insulation usually plays a significant role in the amount of load presented by exposed surfaces. Money invested in insulation will generally pay for itself over time. Other tables are provided so that the estimator can determine the insulation values of various construction materials.

# Table 3A
## MJ8$_{AE}$ Procedure: Cooling HTM for Windows and Glass Doors
### No External Sun Screen, Clear Glass

Cooling Load (Btuh) = HTM × Reference Area
Bay Window HTM = 1.15 × Table HTM
Garden Window HTM = 2.00 × Table HTM
French Door HTM = 0.70 × Table HTM
Use Single Pane, Clear Glass for Jalousie Window

Default indoor design temperature = 75°F.
Outdoor design temperature provided by Table 1.
Load area appears above HTM values.
Table 3, note 2 specifies the order of
application of the HTM adjustment procedures.

**Recommended Adjustments**

1) Full outdoor insect screen = 0.80 × Table HTM
   Half outdoor insect screen = 0.90 × Table HTM
   Full indoor insect screen = 0.90 × Table HTM
   Half indoor insect screen = 0.95 × Table HTM
2) Shade by external overhang (See Table 3E-1)

**No Optional Adjustments**

### Table 3A — Clear Glass

#### No Internal Shade

| Default Assembly Performance | Single Pane | | | | | | Double Pane | | | | | | Triple Pane | | | | | |
|---|---|---|---|---|---|---|---|---|---|---|---|---|---|---|---|---|---|---|
| U-Value | 0.98 | | | | | | 0.56 | | | | | | 0.42 | | | | | |
| SC | 0.85 | | | | | | 0.75 | | | | | | 0.70 | | | | | |
| SHGC | 0.74 | | | | | | 0.65 | | | | | | 0.61 | | | | | |
| Design CTD | 10 | 15 | 20 | 25 | 30 | 35 | 10 | 15 | 20 | 25 | 30 | 35 | 10 | 15 | 20 | 25 | 30 | 35 |
| **Exposure** | HTM for Rough Opening | | | | | | HTM for Rough Opening | | | | | | HTM for Rough Opening | | | | | |
| North | 24 | 29 | 34 | 39 | 44 | 49 | 18 | 21 | 24 | 27 | 29 | 32 | 16 | 18 | 20 | 22 | 24 | 26 |
| NE or NW | 56 | 61 | 66 | 70 | 75 | 80 | 46 | 49 | 52 | 55 | 57 | 60 | 42 | 44 | 46 | 48 | 50 | 53 |
| East or West | 80 | 84 | 89 | 94 | 99 | 104 | 67 | 70 | 73 | 76 | 78 | 81 | 62 | 64 | 66 | 68 | 70 | 72 |
| SE or SW | 68 | 73 | 78 | 83 | 88 | 93 | 57 | 60 | 63 | 65 | 68 | 71 | 52 | 54 | 56 | 59 | 61 | 63 |
| South | 40 | 45 | 50 | 55 | 60 | 65 | 32 | 35 | 38 | 41 | 44 | 46 | 29 | 31 | 33 | 36 | 38 | 40 |

#### Vertical or Horizontal Blinds with Slats At 45 Degrees

| Default Assembly Performance | Single Pane | | | | | | Double Pane | | | | | | Triple Pane | | | | | |
|---|---|---|---|---|---|---|---|---|---|---|---|---|---|---|---|---|---|---|
| U-Value | 0.98 | | | | | | 0.56 | | | | | | 0.42 | | | | | |
| SC | 0.60 | | | | | | 0.50 | | | | | | 0.45 | | | | | |
| SHGC | 0.52 | | | | | | 0.44 | | | | | | 0.39 | | | | | |
| Design CTD | 10 | 15 | 20 | 25 | 30 | 35 | 10 | 15 | 20 | 25 | 30 | 35 | 10 | 15 | 20 | 25 | 30 | 35 |
| **Exposure** | HTM for Rough Opening | | | | | | HTM for Rough Opening | | | | | | HTM for Rough Opening | | | | | |
| North | 16 | 21 | 26 | 31 | 35 | 40 | 11 | 13 | 16 | 19 | 22 | 25 | 9 | 11 | 13 | 15 | 17 | 19 |
| NE or NW | 36 | 41 | 46 | 50 | 55 | 60 | 27 | 30 | 33 | 36 | 38 | 41 | 24 | 26 | 28 | 30 | 32 | 34 |
| East or West | 51 | 56 | 61 | 66 | 71 | 76 | 40 | 43 | 46 | 49 | 51 | 54 | 35 | 37 | 40 | 42 | 44 | 46 |
| SE or SW | 44 | 49 | 54 | 59 | 64 | 68 | 34 | 37 | 40 | 42 | 45 | 48 | 30 | 32 | 34 | 36 | 38 | 40 |
| South | 25 | 30 | 35 | 40 | 45 | 49 | 18 | 21 | 24 | 27 | 29 | 32 | 16 | 18 | 20 | 22 | 24 | 26 |

#### Drape or Roller Shade Half Drawn

| Default Assembly Performance | Single Pane | | | | | | Double Pane | | | | | | Triple Pane | | | | | |
|---|---|---|---|---|---|---|---|---|---|---|---|---|---|---|---|---|---|---|
| U-Value | 0.98 | | | | | | 0.56 | | | | | | 0.42 | | | | | |
| SC | 0.70 | | | | | | 0.60 | | | | | | 0.55 | | | | | |
| SHGC | 0.61 | | | | | | 0.52 | | | | | | 0.48 | | | | | |
| Design CTD | 10 | 15 | 20 | 25 | 30 | 35 | 10 | 15 | 20 | 25 | 30 | 35 | 10 | 15 | 20 | 25 | 30 | 35 |
| **Exposure** | HTM for Rough Opening | | | | | | HTM for Rough Opening | | | | | | HTM for Rough Opening | | | | | |
| North | 17 | 22 | 27 | 32 | 37 | 41 | 12 | 14 | 17 | 20 | 23 | 26 | 10 | 12 | 14 | 16 | 18 | 20 |
| NE or NW | 40 | 45 | 50 | 55 | 60 | 65 | 32 | 34 | 37 | 40 | 43 | 46 | 28 | 30 | 32 | 34 | 36 | 38 |
| East or West | 58 | 63 | 68 | 73 | 78 | 83 | 47 | 50 | 53 | 55 | 58 | 61 | 42 | 44 | 46 | 49 | 51 | 53 |
| SE or SW | 50 | 54 | 59 | 64 | 69 | 74 | 40 | 43 | 45 | 48 | 51 | 54 | 35 | 38 | 40 | 42 | 44 | 46 |
| South | 28 | 32 | 37 | 42 | 47 | 52 | 21 | 24 | 26 | 29 | 32 | 35 | 18 | 20 | 22 | 24 | 27 | 29 |

#### Drape or Roller Shade Fully Drawn

| Default Assembly Performance | Single Pane | | | | | | Double Pane | | | | | | Triple Pane | | | | | |
|---|---|---|---|---|---|---|---|---|---|---|---|---|---|---|---|---|---|---|
| U-Value | 0.98 | | | | | | 0.56 | | | | | | 0.42 | | | | | |
| SC | 0.50 | | | | | | 0.45 | | | | | | 0.40 | | | | | |
| SHGC | 0.44 | | | | | | 0.39 | | | | | | 0.35 | | | | | |
| Design CTD | 10 | 15 | 20 | 25 | 30 | 35 | 10 | 15 | 20 | 25 | 30 | 35 | 10 | 15 | 20 | 25 | 30 | 35 |
| **Exposure** | HTM for Rough Opening | | | | | | HTM for Rough Opening | | | | | | HTM for Rough Opening | | | | | |
| North | 15 | 20 | 25 | 30 | 34 | 39 | 10 | 13 | 16 | 19 | 21 | 24 | 8 | 10 | 12 | 15 | 17 | 19 |
| NE or NW | 31 | 36 | 41 | 46 | 51 | 56 | 25 | 28 | 31 | 33 | 36 | 39 | 21 | 24 | 26 | 28 | 30 | 32 |
| East or West | 44 | 49 | 54 | 59 | 64 | 69 | 37 | 40 | 42 | 45 | 48 | 51 | 32 | 34 | 35 | 38 | 40 | 42 |
| SE or SW | 38 | 43 | 48 | 53 | 58 | 63 | 31 | 34 | 37 | 40 | 42 | 45 | 27 | 29 | 31 | 33 | 35 | 37 |
| South | 22 | 27 | 32 | 37 | 42 | 47 | 17 | 20 | 23 | 25 | 28 | 31 | 14 | 16 | 19 | 21 | 23 | 25 |

*Figure 14* Example of Manual J glass HTMs.

# Table 4A
## Heating and Cooling Performance for Opaque Panels
## U-Values and Group Numbers or CLTD Values

**Construction Number 12**
**Frame Walls and Partitions**

Wall or partition with brick veneer, plus interior finish (40 to 50 Lb / SqFt)
Wall with siding or stucco, or light partition, plus interior finish (7 to 20 Lb / SqFt)
Exterior finish code: b = brick veneer; s = stucco or siding
Framing code: w = wood, m = metal  (studs 16 Inches on center, 75% cavity, 25% framing)
Reference Area = Gross Wall Area - Area of Window and Door Openings

| Construction Number | Insulation R-Values | Description of Construction | Exterior Finish | U-Value with Wood Studs | U-Value with Metal Studs | Group Number |
|---|---|---|---|---|---|---|
| **12A — No Insulation In Stud Cavity** | | | | | | |
| 12A-0b w/m 12A-0s w/m | Cavity: None Board: None | Frame construction, no cavity insulation, no board insulation, wood sheathing | Brick Siding | 0.253 0.240 | 0.315 0.295 | E A |
| 12A-2b w/m 12A-2s w/m | Cavity: None Board: R-2 | Frame construction, no cavity insulation, R-2 board insulation | Brick Siding | 0.194 0.186 | 0.230 0.219 | E A |
| 12A-3b w/m 12A-3s w/m | Cavity: None Board: R-3 | Frame construction, no cavity insulation, R-3 board insulation | Brick Siding | 0.162 0.157 | 0.187 0.180 | F B |
| 12A-4b w/m 12A-4s w/m | Cavity: None Board: R-4 | Frame construction, no cavity insulation, R-4 board insulation | Brick Siding | 0.139 0.135 | 0.157 0.152 | F B |
| 12A-5b w/m 12A-5s w/m | Cavity: None Board: R-5 | Frame construction, no cavity insulation, R-5 board insulation | Brick Siding | 0.122 0.119 | 0.136 0.132 | G C |
| 12A-6b w/m 12A-6s w/m | Cavity: None Board: R-6 | Frame construction, no cavity insulation, R-6 board insulation | Brick Siding | 0.109 0.106 | 0.120 0.117 | G C |
| **12B — R-11 Insulation In 2 x 4 Stud Cavity** | | | | | | |
| 12B-0b w/m 12B-0s w/m | Cavity: R-11 Board: None | Frame construction, R-11 cavity insulation, no board insulation, wood sheathing | Brick Siding | 0.097 | 0.122 | H B |
| 12B-2b w/m 12B-2s w/m | Cavity: R-11 Board: R-2 | Frame construction, R-11 cavity insulation, R-2 board insulation | Brick Siding | 0.086 | 0.106 | I C |
| 12B-3b w/m 12B-3s w/m | Cavity: R-11 Board: R-3 | Frame construction, R-11 cavity insulation, R-3 board insulation | Brick Siding | 0.079 | 0.096 | J D |
| 12B-4b w/m 12B-4s w/m | Cavity: R-11 Board: R-4 | Frame construction, R-11 cavity insulation, R-4 board insulation | Brick Siding | 0.073 | 0.088 | J D |
| 12B-5b w/m 12B-5s w/m | Cavity: R-11 Board: R-5 | Frame construction, R-11 cavity insulation, R-5 board insulation | Brick Siding | 0.068 | 0.081 | K E |
| 12B-6b w/m 12B-6s w/m | Cavity: R-11 Board: R-6 | Frame construction, R-11 cavity insulation, R-6 board insulation | Brick Siding | 0.064 | 0.075 | K F |
| **12C — R-13 Insulation In 2 x 4 Stud Cavity** | | | | | | |
| 12C-0b w/m 12C-0s w/m | Cavity: R-13 Board: None | Frame construction, R-13 cavity insulation, no board insulation, wood sheathing | Brick Siding | 0.091 | 0.115 | I C |
| 12C-2b w/m 12C-2s w/m | Cavity: R-13 Board: R-2 | Frame construction, R-13 cavity insulation, R-2 board insulation | Brick Siding | 0.081 | 0.101 | J D |
| 12C-3b w/m 12C-3s w/m | Cavity: R-13 Board: R-3 | Frame construction, R-13 cavity insulation, R-3 board insulation | Brick Siding | 0.075 | 0.092 | K E |
| 12C-4b w/m 12C-4s w/m | Cavity: R-13 Board: R-4 | Frame construction, R-13 cavity insulation, R-4 board insulation | Brick Siding | 0.069 | 0.084 | K E |
| 12C-5b w/m 12C-5s w/m | Cavity: R-13 Board: R-5 | Frame construction, R-13 cavity insulation, R-5 board insulation | Brick Siding | 0.064 | 0.078 | K F |
| 12C-6b w/m 12C-6s w/m | Cavity: R-13 Board: R-6 | Frame construction, R-13 cavity insulation, R-6 board insulation | Brick Siding | 0.060 | 0.072 | K G |

*Figure 15* Manual J U-values for frame walls and partitions.

## Energy-Efficient Windows

A single pane of glass provides very little insulation. It has an R-value (insulating value) of less than 1. Remember that the greater the R-value, the greater the insulating value. Adding another pane with a ½" air space more than doubles the R-value. The air space between the panes of glass acts as an insulator. Windows are commonly designed with two or three layers of glass separated by ¾" to 1" in order to improve insulation quality.

To obtain even more insulating value, the space between panes in some windows is filled with argon gas, which conducts heat at a lower rate than air. Where single-pane glass is used, it is common to add storm windows.

A special type of glass known as low-e, for low emissivity, provides even greater insulating quality. Emissivity is the ability of a material to absorb or radiate heat. Low-e glass is coated with a very thin metallic substance on the inside of the inner pane of a double-pane window. In cold weather, radiated heat from walls, floors, and furniture reflects back into the room by the low-e coating instead of escaping through the windows. This reduces the heat loss, which in turn saves heating costs. In summer months, radiated heat from outdoor sources, such as the sun, roads, and parking lots, is reflected away from the building by the low-e coating. Although windows with low-e glass are considerably more expensive than standard windows, they usually pay for themselves in reduced heating and cooling costs within three or four years.

 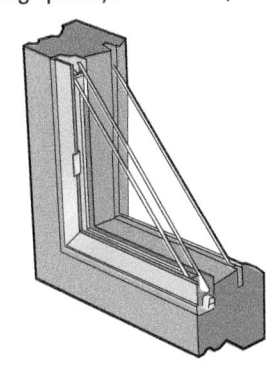

When calculating heating loads, special consideration must be given to walls, floors, and ceilings that adjoin unconditioned spaces such as garages, crawl spaces, and attics. Even with insulation, these surfaces can create as much load as insulated, exposed walls.

Basements also require special treatment when calculating a heating load. The load created by an uninsulated block or brick wall above grade is about equal to that of double-pane glass. Even with insulation, the load is substantial. The portion of the wall that is below grade has significantly less heat loss. Therefore, the portions above and below grade must be treated separately, like partially shaded windows. Basement floors must also be calculated separately. If the first floor is above an unheated basement or crawl space, it can represent a significant heating load, especially if there is no insulation.

Roof area can have a major effect on cooling load. In *Manual J* and other load-estimating methods, the reflective qualities of the roof are important. As discussed earlier, light-colored roofing material will reflect radiant heat better than dark material; light-colored material therefore creates less of a load. Again, insulation is an important factor in the load presented by the roof.

### 2.2.3 Infiltration

Infiltration affects both heating and cooling loads. Air can enter a building through cracks around windows and doors and other construction joints. Fireplaces and vents without dampers are another source of infiltration. Infiltration contributes more to the heating load than the cooling load. This is because of the greater temperature difference between the conditioned inside air and the infiltrating outside air. Caulking and weather-stripping can make a dramatic difference, especially in cold climates.

Infiltration must be factored into the load estimate. The amount of infiltration is a function of the building's tightness. Tightness is a function of several factors, as stated in the ACCA manual:

- *Best* – Continuous infiltration barrier, all cracks and penetrations sealed; tested leakage of windows and doors less than 0.25 cfm per running foot of crack; vents and exhaust fans have dampers; recessed ceiling lights gasketed or taped; no combustion air required or combustion air from outdoors; no duct leakage.
- *Average* – Plastic vapor barrier, major cracks and penetrations sealed; tested leakage of windows and doors between 0.25 and 0.50 cfm per running foot of crack; electrical fixtures that penetrate the envelope not taped or gasketed; vents and exhaust fans dampered; combustion air from indoors; intermittent ignition and flue damper; some duct leakage to unconditioned space.
- *Poor* – No infiltration barrier or plastic vapor barrier; no attempt to seal cracks and penetrations; tested leakage of windows and doors greater than 0.50 cfm per running foot of crack;

vents and exhaust fans not dampered; combustion air from indoors; standing pilot; no flue damper; considerable duct leakage to unconditioned space.

For example, a fireplace evaluation would fall into one of the following categories:

- *Best* – Combustion air from outdoors, tight glass doors and damper.
- *Average* – Combustion air from indoors, tight glass doors or damper.
- *Poor* – Combustion air from indoors, no glass doors or damper.

## Energy Performance Ratings

*GOING GREEN*

The National Fenestration Rating Council (NFRC) is a nonprofit organization created by the window, door, and skylight industry that includes manufacturers, architects, code officials, government agencies, and others as members. The primary goal of the organization is to provide accurate information to measure and compare the energy performance of windows, doors, and skylights. NFRC has established a voluntary energy performance rating and labeling system for fenestration (window and door) products.

The NFRC label contains vital information such as U-factor, solar heat gain coefficient, visible transmittance, and air leakage. Manufacturers can also ask for the entry of a condensation rating on their product labels.

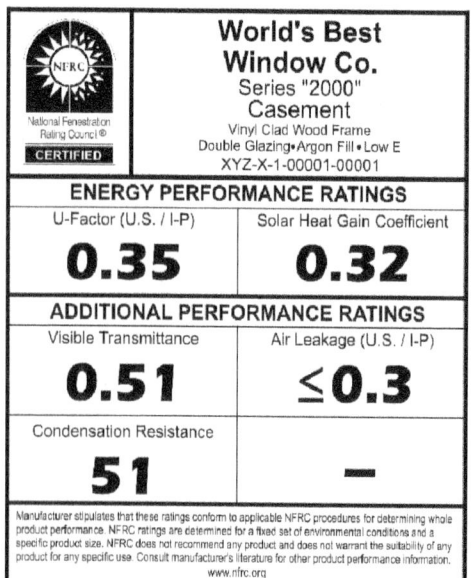

**Figure Credit:** National Fenestration Rating Council (NFRC)

*Figure 16* shows how the effect of infiltration is determined. Infiltration is apportioned to the load estimate based on the square footage of the home. Some load calculation software packages may approach the calculation a different way. Infiltration is a function of the volume of uncontrolled air entering the building. *Figure 16* shows the number of air changes per hour (ACH) for heating and cooling that result from infiltration. One ACH means that infiltration allows all the air within the home to be replaced by outside air once per hour. Of course, when it is very hot or cold outdoors, this represents a major load. Therefore, infiltration must be minimized.

You must convert the ACH to infiltration cubic feet per minute units (ICFM), which is then used to determine the volume of outside air that must be heated and cooled. Note that the *Manual J* table offers information depending on construction, such as single-story or two-story construction.

### 2.2.4 Duct Losses

Duct losses affect both heating and cooling loads. However, it has a greater effect on heating loads. The U.S. Department of Energy states that typical residential duct systems can lose 25 to 40 percent of the heating or cooling energy passing through, due to both leakage and heat transfer. Adding insulation and sealing duct leaks can significantly reduce those losses. *Figure 17* shows heat loss, heat gain, and latent gain factors assigned to duct systems based on outside air temperature (OAT), the square footage of the home, and R6 insulation. Correction factors are provided for different insulation values. There are a number of these tables in *Manual J*, each offered for specific duct system configurations and the location of the duct (i.e. attic, crawl space, or basement).

### 2.2.5 Cooling Load Factors

There are a few load factors that affect only cooling. Temperature swing and internal factors contribute to the cooling load but do not contribute to the heating load. In fact, internal factors typically reduce the heating load, since internal factors are items that produce heat.

People and appliances are internal factors that affect the cooling load. People give off body heat, so the more occupants in the building, the greater the cooling load. If the occupants entertain large groups, that must also be factored into the load estimate. In the absence of the specific number of occupants provided by the owner, a rule of thumb is to consider the load generated by one person per bedroom, plus one.

## Table 5A
### Infiltration Air Change Values for Three or Four Exposures

#### Default Air Change Values for Single Story Construction

| Construction | Air Changes per Hour — Heating | | | | | Infiltration[1] Cfm for One Fireplace |
| | Floor Area of Heated Space (SqFt) | | | | | |
| | 900 Or Less | 901 to 1500 | 1501 to 2000 | 2001 to 3000 | 3001 or More | |
|---|---|---|---|---|---|---|
| Tight | 0.21 | 0.16 | 0.14 | 0.11 | 0.10 | 0 |
| Semi-Tight | 0.41 | 0.31 | 0.26 | 0.22 | 0.19 | 13 |
| Average | 0.61 | 0.45 | 0.38 | 0.32 | 0.28 | 20 |
| Semi-Loose | 0.95 | 0.70 | 0.59 | 0.49 | 0.43 | 27 |
| Loose | 1.29 | 0.94 | 0.80 | 0.66 | 0.58 | 33 |

1)  For one additional fireplace, add 7 Cfm to the above fireplace values. For two or more additional fireplaces, add 10 Cfm (total) to the above.

| Construction | Air Changes per Hour — Cooling | | | | | Fireplace Infiltration Cfm |
| | Floor Area of Air Conditioned Space (SqFt) | | | | | |
| | 900 Or Less | 901 to 1500 | 1501 to 2000 | 2001 to 3000 | More than 3000 | |
|---|---|---|---|---|---|---|
| Tight | 0.11 | 0.08 | 0.07 | 0.06 | 0.05 | |
| Semi-Tight | 0.22 | 0.16 | 0.14 | 0.11 | 0.10 | |
| Average | 0.32 | 0.23 | 0.20 | 0.16 | 0.15 | 0 |
| Semi-Loose | 0.50 | 0.36 | 0.31 | 0.25 | 0.23 | |
| Loose | 0.67 | 0.49 | 0.42 | 0.34 | 0.30 | |

#### Default Air Change Values for Two Story Construction

| Construction | Air Changes per Hour — Heating | | | | | Infiltration[1] Cfm for One Fireplace |
| | Floor Area of Heated Space (SqFt) | | | | | |
| | 900 Or Less | 901 to 1500 | 1501 to 2000 | 2001 to 3000 | 3001 or More | |
|---|---|---|---|---|---|---|
| Tight | 0.27 | 0.20 | 0.18 | 0.15 | 0.13 | 0 |
| Semi-Tight | 0.53 | 0.39 | 0.34 | 0.28 | 0.25 | 13 |
| Average | 0.79 | 0.58 | 0.50 | 0.41 | 0.37 | 20 |
| Semi-Loose | 1.23 | 0.90 | 0.77 | 0.63 | 0.56 | 27 |
| Loose | 1.67 | 1.22 | 1.04 | 0.85 | 0.75 | 33 |

1)  For one additional fireplace, add 7 Cfm to the above fireplace values. For two or more additional fireplaces, add 10 Cfm (total) to the above.

| Construction | Air Changes per Hour — Cooling | | | | | Fireplace Infiltration Cfm |
| | Floor Area of Air Conditioned Space (SqFt) | | | | | |
| | 900 Or Less | 901 to 1500 | 1501 to 2000 | 2001 to 3000 | More than 3000 | |
|---|---|---|---|---|---|---|
| Tight | 0.14 | 0.11 | 0.09 | 0.08 | 0.07 | |
| Semi-Tight | 0.28 | 0.21 | 0.18 | 0.15 | 0.13 | |
| Average | 0.41 | 0.30 | 0.26 | 0.21 | 0.19 | 0 |
| Semi-Loose | 0.64 | 0.47 | 0.40 | 0.33 | 0.29 | |
| Loose | 0.87 | 0.64 | 0.54 | 0.44 | 0.39 | |

*Figure 16* Infiltration air-change values.

**Table 7C-T — Trunk and Branch Supply and Return System in 16C, 16CR or 16CF Attic**

| 7C-T | Ambient dry-bulb temperature = Outdoor db + 11 (heating) and Outdoor db + 25 (cooling) | Supply location = Center of room ceilings | Nominal return Cfm = 400 | Return location = Close to air handler |
|---|---|---|---|---|
| | Duct leakage Cfm per SqFt of duct surface area (supply / return) = 0.06/0.06; 0.09/0.15; 0.12/0.24; 0.24/0.47; 0.35/0.70 | | Duct wall insulation R-value = 2, 4, 6 and 8 | |

**Base Case Heat Loss Factor (BHLF)**
R6 Insulation, ASHRAE Sealed (Supply = 0.12, Return = 0.24)

| Table 1 | Square Feet of Floor Area | | | | |
|---|---|---|---|---|---|
| OAT | 1,000 | 1,500 | 2,000 | 2,500 | 3,000 |
| -10 | 0.220 | 0.240 | 0.260 | 0.280 | 0.310 |
| 0 | 0.200 | 0.220 | 0.240 | 0.260 | 0.290 |
| 10 | 0.180 | 0.200 | 0.220 | 0.240 | 0.260 |
| 20 | 0.170 | 0.180 | 0.200 | 0.220 | 0.240 |
| 30 | 0.140 | 0.170 | 0.180 | 0.200 | 0.220 |
| 40 | 0.110 | 0.140 | 0.160 | 0.180 | 0.200 |

Interpolate or extrapolate for OAT and Floor Area.

| R-Value Correction (WIF—Heat Loss) | R2 | R4 | R6 | R8 |
|---|---|---|---|---|
| | 1.85 | 1.24 | 1.00 | 0.85 |

**Leakage Correction (LCF) for Heat Loss**

| Leakage | R2 | R4 | R6 | R8 |
|---|---|---|---|---|
| 0.06 / 0.06 | 0.84 | 0.77 | 0.71 | 0.67 |
| 0.09 / 0.15 | 0.92 | 0.89 | 0.85 | 0.83 |
| 0.12 / 0.24 | 1.00 | 1.00 | 1.00 | 1.00 |
| 0.24 / 0.47 | 1.27 | 1.37 | 1.44 | 1.52 |
| 0.35 / 0.70 | 1.52 | 1.76 | 1.90 | 2.04 |

**Base Case Sensible Gain Factor (BSGF)**
R6 Insulation, ASHRAE Sealed (Supply = 0.12, Return = 0.24)

| Table 1 | Square Feet of Floor Area | | | | |
|---|---|---|---|---|---|
| OAT | 1,000 | 1,500 | 2,000 | 2,500 | 3,000 |
| 85 | 0.190 | 0.240 | 0.290 | 0.320 | 0.350 |
| 90 | 0.210 | 0.260 | 0.290 | 0.320 | 0.350 |
| 95 | 0.230 | 0.260 | 0.290 | 0.320 | 0.350 |
| 100 | 0.240 | 0.260 | 0.290 | 0.320 | 0.360 |
| 105 | 0.240 | 0.260 | 0.300 | 0.330 | 0.370 |

Interpolate or extrapolate for OAT and Floor Area.

| R-Value Correction (WIF—Sensible Gain) | R2 | R4 | R6 | R8 |
|---|---|---|---|---|
| | 2.00 | 1.28 | 1.00 | 0.86 |

**Leakage Correction (LCF) for Sensible Gain**

| Leakage | R2 | R4 | R6 | R8 |
|---|---|---|---|---|
| 0.06 / 0.06 | 0.80 | 0.74 | 0.69 | 0.64 |
| 0.09 / 0.15 | 0.90 | 0.87 | 0.84 | 0.82 |
| 0.12 / 0.24 | 1.00 | 1.00 | 1.00 | 1.00 |
| 0.24 / 0.47 | 1.34 | 1.45 | 1.52 | 1.59 |
| 0.35 / 0.70 | 1.71 | 1.97 | 2.14 | 2.27 |

**Base Case Latent Gain (BLG)**
R6 Insulation, ASHRAE Sealed (Supply = 0.12, Return = 0.24)

| Table 1 | Square Feet of Floor Area | | | | |
|---|---|---|---|---|---|
| Grains | 1,000 | 1,500 | 2,000 | 2,500 | 3,000 |
| 10 | 191 | 265 | 317 | 363 | 387 |
| 20 | 303 | 421 | 503 | 576 | 614 |
| 30 | 420 | 583 | 697 | 798 | 850 |
| 40 | 541 | 751 | 897 | 1,027 | 1,094 |
| 50 | 666 | 925 | 1,105 | 1,265 | 1,347 |
| 60 | 795 | 1,104 | 1,319 | 1,510 | 1,608 |
| 70 | 929 | 1,290 | 1,541 | 1,764 | 1,879 |

1) The latent gain value is zero if the Table 1 grains value is zero.
2) Interpolate or extrapolate for Grains.
3) Negative Table 1 grains values produce negative latent loads.

**Leakage Correction (LCF) for Latent Gain**

| Leakage | Any R-Value |
|---|---|
| 0.06 / 0.06 | 0.28 |
| 0.09 / 0.15 | 0.63 |
| 0.12 / 0.24 | 1.00 |
| 0.24 / 0.47 | 2.17 |
| 0.35 / 0.70 | 3.58 |

**Surface Area Factors**

| Leakage | Ks | Kr |
|---|---|---|
| 0.06 / 0.06 | 0.694 | 0.306 |
| 0.09 / 0.15 | 0.660 | 0.340 |
| 0.12 / 0.24 | 0.643 | 0.357 |
| 0.24 / 0.47 | 0.624 | 0.376 |
| 0.35 / 0.70 | 0.606 | 0.394 |

**Default Duct Wall Surface Area (SqFt)**

| | Floor Area Look-Up Value | | | | | | | | |
|---|---|---|---|---|---|---|---|---|---|
| 1,000 SqFt | | 1,500 SqFt | | 2,000 SqFt | | 2,500 SqFt | | 3,000 SqFt | |
| Supply | Return | Supply | Return | Supply | Return | Supply | Return | Supply | Return |
| 202 | 91 | 302 | 132 | 388 | 168 | 471 | 200 | 527 | 227 |

See Sections 6-8 and 23-6 for instruction for determining the floor area look-up value.

**Surface Area Adjustment Factor (SAA) for Heat Loss or Sensible Gain**

SAA = Ks x (Actual supply area / Default supply area) + Kr x (Actual return area / Default return area)
See Table 7A-AE, 7B-AE, 7C-AE or 7D-AE for an example application of this equation.

**Surface Area Adjustment (LGA) for Latent Gain**

LGA = Actual return-side area / Default return-side area
See Table 7A-AE, 7B-AE, 7C-AE or 7D-AE for an example application of this equation.

**Procedure for Heat Loss and Sensible Gain Factor Adjustment**
Step 1: Select the base case heat loss factor or base case sensible gain factor.
Step 2: Apply R-value correction to default value.
Step 3: Apply leakage correction to Step 2 value.
Step 4: Apply the surface area adjustment to the Step 3 value.

**Procedure for Latent Gain Adjustment**
Step 1: Select the base case latent gain factor.
Step 2: Apply leakage correction to Step 1 value.
Step 3: Apply surface area adjustment to the Step 2 value.

**Notes**
1) This table provides load factors for systems that feature multiple returns (approximately 400 Cfm per return).
2) Multiple returns improve comfort and room air motion, stabilize room pressures and blower Cfm (as interior doors open and close) and reduce the noise level in the conditioned space.
3) The load factors in this table are compatible with duct systems that are designed according to Manual J, Manual S and Manual D procedures.
4) Duct systems designed by other procedures may not provide adequate air distribution (surface area adjustment does produce an acceptable duct load estimate for such systems).
5) ACCA recommends sealing duct systems that have leakage rates greater than the 0.12 / 0.24 scenario. Use the data for leakier scenarios to evaluate the benefit of the sealing work.

*Figure 17* Manual J duct heat gain/loss factors.

*Figure 18* shows some default scenarios and load values for both appliances and people. A separate *Manual J* table provides more specific information for a variety of appliances. Note that the table uses 230 Btuh for the sensible load from people, plus 200 Btuh to represent the latent load (evaporation of moisture from the skin). Of course, if people are exercising or working hard, their heat output rises sharply. In a room-by-room load estimate, the people are apportioned to the rooms they would normally occupy during the peak load hours, generally the living room, family room, or dining room/kitchen.

Stoves, water heaters, baths, showers, washers, dryers, and lights also add heat to a structure. A default value of 1,200 Btuh is shown in *Figure 18* to cover the sensible load from a refrigerator and range hood. However, there are very few homes today that have such a low appliance-generated internal load. The number of internal loads continues to grow in the U.S. household, although the energy efficiency of many appliances continues to improve. Better energy efficiency for an appliance typically means less heat added to the space.

Part of the load created by people, appliances, and infiltration is latent heat from moisture being vaporized, then re-condensed as liquid condensate by the cooling system. The latent heat load must be accounted for in selecting the equipment. Note that dishwashers impose a much larger latent load on the cooling system than sensible load.

## Table 6A
### Default Scenarios and Values for Internal Loads

| Appliance, Equipment and Lighting Loads | | | |
|---|---|---|---|
| **Default Appliance Load** | **Sensible Btuh** | **Latent Btuh** | **Notes** |
| Refrigerator and range with vented hood. | 1,200 | ~ | 1,200 Btuh applied to the kitchen. |

**Scenario Options**

| | | | |
|---|---|---|---|
| 1) Refrigerator, range with vented hood, dish washer, clothes washer and vented clothes dryer, electronic equipment and lighting allowance. | 2,400 | ~ | 1,000 Btuh for the kitchen, 500 Btuh for the utility room, 900 Btuh allowance for a TV or computer and a few lighting fixtures. |
| 2) Two refrigerators or one refrigerator and one freezer, dish washer, range with vented hood, clothes washer and vented clothes dryer, electronic equipment and lighting allowance. | 3,400 | ~ | 2,000 Btuh for the room or rooms equipped with a refrigerator, 500 Btuh for the utility room, 900 Btuh allowance for a TV or computer and a few lighting fixtures. |

**Adjustment Options**

| | | | |
|---|---|---|---|
| A) Cooking range not equipped with a hood that is vented to outdoors, or an unvented dishwasher operating during the late afternoon in mid summer, or simultaneous use of unvented range and dishwasher. | + 850 | + 600 | Light duty cooking, 25 percent of the available range capacity used for 15 to 20 minutes. One dishwasher cycle load spread over a two hour recovery period. |
| B) Water bed heater (400 Watts, 33 percent duty cycle). | + 450 | ~ | Apply to each bedroom equipped with a water bed . |
| C) Ceiling fan (75 Watts). | + 250 | ~ | Apply to the room where the fan is located. |
| D) Large family using TV's, stereos, computers and laundry room during the late afternoon in mid summer. | + 1,400 | ~ | This is an additional 410 Watt allowance for A/V or computer equipment. |
| E) Allowance for above average lighting load. | + 1,705 | ~ | Five 100 watt lights. |
| F) Unvented clothes dryer. | Clothes dryers must be vented for air quality and efficiency. | | |

**Assignment**

For room cooling load estimates, correlate the equipment loads with the rooms that contain the equipment (existing construction) or which will logically contain the equipment (working from drawings).

### Internal Loads for Full Time Occupants, Guests and Plants

| Item | Sensible Btuh | Latent Btuh | Notes |
|---|---|---|---|
| Full Time Occupant | 230 | 200 | Occupancy = Number Bedrooms + 1 |
| Guest | 230 | 180 | Number of guests specified by owner |
| Each Small Plant | ~ | 10 | Less than 12 Inches high |
| Each Medium Plant | ~ | 20 | 12 inches to 24 inches high |
| Each Large Plant | ~ | 30 | More than 24 inches high |

*Figure 18* Default scenarios and load values for internal loads.

## Additional Resources

*Manual J, Residential Load Calculation.* Washington, DC: Air Conditioning Contractors of America (ACCA).

*Manual N, Commercial Load Calculation.* Washington, DC: Air Conditioning Contractors of America (ACCA).

## 2.0.0 Section Review

1. What effect does radiant energy have on materials?

    a. It warms the air.
    b. It heats glass.
    c. It passes through opaque objects.
    d. It warms dark objects more than light ones.

2. When calculating the heating load of basement walls, how are portions of the wall above grade and portions below grade treated?

    a. They are both treated as insulated walls.
    b. They are each treated differently.
    c. They are both treated as above grade.
    d. They are both treated as below grade.

## SECTION THREE

### 3.0.0 EQUIPMENT SELECTION AND SUPPORT SYSTEMS

### Objective

Explain how to select equipment based on the load estimate and describe common support systems to be considered.

    a. Explain how to select cooling equipment.
    b. Explain how to select heating equipment.
    c. Explain how to select heat pumps.
    d. Describe common support systems to be considered.

### Performance Tasks

    2.  Use manufacturer's product data to select the appropriate heating and cooling equipment based on a load estimate and airflow requirements.

The following four major items of information are necessary to select the equipment for a particular application:

- The type of system
- The sensible heating and/or cooling load
- The latent heat load (for cooling only; not a factor in heating equipment selection)
- The cooling and/or heating airflow needed

Detailed load estimates are performed to identify the equipment that comes as close as possible to matching the true load. In general, the selected equipment should range from no more than 5 percent undersized to 15 percent oversized. A few designers do consider 20 percent oversized to be acceptable.

Undersized equipment will not be able to handle peak loads and will take longer to return the space to the comfort level following temperature setbacks/setups. Oversized equipment will have higher first costs and operating costs, and will cycle more often. Oversized cooling equipment will pull down the load more quickly, which is a slight benefit when owners save energy through temperature setup or setback. However, it may not run long enough to sufficiently reduce the humidity, especially on design or near-design days. Thus, the comfort level will not be satisfactory.

Oversized heating equipment will not be efficient and will not provide good comfort. Oversizing an induced-draft, non-condensing furnace may allow condensation to form in the vent or heat exchanger, which can cause corrosion and failure. In summary, HVAC equipment should be selected carefully with great attention to the estimated loads as well as practical matters of the application.

Residential cooling and heat pump equipment is generally sized in half-ton increments in smaller sizes and 1-ton increments in larger sizes. For example, most manufacturers provide sizes of $1\frac{1}{2}$, 2, $2\frac{1}{2}$, 3, $3\frac{1}{2}$, 4, and 5 tons. With this selection, first determine the heat gain, and then select the size that best matches it. If a 32,000 Btuh heat gain were calculated, a 3-ton (36,000 Btuh) unit would be an appropriate match. Furnace size increments vary by 20,000 Btuh or more in output. For example, a manufacturer might offer 40,000, 60,000, 80,000, and 100,000 Btuh sizes. After determining the heat loss, select an output size that matches or is slightly higher than the heat loss.

The airflow requirement must be calculated separately for cooling and heating. In order to make a preliminary selection of equipment, it is necessary to approximate the heating and/or cooling air volume. However, the final determination on fan size and speed is based on the duct design process, which is discussed later in this module. Cooling volume is generally higher than the heating volume. One method of estimating air volume uses the following formula:

*Cooling*:

$$\text{Air volume (cfm)} = \frac{\text{sensible load (Btuh)}}{1.08 \times (T_1 - T_2)}$$

*Heating*:

$$\text{Air volume (cfm)} = \frac{\text{sensible load (Btuh)}}{1.08 \times (T_2 - T_1)}$$

*Where*:

    $T_1$ = outdoor temperature
    $T_2$ = indoor temperature

### 3.1.0 Cooling Equipment Selection

In many parts of the country, split systems are used to satisfy residential cooling requirements. With a split system, only the small refrigerant lines penetrate the wall. Openings for the refrigerant lines can be easily drilled in most buildings.

## The Competitive Advantage

Installing oversized equipment costs more. The higher cost of larger-capacity equipment is just the first of many added costs. Add to that the cost of larger ducts to handle the increased airflow, a more expensive electrical service to handle higher current, and costly larger-diameter copper refrigerant lines. It all adds up. The contractor that sizes equipment properly often ends up proposing smaller equipment than a competitor who often relies on outdated or rule-of-thumb load-estimating procedures.

*Figure 19* Condensing units.

The condensing unit is placed outside the building, as close to the evaporator as possible, but generally not more than 50' (15 m) apart. Consult the manufacturer for any equipment modifications that are needed if the two parts of a split system require more than 50' of refrigerant piping.

The condensing unit (*Figure 19*) contains the controls, compressor, condensing coil, and fan. The cost of the condensing unit is usually a function of its efficiency. A high-SEER unit may contain a high-efficiency or two-speed compressor and a larger condensing coil. A deluxe model, which is usually associated with a high SEER value as well, may also contain high-pressure and low-pressure safety switches, sound-deadening shields, start-assist gear, a crankcase heater, a filter-drier, and other options. The high-efficiency components add to the initial cost, but will pay back the extra expense over time in reduced energy consumption. The other extra cost devices provide comfort, convenience, and equipment safety.

The indoor unit of a split cooling system must provide a cooling coil, blower, and metering device. When the blower is part of a forced-air furnace, an indoor coil is added to the furnace (*Figure 20*).

Selection of the outdoor unit is made from manufacturer's product data based on capacity in Btuh. Cooling equipment must be selected to match both the sensible and latent heat loads. The indoor unit is selected from a physically compatible group based on its capacity, ability to meet the air volume requirement, and efficiency rating. The components of a split system must be closely matched. They are best made by the same company and designed to work together, but it is not a requirement. When fixed-restrictor

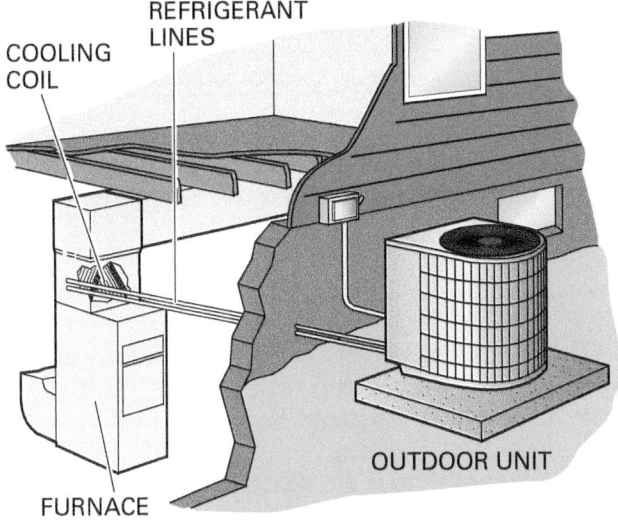

*Figure 20* Furnace with cooling coil.

metering devices are used, the correct piston may be shipped with the condensing unit. The installer then installs the correct piston in the indoor section.

In areas with low humidity and high dry-bulb temperatures, such as in the southwest, it is not uncommon to see an indoor coil with a higher rated capacity than the outdoor unit. This approach, which is sometimes called mix-matching, trades less humidity control for increased efficiency and slightly more cooling capacity. Typically, the indoor unit capacity will be about a half-ton (6,000 Btuh) greater than that of the condensing unit. In desert-like conditions, the evaporator may have as much as one full ton greater capacity than the condensing unit.

Packaged cooling units are popular in many parts of the United States, especially in the south and southwest. Packaged units contain many of the same features designed to enhance equipment

# Mix-Matching

Mix-matching equipment by using an evaporator with a higher rated capacity than the condensing unit can have some positive benefits. The total cooling capacity of the system is increased, due to the additional surface area of the indoor coil. However, the extra load from the larger coil tends to raise the saturated suction temperature and pressure. Since the coil is slightly warmer, less dehumidification occurs.

Although mix-matching can be done using a coil with a fixed-orifice metering device, the positive results are limited since it cannot react to changing conditions. Mix-matched systems perform at their best with a TXV, just as other systems do. Of course, mix-matching is never done using an evaporator with less capacity than that of the condensing unit.

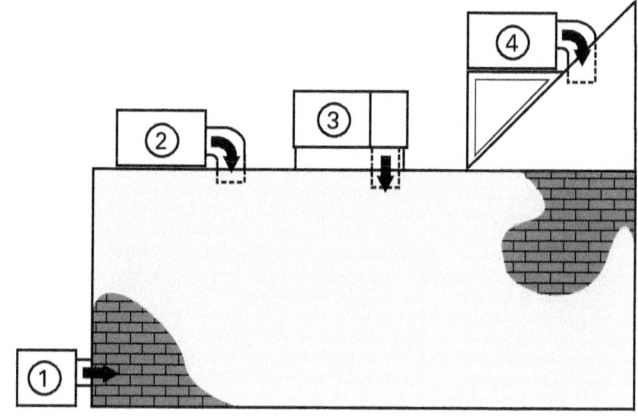

(1) SLAB ON GROUND  (3) PLENUM AND CURB
(2) FLAT ROOF  (4) ROOF JACK

*Figure 22* Packaged unit installation locations.

## 3.2.0 Heating Equipment Selection

There are many ways to deliver heat including gas, oil, and electric furnaces, along with heat pumps. The selection of a heating fuel depends on the local climate as well as the availability and cost of the different forms of energy. Factors that influence heating fuel selection include the following:

- Local traditions
- Type of construction
- Fuel availability
- Fuel costs

operation and safety as are found in split-system products. While straight cooling packaged units are available, most packaged units (*Figure 21*) can heat as well as cool. Heating options available include electric resistance heat, natural gas or propane, or a heat pump combined with any of these options.

Packaged units are installed where indoor space is limited, or the structure is built on a slab or over a crawlspace. Packaged units can be mounted on the ground with the ducts penetrating a sidewall. They are also commonly installed on the roof in commercial as well as residential applications (*Figure 22*).

Local traditions often have a bearing on fuel and equipment selection. In other words, certain practices have gone on for so long, they are accepted as the preferred way to do it. For example, for some years, packaged rooftop units were installed on top of homes in the southwest. Part of the reason they were installed on roofs was that evaporative coolers were traditionally installed there (*Figure 23*), so duct connections were placed there. Today in the southwest, split-systems are now widely installed in new homes. Manufactured homes with a crawlspace often use packaged units that are installed on a slab and ducted through the foundation wall. In homes with basements, upflow forced-air furnaces are widely used.

The availability of a fuel also helps determine the selection. In areas served by gas pipelines, natural gas is generally preferred. Fuel oil and propane are widely used in rural areas, where natural gas may not be readily available. The cost of each form of heat (cost per Btu) is also a factor.

*Figure 21* Packaged gas heat/electric cooling unit.

*Figure 23* Evaporative cooler on a rooftop.

Where fuel oil is costly, a heat pump is a viable alternative. Better yet is a dual-fuel heat pump that takes advantage of the economics of fuel oil and electricity. At higher outdoor temperatures, it is more cost-effective to heat with the heat pump. At lower outdoor temperatures, heating with fuel oil may be more economical and comfortable. Natural gas tends to be the preferred heating fuel overall, and can be used in warm air or hydronic systems (*Figure 24*). Geothermal heat pumps are

somewhat expensive to install and use electricity. However, their high efficiency results in low operating costs, which help offset the high installation costs.

Furnace manufacturers offer standard furnaces with cabinets sized to handle most split-system cooling and heating applications. They also offer high-air models. High-air models are meant to be used in areas where the heating load is moderate to light, while the cooling load is heavy. Areas such as the desert southwest are examples. A home in that area may have a heat loss as low as 40,000 Btuh but have a heat gain of 60,000 Btuh. High-air furnaces are characterized by wider-than-normal cabinets that can accommodate larger cooling coils. They also have a larger blower and a larger blower motor. Typically, a standard furnace of that heating capacity (40,000 Btuh) would be able to provide no more than 2 to 2½ tons of cooling airflow.

## 3.3.0 Heat Pump Selection

A heat pump is an excellent choice when electric heat is the only option available, or where the cost and availability of other fuels is questionable. For design heating days, a heat pump can be supple-

*Figure 24* Gas-fired hydronic heating equipment with condensing boilers.

mented with auxiliary electric heaters, or it can be a dual-fuel system that takes advantage of the economic benefits of both heat sources.

For standard systems, the heat loss of the structure drives the furnace selection process. The cooling equipment is selected based on the heat gain of the structure. With a heat pump, only a single primary piece of equipment is being selected. If the heat pump is sized to the heating load in a cold northern application, there will likely be too much cooling capacity. This will result in inefficiency and poor dehumidification. To prevent this problem, the heat pump is sized to the cooling load and the additional heating capacity needed is made up with auxiliary heat. Of course, a dual-fuel system is also an option. When a dual-fuel system is used, the furnace takes over all heating responsibilities below the balance point, since the heat pump alone cannot keep up. The two units cannot generally run together.

Another basic rule is that the heat pump should be sized to provide the lowest possible balance point in the heating mode without grossly over-sizing it for the cooling load. In some areas, the utility or the building codes specify the minimum balance point. This approach limits the amount of electric resistance heat that can be used. In some cases, it can force the use of a larger heat pump. The staging of resistance heaters based on outdoor temperature to improve energy efficiency and reduce power surges may also be specified by the local utility or building codes. Staging electric heaters when multiple heating elements are installed is always the preferred method.

When electric resistance heaters are used, the combined capacity of the heat pump and the supplemental electric heaters should not exceed 115 percent of the calculated heating load. Electric resistance heaters are also used to provide emergency heat in the event of a heat pump compressor failure. In such cases, the heaters should be sized to provide 80 percent of the heating load. The auxiliary electric heaters can then provide most or all the emergency heat, as long as the outdoor temperatures do not fall below design. Emergency heat requirements may be specified by the local utility.

ACCA has specified the following standard guidelines for selecting heating equipment. These guidelines are based on winter design conditions.

- The capacity of a gas or oil furnace should not be less than the calculated load. It should not exceed the heating design load by more than the next larger size available from the manufacturer.

- The capacity of electric resistance heating equipment should also not be less than the calculated load. It should not exceed the calculated load by more than 10 percent. Note that this pertains to electric resistance heating equipment that is used as the primary source of heat.

### 3.4.0 Support Systems

This section discusses various support systems associated with heating and cooling systems that must be considered during the design and selection process.

### 3.4.1 Refrigerant Piping

A comfort air conditioning system has three primary refrigerant-piping sections:

- The suction line conveys low-temperature, low-pressure, superheated vapor refrigerant from the evaporator outlet to the compressor inlet. It is referred to more accurately as the vapor line in heat pump systems.
- The hot-gas (discharge) line conveys hot, high-pressure vapor refrigerant from the compressor to the condenser.
- The liquid line conveys high-pressure liquid refrigerant from the condenser to the expansion device.

In split-system applications, only the suction and liquid lines, which connect the indoor and outdoor units, are installed at the site. The hot gas line is inside the condensing unit at the factory, and is very short. The proper suction and liquid line sizes are specified in the manufacturer's product data. Pipe sizes can vary for different refrigerants. Do not deviate from the recommended sizes. Undersized lines can cause noisy operation, evaporator starvation, and excessive pressure drop. Oversized lines require a larger refrigerant charge and possible oil return problems, at a minimum.

The diameter and length of the refrigerant lines affect the pressure drop. The smaller the diameter and the longer the line, the greater the pressure drop. In their product data, manufacturers will specify larger suction line diameters if the standard diameter line exceeds a certain length.

Oil return is also important. The compressor discharges a small amount of oil with the refrigerant. During cold startups, larger amounts of oil may be discharged. This oil must be returned to the compressor for the compressor to have proper lubrication.

This is not a problem in the liquid line because oil mixes easily with liquid refrigerant. In the suction (vapor) line, however, it can be a major problem. If this line is oversized, there will not be sufficient refrigerant velocity to move the oil up vertical piping sections (risers), and the oil will settle in low spots. In general, lines should be pitched in the direction of flow. This will help maintain oil flow and avoid backward flow during shutdown.

The location of the indoor coil relative to the compressor in the outdoor unit must be considered when running refrigerant lines (*Figure 25*). If the indoor coil is located above the outdoor unit, the suction line should loop up to the height of the indoor coil (A). This helps prevent liquid refrigerant from migrating to the compressor in the outdoor unit during the compressor off cycle. When the indoor coil and the outdoor unit are at the same level, the suction line should pitch toward the compressor in the outdoor unit (B) with no sags or low spots in any straight run. When the indoor unit is below the outdoor unit, oil must be returned to the compressor in the outdoor unit via a vertical riser in the suction line (C). An oil trap should be installed at the start of the riser to collect oil for feedback to the compressor. Note that these traps are best kept shallow – a preformed P-trap is generally too deep and will hold too much oil without any operational benefit. The horizontal run to the outdoor unit should be pitched slightly toward the outdoor unit. The suction line should always be adequately insulated.

An undersized liquid line may cause enough pressure drop for flash gas to form in the line. Essentially, an undersized liquid line becomes a big capillary tube—a metering device. An oversized liquid line will create the need for extra refrigerant charge. The liquid line should be insulated wherever it is exposed to a great deal of excessive heat, such as in an attic. Heat causes refrigerant to lose subcooling or vaporize, resulting in a loss of capacity. For vertical runs of more than 25' (8 m), refer to the manufacturer's instructions.

### 3.4.2 Condensate Piping

Condensate forms on the evaporator coil of an indoor unit and will accumulate in the condensate pan beneath the coil. The condensate must be removed to a proper drain, or it will overflow and cause damage to the unit or the building. A pipe connected to the pan drains the condensate to an indoor drain or to the outside of the building. Condensate drainage is especially important where units are installed in attics or above drop ceilings.

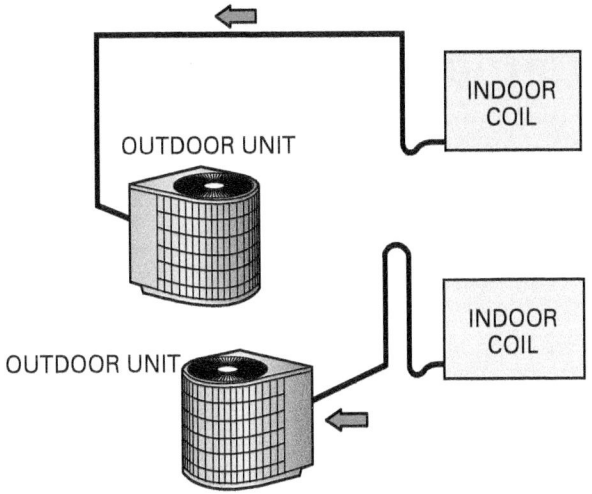

**(A) VAPOR LINE WITH INDOOR COIL ABOVE THE OUTDOOR UNIT**

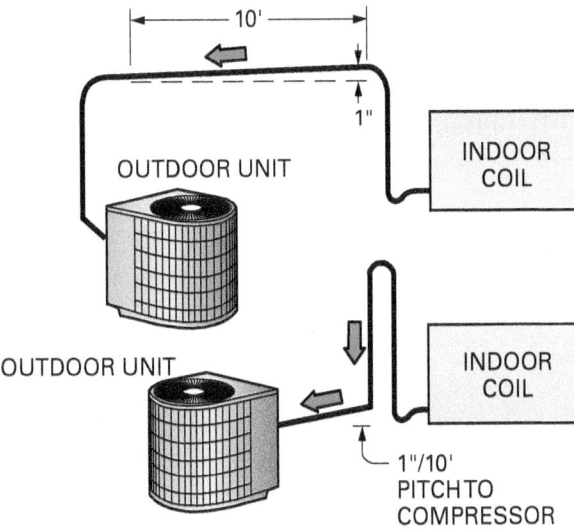

**(B) VAPOR LINE WITH INDOOR COIL AT THE SAME LEVEL AS THE OUTDOOR UNIT**

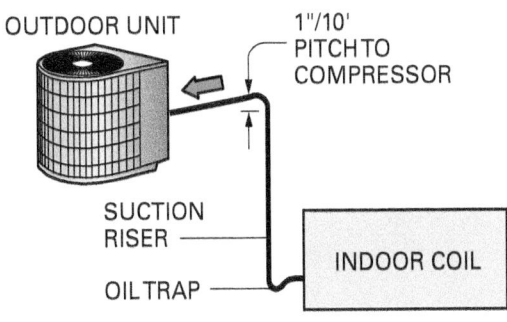

**(C) VAPOR LINE WITH INDOOR COIL BELOW OUTDOOR UNIT**

*Figure 25* Suction (vapor) line design.

# Refrigerant Lines

Field-fabricated refrigerant lines used with split systems should always be constructed of Type ACR copper tubing. Never use copper pipe or tubing meant for use in general plumbing applications. Equipment manufacturers offer soft-copper line sets of different lengths that are intended for use with their equipment. These line sets are evacuated and contain a holding charge of dry nitrogen. Some manufacturers offer line sets with special fittings on the ends of the refrigerant lines that match similar fittings on their equipment.

In such cases, most local codes require the placement of a separate secondary condensate pan (*Figure 26*) under the entire air handler. This pan catches and removes the condensate in the event the evaporator coil drain pan is plugged or damaged. The pan is usually fitted with its own independent drain line, which exits the building on its own. Homeowners are then informed that drainage seen from this particular line indicates there is a problem with the main drain.

Another option is to install a drain pan switch that shuts down the cooling equipment if water is sensed in the secondary pan (*Figure 27*). Using both approaches together is also not out of the question.

The condensate drain typically requires a trap to prevent water from being held in the drain pan by the blower. Traps are especially important when the drain line is under a negative pressure. This happens when the coil and pan are on the return-air (negative pressure) side of the blower. Other system types, such as a ductless indoor unit, may not require a drain trap. Always check the manufacturer's literature for guidance. Drain lines, regardless of the presence of a trap, must be pitched toward the drain at a minimum of a quarter inch per foot.

Condensate drainage is also required for condensing furnaces. A condensate pipe is usually run to a floor drain from the drain connection on the condensing heat exchanger. Where condensate cannot be disposed of by gravity flow, a condensate pump is used (*Figure 28*). Most condensate pumps contain an overflow float switch

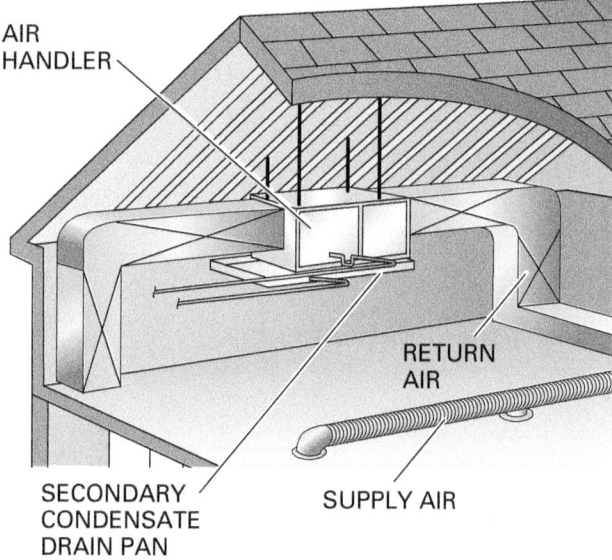

*Figure 26* Secondary drain-pan installation.

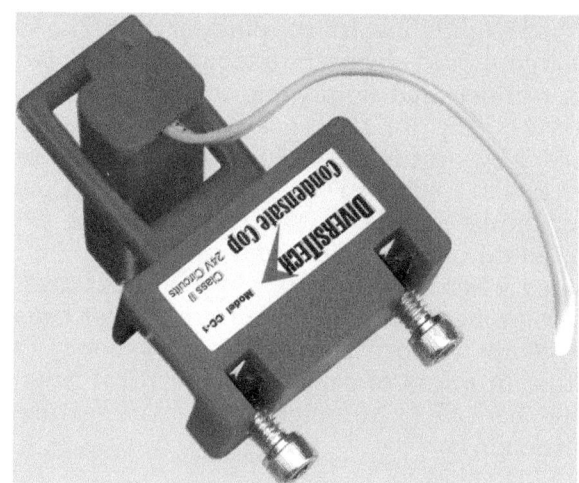

*Figure 27* Float switch for a condensate drain pan.

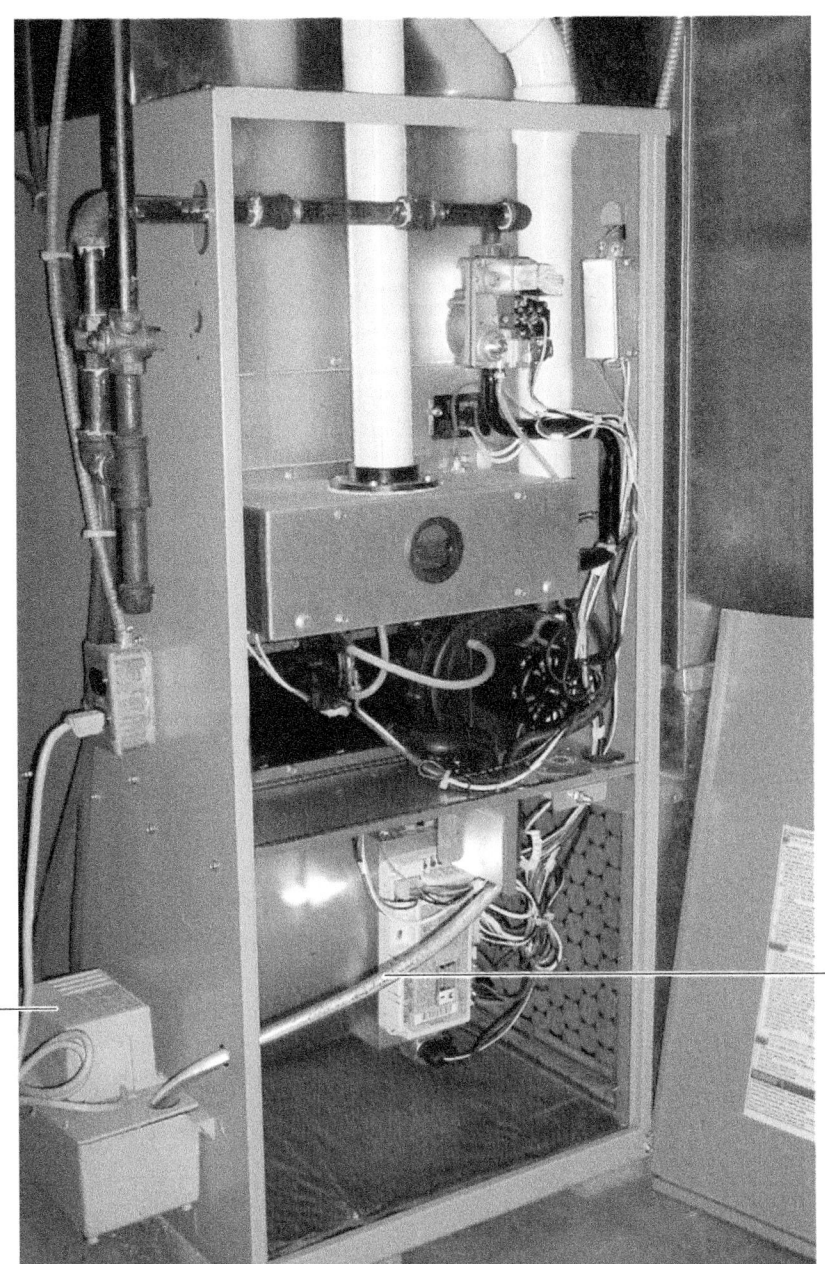

FURNACE
CONDENSATE
PUMP

SECONDARY HEAT
EXCHANGER
CONDENSATE
DRAIN

*Figure 28* Condensate pump installed with a condensing furnace.

that will shut off the heating and cooling equipment if the condensate pump fails. This feature prevents damaging water overflow. The overflow switch, whether it is part of the pump or attached to a secondary drain pan, is wired in series in the system control circuit. The following lists the circuits in which the overflow switch is typically connected:

- *Cooling only air handler* – In series in the Y circuit.
- *Condensing furnace heating only* – In series in the W circuit.

- *Condensing furnace heat/cool operation* – In series in the R circuit. However, note that breaking the R circuit before the thermostat will also result in the system blower being shut down. This can have a significant effect on a drainage situation. The blower generally creates a negative pressure inside the air handling unit cabinet. If the cabinet fills with water, which eventually overflows into the secondary pan, a great deal of water may suddenly leave the air handling unit when the fan shuts down. To avoid this, the switch can be wired to operate a DPDT relay, and the two separate sets of NC contacts can break the Y and W circuits when the relay is energized.

### 3.4.3 Electrical Service

Electrical service is not usually an issue in new construction because the electrical service should be designed to handle the load created by the planned heating and cooling equipment. It is likely to be an issue in cases where air conditioning is being added and/or replaced.

Some older homes may only be equipped with a 100-amp electrical service. Depending on the number of appliances and other electrical loads, that capacity can be quickly used. This leaves no room for additional loads such as air conditioners. This is even more of a concern when electric resistance heaters are being used as the primary heat source or as a supplemental heat source for heat pumps. In these cases, a heavier electrical service might have to be installed. It is easy to estimate the existing electrical load by totaling the current draw in amps (A) for all the appliances and lighting in the building. The following are some examples:

- Electric dryer: 25A
- Washer: 5A to 10A
- Electric oven/range: 45A

- Electric oven/range (self-cleaning): 60A
- Microwave oven: 10A
- Dishwasher: 10A
- Small appliances (combined): 10A to 20A
- Electric hot water heater: 15A to 40A
- Garbage disposal: 5A to 10A
- Lighting (combined): 10A to 15A

A new comfort-cooling system can add anywhere from 15 to 60 amps, depending on the capacity and type of system. A 5-ton heat pump with electric heaters, for example, can add more than 50 amps to the load.

In some homes, that can be enough to exceed the capacity of the existing electrical service. It may seem reasonable to assume that all the loads will not be operating at once, and that there may be a difference between rated load and actual load. Always consult a qualified master electrician before reaching a decision. Although HVAC personnel should understand these issues, they should not make the final decision regarding the size of the electrical service.

## Electrical Service

*GOING GREEN*

When replacing an older HVAC unit with a newer unit, the electrical service is often more than adequate because newer equipment is much more energy-efficient than older equipment, especially equipment that is more than 10 years old. This means it probably draws significantly less current than an older unit of the same capacity. It may, therefore, be possible to increase capacity with a newer unit and still draw no more than, or even less current than, the lower capacity unit it replaces.

## Additional Resources

*Manual S, Residential Equipment Selection.* Washington, DC: Air Conditioning Contractors of America (ACCA).

## 3.0.0 Section Review

1. In a desert setting, the indoor coil of a split-system cooling unit is often oversized by _____.

   a. 1,000 Btuh
   b. 6,000 Btuh
   c. 12,000 Btuh
   d. 18,000 Btuh

2. Which of the following is *not* typically a factor that affects the selection of a heating fuel?

   a. Local customs
   b. Fuel availability
   c. Structure heat loss
   d. Fuel cost

3. With a standard dual-fuel heat pump, the heat pump should provide the heat _____.

   a. when the outdoor temperature is below the balance point
   b. when the outdoor temperature is above the balance point
   c. during all daylight hours, but not at night
   d. if the thermostat is operating in its setback mode

4. When installing a condensate pump on an air handler used only for cooling, you should connect the overflow switch in series with the _____.

   a. Y control circuit
   b. condensate pump motor
   c. W control circuit
   d. condensate flow switch

### 4.0.0 AIR DISTRIBUTION SYSTEM DESIGN

## Objective

Explain how to design air distribution systems.
a. Identify basic duct design considerations.
b. Identify various duct system layouts.
c. Identify various duct system components.
d. Describe how to design and size duct systems.
e. Identify system design factors unique to commercial buildings.

## Performance Tasks

3. Determine the number, location, and sizes of supply outlets and return inlets needed in a building.
4. Use standard duct sizing tables, duct design calculator, or software application to size the trunk and branch ducts for a selected low-pressure air distribution system.
5. Calculate the total system friction loss (external static pressure) for a selected air distribution system.

## Trade Terms

**Brake horsepower:** The total power needed to drive a fan to deliver the required volume of air through a duct system. It is greater than the expected power needed to deliver the air because it includes losses due to turbulence, inefficiencies in the fan, and bearing losses.

**Equal-friction method:** A method of sizing air distribution systems by designing for a consistent amount of pressure loss per unit length of duct (usually 100 feet).

**Static-regain method:** A method of sizing ducts such that the regain in static pressure due to a decreasing velocity between two points fully or partially makes up for the frictional resistance between the two points.

**Velocity-reduction method:** A method of sizing ducts such that the desired velocities occur in specific duct lengths.

Knowing the basics of air distribution system design will help when installing an HVAC system. Knowledge of duct design also helps to recognize and solve duct system problems, such as noise, vibration, or incorrect air distribution. The first part of this section reviews some basic air system operating principles. The second part covers factors relating to air system design. The desired design goals for all air distribution systems include the following:

- Supply the right quantity of air to each conditioned space.
- Supply the air in each space so that stratification is minimized and air motion is adequate but not drafty.
- Condition the air to maintain the proper comfort zones or the necessary conditions for a commercial or manufacturing process.
- Provide for the return of air from all conditioned areas to the air handler.
- Operate efficiently without excessive power consumption or noise.
- Operate with minimum maintenance.

### 4.1.0 Duct Design Basics

The resistance to airflow caused by the components of a duct system is overcome by the system fan. The fan supplies the energy needed to overcome the duct resistance and maintain the necessary airflow. Air can be moved by creating areas of differing pressures, causing it to move from the higher pressure towards the lower pressure. All fans produce both high- and low-pressure areas. The air inlet to a fan is below atmospheric pressure, while the outlet of the fan is above atmospheric pressure.

As shown in *Figure 29*, normal atmospheric pressure exists initially at the return grille and supply diffuser. With the blower running, the pressure at the face of the return air grille is slightly lower than atmospheric pressure; therefore, air moves into the duct. The pressure decreases to its lowest point at the input to the blower. Through the action of the fan, the air pressure is increased to its highest level at the blower discharge. From there, the air resumes its tendency to flow from the area of higher pressure at the fan discharge, to the area of lower pressure.

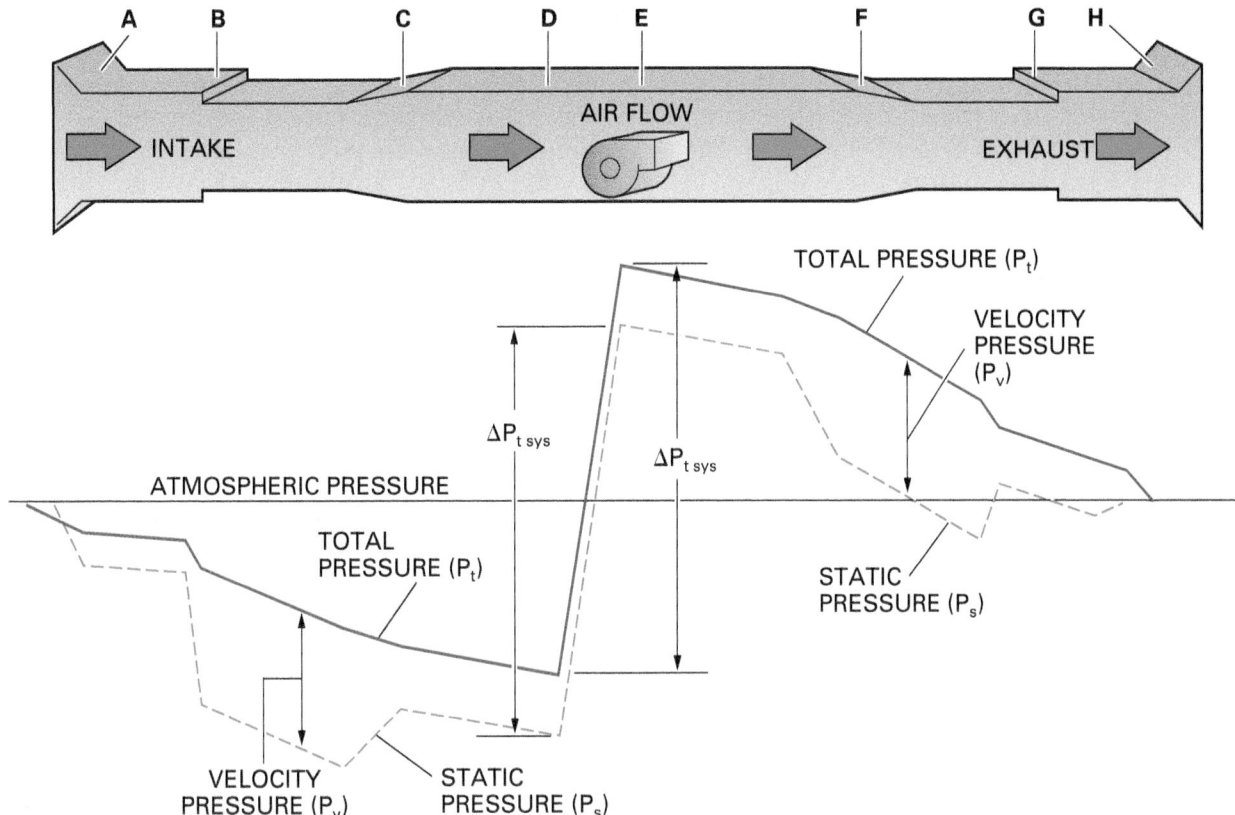

*Figure 29* Pressure relationships in a duct.

The amount of pressure difference needed to move air through a duct system depends on the velocity and volume of air, the cross-sectional area of the duct, and the length of the duct. Velocity reflects how fast the air is moving, and is usually measured in feet per minute (fpm) or meters per second (mps). Volume flow rate is the amount of air, in cubic feet (ft³) or cubic meters (m³), that flows past a point in one minute. Cubic feet per minute is usually abbreviated as cfm, while cubic meters per second is usually abbreviated as cms. The air volume, in cfm, can be calculated by multiplying the air velocity, in fpm, by the area (in square feet) of the duct it is moving through, as follows:

Volume flow rate (cfm) =
duct area (ft²) × velocity (fmp)

This formula can be rearranged as follows:

$$\text{Velocity (fpm)} = \frac{\text{Volume flow rate (cfm)}}{\text{Area (ft}^2)}$$

$$\text{Area (ft}^2) = \frac{\text{Volume flow rate (cfm)}}{\text{Velocity (fpm)}}$$

Similarly, the volume using metric units is calculated as follows:

Volume flow rate (cms)
= area (square meters) × velocity (mps)

The flow rate can, of course, be measured directly with various test instruments.

### 4.1.1 Pressure Relationships Within a Duct

Three pressures exist in an operating duct system: total pressure, static pressure, and velocity pressure. Total pressure determines how much energy must be supplied to the system by the fan to maintain airflow. Total pressure always decreases in the direction of airflow. For any cross-section of the duct, the total pressure is the sum of the static pressure and the velocity pressure. Static pressure and velocity pressure are present when the duct system is in operation. Static pressure is a stationary air pressure that is exerted uniformly in all directions within the duct. It is the same kind of pressure that is applied equally on the internal walls of a balloon or inflated tire.

Velocity pressure is the pressure caused by the velocity and weight of the moving air. It acts in the direction of airflow only. Mathematically, it is the difference between the total pressure and the static pressure. In other words, static and velocity pressure together comprise the total pressure. Static and velocity pressures can either increase or decrease in the direction of airflow.

The magnitudes of the total, static, and velocity pressures can be calculated as follows:

$$P_t = P_s + P_v$$
$$P_s = P_t - P_v$$
$$P_v = P_t - P_s$$

*Where:*

$P_t$ = total pressure
$P_s$ = static pressure
$P_v$ = velocity pressure

The levels of the static, velocity, and total pressures in a duct system are relatively small. The unit of measurement used, therefore, needs to be small as well. Inches of water column (in. w.c.) or inches of water gauge (in. w.g.) are the units of measure commonly used to express pressure in a duct system. The terms *in. w.c.* and *in. w.g.* are interchangeable. Inches of water column is the height, in inches, to which the pressure will lift a column of water. There are several metric units used for low pressures. Centimeters of water (cm w.c.) is a popular unit for this purpose.

The instrument used to measure pressures in a duct system is the manometer. Manometers can be calibrated or manufactured in a variety of units.

## 4.1.2 Friction Losses

The inside surface of the duct offers resistance to the flow of air. The velocity of airflow within a duct is not uniform. It varies from zero at the duct walls to a maximum at the center of the duct. This variation in velocity is caused by the resistance encountered by the air molecules as they are dragged over the duct surfaces. The resistance to airflow or velocity in straight duct sections is called friction loss. Pressure drop in a straight duct is caused by surface friction. It varies with the air velocity, the duct size and length, and the interior surface roughness. The amount of friction loss in straight duct is normally found by using air friction charts. These charts are reviewed later in this section.

As the cross-sectional area of a straight duct section becomes smaller, as shown in duct sections BC and FG in *Figure 29*, an increase in the airflow velocity and velocity pressure occurs.

As a result, the static and total pressure lines drop more rapidly than in the larger cross-sectional area ducts. This drop occurs because the pressure losses increase as the square of the velocity increases. As the air moves within a constant-area straight duct section, the static pressure and total pressure losses increase at the same rate. This is because the velocity and velocity pressure ($P_v$) of the air flowing within the duct are constant. *Figure 30* is an example of a chart used to determine friction loss in flexible ducts.

## 4.1.3 Dynamic Losses

Dynamic losses occur when there are changes in the direction or velocity of the air, such as at transitions, elbows, and other fittings. Dynamic losses also occur at obstructions such as dampers. When duct cross-sectional areas transition to a smaller size, whether abruptly or gradually, some turbulence results. This is because of the sudden change in airflow velocity. Both the velocity and velocity pressure increase in the direction of airflow, while the absolute values for both the total and static pressures decrease. The result is that a greater loss in total pressure takes place than would occur in a steady flow through an equal length of straight duct with a uniform cross-section.

**CHART 3 – THERMAFLEX M-KC, S-LP-10, S-TL**
FLEXIBLE DUCT – STRAIGHT RUN
FRICTION LOSS PER 100 FT

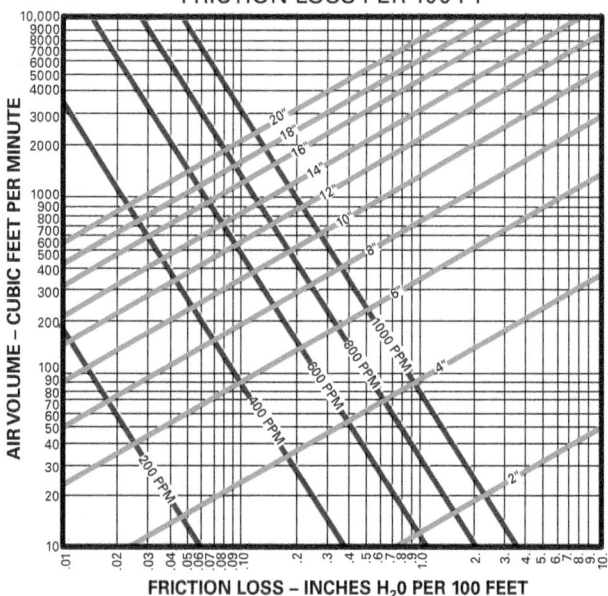

*Figure 30* Friction loss chart for flexible round duct.

The amount of loss in excess of the straight-duct friction is called the dynamic loss. In *Figure 29*, Points A, B, C, F, G, and H represent changes in the duct that will cause dynamic losses. Dynamic pressure losses can be expressed as a loss coefficient value or C-factor. A loss coefficient value is a known, dimensionless value assigned to each type of duct elbow, transition, or other fitting. To determine the pressure loss through a specific kind of duct fitting, the C-factor is multiplied by the velocity pressure in the fitting.

The dynamic pressure loss of a specific type of fitting can also be expressed by its equivalent length of straight duct value. This is usually a more convenient value with which to work. For example, a radius elbow has the same resistance as 25' (7.6 m) of straight duct of the same size. C-factors are more accurate values of pressure loss. The accuracy is typically more important when designing large and complex duct systems. The equivalent length of straight duct values is commonly used when designing residential or light commercial duct systems.

Both C-factors and equivalent length of straight duct values for the various kinds of fittings are found using published tables. These tables are available from the Sheet Metal and Air Conditioning Contractors' National Association (SMACNA), ASHRAE, and most duct component manufacturers. *Figure 31* shows examples of the loss coefficient data and equivalent length of straight duct data given in a table for a rectangular elbow. A deeper discussion of equivalent length is provided later in this module.

### 4.1.4 Static Regain

Abrupt or gradual increases in duct cross-sectional area cause a decrease in airflow velocity and velocity pressure. There is also a decrease in total pressure accompanied by an increase in static pressure. This increase in static pressure is called static regain. It is caused by the conversion of velocity pressure to static pressure.

### 4.1.5 Duct System External Static Pressure and Supply Fan Relationship

The total pressure that the system fan must supply is the sum of the friction losses in the supply duct system, return duct system, and all the components not included in the fan rating (*Figure 32*). These include all straight duct sections; the dynamic losses of each duct fitting or obstruction, such as registers and grilles; and the pressure loss

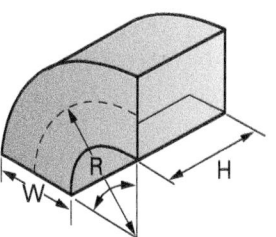

COEFFICIENTS FOR 90°:

| R/W | COEFFICIENTS C | | | | | | | | |
|-----|-----|-----|-----|-----|-----|-----|-----|-----|-----|
| | H/W | | | | | | | | |
| | 0.25 | 0.5 | 0.75 | 1.0 | 1.5 | 2.0 | 4.0 | 6.0 | 8.0 |
| 0.5 | 1.5 | 1.4 | 1.3 | 1.2 | 1.0 | 1.0 | 1.1 | 1.2 | 1.2 |
| 0.75 | 0.57 | 0.52 | 0.48 | 0.44 | 0.40 | 0.39 | 0.40 | 0.43 | 0.44 |
| 1.0 | 0.27 | 0.25 | 0.23 | 0.21 | 0.19 | 0.18 | 0.19 | 0.27 | 0.21 |
| 1.5 | 0.22 | 0.20 | 0.19 | 0.17 | 0.15 | 0.14 | 0.15 | 0.17 | 0.17 |
| 2.0 | 0.20 | 0.18 | 0.16 | 0.15 | 0.14 | 0.13 | 0.14 | 0.15 | 0.15 |

ELBOW, RECTANGLE, SMOOTH RADIUS
WITHOUT VANES FITTING LOSS (P$_t$) = C × V
USE THE VELOCITY PRESSURE V OF THE
UPSTREAM SECTION

**LOSS COEFFICIENT (C-FACTOR)**

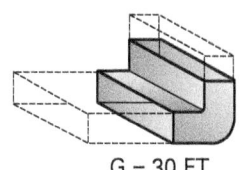

G = 30 FT
**EQUIVALENT LENGTH OF DUCT**

*Figure 31* Examples of loss coefficient and equivalent length of straight duct data.

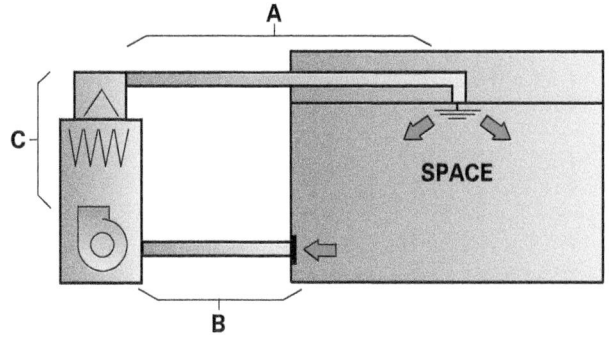

**TOTAL FRICTION LOSS = A + B + C**
A = SUPPLY SYSTEM LOSS
B = RETURN SYSTEM LOSS
C = COMPONENTS NOT INCLUDED IN FAN RATING

*Figure 32* System friction losses.

of each duct component, such as coils, filters, and dampers, in the system. The room itself is part of the air distribution system too, but doesn't represent significant resistance to flow. The velocity is very low, making the friction loss very low.

The pressure loss or drop for a given system component is available from the equipment manufacturer. The chart in *Figure 33* lists pressure drop values for a specific ceiling diffuser. The total pressure loss of the duct system components external to the fan assembly is called the *external static pressure* (*ESP*). The sizing of the system fan is based on the external pressure losses resulting from the system ductwork and its components. Resistances internal to the fan assembly (not contributing to the ESP) result from losses in the fan assembly accounted for by the manufacturer.

In *Figure 33*, you will see the following footnote: "Terminal velocity of 50 fpm." This means that the distance of throw is based on the air slowing to a velocity of 50 fpm. In other words, the measurement of the throw distance stops when the air leaving the diffuser has slowed to that speed.

The fan, or blower (*Figure 34*), provides the pressure difference that forces the air into the supply duct system, through the grilles and registers, and into the conditioned space. It also must overcome the pressure loss involved in the return of the air as it flows into the return air grilles and through the return ductwork system back to the fan. Brake horsepower is the total power required to drive the fan when delivering the required volume of air through a duct system. It is greater than the power to deliver the air alone because it includes losses due to turbulence and other inefficiencies in the fan, plus bearing friction.

The performance of all fans and blowers is governed by three rules commonly known as the *Fan Laws*. The volume of air movement, fan speed, external static pressure, and horsepower (hp) are all closely related. There are three fan laws:

- *Fan Law 1* – The volume of air delivered by a fa=n varies directly with the speed of the fan:

$$\text{New cfm} = \frac{\text{New rpm} \times \text{existing cfm}}{\text{Existing rpm}}$$

*Or:*

$$\text{New rpm} = \frac{\text{New cfm} \times \text{existing rpm}}{\text{Existing cfm}}$$

- *Fan Law 2* – The static pressure of a system varies directly with the square of the ratio of the fan speeds:

$$\text{New } P_s = \text{Existing } P_s \times \left(\frac{\text{New rpm}}{\text{Existing rpm}}\right)^2$$

- *Fan Law 3* – The horsepower varies directly with the cube of the ratio of the fan speeds:

$$\text{New hp} = \text{existing hp} \times \left(\frac{\text{new rpm}}{\text{existing rpm}}\right)^3$$

Fan curve charts or fan performance charts are normally used to find the relationships that exist for a set of system conditions involving $P_s$, fan rpm, fan bhp, and cfm. These charts are produced by equipment manufacturers for the specific model of equipment being used. Such charts are based on these fan laws.

| 24" SQUARE CEILING DIFFUSER | | | | | | | | | |
|---|---|---|---|---|---|---|---|---|---|
| FACE VELOCITY | | 300 | 400 | 500 | 600 | 700 | 800 | 900 | 1000 |
| PRESSURE LOSS | | 0.006 | 0.010 | 0.016 | 0.022 | 0.031 | 0.040 | 0.050 | 0.062 |
| 6 Ak 0.165 | cfm | 50 | 65 | 85 | 100 | 115 | 130 | 150 | 165 |
| | Throw | 3.5 | 4.5 | 5.5 | 6.5 | 8 | 9 | 10 | 11 |
| 8 Ak 0.280 | cfm | 85 | 110 | 140 | 170 | 195 | 225 | 250 | 280 |
| | Throw | 4.5 | 5.5 | 7 | 8.5 | 10 | 11 | 12 | 14 |
| 10 Ak 0.420 | cfm | 125 | 170 | 210 | 250 | 295 | 335 | 380 | 420 |
| | Throw | 5 | 6.5 | 8 | 9.5 | 11.5 | 13 | 15 | 16 |
| 12 Ak 0.595 | cfm | 180 | 240 | 300 | 355 | 415 | 475 | 535 | 595 |
| | Throw | 6 | 8 | 10 | 11.5 | 13.5 | 15.5 | 17.5 | 19 |
| 14 Ak 0.820 | cfm | 245 | 330 | 410 | 490 | 575 | 655 | 740 | 820 |
| | Throw | 7 | 9 | 11.5 | 13.5 | 16 | 18 | 20 | 22.5 |
| 16 Ak 1.03 | cfm | 310 | 410 | 515 | 620 | 720 | 825 | 925 | 1030 |
| | Throw | 7.5 | 10 | 12.5 | 15 | 18 | 20 | 22 | 25 |
| 18 Ak 1.33 | cfm | 400 | 530 | 665 | 800 | 930 | 1065 | 1200 | 1330 |
| | Throw | 8.5 | 11 | 14 | 17 | 20 | 23 | 26 | 28 |
| 20 Ak 1.60 | cfm | 480 | 640 | 800 | 960 | 1120 | 12801 | 1440 | 1600 |
| | Throw | 9.5 | 12 | 16 | 18 | 22 | 25 | 28 | 31 |
| 22 Ak 1.90 | cfm | 570 | 760 | 950 | 1140 | 1330 | 1520 | 1710 | 1900 |
| | Throw | 10.5 | 13.5 | 17 | 19 | 24 | 27 | 30 | 33 |
| 24 Ak 2.30 | cfm | 690 | 920 | 1150 | 1380 | 1610 | 1840 | 2070 | 2300 |
| | Throw | 11 | 14.5 | 18.5 | 22 | 26 | 30 | 33 | 36 |

Terminal velocity of 50 fpm

*Figure 33* Example of a ceiling diffuser pressure-drop chart.

### 4.1.6 Airflow in a Typical System

*Figure 35* shows a simplified building air distribution system. It will be used to discuss the airflow through a system, and review some of the concepts and pressure relationships. The volume of air that an air distribution system must deliver is normally based on the mode of operation (cooling or heating) that needs the most airflow. Generally, this is the cooling mode. But there are exceptions.

In *Figure 35*, the airflow shown is for the cooling mode related to a 3-ton cooling system. Therefore, the system fan must be able to supply the volume of air needed for 3 tons of cooling. Since cooling requires roughly 400 cfm of air per ton, the blower in the example system must supply 1,200 cfm (3 x 400 cfm) or more of air.

Further, the external static pressure of the duct system that the fan must work against is 0.48 in. w.c. This is derived by adding the absolute values of the supply and return external static pressures of 0.29 in. w.c. and –0.19 in. w.c., respectively.

**(A) BELT-DRIVE**

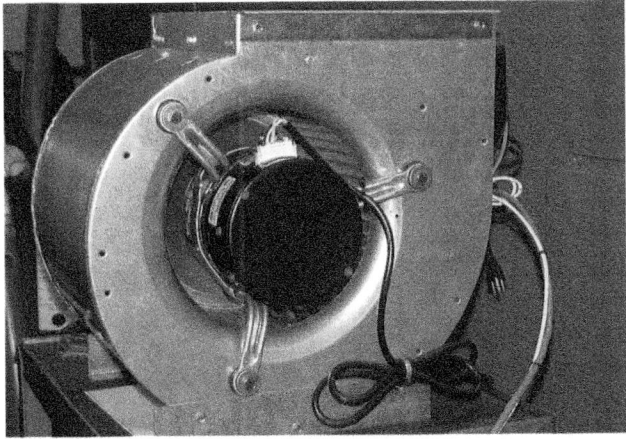

**(B) DIRECT-DRIVE**

*Figure 34* Blower assemblies.

The system shown has 12 air supply outlets: two delivering 150 cfm, eight delivering 100 cfm, and two delivering 50 cfm. The return air is taken into the system through two centrally located grilles.

While studying *Figure 35*, consider the structure as part of the system. The supply air leaves all the supply registers and sweeps the walls of the building. Then, it travels through the conditioned spaces within the building as it flows toward the return air grilles. The duct system begins at the two return air grilles. Relative to the atmospheric pressure of the rooms, there is a slight negative pressure at the grilles. The pressure on the blower side of the return air filter is lower than the pressure in the rooms. This results in the higher room pressures pushing the air through the return air grilles and filter and into the return duct.

As the air flows down the return duct towards the blower, the pressure continues to decrease because of friction losses in the duct. At the inlet to the blower wheel, the air pressure is at its lowest pressure in the system. The difference in static pressure between the input and output of the blower is 0.48 in. w.c.

The air at the blower outlet is pushed through the cooling coil, where it encounters a pressure drop of 0.08 in. w.c. After the air enters the supply duct, it undergoes a slight additional pressure drop at the tee where the duct is split into two reducing trunks, one to feed each end of the building. Some pressure drop is occurring throughout the duct system as the air travels through, scrubbing along the duct walls.

Each initial section of the supply duct on either side of the tee must handle 600 cfm of air. The supply air distribution is virtually the same on each side. Looking at the supply air ducts to the right of the tee, two supply air duct branches are the first connections. One requires 150 cfm, while the other requires only 50 cfm. This reduces the quantity of air that needs to flow into the next section of the trunk to 400 cfm. Since there is a significant reduction in air volume, the duct must be made smaller to keep the static pressure and air velocity roughly equal throughout the system.

Immediately after these two branch connections, there is a transition shown. The next section supplies air to two 100 cfm outlets. The duct is again transitioned to a smaller size after these two branch line connections. The final section of trunk duct serves the last two 100 cfm outlets. Normally, dampers are installed at each branch take-off to balance the quantity of air supplied to each room.

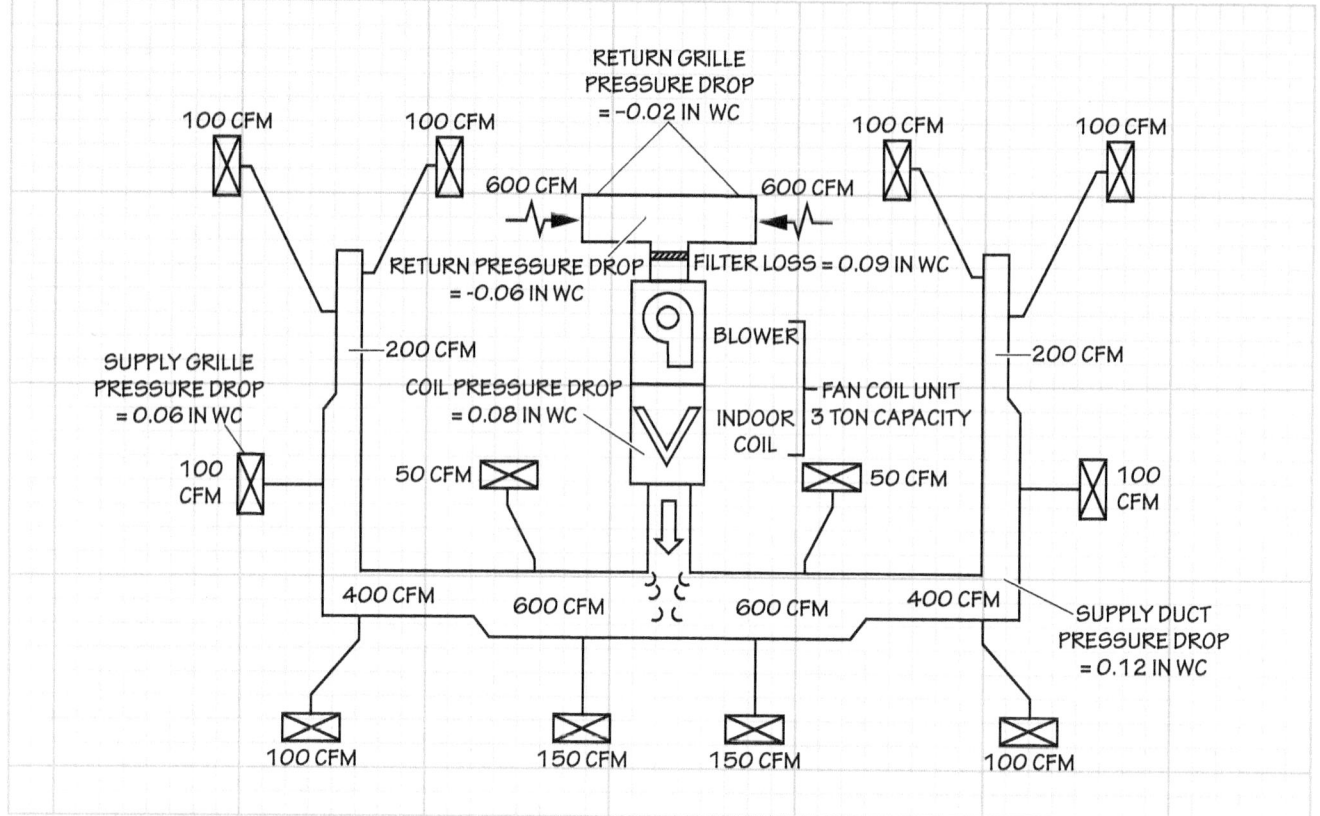

- SYSTEM COOLING CAPACITY = 3 TONS
- AT 400 CFM PER TON, TOTAL AIRFLOW REQUIRED = 1,200 CFM
- TOTAL RETURN DUCT STATIC PRESSURE AT 1,200 CFM = –0.19 IN WC
    - –0.06 IN WC RETURN DUCT STATIC PRESSURE LOSS
    - –0.09 IN WC FILTER STATIC PRESSURE LOSS
    - –0.04 IN WC TOTAL RETURN AIR GRILLE LOSS
- TOTAL SUPPLY DUCT SYSTEM STATIC PRESSURE AT 1,200 CFM = 0.29 IN WC
    - 0.12 IN WC SUPPLY DUCT STATIC PRESSURE LOSS
    - 0.17 IN WC TOTAL SUPPLY GRILLE STATIC PRESSURE LOSS
- TOTAL EXTERNAL STATIC PRESSURE = 0.48 IN WC

*Figure 35* Simplified air distribution system.

# Duct Leakage

All duct system designs assume a tight system with no losses due to leakage of conditioned air. However, most duct systems, especially sheet metal systems, are inherently leaky. According to the US Department of Energy, 25 to 40 percent of the energy in the conditioned air in ducts is lost through leakage or lack of insulation. Much of the leakage can be attributed to poor workmanship when assembling the system. The joints in sheet metal systems will generally have some leakage, even if properly assembled. To correct this problem, the system designer should specify that all duct joints and connections be sealed at the time of installation. The sealing method will vary according to the duct material used. Fiberglass ductboard and flexible duct joints can be sealed with aluminum foil tape. Sheet metal ducts can be sealed with aluminum foil tape or with mastic.

Today, most codes require that all duct systems be sealed. Those same codes may also dictate the type of material to be used for duct sealing.

## 4.2.0 Air Distribution System Layouts

There are many air distribution system design approaches. Most of them consist of two independent duct systems: the supply duct system and the return duct system. The supply duct system receives air from the blower and distributes it through the registers and diffusers into the conditioned space. The return duct system collects air from the conditioned space and returns it to the input of the blower.

Some large commercial systems use a return air fan to overcome friction loss in the return ducts. The design of air distribution systems for commercial and industrial structures varies widely, depending on the structure and its intended use. Because the designs of air distribution systems for residential applications are simpler and more uniform, they will be used as the basis for discussion in the remainder of this section. Although the size of the heating and cooling loads, the system layout, and the physical size of the system components will vary, the principles of operation and types of components are basically the same in all duct systems.

### 4.2.1 Duct Systems Used in Cold Climates

The type of duct system used in a building is mainly determined by the climate. In cold climates, most buildings use perimeter duct systems, which have floor or baseboard supply diffusers along the exterior walls of the building. Use of floor or baseboard supply diffusers provides a good tradeoff for heating and cooling performance.

In winter, the warm air supplied by the furnace blankets the outside walls and windows. This compensates for the cold downdrafts that tend to develop at the outside walls, windows, and doors. The return air grilles are located on the interior partition walls, at or near the floor. Central returns may be used, or for better performance, individual returns can be installed in each room. Locating return grilles on the interior walls near floor level helps remove any cool air from the floor where it tends to collect or stratify.

*Figure 36* shows the pattern of room airflow during the heating and cooling modes of system operation. During the heating mode, the heated air blankets the outside walls and windows. Because it is warmer and lighter than the room air, it spreads across the ceiling and down the inside wall. Room air is drawn (induced) into the warm airstream and mixes with it. A resulting stratified zone of cool air tends to collect near the floor then leaves the room through a low sidewall return.

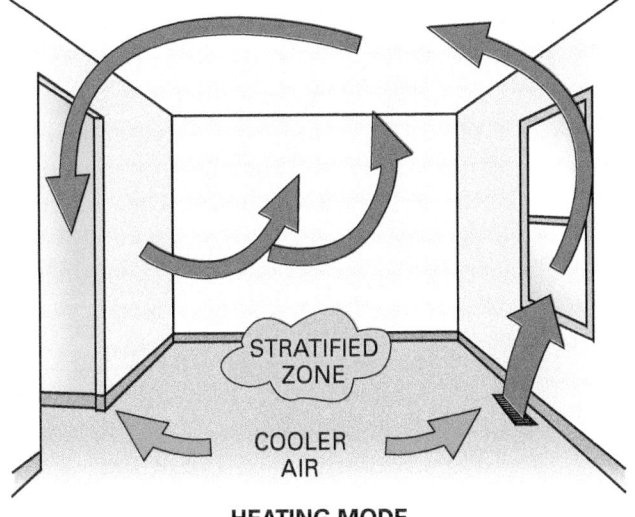

**HEATING MODE**

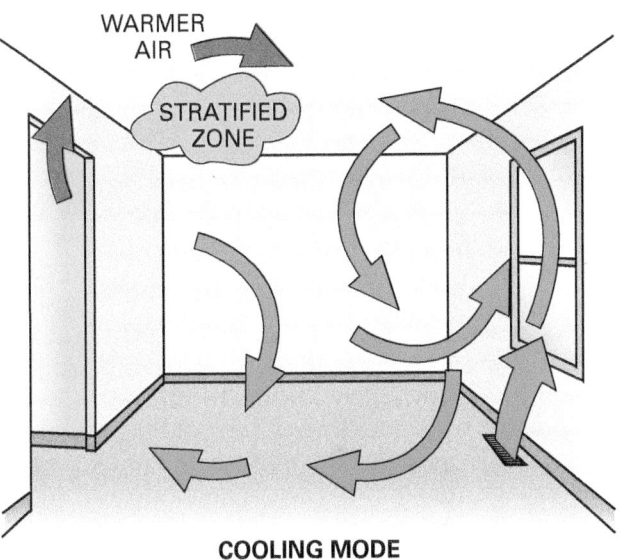

**COOLING MODE**

*Figure 36* Room air distribution patterns for a perimeter duct system.

During the cooling mode, cold supply air travels up the outside wall and windows and strikes the ceiling. Because it is cooler and heavier than the room air, it travels a short distance along the ceiling and then drops back down into the room. The cold air mixes with the room air, leaving only a small stratified layer of warm air near the ceiling. High sidewall returns would minimize this problem, but would result in a loss of heating performance. In this situation, the use of a slow-rotating ceiling fan during the heating and cooling seasons would help break up the stratified air, resulting in better indoor comfort.

Perimeter systems can have various layouts. The common ones include the following:

- Loop perimeter
- Extended plenum
- Reducing plenum

Loop perimeter duct systems are seen in structures built on concrete slabs in colder climates (*Figure 37*). They are easily used with centrally located, down-flow air handlers. The perimeter loop is a continuous round duct of constant size imbedded in the slab. It runs close to the outer walls, with the outlets next to the wall. The perimeter loop is fed by several branches from the plenum. When the furnace fan is running, warm air is in the whole loop, which helps to keep the slab at an even temperature. Heat loss to the outside is minimized by insulation around the slab. The loop has a constant pressure and provides the same pressure to all outlets.

The extended-plenum duct system (*Figure 38*) uses rectangular trunks for the main supply and return ducts. The supply and return trunk ducts are a consistent size over the full length. This is the reason it is called an extended-plenum system. These systems are commonly used in below-floor (basement or crawl space) or ceiling (attic) installations. This type of system does not perform well in large structures and those with many outlets. Since there are no transitions, the pressure and velocity of air in the duct becomes low at the far end of a long duct, resulting in limited airflow.

Separate branch ducts run from the trunk duct to each supply outlet. The extended-plenum system works best when the air handler is located at the center of the main duct. However, it can be routed in one direction. The trunk ducts are normally installed near the center line of the building, and their dimensions are constant over their entire length. The branch ducts are normally round but can be rectangular. An air volume damper is usually installed in each branch duct near the trunk. This allows the airflow to be balanced with all supply air outlets fully open. Recommendations for the design of an extended plenum duct system are as follows:

- The supply and return ducts should extend no more than 24' (7.3 m) from the air handler.

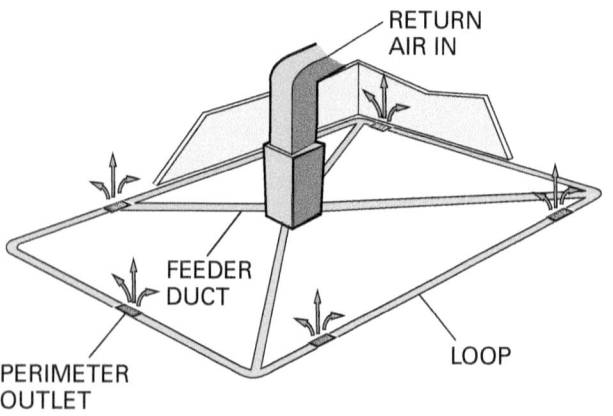

*Figure 37* Loop perimeter duct system.

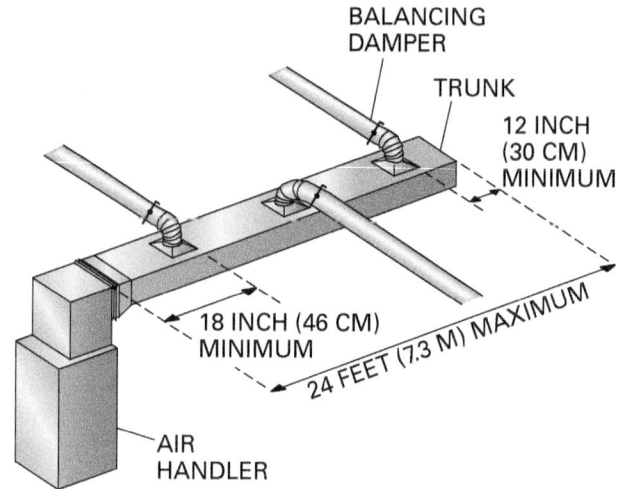

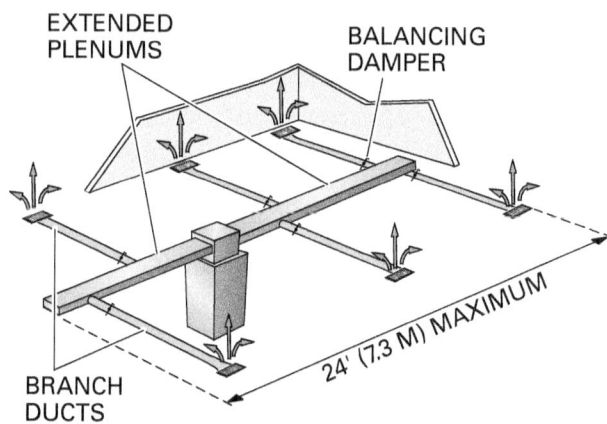

*Figure 38* Extended-plenum duct system.

- The first branch should be at least 18" (46 cm) from the beginning of the main duct.
- The main trunk should extend at least 12" (30 cm) past the last branch takeoff.
- Balancing dampers should be installed at each branch takeoff.

*Figure 39* shows an example of a reducing-trunk system. A reducing-trunk system provides the best performance but at a slightly higher cost. It works well in larger buildings that require longer duct runs. Remember that extended-plenum systems are limited to 24' (7.3 m). It is a better choice for systems where the air handler is installed on one end of the main trunk duct rather than in the middle. Indeed, it is a better choice for most applications. When properly sized, the same pressure drop is maintained from one end of the duct system to the other. This allows all branches to operate at roughly the same pressure. When it is properly designed and installed, it requires less effort to balance. The following are some recommended practices for laying out a reducing-trunk duct system:

- The first main duct section should be no longer than 20' (6 m).

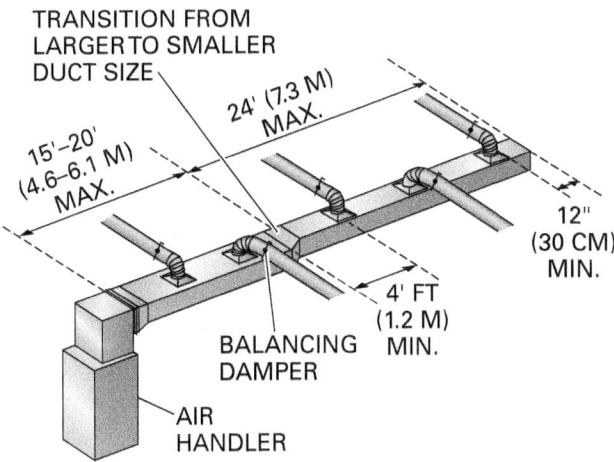

*Figure 39* Reducing-trunk system.

- The length of each reducing section after that should not exceed 24' (7.3 m).
- The first branch duct connection down from a single-taper transition should be at least 4' (1.2 m) from the beginning of the transition fitting. This distance allows the air turbulence caused by the fitting to subside before the air is sent into the next branch duct. If the distance is less than 4', the branch ducts near the transition can be hard to balance as the air may tend to bypass them.
- The trunk duct should extend at least 12" (30 cm) from the last branch duct.

### 4.2.2 Duct Systems Used in Warm Climates

In warm climates, buildings should have duct systems that favor cooling performance over heating performance. Perimeter systems like those used in cold climates can work reasonably well in some warm areas. However, their use is normally limited to buildings constructed over a basement or crawl space. Because cold floors and downdrafts from the outside walls are not normally a problem in warm climates, the air supply outlets do not need to be located at the building perimeter. In warm climates, supply and return air openings can be mounted high on the interior walls or in the ceiling to intensify cooling.

*Figure 40* shows the room airflow with high-sidewall outlets. In the cooling mode, cool air moves across the ceiling and washes the far wall. The room air mixes well with the supply air, and almost no stratification occurs. Air motion throughout the room is good. In the heating mode, the supply air remains near the ceiling and moves partway down the outside wall. Because of its buoyancy, the warm air does not descend the wall very far. This causes a stratified area near the floor where cool air tends to build up.

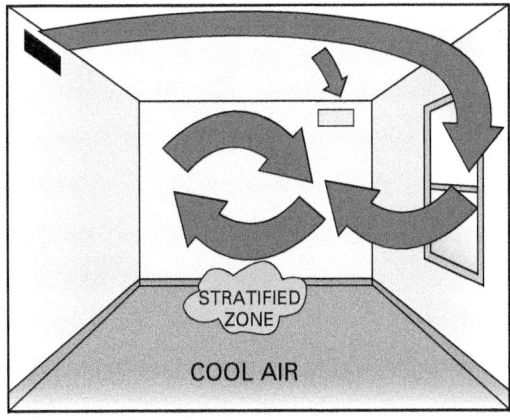

**HEATING MODE**

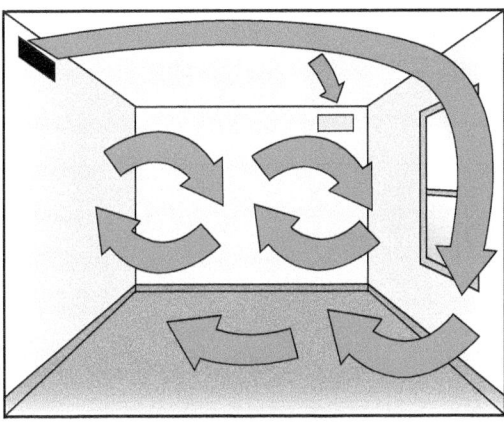

**COOLING MODE**

*Figure 40* Room air distribution patterns for high-sidewall outlets.

The returns in a high-sidewall outlet system can vary. Some designers choose a high-sidewall return to help reduce warm air stratification near the ceiling in both modes. A central ceiling return (usually located in a central hallway) is commonly seen, especially if the furnace or air handler is located above the ceiling. If a closet-mounted furnace or air handler is used, the appliance is typically mounted on a platform with one or two return grilles near the floor tied into the area under the platform that serves as a return plenum.

Ceiling diffusers are one of the best air supply methods used for cooling, but they are not as effective for heating in perimeter areas. In the cooling mode, supply air from the diffuser mixes well with the room air (*Figure 41*). Air motion in the room is good with no stagnant areas. In the heating mode, warm air tends to rise toward to the ceiling. Very little of it reaches the lower portions of the occupied space. Also, because of convection currents moving down the cold outside wall, cool air is drawn across the floor and up the inside wall.

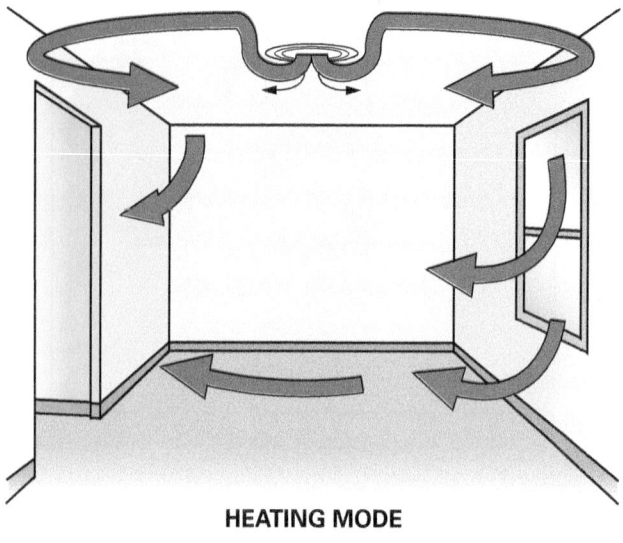

**HEATING MODE**

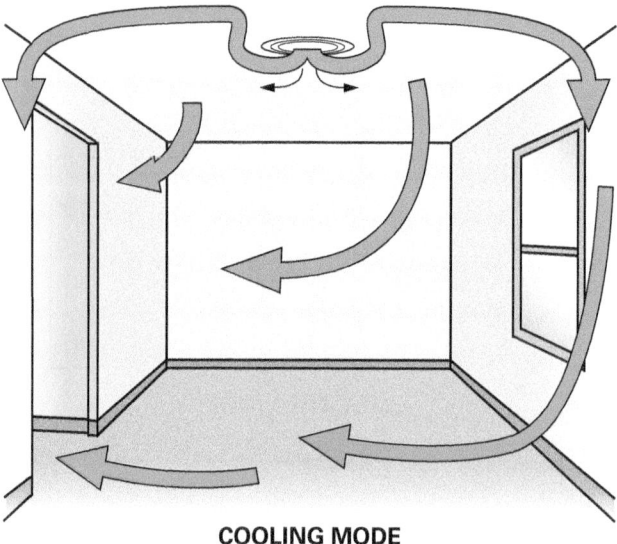

**COOLING MODE**

*Figure 41* Room air distribution patterns for ceiling outlets.

Duct systems used in warm climates have various layouts. The three typical layouts used in buildings on concrete slabs are overhead trunk, overhead or attic radial, and overhead extended-plenum.

Regardless of the type of system, the ductwork should be insulated if it is installed in an attic. If the ductwork is installed above a suspended ceiling, it may or may not be insulated, depending on whether the space above the suspended ceiling is considered conditioned space. Overhead trunk and overhead extended-plenum systems should be laid out as those shown in *Figure 38* and *Figure 39*.

Trunk and extended plenum types of systems are designed and laid out the same, whether installed in a basement, crawl space, attic, or above a ceiling. In the overhead radial system, separate branch ducts are routed from a common supply

air plenum to each outlet. The branch ducts are metal, ductboard, or flexible duct and are usually insulated, depending on whether they are in the conditioned space. Overhead radial systems often use a central return.

## 4.3.0 Duct System Components

Building code requirements pertaining to the design of air distribution systems vary widely across the country. Almost all localities have minimum standards or codes that determine the type of materials and methods that must be used. Also, the materials and methods used in residential, commercial, and industrial air distribution systems vary. Become familiar with and follow all codes that apply in your specific area.

Areas that experience earthquakes have stringent local codes that cover all aspects of building construction, including the HVAC equipment installation. System designers must be aware of those additional code requirements and comply with them. On a national level, the Federal Emergency Management Agency (FEMA) publishes *FEMA 412 Installing Seismic Restraints for Mechanical Equipment*. At the industry level, SMACNA has a 200-page publication entitled *Seismic Restraint Manual: Guidelines for Mechanical Systems*. This publication is also endorsed by the American National Standards Institute (ANSI). The manual shows designers and contractors how to select and apply quake-resistant restraints for sheet metal ducts, piping, and conduit. ASHRAE also devotes a chapter of their handbook series to seismic and wind restraint design.

# Seismic Considerations

HVACR systems designed for use in earthquake-prone areas, such as the western United States and Alaska, have special requirements. Earthquakes place unusual stresses on all types of structures and the HVACR equipment within them. If not properly compensated for, those stresses can cause HVACR equipment to move or fall, with potentially disastrous results. For example, if a furnace gas line breaks or shears, the leaking gas becomes an immediate fire and explosion hazard. For that reason, flexible gas line connectors are used in those areas. Falling or moving equipment can further damage the structure and may cause injury to anyone nearby.

Duct systems are classified by their use and static pressure. *Table 2* lists the pressure levels normally used in residential, commercial, and industrial systems. The pressure in residential systems does not exceed 1 in. w.c. In fact, residential duct pressure rarely approaches this level. Duct systems in public assembly, business, educational, general factory, and mercantile buildings are normally classified as commercial systems. The highest design pressure used in commercial applications is 10 in. w.c. Commercial systems that operate at this pressure are extremely rare. Industrial duct systems include air pollution control systems and industrial exhaust systems, as well as HVAC systems. Industrial systems may be designed for a wide variety of pressures, including those found in commercial projects and higher.

Duct systems are also classified as low-velocity or high-velocity systems. *Table 3* lists the duct velocity levels and supply outlet pressure levels used with low-velocity and high-velocity systems. When space is available, it is recommended that larger, low-velocity duct systems be used. This is because they generally have much lower system noise levels. High-velocity duct systems have higher noise levels and are used mainly where space limitations prevent the use of larger ducts. Some installations may have areas with space restrictions that require the use of smaller duct operating at higher velocities. As soon as these areas are passed and space is available, the duct size should be increased so that the velocity rate drops sharply and is gradually reduced toward the end of the duct system.

The remainder of this section will focus on the low-velocity duct system and components commonly used in residential and light commercial applications. *Table 4* lists the recommended maximum duct velocities for common low-velocity applications.

**Table 2** Duct System Classification by Pressure

| System | Pressure Level |
| --- | --- |
| Residential | ±0.5 in wc |
| | ±1 in wc |
| Commercial | ±0.5 in wc |
| | ±1 in wc |
| | ±2 in wc |
| | ±3 in wc |
| | ±4 in wc |
| | ±6 in wc |
| | ±10 in wc |
| Industrial | Pressure varies with application |

**Table 3** Duct System/Supply Outlet Classifications

| Duct Velocities | Velocity |
| --- | --- |
| Low-velocity duct | Main duct 1,000 to 2,400 fpm<br>Branches 600 to 1,600 fpm |
| High-velocity duct | Main duct 2,500 to 4,500 fpm<br>Branches 2,000 to 4,000 fpm |
| Air Outlet Pressures | Pressure |
| Low-pressure outlets | Static pressure 0.1 to 0.5 in wc |
| High-pressure outlets | Static pressure 1 to 3 in wc |

The main components of an air distribution duct system are as follows:

- Main trunk and branch ducts
- Fittings and transitions
- Supply air outlets and return air inlets
- Dampers
- Insulation and vapor barriers

### 4.3.1 Main Trunk and Branch Ducts

Duct systems can be installed in basements, crawl spaces, attics, and concrete floors (slabs). In commercial jobs, they are often installed in open areas, such as warehouses and garages. Ductwork can be made from metal, fiberglass ductboard, plastic, or even ceramic materials. Galvanized sheet metal or fiberboard ducts are typically used for heating/cooling air distribution. When installed in a concrete slab, ducts are usually made of metal, plastic, or ceramic. Spiral metal and flexible ducts are also commonly used. Where weight is a factor, aluminum duct is sometimes used.

Galvanized steel duct can be round, square, or rectangular. Popular sizes of round and rectangular steel ducts, along with an assortment of standard fittings, can be obtained from HVACR supply houses. *Figure 42* and *Figure 43* show typical steel rectangular and round duct systems and components, respectively. For large commercial jobs involving customized ductwork, the ducts and fittings are often made separately in a metal shop or are fabricated at the job site. Because sheet metal duct is rigid, the layout must be well planned, and all the pieces exactly cut, or the duct system will not fit together.

The thickness of galvanized steel and other metal duct is expressed as a gauge thickness. When a duct is made of 28-gauge sheet metal, this means that the thickness of the duct walls is $\frac{1}{28}$". Likewise, a sheet metal duct made from 24-gauge metal has a wall thickness of $\frac{1}{24}$", and so on. Larger ducts are made from thicker metal and are more rigid than smaller ducts. This helps prevent them from swelling and making popping noises

**Table 4** Recommended Maximum Duct Velocities for Common Low-Velocity Applications

| Main Duct | Branch Ducts | | | |
|---|---|---|---|---|
| Application | Supply (fpm) | Return (fpm) | Supply (fpm) | Return (fpm) |
| Apartments, residences | 1,000 | 800 | 600 | 600 |
| Auditoriums, theaters | 1,300 | 1,100 | 1,000 | 800 |
| Banks, meeting rooms, libraries, offices, restaurants, retail stores | 2,000 | 1,500 | 1,600 | 1,200 |
| Hospital rooms, hotel rooms | 1,500 | 1,300 | 1,200 | 1,000 |

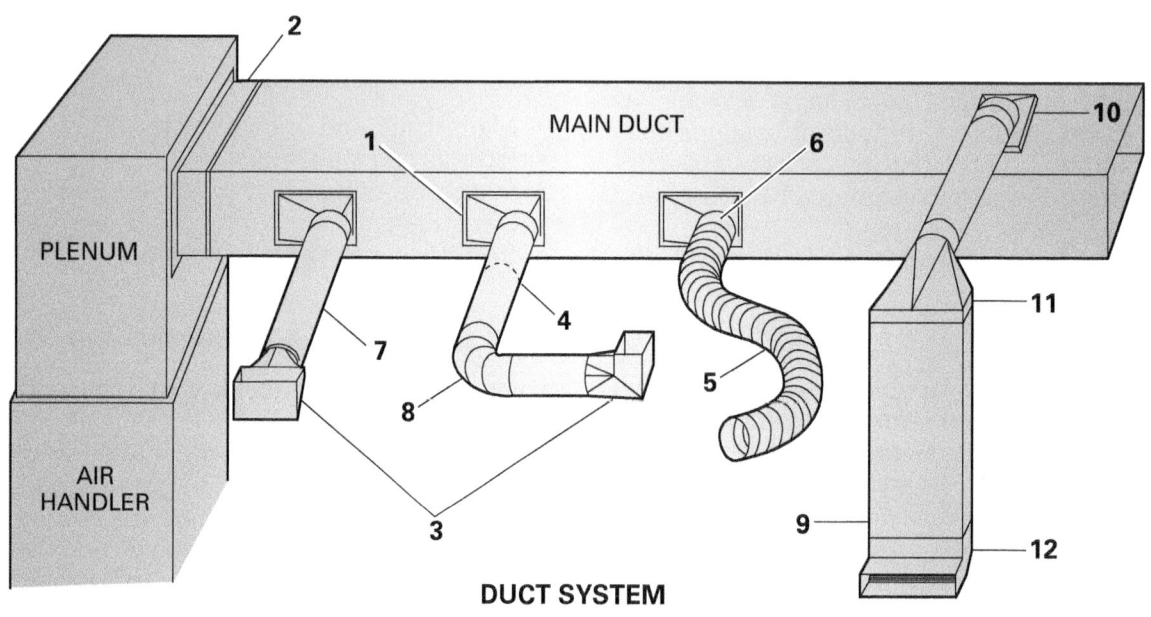

DUCT SYSTEM

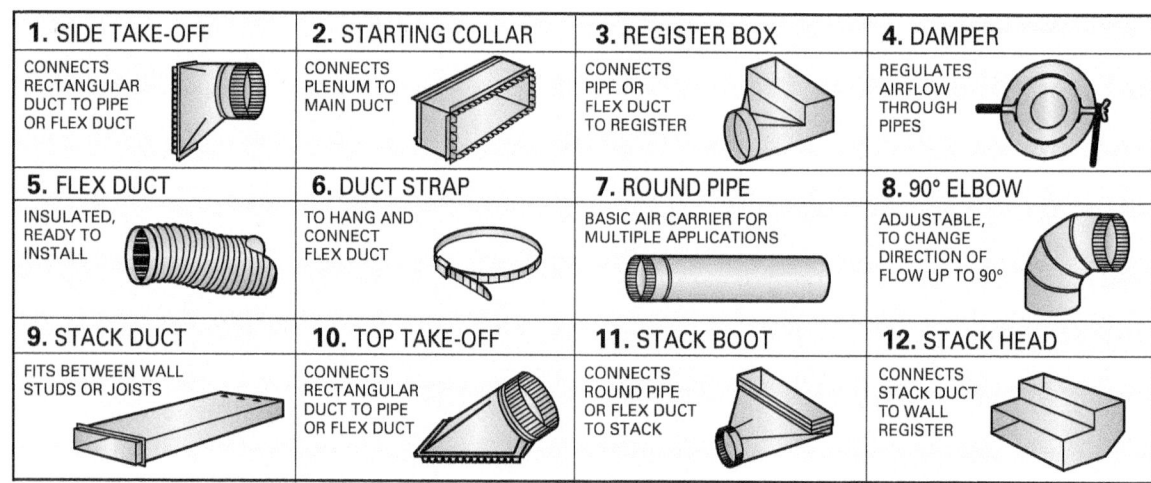

FITTINGS AND TRANSITIONS

*Figure 42* Typical rectangular duct system and components.

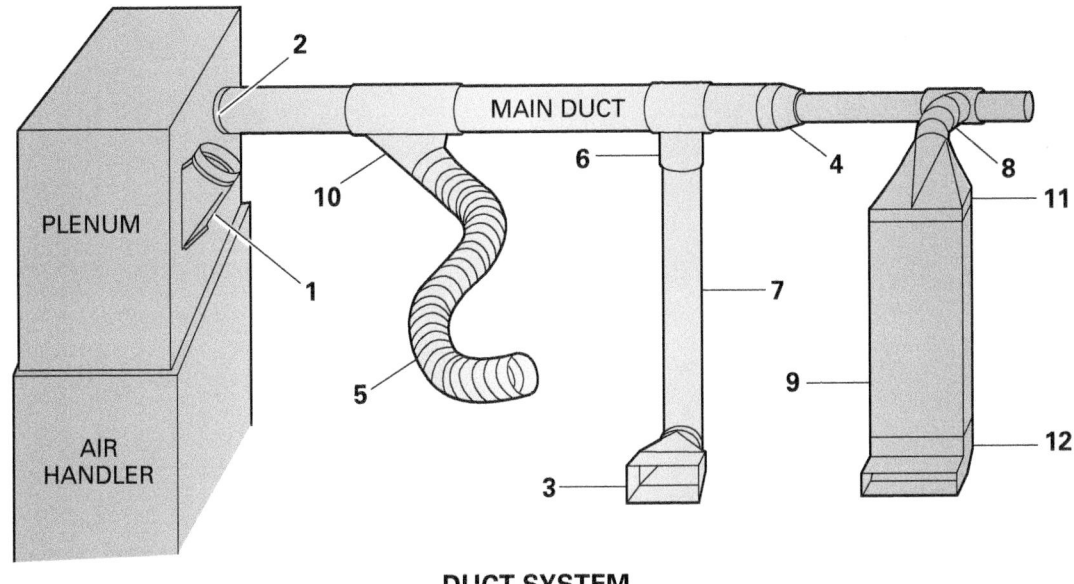

**DUCT SYSTEM**

**FITTINGS AND TRANSITIONS**

| **1.** TAKE-OFF | **2.** STARTING COLLAR | **3.** REGISTER BOX | **4.** REDUCER/INCREASER |
|---|---|---|---|
| CONNECTS RECTANGULAR DUCT TO PIPE OR FLEX DUCT | CONNECTS PLENUM TO MAIN DUCT | CONNECTS PIPE OR FLEX DUCT TO REGISTER | TO REDUCE OR INCREASE BETWEEN DIFFERENT DIAMETER PIPES |
| **5.** FLEX DUCT | **6.** TEE JOINT | **7.** ROUND PIPE | **8.** 90° ELBOW |
| INSULATED, READY TO INSTALL | CONNECTS MAIN DUCT TO BRANCH | BASIC AIR CARRIER FOR MULTIPLE APPLICATIONS | ADJUSTABLE, TO CHANGE DIRECTION OF FLOW UP TO 90° |
| **9.** STACK DUCT | **10.** WYE BRANCH | **11.** STACK BOOT | **12.** STACK HEAD |
| FITS BETWEEN WALL STUDS OR JOISTS | CONNECTS MAIN DUCT TO BRANCH | CONNECTS ROUND PIPE OR FLEX DUCT TO STACK | CONNECTS STACK DUCT TO WALL REGISTER |

*Figure 43* Typical round duct system and components.

(called oil-canning or drumming) when the system blower starts and stops. Also, lines or ridges called cross-breaks (*Figure 44*) are used on large sheet metal panels or ducts to make them more rigid. *Figure 45* shows some typical metal gauges used for rectangular and round metal ducts.

The aspect ratio of a duct is often used to classify a duct size and estimate its fabrication cost. Aspect ratio is the ratio of the duct's width to its height. For example, if a duct is 18" wide and 6" high (45 x 15 cm), the aspect ratio is 18 ÷ 6 (45 ÷ 15), or 3:1.

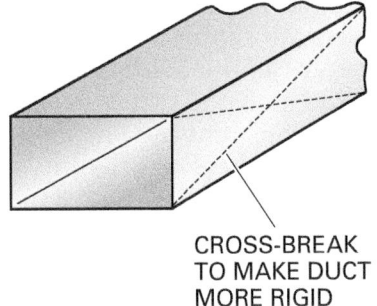

CROSS-BREAK TO MAKE DUCT MORE RIGID

*Figure 44* Cross-break on metal duct.

| RECTANGULAR DUCT WIDTH IN INCHES | COMMERCIAL | | RESIDENTIAL |
|---|---|---|---|
| | SHEET METAL GALVANIZED | ALUMINUM | SHEET METAL GALVANIZED |
| UP TO 12 | 26 | 0.020 | 28 |
| 13–23 | 24 | 0.025 | 26 |
| 24–30 | 24 | 0.025 | 24 |
| 31–42 | 22 | 0.032 | – |
| 43–54 | 22 | 0.032 | – |
| 55–60 | 20 | 0.040 | – |
| 61–84 | 20 | 0.040 | – |
| 85–96 | 18 | 0.050 | – |
| OVER 96 | 18 | 0.050 | – |

**RECTANGULAR DUCT**

| ROUND DUCT DIAMETER IN INCHES | COMMERCIAL SHEET STEEL GALVANIZED GAUGE | RESIDENTIAL SHEET STEEL GALVANIZED GAUGE |
|---|---|---|
| UP TO 12 | 26 | 30 |
| 13–18 | 24 | 28 |
| 19–28 | 22 | – |
| 27–36 | 20 | – |
| 35–52 | 18 | – |

**ROUND DUCT**

*Figure 45* Typical metal-duct gauges.

### 4.3.2 Duct Assembly, Insulation, and Support

Sections of square or rectangular duct are assembled using one of several available fasteners. Several different connection types are shown in *Figure 46*. Typically, S-slip and drive connectors, and snap-lock connections (often provided on prefabricated duct) are used with smaller duct sizes. Round duct sections are normally fastened together with self-tapping sheet metal screws. These fasteners make a mechanically strong connection, but also one that is not air-tight. For further sealing, the joint can be taped using aluminum foil duct tape. Duct sealing mastic that can be brushed on or applied with a caulking gun is also available. Leaking joints cut down on the amount of air available for delivery to the outlets and waste significant amounts of energy.

Rectangular metal duct is often lined with fiberglass duct liner to provide insulation and to deaden sound. When using lined metal duct, keep in mind that the lining reduces the internal cross-sectional area of the duct, making it smaller. For example, lining a section of 18" × 8" (46 × 20 cm) metal duct with a ½" (1.3 cm) duct liner effectively reduces the duct size to 17" × 7" (43 × 18). If using lined metal duct, a larger duct will have to be used to get the same internal size of unlined duct. These duct liners also increase resistance to airflow because of their rough surfaces.

A duct system must be well supported so that it does not move. If it is not properly supported, movement can occur when the fan starts, or from expansion and contraction. Sheet-metal duct systems also transmit vibrations from the air handling equipment. This type of movement or vibration can be minimized by using flexible or fabric joints at key points in the system. *Figure 47* shows some typical duct noise and vibration control devices.

## Duct Coatings

As a means of improving indoor air quality, some manufacturers are producing steel covered with an antimicrobial coating for use in the construction of ductwork. Its purpose is to deter mold, mildew, fungus, and bacteria that might enter the HVAC airstream.

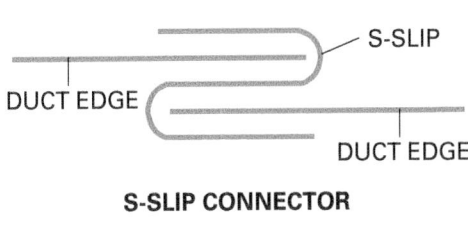

S-SLIP CONNECTOR

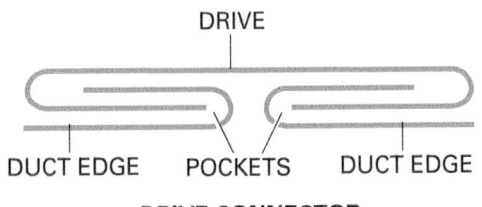

DRIVE CONNECTOR

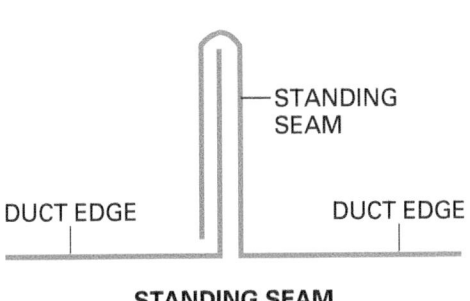

STANDING SEAM

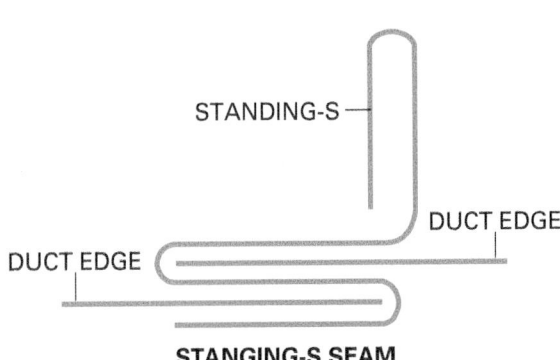

STANGING-S SEAM

*Figure 46* Common rectangular metal-duct connection methods.

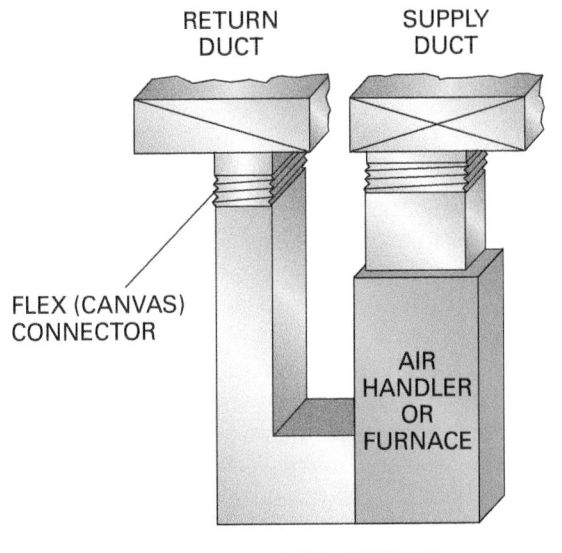

VIBRATION/NOISE CONTROL
AT AIR HANDLER

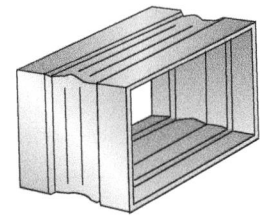

VIBRATION/NOISE/MOVEMENT
CONTROL IN DUCT RUNS

*Figure 47* Duct vibration and noise control devices.

### 4.3.3 Fiberglass Ductboard

Fiberglass ductboard can be used almost anywhere that metal duct is used in applications involving velocities up to 2,400 fpm and pressures of ±2 in. w.c. Molded round fiberglass ducts are available to handle higher pressures. Fiberglass duct has more friction loss than metal duct, but it is quieter because the ductboard absorbs blower and air noises better. Fiberglass duct is available in flat sheets for fabrication or as prefabricated round duct. Fiberglass duct is normally 1" (2.5 cm) thick with an aluminum foil backing. This backing is reinforced with fiber to make it strong. The inside surface of the ductboard is coated with plastic or a similar substance to prevent the erosion of duct fibers into the supply air and to provide smoother airflow.

Ducts are made from folded sheets of fiberglass ductboard using special knives or cutting machines. When two pieces are fastened together, an overlap of foil is left so that one piece can be stapled to the other using special staples (*Figure 48*). The joint is then taped to make it airtight. Round fiberglass duct is also easy to install because it can be cut to size with a knife. Fiberglass duct systems must be well supported, or they will sag over long runs. Proper hangers for ductboard make contact over a reasonable surface area so that the hanger does not dig into the duct wall.

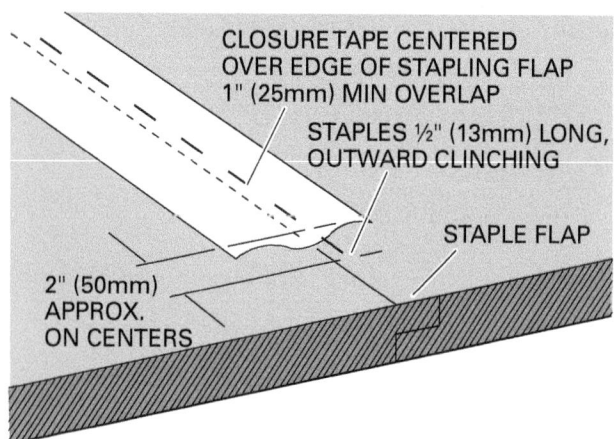

*Figure 48* Fiberglass ductboard joint.

### 4.3.4 Flexible Round Duct

Flexible round duct (*Figure 49*) is typically available in sizes of up to 24" (61 cm) in diameter. Flexible duct is available with only a reinforced aluminum-foil backing for use in conditioned areas. It also comes pre-insulated and protected by a vapor barrier made of vinyl or foil for use in unconditioned areas.

Flexible duct is typically used in spaces where obstructions make the use of rigid duct difficult or impossible. Flexible duct is easy to route around corners and other bends. Duct runs should be kept as short and straight as possible. Gradual bends should be used because tight turns greatly reduce the airflow and can cause the duct to collapse.

Long runs of flexible duct are not recommended unless the friction loss is taken into account. Even when properly installed, most flex ducts cause at least two to four times as much resistance to airflow as the same diameter sheet metal duct. To avoid sags in the run, flexible duct should be amply supported with one-inch wide or wider metal bands to keep the duct from collapsing and reducing the inside dimension (*Figure 50*). Some flexible duct comes with built-in eyelet holes for hanging the duct.

> **NOTE**
>
> Flexible duct runs are often limited to 6' (1.8 m) in length. However, local codes vary. Check the codes in your area for specific flexible-duct installation requirements.

### 4.3.5 Fittings and Pressure Loss

Fittings in duct systems, such as elbows, angles, takeoffs, and boots, change the direction of airflow or change its velocity. Transitions are typically used to change from one size or shape of duct to another.

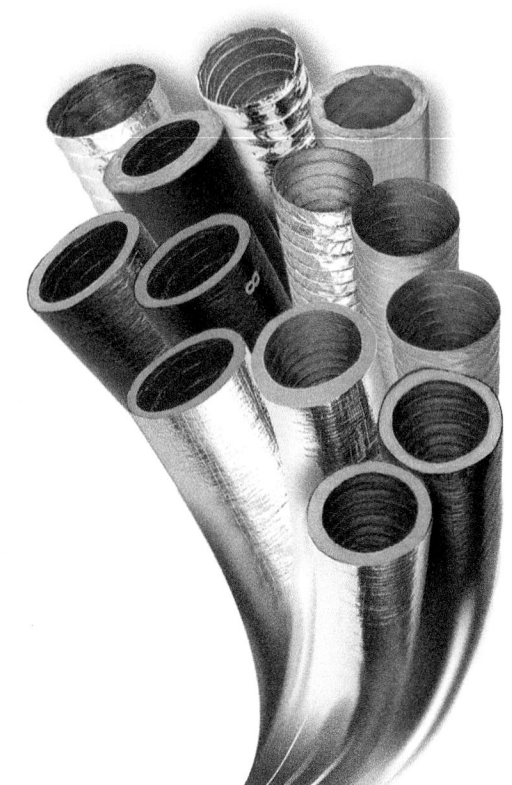

*Figure 49* Flexible duct.

Air moving in a duct has inertia that makes it tend to flow in a straight line. Each fitting in a duct run adds friction and decreases the quantity of air the duct can carry. It takes energy to overcome the resistance (friction) of a fitting. Adding fittings to a duct run has the same effect on the pressure loss of a duct as increasing its overall length. Therefore, duct runs should be made as straight as possible. It is also why using unnecessary fittings or ones not best suited for the job must be avoided.

The pressure loss that results from using a specific type of fitting can be found by multiplying its loss coefficient value (C-factor) by the velocity pressure in the fitting. C-factors give more accurate values of pressure loss and are typically used when designing very complex duct systems. C-factors for the various kinds of fittings are found in published tables readily available in SMACNA, ASHRAE, and duct component manufacturers' published literature.

The equivalent-length-of-straight-duct values assigned to a fitting are more commonly used when designing residential or light commercial duct systems. This means that a specific fitting produces a pressure drop equal to a certain number of feet of straight duct length of the same size. For each standard type of fitting, the pressure drop is known and has been converted to the equivalent feet of duct length. This information

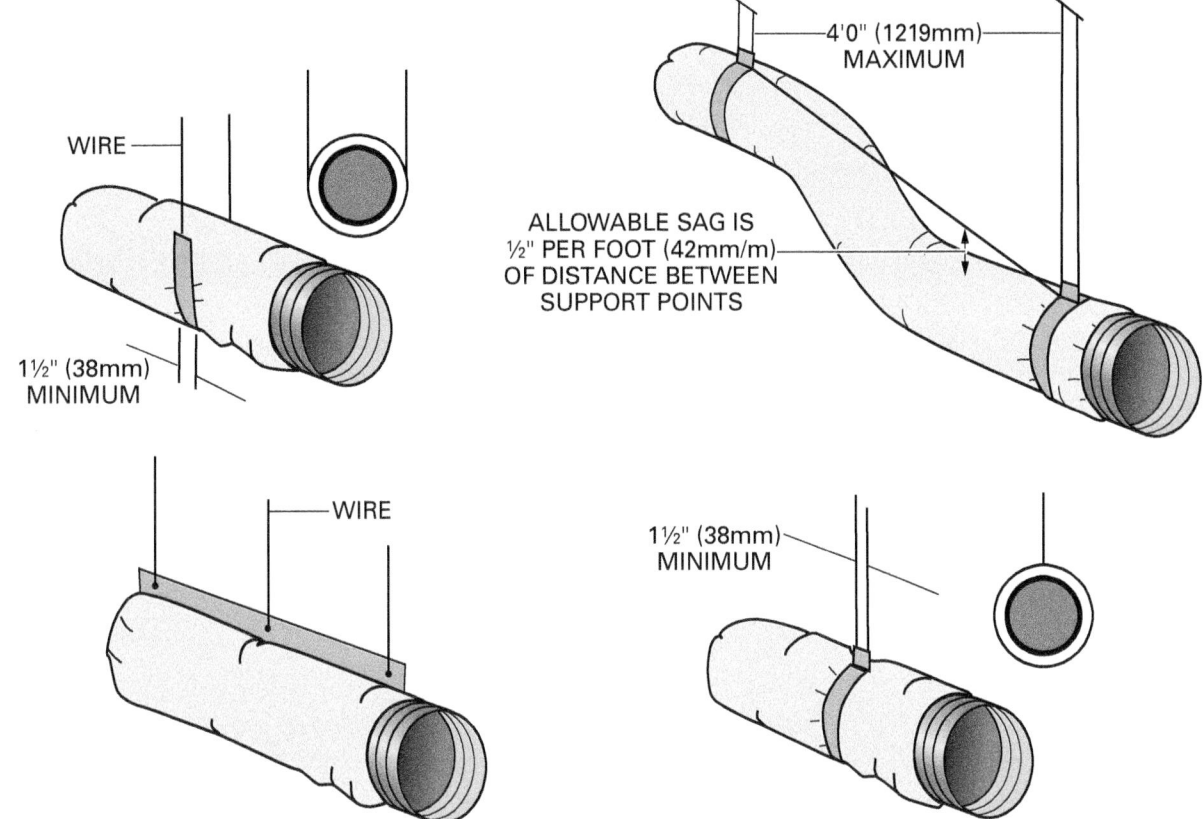

WIRE

1½" (38mm)
MINIMUM

4'0" (1219mm)
MAXIMUM

ALLOWABLE SAG IS
½" PER FOOT (42mm/m)
OF DISTANCE BETWEEN
SUPPORT POINTS

WIRE

1½" (38mm)
MINIMUM

*Figure 50* Supporting flexible duct.

is available in a set of charts published in air duct system-related SMACNA and ASHRAE documents and in some duct component manufacturers' product literature. These charts show the standard types of fittings and/or transitions and the value for the equivalent feet of length used for each one. The total equivalent feet of length for a duct run is then calculated by adding all the equivalent lengths for fittings in the run to the actual length of straight duct used. Each type of fitting or transition presented in the charts is identified by a letter and a number.

The letter identifies the type of fitting and the number indicates the equivalent length of straight duct, in feet, for that fitting. In the example shown in *Figure 51*, Elbow G makes a sharp change in direction, resulting in greater losses. Note that it has a short radius on the outside of the bend, and a square throat. This single elbow adds 30' (9.1 m) of equivalent length to the 100' (30.5 m) straight duct. The resulting pressure drop is therefore the same as that of 130' (39.6 m) of straight duct. On the other hand, Elbow F makes a longer and smoother change in direction and has fewer losses. The throat of the elbow also has a radius to enhance flow. Its equivalent length is only 10' (3.1 m) as a result. The required turn in the duct would be made, but with a lower pressure drop.

Supply air outlets control the air pattern within a conditioned area to obtain proper air motion and to blend the supply air with the room air so that the room is comfortable without excess noise or drafts. Supply air outlets are selected for each area in a building based on the volume of air required, the distance available for throw or radius of diffusion, the building structure, and the decor. These components were discussed in detail in an earlier module.

Return air inlets are located to be compatible with supply outlets and ductwork. Generally, return inlets are positioned to return room air of the greatest temperature differential that collects in stagnant areas. This results in the warmest air being returned during cooling and the coolest air being returned during heating. In general, the same types of grilles and diffusers used for supplying air are used for returning air. Normally, they do not have any deflection or volume control devices.

Volume dampers are used to control the volume of airflow through the various sections of an air distribution system. They do this by introducing a resistance to airflow in the system. Volume dampers are normally installed in each branch duct serving a conditioned area. Without volume dampers, air distribution systems cannot be properly balanced, causing some rooms to receive

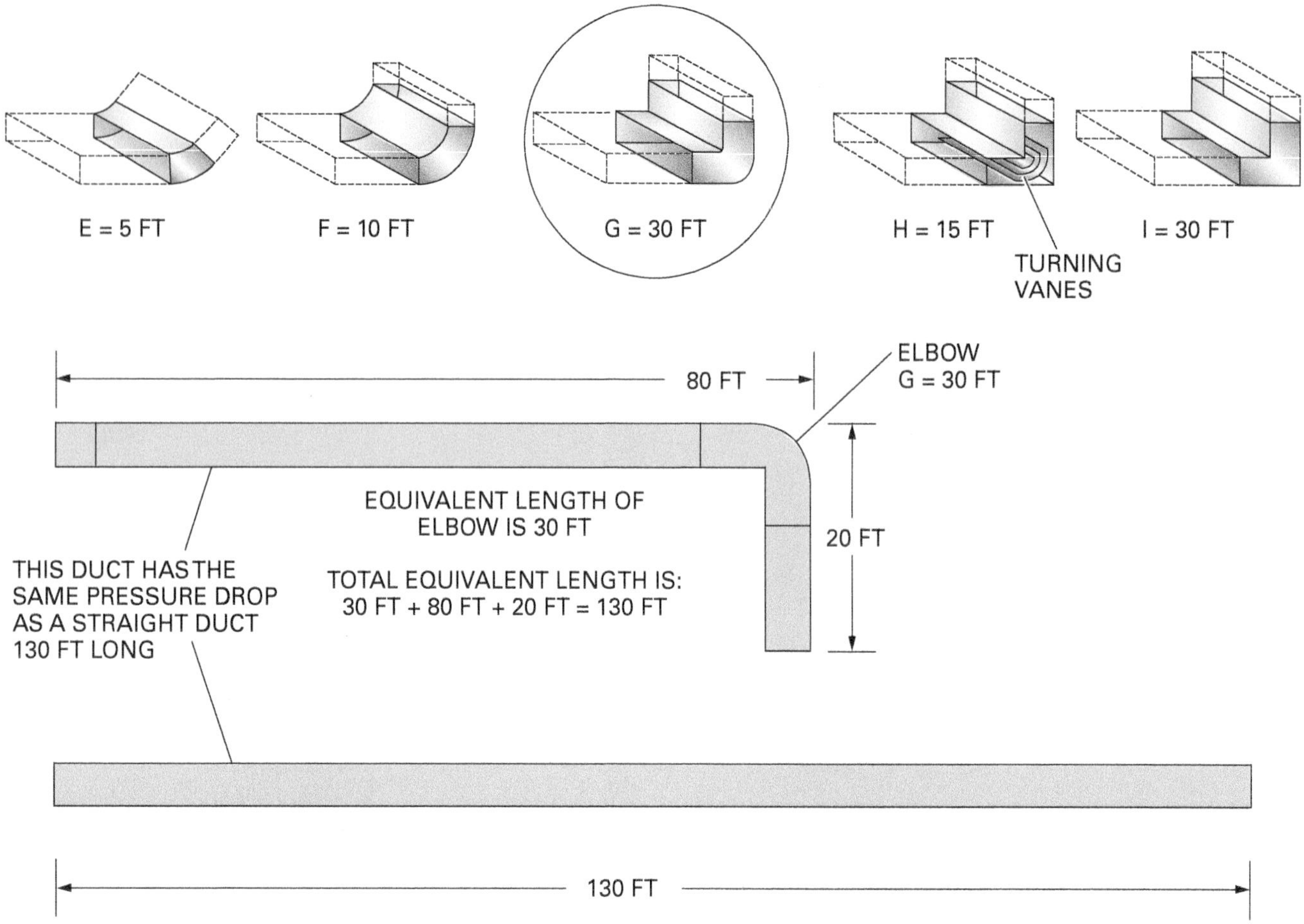

E = 5 FT    F = 10 FT    G = 30 FT    H = 15 FT    I = 30 FT

TURNING VANES

ELBOW
G = 30 FT

80 FT

20 FT

EQUIVALENT LENGTH OF
ELBOW IS 30 FT

TOTAL EQUIVALENT LENGTH IS:
30 FT + 80 FT + 20 FT = 130 FT

THIS DUCT HAS THE
SAME PRESSURE DROP
AS A STRAIGHT DUCT
130 FT LONG

130 FT

*Figure 51* Example of equivalent length.

too much air and others not enough. In high-pressure systems, volume dampers may actually be referred to as pressure-reducing valves. They are usually installed between a high-pressure trunk duct and a low-pressure branch duct.

Dampers are usually manually adjustable. When used in electronically controlled zoned systems, however, dampers are automatically controlled. Sometimes dampers are used to mix two airflows, such as fresh and recirculated air. By code requirements, commercial and industrial buildings normally have automatic fire dampers installed in all the vertical duct runs. All ducts, especially vertical ducts, have the capacity to carry smoke and flames.

A damper used to balance a system should be installed in an accessible place in each branch supply duct. The closer the dampers are to the main duct or supply air plenum, the better. They should be tight-fitting with minimum leakage. The built-in damper that is part of a supply diffuser or register should not be used to balance an air system. It should only be used to apply or shut off airflow to a room. When partially closed, it disrupts the performance of the register and often creates noise.

Generally, single-vane volume dampers are used in branch ducts where the total static pressure is less than 0.5 in. w.c. Multi-vane or multi-blade volume dampers are normally used in ductwork where the total static pressure exceeds 0.5 in. w.c.

### 4.4.0 Duct System Sizing and Design

The design of an air distribution duct system is based on the heat loss and/or heat gain for the entire building and for each of the rooms or areas within the building. To make an initial equipment selection, the required airflow is determined. This value is further refined in designing the duct system or evaluating an existing system. It is possible that the initial equipment selection will need to be changed somewhat, as it may be determined that it cannot overcome the static-pressure losses in the duct system. In other words, what is discovered during the air distribution system design process can affect the final equipment selection. This is far more common in commercial/industrial applications than in residential.

### 4.4.1 General Design Procedure

Once the total building air volume requirement and the heat gain and losses for each area in the building are obtained, the duct design process can begin:

*Step 1* Study the building plans, and locate the supply outlets and return inlets to provide proper distribution and circulation of the air within each space.

*Step 2* Calculate the air volume to be delivered by each supply outlet and returned by each return inlet. Size the supply outlets and return inlets using the air volume and the grille manufacturer's product data sheets.

*Step 3* Sketch the duct system, connecting the supply outlets and return inlets with the air handling units/air conditioners. Note that the physical layout of the duct system is driven by the building construction and the placement of other building system components.

*Step 4* Divide the duct system layout into sections and identify each of the sections. The duct system should be divided at all points where flow direction, size, or shape changes.

*Step 5* Size the main and branch ducts by the selected method. Methods commonly used are the equal-friction method, static-regain method, and velocity-reduction method. There are a variety of apps available for this purpose, in addition to manual duct-design calculators produced by SMACNA and major HVACR equipment manufacturers.

*Step 6* Calculate total system pressure loss and then ensure the selected fan can provide the needed airflow against this calculated resistance.

*Step 7* Lay out the system in detail. It often takes several attempts to find a duct and air distribution layout that fits the building's construction and functions well. If duct routing and fittings vary greatly from the original plan, it may be necessary to recalculate the pressure losses and resize the duct. Load-estimating software that includes a duct system design feature can simplify the duct system design process. *Figure 52* shows a screen from a popular load-estimating program that has this feature.

### 4.4.2 Selecting Supply Outlets and Return Inlets

Because the heating and cooling loads are not distributed evenly throughout a building, the number and location of supply outlets and return inlets must be accurately determined for each conditioned area; otherwise, the air system will perform poorly. Once installed, it can be difficult and costly to make corrections. One manufacturer recommends that at least one outlet be used for each 4,000 Btuh heat gain, or 8,000 Btuh heat loss, whichever is greater, in any room or area. The 4,000 Btuh value equates to roughly 130 cfm of airflow. However, commercial air distribution systems often have larger registers and diffusers that provide far more airflow than this.

A good duct design provides for as much capacity for return air as it does for supply air. Individual return grilles can be placed in rooms for good air circulation and to avoid stratification. Normally, this approach provides better performance than a central return. However, it is also a costlier approach that is not often found in residential systems.

Return air inlets should not be located directly in the primary airstream from supply outlets. If this occurs, the supply air can be short circuited back into the return air system without mixing well with the room air. Also, return inlets should not be installed in areas such as bathrooms that can have undesirable odors. Otherwise, the odors will spread throughout the building by way of the duct system. When a return air path is not placed in spaces that can be isolated, the entry door can be undercut to allow air to escape the room. Alternatively, transfer grilles and ducts can be used that allow air to leave the room. A transfer duct is simply a short section of duct with a grille on each end, allowing air to exit one room and enter another.

### 4.4.3 Calculating Supply Outlet Volume and Size

To determine the size of the supply outlets for each area, it is first necessary to calculate and record the air volume needed for each room. Calculate the air volume for both cooling and heating. As discussed earlier in this module, the air volumes for cooling and heating are calculated using the following formulas:

*Cooling*:

$$\text{Air volume flow rate (cfm)} = \frac{\text{sensible load (Btuh)}}{1.08 \times (T_1 - T_2)}$$

---

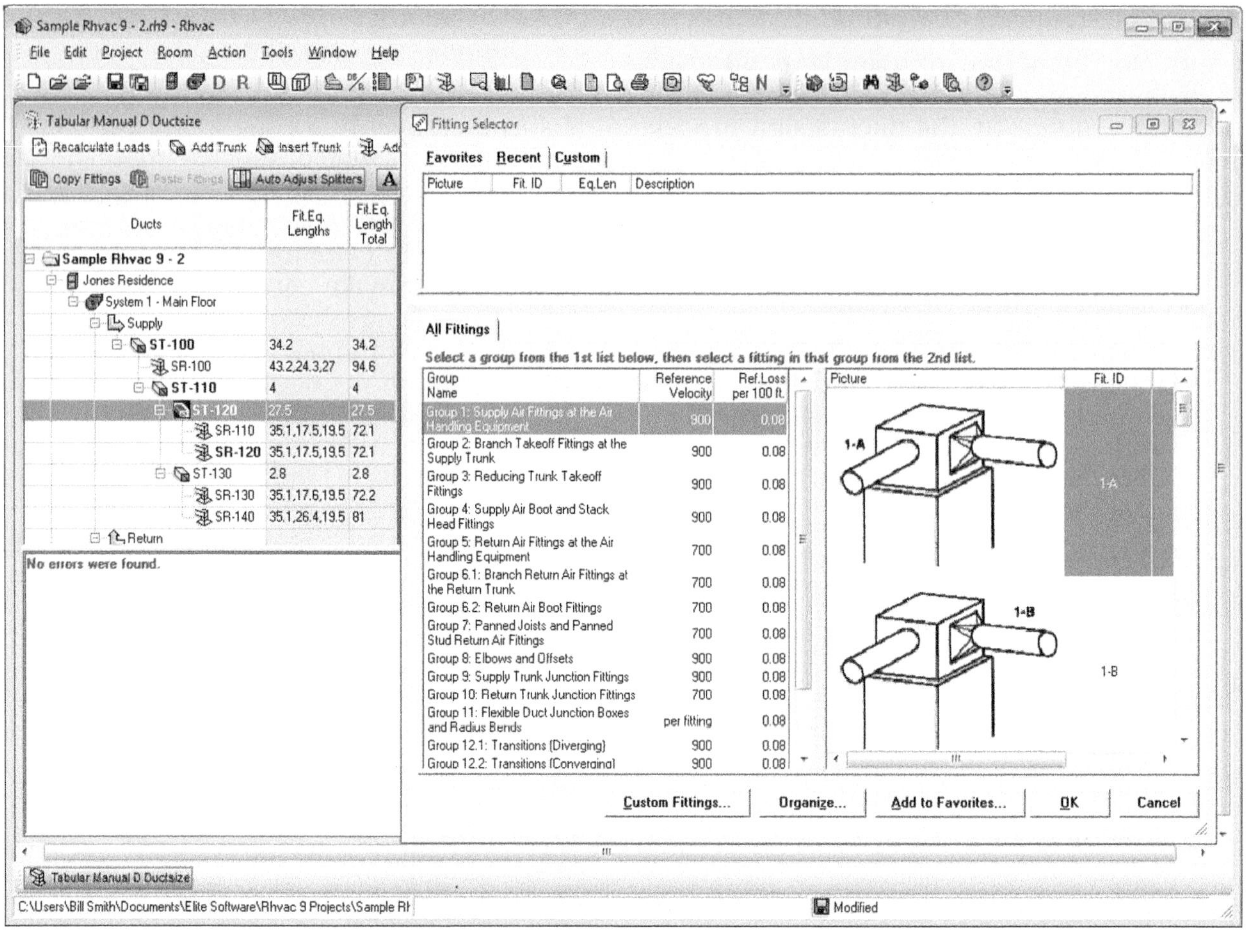

**Figure 52** Example of a data screen from duct design software.

*Heating:*

$$\text{Air volume flow rate (cfm)} = \frac{\text{sensible load (Btuh)}}{1.08 \times (T_2 - T_1)}$$

*Where:*

Sensible load = value in Btuh for the heat gain or loss as determined by a room-by-room load estimate

$T_1$ = design outdoor temperature

$T_2$ = design indoor temperature

Once the cooling and heating air volume for each room have been calculated, the larger of the two values (cooling or heating) is used to size the supply outlets for that room. The required flow rate for cooling is almost always the larger of the two, but the heating requirement may be larger in some areas. If more than one outlet is being used to supply a room, the total room air volume must be divided by the number of outlets to determine the size of the individual outlets. For example, if a room requiring 184 cfm is supplied by two outlets, the air that must be delivered by each outlet would be 92 cfm.

Once the air volume for each supply outlet is known, the types of diffusers and their sizes can be selected using the manufacturer's product data. *Figure 53* shows an example of the air volume requirements for various rooms in a house and the selected sizes of supply outlets. *Appendix A* lists the air volume ratings and sizes for the most common types of ductwork and supply/return grilles.

Compromises are almost always made when sizing supply outlets. In the example shown, each of the room outlets could have been custom-sized to match its air volume more closely. However, to give a uniform appearance in the rooms, a standard size is commonly used for most, if not all, outlets.

In the example shown, both the $2\frac{1}{2} \times 14$ and $4 \times 10$ floor diffusers can supply the needed air volume for all outlets. However, the $4 \times 10$ outlet size was selected because the $2\frac{1}{2} \times 14$ floor diffusers are more difficult to fit between the floor joists. Also, the $4 \times 10$ outlet is a standard size stocked by most suppliers.

# AIR SYSTEM DESIGN WORKSHEET

Job Name **EXAMPLE**

Job No.

Date

Address

Estimator

Salesman

## SUPPLY SIZING

| ROOM NAME | CFM FROM LOAD ESTIMATE | | GREATER CFM** | CFM PER OUTLET | SUPPLY OUTLET TYPE | OUTLET NO. | ACTUAL SUPPLY OUTLET SIZE (IN) | SLECTED SUPPLY OUTLET SIZE (IN) | BRANCH DUCT SIZE | REMARKS |
|---|---|---|---|---|---|---|---|---|---|---|
| | HEAT | COOL | | | | | | | | |
| L.R. | 102 | 154 | 154 | 77 | FLOOR | 2, 3 | 2 1/4 x 12 | 4 x 10 | 6" | |
| KIT. | 155 | 184 | 184 | 92 | FLOOR | 11, 12 | 2 1/4 x 14 | 4 x 10 | 6" | |
| D.R. | 73 | 52 | 73 | 73 | FLOOR | 1 | 2 1/4 x 12 | 4 x 10 | 6" | |
| LAV. | 13 | 14 | 14 | 14 | FLOOR | 9 | 2 1/4 x 10 | 4 x 10 | 6" | |
| BATH | 32 | 23 | 32 | 32 | FLOOR | 5 | 2 1/4 x 10 | 4 x 10 | 6" | |
| BR #1 | 90 | 71 | 90 | 90 | FLOOR | 6 | 2 1/4 x 14 | 4 x 10 | 6" | |
| BR #2 | 80 | 65 | 80 | 80 | FLOOR | 7 | 2 1/4 x 12 | 4 x 10 | 6" | |
| BR #3 | 52 | 58 | 58 | 58 | FLOOR | 8 | 2 1/4 x 10 | 4 x 10 | 6" | |
| FOYER | 30 | 36 | 36 | 36 | FLOOR | 4 | 2 1/4 x 10 | 4 x 10 | 6" | |
| B.FOYER | 53 | 63 | 63 | 63 | FLOOR | 10 | 2 1/4 x 10 | 4 x 10 | 6" | |
| | | | 784 | | | | | | | |

## RETURN SIZING

| RETURN LOCATION | RETURN NO. | CFM** | GRILLE SIZE (IN) | NO. STUD SPACES REQ'D | NO. PANNED JOIST REQ'D | DUCT SIZE | REMARKS |
|---|---|---|---|---|---|---|---|
| BR #1 | | 90 | 12 x 6 | 1 | 1 | PANNED | LOW WALL |
| BR #2 | | 80 | 12 x 6 | 1 | 1 | | RETURNS |
| BR #3 | | 58 | 12 x 6 | 1 | 1 | | |
| HALL S | | 111 | 12 x 6 | 1 | 1 | | |
| HALL N | | 111 | 12 x 6 | 1 | 1 | ↓ | ↓ |
| D.R. | | 334 | 30 x 8 | 2 | 2 | | |
| | | 784 | | | | | |

* The larger of the heating or cooling cfm is used in this column.
** Return sizing should be based on supply cfm.

*Figure 53* Example of room-by-room worksheet.

### 4.4.4 Calculating Return Inlet Volume and Size

A good design provides as much capacity for return air as it does for supply air. Many well-designed duct systems fail to deliver the correct amount of air because the returns feeding the air handler are too small or have insufficient or small return grilles. Pay as much attention to the return duct as you do to the supply duct. Sizing return grilles is done the same way as sizing supply outlets. *Figure 53* also shows the return air volume requirements for the various rooms used in the previous supply outlet example and the selected sizes of return inlets and grilles. As with the supply outlets, the return air grille sizes were also standardized.

When the air handler is installed in a crawlspace, attic, or other limited-access location, it is common to include filters in the return air grilles. This replaces the need for a filter at the return side of the air handler. If very high efficiency filters, such as high-efficiency particulate arrestance (HEPA) filters, are used, it can result in extreme pressure drop through the filter. A common residential blower for example, cannot move air effectively through a HEPA filter.

The pressure drop of unusual filters must be taken into account during the design process. When filtration is provided at the return grille, designers must consider the velocity of air passing through the filter for it to work properly without causing a significant restriction. Filter grilles should generally be sized larger than return grilles that do not house a filter.

### 4.4.5 Sizing the Ducts

There are several acceptable methods used to size ductwork. This module uses a method called the equal-friction method. This method was selected because it is the most common means of sizing low-pressure supply, return, and exhaust duct systems. Also, most of the manufacturers' duct-design calculators are based on this method. Equal friction does not mean that the total friction remains constant throughout the system. It means that a specific friction loss or static pressure loss per 100' of duct is selected before the ductwork is laid out. Then, this loss value per 100' is used consistently to size the duct. The equal-friction method automatically reduces the air velocities in the direction of airflow such that, by using a reasonable initial airflow velocity, the chances of introducing noise are reduced or eliminated.

The total equivalent length of fittings in a duct run is the sum of all the individual equivalent lengths. *Figure 54* shows an example of a simplified duct system. The method used to find the equivalent length of fittings in the duct run to outlet S5 is explained in this section. To aid in understanding this example, the types of fittings in the duct run to outlet S5 have been marked on the figure. For each duct fitting, the capital letter identifies the type of fitting and the number indicates the equivalent length of straight duct in feet for that fitting.

As shown, the fittings in the path from the air handling unit to outlet S5 (not including the length of straight duct involved) has an equivalent length of 115', derived as follows:

- (1) G30 Floor diffuser box = 30'
- (3) E10 90-degree adjustable elbows = 30'
- (1) B15 Side takeoff = 15'
- (1) E5 Reduction = 5'
- Additional loss for one of the first three takeoff fittings after the E5 transition = 25'
- (1) B10 Plenum takeoff = 10'

$$30' + 30' + 15' + 5' + 25' + 10' = 115'$$

When sizing supply outlets, it is important to remember that the capacity of a supply diffuser or grille is no greater than that of the branch duct connected to it. This can be an issue when using wall vertical-stack ducts installed inside the building partition walls to feed the supply outlets on the upper floors of a building. For example, a 4 x 10 diffuser with a capacity of 110 cfm can only deliver about 100 cfm if fed by a $3\frac{1}{4} \times 10$ vertical-stack duct. If more than one outlet is connected to the duct, they duct will limit their capacity.

To size the ducts, it is necessary to find the supply and return runs that have the longest total equivalent length. The total equivalent length of a run is the measured horizontal and vertical straight lengths of duct in the path, plus the equivalent length of all fittings in the path. Using the example shown in *Figure 54*, the supply run for outlet S5 has a total equivalent length of 157' (115' + 42'). Note that the duct run to outlet S4 has the longest total equivalent length of 179' (130' + 49') for the example system.

This comparison between the S5 and S4 duct runs is presented to emphasize that if the duct run with the greatest resistance is not obvious, it may be necessary to calculate the resistance of several or all of the runs in order to make certain that the longest equivalent length is used to establish the system pressure loss.

The sizing of trunk and branch ducts can be done using different methods. There are various friction charts to choose from, as well as simple manual calculators (*Figure 55*) and apps (*Figure 56*). All come with a set of instructions describing

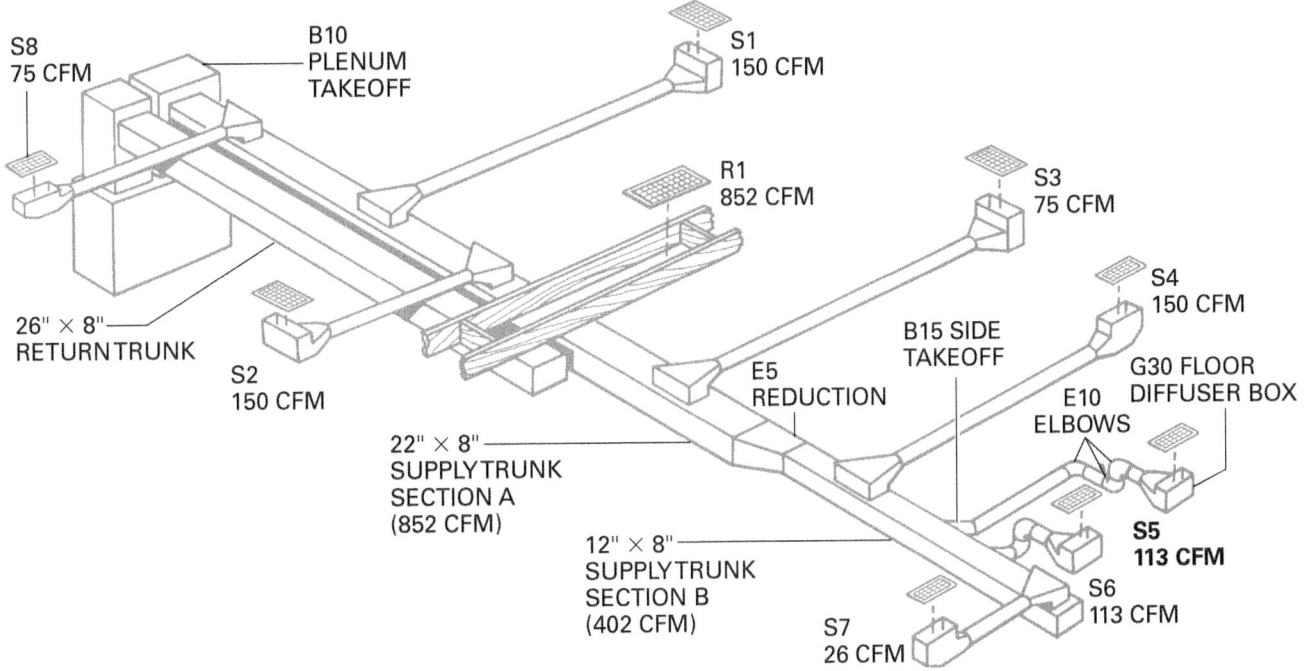

| OUTLET | EQUIVALENT LENGTH OF FITTINGS (FT) | MEASURED LENGTH OF DUCT (FT) | TOTAL EQUIVALENT LENGTH OF RUN (FT) | BRANCH DUCT SIZE (ROUND) |
|---|---|---|---|---|
| S1 | 105 | 29 | 134 | 7" |
| S2 | 105 | 19 | 124 | 7" |
| S3 | 80 | 43 | 123 | 6" |
| S4 | 130 | 49 | 179 | 7" |
| S5 | 115 | 42 | 157 | 6" |
| S6 | 105 | 42 | 147 | 6" |
| S7 | 105 | 42 | 147 | 6" |
| S8 | 125 | 12 | 137 | 6" |
| RETURN R1 | 140 | 24 | 164 | |

*Figure 54* Example duct system.

how to use them to size duct. Most, with the appropriate knowledge of the process, can be used to design a system without additional references. The apps are quite versatile and are available for smartphones and tablets as well as for computers.

The app allows the user to input the desired air volume or velocity, then use sliders or a keyboard to change the duct size. As the duct size changes, the resulting information is displayed for the chosen air volume or velocity. Refer to *Figure 57*. An air volume value of 800 cfm has been entered, and the duct size set to 16" × 10". The equivalent round-duct diameter of 13.7" is offered, along with a friction loss of 0.064 in. w.c. per 100' of duct. The

velocity is shown to be 777.7 fpm. The designer can adjust the duct size accordingly if the values do not fall within the desired parameters.

Let's assume the designer is working with a design static pressure per 100' of duct of 0.08 in. w.c. In *Figure 58*, we see that the friction in the Exact column is 0.064 in. w.c. To see if it is advantageous to change the duct size, the slider for duct height in *Figure 57* has been reduced until the friction per 100' of duct equals 0.08 in. w.c. The duct size is now 16" × 9.2". Many shops prefer to work only with whole numbers, so the final size chosen would likely be 16" × 9".

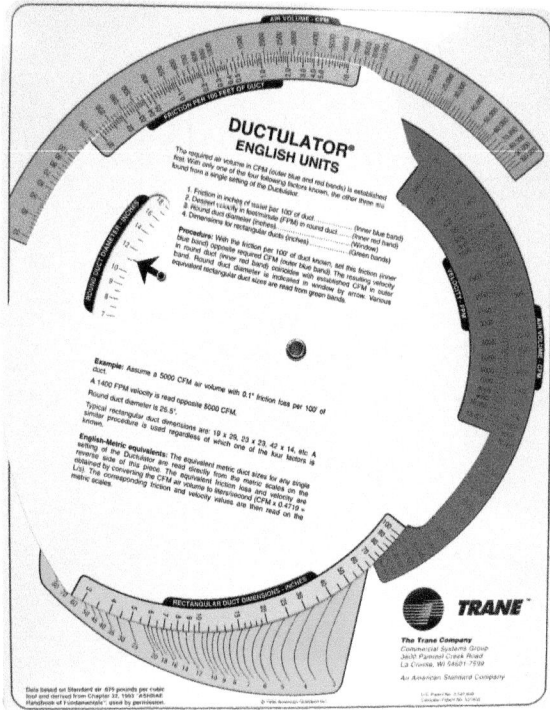

*Figure 55* Duct-sizing calculator.

One important point to remember is that the friction values are based on losses per 100' of duct. When using the app to size the duct for a run with a total equivalent length that is greater or less than 100', the friction loss must be corrected for the actual equivalent length of the duct. The corrected value is easily calculated as follows:

Corrected friction =
$$\text{total equivalent} \atop \text{duct length} \times \left( \frac{\text{design friction per 100'}}{100} \right)$$

For example, here is the corrected friction loss for a duct with a total equivalent length of 35', sized at a design friction loss of 0.08 in. w.c. per 100':

Corrected friction =
$$35 \times \left( \frac{0.08 \text{ in. w.c.}}{100} \right) = 0.028 \text{ in. w.c.}$$

The actual friction loss of 0.028 in. w.c. for this 35' section of duct is then added to the rest of the losses as part of the total system friction loss. Failing to correct the friction for the actual length of duct can lead to an incorrect blower selection and more or less airflow than expected. Most apps used allow for the entry of the duct length, and the total loss for the entire length is calculated automatically.

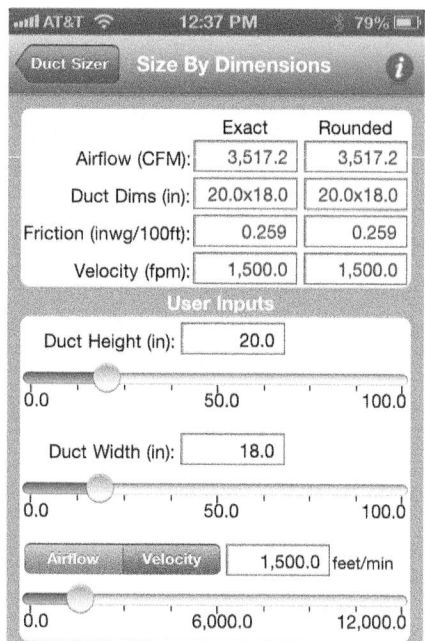

*Figure 56* Duct-sizing app.

Simple low-velocity systems can also be sized using tables. They typically show the specific round and/or rectangular duct sizes used for various air volumes. Such charts are often based on a predetermined design supply duct friction loss of 0.08 in. w.c. per 100'. This means that the duct system will create 0.08 in. w.c. of friction loss for each 100' of equivalent length that the air travels through. A predetermined design friction loss of 0.05 in. w.c. per 100' is often used for the return duct sizing on charts. These values are shown in footnotes to inform the user of the basis for the information.

Ductwork is most often sized starting at the air handler or furnace and working toward the end of the system. Refer to the duct system example in *Figure 54*. Section A of the supply trunk must carry the total system air quantity of 852 cfm to feed outlets S1 through S8. The duct size selected is 22" × 8". Section B of the supply trunk is sized as 12" × 8". This is based on the need to deliver the 402 cfm of air used to feed outlets S4 through S7. In the Figure 54 example, the design static pressure for the supply duct was 0.06 in. w.c. per 100'.

The return duct is sized at 26" × 8". It must return 852 cfm to the system, and is sized for a friction loss of roughly 0.05 in. w.c. per 100'. As shown, the branch ducts are custom-sized using round duct based on their air volume.

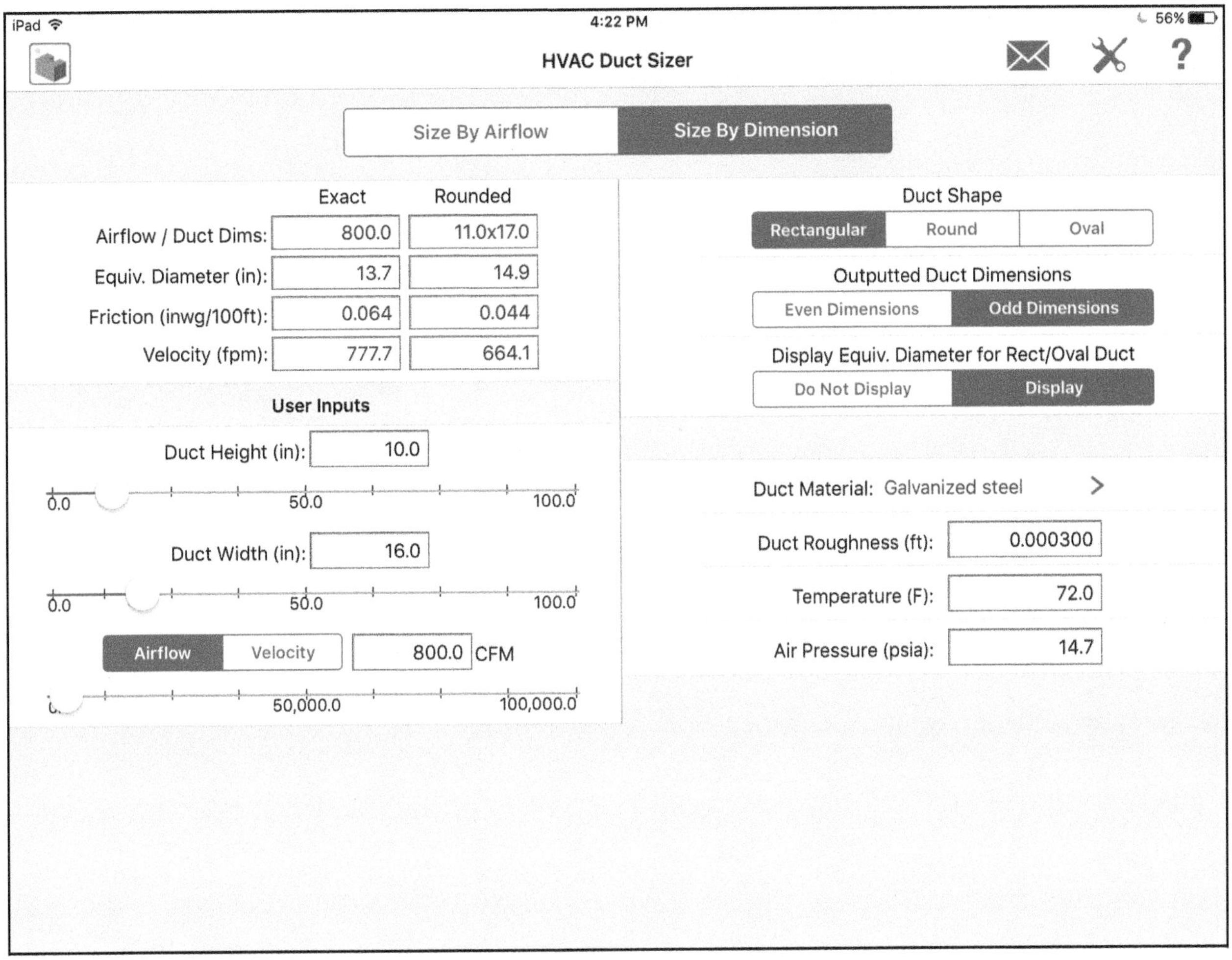

| | Size By Airflow | Size By Dimension |

| | Exact | Rounded |
| --- | --- | --- |
| Airflow / Duct Dims: | 800.0 | 11.0x17.0 |
| Equiv. Diameter (in): | 13.7 | 14.9 |
| Friction (inwg/100ft): | 0.064 | 0.044 |
| Velocity (fpm): | 777.7 | 664.1 |

**Duct Shape**

| Rectangular | Round | Oval |

**Outputted Duct Dimensions**

| Even Dimensions | Odd Dimensions |

**Display Equiv. Diameter for Rect/Oval Duct**

| Do Not Display | Display |

**User Inputs**

Duct Height (in): 10.0

0.0     50.0     100.0

Duct Width (in): 16.0

0.0     50.0     100.0

| Airflow | Velocity | 800.0 CFM |

50,000.0     100,000.0

Duct Material: Galvanized steel   >

Duct Roughness (ft): 0.000300

Temperature (F): 72.0

Air Pressure (psia): 14.7

*Figure 57* Example using an app for duct sizing.

Depending on the application, compromises are sometimes made when sizing branch ducts. For example, outlets S5 and S6 are shown supplied with 6" round duct, even though they must carry 113 cfm. This was done for the sake of duct-size standardization. A designer may prefer a 7" diameter duct to reduce friction loss, velocity, and noise. However, 7" round duct is very rare in the marketplace. Common sizes and the cost of customizing duct sizes must be considered carefully.

Flexible insulated duct is widely used, and has a much higher resistance to airflow than sheet metal duct. Therefore, a larger diameter of flex duct may be required to get the same performance. The information found in *Table 5* was derived from software to provide a quick reference, and is based on a friction loss of 0.08 in. w.c. per 100'. It shows that a 6" sheet metal duct can carry 100 cfm, while a 6" flex duct can carry only 55 cfm at the same friction loss. To carry 100 cfm through a flex duct, an 8" diameter would be required to avoid exceeding the desired friction loss.

The total friction loss of the system needs to be calculated to ensure the system blower can provide the needed performance. As previously shown, the total friction loss for an air system is the sum of the supply system loss, return system loss, and components not included in the system fan rating. The friction loss for the supply system ductwork is found by taking the equivalent length of the longest run and multiplying it by the system design friction rate. In the example system in *Figure 54*, the run feeding outlet S4 is the longest run, with a total equivalent length of 179'. The design friction loss used for duct sizing was 0.06 in. w.c. per 100'. Therefore, the supply-duct friction loss for the example system is determined as follows:

Friction loss for longest run =
   179' × 0.06 in. w.c./100'
Friction loss for longest run = 179' × 0.0006
Friction loss for longest run = 0.143 in. w.c.

The pressure loss of the supply diffuser at the end of the run must also be added to the supply duct loss. This value can be found using the manufacturer's product data sheets for the grille. A common friction loss for a properly sized diffuser

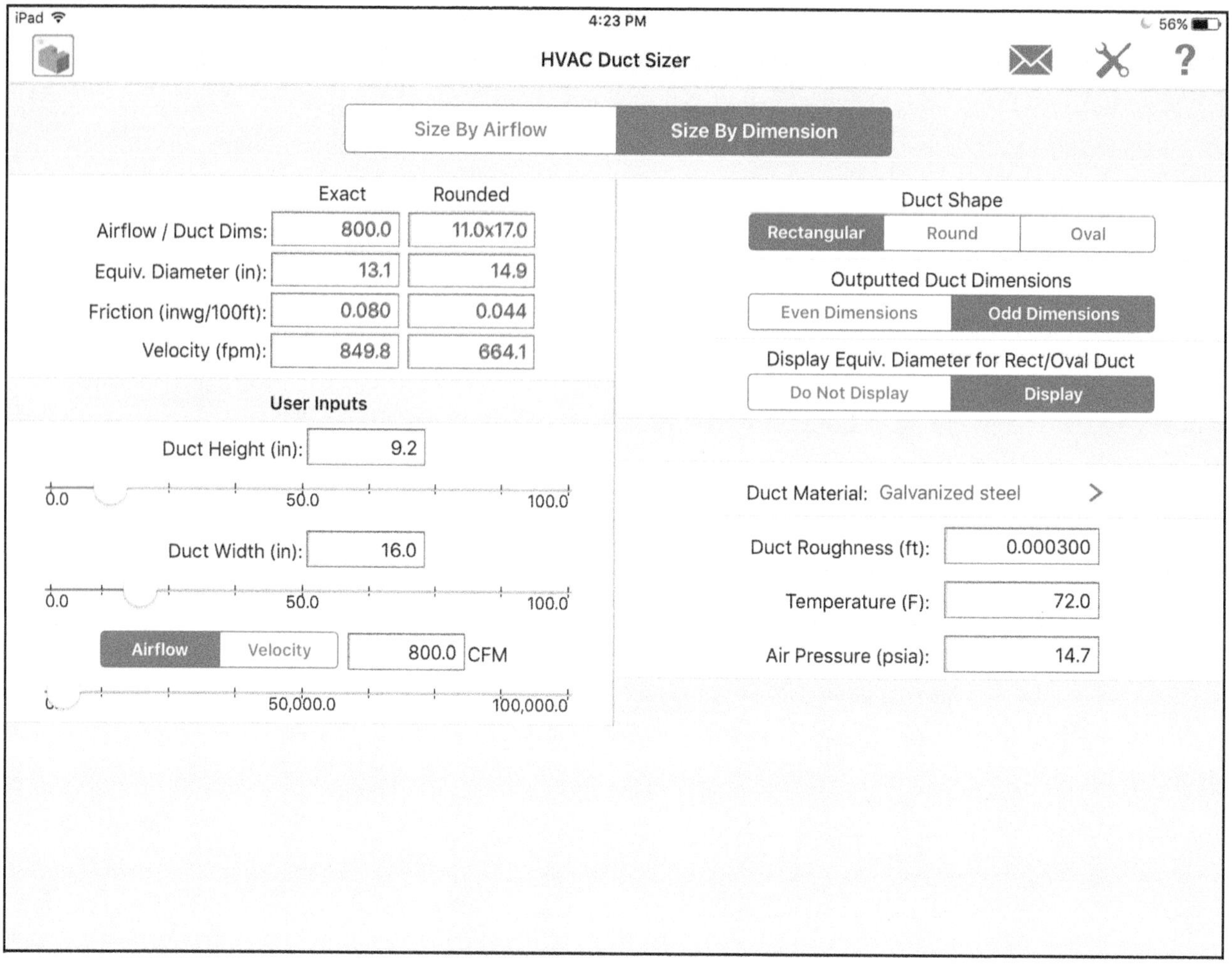

**Figure 58** Adjusting duct size to achieve a target friction loss.

is 0.05 in. w.c. Using that value for our example and adding it the supply duct loss, the total supply loss is 0.193 in. w.c.

The friction loss for the return system duct is found the same way as for the supply system duct. Take the equivalent length of the longest return run and multiply it by the return system design friction rate. In the example system in *Figure 54*, the duct run for return R1 is the only return run. It has a total equivalent length of 164'. Also, the design friction rate used for the return duct system was 0.05 in. w.c. per 100' of duct. Therefore, the return friction loss for the example system is:

Return duct friction loss =
164' × 0.05 in. w.c./100'
Return duct friction loss = 164' × 0.0005
Return duct friction loss = 0.082 in. w.c.

Remember that the pressure loss of the return grille must be added to the return duct loss. This information can be found using the manufacturer's product data sheets for the specific grille. However, for this example, use the same value of 0.05 in. w.c. that was used for the supply diffuser.

The total resulting return duct friction loss is 0.132 in. w.c.

The last component of the system friction loss accounts for the losses for components not included in the system fan rating, such as a wet evaporator coil or an electronic air cleaner. The friction loss for each of these is normally provided in the equipment product literature. A friction loss of 0.25 in. w.c. is within the common range of friction losses for these items, and will be used to complete the calculation for the total system loss of the example system:

Total system friction loss =
0.139" supply + 0.132" return + 0.25" equipment =
0.575 in. w.c.

For any system, there should be an operating balance between the energy produced by the fan and the energy consumed by the duct system. For the example system, at the design airflow of 852 cfm, the total friction loss against which the fan must operate is 0.575 in. w.c. Once the system external static pressure is known, the next step is to refer to the manufacturer's product data sheets

NCCER – *HVAC Level Four*   03407

**Table 5** Maximum CFM Through Runout Ductwork

| Runout Size | Supply CFM | Return CFM |
|---|---|---|
| Sheet Metal or Ductboard | | |
| 5" diameter | 60 | 45 |
| 6" diameter | 100 | 75 |
| 7" diameter | 140 | 110 |
| 8" diameter | 210 | 160 |
| 3¼" × 8" stack | 70 | 55 |
| 3¼" × 10" stack | 100 | 75 |
| 3¼" × 14" stack | 140 | 110 |
| 2¼" × 12" stack | 70 | 55 |
| 2¼" × 14" stack | 90 | 70 |
| Flex Duct* | | |
| 6" diameter | 55 | 40 |
| 8" diameter | 120 | 90 |
| 10" diameter | 200 | 160 |
| 12" diameter | 320 | 250 |
| 14" diameter | 480 | 375 |
| 16" diameter | 660 | 530 |
| 18" diameter | 880 | 680 |
| 20" diameter | 1,200 | 900 |

*The maximum duct capacity varies depending upon length, bends, and sags. The numbers shown assume straight runs cut to the proper length.

for the selected equipment to verify that its fan can deliver the needed volume of air at this resistance, and what fan speed may be required to do so.

*Figure 59* shows an example of a blower performance chart for a furnace. Since an external static pressure of 0.6 in. w.c. is the closest to our condition of 0.575 in. w.c. of friction loss, use this column of the chart. In this case, all the available furnace models in the chart can meet the airflow requirement in low or medium-low speed.

### 4.4.6 Existing Duct and Supply Outlet Capacity

When sizing equipment for an add-on or replacement job, it is necessary to evaluate the existing duct system to see if it has the capacity to handle the increase in required air volume. One method of determining this is to calculate the capacity ratio of the new system air volume requirements to total capacity of the existing duct system using the formula provided here:

$$\text{Duct capacity ratio} = \frac{\text{New supply cfm}}{\text{Existing supply-air outlet capacity}}$$

For an existing air distribution system, the new air volume needed should not exceed the duct's existing capacity by more than 10 percent. It should be considered unacceptable if it exceeds it by more than 15 percent.

The supply outlets and return grilles should also be checked for the following:

- Check the type of existing supply outlets to make sure they are suitable for the application and climate.
- Compare the airflow capacities of the existing supply outlets and return inlets with the air volume requirement for the upgraded system. The capacity ratio for each supply outlet or return grille should be calculated using the formula below:

$$\text{Outlet capacity ratio} = \frac{\text{Required supply cfm}}{\text{Total existing outlet capacity (cfm)}}$$

The required air volume should not exceed the existing grill/diffuser capacity by more than 15 percent. If it does, additional or different grilles and diffusers must be added to the project.

### 4.4.7 Insulation and Vapor Barriers

When ducts pass through an unconditioned space, such as an attic or crawl space, heat transfer takes place between the air in the duct and the air in the unconditioned space. Insulation should be applied to the duct if it passes through an unconditioned space. Many installations use ductboard for the main supply and return ducts, which does not require added insulation.

| FAN PERFORMANCE – HIGH-EFFICIENCY FURNACE AIR DELIVERY - CFM (WITH FILTER) | | | | | | | | |
|---|---|---|---|---|---|---|---|---|
| FURNACE | | EXTERNAL STATIC PRESSURE (IN. W.G.) | | | | | | |
| SIZE | SPEED | 0.1 | 0.2 | 0.3 | 0.4 | 0.5 | 0.6 | 0.7 |
| 40-A | High | 1425 | 1370 | 1320 | 1265 | 1195 | 1125 | 1060 |
| | Med-High | 1315 | 1275 | 1230 | 1180 | 1125 | 1070 | 1000 |
| | Med-Low | 1145 | 1115 | 1090 | 1045 | 1010 | 955 | 880 |
| | Low | 1035 | 1000 | 970 | 930 | 900 | 850 | 800 |
| 40-B | High | 1515 | 1465 | 1405 | 1345 | 1275 | 1210 | 1115 |
| | Med-High | 1350 | 1310 | 1275 | 1220 | 1160 | 1100 | 995 |
| | Med-Low | 1155 | 1140 | 1110 | 1070 | 1020 | 945 | 825 |
| | Low | 975 | 960 | 945 | 915 | 870 | 780 | 710 |
| 60 | High | 1510 | 1450 | 1380 | 1320 | 1230 | 1150 | 1045 |
| | Med-High | 1350 | 1300 | 1245 | 1190 | 1115 | 1030 | 945 |
| | Med-Low | 1165 | 1125 | 1090 | 1045 | 980 | 910 | 775 |
| | Low | 965 | 945 | 915 | 880 | 830 | 735 | 640 |
| 80 | High | 1590 | 1530 | 1470 | 1390 | 1315 | 1225 | 1140 |
| | Med-High | 1425 | 1390 | 1345 | 1285 | 1200 | 1135 | 1040 |
| | Med-Low | 1250 | 1220 | 1185 | 1135 | 1085 | 1010 | 930 |
| | Low | 1060 | 1055 | 1030 | 980 | 930 | 870 | 805 |

*Figure 59* Blower performance chart example.

Metal duct can be insulated in three ways: on the outside, on the inside, or both. Insulation on both sides is rarely needed, but is sometimes used. Insulation inside the duct is installed by the duct fabricator. Insulation with a vapor barrier can be wrapped around the outside of the duct after it has been installed. Always use a vapor barrier on duct insulation. Local code requirements may dictate the insulation thickness. Once installed, all insulation joints must be properly sealed. To avoid condensation damage, all punctures, seams, and slits in the vapor barrier must be sealed with tape or mastic.

Sheet metal duct with exterior insulation has a lower pressure loss, and is therefore more efficient than sheet metal lined on the inside. The surface of the liner creates more friction loss than the smooth surface of the metal duct. Another advantage of exterior insulation is that the cost of the unlined metal duct is lower, since the duct doesn't have to be larger to accommodate the liner thickness. A duct with 1" of internal insulation must have width and height dimensions that are 2" larger to deliver the same amount of air. In fact, since the liner creates greater friction loss, the duct may need to be even larger. The disadvantages of using duct with external insulation are that it requires field labor to install; there is a greater chance of damage to the material; and it tends to be noisier than a lined system.

In attic installations, ducts must be insulated to maintain proper cooling and heating in the conditioned rooms of the building. *ASHRAE Standard 90-80* specifies the minimum acceptable value (R-value) of insulation. *Figure 60* shows an example using *ASHRAE Standard 90-80* to calculate the R-value for insulation related to the cooling mode.

The R-value must be calculated for the heating mode in the same manner. The amount of insulation eventually used is determined by the mode with the greatest need. Because attic systems are more common in warm climates, it is the cooling mode that usually determines the amount of insulation required there. Local code requirements may, however, call for even thicker insulation.

## 4.5.0 Commercial Building Design Factors

*ACCA Manual N* provides load-estimating data and instructions for small to medium commercial buildings using packaged or split systems, including heat pumps. Neither method is intended for large buildings that use built-up systems or central station equipment. Such designs must be done by qualified engineers and architects using information and procedures from the latest ASHRAE handbook.

Although some of the load factors and calculations for commercial systems are similar to those for residential systems, there are several major differences. Some examples of how commercial estimates differ from residential estimates are as follows:

- Commercial buildings use different construction materials. While homes lean toward wood frame or masonry construction, commercial buildings often use facades of metal, masonry, and glass.
- It is often necessary to deal individually with core and exterior zones. Because of the effect of radiation and infiltration on the perimeter of a building, the loads at the perimeter and core can vary widely. It is not unusual to have situations during occupied periods when perimeter zones need cooling and the core zone needs heating or just ventilating.
- A commercial building is more likely to have a large, open area, such as a lobby or atrium with vaulted ceilings, and a lot of traffic in and out of the building with its resulting infiltration.
- The load may vary significantly as the day progresses. Solar gain can be very high on a large structure with a great deal of glass. Since the sun moves throughout the day, the load imposed by it moves as well. Unlike *Manual J*, the calculations in *Manual N* are made for different times of day because the load varies greatly between those times. Note especially how different the solar radiation and transmission loads are from noon to 6 pm. Also note the change in sensible cooling load.
- The people load is a critical factor in commercial system design. Shopping centers, theaters, and office buildings have some periods when they are heavily occupied and other periods when they may be unoccupied. The load varies widely between these periods. A lot of human traffic also means many door openings and closings.
- Loads can vary radically from one area to another within the building. While this is true to some extent in residential work, it is less of a factor because residential buildings are generally much smaller and use different construction materials. In a southern-exposure room with a lot of glass at mid-afternoon, the load might be three to four times that of a northern exposure room and two to three times that of a southern exposure room with no glass.

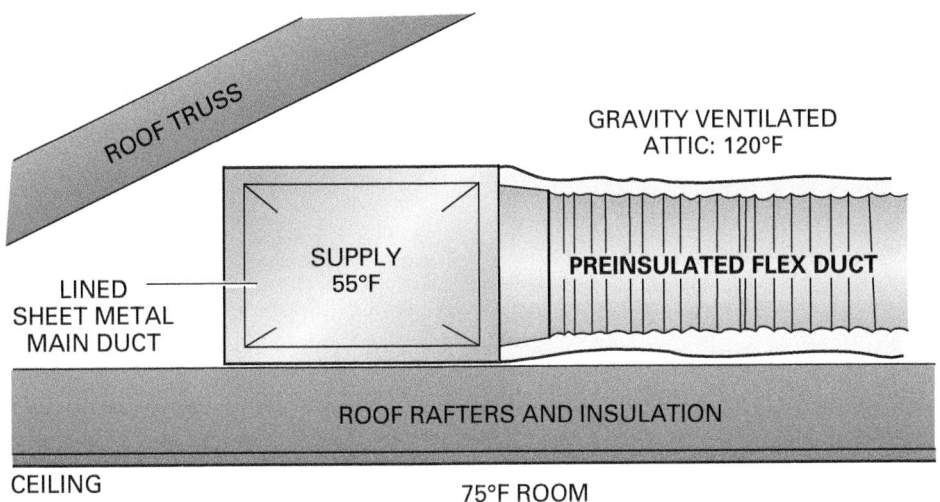

$$
\begin{array}{lll}
\mathbf{TD}_{DUCT} = 120 - 55 & R = TD/15 & 1'' = \text{ABOUT} \\
\quad = 65°F & \quad = 65/15 & \quad R4 \\
& \quad = 4.3 &
\end{array}
$$

$$\mathbf{R = TD/15}$$

WHERE:

  R = THERMAL RESISTANCE TO HEAT FLOW
    OF THE DUCT WITH INSULATION
    (R-VALUE)

  TD = TEMPERATURE DIFFERENCE BETWEEN
    INSIDE AND OUTSIDE DUCT

*Figure 60* Example of R-value calculation using *ASHRAE Standard 90-80.*

• The interiors of commercial buildings are often rearranged to accommodate tenants. The system design must therefore be flexible enough to deal with changing load patterns.

Commercial buildings have many of the same heating and cooling loads as residential buildings. They also have other cooling-load considerations such as:

• Large latent loads
• Large lighting loads
• Computers
• Printers and copiers

## Additional Resources

*FEMA 412, Installing Seismic Restraints for Mechanical Equipment.* Washington, DC: Federal Emergency Management Agency.

*HVAC Duct Construction Standards—Metal and Flexible.* Chantilly, VA: Sheet Metal and Air Conditioning Contractors National Association (SMACNA).

*Manual J, Residential Load Calculation.* Washington, DC: Air Conditioning Contractors of America (ACCA).

*Manual N, Commercial Load Calculation.* Washington, DC: Air Conditioning Contractors of America (ACCA).

*Manual D, Residential Duct Design.* Washington, DC: Air Conditioning Contractors of America (ACCA).

*Manual G, Selection of Distribution Systems.* Washington, DC: Air Conditioning Contractors of America (ACCA).

### 4.0.0 Section Review

1. The loss coefficient (C-factor) is used to calculate the _____.

   a. pressure loss through a straight duct section
   b. pressure loss through a specific type of fitting
   c. size of a rectangular duct
   d. equivalent length of a specific type of fitting

2. Where is a loop-perimeter duct system likely to be found?

   a. A slab-built home in Florida
   b. A home with a basement in Ohio
   c. A home with a basement in Texas
   d. A slab-built home in North Dakota

3. Why are low-velocity ducts often recommended for comfort cooling?

   a. Lower cost
   b. Quieter operation
   c. Smaller duct size
   d. No insulation required

4. The R-value chosen for duct insulation in an attic is based on the _____.

   a. mode of operation with the greatest temperature difference
   b. R-value calculated for the cooling mode
   c. R-value calculated for the heating mode
   d. mode of operation used the most

5. Which statement about commercial building HVAC system design is *incorrect*?

   a. *ACCA Manual N* is used for commercial-building load calculation.
   b. Large air distribution systems require the expertise of an engineer.
   c. A commercial building load is consistent throughout the day.
   d. Large latent loads are common in commercial buildings.

## SUMMARY

Improper sizing of heating and cooling equipment can lead to comfort problems. If the equipment is undersized or oversized, the building occupants will be uncomfortable, especially on days when design conditions occur. If the air distribution system is improperly designed, similar problems as well as noisy operation can occur. Even if the original system was properly designed, later additions or modifications to the building or system can cause problems.

It is often up to the service technician to spot design or installation problems that could lead to unsatisfactory operation. By understanding the basic principles of system design, technicians will be better equipped to recognize these issues.

1. Below-grade (underground) walls are a factor in calculating _____.

   a. heating and cooling loads
   b. heating loads only
   c. cooling loads only
   d. heat gains

2. The term convection refers to heat that is transferred by the _____.

   a. flow of heat through a substance
   b. light shining off reflective objects
   c. movement of a fluid
   d. contact of materials or objects

3. All these factors affect heat flow through a wall *except* _____.

   a. relative humidity
   b. the area of the wall
   c. the indoor and outdoor temperatures
   d. the heat-conducting properties of the wall

4. The U-factor represents the _____.

   a. amount of below-grade wall
   b. amount of area on the upper floors of a building
   c. rate at which heat will flow through a structure
   d. amount of ultraviolet light to which an area is exposed

5. When the area of a window is multiplied by the HTM, the result represents the _____.

   a. heat gain for the window
   b. heat loss for the window
   c. heating or cooling load in Btuh
   d. heating or cooling load in Btus per sq ft

6. Which of these factors affects both the heating and cooling load calculations?

   a. Floors
   b. Infiltration
   c. Swing factor
   d. Internal factors

7. Latent heat is created by all the following *except* _____.

   a. sunlight
   b. people
   c. appliances
   d. infiltration

8. All the following factors must be considered in the selection of a furnace for a heating-only application *except* _____.

   a. latent heat load
   b. sensible heat load
   c. air volume (cfm)
   d. type of system

9. The indoor unit of a split system contains the _____.

   a. cooling coil, metering device, and blower
   b. cooling coil, condenser, and blower
   c. compressor, condenser, and blower
   d. condenser, metering device, and blower

10. What is the distance apart that the two components of a split system can typically be placed without any special modifications?

    a. 30' (9 m)
    b. 50' (15 m)
    c. 70' (21 m)
    d. 90' (27 m)

11. In an area where the cooling load exceeds the heating load, the furnace should be _____.

    a. gas-fired
    b. oil-fired
    c. a high-air model
    d. outdoors

12. A suction line carries _____.

    a. hot, high-pressure vapor from the compressor to the condenser
    b. low-temperature, low-pressure vapor from the evaporator to the compressor
    c. high-pressure liquid refrigerant from the condenser to the evaporator
    d. high-pressure liquid from the condenser to the metering device

13. When the indoor coil is at a higher level than the outdoor unit, the suction line should _____.

    a. be pitched toward the indoor coil
    b. have a trap at the compressor suction inlet
    c. have a loop that rises to the height of the indoor coil
    d. be no more than 50' in length

14. When liquid lines are being installed, _____.

    a. insulate them if they are exposed to cold weather
    b. insulate them when they are exposed to sunlight or high heat
    c. they are best installed using long vertical runs
    d. oversized lines will cause flash gas, affecting the metering device

15. Airflow velocity is usually measured in _____.

    a. feet per minute (fpm)
    b. cubic feet per minute (cfm)
    c. feet per second (fps)
    d. miles per hour (mph)

16. Within an air distribution system, the highest pressure is found at the _____.

    a. conditioned space
    b. input to the return duct
    c. input to the blower
    d. output of the blower

17. The term *external static pressure* refers to losses caused by _____.

    a. registers and return grilles
    b. conditions external to the duct system
    c. the fan assembly
    d. all components external to the fan assembly

18. Failure to use a corrected friction value when sizing duct runs under 100' long can result in _____.

    a. undersizing the duct
    b. oversizing the duct
    c. incorrect blower selection
    d. incorrect filter selection

| Runout Size | Supply CFM | Return CFM |
|---|---|---|
| **Sheet Metal or Ductboard** | | |
| 5" diameter | 60 | 45 |
| 6" diameter | 100 | 75 |
| 7" diameter | 140 | 110 |
| 8" diameter | 210 | 160 |
| 3¼" × 8" stack | 70 | 55 |
| 3¼" × 10" stack | 100 | 75 |
| 3¼" × 14" stack | 140 | 110 |
| 2¼" × 12" stack | 70 | 55 |
| 2¼" × 14" stack | 90 | 70 |
| **Flex Duct\*** | | |
| 6" diameter | 55 | 40 |
| 8" diameter | 120 | 90 |
| 10" diameter | 200 | 160 |
| 12" diameter | 320 | 250 |
| 14" diameter | 480 | 375 |
| 16" diameter | 660 | 530 |
| 18" diameter | 880 | 680 |
| 20" diameter | 1,200 | 900 |

\* The maximum duct capacity varies depending upon length, bends, and sags. The numbers shown assume straight runs cut to the proper length.

*Figure RQ01*

19. Refer to *Figure RQ01*. What size round flexible duct should be used to carry a supply air quantity of 410 cfm?

    a. 8"
    b. 12"
    c. 14"
    d. 20"

20. Once the total volume of airflow and total external static pressure are both known, the _____.

    a. supply duct material can be chosen
    b. return duct material can be chosen
    c. heat loss and gain can be determined based on these values
    d. blower performance data can be checked to see if it can deliver sufficient airflow

# Trade Terms Introduced in This Module

**Brake horsepower:** The actual total power needed to drive a fan to deliver the required volume of air through a duct system. It is greater than the expected power needed to deliver the air because it includes losses due to turbulence, inefficiencies in the fan, and bearing losses.

**Equal-friction method:** A method of sizing air distribution systems by designing for a consistent amount of pressure loss per unit length of duct (usually 100 feet).

**Homogeneous:** A mixture or material that is uniform in its composition throughout.

**R-value:** The thermal resistance of a given thickness of insulating material.

**Static-regain method:** A method of sizing ducts such that the regain in static pressure due to a decreasing velocity between two points fully or partially makes up for the frictional resistance between the two points.

**Thermal conductivity:** the heat flow per hour (Btuh) through one square foot of one-inch thick homogeneous material for every 1°F of temperature difference between the two surfaces.

**U-factor:** The unit of measure for thermal conductivity.

**Velocity-reduction method:** A method of sizing ducts such that the desired velocities occur in specific duct lengths.

SAMPLE RESIDENTIAL LOAD ESTIMATE

## Load Short Form
### *Entire House*
Bertie Heating & Air Conditioning, Inc.

Job:  6115 NW 117th Place-326...
Date:  Jun 28, 2013
By:  Amy Kauper
Plan:  Creekside Villas Lot 37

1730 NE 23rd Avenue, Gainesville, FL 32609-3904  Phone: 352-331-2005  Fax: 352-371-5770  Email: amy@bertieair.com  Web: www.bertieair.com  License: CAC-058522

### Project Information

For:  Duration Builders, Inc.
11787 NW 61st Terrace, Alachua, FL 32615
Phone: 386-462-0511
Web: www.durationbuilders.com  Email: bjones@durationbuilders.com

### Design Information

|  | Htg | Clg |  | Infiltration |  |
|---|---|---|---|---|---|
| Outside db (°F) | 33 | 92 | Method | | Simplified |
| Inside db (°F) | 70 | 75 | Construction quality | | Average |
| Design TD (°F) | 37 | 17 | Fireplaces | | 0 |
| Daily range | - | M | | | |
| Inside humidity (%) | 50 | 50 | | | |
| Moisture difference (gr/lb) | 32 | 47 | | | |

### HEATING EQUIPMENT

| | |
|---|---|
| Make | Goodman Mfg. |
| Trade | GOODMAN, JANITROL, AMANA DISTI... |
| Model | SSZ140301A* |
| AHRI ref | 4355477 |

| | | |
|---|---|---|
| Efficiency | 8.6 HSPF | |
| Heating input | | |
| Heating output | 28000 | Btuh @ 47°F |
| Temperature rise | 26 | °F |
| Actual air flow | 1000 | cfm |
| Air flow factor | 0.039 | cfm/Btuh |
| Static pressure | 0.51 | in H2O |
| Space thermostat | | |

### COOLING EQUIPMENT

| | |
|---|---|
| Make | Goodman Mfg. |
| Trade | GOODMAN, JANITROL, AMANA DISTI... |
| Cond | SSZ140301A* |
| Coil | ASPF313716E* |
| AHRI ref | 4355477 |

| | | |
|---|---|---|
| Efficiency | 13.0 EER, 15 SEER | |
| Sensible cooling | 21000 | Btuh |
| Latent cooling | 9000 | Btuh |
| Total cooling | 30000 | Btuh |
| Actual air flow | 1000 | cfm |
| Air flow factor | 0.046 | cfm/Btuh |
| Static pressure | 0.51 | in H2O |
| Load sensible heat ratio | 0.77 | |

| ROOM NAME | Area (ft²) | Htg load (Btuh) | Clg load (Btuh) | Htg AVF (cfm) | Clg AVF (cfm) |
|---|---|---|---|---|---|
| Master Suite | 216 | 5621 | 4483 | 222 | 205 |
| Bath | 62 | 1367 | 1125 | 54 | 51 |
| Closet | 44 | 64 | 707 | 3 | 32 |
| Bedroom 2 | 156 | 2165 | 4011 | 86 | 183 |
| Bath 2 | 47 | 688 | 879 | 27 | 40 |
| Bedroom 3 | 140 | 4122 | 2410 | 163 | 110 |
| Living Room | 341 | 7669 | 4695 | 303 | 214 |
| Utility | 57 | 1072 | 1050 | 42 | 48 |
| Kitchen | 120 | 1322 | 1216 | 52 | 55 |
| Foyer/Hall | 100 | 1228 | 1336 | 49 | 61 |

Calculations approved by ACCA to meet all requirements of Manual J 8th Ed.

 **wrightsoft** Right-Suite® Universal 2013 13.0.03 RSU08133
...ders, Inc\Creekside Lot 37\Creekside lot 37.rup  Calc = MJ8  Front Door faces:  N

2013-Sep-06 09:44:32
Page 1

| | | | | | |
|---|---|---|---|---|---|
| Entire House | 1282 | 25317 | 21912 | 1000 | 1000 |
| Other equip loads | | 0 | 0 | | |
| Equip. @   0.97   RSM | | | 21254 | | |
| Latent cooling | | | 6712 | | |
| TOTALS | 1282 | 25317 | 27966 | 1000 | 1000 |

Calculations approved by ACCA to meet all requirements of Manual J 8th Ed.

# Building Analysis
## *Entire House*
### Bertie Heating & Air Conditioning, Inc.

Job: 6115 NW 117th Place-326...
Date: Jun 28, 2013
By: Amy Kauper
Plan: Creekside Villas Lot 37

1730 NE 23rd Avenue, Gainesville, FL 32609-3904  Phone: 352-331-2005  Fax: 352-371-5770  Email: amy@bertieair.com  Web: www.bertieair.com  License: CAC-058522

## Project Information

For:  Duration Builders, Inc.
11787 NW 61st Terrace, Alachua, FL 32615
Phone: 386-462-0511
Web: www.durationbuilders.com  Email: bjones@durationbuilders.com

## Design Conditions

**Location:**
Gainesville Regional AP, FL, US
Elevation: 164 ft
Latitude: 30°N

| Indoor: | Heating | Cooling |
|---|---|---|
| Indoor temperature (°F) | 70 | 75 |
| Design TD (°F) | 37 | 17 |
| Relative humidity (%) | 50 | 50 |
| Moisture difference (gr/lb) | 32.5 | 47.4 |

| Outdoor: | Heating | Cooling |
|---|---|---|
| Dry bulb (°F) | 33 | 92 |
| Daily range (°F) | - | 18 ( M ) |
| Wet bulb (°F) | - | 76 |
| Wind speed (mph) | 15.0 | 7.5 |

**Infiltration:**
| | |
|---|---|
| Method | Simplified |
| Construction quality | Average |
| Fireplaces | 0 |

## Heating

| Component | Btuh/ft² | Btuh | % of load |
|---|---|---|---|
| Walls | 3.3 | 4526 | 17.9 |
| Glazing | 37.0 | 4585 | 18.1 |
| Doors | 12.8 | 239 | 0.9 |
| Ceilings | 1.2 | 1501 | 5.9 |
| Floors | 4.8 | 6125 | 24.2 |
| Infiltration | 2.7 | 3548 | 14.0 |
| Ducts | | 4792 | 18.9 |
| Piping | | 0 | 0 |
| Humidification | | 0 | 0 |
| Ventilation | | 0 | 0 |
| Adjustments | | 0 | |
| **Total** | | **25317** | **100.0** |

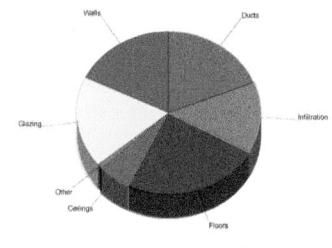

## Cooling

| Component | Btuh/ft² | Btuh | % of load |
|---|---|---|---|
| Walls | 2.1 | 2840 | 13.0 |
| Glazing | 41.9 | 5191 | 23.7 |
| Doors | 10.3 | 192 | 0.9 |
| Ceilings | 1.7 | 2168 | 9.9 |
| Floors | 0 | 0 | 0 |
| Infiltration | 0.6 | 842 | 3.8 |
| Ducts | | 6080 | 27.7 |
| Ventilation | | 0 | 0 |
| Internal gains | | 4600 | 21.0 |
| Blower | | 0 | 0 |
| Adjustments | | 0 | |
| **Total** | | **21912** | **100.0** |

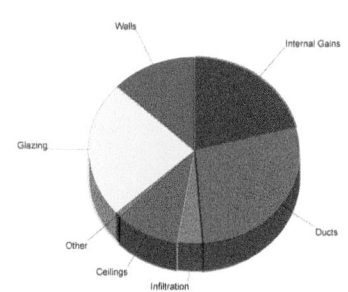

Latent Cooling Load = 6712 Btuh
Overall U-value = 0.114 Btuh/ft²-°F

Data entries checked.

**wrightsoft**
Right-Suite® Universal 2013 13.0.03 RSU08133
...ders, Inc\Creekside Lot 37\Creekside lot 37.rup  Calc = MJ8  Front Door faces: N

2013-Sep-06 09:44:32
Page 1

# Project Summary
## *Entire House*
### Bertie Heating & Air Conditioning, Inc.

**Job:** 6115 NW 117th Place-326...
**Date:** Jun 28, 2013
**By:** Amy Kauper
**Plan:** Creekside Villas Lot 37

1730 NE 23rd Avenue, Gainesville, FL 32609-3904 Phone: 352-331-2005 Fax: 352-371-5770  Email: amy@bertieair.com Web: www.bertieair.com License: CAC-058522

## Project Information

**For:**  Duration Builders, Inc.
11787 NW 61st Terrace, Alachua, FL 32615
Phone: 386-462-0511
Web: www.durationbuilders.com  Email: bjones@durationbuilders.com

**Notes:**

## Design Information

**Weather:**  Gainesville Regional AP, FL, US

### Winter Design Conditions

| | | |
|---|---|---|
| Outside db | 33 | °F |
| Inside db | 70 | °F |
| Design TD | 37 | °F |

### Summer Design Conditions

| | | |
|---|---|---|
| Outside db | 92 | °F |
| Inside db | 75 | °F |
| Design TD | 17 | °F |
| Daily range | M | |
| Relative humidity | 50 | % |
| Moisture difference | 47 | gr/lb |

### Heating Summary

| | | |
|---|---|---|
| Structure | 20525 | Btuh |
| Ducts | 4792 | Btuh |
| Central vent (0 cfm) | 0 | Btuh |
| Humidification | 0 | Btuh |
| Piping | 0 | Btuh |
| Equipment load | 25317 | Btuh |

### Sensible Cooling Equipment Load Sizing

| | | |
|---|---|---|
| Structure | 15832 | Btuh |
| Ducts | 6080 | Btuh |
| Central vent (0 cfm) | 0 | Btuh |
| Blower | 0 | Btuh |
| Use manufacturer's data | n | |
| Rate/swing multiplier | 0.97 | |
| Equipment sensible load | 21254 | Btuh |

### Infiltration

| | | |
|---|---|---|
| Method | | Simplified |
| Construction quality | | Average |
| Fireplaces | | 0 |

| | Heating | Cooling |
|---|---|---|
| Area (ft²) | 1282 | 1282 |
| Volume (ft³) | 11819 | 11819 |
| Air changes/hour | 0.45 | 0.23 |
| Equiv. AVF (cfm) | 89 | 45 |

### Latent Cooling Equipment Load Sizing

| | | |
|---|---|---|
| Structure | 5452 | Btuh |
| Ducts | 1260 | Btuh |
| Central vent (0 cfm) | 0 | Btuh |
| Equipment latent load | 6712 | Btuh |
| Equipment total load | 27966 | Btuh |
| Req. total capacity at 0.70 SHR | 2.5 | ton |

### Heating Equipment Summary

| | |
|---|---|
| Make | Goodman Mfg. |
| Trade | GOODMAN, JANITROL, AMANA DISTI... |
| Model | SSZ140301A* |
| AHRI ref | 4355477 |

| | | |
|---|---|---|
| Efficiency | 8.6 HSPF | |
| Heating input | | |
| Heating output | 28000 | Btuh @ 47°F |
| Temperature rise | 26 | °F |
| Actual air flow | 1000 | cfm |
| Air flow factor | 0.039 | cfm/Btuh |
| Static pressure | 0.51 | in H2O |
| Space thermostat | | |

### Cooling Equipment Summary

| | |
|---|---|
| Make | Goodman Mfg. |
| Trade | GOODMAN, JANITROL, AMANA DISTI... |
| Cond | SSZ140301A* |
| Coil | ASPF313716E* |
| AHRI ref | 4355477 |

| | | |
|---|---|---|
| Efficiency | 13.0 EER, 15 SEER | |
| Sensible cooling | 21000 | Btuh |
| Latent cooling | 9000 | Btuh |
| Total cooling | 30000 | Btuh |
| Actual air flow | 1000 | cfm |
| Air flow factor | 0.046 | cfm/Btuh |
| Static pressure | 0.51 | in H2O |
| Load sensible heat ratio | 0.77 | |

Calculations approved by ACCA to meet all requirements of Manual J 8th Ed.

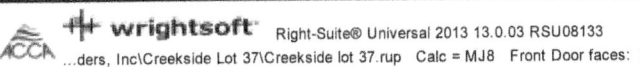
**wrightsoft**  Right-Suite® Universal 2013 13.0.03 RSU08133
...ders, Inc\Creekside Lot 37\Creekside lot 37.rup  Calc = MJ8  Front Door faces: N

2013-Sep-06 09:44:32
Page 1

 03407  Heating and Cooling System Design

Module Eight  73

**Level 1**

Living Room

151 cfm

151 cfm

111 cfm

Master Suite

111 cfm

414 cfm

32 cfm

Closet

Bath

54 cfm

Kitchen

55 cfm

190 cfm

Bedroom 2

183 cfm

Utility 48 cfm

282 cfm

Bath 2

40 cfm

Foyer/Hall

61 cfm

239 cfm

Garage

Bedroom 3

163 cfm

| **Job #: 6115 NW 117th Place-32615** | **Bertie Heating & Air Conditioning, I...** | Scale: 1 : 87 |
|---|---|---|
| **Performed by Amy Kauper for:** | | Page 1 |
| Duration Builders, Inc. | 1730 NE 23rd Avenue | Right-Suite® Universal 2013 |
| 11787 NW 61st Terrace | Gainesville, FL 32609-3904 | 13.0.03 RSU08133 |
| Alachua, FL 32615 | Phone: 352-331-2005 Fax: 352-371-5770 | 2013-Sep-06 09:44:33 |
| Phone: 386-462-0511 | www.bertieair.com  amy@bertieair.com | ...side Lot 37\Creekside lot 37.rup |
| www.durationbuilders.com  bjones@durationbuilders.com | | |

# Manual S Compliance Report
## *Entire House*

### Bertie Heating & Air Conditioning, Inc.

Job: 6115 NW 117th Place-326...
Date: Jun 28, 2013
By: Amy Kauper
Plan: Creekside Villas Lot 37

1730 NE 23rd Avenue, Gainesville, FL 32609-3904 Phone: 352-331-2005 Fax: 352-371-5770 Email: amy@bertieair.com Web: www.bertieair.com License: CAC-058522

## Project Information

For: Duration Builders, Inc.
11787 NW 61st Terrace, Alachua, FL 32615
Phone: 386-462-0511
Web: www.durationbuilders.com Email: bjones@durationbuilders.com

## Cooling Equipment

### Design Conditions

| | | | | | | |
|---|---|---|---|---|---|---|
| Outdoor design DB: | 92.0°F | Sensible gain: | 21912 | Btuh | Entering coil DB: | 77.0°F |
| Outdoor design WB: | 76.3°F | Latent gain: | 6712 | Btuh | Entering coil WB: | 63.5°F |
| Indoor design DB: | 75.0°F | Total gain: | 28624 | Btuh | | |
| Indoor RH: | 50% | Estimated airflow: | 1000 | cfm | | |

### Manufacturer's Performance Data at Actual Design Conditions

Equipment type: Split ASHP
Manufacturer: Goodman Mfg.       Model: SSZ140301A*+ASPF313716E*
Actual airflow:   1000  cfm
Sensible capacity: 23031  Btuh   105% of load
Latent capacity:   3416  Btuh    51% of load
Total capacity:    26447  Btuh   92% of load  SHR: 87%

## Heating Equipment

### Design Conditions

| | | | | | | |
|---|---|---|---|---|---|---|
| Outdoor design DB: | 33.4°F | Heat loss: | 25317 | Btuh | Entering coil DB: | 69.0°F |
| Indoor design DB: | 70.0°F | | | | | |

### Manufacturer's Performance Data at Actual Design Conditions

Equipment type: Split ASHP
Manufacturer: Goodman Mfg.       Model: SSZ140301A*+ASPF313716E*
Actual airflow:   1000  cfm
Output capacity:  27000  Btuh   107% of load        Capacity balance:  37  °F
Supplemental heat required:    0  Btuh              Economic balance: -99  °F

Backup equipment type: Elec strip
Manufacturer: Generic          Model: AFUE 100
Actual airflow:   1000  cfm
Output capacity:     0  kW    0% of load   Temp. rise:   0  °F

The above equipment was selected in accordance with ACCA Manual S.

---

**wrightsoft** Right-Suite® Universal 2013 13.0.03 RSU08133
...ders, Inc\Creekside Lot 37\Creekside lot 37.rup  Calc = MJ8  Front Door faces: N

2013-Sep-06 09:44:32
Page 1

---

03407 Heating and Cooling System Design                    Module Eight  75

| | | Residential Plans Examiner Review Form for HVAC System Design (Loads, Equipment, Ducts) | Form RPER 1 15 Mar 09 |
|---|---|---|---|

## City of Alachua
### Header Information

Contractor: Bertie Heating & Air Conditioning, Inc.

Mechanical license: CAC-058522

Building plan #: Creekside
Villas Lot 37

Home address (Street or Lot#, Block, Subdivision): 11787 NW 61st Terrace, Entire House

| REQUIRED ATTACHMENTS | ATTACHED | |
|---|---|---|
| | Yes | No |
| Manual J1 Form (and supporting worksheets): | ☐ | ☐ |
| or MJ1AE Form* (and supporting worksheets): | ☐ | ☐ |
| OEM performance data (heating, cooling, blower): | ☐ | ☐ |
| Manual D Friction Rate Worksheet: | ☐ | ☐ |
| Duct distribution sketch: | ☐ | ☐ |

## HVAC LOAD CALCULATION (IRC M1401.3)

### Design Conditions

**Winter Design Conditions**
| | | |
|---|---|---|
| Outdoor temperature: | 33 | °F |
| Indoor temperature: | 70 | °F |
| Total heat loss: | 25317 | Btuh |

**Summer Design Conditions**
| | | |
|---|---|---|
| Outdoor temperature: | 92 | °F |
| Indoor temperature: | 75 | °F |
| Grains difference: | 47 gr/lb @50% RH | |
| Sensible heat gain: | 21912 | Btuh |
| Latent heat gain: | 6712 | Btuh |
| Total heat gain: | 28624 | Btuh |

### Building Construction Information

**Building**
| | |
|---|---|
| Orientation: | Front Door faces North |
| North, East, West, South, Northeast, Northwest, Southeast, Southwest | |
| Number of bedrooms: | 3 |
| Conditioned floor area: | 1282 ft² |
| Number of occupants: | 20 |

**Windows**
| | |
|---|---|
| Eave overhang depth: | 1.5 ft |
| Internal shade: | none |
| Blinds, drapes, etc. | |
| Number of skylights: | 0 |

## HVAC EQUIPMENT SELECTION (IRC M1401.3)

### Heating Equipment Data

| | |
|---|---|
| Equipment type: | Split ASHP |
| Furnace, Heat pump, Boiler, etc. | |
| Model: | Goodman Mfg. |
| | SSZ140301A*+ASPF313716E* |
| Heating output capacity: | 27000 Btuh |
| Heat pumps - capacity at winter design outdoor conditions | |
| Aux. heating output capacity: | 0 Btuh |

### Cooling Equipment Data

| | |
|---|---|
| Equipment type: | Split ASHP |
| Air Conditioner, Heat pump, etc. | |
| Model: | Goodman Mfg. |
| | SSZ140301A*+ASPF313716E* |
| Total cooling capacity: | 26447 Btuh |
| Sensible cooling capacity: | 23031 Btuh |
| Latent cooling capacity: | 3416 Btuh |

### Blower Data

| | |
|---|---|
| Heating cfm: | 1000 |
| Cooling cfm: | 1000 |
| Static pressure: | 0.51 in H2O |
| Fan's rated external static pressure for design airflow | |

## HVAC DUCT DISTRIBUTION SYSTEM DESIGN (IRC M1601.1)

| | | | | | | | |
|---|---|---|---|---|---|---|---|
| Design airflow: | 1000 | cfm | Longest supply duct: | 0 | ft | Duct Materials Used | |
| Equipment design ESP: | 0.51 | in H2O | Longest return duct: | 0 | ft | Trunk duct: | |
| Total device pressure losses: | -0.1 | in H2O | **Total effective length (TEL):** | 0 | ft | | |
| **Available static pressure (ASP):** | 0.36 | in H2O | **Friction rate:** | 0.880 | in/100ft | Branch duct: | Round flex vinyl |
| | | | Friction Rate = ASP ÷ (TEL x 100) | | | | |

I declare the load calculation, equipment, equipment selection and duct design were rigorously performed based on the building plan listed above. I understand the claims made on these forms will be subject to review and verification.

Contractor's printed name: _____

Contractor's signature: _____  Date: _____

Reserved for County, Town Municipality or Authority having jurisdiction use.

*Home qualifies for MJ1AE Form based on Abridged Edition Checklist

**⊞ wrightsoft** Right-Suite® Universal 2013 13.0.03 RSU08133

# Additional Resources

This module presents thorough resources for task training. The following resource material is suggested for further study.

*FEMA 412, Installing Seismic Restraints for Mechanical Equipment*. Washington, DC: Federal Emergency Management Agency.

*HVAC Duct Construction Standards—Metal and Flexible*. Chantilly, VA: Sheet Metal and Air Conditioning Contractors National Association (SMACNA).

*Manual J, Residential Load Calculation*. Washington, DC: Air Conditioning Contractors of America (ACCA).

*Manual N, Commercial Load Calculation*. Washington, DC: Air Conditioning Contractors of America (ACCA).

*Manual D, Residential Duct Design*. Washington, DC: Air Conditioning Contractors of America (ACCA).

*Manual G, Selection of Distribution Systems*. Washington, DC: Air Conditioning Contractors of America (ACCA).

*Manual S, Residential Equipment Selection*. Washington, DC: Air Conditioning Contractors of America (ACCA).

# Figure Credits

| Answer | Section Reference | Objective |
|---|---|---|
| **Section One** | | |
| 1. b | 1.1.0 | 1a |
| 2. d | 1.2.0 | 1b |
| **Section Two** | | |
| 1. d | 2.1.2 | 2a |
| 2. b | 2.2.2 | 2b |
| **Section Three** | | |
| 1. c | 3.1.0 | 3a |
| 2. c | 3.2.0 | 3b |
| 3. b | 3.3.0 | 3c |
| 4. a | 3.4.2 | 3d |
| **Section Four** | | |
| 1. b | 4.1.3 | 4a |
| 2. d | 4.2.1 | 4b |
| 3. b | 4.3.0 | 4c |
| 4. a | 4.4.7 | 4d |
| 5. c | 4.5.0 | 4e |

# NCCER CURRICULA — USER UPDATE

NCCER makes every effort to keep its textbooks up-to-date and free of technical errors. We appreciate your help in this process. If you find an error, a typographical mistake, or an inaccuracy in NCCER's curricula, please fill out this form (or a photocopy), or complete the online form at **www.nccer.org/olf**. Be sure to include the exact module ID number, page number, a detailed description, and your recommended correction. Your input will be brought to the attention of the Authoring Team. Thank you for your assistance.

*Instructors* – If you have an idea for improving this textbook, or have found that additional materials were necessary to teach this module effectively, please let us know so that we may present your suggestions to the Authoring Team.

### NCCER Product Development and Revision
13614 Progress Blvd., Alachua, FL 32615

**Email:** curriculum@nccer.org
**Online:** www.nccer.org/olf

❏ Trainee Guide    ❏ Lesson Plans    ❏ Exam    ❏ PowerPoints    Other _____

Craft / Level: _____    Copyright Date: _____

Module ID Number / Title: _____

Section Number(s): _____

Description: _____

_____

_____

_____

Recommended Correction: _____

_____

_____

_____

Your Name: _____

Address: _____

Email: _____    Phone: _____

# Commercial/Industrial Refrigeration

## OVERVIEW

Supermarkets, warehouses, and commercial food industries require much larger and more complex refrigeration systems than those used in small retail operations. The commercial and industrial applications are also likely to be custom engineered for the structure and can involve highly complex control systems. This module examines the equipment, control systems, and refrigerants used in commercial and industrial refrigeration applications.

## Module 03408

Trainees with successful module completions may be eligible for credentialing through NCCER's National Registry. To learn more, go to **www.nccer.org** or contact us at **1.888.622.3720**. Our website has information on the latest product releases and training, as well as online versions of our *Cornerstone* magazine and Pearson's product catalog.

Your feedback is welcome. You may email your comments to **curriculum@nccer.org,** send general comments and inquiries to **info@nccer.org**, or fill in the User Update form at the back of this module.

This information is general in nature and intended for training purposes only. Actual performance of activities described in this manual requires compliance with all applicable operating, service, maintenance, and safety procedures under the direction of qualified personnel. References in this manual to patented or proprietary devices do not constitute a recommendation of their use.

## Objectives

When you have completed this module, you will be able to do the following:

1. Describe methods used to freeze, store, and transport food products.
   a. Describe methods used to freeze food products.
   b. Describe methods used to store food products.
   c. Describe methods used to transport refrigerated food products.
2. Identify and describe various commercial and industrial refrigeration system components.
   a. Identify and describe various compressor configurations.
   b. Describe the application, control, and installation of air-cooled condensers.
   c. Identify and describe various evaporator and display case configurations.
   d. Identify and describe various refrigeration system accessories.
   e. Identify and describe various refrigerant control devices.
3. Identify and describe various types of defrost systems.
   a. Identify and describe off-cycle defrost systems.
   b. Identify and describe electric defrost systems.
   c. Identify and describe hot-gas defrost systems.
4. Describe the main characteristics of ammonia-based refrigeration systems.
   a. Describe the properties and safety considerations of ammonia as a refrigerant.
   b. Describe ammonia systems and the basic components.

## Performance Tasks

Under the supervision of your instructor, you should be able to do the following:

1. Install or make repairs to a packaged refrigeration condensing unit.
2. Install or make repairs to a packaged unit cooler in a refrigeration system.
3. Identify at least three of the following devices (selection provided by the instructor) commonly used in refrigeration systems:
   - Crankcase pressure regulator
   - Evaporator pressure regulator
   - Condenser head pressure regulator
   - Hot gas bypass regulator
   - Pressure-controlled cylinder unloader
   - Solenoid-controlled cylinder unloader

## Trade Terms

| | | | |
|---|---|---|---|
| Anhydrous | Deck | Queen valve | Total heat rejection |
| Commodities | Immiscible | Satellite compressor | (THR) value |
| Cooler | King valve | Secondary coolant | Unit cooler |
| Cryogenics | Latent-heat defrost | Sublimation | |
| Cryogenic fluid | Lyophilization | | |

## Industry-Recognized Credentials

If you are training through an NCCER-accredited sponsor, you may be eligible for credentials from NCCER's Registry. The ID number for this module is 03408. Note that this module may have been used in other NCCER curricula and may apply to other level completions. Contact NCCER's Registry at 888.622.3720 or go to **www.nccer.org** for more information.

# Contents

*Topics to be presented in this module include:*

## Figures and Tables

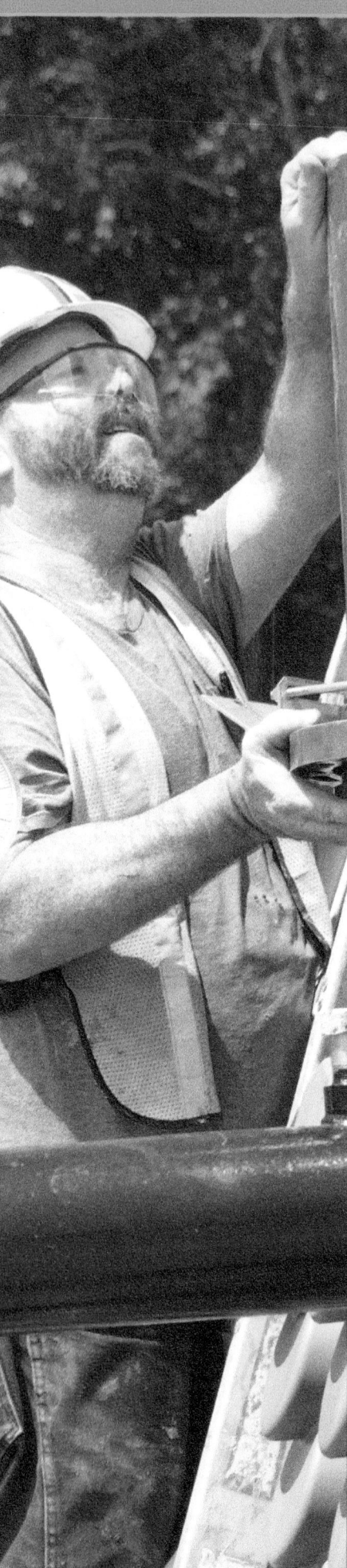

## Figures and Tables (continued)

## SECTION ONE

### 1.0.0 FOOD PRODUCT REFRIGERATION

#### Objectives

Describe methods used to freeze, store, and transport food products.

    a.  Describe methods used to freeze food products.

    b.  Describe methods used to store food products.

    c.  Describe methods used to transport refrigerated food products.

#### Trade Terms

**Commodities**: Commercial items such as merchandise, wares, goods, and produce.

**Cooler**: A refrigerated storage device that protects commodities at temperatures above 32°F.

**Cryogenics**: Refrigeration that deals with producing temperatures of –250°F and below.

**Cryogenic fluid**: A substance that exists as a liquid or gas at ultra-low temperatures of –250°F and below.

**Lyophilization**: A dehydration process which first incorporates freezing of a material, then a reduction of the surrounding pressure with simultaneous addition of heat to remove moisture. Moisture leaves the material through sublimation. Also known as freeze-drying.

**Sublimation**: The change in state directly from a solid to a gas, such as the changing of ice to water vapor, without passing through the liquid state at any point.

The common purpose of commercial and industrial refrigeration systems and comfort cooling systems are to remove heat. Refrigeration systems are mainly used to remove heat from substances for the food, chemical, and manufacturing industries. Air-conditioning systems are used to remove heat from buildings and vehicles to provide comfort control. Excluding their applications and the obvious differences in physical layout and packaging, both types of systems are basically constructed of common hardware: compressors, heat exchangers (evaporators and condensers), fans, pumps, pipe, duct, and controls. Their main working elements are air, water, and refrigerants.

Most commercial refrigeration involves the refrigeration and freezing of food and other perishable commodities. This module focuses on the types of refrigeration equipment used for this purpose. The first part of this module briefly describes the methods used to preserve food. It also covers the various kinds of commercial and industrial refrigeration equipment, such as refrigerated warehouses, walk-in coolers, and display cases. The second part of the module covers the refrigeration methods and components used in commercial refrigeration systems that are unique or different from the components used in comfort cooling systems. Also covered are some special refrigeration applications and systems.

The field of food processing starts at the point of harvest and extends to consumption (*Figure 1*). Throughout this process, the food must be properly prepared and stored in order to maintain its freshness and quality. Refrigerated storage is defined as any space where refrigeration is used to provide controlled storage conditions. It can be an entire building such as a warehouse, a section of a building dedicated to storage, or a single standalone unit. Refrigerated storage areas are often divided into several sections, with each section being cooled at a different temperature. The specific temperature used is determined by the safe preservation temperature of the commodities involved.

### 1.1.0 Commercial Freezing Methods

Some cold storage plants and warehouses are used to both process and freeze foods. The process of freezing reduces the temperature of a food product from the ambient level to the storage level and changes most of the water in the product to ice. As shown in *Figure 2*, the freezing process has three phases. The first phase removes the product's sensible heat. During the second phase, the product's latent heat of fusion is removed and the water in the product is changed to ice crystals. In phase three, continued cooling of the product removes additional sensible heat and reduces the temperature to the required frozen storage temperature.

Freezing is a time-temperature related process. The longest part of the process is normally the latent heat of fusion as the water turns to ice.

Several methods are used for freezing food products. These methods are briefly described in the following sections.

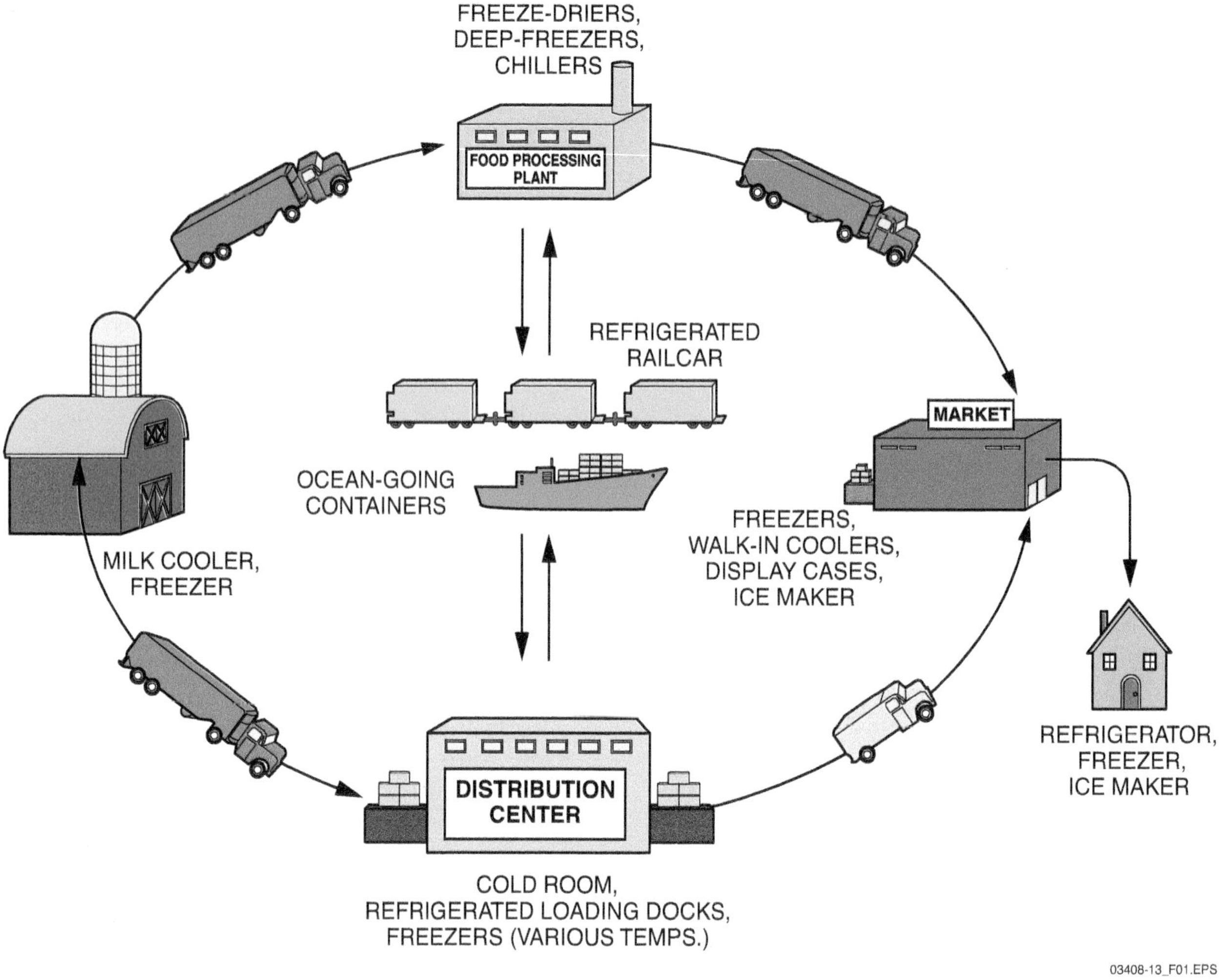

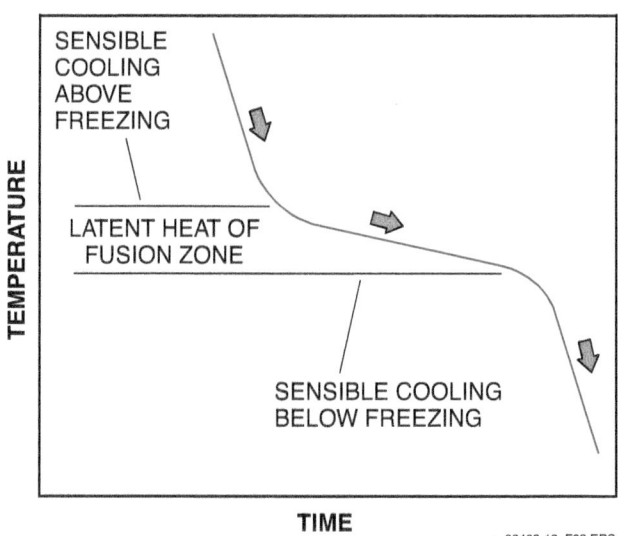

Figure 1 Food processing.

## 1.1.1 Air Blast Freezing

Air blast freezing employs cold air circulated around the product at high velocities, maximizing the heat exchange process. The air removes heat from the product via convection and releases it to an air-to-refrigerant heat exchanger coil. One advantage of the blast freezing approach is the ability to freeze objects of inconsistent or irregular size and shape. The air discharge temperatures of a blast freezing evaporator coil are typically –20°F to –40°F, but varies depending on the product.

The two main types of blast freezing applications are batch freezing and continuous freezing. In batch freezing applications, product is loaded into the freezer and allowed to remain in place until the process is complete. The product is often palletized or placed in moveable racks, then placed in a room equipped with the necessary evaporator coils and high-velocity fans. To ensure that the air makes adequate contact with the product, the individual product packages are often separated on the top and sides to allow free air flow. As shown in *Figure 3*, the evaporator coils

Figure 2 Typical freezing curve.

are often separated from the product by a horizontal partition, directing discharge air from the coils through the product load and back to the coil inlet. The arrangement can be series flow, with air flowing along the length of the load, or cross-flow, with the air moving perpendicular to the length of the load. In continuous freezing, product moves by conveyor or other means through the refrigerated enclosure at a constant speed appropriate for the product type.

## 1.1.2 Contact Freezing

Contact freezing produces packaged or unpackaged frozen products by pressing them between cold metal plates or by contact with a moving belt exposed to the refrigerated surface. Contact freezing is typically used for products of consistent size and shape, to ensure that proper contact is made with individual pieces. Heat is extracted by direct conduction through the plate surfaces and into the refrigerant circulating through channels or tubes integrated into the plate. Contact freezing is considered to be one of the most efficient approaches because contact transfers heat more rapidly. One other advantage is that there is little or no loss of moisture from the product, as opposed to freezing methods using high volumes of air. Moisture loss means product weight loss and possible quality degradation, significantly impacting the value of the end product.

*Figure 4* shows one example of a horizontal plate freezer. Product such as fish is loaded between the plates. Once loaded, the plates are pressed down on the product and the freezing process begins. Plate freezers can also use vertical plate arrangements.

Belt freezers, as shown in *Figure 5*, are generally used for unpackaged products. The film is pulled with rollers, moving the product along the surface. Used film is disposed of, ensuring good hygiene. Soft, sticky and wet products (scallops, for example) can be a problem with contact freezers, as product can adhere to the refrigerated surface. The special and extremely thin film used with belt freezers like the one shown does not allow

product adhesion, and does not significantly impede heat transfer.

### 1.1.3 Cryogenic Freezing

In the refrigeration industry, cryogenics is generally accepted to define the freezing of product using fluid (refrigerant) temperatures below –250°F. The boiling points of common cryogenic fluids, such as carbon dioxide, nitrogen, and helium, which are normally in their gaseous state, are below this temperature. The boiling points of common refrigerants such as HFCs and HCFCs are above this temperature. Liquid nitrogen is most commonly used, while liquid helium provides the lowest temperature.

Cryogenic freezing is generally done by injecting the fluid into the freezing chamber. In some cases, only the surface is to be frozen cryogenically. The product is then transferred to a blast freezer to complete the process. The frozen surface reduces moisture loss from the product. This process is known as cryomechanical freezing.

*Figure 6* shows one example of a cryogenic freezing unit. Cryogenics can be used in spiral-style or tunnel-belt freezers, as well as batch freezers. In other cases, the product is immersed in the cryogenic fluid, floating in a tank maintained at a specific temperature and pressure. In all cases, the cryogenic fluid absorbs heat from the product, causing the fluid to vaporize. This change of state, a latent heat process, creates the highest heat ex-

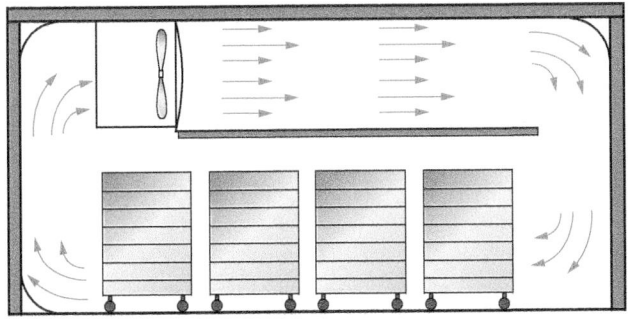

TYPICAL BLAST FREEZER LAYOUT

03408-13_F03.EPS

*Figure 3* Blast freezer layout.

03408-13_F04.EPS

*Figure 4* Horizontal plate freezer.

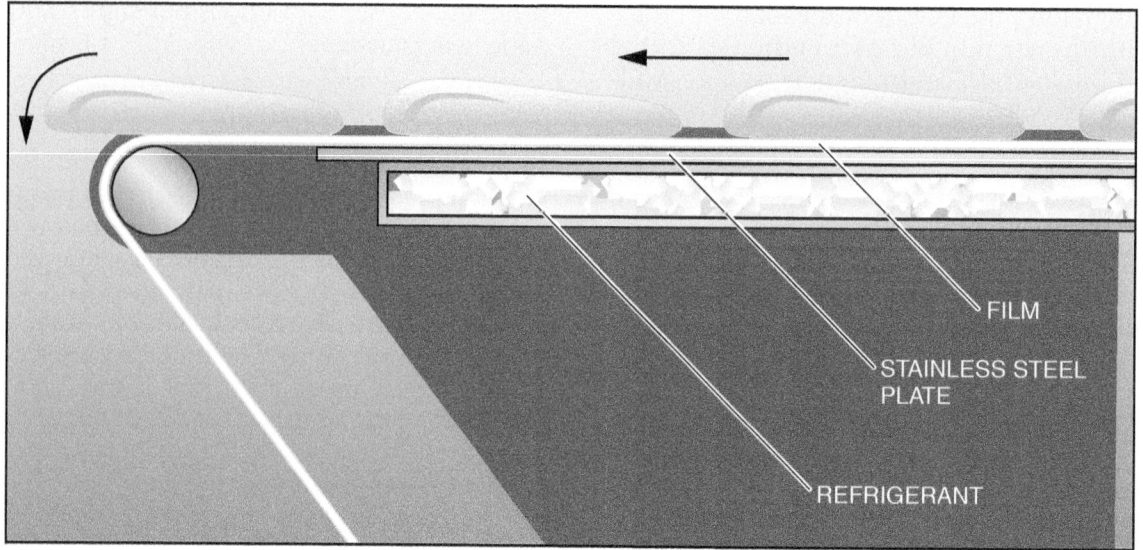

Figure 5 Belt freezer design.

change rate. This is the case in all common refrigeration applications regardless of the refrigerant used.

### 1.1.4 Freeze-Drying

Freeze-drying, also known as lyophilization, is generally a cryogenic process, relying first on the extremely rapid freezing action of this method. The product is then subjected to high levels of vacuum and sufficient heat to dehydrate the product, removing all moisture through sublimation. The result is a product that no longer requires refrigeration for long-term preservation, is much lighter in weight, and can be returned to its original state by simply adding water. *Figure 7* is an example of a typical commercial unit.

### 1.2.0 Food Storage

Cold storage is used to preserve perishable food in its fresh, wholesome state for extended periods. This is done by controlling the food's temperature and humidity during storage. Cold storage conditions are determined by the type of food being stored. They are also determined by the length of time the food is to be held in storage.

Foods placed in cold storage can be divided into two broad groups based on their required temperature range: perishable or chilled temperature range and frozen temperature range.

- *Perishable or chilled temperature range* – This group includes commodities that must be refrigerated below a specified temperature but not allowed to freeze. For some products in this category, such as fruits, vegetables, and

Figure 6 Cryogenic spiral freezer.

Figure 7 Freeze-drying unit.

flowers, living processes (such as the growth of shoots or ripening) continue during storage. Exposure to cold temperatures only retards these processes, with the degree of retardation depending on the temperature level being maintained. Airflow is crucial to removing the heat generated during the respiratory cycle of many products in the perishable temperature range. With high air velocity also comes the potential for higher evaporation rates from many types of produce. This can cause premature dehydration. Avoiding dehydration requires that humidity be maintained at the proper level for each type of product. For example, celery is best stored at temperatures between 32°F and 36°F, with a relative humidity of 98 to 100 percent. Pumpkins, on the other hand, are best stored at temperatures between 50°F and 55°F, and a relative humidity of 50 to 70 percent. The storage temperatures and humidity requirements for the different kinds of produce are so varied that no generalizations can be made. Some

require a curing period before storage. Other products, such as certain kinds of potatoes, require different storage conditions depending on their intended use.

- *Frozen temperature range* – This group includes those products whose temperature must not rise above freezing, such as ice cream and frozen prepared foods, meats, meat products, and fish. Airflow is not as critical for frozen products as it is for non-frozen products.

Cold storage plants (*Figure 8*) are used to process and/or store large quantities of foods intended for distribution to supermarkets and restaurants. The plant or warehouse food handling methods and storage requirements determine the building design. Refrigerated storage plants and warehouses that handle a variety of foods are divided as required into several cooler rooms (chill rooms) and freezer rooms to preserve the foods involved. Each refrigerated room is separated from the other rooms based on the storage

## Other Uses of Freeze Drying

In addition to foods, freeze-drying is used to manufacture pharmaceuticals and biological samples, as well as in the recovery of valuable water-damaged documents.

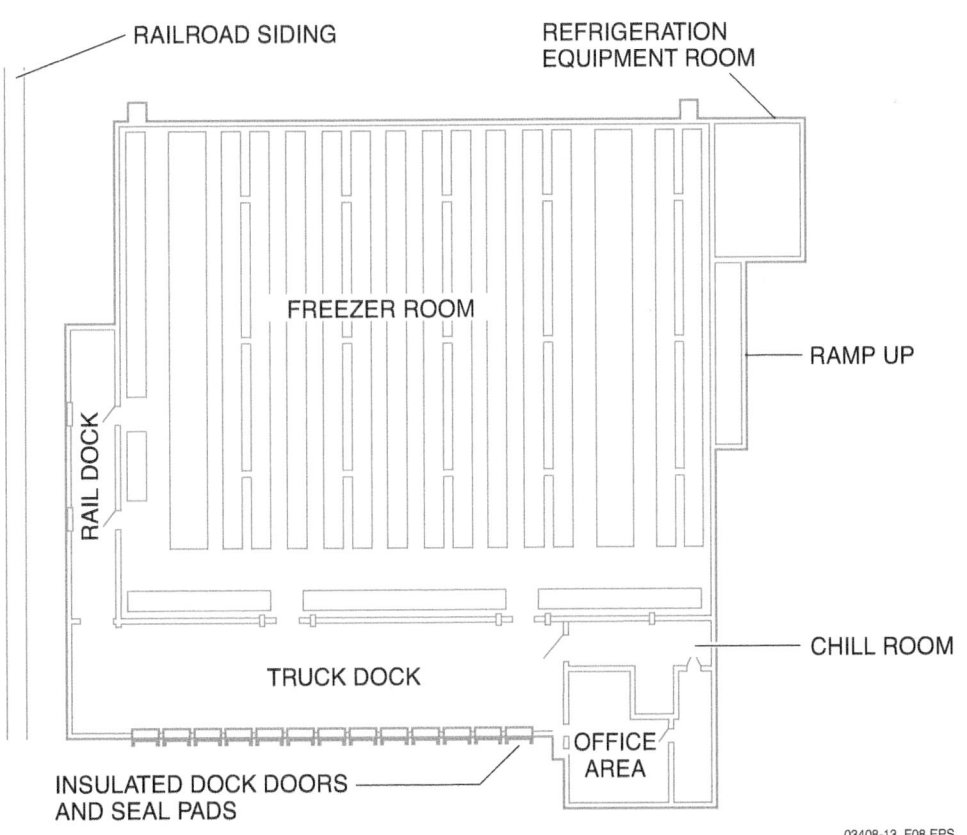

*Figure 8* Floor plan for a typical refrigerated warehouse.

temperature needed. Typically, the temperatures in a warehouse can range from 35°F to 60°F with humidity control, and to –20°F without humidity control. Generally, a custom-built refrigeration system is used to cool the entire warehouse and maintain the individual temperatures within the refrigerated rooms.

### 1.2.1 Chill Rooms

Chill rooms are used for temporary storage of foods. Fruits and vegetables are the main foods stored in chill rooms. They are normally placed in the chill rooms in their freshly harvested condition, after the necessary sorting and cleaning. Medium-range temperatures from 32°F to 60°F are used in chill rooms. The specific temperature maintained in any one room is determined by the storage temperature and humidity level needed for the type of food being stored. The list of specific temperatures and relative humidity required for the various fruits and vegetables is extensive.

Foods stored in chill rooms are typically kept at temperatures just above their freezing point to maintain top product quality. The required temperature must be precisely maintained. In some cases, variations as small as 2°F to 3°F can result in loss of product quality. Most chill rooms also have a means of controlling the relative humidity to within 3 to 5 percent of the desired level. This is often done by controlling the temperature of the evaporators using spray nozzles or by using precise humidifier controls. Air is circulated within the room to maintain an even temperature.

### 1.2.2 Freezer Rooms

Held at temperatures below freezing, freezer rooms are used for the prolonged storage of frozen food. This allows the storage time to be lengthened from days or weeks to months. These temperatures are typically maintained at levels ranging between 10°F and –10°F whereas rooms used to store ice cream are often maintained at a temperature of –20°F.

Frozen foods can deteriorate during the period between production and consumption. The most important factors contributing to the deterioration of frozen foods are the storage temperature and storage time.

## 1.3.0 Transporting Refrigerated Food Products

In addition to storage, the movement of products from one point in the food processing chain to the next requires refrigerated transport. Products intended for export can spend from one to three weeks in some type of refrigerated transport unit.

Refrigerated transport units maintain product temperature during shipment, thereby keeping products fresh until they reach distant markets. Without the ability to control product temperature in transit, consumers would not be able to enjoy many of the foods now found in local stores. Refrigerated transport equipment can be grouped into three methods of shipping: refrigerated shipboard containers, refrigerated truck and trailer units, and refrigerated railcars.

### 1.3.1 Trailer and Truck Units

At some point in the supply chain, refrigerated products will be transported by refrigerated trucks. Trucks used to haul perishable and frozen loads come in a variety of sizes and shapes. The most common are over-the-road refrigerated trailers, trucks, and direct-drive delivery vans.

Refrigerated trailers are used to haul just about every type of temperature-sensitive cargo. These units are used to transport products over long distances, typically across the state or even across the country. Unlike a standard trailer, these units are insulated, have swing doors with seals, and have a refrigeration unit mounted onto the front.

The refrigeration equipment is a packaged unit that contains all the components necessary to remove heat from the cargo area (*Figure 9*). These components include the compressor, condenser, metering device, evaporator, and all applicable controls. The trailer's refrigeration unit is heavy-duty, allowing it to endure the constant bumps and vibration present during travel. Trailer units can use electric heaters for heat and defrost, but in most cases, heat and defrost come from the use of hot gas. The trailer unit has an integral power supply. Power is provided to an open-drive compressor driven by a small diesel engine connected via a driveshaft. The engine also turns an alternator to supply DC voltage to the unit's electrical components. Fans can be belt driven or direct

## Distribution

Refrigerated warehouses often function as storage and pickup/distribution points for produce, meats, and frozen foods supplied from several sources. These warehouses are highly automated to allow efficient handling, storage, and movement of the food within the building and to the building loading docks.

driven. The unit's controls monitor and protect the cargo, and also protect and control the engine.

Cargo placement within a trailer plays an important role in removing heat from the inside of the trailer. A trailer unit supplies cold air out of the top of the refrigerated space. Sometimes a chute or duct is used to evenly distribute the air along the length of the trailer. The fan forces the air into the spaces around the cargo, where it picks up heat and is drawn back along the floor to the evaporator. Some trailers are equipped with a T-slotted floor, but most use pallets and depend on a clear path through the open pallets. If the cargo is loaded too tightly, the air will not be able to circulate through the load and absorb the heat. For produce with a high rate of respiration, this condition could cause the cargo to spoil.

The primary difference between a refrigerated truck (*Figure 10*) and a refrigerated trailer is that a trailer requires a separate drive vehicle (cab) to tow it from place to place, while a truck is self-contained. Truck units also differ in size and cooling capacity. Trucks are not as large as trailers and are typically used for local delivery of cargo in cities or between adjacent cities. Truck units have a packaged refrigeration unit mounted on the front of the cargo area, just above the driver's cab.

Both refrigerated trucks and trailers can be configured with multi-temperature capabilities. A multi-temperature truck or trailer consists of a refrigeration unit supplying two or more evaporator sections connected in parallel.

Direct-drive refrigeration units are used in trucks of lower capacity used for local delivery. They differ from the standard truck unit in that their compressor is a much smaller automotive-type compressor located in the engine compartment instead of being packaged in the refrigeration unit. These units carry smaller loads or operate with higher cargo temperature ranges. Power to drive the compressor is transferred from the engine by a belt. Electrical power is supplied by the vehicle's alternator. Direct-drive systems can combine the evaporator and condenser section into one frame or they can use remote evaporators.

### 1.3.2 Refrigerated Shipboard Containers

A large variety of products can be transported in containers. The unit must be capable of controlling a very wide temperature range, from 90°F down to –25°F. The refrigeration cycle is similar to that of a walk-in cooler, but the unit itself must be

03408-13_F09.EPS

*Figure 9* Trailer refrigeration unit.

03408-13_F10.EPS

*Figure 10* Refrigerated truck.

## Breathing Produce

Fresh products, such as apples and potatoes, continue to ripen long after they are harvested. The respiration associated with the ripening process creates heat and ethylene gas. If proper airflow isn't maintained, the buildup of heat and ethylene gas will speed up the ripening process, causing the produce to prematurely spoil.

extra heavy duty in order to withstand the abuse of traveling on a flatbed, being moved by a fork-lift, and being lifted on and off the container ship by a crane. Corrosion resistance is also important to the unit's design as saltwater spray will be present for most of its operational life. Exported commodities are usually shipped in container units. These heavily insulated steel or aluminum boxes are typically 8' wide by 8' high by 20' or 40' long, with a refrigeration unit installed at one end. The refrigeration equipment is a packaged unit that contains all the components necessary to remove heat from the cargo area. This includes the compressor, condenser, metering device, evaporator, and all electrical controls. *Figure 11* shows a typical refrigerated container.

The uniform size of containers allows them to be stacked aboard ship and on the docks. The loaded containers can be transported to and from the docks via flatbed trailers or railcars. When they arrive, they are either immediately placed on board the ship or held for future shipment in the dock's staging area.

Most container units use hermetic or semi-hermetic compressors. These units plug into the electrical power panel of the ship or dock facility while they are stationary. When a container is being transported to and from the dock, a portable generator set provides power so the unit can maintain the proper temperature. *Figure 12* shows nose-mount and under-mount generator sets.

Microprocessor controls allow the refrigeration unit to monitor system conditions. This provides the best settings for temperature control and energy efficiency. Heating and evaporator defrosting is done with electric heating elements. Although the condenser is typically air cooled, most manufacturers offer an optional water-cooled condenser. Water-cooled condensers allow the heat from the inside of the container to be removed, even when stacked tightly on the container ship. Air-cooled condensers can have problems with

UNDER-MOUNT GENERATOR SET

NOSE-MOUNT GENERATOR SET

03408-13_F12.EPS

*Figure 12* Under-mount and nose-mount generator sets.

03408-13_F11.EPS

*Figure 11* Refrigerated container.

## Food Safety

The amount of protection provided by the package containing the frozen food is also important. Some of the bacteria in frozen foods can be killed during freezing and frozen storage, but all the bacteria are never completely destroyed. When defrosting, foods are still subject to bacterial decomposition. This is why it is wise to defrost foods in the refrigerator, not the warm environment of a countertop.

high ambient temperatures and lack of airflow. Of course, water-cooled systems must be provided with a water source and a means of rejecting the collected heat.

Proper placement of the product in the container is vital if the refrigeration unit is to operate correctly. Air must be able to circulate around the product in order to absorb the heat and return to the refrigeration unit. The unit supplies cool air to the product through grooves or T-slots in the floor. The air warmed by the product rises to the top of the container and is drawn into the evaporator by a fan.

Advanced options for containers include humidity control and controlled atmosphere (CA). Humidity controls add moisture to the supply air to protect produce from drying out. CA is a process by which the oxygen, nitrogen, and carbon dioxide levels inside the container are adjusted to minimize ripening, extending the cargo's shelf life.

### 1.3.3 Refrigerated Railcars

Improvements to refrigerated containers and trailer units have had a direct impact on railcars. In many cases, the same refrigerated container used to transport cargo in a ship can be placed on a flatbed railcar to continue its trip inland. These containers either plug into a power source built into the railcar or use a portable generator set mounted onto the nose of the container.

Refrigerated railcars also use technological advances from the truck/trailer industry. By mounting a high-capacity trailer refrigeration unit onto a heavily insulated boxcar, larger loads can be delivered at reduced delivery times. Refrigerated railcars of this kind are capable of protecting 130,000 pounds of cargo with a temperature range of –20°F to 87°F. Another way of using railroads to transport perishable products is to load refrigerated trailers onto flatbed rail cars (*Figure 13*). The trailers draw power from the train while in transit.

> **WARNING!**
>
> Never enter a controlled atmosphere container right immediately after opening the door. The low levels of oxygen required to preserve produce can be extremely harmful to humans. Before entering the container:
>
> - Turn off the CA option.
> - Open the doors fully.
> - Wait 30 to 45 minutes to allow oxygen levels to return to normal before entering.

03408-13_F13.EPS

*Figure 13* Refrigerated trailers on railcars.

## Break Bulk Cargo

Before the introduction of container ships, produce was loaded on board by placing the crates in large cargo nets that were swung up to the ship and into the hold. The crates were then stacked inside the hold. This process took a long time, and much of the produce spoiled from the heat and rough handling. Goods loaded in this manner were called break bulk cargo.

## Help from Outer Space

Some refrigerated railcars and trucks are equipped with global positioning units that provide instantaneous tracking and monitoring information. Global positioning tracking allows dispatchers to confirm the location and direction of cargo wherever it goes. This is two-way communication, allowing a technician to monitor load temperature, refrigerant pressures, and fuel level, as well as providing the capability to change the unit setpoint, initiate a defrost, or download data.

## Did You Know?
## Controlled Atmosphere Warning Sign

When working with a controlled atmosphere container, you may encounter a sign simllar to this. The container doors should be fully opened for 30 to 45 minutes for the container's oxygen levels to return to normal prior to being entered by a human.

⚠ WARNING

HAZARDOUS ATMOSPHERE INSIDE.
VENTILATE CONTAINER BEFORE ACCESS.

03408-13_SA02.EPS

## Multiple Cargos, One Load

Multi-temperature trucks and trailers are useful for making deliveries to smaller rural grocery stores. These smaller stores often do not require a full load of dry goods, a full load of perishable items, or a full load of frozen products. By using a truck with multi-temperature capability, the driver can deliver smaller amounts of each product and still haul full loads.

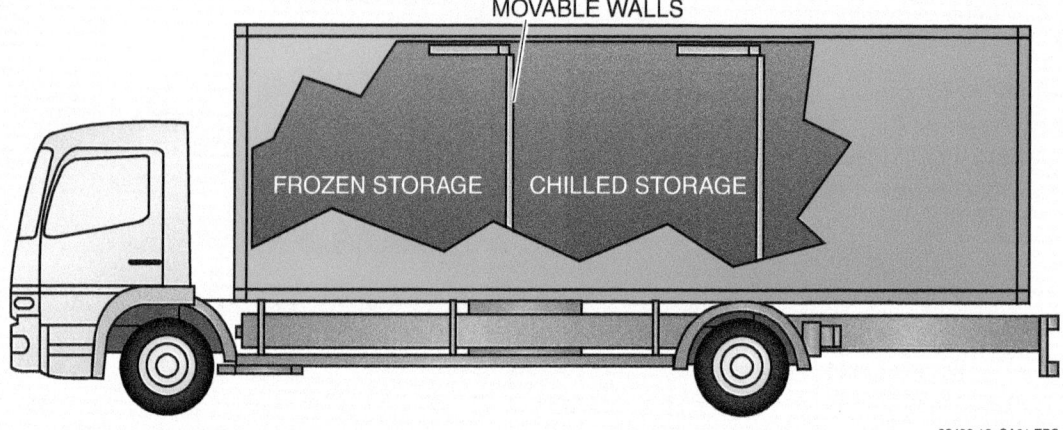

MOVABLE WALLS

FROZEN STORAGE    CHILLED STORAGE

03408-13_SA01.EPS

## Additional Resources

*ASHRAE Handbook, Refrigeration Systems and Applications*, Latest Edition. Atlanta, GA: American Society of Heating, Refrigerating, and Air Conditioning Engineers.

*ASHRAE Handbook, HVAC Systems and Equipment*, Latest Edition. Atlanta, GA: American Society of Heating, Refrigerating, and Air Conditioning Engineers.

## 1.0.0 Section Review

1. During air blast freezing, what are the typical coil air discharge temperatures?
   a. −10°F to −30°F
   b. −15°F to −35°F
   c. −20°F to −40°F
   d. −30°F to −50°F

2. Cold storage conditions are determined by the _____.
   a. way foods are packaged
   b. type of food being stored
   c. amount being stored
   d. type of storage building

3. Why are air-cooled condensers *not* commonly used on container ships?
   a. Low ambient temperatures and impeded airflow
   b. High ambient temperatures and difficulty controlling humidity levels
   c. Collected heat rejection and temperature monitoring is difficult
   d. High ambient temperatures and poor airflow

### 2.0.0 REFRIGERATION SYSTEM COMPONENTS

### Objectives

Identify and describe various commercial and industrial refrigeration system components.

  a. Identify and describe various compressor configurations.
  b. Describe the application, control, and installation of air-cooled condensers.
  c. Identify and describe various evaporator and display case configurations.
  d. Identify and describe various refrigeration system accessories.
  e. Identify and describe various refrigerant control devices.

### Performance Tasks

1. Install or make repairs to a packaged refrigeration condensing unit.
2. Install or make repairs to a packaged unit cooler in a refrigeration system.
3. Identify at least three of the following devices (selection provided by the instructor) commonly used in refrigeration systems:
   - Crankcase pressure regulator
   - Evaporator pressure regulator
   - Condenser head pressure regulator
   - Hot gas bypass regulator
   - Pressure-controlled cylinder unloader
   - Solenoid-controlled cylinder unloader

### Trade Terms

**Deck**: A refrigeration industry trade term that refers to a shelf, pan, or rack that supports the refrigerated items stored in coolers and display cases.

**King valve**: The manual refrigerant shut-off valve located at the outlet of the receiver.

**Queen valve**: The manual refrigerant shut-off valve located at the inlet of the receiver.

**Satellite compressor**: A separate compressor that uses the same source of refrigerant as those used by a related group of parallel compressors. However, the satellite compressor functions as an independent compressor connected to a cooling area that requires a lower temperature level than that being maintained by the related compressors.

**Secondary coolant**: Any cooling liquid that is used as a heat transfer fluid. It changes temperature without changing state as it gains or loses heat.

**Total heat rejection (THR) value**: A value used to rate condensers. It represents the total heat removed in desuperheating, condensing, and subcooling a refrigerant as it flows through the condenser.

**Unit cooler**: A packaged refrigeration system assembly containing the evaporator, expansion device, and fans. It is commonly used in chill rooms and walk-in coolers.

Many types of mechanical refrigeration systems are used to cool or freeze foods and other substances. The operation of and components in a refrigeration system used to cool food and other products are basically the same as those used in comfort cooling systems. *Figure 14* shows a basic refrigeration system. Redundant, or back-up, equipment is commonly added to prevent product and financial losses if the primary refrigeration system fails.

The basic refrigeration system components are the compressor, evaporator, condenser, and expansion device. What changes from system to system is the type of refrigerant, the system operating temperatures, the size and style of the components, and the installed locations of the four components and the lines. The events that take place within the mechanical refrigeration system happen again and again, in the same order. This series of events is called the refrigeration cycle. For review, the operation of the basic refrigeration system and its components are described in detail in the HVAC Level One module, *Introduction to Cooling*.

The following are the functions performed by each component:

- *Evaporator* – This is a heat exchanger in which the heat from the area or item being cooled is transferred to the refrigerant.
- *Compressor* – This creates the pressure differences in the system needed to make the refrigerant flow and the refrigeration cycle work.
- *Condenser* – This is a heat exchanger in which the heat absorbed by the refrigerant is transferred from the refrigerant to the cooler outdoor air or another cooler substance, such as water.
- *Expansion device* – This provides a pressure drop that lowers the boiling point of the refrigerant just before it enters the evaporator. The expansion device is also commonly called the metering device.

*Figure 14* also shows the piping used to connect the basic components in order to provide the path for refrigerant flow. Together, the components, accessories, and lines form a closed refrigeration system. The tubing lines are as follows:

- *Suction line* – This carries heat-laden refrigerant gas from the evaporator to the compressor.
- *Hot gas line* – This carries hot refrigerant gas from the compressor to the condenser. It is also referred to as the discharge line.

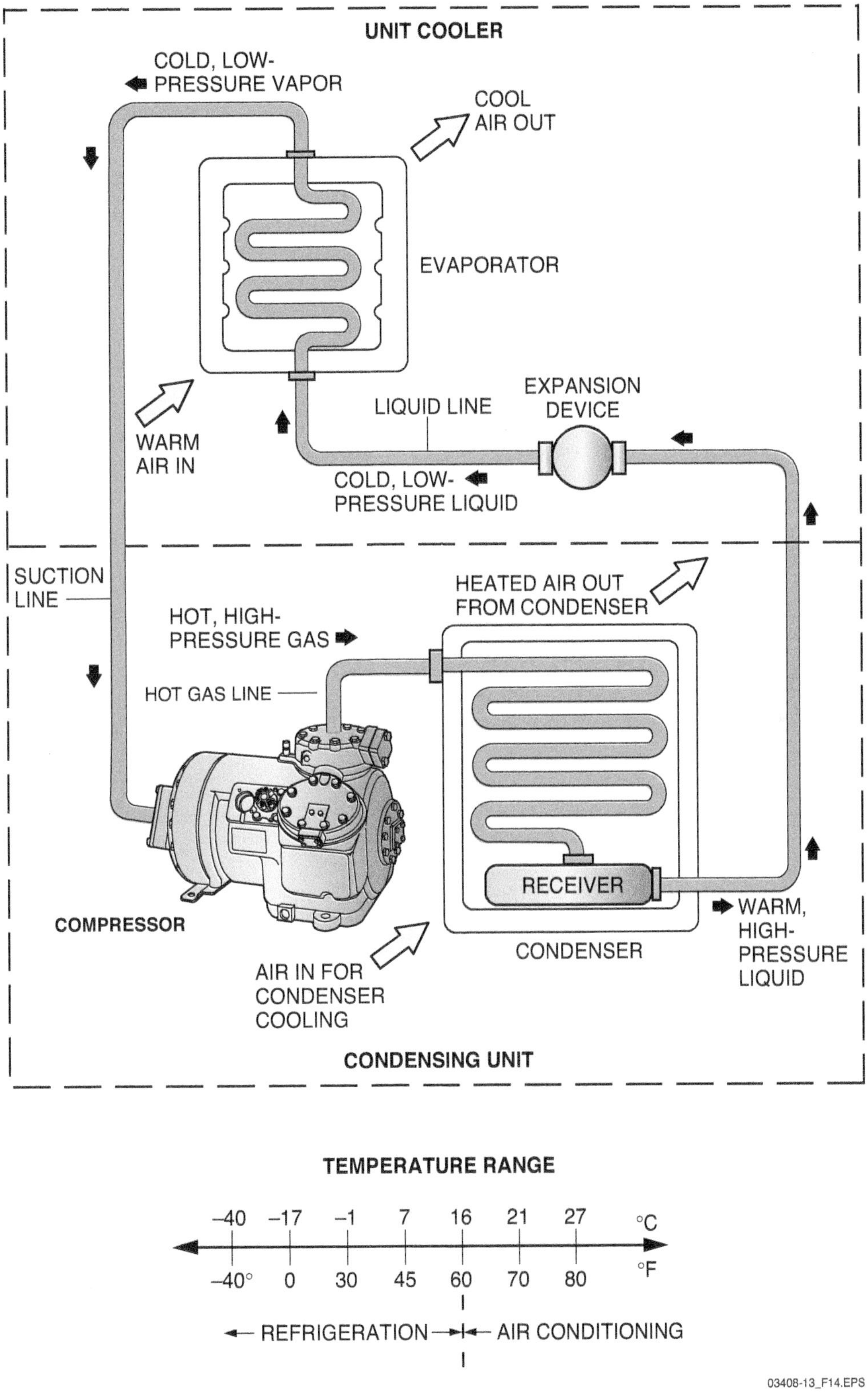

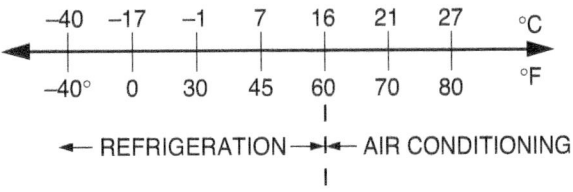

*Figure 14* Basic refrigeration system.

- *Liquid line* – This carries liquid refrigerant, formed in the condenser, to the expansion device.

In refrigeration systems used to cool warehouses, cold storage (chill) rooms, and walk-in coolers, the compressor, condenser, and related components are commonly packaged into an assembly called a condensing unit (*Figure 15*). In practice, condensing units can be either air cooled or water cooled. Air-cooled units can contain one or more compressors and/or fans. Condensing units can be installed in equipment or mechanical rooms, mounted on the roof or other outdoor location, or both. The evaporator, expansion device, and fans are also packaged into an assembly that is installed in the individual chill room or walk-in cooler. This assembly is commonly called a unit cooler. Some large refrigeration systems can consist of individual components piped together at the job site into a customized system. Self-contained reach-in coolers and display cases have all the system components packaged into a unit that is enclosed in the cabinet. In-depth coverage of systems used in the retail sector can be found in the HVAC Level Three module, *Retail Refrigeration Systems*.

Refrigeration equipment manufacturers classify equipment according to its cooling class, which is based on the evaporator operating temperature. Within each classification, there are a variety of models to cover the capacity requirements of various applications. *Table 1* shows common refrigeration applications and cooling classes assigned by one manufacturer.

Refrigeration systems used in supermarkets usually have the most demanding cooling requirements (*Figure 16*). A typical supermarket might use one or more medium-temperature refrigeration systems with parallel compressors for the meat, deli, dairy, and produce refrigerators and walk-in coolers. The system can have a separate compressor for the meat or deli refrigerators, or all units can have a single compressor. The low-temperature refrigerators and coolers can be

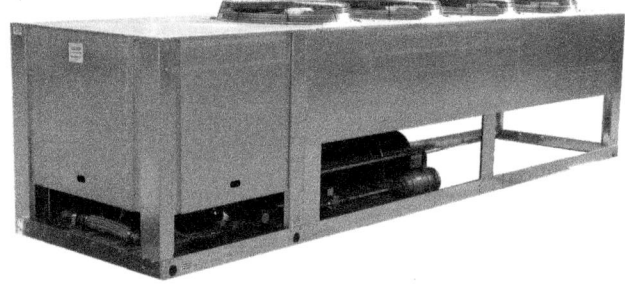

03408-13_F15.EPS

*Figure 15* Commercial refrigeration condensing unit and unit cooler.

**Table 1** Cooling Applications and Temperature Class

| Application | Storage Temp. Range | Cooling (Evaporator Temp. Class) | | |
| --- | --- | --- | --- | --- |
| | | Low −40°F to −10°F | Low −10°F to +30°F | High +30°F to +50°F |
| Meat cutting and packaging, fresh fruit and vegetable storage, floral boxes | +35°F to +60°F | | | X |
| Meat storage rooms and extended storage of many fruits and vegetables | +28°F to +34°F | | | X |
| Low ceiling coolers for all applications | +35°F and above | | | X |
| Low ceiling freezers for all applications | −20°F to +34°F | X | X | |
| Walk-in coolers and warehouses for meat, dairy, and produce | +35°F and above | | | X |
| Walk-in freezers | −30°F to +34°F | X | X | |

03408-13_T01.EPS

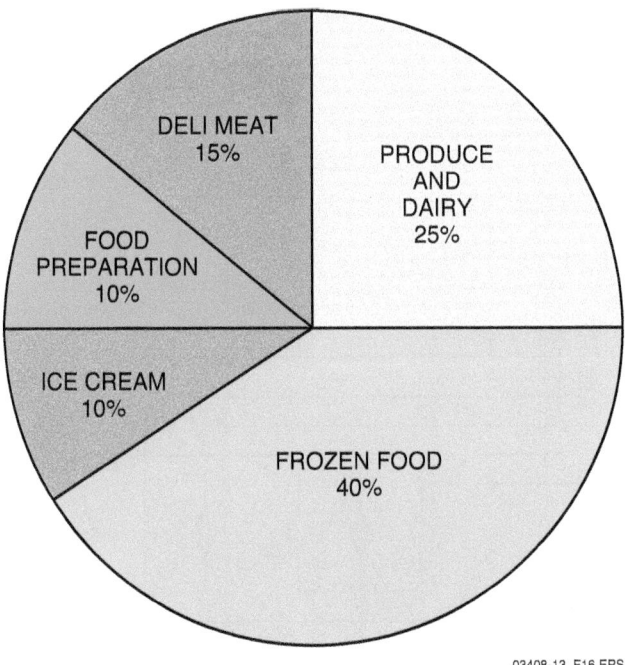

*Figure 16* Distribution of cooling requirements in a typical supermarket.

grouped on one or more systems with parallel compressors. The ice cream refrigerators might be on a satellite compressor or single compressor. Meat cutting and food preparation rooms are usually placed on a single unit.

An important reason to group the various refrigeration circuits with similar temperature requirements is to improve compressor efficiency. The lower the required system temperature, the lower the compressor energy efficiency ratio (EER), thus the more expensive it is to operate. EERs for systems operating at extremely low temperatures can be as low as 3, as opposed to ratings at or above 14 with systems operating at or above 40°F evaporator temperatures.

## 2.1.0 Compressors

Reciprocating, rotary, scroll, and screw compressors are all used in refrigeration systems. The particular type used is determined by the refrigeration load and application. Hermetic and semi-hermetic reciprocating compressors are the most widely used for commercial applications, with the semi-hermetic being the most common. Direct-drive and/or belt-drive open reciprocat-

ing compressors are used in mobile refrigeration units and on older stationary refrigeration systems. For review, the operating principles for these compressors are described in detail in the HVAC Level One module *Introduction to Cooling* and HVAC Level Two module *Compressors*.

### 2.1.1 Single-Compressor Applications

Single compressors are found in about half of the refrigeration systems in use. The compressor sizes commonly range between 0.5 and 30 horsepower (hp). A single compressor is often used to cool multiple evaporators in a line of display cases or coolers (*Figure 17*) or in multi-temperature mobile refrigeration units.

The advantages of using a single compressor to supply multiple evaporators are two-fold. One is the relatively low cost and the second is the heat from the equipment can be easily captured in a heat recovery unit, which is then used to heat the store or its domestic hot water supply.

With multiple evaporators connected to a single compressor, and each evaporator experiencing different loads throughout the day, the use of compressor unloaders is generally recommended. This approach allows the compressor to adjust its capacity as necessary to accommodate the load at any given time, greatly reducing the number of start-stop cycles. A single compressor can also be used to serve multiple evaporators operating at different temperatures. In this application, as shown in *Figure 17*, the compressor must be set to operate at the suction pressure and corresponding evaporator temperature needed by the coldest evaporator in the system. To prevent the other evaporators operating at higher temperatures from falling to this pressure and temperature, evaporator pressure regulators (EPRs) are installed on only those suction lines. EPRs are examined later in this module. In more sophisticated applications, the compressor's suction pressure setpoint can be automatically controlled and adjusted as evaporator loads are satisfied, preventing it from operating at lower setpoints than are necessary.

### 2.1.2 Two-Stage Compressors

Conventional compressors use a single-stage process to compress the refrigerant gas in one or more

---

# Energy Efficient Ratios (EERs)

Energy efficiency ratios (EERs) were discussed in earlier HVAC training modules. The EER is the number of Btus produced by a system per watt of energy consumed.

---

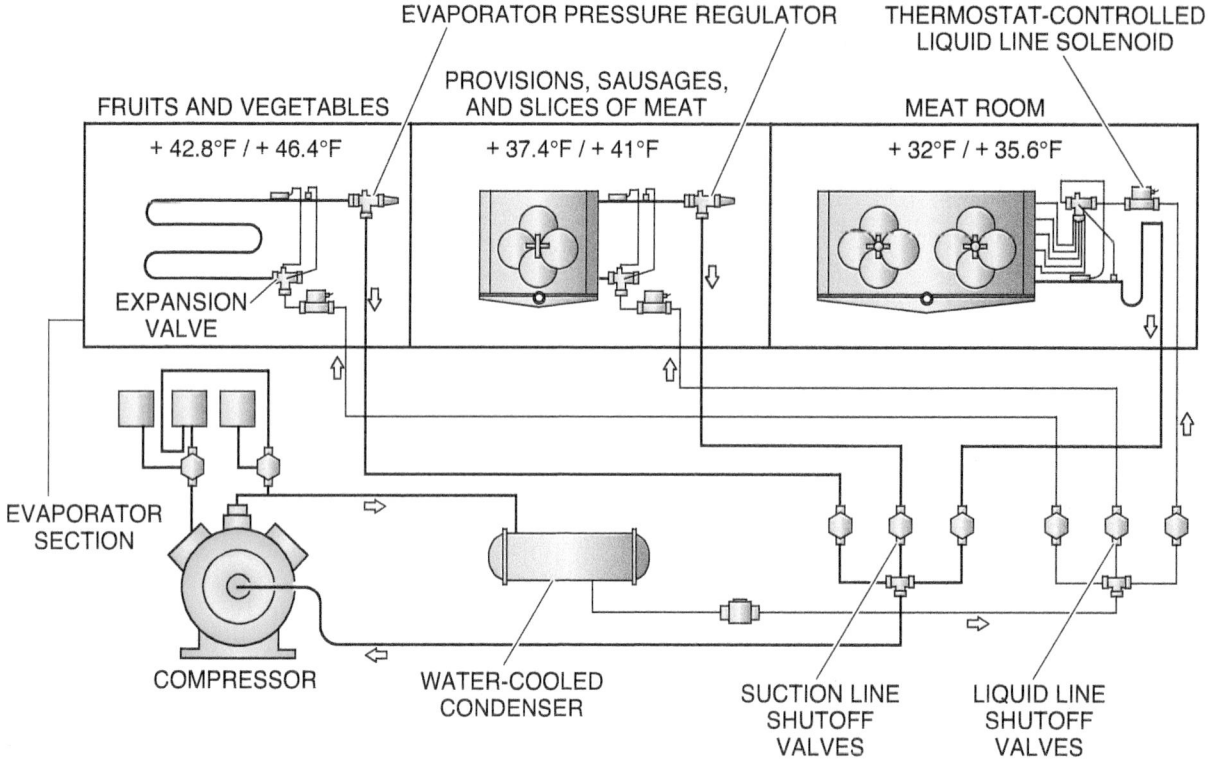

EVAPORATOR PRESSURE REGULATOR

THERMOSTAT-CONTROLLED
LIQUID LINE SOLENOID

PROVISIONS, SAUSAGES,
AND SLICES OF MEAT

FRUITS AND VEGETABLES

MEAT ROOM

+ 42.8°F / + 46.4°F

+ 37.4°F / + 41°F

+ 32°F / + 35.6°F

EXPANSION
VALVE

EVAPORATOR
SECTION

COMPRESSOR

WATER-COOLED
CONDENSER

SUCTION LINE
SHUTOFF
VALVES

LIQUID LINE
SHUTOFF
VALVES

03408-13_F17.EPS

*Figure 17* Simplified single-compressor system with multiple evaporators.

of its cylinders. Single-stage compression means the refrigerant gas is drawn into the suction valve of the compressor cylinder(s) on the piston downstroke, and then is compressed to the required pressure on the compression upstroke of the piston. The compressed gas is then discharged from the compressor cylinder(s) for routing to the condenser.

Two-stage compressors are used in ultra low-temperature systems to pump very low-pressure suction line vapor up to the condensing pressure and temperature conditions. They are also used in some refrigeration applications where the compression ratio of a single-stage compressor would exceed 10 to 1. Compression ratio is calculated as absolute discharge pressure divided by absolute suction pressure. The absolute pressure (psia) is obtained by adding gauge pressure (psig) to atmospheric pressure (14.7 psi, sometimes rounded to 15 for simplicity).

Two-stage compressors (*Figure 18*) compress the refrigerant gas in a two-step process involving two sequential cylinders or compressors. The system suction gas is applied to the first stage, usually a large cylinder, where it is compressed to about the midpoint of the compression curve. The gas is then discharged into the suction valve of a second cylinder. In the second cylinder, the refrigerant gas is compressed again and raised to the final condensing pressure and temperature

conditions. It is then discharged for routing to the condenser. As shown, a typical inter-stage pressure between the first and second cylinders (suction pressure for the second cylinder) is 50 psig. The use of a two-stage compressor results in a lower compression ratio per stage than that with a single-stage compressor. In the example shown, the compression ratio of the first stage is 4.4 to 1, and for the second stage, it is 3.3 to 1. If a single-stage compressor were used with the same suction and discharge pressures, the compression ratio would be 14.6 to 1.

The compression ratio for a two-stage compressor is calculated as follows:

$$\text{Compression ratio} =$$

$$\frac{\text{gauge discharge pressure} + 14.7}{\text{gauge suction pressure} + 14.7}$$

$$\text{Stage 1} = \frac{50 + 14.7}{0 + 14.7} = 4.4$$

For a single-stage compressor:

$$\frac{200 + 14.7}{0 + 14.7} = 14.6$$

One disadvantage of using two-stage compression is the high temperatures developed between the first and second stages as a result of compressing the refrigerant in the first stage. One method

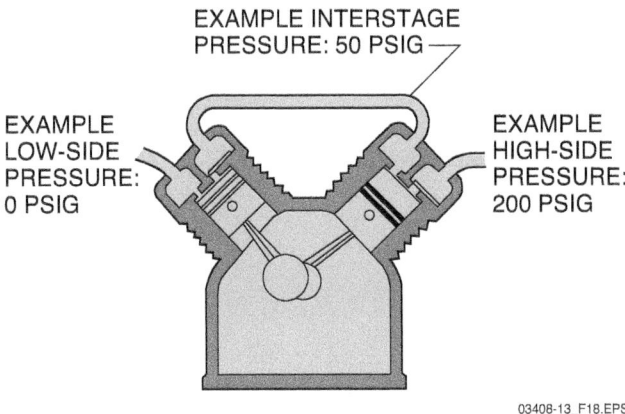

EXAMPLE INTERSTAGE
PRESSURE: 50 PSIG

EXAMPLE
LOW-SIDE
PRESSURE:
0 PSIG

EXAMPLE
HIGH-SIDE
PRESSURE:
200 PSIG

03408-13_F18.EPS

*Figure 18* Two-stage compressor.

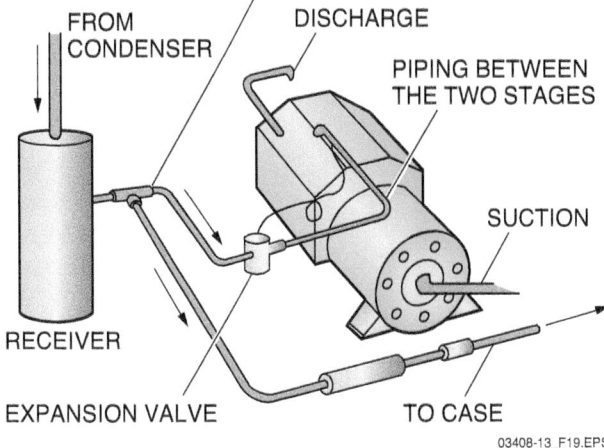

SMALL AMOUNT OF LIQUID REFRIGERANT DIVERTED
FOR THE PURPOSE OF DESUPERHEATING THE GAS
IN THE INTERSTAGE PIPING

FROM
CONDENSER

DISCHARGE

PIPING BETWEEN
THE TWO STAGES

SUCTION

RECEIVER

EXPANSION VALVE

TO CASE

03408-13_F19.EPS

*Figure 19* Desuperheating method used in a two-stage compressor.

used to eliminate this problem is to desuperheat the gas. This is done by allowing a certain amount of liquid refrigerant from the system receiver to be metered through an expansion valve into the compressor inter-stage piping (*Figure 19*). As a result, the refrigerant flash gas acts to cool the discharge gas from the first stage before it is applied to the second stage. This is the same approach used in some hot gas bypass arrangements to desuperheat the suction gas to prevent compressor overheating.

### 2.1.3 Satellite Compressors

A satellite compressor is a separate compressor that uses the same source of refrigerant as that used by a related group of parallel compressors. However, the satellite compressor functions as an independent compressor connected to a cooling area. This requires a lower temperature level than that being maintained by the related parallel compressors. Systems operating at a lower temperature run lower suction pressures and are less efficient. The use of a satellite compressor allows the parallel compressors in the same system to operate at a higher, more efficient suction pressure. The satellite compressor can be a remote unit, or it can be installed on the same rack as a group of parallel compressors (*Figure 20*). Satellite compressors can be used in both medium- and low-temperature applications. For example, it is common for a low-temperature frozen food cooling system to use a satellite compressor to cool the lower-temperature ice cream freezers.

### 2.1.4 Multiple-Compressor Applications

The use of multiple compressors connected in parallel allows greater system capacities and the ability to meet varying load conditions more effectively. Connecting two or three smaller compressors in parallel normally results in a higher

Btu-per-horsepower capacity than when one larger compressor is used. Compressors in the 5hp, 7.5hp, and 10hp size are typically used for this purpose. Compressors connected in parallel can also provide a system backup in the event that one of the other compressors breaks down.

Parallel compressors are normally operated with a large receiver and multi-tapped liquid, suction, and discharge (hot) gas manifolds. These manifolds or headers provide the point where the evaporators located in each of the various display cases and/or coolers are connected to the system. The condenser used with parallel compressors can be part of the compressor assembly or rack, or it can be remotely located. *Figure 21* shows a basic system using two compressors connected in parallel.

The compressors can be the same or different sizes. In parallel compressor systems, the compressors can be cycled on and off as needed for capacity control and can be controlled or staged based on a drop in the system suction pressure. If the compressors are of equal size, one or more mechanical or electronic methods of capacity control can be used to cycle the compressors on and off while the system maintains one economical pressure range. Parallel compressors of different sizes can be staged to obtain more steps of capacity than the same number of equally sized compressors. *Figure 22* shows an example of capacity control that can be obtained by using three parallel compressors of 5hp, 7.5hp, and 10hp.

In systems using multiple compressors, attention must be given to the return of the oil to the compressors, compressor protection, and the effect on the remainder of the system when some of the compressors are in the Off cycle. Oil must re-

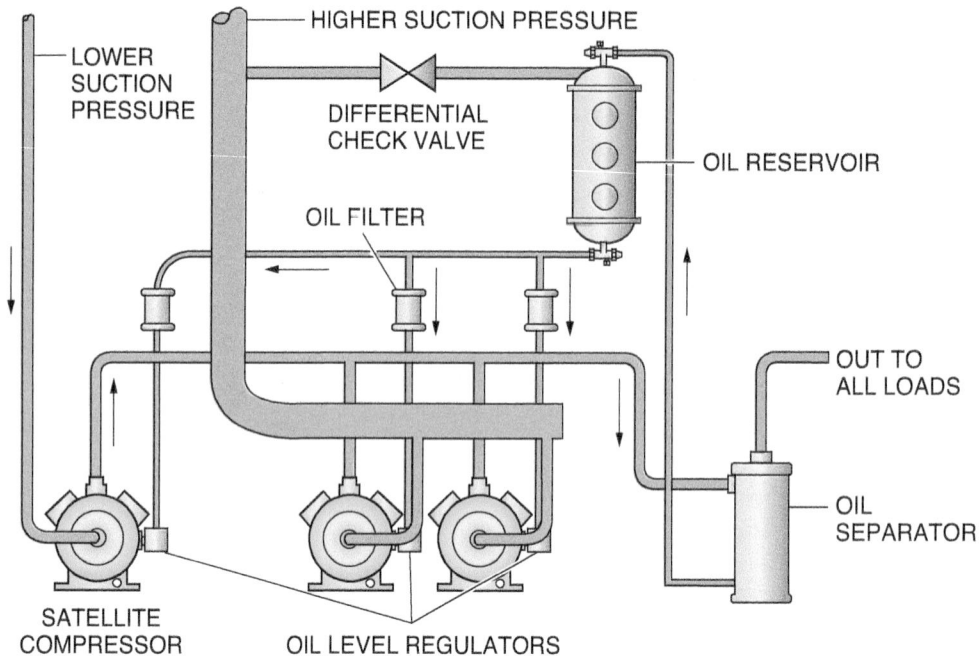

LOWER SUCTION PRESSURE

HIGHER SUCTION PRESSURE

DIFFERENTIAL CHECK VALVE

OIL FILTER

OIL RESERVOIR

OUT TO ALL LOADS

OIL SEPARATOR

SATELLITE COMPRESSOR

OIL LEVEL REGULATORS

03408-13_F20.EPS

*Figure 20* Satellite compressor.

turn to the compressors under all operating conditions. The use of multiple compressors introduces more places to trap and hold the oil. The use of multiple condensers and/or multiple evaporators with multiple compressors further increases the potential for trapping and storing both liquid refrigerants and oil. To the extent possible, piping layouts must avoid oil accumulation in the portions of the system that are not operating. Compressors connected in parallel require crankcase and hot gas line equalization to provide adequate protection. This is because unequal pressures in multiple compressors can cause oil to be blown or drawn from one compressor crankcase to another, creating mechanical problems.

Other guidelines that should be observed when selecting and installing parallel compressors are as follows:

- If possible, standardize on 5hp, 7.5hp, and 10hp compressors. Equip each compressor with pressure controls that satisfy all refrigeration requirements.
- Group similar types of equipment that operate at approximately the same suction pressure on the same circuit. For example, group frozen food cases together and ice cream cases together. Never combine medium-temperature or high-temperature equipment designed to cool non-frozen foods with low-temperature equipment designed to maintain frozen foods.

## Supermarket Parallel Rack Systems

Packaged racks of compressors, complete with controls and safety device, are readily available for installation. Features and available options include integral oil reservoirs with all required oil management devices, a wide selection of compressor types, and alarm systems that visually and audibly indicate a loss of refrigerant or oil as well as mechanical component failure.

03408-13_SA03.EPS

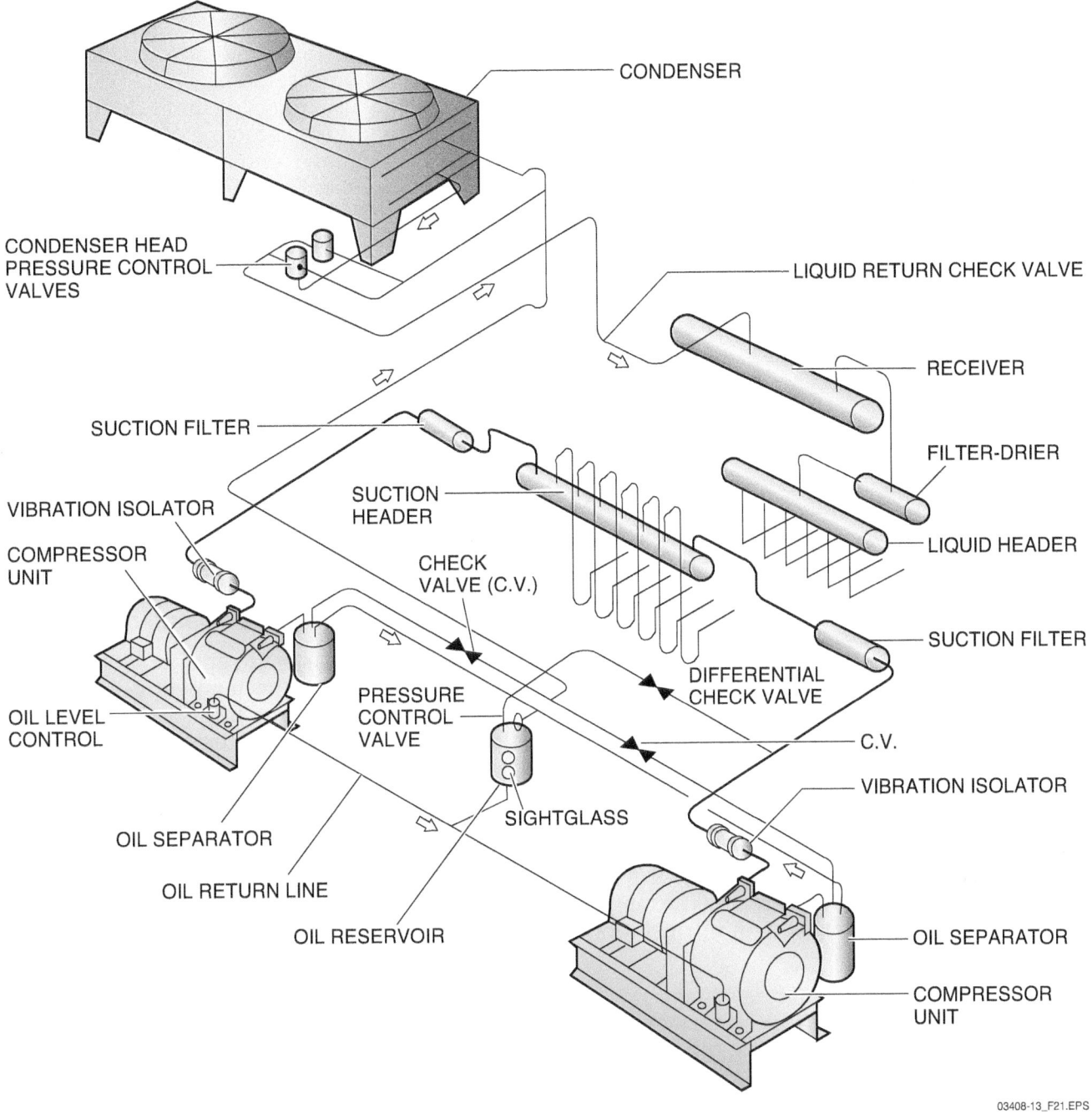

CONDENSER

CONDENSER HEAD
PRESSURE CONTROL
VALVES

LIQUID RETURN CHECK VALVE

RECEIVER

FILTER-DRIER

SUCTION FILTER

VIBRATION ISOLATOR

COMPRESSOR
UNIT

SUCTION
HEADER

LIQUID HEADER

CHECK
VALVE (C.V.)

SUCTION FILTER

DIFFERENTIAL
CHECK VALVE

OIL LEVEL
CONTROL

PRESSURE
CONTROL
VALVE

C.V.

OIL SEPARATOR

SIGHTGLASS

VIBRATION ISOLATOR

OIL RETURN LINE

OIL RESERVOIR

OIL SEPARATOR

COMPRESSOR
UNIT

03408-13_F21.EPS

*Figure 21* Parallel-connected compressor system with multiple evaporators.

- If the compressor size needed for a group of ice cream cases is less than 5hp, the ice cream cases should be combined with frozen food cases that can be defrosted at the same time.
- Combine walk-in coolers. If necessary, produce cases and cutting room loads can be added to the walk-in cooler circuit in order to make use of 5hp to 10hp compressors.
- Combine multi-shelf high-temperature and medium-temperature cases. Produce and dairy coolers can be combined with these units to optimize compressor sizing.

- Put open meat cases on their own compressor. If they must be combined with other equipment, use individual EPR valves for each case.
- Use EPR valves instead of solenoids and thermostats whenever possible. Use solenoids and multiple circuit time clocks only when combining systems that have different defrost requirements.

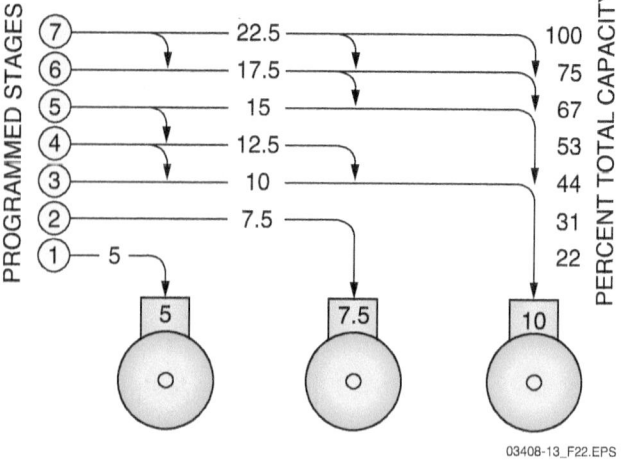

ARROWS INDICATE RUNNING COMPRESSORS

*Figure 22* Capacity control by staging unequally sized compressors.

## 2.2.0 Condensers

Some commercial and industrial refrigeration systems use evaporative condensers or water-cooled condensers with cooling towers. However, the most common condensing apparatus is the air-cooled condenser. A refrigeration system air-cooled condenser can be an integral part of an indoor or outdoor condensing unit assembly, or it can be a separate unit installed at a remote location from the compressor(s) (*Figure 23*). The remote condenser can be placed outdoors, or it can be placed indoors to heat portions of the building in winter. Both single-circuit and multiple-circuit condensers are used.

The typical temperature difference (TD) ranges used for air-cooled condensers in refrigeration systems are as follows:

- 10°F to 15°F for low-temperature systems
- 15°F to 20°F for medium-temperature systems
- 15°F to 30°F for high-temperature systems

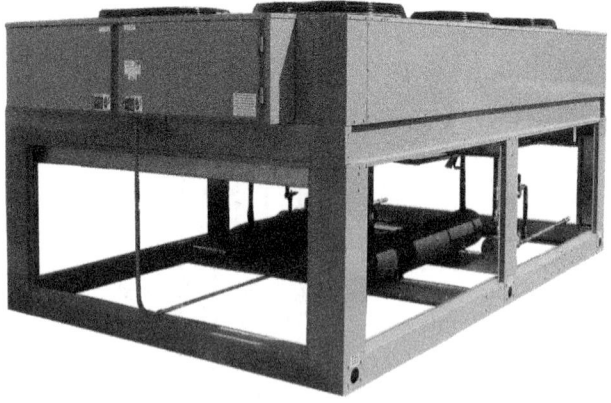

*Figure 23* Remote air-cooled condenser.

### 2.2.1 Condenser Ratings

Condensers used in refrigeration systems are rated by their total heat rejection (THR) value. THR is the total heat removed in the desuperheating, condensing, and subcooling of the refrigerant as it flows through the condenser. The rating of an air-cooled condenser is based on the TD between the dry-bulb temperature of the air entering the condenser coil and the saturated condensing temperature corresponding to the pressure at its input. Normally, the condenser picked to work with a particular compressor is selected by matching their THR ratings. The THR capacity of a condenser is considered proportional to its TD. For example, the capacity of a condenser at 30°F is about 50 percent greater than that of the same condenser at a TD of 15°F. Condensers can also be rated in terms of net refrigeration effect (NRE), also known as net heat rejection (NHR). NHR is the total heat rejection minus the heat of compression added to the refrigerant in the compressor.

### 2.2.2 Control of Air-Cooled Condensers

For a refrigeration system to operate properly, the condensing pressure and temperature must be maintained within certain limits. An increase in condensing temperature causes a loss in capacity. If extreme enough, it can overload the compressor. Low condensing pressures can reduce the flow of refrigerant through conventional expansion valves, resulting in a starved evaporator. Some medium-temperature and low-temperature refrigeration systems use electronically controlled, low-pressure drop expansion valves. These valves are used to ensure that a sufficient supply of refrigerant is fed to the evaporator even when condensing pressures are very low. Conventional thermostatic expansion valves (TXVs) generally require a specific pressure drop to operate at their rated capacity, based on the refrigerant used. For example, HCFC-22 expansion valves are manufacturer-rated at a 100-psi pressure drop. With conventional TXVs, the discharge pressure must be maintained high enough to produce a pressure drop in this range in order to maintain full refrigeration capacity.

To prevent excessively low head pressures that can occur during winter operation, refrigerant-side or air-side control methods are commonly used with condensers. Refrigerant-side control can be done by adjusting the amount of active condensing surface used in the condenser coil. This is done by the controlled flooding of the condenser coil with liquid refrigerant. This method requires the use of a receiver and a larger charge of refrigerant. It also involves the use of temperature-actuated and/or pressure-actuated valves to meter the proper amount of refrigerant needed to flood the condenser in response to the variable loads.

Another common method of refrigerant-side control is to use a condenser with two equal parallel sections, each handling 50 percent of the load during normal summer operation. During the winter, only half of the condenser is used. Solenoid or three-way valves are used to isolate one of the condenser sections as well as any pump-down circuits and fans.

Due to its low initial cost, air-side control is one of the most common methods of head pressure control. It is accomplished by using one of three methods, or a combination of the three: fan cycling, modulating condensing unit dampers, or fan speed control.

Condenser fan cycling can be used even on condensing units with a single fan. As head pressure falls, a pressure switch opens and cycles the fan off. Once the head pressure rises to a predetermined setpoint, the switch closes and the fan restarts. Although simple and inexpensive, this method often results in constant and rapid head pressure changes. This, in turn, causes the expansion valve to constantly change its position in an attempt to regulate the flow of refrigerant and cope with the changing pressure differential. Fan cycling provides better performance when used with larger condensers having multiple fans. Although the head pressure will continue to change up and down between the pressure settings, it is generally a much slower process and the expansion valve responds to the changes more smoothly. With condensers using multiple fans, one fan is often left to run constantly while the others can be cycled as needed.

Some condensers are fitted with dampers on the discharge side of the fans. The damper position is usually controlled directly by refrigerant pressure, with the damper modulating toward the closed position as the head pressure falls. Although this generally results in a more consistent head pressure, the refrigerant-operated damper actuators expose the system to a greater potential for refrigerant leakage. Systems using electronic monitoring of head pressure and positioning of an electric damper actuator are available, but are more expensive. With either approach, air flow through the condenser coil can be modulated from 0 percent to 100 percent. This method is sometimes combined with fan cycling; one fan is fitted with a damper while the other fans are cycled on and off.

The use of variable-speed fans is another method of head pressure control that provides consistent pressures. This method has gained popularity as the cost and reliability of solid-state motor speed controls has improved. As shown in *Figure 24*, one type of control monitors liquid line temperature and uses this input to determine fan speed. As a general rule, fan motors using ball bearings generally perform better than those using impregnated sleeve bearings in speed control applications. This is due to lubrication issues, because sleeve bearings often do not produce proper lubrication at low speeds. This type of control allows for use of sleeve-bearing motors by incorporating a minimum speed setpoint. It also has the advantage of simple installation, since the refrigerant circuit is not accessed. *Figure 25* shows an example of a condenser fan speed control that accepts a pressure input from a transducer. This provides a more direct method of speed control than monitoring liquid line temperature. The control pictured can actually accept a temperature or a pressure input.

### 2.2.3 Multiple Condensers

Hot gas line connections to multiple air-cooled condensers must maintain roughly equal pressure drop to each condenser. Otherwise, there is a tendency for liquid refrigerant to back up in the condenser with the lower pressure. One piping method used to help achieve equal pressures uses a slightly oversized hot gas line or header to supply the condensers. This oversized header is

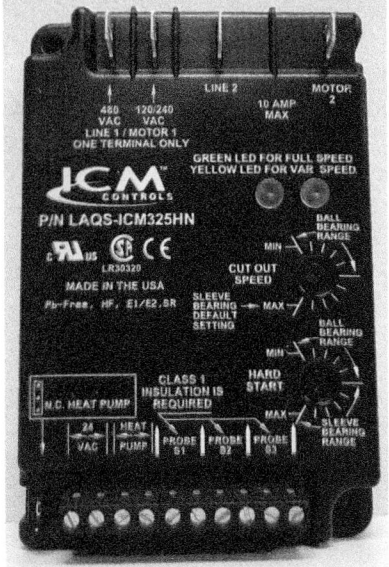

03408-13_F24.EPS

*Figure 24* Temperature-controlled condenser fan speed control.

03408-13_F25.EPS

*Figure 25* Pressure-controlled condenser fan speed control.

essentially frictionless. This allows the takeoffs feeding each of the condensers to be individually sized to provide the required pressure drop or friction loss across the related condenser.

### 2.2.4 *Increasing Liquid Line Refrigerant Subcooling*

Subcooling is used in low-temperature and other refrigeration systems to reduce the refrigerant temperature in the liquid line below the saturation temperature. Remember that there is a loss of capacity in most systems at the outlet of the metering device. This is due to the volume of refrigerant that is vaporized to reduce the temperature of the remaining refrigerant to the saturation temperature corresponding to the evaporator pressure. Refrigerant that is vaporized at the outlet of the metering device cannot contribute its latent heat capacity to the cooling process in the evaporator. Subcooling the liquid refrigerant reduces the temperature difference across the metering device, and less refrigerant is vaporized there as a result. This can result in capacity increases of roughly 0.5 percent per 1°F of additional subcooling provided at the same suction and discharge pressures.

One method used to increase subcooling is to add a smaller coil section in the air entry side of the system condenser. With this arrangement, the liquid refrigerant that leaves the main condenser coil is routed to the receiver as in the basic system. However, from the receiver, the liquid refrigerant is routed back through the small subcooling coil in the condenser, where it receives additional subcooling before being applied to the system expansion valve(s). Subcooling can also be increased by integrating a subcooling section within the con-

denser's main coil assembly. There, the coil's circuitry at the liquid area of the coil is constructed to provide more passes through fewer tube circuits.

Additional subcooling can also be provided by using a water-cooled condenser in addition to the air-cooled condenser. The water becomes a secondary coolant. The best approaches for this in the commercial/industrial sector use sufficiently cool wastewater from other processes. This method uses water that is otherwise wasted to reduce energy consumption and increase refrigeration capacity.

Suction-to-liquid heat exchangers (*Figure 26*) can also be added to systems to increase subcooling, although the value can be questionable. These are simple double-walled heat exchangers, with suction gas flowing through one side and liquid refrigerant flowing through the other. There is no question that the liquid refrigerant does experience additional subcooling from the cool suction vapor, but it is argued that the heat energy never leaves the system since the heat is absorbed by the suction gas. The increased pressure drop in both refrigerant streams can also add to system energy consumption. An added advantage is that the process increases the superheat of the suction vapor, helping to ensure that liquid refrigerant droplets do not return to the compressor. This is an important consideration because commercial and industrial refrigeration systems generally operate at lower superheat values than comfort cooling systems. A variety of studies indicate that the overall effect is positive with some refrigerants and negative with others, so it is suggested that some research be done before applying suction-to-liquid heat exchangers.

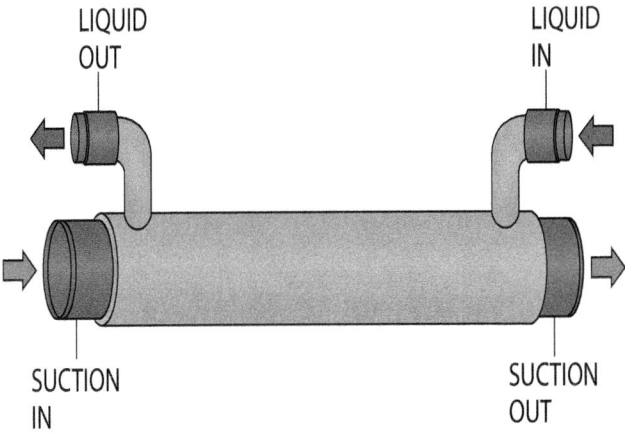

LIQUID OUT    LIQUID IN

SUCTION IN    SUCTION OUT

*Figure 26* Suction-to-liquid heat exchangers.

Another common method used to increase subcooling in the liquid line does not involve the condenser. This method uses a higher-temperature refrigeration circuit to cool the liquid line of a separate lower-temperature system. This subcools the liquid refrigerant flowing in the low-temperature system liquid line. This method is popular because a high-temperature system can remove heat with much more efficiency than could be achieved using one of the condenser-related subcooling methods described earlier.

## 2.2.5 Guidelines for Installing Remote Air-Cooled Condensers

The most important factor to take into account when installing air-cooled condensers or condensing units is the need for a supply of ambient air to the condenser and the removal of heated air from the condensing unit or remote condenser area. Always follow the manufacturer's recommendations concerning clearances from walls and obstructions that should be maintained around the top, bottom, and sides of the unit. If the manufacturer's requirements are not followed, higher system head pressures can result. This can cause poor system operation and possible failure of the equipment. It is also important that adequate space be provided around the unit so that the components can be accessed for removal and/or servicing. Condenser units must not be located in the vicinity of steam, hot air, or exhaust fumes.

Ideally, condenser units should be mounted over corridors, utility areas, restrooms, and other areas where high levels of sound are not important. Roof-mounted units should be installed level on steel channels or I-beam frames capable of supporting the weight of the unit. Mount the condenser over building columns or loadbearing walls. The condenser should be mounted away

## Airflow

Heat from the refrigerated commodity cannot be removed unless the cool refrigerated air can circulate freely around it and return to the evaporator.

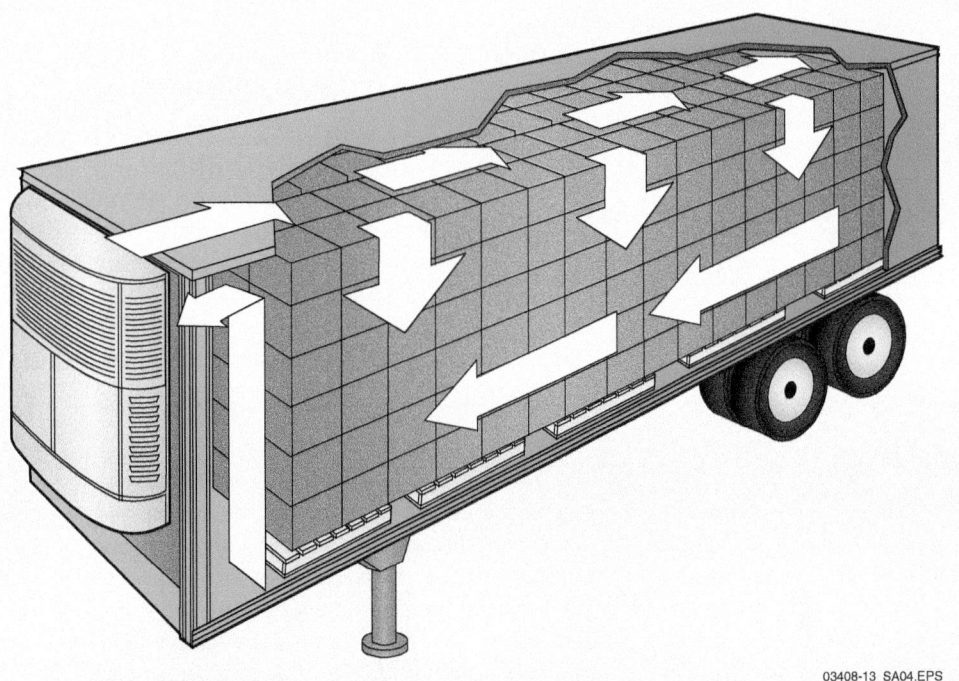

03408-13_SA04.EPS

from noise-sensitive areas and have adequate support to avoid vibration and the transmission of noise into the building. Vibration-absorbing pads or springs should be installed between the condensing unit legs or frame and the roof mounting assembly. Some additional guidelines that should be observed when installing remote air-cooled condensers are as follows:

- Never use the condenser coil manifolds, control panels, or return piping bends to lift or move the condenser unit.
- Use the building plans or condenser manufacturer's specifications to identify which set of condenser circuits should be connected to a specific compressor.
- Route and support all piping so that vibration and the stress caused by thermal expansion and building settling are minimized.
- Construct a weatherproof enclosure over all the piping access openings in the roof.
- Route all discharge lines away from the control panel.
- Route all liquid return lines so that they do not protrude more than a few inches in front of the condenser before dropping to the roof level.
- Insulate all discharge piping near the control box to protect workers from burns or electrical shock.
- Build a wooden platform or other nonconductive structure for workers to stand on when servicing electrical devices.

### 2.3.0 Evaporators and Display Cases

Many kinds of evaporators are used in commercial refrigeration. They can be divided into two main groups. One group includes the evaporators used for cooling air, which in turn cools the contents of coolers or display cases. The other group includes liquid-cooling evaporators that are used to cool liquids such as water, soft drinks, other beverages, and brines.

### 2.3.1 Natural Convection Evaporators

Natural convection evaporators are used in many types of open display cases. Air circulation in these evaporators depends on thermal circulation, where warm air rises and cool air descends. These evaporators can be grouped into three classes: frosting, defrosting, and non-frosting.

Frosting evaporators are commonly used in frozen food storage cases. They continuously build up frost from the moisture in the air. The refrigeration system must be shut down periodically to allow the evaporator to defrost. The defrost cycle can be either automatic or manual. If the refrigerant temperature is below 28°F, heat energy of some kind must be used to defrost the coil; otherwise, the evaporator must be turned off longer than a normal Off cycle.

Defrosting evaporators run on a defrosting cycle. While the system is operating and the condensing unit is running, the temperature of the evaporator is low. This results in a buildup of frost on the evaporator. When the compressor shuts off, the coil warms above 32°F and the frost melts. This defrosting process is called air defrosting. One disadvantage with some defrosting evaporators is that when the top of the evaporator defrosts, moisture can flow down the evaporator surface. If this happens before it can properly drain, the moisture can freeze around the lower parts of the evaporator, eventually blocking air circulation and interfering with proper refrigeration.

Non-frosting evaporators operate at temperatures just at or above freezing. When operating properly, frost does not form on these evaporators.

### 2.3.2 Forced Convection Evaporators (Unit Coolers)

The unit coolers (*Figure 27*) used in food preparation rooms, chill rooms, and walk-in coolers incorporate a forced convection evaporator. It is a packaged assembly usually consisting of the evaporator coil, blower fan, expansion valve, and defrost mechanism. The size and shape of the storage area generally determines the type and number of unit coolers used and their location(s).

## Subcooling

The high-temperature, high-pressure liquid refrigerant entering the metering device carries some sensible heat. This sensible heat, along with the latent heat needed to change states, is instantly absorbed by the expanding refrigerant. Some of the refrigerant boils off into vapor to provide a heat sink. The vapor created in this manner is called flash gas. Flash gas cannot absorb any significant heat in the evaporator and therefore reduces system capacity. More subcooling of the liquid refrigerant means less flash gas is produced, which results in more heat being removed from the refrigerated space. More subcooling equals higher evaporator efficiency.

Forced convection cooling of the room results from the fan(s) forcing air over and through the evaporator coil.

Unit coolers can cool a refrigerated room or cabinet quickly; however, they have a tendency to cause rapid dehydration or drying of the foods unless care is taken to match the unit with its intended use. When storing many varieties of fresh produce, the humidity in the storage area should be kept high to maintain product quality. Conversely, the humidity should be kept low in a room used to handle and chill warm, red meat carcasses. This is necessary to avoid the formation of fog and to prevent condensing moisture from dripping onto the meat.

Unit coolers used in areas requiring high humidity levels normally have large coils operating with small air-to-refrigerant temperature differences (4°F to 8°F). The rate of airflow across the evaporator coil is high to provide the required refrigeration with the small drop in temperature. When unit coolers are used in areas requiring low humidity, the evaporator coil is small and operates with higher air-to-refrigerant temperature differences (20°F to 30°F). The rate of airflow across the evaporator is low. The evaporator coil(s) in unit coolers that operate at saturated suction temperatures below 30°F must be defrosted periodically.

When installing unit coolers, the most important factor to take into account is the need for adequate airflow across the evaporator. Always follow the manufacturer's recommendations concerning clearances from walls and obstructions around the top, bottom, and sides of the unit. Some guidelines that should be observed when installing unit coolers are as follows:

- Make sure that the air pattern covers the entire room or area.
- Avoid installing the unit directly above doors and door openings in low-temperature and medium-temperature applications.
- Allow sufficient space between the rear of the unit cooler and the wall to permit free air movement.

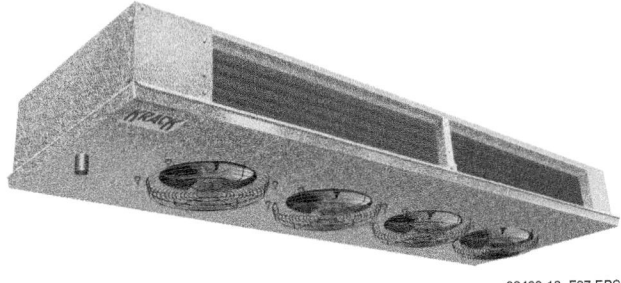

03408-13_F27.EPS

*Figure 27* Typical unit cooler.

- Always trap drain lines individually to prevent vapor migration.
- Traps on low-temperature units must be outside of the refrigerated enclosure. Traps subject to freezing temperatures must be wrapped with heat tape and insulated.
- In coolers or freezers with glass display doors, be sure that the cooler air discharge blows above, not directly at, the doors. If the doors extend above the blower level, use a baffle to properly direct the air.

### 2.3.3 Multiple Evaporators

Many different configurations can exist in a multiple evaporator system. Each coil has its own metering device. Frequently, each coil is also controlled by a liquid solenoid shutoff valve. While each coil acts more or less as an independent coil, be sure that each expansion valve is sensitive to the condition of the coil it controls and is not affected by the conditions in the other coils. Also ensure that the control bulb of the lower expansion valve is mounted at a location that is not affected by refrigerant slop-over from the coil above it. Because of the numerous conditions that can be encountered with expansion valves when installed with multiple evaporators, it is important to install the expansion valves as recommended by the manufacturer.

The physical location of the suction outlets of different evaporator coils also varies widely. Many are at the bottom of the coil, thus creating a natural self-draining condition for any oil in the coil. Also, many coils have headers, thus creating a potential trap at the bottom of the header unless a drain is provided to drain the oil. If drains are not provided by the coil manufacturer, special oil drain lines must be built into the system when it is installed. Typically, these oil drain lines are positioned so that the oil flows by gravity into the refrigerant suction line or the TXV equalizing line. Because of the many conditions that can be encountered, it is important to always install the evaporators and related suction and liquid line piping as recommended by the evaporator manufacturer.

### 2.3.4 Eutectic Plate Evaporators

A eutectic plate evaporator, commonly called a holding plate, is a type of evaporator used in mobile land-based and marine refrigeration systems where continuous power is not readily available to run the refrigeration system on a 24-hour basis. The holding plate is a box-like assembly that houses coiled tubing through which the system refrigerant gas is passed, surrounded by a chamber filled with a eutectic solution. A eutectic solution is one that maintains a constant temperature while changing state from a solid to a liquid (thawing) or from a liquid to a solid (freezing). It is known from previous training that this is a common characteristic of all substances that change states. However, choosing the substance requires study. The particular eutectic solution used must be designed to freeze at a temperature suitable for the refrigeration application. For example, 0°F to 10°F can be required for a freezer unit or 24°F to 26°F for a refrigerator unit. A holding plate is installed in or on the interior walls or ceiling of the refrigerated storage container. Refrigeration systems that use holding plates instead of standard evaporators are called holdover refrigeration systems.

The use of a holdover refrigeration system in a truck, railcar, or boat enables the refrigeration system equipment to run for only part of the day and still maintain stable refrigeration cold box temperatures for up to 24 hours or more. For example, in a truck holdover refrigeration system, the system compressor is typically energized and allowed to run at night. This allows the refrigerated box to be precooled and the eutectic solution contained in the holding plate to completely freeze. This is done by circulating the system refrigerant gas through the evaporator tubing inside the holding plate. When the compressor is energized, the cycle time is controlled by a temperature probe that monitors the temperature of the holding plate and cycles the system as required. The compressor in these systems can be driven in one of several ways, such as directly from an engine or by an AC or DC electric motor.

The holdover refrigeration system compressor remains de-energized while the truck is being used to make delivery runs during the day. During this interval, called the holdover period, the frozen eutectic solution in the holding plate absorbs the heat in the refrigerated container, maintaining the loaded chilled or frozen cargo at the proper temperature. This is accomplished because the frozen solution in the holding plate acts like a block of ice to maintain the temperature of the container for the required holdover period. Note that the surface area of the holding plate and number of holding plates used determines the time available for the holdover period. This depends on the maximum heat load. When the eutectic solution in the holding plate completely thaws, the system must run again to refreeze the eutectic solution and start the holdover cycle again.

### 2.3.5 Display Cases (Refrigerators)

Open, self-service display cases (*Figure 28*) that operate at medium and low temperatures are used extensively in food markets. Display cases are used to maintain the temperature of a prod-

## Suction Line Accumulators

Suction line accumulators have a U-shaped tube inside the shell that ensures only vapor is returned to the compressor. A small hole is drilled in the bottom of the U-tube to allow the trapped oil to return to the compressor.

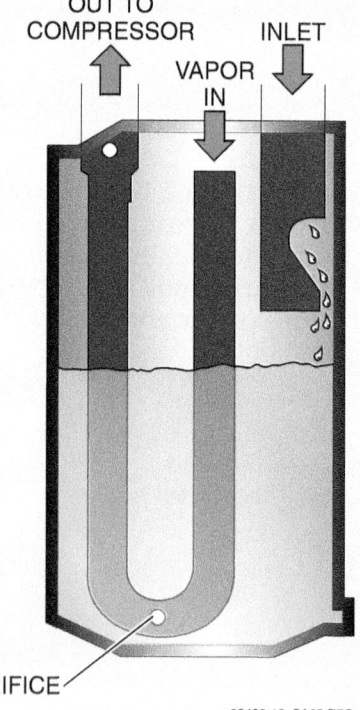

03408-13_SA05.EPS

# Moisture Removal

For every pint of moisture condensed in a self-service display case, about 1,000 Btuh of heat must be removed. This process occurs continuously, requiring that the condensate be removed from the case through a drain.

uct that has previously been cooled to the proper temperature.

Display case operation is significantly affected by temperature, humidity, and movement of the surrounding air. The refrigeration load for food store display cases is normally rated by the manufacturer based on ambient summer design conditions of 75°F and 55 percent relative humidity. Satisfactory operation and efficiency of display case refrigerators will vary when the store conditions are different than the rated conditions.

Sources of radiant heat, such as display lighting, ceiling temperatures, product packaging, and density of display case loading, all affect display case operation. If not controlled properly, all of these heat sources can raise the surface temperatures of the products being displayed, thereby shortening their shelf life. For example, an 80°F ceiling temperature can raise the surface temperature of meat in a display case 3°F to 5°F. At 100°F, the surface temperature of meat can be raised 4°F to 8°F. The surface temperature of a loosely wrapped package of meat with an air space between the film and surface can be 2°F to 4°F above the ambient temperature because of the greenhouse effect within the packaging.

Display cases used in market installations are normally an endless construction type, which allows a group of individual displays to be joined together to form a continuous display bank. Separate end sections are provided for the first and last units in the display bank. All the display cases are constructed with surface zones of transition between the refrigerated area and the room atmosphere. Thermal breaks separate the zones to minimize the amount of surface on the refrigerator below the dew point. The evaporators, air distribution system, and method of defrost built into each type of display case are highly specialized and designed to produce the particular display results desired.

A deck is a shelf, pan, or rack that supports the refrigerated item. Display cases are often classified by the number of decks they contain:

*Figure 28* Open, self-service display cases.

- Multiple-deck dairy display
- Multiple-deck produce display
- Top-discharge produce display
- Multiple-deck fruit and produce display
- Multiple-deck frozen food display
- Back-to-back, single-deck frozen food/ice cream display

The display cases listed above can all be classified as open, self-service display cases. The open display case refrigerates the food using a blanket of cold air. The foods are stored below the top level of the cold blanket of air and are usually wrapped in clear plastic to keep the food and case sanitary. The evaporator in an open display case is normally located in the bottom of the cabinet with a fan used to circulate the air over the coil and through the case.

When compared to closed coolers, open, self-service display cases need additional refrigeration capacity because of their higher refrigeration loss. The refrigeration losses in an open display case are great enough that about twice the amount of refrigeration capacity is needed to cool a product than would be needed to cool the same product in a closed case. Warm, moist air from outside the cabinet mixes with the cold, dry air in the case. Customers reaching into the case to remove stored products cause air movement and additional mixing of warm and cold air. This adds a substantial load to the refrigeration cycle because it must cool the additional air and condense the additional moisture. To minimize the mixing of cold and warm air, open display cases should be located away from all externally induced air circulation devices, such as fans, entrance doors, heating duct openings, and air blasts from unit heaters.

Open frozen food display cases are used to display a variety of frozen foods. Because they operate at temperatures of 0°F and below, a substantial difference exists between the room air and the cold air within the case. Because of this greater temperature difference, more moisture-laden air infiltrates the case, increasing the refrigeration losses. These losses, coupled with the lower operating temperatures, require that the condenser used with a frozen food case be larger than that used with a medium-temperature unit.

It is important to note that the dramatic refrigeration losses described here do have a positive effect elsewhere. Although the losses create higher refrigeration loads, the comfort cooling load of the facility itself can be significantly reduced. As a result, many supermarkets have significantly less comfort cooling equipment installed, helping to offset the investment in refrigeration apparatus. In some cooler geographic areas, the cooling effect of the open refrigeration cases, offset by the heat gain of the facility, can nearly or completely eliminate the need for comfort cooling systems. Conversely, refrigeration losses to the space can increase the winter heating load significantly, resulting in substantial energy costs during the winter months, as the two systems are working against each other. The process of calculating heat gains and losses for food markets must therefore be done with great care and attention to detail to ensure accuracy and proper performance of all mechanical systems as a combined unit.

Some open frozen food cases use vertical plate evaporators. These evaporators are placed about 12" apart along the full length of the storage compartment. The packaged frozen foods are stored between the plates. In some cases, frozen foods cannot be stored in the upper 2" or 3" of the plate compartments due to the warm air that penetrates through that top layer of cold air. The evaporator plates in these display cases are marked to show the level beyond which the products should not be stored. Some display cases use a method of forcing a cold air blanket over the top of the evaporator plates to permit storing food up to the top edge of the plates. This is done by using a forced-air fan, an auxiliary coil built into the structure above the case, or an extra plate evaporator placed lengthwise above the top edge of the plates in the freezing compartment. A blanket of cold air at least 2" or 3" thick must separate the frozen food and the warm, moist air above the storage area in the display case.

### 2.4.0 Refrigeration System Accessories

Accessories (*Figure 29*) are added to the basic refrigeration system in order to improve safety, endurance, efficiency, or servicing. This is especially

## Displaying Produce

Display cases must be loaded properly. The product on display should never be piled so high that it is out of the refrigerated zone. The load line recommended by the display case manufacturer should always be observed to maintain proper cooling. Also, the product should not be stacked so that circulation of the refrigerated air is blocked.

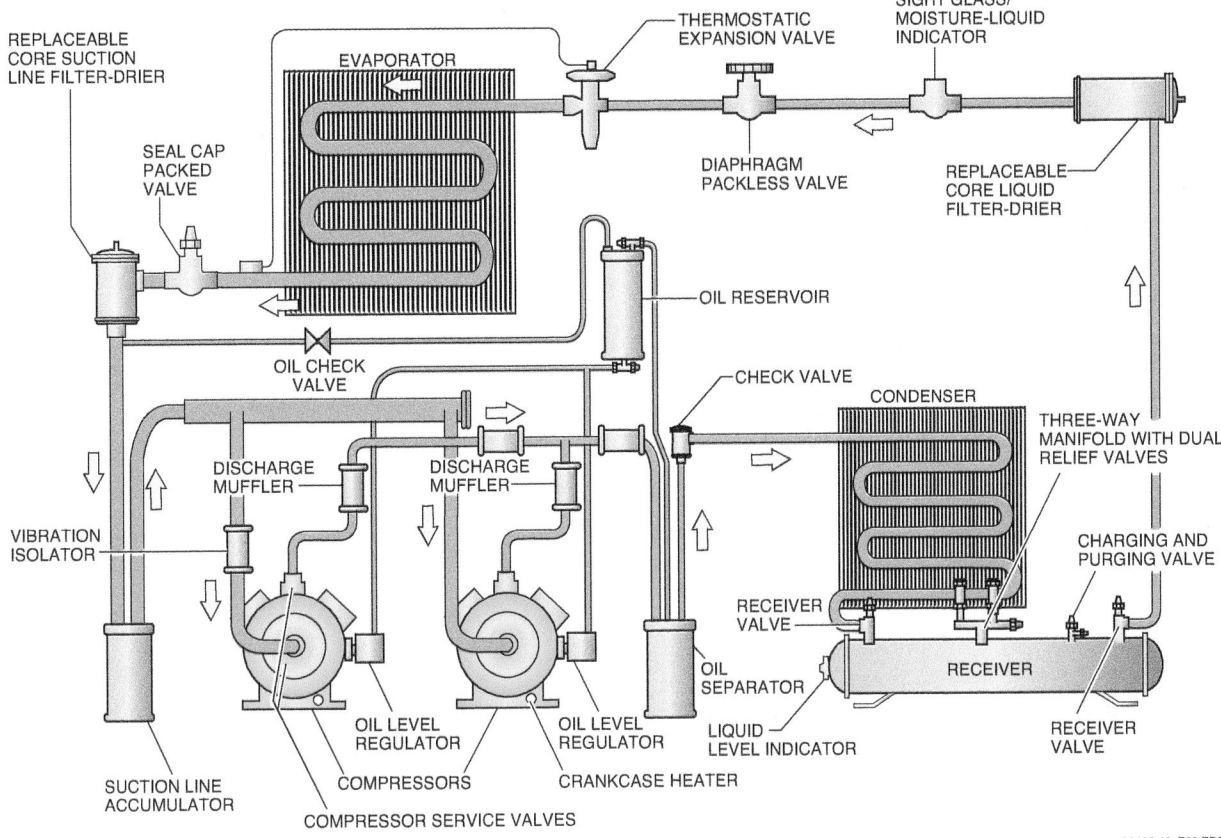

*Figure 29* Parallel compressor refrigeration system with accessories.

Labels in figure:
- REPLACEABLE CORE SUCTION LINE FILTER-DRIER
- SEAL CAP PACKED VALVE
- EVAPORATOR
- THERMOSTATIC EXPANSION VALVE
- SIGHT GLASS/ MOISTURE-LIQUID INDICATOR
- DIAPHRAGM PACKLESS VALVE
- REPLACEABLE CORE LIQUID FILTER-DRIER
- OIL RESERVOIR
- OIL CHECK VALVE
- CHECK VALVE
- CONDENSER
- THREE-WAY MANIFOLD WITH DUAL RELIEF VALVES
- DISCHARGE MUFFLER
- DISCHARGE MUFFLER
- VIBRATION ISOLATOR
- CHARGING AND PURGING VALVE
- RECEIVER VALVE
- OIL SEPARATOR
- RECEIVER
- LIQUID LEVEL INDICATOR
- RECEIVER VALVE
- SUCTION LINE ACCUMULATOR
- OIL LEVEL REGULATOR
- COMPRESSORS
- OIL LEVEL REGULATOR
- CRANKCASE HEATER
- COMPRESSOR SERVICE VALVES

03408-13_F29.EPS

true in commercial refrigeration systems. Some of these components are factory installed, while others can be installed in the field. This section briefly describes the most common accessories. They include the following:

- Filter-driers
- Sight glass/moisture-liquid indicators
- Suction line accumulators
- Crankcase heaters
- Oil separators and oil control systems
- Receivers
- Manual shutoff and service valves
- Fusible plugs, relief valves, and relief manifolds
- Check valves
- Compressor mufflers
- Vibration isolators

### 2.4.1 Filter-Driers

No matter how many precautions are taken during the assembly, installation, and service of a refrigeration system, some moisture and other contaminants will enter the system. Leaks in the system are the main cause of such contamination. Water vapor and foreign matter in sufficient quantities will cause problems in any refrigeration system. Synthetic polyolester (POE) oils increase the chances of a system moisture problem. POE oils attract and absorb moisture from the ambient surroundings faster and in greater quantities than conventional mineral oils. Because they absorb moisture, using POE oils increases the amount of moisture in the system. Moisture can freeze within the orifice of the expansion valve, cause corrosion of metal parts, and/or wet the motor windings of hermetic and semi-hermetic compressors. In the presence of system heat, the moisture will react with the system oil and refrigerant to produce corrosive acid. Foreign matter can also contaminate the compressor oil and become lodged in the various mechanical parts of the system.

Filter-driers (*Figure 30*) are used to remove water vapor and foreign matter from the refrigerant stream. A filter-drier combines the functions of a refrigerant filter and a refrigerant drier in one device. Its use protects the metering device, evaporator, and compressor from foreign matter such as dirt, scale, or rust. The drier portion removes moisture from the system and traps it. Filter-driers are installed downstream of the condenser or receiver outlet service valve and upstream of the liquid line solenoid valve (if so equipped). On some occasions, especially in the suction line, a simple filter is used

---

(A) CROSS SECTION SHOWING DESICCANT
BALLS AND SOLID-PARTICLE SCREEN

(B) CROSS SECTION SHOWING
SOLID CORE

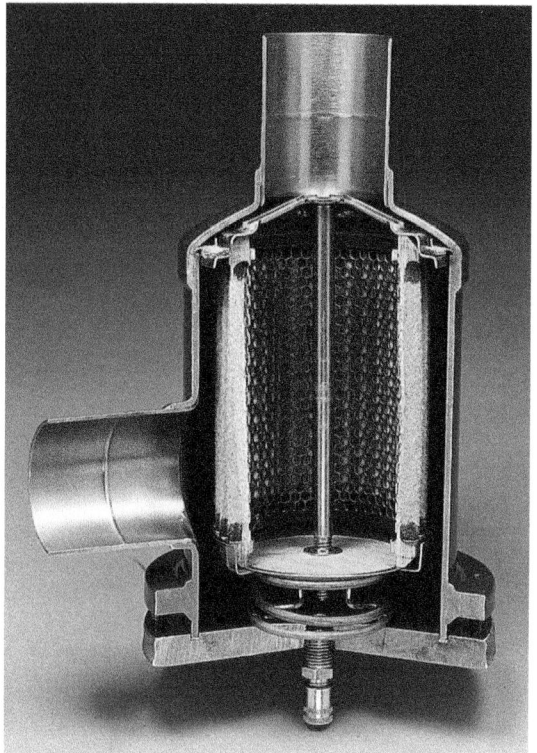

(C) CROSS SECTION OF FILTER
WITH REPLACEABLE CORE

(D) REPLACEABLE CORE

03408-13_F30.EPS

*Figure 30* Filter-driers.

in lieu of the filter-drier. Suction line filter-driers are installed upstream of the compressor service valve, accumulator, or other accessories that can be installed. This is appropriate when moisture is not considered to be an issue. Filters often cause lower pressure drops than filter-driers.

Filter-driers are available both as sealed (throw-away) units and with replaceable cores. Some are equipped with Schrader-type access valves that

## Filter-Drier

A filter-drier is similar to a dried-out sponge. Its drying media attracts and holds water until it can hold no more. Once the filter-drier has absorbed all the moisture it can hold, any remaining water flows right through it, just like water poured onto a saturated sponge. If a filter-drier has a restriction, it is because there are solid particles trapped inside it, not because it is saturated with water.

allow measurement of the pressure drop across the filter. This allows restricted filters or cores to be identified quickly so they can be replaced when the pressure drop is excessive. Normally, filter-driers are replaced immediately after a major system repair.

It is impossible to determine the volume of contamination that is in a system. For this reason, when selecting a replacement unit, choose the largest size filter-drier that fits the available space. Use manufacturer-provided tables to determine the pressure drop that the filter-drier adds based on a given rate of flow or system tonnage.

### 2.4.2 Sight Glass/Moisture-Liquid Indicators

The sight glass (*Figure 31*) is like a window that allows you to view the condition of the system refrigerant at the sight glass location. It is typically used when checking the refrigerant charge. The normal location for the sight glass is in the liquid line, downstream of the condenser or receiver outlet and the filter-drier, and immediately ahead of the expansion valve. The appearance of bubbles in the sight glass generally indicates a refrigerant shortage, but it can also signify a restriction in the liquid line. Frequently, the sight glass also serves as a moisture indicator. This type of sight glass indicates the presence of moisture in the system by the color of the moisture-sensing element. Typically, blue indicates a dry and safe system, light violet cautions that some moisture is present, and a darker violet indicates that the amount of moisture in the system is at a dangerous level. Sight glass/moisture-liquid indicators

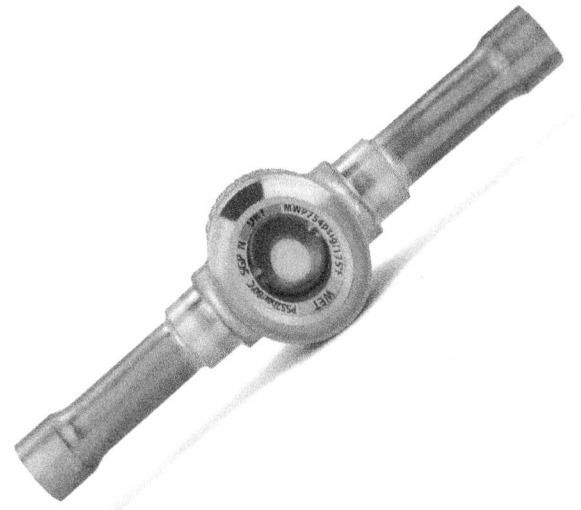

*Figure 31* Sight glass/moisture-liquid indicator.

normally require no service. However, in cases of extreme acid formation in a system after a compressor burnout, the acid can damage the sensing element or etch the glass. This requires that the device be replaced.

### 2.4.3 Suction Line Accumulators

The suction line accumulator is a trap that prevents the compressor from taking in slugs of liquid refrigerant or compressor oil. If liquid refrigerant is allowed to enter the compressor, noisy operation, high consumption, and compressor damage can result. Accumulators are installed in the suction line as close to the compressor suction inlet as possible. At this location, any quantities of liquid refrigerant or oil will be trapped temporarily in the accumulator. The trapped refrigerant remains in the accumulator until it is evaporated. Some accumulators have heaters that help to vaporize the refrigerant liquid.

### 2.4.4 Crankcase Heaters

Crankcase heaters are installed on most compressors to prevent refrigerant from migrating into the system and mixing with oil when the compressor is off. All heaters evaporate refrigerant from the oil. Heaters are typically fastened to the bottom of the crankcase or inserted directly into the compressor crankcase (immersion type). Band-type heaters that encircle the outside shell of welded hermetic compressors are frequently used. Some of these use positive temperature coefficient (PTC) technology to regulate the output of the heater; as the compressor warms, the heater produces less heat. Band-type heaters are inappropriate for use with semi-hermetic and open drive compressors. *Figure 32* shows a variety of clamp-on and insertion type crankcase heaters.

### 2.4.5 Oil Separators and Oil Control Systems

Oil coats the inside of every component through which it passes. It reduces the heat transferability and efficiency of the evaporator and condenser. Oil separators minimize the amount of oil that circulates through a refrigeration system. They also slow down the accumulation of oil in places from which it is difficult to return. The oil separator (*Figure 33*) is one of the primary components of any oil control system. They are typically installed in the hot gas discharge line as close to the compressor as practical. Separators usually have a small sump to collect the trapped oil. A float valve in the sump maintains a seal between the high-pressure and low-pressure sides of the system. As

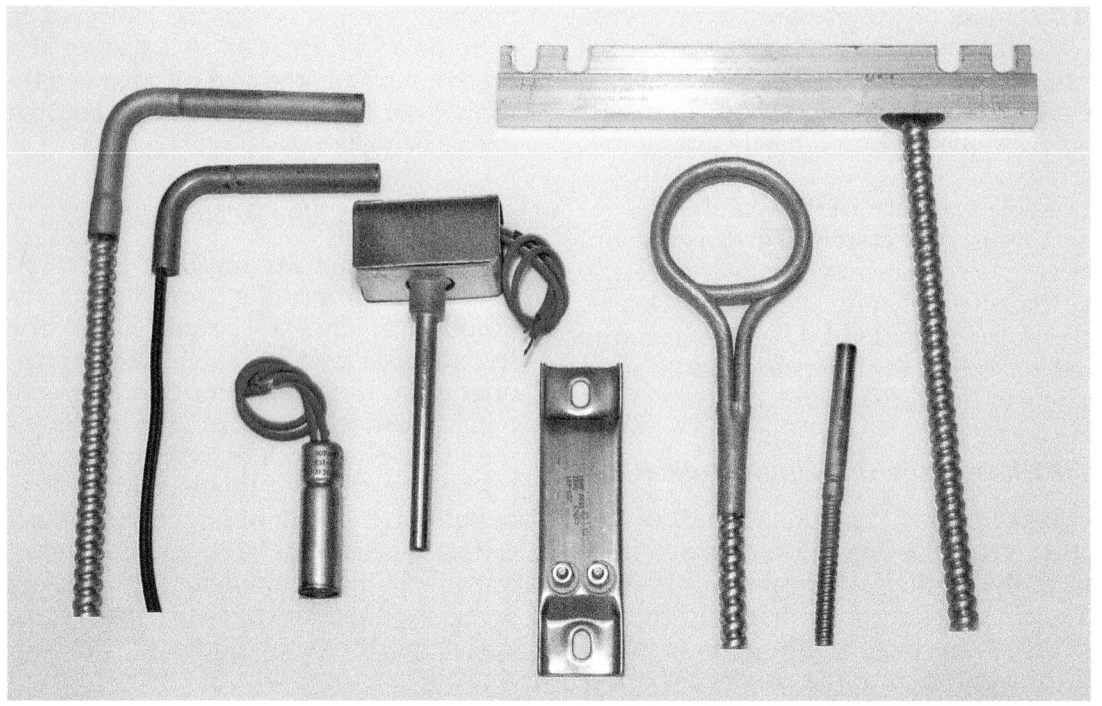

Figure 32 Crankcase heaters.

03408-13_F33.EPS

Figure 33 Oil separator.

the oil level in the sump rises, the float raises and allows high-pressure refrigerant vapor to push the oil back to the low-pressure crankcase of the compressor. When used with a single compressor, the oil separator generally functions without additional system components except for an oil filter (*Figure 34*), which captures any debris.

In parallel compressor systems, the correct oil level must be maintained in each of the compressors, regardless of operating conditions. The amount of oil leaving the individual compressors through the discharge line can vary dramatically due to age, wear, and operating time. Additional components are included, along with the oil separator, to maintain the oil level. The oil separator sends the accumulated oil to an oil reservoir (*Figure 35*), since it does not have a large holding capacity. To reduce the pressure in the reservoir to a level slightly higher than the pressure in the compressor crankcase, high pressure is relieved through the oil differential check valve (*Figure 36*) back to the low-pressure side of the system. The valve shown is preset to an appropriate value to ensure that adequate pressure is maintained in the reservoir to push oil into the crankcase and relieve any pressure in excess of this value.

Oil is delivered back to the individual compressor crankcases through the oil level control (*Figure 37*). One of the flanged ports shown on the back of the control is used to mount it to the compressor crankcase, while the other is covered with a sight glass for viewing. As the oil level drops in the crankcase, the internal float of the control allows oil to enter. The oil level is adjustable in this particular device from ¼ sight glass to ½ sight glass. *Figure 38* is an example of a complete oil level control system.

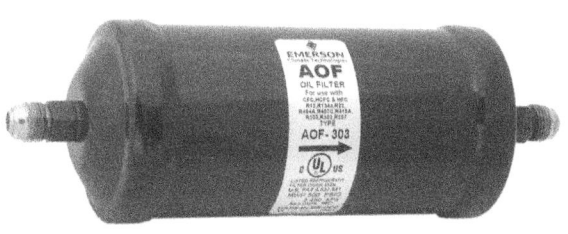

*Figure 34* Oil filter.

03408-13_F34.EPS

03408-13_F35.EPS

*Figure 35* Oil reservoir.

03408-13_F36.EPS

*Figure 36* Oil differential check valve.

## 2.4.6 Receivers

The liquid receiver is a tank or container used to store liquid refrigerant in the refrigeration system. This storage is needed in air-cooled refrigeration systems that have large load variations where the system is not required to remove heat at a constant rate. These variations in load allow the refrigerant in the condenser to freely drain into and accumulate in the receiver for temporary storage. The receiver can also be used to store the system charge during system service procedures or prolonged shutdown periods. Note that most water-cooled refrigeration systems do not use a receiver because it is an integral part of the water-cooled condenser. The receiver is installed in the liquid line between the condenser and the metering device. Receivers have inlet and outlet connections and a relief port. They normally contain a sight glass or test cocks that can be used to determine the amount of liquid refrigerant in the receiver.

03408-13_F37.EPS

*Figure 37* Oil level control.

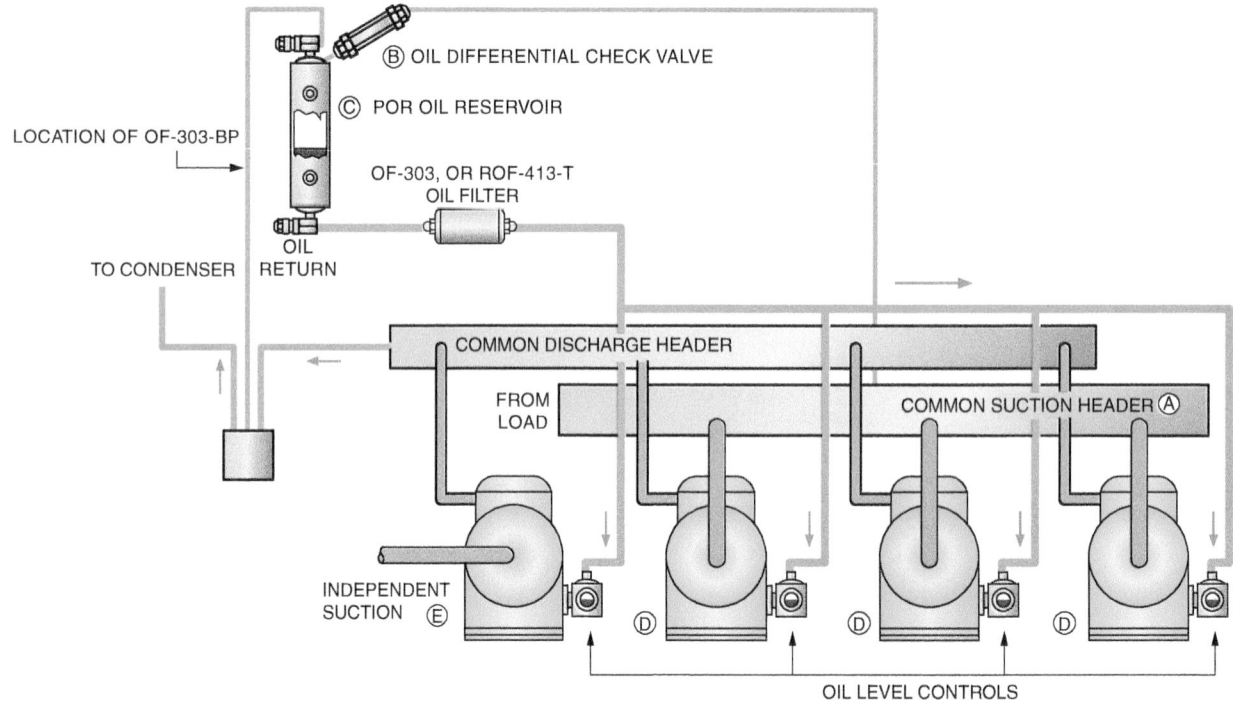

*Figure 38* Oil level control system.

03408-13_F38.EPS

### 2.4.7 Manual Shutoff and Service Valves

Various types of manual shutoff and service valves are installed in refrigeration systems. They enable technicians to seal off parts of the system while installing gauges. They also provide access to the closed system for servicing. Service valves can be installed in any line that may need to be valved off, usually at a place that is easily accessible to the technician. Common valves include charging valves installed in the liquid line or the downstream side of a receiver, receiver outlet and inlet shutoff valves, and compressor shutoff and service valves.

Although all refrigeration valves are important in one situation or another, there are two that are a bit more significant. What is likely considered to be the most important service valve in a system is referred to as the king valve. This is the valve found on the outlet of a receiver. The king valve is used to pump a system down. When it is closed while the system continues to run, the compressor pumps the refrigerant out of all lines and components downstream of it, and the refrigerant accumulates in the receiver. Another

important valve is sometimes referred to as the queen valve. The queen valve is located at the inlet to the receiver. Together, the two valves can be used to trap virtually all of the refrigerant charge of a large system in the receiver. This is usually done for maintenance or repair purposes. Once a pump-down cycle is complete, the compressor is shut down and the queen valve is closed. Only relatively small amounts of refrigerant should then remain in the system.

Valves can have packing around their valve stems, or they can be packless. Many are provided with back-seating construction; that is, when the valve is fully opened, the valve disc seats against a second seat. This arrangement seals the packing from the system pressure, preventing the leakage of refrigerant between the packing and the valve stem. Back-seating valves frequently have a port located between the packing and the back seat of the valve. This service port can be used for refrigerant charging or as a connection point for the gauge manifold set when measuring system pressures. The port is opened to the system by

## Phantom Leak

When a system is diagnosed as being low on charge, the source of the leak must be found and corrected before additional refrigerant is added. If no leak is detected, examine the relief valve. This is especially true for systems that are repeatedly low on charge with no detectable leak. Once a relief valve has opened, it cannot reseat properly when it resets, or it can open at a lower pressure than that for which it was originally designed.

closing (front-seating) the valve or by placing the valve in any intermediate position between open and closed. It is shut off from the system by fully opening (back-seating) the valve. Sealed valve caps are used with packed valves as a further precaution against leakage. To prevent leaks, these should always be in place, unless removed in order to adjust the valve position. *Figure 39* shows the three service valve positions.

Schrader valves are used in some systems that are not equipped with service shutoff valves. A Schrader valve uses the same type of valve as those used to pump air into an automobile tire. Schrader valves are self-sealing and have a spring-loaded core that closes when released and opens when depressed. They provide access to the system when the core is depressed.

### 2.4.8 Fusible Plugs, Relief Valves, and Relief Valve Manifolds

The fusible plug is a device related to temperature and pressure that melts when a specific temperature that corresponds to a given refrigerant pressure is reached. Once the plug melts, the entire refrigerant charge is released. For this reason, the use of fusible plugs is being eliminated in newer systems.

Relief valves are safety devices designed to relieve pressure, preventing the buildup of excessive refrigerant pressure within a system. They are normally installed in the condenser or liquid receiver. In addition, some codes require that a relief valve be installed on the low-pressure side of the system. Relief valves reset after the excessive pressure has been released, trapping the remaining refrigerant charge.

Relief valve manifolds enable the installation of two relief valves at the same piping position. This allows for one of the two relief valves to be safely isolated for service or replacement, while the second valve remains active.

### 2.4.9 Check Valves

Check valves allow refrigerant to flow in one direction only. They are used to prevent the refrigerant in the system from flowing into system components or piping where it is not wanted. A typical application is when a compressor is located in a much cooler place than the condenser. With this condition, it is possible for refrigerant to migrate from the condenser and condense at the compressor discharge connection during periods of shutdown. If the compressor valves are not tight, the refrigerant can enter the cylinders where it will cause severe slugging upon compressor restart. The installation of a check valve in the compressor discharge line prevents this problem.

### 2.4.10 Compressor Mufflers

Mufflers (*Figure 40*) are used mainly in systems with open or semi-hermetic reciprocating compressors. Reciprocating compressors generate audible pulsations that transmit into and travel along the system piping. A muffler installed in the discharge line, as near the compressor as practical, is used to remove or dampen these

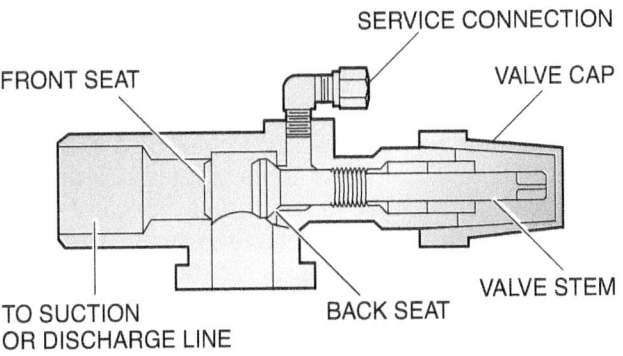

**BACK-SEATED POSITION**

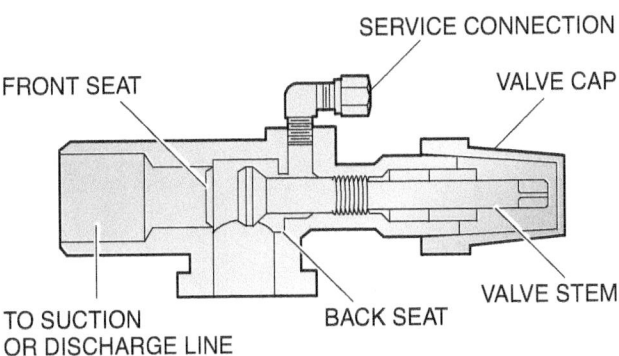

**MID-SEATED POSITION**

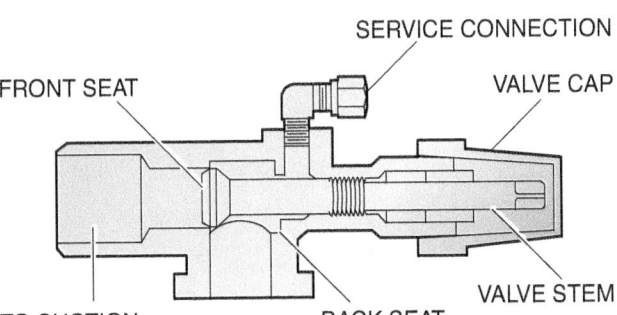

**FRONT-SEATED POSITION**

03408-13_F39.EPS

*Figure 39* Service valve positions.

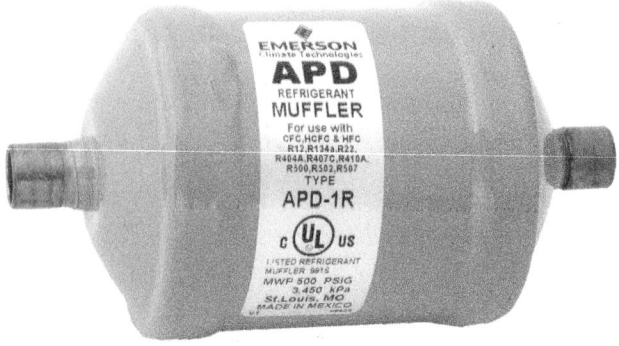

*Figure 40* Compressor muffler.

pulsations. Mufflers lower the system noise and prevent possible damage to piping and components caused by vibration. Mufflers should add only a minimal pressure drop when in the discharge line.

## 2.5.0 Refrigeration System Control Devices

Control devices used to regulate the operation of a refrigeration system can be either manual or automatic and respond to temperature and/or pressure.

### 2.5.1 Crankcase Pressure Regulating Valves

Crankcase pressure regulating (CPR) valves are installed in the suction line ahead of the compressor (*Figure 41*). The valve controls the maximum pressure at the compressor suction line and provides overload protection for the compressor motor. The valve is adjusted for the maximum pressure specified by the manufacturer of the condensing unit when available.

Compressors in commercial and industrial refrigeration systems commonly experience an overload during two conditions. One condition occurs during initial startup or a restart of the system after an extended shutdown, especially in low-temperature systems. At this time, the compressor is likely operating well outside of its selected operating range, and the load must be managed to prevent compressor overheating and/ or motor damage. The other condition follows a defrost period, when the warm evaporator places an additional load on the compressor for a short period of time. When an overload occurs, the valve modulates to prevent suction gas pressures from exceeding the pressure for which the valve is adjusted. When the overload passes and the pressure drops below the valve setting, the valve opens wide.

The valve setting is determined by a pressure spring and the valve closes on a rise in suction line pressure. One common field adjustment method is to operate the system under what is perceived to be an excessive load while monitoring the compressor amperage with an ammeter. The valve is then adjusted toward the closed position until the compressor amperage falls just below the compressor rated load amps (RLA) as stated on its data plate.

### 2.5.2 Evaporator Pressure Regulating Valves

Evaporator pressure regulating (EPR) valves are installed in the suction line between an evaporator and the compressor (*Figure 42*). They are designed to maintain the desired minimum pressures, and therefore temperatures, within close limits. These valves are used to control the evaporator temperature on systems that use multiple evaporators operating at different temperatures or on systems where the evaporating temperature cannot be allowed to fall below a predetermined level.

The EPR valve responds to the inlet pressure of the evaporator and opens when the inlet pressure equals or exceeds the opening setpoint. As the evaporator temperature continues to rise, the valve opening increases, permitting an increased

THERMOSTATIC EXPANSION VALVE (TXV)

*Figure 41* Crankcase pressure regulating (CPR) valve.

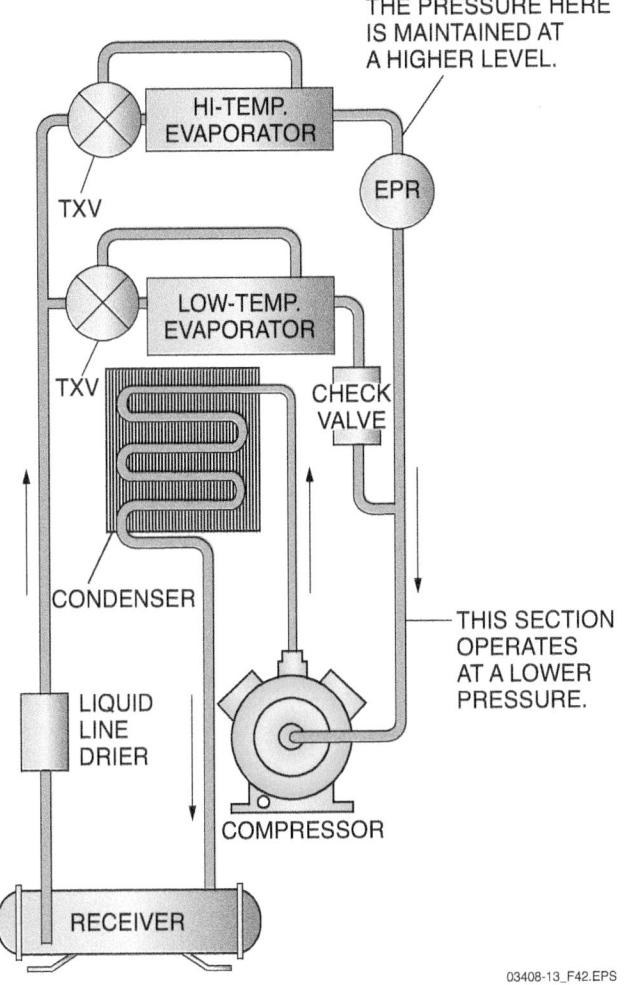

THE PRESSURE HERE IS MAINTAINED AT A HIGHER LEVEL.

HI-TEMP. EVAPORATOR

EPR

TXV

LOW-TEMP. EVAPORATOR

TXV

CHECK VALVE

CONDENSER

THIS SECTION OPERATES AT A LOWER PRESSURE.

LIQUID LINE DRIER

COMPRESSOR

RECEIVER

03408-13_F42.EPS

*Figure 42* Evaporator pressure regulator.

flow of refrigerant. The valve closes off when the inlet pressure drops below the setpoint. Consequently, the evaporator temperature does not drop below this established point.

On multiple evaporator systems where different evaporator temperatures are required, EPR valves are used to hold the saturation pressure at the required setpoint above the common system suction pressure. The EPRs prevent the lowering of temperatures in the warmer evaporators below that desired, while the compressor continues operating to satisfy the temperature requirements needed for the colder evaporators.

### 2.5.3 Air-Cooled Condenser Pressure Regulator

The condenser pressure regulator (head pressure control) maintains proper condenser pressures when the ambient temperature of the outdoor air is low. The head pressure control is a three-way modulating valve (*Figure 43*). It is controlled by the discharge (head) pressure. When the outdoor air temperatures are high enough to cause the compressor discharge pressure to be above the head pressure control valve setpoint, refrigerant flows through the system in the normal manner. It flows from the compressor through the condenser and valve ports to the receiver, then through the metering device and into the evaporator.

As the ambient outdoor air temperatures fall, there is a corresponding decrease in the discharge pressure. When the discharge pressure falls below the head pressure control valve setpoint, some of the compressor discharge gas is routed through the valve into the receiver, creating a higher pressure at the condenser outlet. Note that the valve does not bypass all refrigerant around the condenser; a portion of the refrigerant flows into the condenser. As it exits, it mixes with the hot gas stream at the valve. The higher pressure at the condenser outlet reduces

## Supermarket Energy and Global Warming

GOING GREEN

Supermarkets are among the highest energy-consuming types of commercial facilities in the world, with the refrigeration systems consuming the lion's share at 50 percent or more. Huge amounts of heat must be rejected, and proportionally large amounts of refrigerant are present in the facility due to the required refrigeration capacity and lengthy piping runs. As a result, the Total Equivalent Warming Impact (TEWI) of grocery operations is very high.

Numerous studies have been done on TEWI reduction in supermarket operations and many strategies are employed to reach this goal. For example, rejected heat is captured from refrigerants and used to heat water for food preparation and cleaning activities, as well for use in comfort heating. A reduction of refrigerant charge results in a reduction of financial risk associated with refrigerant loss, and a reduction in the potential for environmental damage when leakage does occur. One method used to reduce the total refrigerant charge is to use secondary coolants. These are other fluids that have no significant environmental impact and function as a heat transfer agent. These coolants can be chilled by the system inside the mechanical room environment, and then pumped to the various refrigeration loads throughout the store. This strategy can reduce the required primary refrigerant charge by thousands of pounds in a single facility.

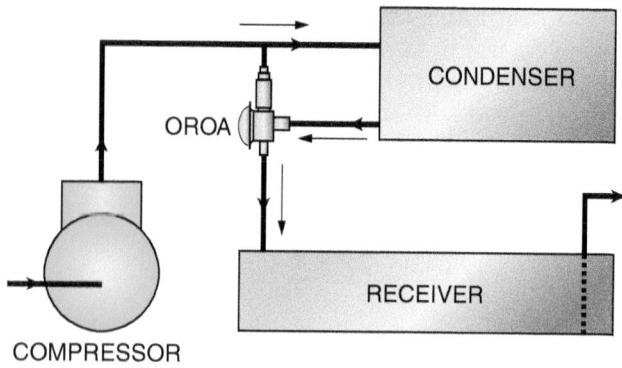

**(A) NORMAL FLOW**

CONDENSED REFRIGERANT MIXES
WITH HOT GAS AT THE VALVE

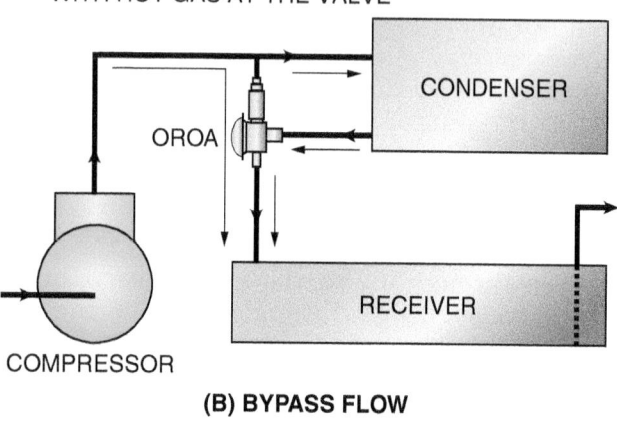

**(B) BYPASS FLOW**

03408-13_F43.EPS

*Figure 43* Condenser pressure regulator (three-way valve).

the flow and causes the level of condensed liquid to rise in the condenser. This flooding of the condenser with liquid refrigerant reduces the available condensing surface. The result is to increase the pressure in the condenser and maintain an adequate high-side pressure. Although the three-way valve is popular for small systems, its capacity is limited. Larger systems accomplish the same task by using by using two separate valves to control refrigerant flow and maintain head pressure. *Figure 44* provides an example of this arrangement. The desired head pressure is adjustable at the upper valve. When using the single valve, the pressure setting is fixed at the factory and cannot be adjusted.

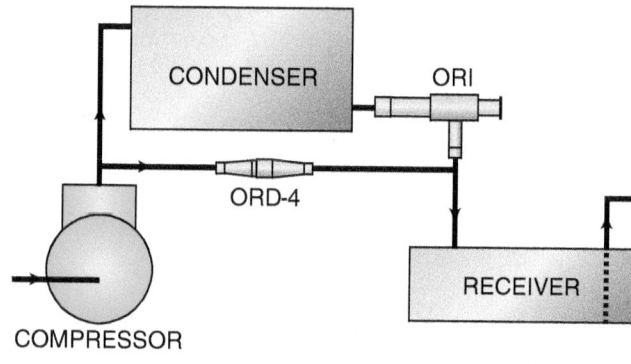

03408-13_F44.EPS

*Figure 44* Condenser pressure regulation (two valves).

### 2.5.4 Bypass Control Valves

Bypass control valves for air-cooled condensers function as condenser bypass devices on air-cooled condensers with winter start control. These types of valves are installed in the condenser bypass line between the compressor discharge and the liquid line receiver (*Figure 45*). These devices operate in response to receiver pressure. The bypass valve opens when the receiver pressure is below the valve setpoint.

As each compressor starts on cycle, high-pressure discharge gas is admitted directly to the receiver through this valve. The pressure quickly builds up in the receiver; when the valve operating pressure is reached, it closes and the compressor discharges normally into the condenser. Without a bypass control, some delay would occur before adequate receiver pressure is reached. This is due to the time required for the larger condenser mass to reach operating temperatures.

### 2.5.5 Capacity Control Devices

Capacity control refers to the method used to control a system so that its heat removal ability (capacity) matches changing system load conditions. In commercial refrigeration systems, the load on the evaporator can vary widely, ranging from full load to a small percentage of full load. Differences in the load can be caused by the type of product being cooled, the quantity of product being cooled, and the ambient temperature. At system loads less than full load, less refrigerant is boiled

## Compressor Failures

Failure analysis has determined that most compressors failing in the first week of operation do so because of a lack of lubrication. The most likely cause of this problem is failure to properly connect the crankcase heater. Without a crankcase heater, liquid accumulates in the compressor during the Off cycle. When the compressor starts up, the liquid refrigerant flushes the oil out of the lubrication system. The result is usually a broken rod, piston, or crankshaft.

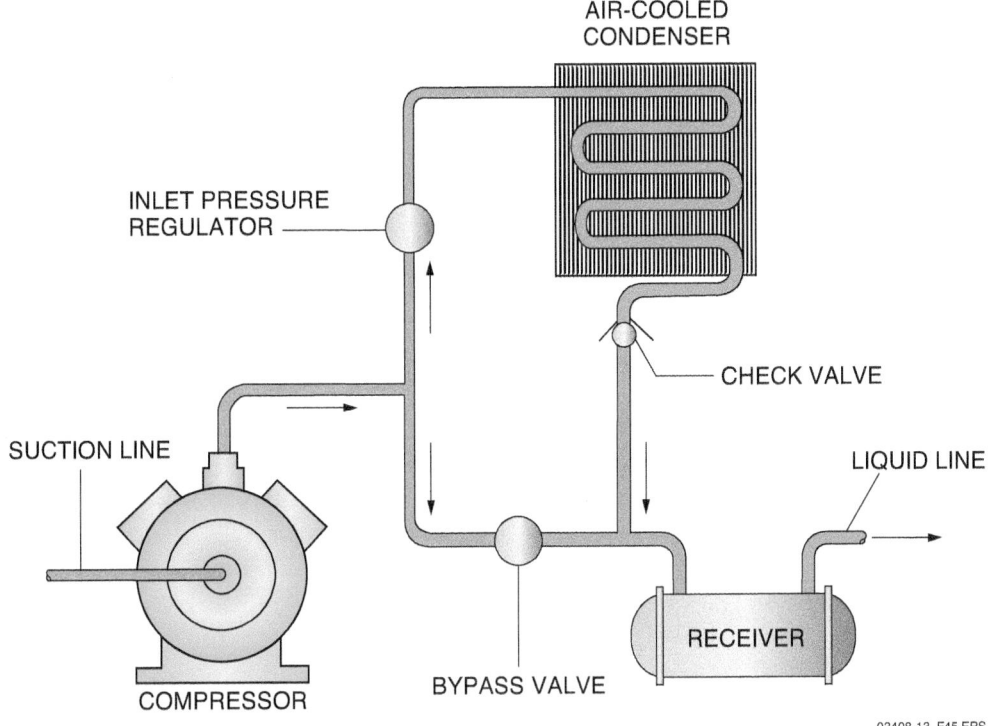

AIR-COOLED
CONDENSER

INLET PRESSURE
REGULATOR

CHECK VALVE

SUCTION LINE

LIQUID LINE

BYPASS VALVE

RECEIVER

COMPRESSOR

03408-13_F45.EPS

*Figure 45* Bypass control valve.

off within the evaporator, causing the compressor suction temperature and pressure to reduce. When a system operates at or near minimum load conditions for an extended interval, problems can occur that affect the operating life of the compressor. Low suction pressures and temperatures result in reduced refrigerant gas flow, which can cause the motor in hermetic compressors to overheat. It can also result in poor oil return, which can cause compressor lubrication problems.

Various forms of capacity control have been developed to regulate the capacity of the system compressor to match changing load conditions. Some very common methods of capacity control include compressor on-off, compressor cylinder unloading, and multi-speed compressors. As these methods have been discussed in previous modules and are directly related to the compressor itself, several other methods related primarily to the refrigerant side of the system will be examined in this module. These methods include hot gas bypass and suction modulation.

In cases where short cycling the compressor or using cylinder unloaders is not satisfactory, or where energy management is not a factor, compressor capacity can be controlled by using a hot gas bypass. The hot gas bypass system is simply a method of bypassing the condenser with the compressor discharge gas to prevent the compressor suction pressure from falling below a predetermined setpoint, thereby limiting the saturated

suction temperature of the evaporator. This prevents the system from reducing the temperature of the refrigerated space any further, since the evaporator coil temperature cannot be reduced further.

All hot gas bypass valves (*Figure 46*) open in response to a decrease in pressure on the low side of the system. They modulate as necessary to maintain the predetermined low-side pressure, admitting as much high-pressure vapor as needed. A solenoid valve is typically installed upstream of the regulator, or incorporated into the valve itself, to allow this part of the circuit to be disabled, such as when the load is satisfied and the compressor enters a pump-down mode. The solenoid also prevents any refrigerant migration during the Off cycle.

On close-connected systems using a single evaporator, the hot gas is bypassed into the evaporator immediately after the expansion valve. *Figure 47* provides one example of this application. Special tee fittings (*Figure 48*) are available from the manufacturers of hot gas bypass regulators and accessories to introduce the hot gas between the expansion valve and the refrigerant distributor assembly. Some distributors and tee fittings are also available as a single assembly. This special fitting allows the hot gas to enter the refrigerant stream without disturbing the operation of the expansion valve itself. Its design also ensures good mixing of the hot gas and the liquid/vapor

*Figure 46* Hot gas bypass valves.

03408-13_F46.EPS

refrigerant mixture for smooth flow into the evaporator coil. A common piping tee should not be used for this application. The distinct advantage of this method is that the expansion valve continues to do its job. As superheat begins to rise as a result of the introduction of hot gas, the expansion valve responds by opening further to maintain the proper superheat value. This prevents the

compressor from overheating due to high superheat. The suction pressure and refrigeration capacity remain at the desired level.

On systems where the condensing unit and evaporator are distant from each other, or when multiple evaporators are involved, the hot gas can be introduced into the suction line upstream of the compressor (*Figure 49*). Although functional,

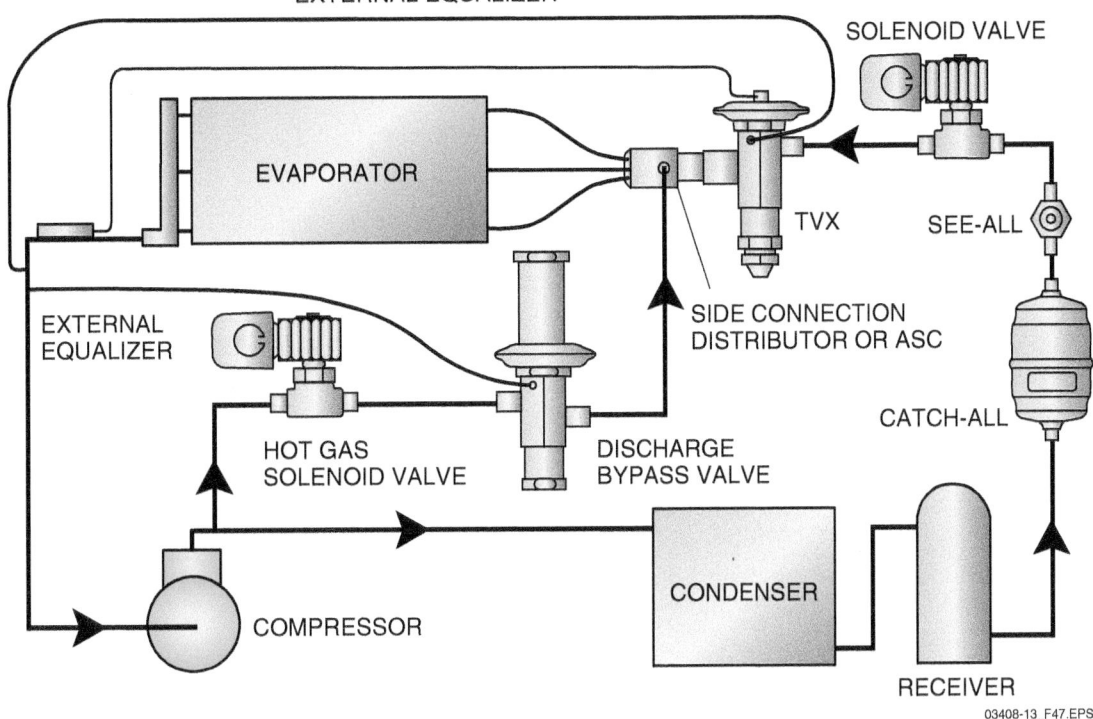

*Figure 47* Hot gas bypass, introduced at the expansion valve.

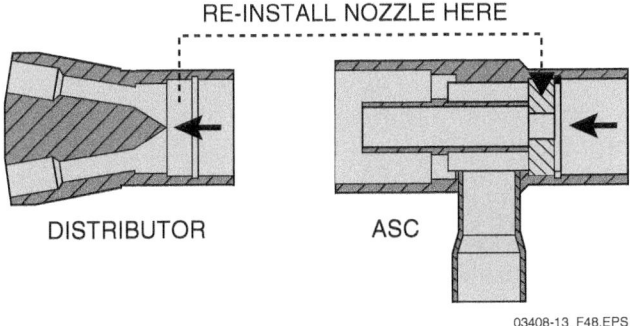

*Figure 48* Typical hot gas connector installation.

this is not the preferred method, as additional steps must be taken to ensure the compressor does not overheat. First, the elevated superheat from the hot gas must be reduced. This is done by using a separate expansion valve known as the desuperheating expansion valve. This valve must be carefully selected by the manufacturer for each application, rather than from a typical catalog. Mixing liquid refrigerant together with hot gas vapor in the suction line, especially near the compressor inlet, can lead to disastrous results. To ensure that the compressor remains protected from liquid refrigerant floodback, and to ensure the hot gas and liquid refrigerant are thoroughly mixed together, a suction accumulator is usually installed downstream of the connection from both devices. This design effectively provides the function of capacity control through hot gas by-

pass, and must be used for other practical reasons on some systems. However, it should be avoided when the first option is feasible.

Suction modulation capacity control is used on equipment with advanced microprocessor controls. It uses a series of sensors to input system pressure and temperature information to the microprocessor on a continuous basis. To control capacity, the microprocessor sends a signal to a drive module that engages a stepper motor. The stepper motor valve is located in the suction line just before the compressor and opens or closes to regulate the flow of refrigerant through the suction line. The suction line is large enough to allow the valve to handle significant demands for cooling such as in deep freeze applications or quick temperature pull-downs. *Figure 50* shows a typical suction modulation valve.

### 2.5.6 *Pump-Down Control*

Refrigeration systems usually equalize during the Off cycle, which means the high-side pressure and the low-side pressure reach equilibrium at some point between the normal operating pressures. This equalization results in a mixture of gas and liquid refrigerant throughout the system in most direct expansion systems. However, refrigerant gas occasionally migrates to the compressor and condenses into a liquid in the compressor crankcase. Another phenomenon somewhat unique to

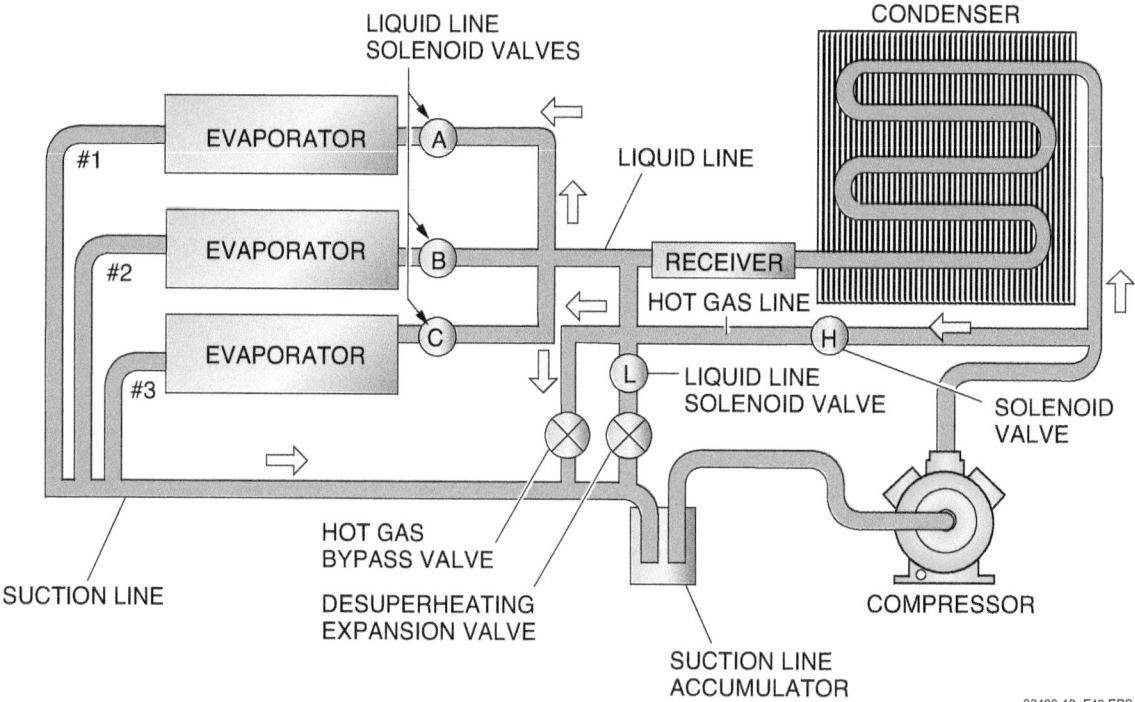

**LIQUID LINE SOLENOID VALVES**

CONDENSER

EVAPORATOR #1

EVAPORATOR #2

EVAPORATOR #3

A

B

C

LIQUID LINE

RECEIVER

HOT GAS LINE

H

L

LIQUID LINE SOLENOID VALVE

SOLENOID VALVE

SUCTION LINE

HOT GAS BYPASS VALVE

DESUPERHEATING EXPANSION VALVE

SUCTION LINE ACCUMULATOR

COMPRESSOR

03408-13_F49.EPS

*Figure 49* Hot gas bypass valve with multiple evaporators.

refrigeration, especially in low-temperature applications, is that a significant amount of refrigerant vapor left motionless in the evaporator during an off cycle quickly condenses to liquid. This will cause flooding back to the compressor at startup. Both conditions can result in lubricating oil loss, scored compressor bearings, broken piston rods, and valve damage. To alleviate this problem, a pump-down cycle can be used.

The advantages of using a pump-down cycle include the following:

- Reduced flooded starts
- Refrigerant pumped into the condenser/receiver during the Off cycle
- Quicker cooling during the On cycle

During the pump-down cycle, the compressor pumps the refrigerant charge into the condenser and liquid line or into a receiver at the end of each operating cycle. This keeps the liquid refrigerant out of the low side of the system, which includes the evaporator, suction line, and compressor crankcase.

## Pilot-Operated EPRs

Normally, EPRs are installed as close to the evaporator as possible. If they must be installed some distance from the evaporator or downstream from a riser, a pilot-operated EPR, like this one, is typically used, and the pilot line is connected to the evaporator coil.

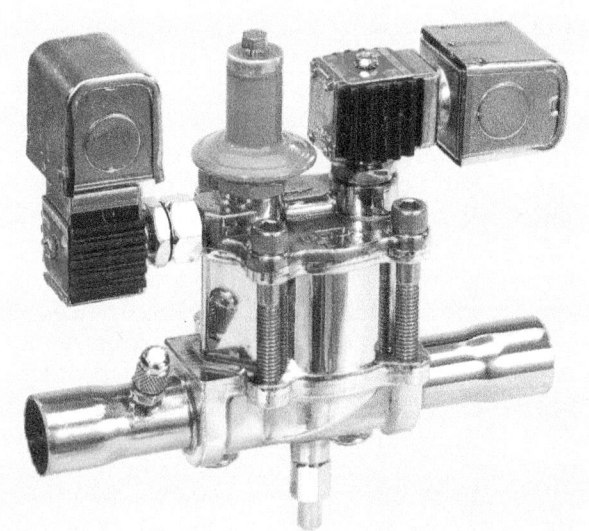

03408-13_SA06.EPS

Crankcase or sump heaters usually prevent refrigerant from condensing in the crankcase during the Off cycle. However, systems with larger-than-normal refrigerant operating charges can have liquid refrigerant in the compressor crankcase after a long Off cycle. A suction line accumulator (*Figure 51*) and a pump-down cycle will normally protect the compressor from liquid slugging during startup after a prolonged Off cycle. The suction line accumulator offers system protection during the running cycle, whereas the pump-down cycle offers protection during the Off cycle.

When a pump-down cycle is used, a thermostat controls a liquid line solenoid valve that opens or closes in response to the thermostat's setpoint. By closing the liquid line, the refrigerant is pumped into the condenser and receiver. The liquid line solenoid valve should be located just upstream from the expansion valve and as close to the evaporator coil inlet as possible. This allows for maximum storage volume of the liquid refrigerant during pump-down and prevents the liquid line from sweating. Compressor operation is controlled by a low-pressure cutout control that is set for a suction pressure below any value expected during normal operation.

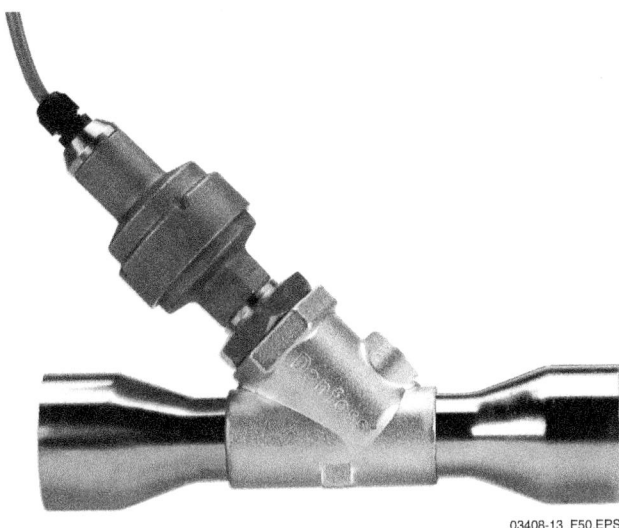

03408-13_F50.EPS

*Figure 50* Suction modulation valve.

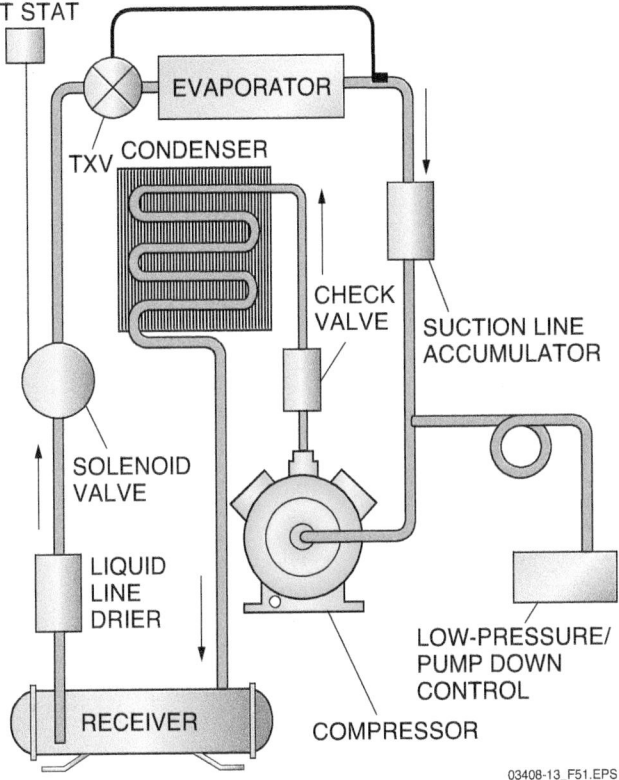

03408-13_F51.EPS

*Figure 51* Pump-down system.

# Head Pressure Control

When using a head pressure control, such as the one shown here, it is essential that enough refrigerant is charged into the system to completely fill the condenser for the lowest ambient temperature condition. Also, the receiver must be large enough to hold the excess refrigerant during all other operating conditions. If it is too small, the refrigerant can back up into the condenser and produce excessively high discharge pressures when the ambient outdoor air temperature is high.

## Additional Resources

*ASHRAE Handbook, Refrigeration Systems and Applications*, Latest Edition. Atlanta, GA: American Society of Heating, Refrigerating, and Air Conditioning Engineers.

*ASHRAE Handbook, HVAC Systems and Equipment*, Latest Edition. Atlanta, GA: American Society of Heating, Refrigerating, and Air Conditioning Engineers.

## 2.0.0 Section Review

1. What are the types of compressors used in refrigeration systems?

   a. Standalone, rotary, screw, and centrifugal
   b. Reciprocating, rotary, scroll, and centrifugal
   c. Reciprocating, rotary, scroll, and screw
   d. Reciprocating, centrifugal, two-stage, and rotary

2. The typical temperature difference (TD) range used for a low-temperature air-cooled refrigeration condenser is _____.

   a. 0°F to 15°F
   b. 5°F to 10°F
   c. 10°F to 15°F
   d. 15°F to 20°F

3. What is the defrost process called when the compressor shuts off and the coil naturally warms above 32°F?

   a. Forced defrosting
   b. Atmospheric defrosting
   c. Air defrosting
   d. Hot-gas defrosting

4. When viewing a sight glass, what color indicates the system has a dangerously high amount of moisture?

   a. Dark violet
   b. Blue
   c. Brown
   d. Red

5. Evaporator pressure regulating (EPR) valves are installed in the suction line between a(n) _____.

   a. compressor and receiver
   b. evaporator and compressor
   c. receiver and evaporator
   d. check valve and evaporator

### 3.0.0 DEFROST SYSTEMS

#### Objectives

Identify and describe various types of defrost systems.

a. Identify and describe off-cycle defrost systems.
b. Identify and describe electric defrost systems.
c. Identify and describe hot-gas defrost systems.

#### Performance Tasks

1. Install or make repairs to a packaged refrigeration condensing unit.
2. Install or make repairs to a packaged unit cooler in a refrigeration system.

#### Trade Term

Latent-heat defrost: A method used in hot-gas defrost applications that uses the heat added in one or more active evaporators as a source of heat to defrost another evaporator coil.

As long as there is a supply of moisture inside the refrigerated space, it will condense on the surface of the evaporator, just as is the case in comfort cooling applications. However, when the coil surface is below freezing, frost and ice will build, impeding and eventually blocking airflow through the coil. The coil must be defrosted for it to resume proper operation.

In commercial and industrial refrigeration applications, the primary defrost approaches are as follows:

- Off-cycle and/or timed defrost
- Electric defrost
- Hot-gas defrost

Regardless of the defrost approach used, there are several common factors regarding frost accumulation on the evaporator coils. Fixtures that were selected and designed to operate at high air-to-refrigerant temperature differences (such as those that are not required to maintain higher humidity levels) will generally build frost more rapidly as their coil temperature will likely be well below the dew point for the box. This encourages greater moisture condensation on the coil. Fixtures that have low air-to-refrigerant temperature differences will not accumulate condensate and frost quite as quickly.

Another factor in frost accumulation is related to the use of the fixture and the infiltration of outside air. Many refrigerated enclosures are expected to experience high consumer or operator traffic. In hot, humid weather, a fixture that is opened many times each day for product access will constantly be exposed to hot, moist air. This significantly increases the volume of airborne moisture in the enclosure. In addition, the infiltration of hot air replacing refrigerated air significantly increases the required refrigeration operating time as well. Conditions such as this can overwhelm and interfere with fixture performance in spite of defrost cycles. Consider the potential difference in performance of an identical refrigeration system installed in two different locations, one in a remotely located store with little consumer traffic in Arizona, and the other in a consistently busy, 24-hour refrigerated distribution center in coastal Florida.

It is important to note that all heat added to the refrigerated enclosure during the defrost cycle, regardless of the source, must again be removed once the process is complete. This fact has a significant impact on the refrigeration load, and the time a system is projected to spend in the defrost mode must be subtracted from the available time for refrigeration. For example, if the total refrigeration load for a freezer were to be calculated as 96,000 Btus over a 24-hour period (4,000 Btuh), but the unit is expected to operate in defrost for 3 hours per day, then the refrigeration equipment must be able generate this capacity in 21 hours in-

---

### GOING GREEN

### Infiltration

Technicians should note that poor, badly worn door gaskets and hinges can seriously impact the performance and energy efficiency of a refrigerated fixture. The constant infiltration of warm and moist outside air through poorly fitted doors and gaskets can prevent even the best fixture from reaching its setpoint, increase frost accumulation reducing coil heat transfer, and increase refrigeration equipment run-time and operating costs as well. The maintenance and replacement of door gaskets and hinges that experience high wear and abuse are as important to the function of a fixture as the proper refrigerant charge.

stead of 24 hours. This would require 4,571 Btuh of actual refrigeration capacity.

## 3.1.0 Off-Cycle Defrost

Off-cycle defrost is the simplest and most passive of the defrost approaches. In fixtures that maintain temperatures at or above 36°F, the coil is simply allowed to defrost during the normal Off cycle. Fixtures in this category will usually operate at a coil/saturated suction temperature of 16°F to 31°F, depending on the desired humidity levels for the stored product and the box design. At these temperatures, the coil will certainly build frost, but should defrost naturally once the refrigeration apparatus has cycled off because the fixture temperature is above freezing.

One assist to this natural defrost approach is to ensure that the fixture temperature is not set lower than needed. Maintaining the box even a few degrees colder than necessary can seriously impede or defeat the natural defrost process, resulting in frozen coils and rising box temperatures. Once this occurs, the fixture will require shutdown for an extended period to rid the coil of accumulated ice.

Some systems that operate at slightly lower temperatures than those using simple off-cycle defrost will require a longer period to clear accumulated ice than a normal Off cycle provides. Fixtures that operate at 32°F to 36°F often benefit from this approach. Typically, a 24-hour defrost timer (*Figure 52*) with normally closed contacts placed in series with the fixture's operating controls is set for a reasonable period of time, yet significantly longer than a normal Off cycle. The fixture temperature can rise a few degrees above normal during the extended period, but will quickly recover. As is true with all defrost scenarios, setting

the timer for a defrost period that coincides with the fixture's lightest period of use (late at night, for example) helps prevent the fixture temperature from rising beyond an acceptable level. Allowing a defrost cycle to occur during heavy use can result in warmer products and possible product loss. Several defrost cycles per day are often required, depending on the operating conditions.

## 3.2.0 Electric Defrost

Electric defrost systems are frequently used in high- and medium-temperature systems that operate at temperatures too cold for off-cycle or timed defrost to be effective, as well as in commercial freezers. Electric defrost components are relatively inexpensive and reliable, and are simple to troubleshoot and repair.

*Figure 53* shows a typical arrangement for resistive defrost heaters, in this case installed on both the face and back of an evaporator coil. For proper operation, the heater must be in good thermal contact with the coil fins to effectively distribute heat and minimize the required defrost period. During the defrost period, a time clock stops the refrigeration equipment and energizes the electric heater. Depending on the unit design, the evaporator fans can continue to operate through the defrost cycle. In many cases, the size and shape of the defrost heater will be unique to a given evaporator coil or manufacturer. This requires access to original parts when replacement becomes necessary. Some coil designs require that the heater be placed deep inside the coil assembly, making repair and/or replacement more challenging, but resulting in a highly effective and rapid defrost cycle.

A few units are equipped with radiant electric heaters (*Figure 54*). The resistance heater wire is spirally wound and enclosed inside a quartz tube, with the wires passed through the ends of the sealed tube cap. Using infrared radiant heat, defrost can be accomplished without the heater being in contact with the coil itself. They are capable of reaching a high temperature very quickly and are corrosion-resistant. The quartz enclosure can be a bit fragile, however.

When electric defrost is required, the condensate removed from the coil must also be maintained above freezing until it has completely drained away from the chilled space. Most condensate collection pans, either as part of a unitary evaporator coil assembly or as a separate collection point inside the fixture, must be equipped with electric heaters such as the one shown in *Figure 55*. Some installations where the drain line is routed through a freezing area can require that

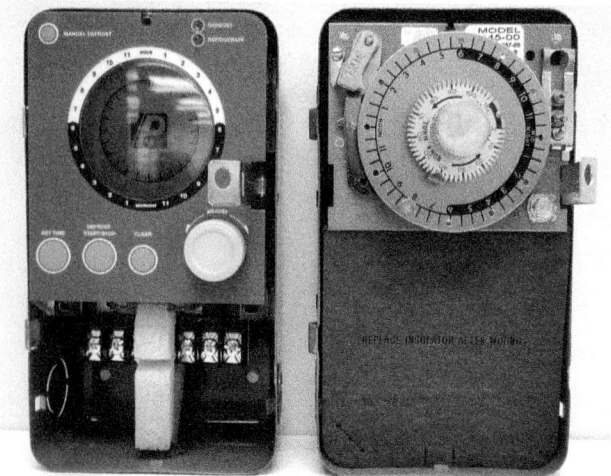

03408-13_F52.EPS

*Figure 52* Typical defrost timers.

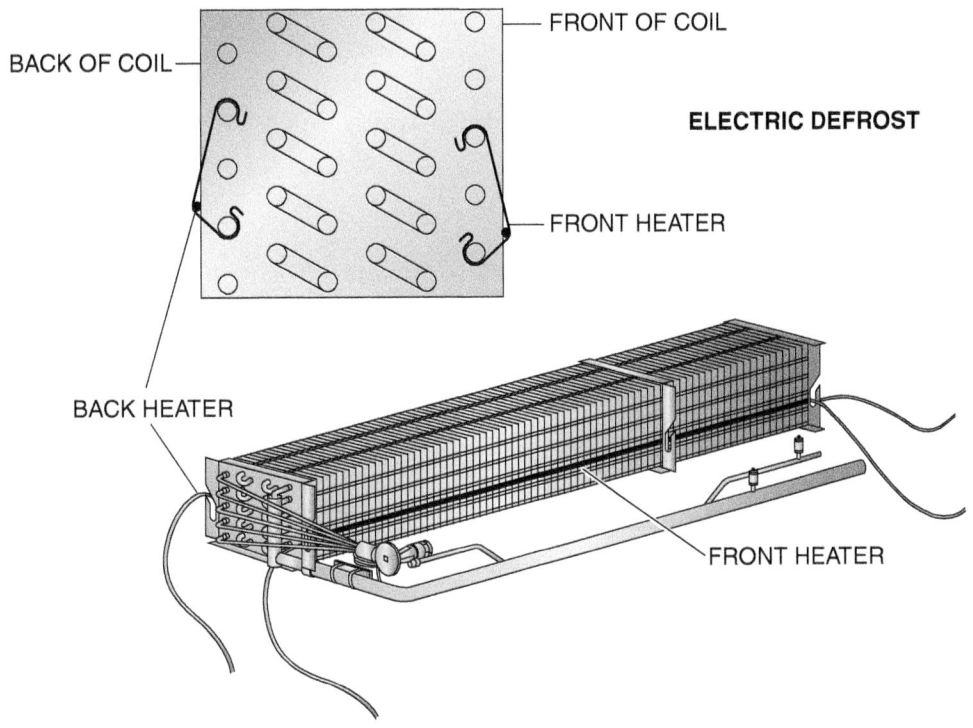

ELECTRIC DEFROST

BACK OF COIL

FRONT OF COIL

FRONT HEATER

BACK HEATER

FRONT HEATER

03408-13_F53.EPS

*Figure 53*  Evaporator electric heat arrangement.

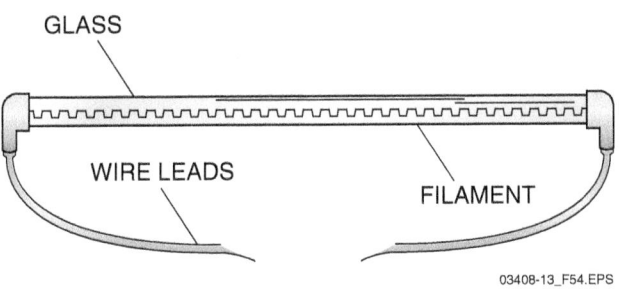

GLASS

WIRE LEADS

FILAMENT

03408-13_F54.EPS

*Figure 54*  Infrared quartz heater assembly.

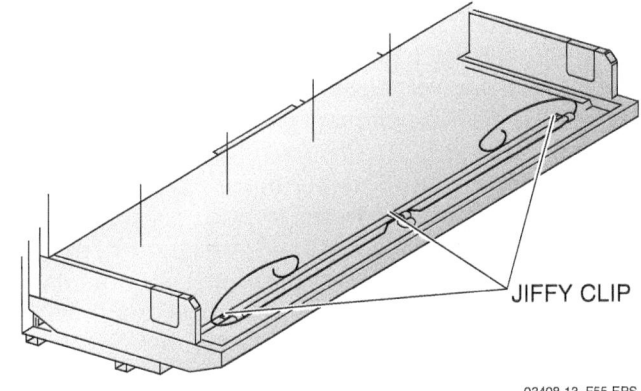

JIFFY CLIP

03408-13_F55.EPS

*Figure 55*  Condensate pan heater.

the condensate drain line be kept above freezing as well. This is usually done by attaching electric pipe heating cable to the outside of the line, or by using an appropriate waterproof heat cable routed on the inside of the condensate drain line.

Electric defrost systems are generally controlled by a timer like those seen in *Figure 52*. Electromechanical clocks have a dial assembly (*Figure 56*), allowing the timer to be configured for the time, duration, and number of defrost periods during a 24-hour period. The inner defrost dial sets the duration of the defrost period. Defrost is terminated after the period has elapsed. The timer contacts then return to their normal position and the refrigeration cycle resumes.

Electric defrost can also be terminated by a temperature sensor or line-mounted thermostat. By monitoring the temperature of the coil, or the refrigerant suction line adjacent to the coil, the thermostat provides an indication that the coil has reached the appropriate temperature. Defrost can then be terminated before the time clock has reached the allowed maximum time. The proper setpoint for temperature termination can differ among products, as it is based on the manufacturer's chosen location and laboratory testing. Line-mounted thermostats are generally nonadjustable, so it is important that the proper device be chosen for replacement of the sensor.

The final choice of defrost control and termination is often an option provided to the end user, either as a standard feature or at an added cost. Defrost termination can be time-only, temperature-only, or a combination of the two.

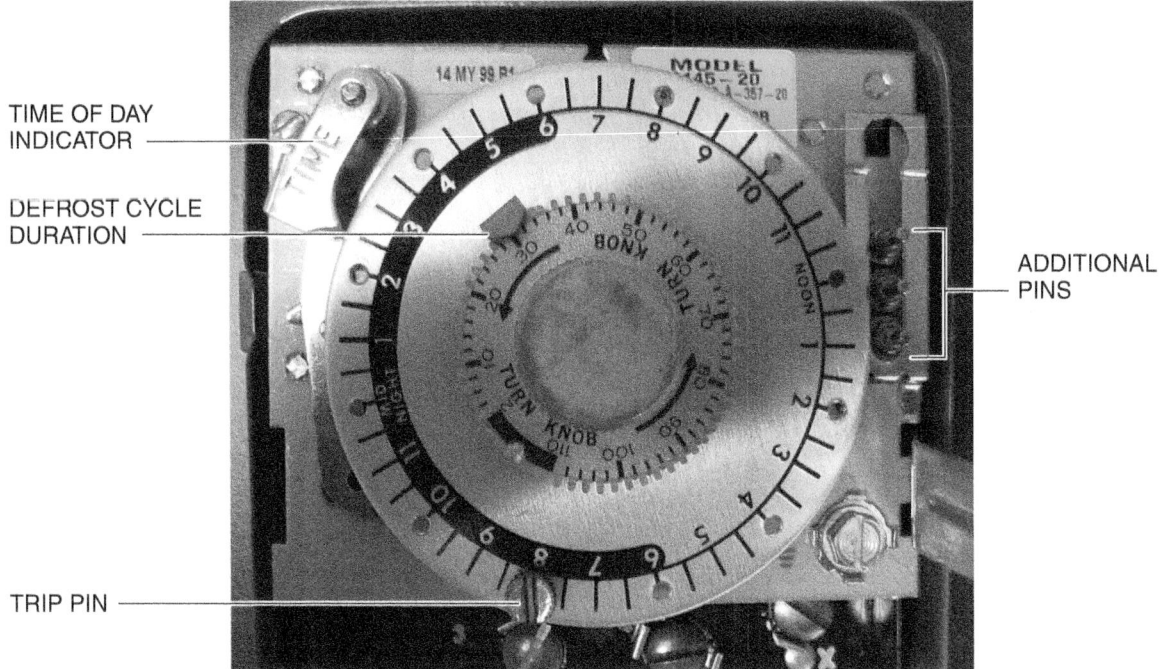

TIME OF DAY INDICATOR

DEFROST CYCLE DURATION

ADDITIONAL PINS

TRIP PIN

03408-13 F56.EPS

*Figure 56* Typical defrost timer dial.

### 3.3.0 Hot-Gas Defrost

Hot-gas defrost is widely considered to be the fastest and most efficient approach. Instead of having only the surface area contact on the coil provided by electric heating elements, the entire surface of the coil tubing is incorporated into the defrost process. However, there are more complex issues in the refrigeration circuit to address with this method. Due to the additional refrigerant circuit components required, and the increased potential for compressor damage when defrost components do not operate properly, great care must be taken in its application.

Hot-gas defrost, in its simplest form, sends discharge gas from the compressor directly to the evaporator, bypassing the metering device through the use of several solenoid valves. The coil temperature is allowed to rise to a predetermined pressure and the refrigerant vapor rapidly condenses in the cold evaporator as the process begins. The latent heat of condensation, not the temperature of the discharge gas, is the primary factor in removing enough heat to bring the coil surface above freezing.

One problem with this approach is that little or no heat is being added to the refrigeration circuit beyond compressor motor heat and the heat of compression. This occurs because the single evaporator is now condensing refrigerant; as a result, the system may not able to develop sufficient heat to complete a defrost cycle. This can be avoided to some degree by more frequent, but short, defrost periods. Another more serious problem is the potential for condensed liquid refrigerant to return to the compressor in significant volume. For hot-gas defrost to work well, any refrigerant leaving the evaporator as a liquid must be re-vaporized fully and a source of heat must be incorporated. This can be done in several ways.

One approach works well with multiple (two or more) evaporators. This method is sometimes referred to as latent-heat defrost. One or more evaporators remain in service, absorbing heat and transferring it to the refrigerant, while one coil is being defrosted. Hot gas is ported to the defrosting coil through a series of solenoid valves.

As shown in *Figure 57*, Evaporator Coil 1 is in the defrosting mode and is serving as the condenser to Evaporator Coil 2. The compressed vapor from the compressor uses a direct flow to the evaporator by opening and closing the applicable valves. At this point, the refrigerant condenses in Evaporator Coil 1 and melts the ice. The refrigerant then flows to, and collects in, the liquid receiver. The liquid refrigerant then flows to the metering device of Evaporator Coil 2. The vapor refrigerant leaves Evaporator Coil 2 and takes its normal path back to the compressor. This approach works especially well on large supermarket systems where many evaporators are connected to a compressor rack system.

Another effective approach, known as a thermal bank, uses a tank of heat exchanging liquid that is warmed by the hot discharge gas during the refrigeration cycle. This method absorbs and stores heat during the normal operating cycle. It also makes use of heat energy that is otherwise transferred to the outside air and wasted. When the defrost cycle is initiated, this stored heat is used to re-evaporate the refrigerant. This results in a very efficient and rapid defrost with limited potential for liquid floodback to occur.

During the normal operating cycle, hot gas refrigerant is sent through a tubing coil submerged inside the thermal bank. The water is heated and the refrigerant is pre-cooled before moving on to the condenser. Once the defrost cycle begins, hot gas is first ported to the evaporator drain pan and on to the evaporator coil, where the ice melts and the refrigerant gas is condensed to a liquid. As this liquid or liquid/vapor refrigerant mixture exits the evaporator, it passes through the thermal bank, absorbing heat from the water and re-evaporating the refrigerant. The water in the bank may even be frozen during the process.

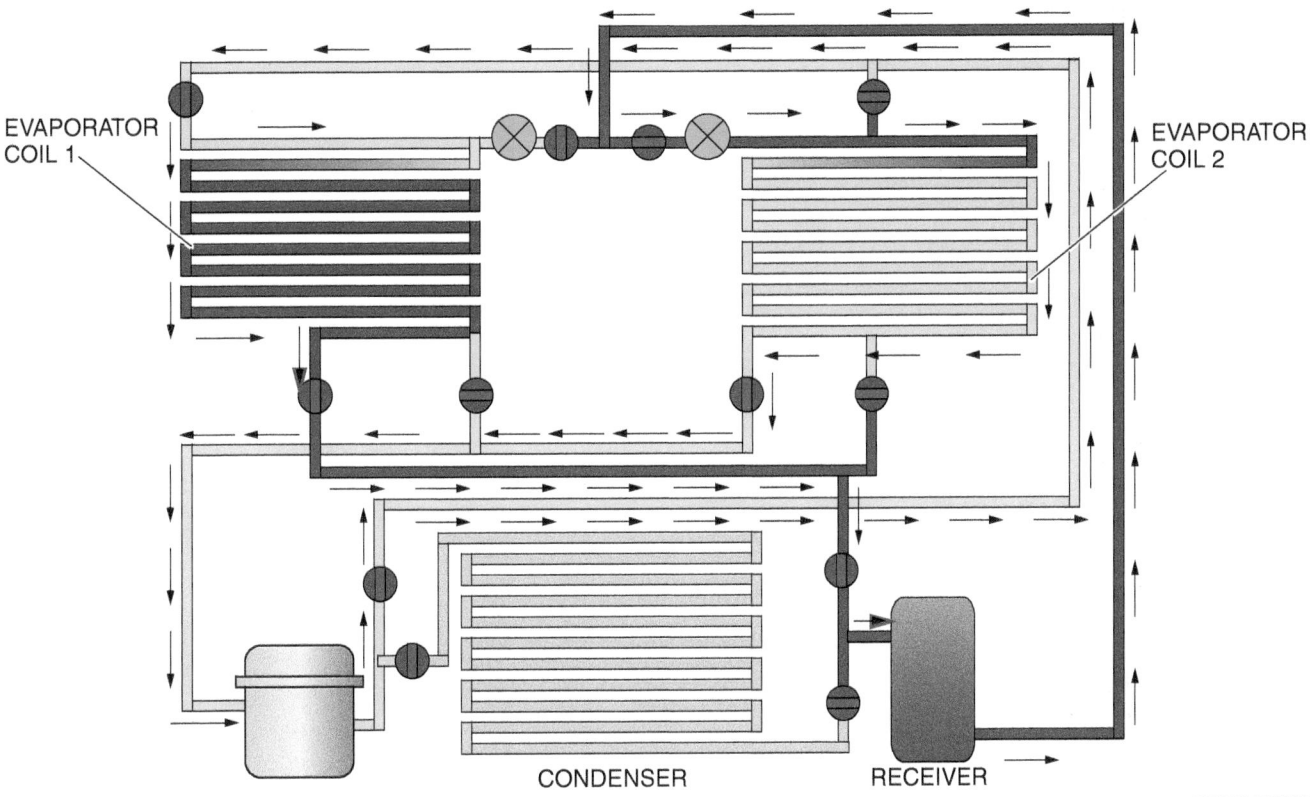

03408-13_F57.EPS

*Figure 57* Dual evaporator system with hot-gas defrost.

1. Systems that make use of off-cycle defrost generally have evaporator coils operating in a temperature range of _____.

   a. 15°F to 30°F
   b. 16°F to 31°F
   c. 16°F to 32°F
   d. 17°F to 32°F

2. What device is used to stop the refrigeration equipment and energize the electric heater to start an electric defrost period?

   a. Time clock
   b. Internal alarm
   c. Frost indicator
   d. Timed alarm

3. One way hot-gas defrost can be accomplished is to use a tank of heat exchanging liquid to store heat, also known as a(n) _____.

   a. thermal bank
   b. thermal blanket
   c. flow bank
   d. air blanket

### 4.0.0 AMMONIA REFRIGERATION

#### Objectives

Describe the main characteristics of ammonia-based refrigeration systems.

a. Describe the properties and safety considerations of ammonia as a refrigerant.
b. Describe ammonia systems and the basic components.

#### Trade Terms

**Anhydrous:** Containing no water.

**Immiscible:** A condition in which a refrigerant does not dissolve in oil and vice versa.

The choice of refrigerant used in low-temperature industrial systems is mainly between HFC-404A and ammonia. Obviously, there is a choice only when the federal, state, and local codes permit the use of ammonia. Codes in many locations restrict the use of ammonia systems. In general, it is only used in applications that are physically separated from public access. Ammonia systems are typically custom designed for a particular application. These systems can span a wide range of evaporating and condensing conditions. They can be used to cool a food freezing plant operating at temperatures from +50°F to –50°F, or to cool a distribution warehouse requiring multiple temperatures for varying commodities. Generally, the larger the application, the more likely it would be that ammonia will be the refrigerant of choice. Ammonia system construction and installation becomes more cost-effective as size increases.

#### 4.1.0 Ammonia (R-717)

Anhydrous ammonia, called refrigerant R-717, has excellent heat transfer qualities and is commonly used in industrial refrigeration systems. Ammonia is a chemical compound of nitrogen and hydrogen ($NH_3$). Refrigerant grade anhydrous ammonia is a clear, colorless liquid or gas, free from visible impurities. It is a minimum of 99.95 percent pure ammonia. Its boiling point at atmospheric pressure is –28°F and its melting point from a solid is –108°F. Because of its low boiling point, refrigeration can occur at temperatures considerably below 0°F without the need for

below atmospheric pressures in the evaporator. It has a latent heat value of 565 Btu/lb at 5°F. This enables large refrigeration results while using relatively small equipment. In the presence of moisture, ammonia chemically reacts with and attacks copper and bronze; therefore, steel piping and fittings are used in ammonia systems.

> **WARNING!**
>
> The term *anhydrous* refers to the lack of water in ammonia for commercial refrigeration use. Ammonia is highly hydroscopic, seeking water from any available source. A volume equal to 1,300 gallons of ammonia vapor can be absorbed into a single gallon of water. Its attraction to water places moist surfaces of the human body, such as eyes, nasal passages, mouth and lungs, at risk when exposed to excessive amounts of ammonia vapor.

Ammonia is an immiscible refrigerant. This means that it does not dissolve in oil. In mechanical (compressor-driven) ammonia refrigeration systems, this causes the system mineral-based oil to separate from the ammonia and collect in the evaporator or low side of the system. This oil must be removed, or it will coat the heat transfer surfaces of the evaporator, reducing system performance. In some systems draining occurs automatically; those that do not must be drained manually.

One straightforward advantage to using ammonia as a refrigerant is that leaks are self-alarming; the odor alone gets immediate attention and alerts system operators to the presence of a leak. Concentrations in air as small as 3 to 5 parts per million (ppm) can be detected by humans. Small leaks can be pinpointed by using small handheld wicks coated with sulfur. The wick is lit to produce a relatively colorless smoke. In the presence of ammonia vapors, the smoke then turns white and very dense. Since ammonia is one of the oldest refrigerants in use today, the simple sulfur stick has been used to locate ammonia leaks for over 100 years.

#### 4.1.1 Ammonia Safety Considerations

Ammonia has a harsh effect on the respiratory system. Only very small quantities can be breathed safely. Five minutes at 50 ppm is the maximum exposure allowed by OSHA. Ammonia is hazardous to life at 5,000 ppm and is flammable at 150,000 to 270,000 ppm. Ammonia has an odor that can be smelled at 3 ppm to 5 ppm. This odor gets very irritating at 15 ppm. While it is dangerous to human health, ammonia is considered environmentally safe.

Rusting pipes and vessels in older systems containing ammonia can create a safety hazard. Also, cold liquid refrigerant should not be confined between closed valves in a pipe where the liquid can warm. The liquid expands as it warms, increasing the pressure and possibly causing the pipe to burst. Slight smells of ammonia sometimes can occur because of leakage at valve packings or pump seals, or venting from recently drained oil or from an air-purging device. The scent of ammonia in the air is quite common near the equipment of a large system.

Any significant escape of ammonia must be prevented from causing personal injury. When an ammonia leak occurs where people are present, the odor provides a warning. In unoccupied areas, such as refrigerated storage rooms or unmanned machine rooms, automatic sensors are used to provide a warning of a potential problem (*Figure 58*). These sensors include detector tubes and solid-state electronic, electrochemical, or infrared devices. The sensitivity of each type of sensor varies with its design and intended use. The solid-state sensor works well for initiating alarms in the range of ammonia concentration of several hundred ppm, but is not very sensitive in the range of 20 ppm to 50 ppm. Electrochemical units provide better protection for occupied areas because many are sensitive to ammonia concentrations in the range of 0.7 ppm to 1,000 ppm.

OSHA requires that maintenance and service personnel who are expected to respond in any way to an ammonia release or other emergency regarding ammonia receive appropriate training in accordance with its regulations. Having experience working with ammonia is not sufficient to satisfy the OSHA requirement, and the training must be appropriate for the duties the individual is expected to perform in an emergency situation.

03408-13_F58.EPS

*Figure 58* Ammonia detection equipment.

End users of ammonia, as well as contractors, have a responsibility to ensure that the proper training has been received and documented. Contractors should take responsibility for general safety and emergency response training. The end user should ensure that all contractor personnel who work at the site, as well as their own maintenance employees, receive site-specific training.

Before working with ammonia, become familiar with the information contained on its MSDS. You should also know the first-aid procedures used for exposure to ammonia. In addition, the safety guidelines listed below should also be observed:

- Wear the proper protective equipment, such as a respirator mask, face shield, and gloves when working with ammonia to avoid inhalation and contact with your skin and eyes.

### Case History

## Ammonia Release

In March of 1997, a cloud containing 2,200 pounds of ammonia was released from the relief valves of a refrigeration system used to produce ice cream. The release occurred after runoff from melting snow leaked from an indoor drainage pipe onto the electrical control panel for the refrigeration system's ammonia compressors. The panel shorted out, and the system shut down. Employees were evacuated from the building with no injuries being reported.

The operators of the facility neglected to inform the Environmental Protection Agency (EPA) until five days after the accident. The company was fined $41,252 for failing to notify the EPA of a serious ammonia leak in a timely manner. The EPA requires immediate notification whenever there is a release of a reportable quantity of a hazardous substance. The reportable quantity for ammonia is 100 pounds.

- If bodily contact with ammonia is made, flush any affected areas with copious amounts of cool water.
- Remove clothing unless it is so saturated with liquid ammonia that it is frozen to the skin.
- Do not wear contact lenses while working with ammonia.
- If a spill has occurred and no one is in the room or in danger, start ventilation equipment in the room before entering it, even if you are wearing a respirator mask.

### 4.2.0 Ammonia Refrigeration Systems and Components

Ammonia systems fall into two classes: mechanical and non-mechanical. In the mechanical system, a compressor functions to lower the pressure and temperature of the ammonia (refrigerant) in the evaporator, allowing it to boil and absorb heat from the item being cooled. The pressure and temperature of the ammonia in the condenser are raised, allowing it to condense and give up its heat to either air or water. The compressor also maintains a pressure difference between the high-pressure and low-pressure sides of the system. This pressure difference causes the ammonia to flow throughout the system.

Non-mechanical ammonia systems are called absorption systems. They do not have a compressor. Instead, they use a chemical cycle to provide cooling. Absorption systems have many applications and are used widely in recreational vehicle refrigeration systems. They are also used in residential refrigeration applications where the use of electrical devices is limited. However, mechanical systems remain the focus of this module.

### 4.2.1 Single-Stage and Compound Ammonia Refrigeration Systems

Mechanical ammonia refrigeration systems can be single-stage or multi-stage (compound) systems. Compound systems compress the gas from the evaporator to the condenser in several stages. Typically, compound systems are used when producing temperatures of –15°F and below. Single-stage ammonia systems consist of the same basic components as used in a single-stage refrigeration system that operates with a halocarbon-based refrigerant.

*Figure 59* shows a basic compound (two-stage) ammonia system. As shown, it has two compressors: a low-stage compressor and a high-stage compressor. It also has two cooling loads: a low-temperature evaporator and a medium-temperature evaporator.

During operation, low-pressure vapor returned from the primary evaporator(s) is applied to the suction input of the low-stage compressor. There, the vapor is compressed, then discharged at an intermediate pressure into an intercooler unit. This unit is located between the two compressors. The intercooler acts to desuperheat the discharge gas from the low-stage compressor to prevent

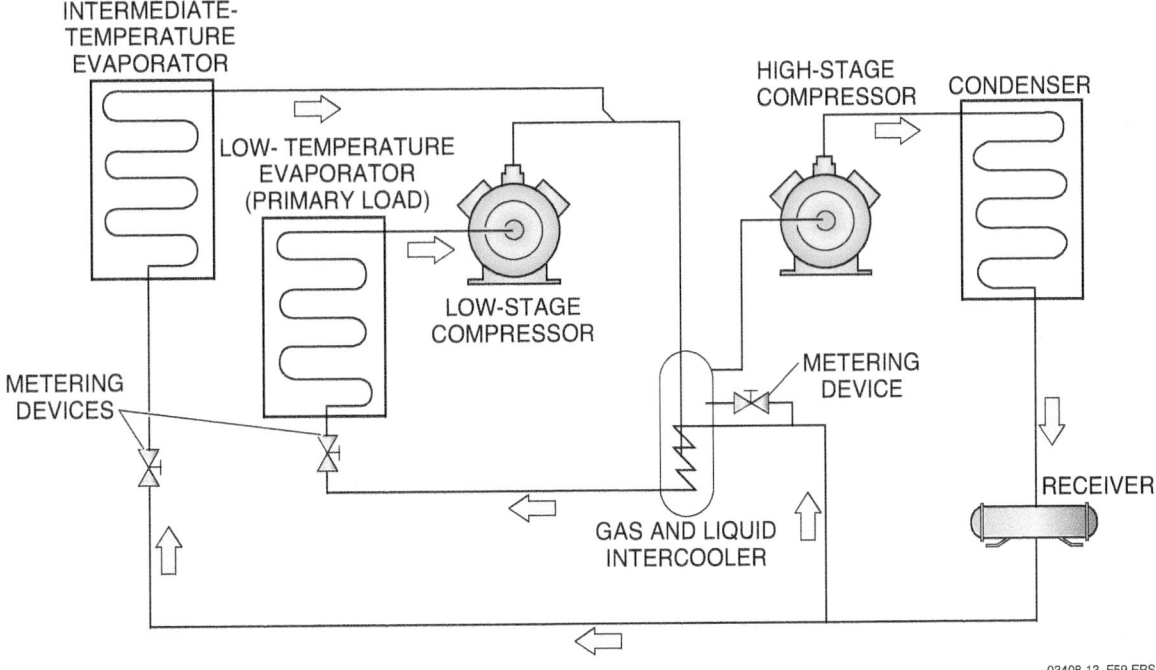

03408-13_F59.EPS

*Figure 59* Basic two-stage mechanical ammonia refrigeration system.

the overheating of the high-stage compressor. In *Figure 59*, note the metering device that feeds refrigerant to the intercooler to desuperheat the discharge gas passing through it. Desuperheating is done in the intercooler by routing the discharged gas from the low-stage compressor through liquid refrigerant contained in the unit. The superheat is removed as the vapor bubbles rise through the liquid.

Another method mixes the liquid refrigerant normally entering the intercooler with the discharge gas. Desuperheated intermediate-pressure vapor flows from the intercooler into the suction input of the high-stage compressor. There, the vapor is compressed and discharged to the system condenser. The hot high-pressure gas cools in the condenser and the resulting liquid refrigerant flows from the condenser into the receiver. From the receiver, the liquid refrigerant flows to and through the evaporator metering devices for use in the evaporators. Liquid refrigerant also flows to the intercooler metering device.

### 4.2.2 Ammonia Liquid Recirculation Systems

Ammonia liquid recirculation systems (*Figure 60*) are also called liquid overfeed systems. In this type of system, ammonia is supplied from a low-pressure receiver to the system evaporators. The low-pressure receiver is a vessel that stores liquid ammonia at low pressure and is used to supply the evaporators with liquid ammonia, circulated by a pump. The compressor suction line is connected to the low-pressure receiver, maintaining the pressure and, in turn, the temperature of the liquid refrigerant being recirculated.

Large suction line accumulators are often incorporated to prevent compressor damage in the event the low-pressure receiver overfills with liquid. At startup, there is also the potential for liquid carryover into the suction line if the pressure is reduced too quickly and the ammonia liquid in the low-pressure receiver boils violently. Therefore, the pressure must be reduced slowly in the low-pressure receiver when starting a warm system. The compressor should not be allowed to immediately reach its full capacity.

The low-pressure receiver also takes the suction gas from the evaporators and separates the gas from the liquid. Because the amount of ammonia fed into the evaporators is usually several times the amount that is actually evaporated there, liquid ammonia is always present in the suction return to the low-pressure receiver. Makeup ammonia is supplied to the low-pressure receiver from the high-pressure receiver through the metering device, which is usually a float valve.

### 4.2.3 Compressors

Reciprocating, rotary vane, and screw compressors are commonly used in ammonia systems. Reciprocating compressors are most often used in smaller single-stage or multiple-stage systems. Screw compressors are typically used in systems over 100hp. Various compressor combinations can

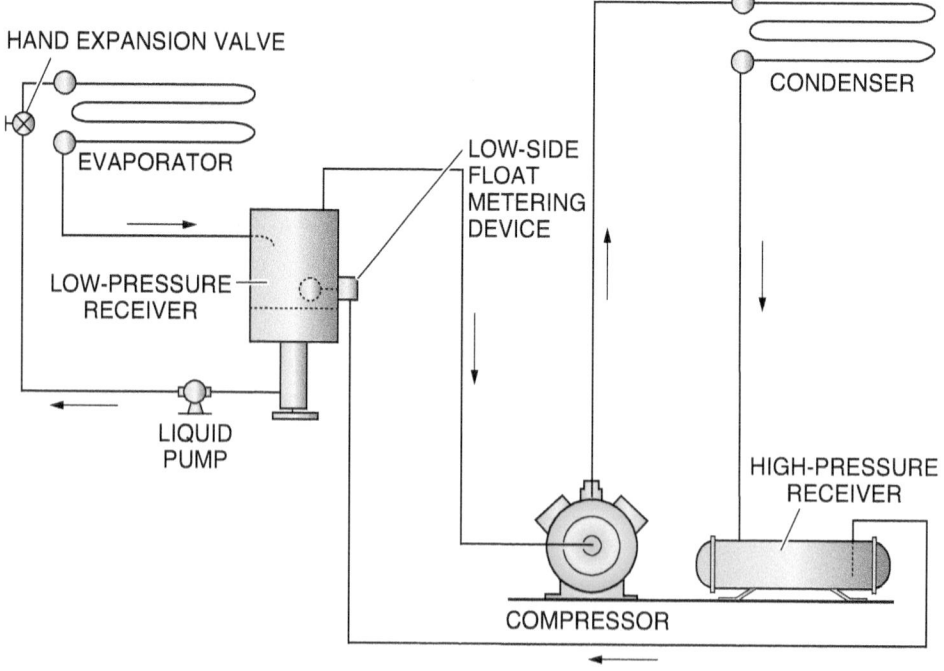

03408-13_F60.EPS

*Figure 60* Simplified ammonia liquid recirculation system.

be used in multi-stage systems. The low-pressure stage can be a rotary vane or screw compressor where large volumes of gas are being moved. The high-pressure stage can be a reciprocating or screw compressor.

When reciprocating compressors that require oil cooling are used, external heat exchangers using a refrigerant or secondary cooling are frequently used to cool the oil.

Oil for screw compressors is usually cooled by either the direct injection of liquid refrigerant into the compressor to cool the discharge gas external cooling by a heat exchanger or external cooling using refrigerant from the condenser in a shell-and-tube heat exchanger.

### 4.2.4 Condensers and Receivers

Water-cooled shell-and-tube and evaporative condensers are commonly used with ammonia systems whereas air-cooled condensers are seldom used. The condenser can be a single unit, or it can be multiple units connected in parallel. When using shell-and-tube condensers, the piping should be arranged so that the condenser tubes are always filled with water. The condenser should also have air vents on its head with hand valves for manual purging. Receivers fed by the condenser should always be mounted below the condenser

so that the condensing surface is not flooded with ammonia. Evaporative condensers must be located where there is sufficient airflow and no obstructions that can cause recirculation of the discharged air. When using a single evaporative condenser, the receiver must always be operated at a lower pressure than the condensing pressure. It also operates at a cooler temperature than the condensing temperature.

### 4.2.5 Evaporators and Evaporator Control

Several types of evaporators can be used in ammonia systems. Fan coil, direct-expansion evaporators can be used, but their use is limited to systems with suction temperatures of 0°F or higher. Flooded shell-and-tube evaporators are often used in ammonia systems where indirect or secondary cooling fluids such as water, brine, or glycol must be cooled. The various metering devices used to control ammonia system evaporators are basically the same kind and operate the same way as the metering devices used in halocarbon-based refrigerant systems. The same thing is true for the various methods used to defrost the evaporators in ammonia systems that operate below 35°F. Ammonia evaporator coils are made of either aluminum or steel tubing.

### 4.0.0 Section Review

1. What is the first warning of an ammonia leak?

   a. Watery eyes
   b. Difficulty breathing
   c. Noticeable odor
   d. Skin feels cold

2. Non-mechanical ammonia systems are called _____.

   a. manual systems
   b. single-stage systems
   c. compound systems
   d. absorption systems

# SUMMARY

While commercial and industrial refrigeration systems are used to cool substances for the food, chemical, and manufacturing industries, the principle use of commercial refrigeration systems involves the refrigeration and freezing of food.

Excluding their applications and the obvious differences in the physical layout and packaging, refrigeration systems operate in the same way as comfort air-conditioning systems. Both types of systems have the same basic hardware; compressors, heat exchangers (evaporators and condensers), fans, pumps, pipe, duct, and controls. Their main working fluids are air, water, and refrigerants. The major difference between commercial/industrial refrigeration systems and comfort air-conditioning systems is the operating temperature range that the refrigeration systems serve. Depending on the specific application, commercial/industrial refrigeration systems can have operating temperatures that range between 60°F and –250°F.

Some other differences include the following:

- The number of evaporators connected to a single condensing unit
- The number of compressors used
- The number of condensers used
- The variety of refrigerants used
- The variety of defrosting methods used
- More than one temperature often provided by a single system

1. The specific temperature used for product preservation is determined by the _____.

   a. type of preparation
   b. type of commodity
   c. type of packaging
   d. type of pre-storage

2. The commercial freezing method by which the product to be frozen is pressed between cold metal plates is called _____.

   a. freeze drying
   b. contact freezing
   c. air blast freezing
   d. cryogenic freezing

3. Lyophilization produces food products that _____.

   a. contain all the moisture originally found in the product
   b. require freezing until the point of consumption
   c. are lightweight and require no refrigeration
   d. must always be cooked before consumption

4. The temperature and humidity maintained in cold storage are determined by the type of food being stored and the _____.

   a. local climate
   b. size of the storage room
   c. number of pallets
   d. length of time it is to be held in storage

5. To maintain top quality, fruits and vegetables stored in chill rooms are normally stored _____.

   a. at temperatures just below their ripening temperature
   b. in airtight bins
   c. at a temperature just above their freezing point
   d. between vertical plate evaporators

6. Freezers used to store ice cream often operate at _____.

   a. −20°F
   b. 0°F
   c. 20°F
   d. 32°F

7. Which of the following are typically used for local delivery trucks?

   a. Direct-drive refrigeration units
   b. Trailer refrigeration units
   c. Cryogenic freezers
   d. Large insulated ice blocks

8. The energy efficiency ratio (EER) of a compressor is the lowest (least efficient) when used to cool _____.

   a. food preparation rooms
   b. meat cutting rooms
   c. meat refrigerators
   d. ice cream freezers

9. Parallel-connected multiple compressors are used mainly to _____.

   a. cool multiple evaporators in a line of display cases
   b. cool multi-temperature mobile refrigeration units
   c. increase system capacity and meet varying load conditions
   d. operate a system at extremely low temperatures

10. A refrigerant-side method to control air-cooled condensers in response to varying loads is _____.

    a. cycling the fans on and off
    b. modulating the unit dampers
    c. changing the fan speed
    d. adjusting the amount of active condensing surface in the condenser coil

11. The refrigeration load for food store display cases is normally rated by the manufacturer based on the _____.

    a. summer design conditions
    b. number of customers per day
    c. thermal breaks between zones
    d. thickness of the cold air blanket

12. A device used to minimize the amount of oil circulating through a refrigeration system is a(n) _____.

    a. oil separator
    b. oil reservoir
    c. oil differential check valve
    d. crankcase pressure regulator

13. Relief valves are usually installed in the _____.

    a. condenser or liquid receiver
    b. evaporator or condenser
    c. condenser or compressor
    d. compressor or liquid receiver

14. One method of preventing compressor overload from excessive refrigeration load is accomplished by using a _____.

    a. crankcase pressure regulator
    b. hot gas bypass regulator
    c. compressor discharge modulator
    d. condenser fan cycling

15. What condition can interfere with the refrigeration system despite the defrost cycles?

    a. Poor air circulation
    b. Dry environments
    c. Humid environments
    d. Cold environments

16. Off-cycle defrost is best used in fixtures that maintain temperatures at or above _____.

    a. 38°F
    b. 34°F
    c. 36°F
    d. 32°F

17. During an off-cycle defrost, what can impede or defeat the natural defrost process?

    a. Colder box temperatures
    b. Warmer fixture temperatures
    c. Cold coil temperatures
    d. Warmer box temperatures

18. A defrost method used with multiple evaporators where one evaporator is being defrosted while the others continue refrigerating is called _____.

    a. latent heat defrost
    b. time defrost
    c. hot gas bypass
    d. off-cycle defrost

19. Ammonia is _____.

    a. a chemical compound of nitrogen and hydrogen
    b. harmful to the environment
    c. a miscible refrigerant
    d. safe for humans

20. Because ammonia is an immiscible refrigerant, the mineral-based oil separates from the ammonia and collects in the _____.

    a. condenser
    b. sight glass
    c. evaporator
    d. compressor

# Trade Terms Introduced in This Module

**Anhydrous:** Containing no water.

**Commodities:** Commercial items such as merchandise, wares, goods, and produce.

**Cooler:** A refrigerated storage device that protects commodities at temperatures above 32°F.

**Cryogenics:** Refrigeration that deals with producing temperatures of –250°F and below.

**Cryogenic fluid:** A substance that exists as a liquid or gas at ultra-low temperatures of –250°F and below.

**Deck:** A refrigeration industry trade term that refers to a shelf, pan, or rack that supports the refrigerated items stored in coolers and display cases.

**Immiscible:** A condition in which a refrigerant does not dissolve in oil and vice versa.

**King valve:** The manual refrigerant shut-off valve located at the outlet of the receiver.

**Latent-heat defrost:** A method used in hot-gas defrost applications that uses the heat added in one or more active evaporators as a source of heat to defrost another evaporator coil.

**Lyophilization:** Also known as freeze-drying. A dehydration process which first incorporates freezing of a material, then a reduction of the surrounding pressure with simultaneous addition of heat to remove moisture. Moisture leaves the material through sublimation.

**Queen valve:** The manual refrigerant shut-off valve located at the inlet of the receiver.

**Satellite compressor:** A separate compressor that uses the same source of refrigerant as those used by a related group of parallel compressors. However, the satellite compressor functions as an independent compressor connected to a cooling area that requires a lower temperature level than that being maintained by the related compressors.

**Secondary coolant:** Any cooling liquid that is used as a heat transfer fluid. It changes temperature without changing state as it gains or loses heat.

**Sublimation:** The change in state directly from a solid to a gas, such as the changing of ice to water vapor, without passing through the liquid state at any point.

**Total heat rejection (THR) value:** A value used to rate condensers. It represents the total heat removed in desuperheating, condensing, and subcooling a refrigerant as it flows through the condenser.

**Unit cooler:** A packaged refrigeration system assembly containing the evaporator, expansion device, and fans. It is commonly used in chill rooms and walk-in coolers.

# Additional Resources

This module presents thorough resources for task training. The following resource material is suggested for further study.

*ASHRAE Handbook, Refrigeration Systems and Applications*, Latest Edition. Atlanta, GA: American Society of Heating, Refrigerating, and Air Conditioning Engineers.

*ASHRAE Handbook, HVAC Systems and Equipment*, Latest Edition. Atlanta, GA: American Society of Heating, Refrigerating, and Air Conditioning Engineers.

# Figure Credits

| Answer | Section Reference | Objective |
|--------|-------------------|-----------|
| **Section One** | | |
| 1.c | 1.1.1 | 1a |
| 2.b | 1.2.0 | 1b |
| 3.d | 1.3.2 | 1c |
| **Section Two** | | |
| 1.c | 2.1.0 | 2a |
| 2.c | 2.2.0 | 2b |
| 3.c | 2.3.1 | 2c |
| 4.a | 2.4.2 | 2d |
| 5.b | 2.5.2 | 2e |
| **Section Three** | | |
| 1.b | 3.1.0 | 3a |
| 2.a | 3.2.0 | 3b |
| 3.a | 3.3.0 | 3c |
| **Section Four** | | |
| 1.c | 4.1.0 | 4a |
| 2.d | 4.2.0 | 4b |

# NCCER CURRICULA — USER UPDATE

NCCER makes every effort to keep its textbooks up-to-date and free of technical errors. We appreciate your help in this process. If you find an error, a typographical mistake, or an inaccuracy in NCCER's curricula, please fill out this form (or a photocopy), or complete the online form at **www.nccer.org/olf**. Be sure to include the exact module ID number, page number, a detailed description, and your recommended correction. Your input will be brought to the attention of the Authoring Team. Thank you for your assistance.

*Instructors* – If you have an idea for improving this textbook, or have found that additional materials were necessary to teach this module effectively, please let us know so that we may present your suggestions to the Authoring Team.

**NCCER Product Development and Revision**
13614 Progress Blvd., Alachua, FL 32615

**Email:** curriculum@nccer.org
**Online:** www.nccer.org/olf

❏ Trainee Guide    ❏ Lesson Plans    ❏ Exam    ❏ PowerPoints    Other _____

Craft / Level: _____ Copyright Date: _____

Module ID Number / Title: _____

Section Number(s): _____

Description: _____

_____

_____

_____

Recommended Correction: _____

_____

_____

_____

Your Name: _____

Address: _____

_____

Email: _____ Phone: _____

# Alternative and Specialized Heating and Cooling Systems

## OVERVIEW

The twenty-first century has brought in higher energy costs and a new awareness of the potential damage that the unrestricted burning of fossil fuels can have on the environment. HVACR equipment consumes significant quantities of electrical power as well as fossil fuels. Finding and using new alternatives to traditional heating and cooling methods makes good sense for a number of reasons. In addition, there are some systems that are used in unique applications that technicians should be familiar with. This module explores some alternative methods of providing comfort, and examines some of the unique applications of HVACR principles.

## Module 03409

Trainees with successful module completions may be eligible for credentialing through NCCER's National Registry. To learn more, go to **www.nccer.org** or contact us at **1.888.622.3720**. Our website has information on the latest product releases and training, as well as online versions of our *Cornerstone* magazine and Pearson's product catalog.

Your feedback is welcome. You may email your comments to **curriculum@nccer.org,** send general comments and inquiries to **info@nccer.org**, or fill in the User Update form at the back of this module.

This information is general in nature and intended for training purposes only. Actual performance of activities described in this manual requires compliance with all applicable operating, service, maintenance, and safety procedures under the direction of qualified personnel. References in this manual to patented or proprietary devices do not constitute a recommendation of their use.

# ALTERNATIVE AND SPECIALIZED HEATING AND COOLING SYSTEMS

## Objectives

When you have completed this module, you will be able to do the following:

1. Identify and describe various alternative heating and cooling systems.
   a. Identify and describe solid-fuel heating equipment.
   b. Identify and describe waste-oil heating equipment.
   c. Identify and describe passive and active solar heating systems.
   d. Identify and describe evaporative coolers.
2. Identify and describe various unique heating and cooling systems and equipment.
   a. Identify and describe direct-fired make-up air units.
   b. Identify and describe computer room cooling systems and equipment.
   c. Identify and describe enclosure- and spot-cooling equipment.
   d. Identify and describe valance and chilled-beam cooling approaches.
   e. Identify and describe air turnover systems.

## Performance Tasks

This is a knowledge-based module; there are no performance tasks.

## Trade Terms

Active solar heating system
Air stratification
Catalytic element
Chilled-beam cooling system
Creosote
Direct-fired make-up air unit
Evaporative cooler
Hot aisle/cold aisle configuration

Indirect solar hydronic heating system
Micro-channel evaporator coil
Passive solar heating system
Thermosiphon system
Type HT vent
Type PL vent
Valance cooling system

## Industry-Recognized Credentials

If you are training through an NCCER-accredited sponsor, you may be eligible for credentials from NCCER's Registry. The ID number for this module is 03409. Note that this module may have been used in other NCCER curricula and may apply to other level completions. Contact NCCER's Registry at 888.622.3720 or go to **www.nccer.org** for more information.

# Contents

# Figures

## 1.0.0 ALTERNATIVE HEATING AND COOLING SYSTEMS

### Objectives

Identify and describe various alternative heating and cooling systems.

a. Identify and describe solid-fuel heating equipment.
b. Identify and describe waste-oil heating equipment.
c. Identify and describe passive and active solar heating systems.
d. Identify and describe evaporative coolers.

### Trade Terms

**Active solar heating system**: A type of solar heating system that uses fluids pumped through collectors to gather solar heat. It is the most complex solar heating system and provides the best temperature control.

**Catalytic element**: A device used in wood-burning stoves that helps reduce smoke emissions.

**Creosote**: A black, sticky combustible byproduct created when wood is burned in a stove. It must be periodically cleaned from within the stove and chimney before it can build up to the point where it can cause a fire.

**Evaporative cooler**: A comfort cooling device that cools air by evaporating water. It is commonly used in hot, dry climates. Cooling effectiveness drops as the relative humidity of the outdoor air increases.

**Indirect solar hydronic heating system**: A type of active solar heating system in which a double-walled heat exchanger is used to prevent toxic antifreeze in the solar collectors from contaminating the potable water system.

**Passive solar heating system**: A type of solar heating system characterized by a lack of moving parts and controls. The sun shining through a window on a winter day is an example of passive solar heat.

**Thermosiphon system**: A type of passive solar heating system in which the difference in fluid temperature in different parts of the system causes the fluids to flow through the system.

**Type HT vent**: A metal vent capable of withstanding temperatures up to 1,000°F. It is commonly used to vent wood-burning stoves and furnaces.

**Type PL vent**: A type of metal vent specifically designed for stoves that burn wood pellets or corn.

Alternative heating and cooling methods and systems can take many forms. The alternative approaches covered in this section include the following:

- *Equipment that burns solid fuel, such as stoves, furnaces, and boilers* – Fuels commonly burned include, but are not limited to, wood, wood pellets, shelled corn, and coal.
- *Waste-oil heaters* – These devices burn waste motor oil, used transmission and hydraulic fluid, waste vegetable oil, and other petroleum-based fluids.
- *Solar heating systems* – These systems can be passive or active. Passive systems have few or no moving parts or controls. Active systems employ collectors, circulating pumps, and sophisticated control systems.
- *Evaporative coolers* – This form of cooling relies solely on the cooling effect that occurs when water is evaporated. In some climates, it can provide adequate cooling for a structure without additional assistance from a mechanical refrigeration system.

### 1.1.0 Solid-Fuel Equipment

Solid fuels such as wood and coal have been used to provide heat since prehistoric times. In the United States, coal and wood were widely used for home heating until the middle of the twentieth century. At that time, solid fuels were replaced by cleaner and more convenient-to-use fuels, such as natural gas and fuel oil. By the 1970s, rising energy costs forced people to take another look at solid fuels as an energy source. As a result, solid-fuel burning stoves, furnaces, and boilers regained some popularity, especially in rural areas where this type of fuel is plentiful and inexpensive.

#### 1.1.1 Wood-Burning Stoves

Wood-burning stoves (*Figure 1*) and fireplace inserts are used to provide area heating or to supplement a conventional heating system. Wood-burning stoves and similar equipment can burn chunks of wood, wood pellets, or shelled corn. If so designed, coal may be burned in some wood-burning stoves. Stoves are typically constructed of welded steel plates or cast iron. The combustion chamber may be lined with firebrick for enhanced combustion and heat retention. Modern stoves are designed to burn the fuel through controlled combustion. This can be accomplished using a manual air damper, a thermally activated air damper, or by a thermostatically controlled, electrically operated combustion air blower or air damper.

Wood stoves and fireplace inserts sold in the United States since 1988 must meet EPA smoke emission standards. If the stove is equipped with a catalytic element, smoke emissions cannot exceed 4.1 grams per hour per EPA guidelines. The catalytic element is basically a coated ceramic honeycomb that captures smoke and particles and provides them a place to re-burn. The elements do require replacement as they degrade, and equipped stoves require more effort to properly fire and start. Non-catalytic stoves use a number of construction techniques such as internal baffles to enhance combustion and reduce smoke emissions. Non-catalytic stoves cannot emit more than 7.5 grams of smoke per hour. Wood stoves that meet these emission standards, catalytic or non-catalytic, can become EPA-certified appliances (*Figure 2*). Note however, that some states, such as Washington, have enacted more stringent standards. It is important to be aware of any such standards in your area.

Stoves provide heat through radiation. Some stoves can be fitted with an optional circulating

03409-13_F01.EPS

*Figure 1* Installed wood stove.

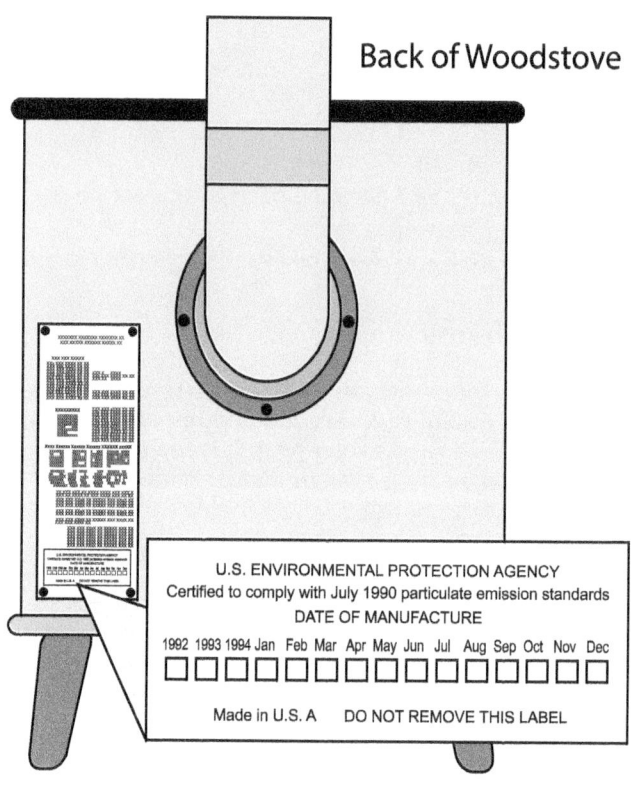

U.S. ENVIRONMENTAL PROTECTION AGENCY
Certified to comply with July 1990 particulate emission standards
DATE OF MANUFACTURE

1992 1993 1994 Jan Feb Mar Apr May Jun Jul Aug Sep Oct Nov Dec
☐ ☐ ☐ ☐ ☐ ☐ ☐ ☐ ☐ ☐ ☐ ☐ ☐ ☐ ☐

Made in U.S.A    DO NOT REMOVE THIS LABEL

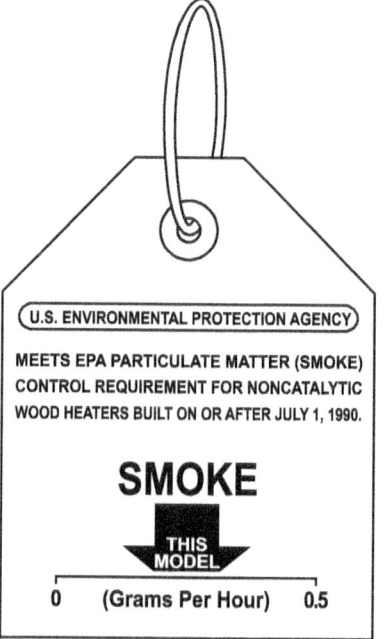

U.S. ENVIRONMENTAL PROTECTION AGENCY

MEETS EPA PARTICULATE MATTER (SMOKE)
CONTROL REQUIREMENT FOR NONCATALYTIC
WOOD HEATERS BUILT ON OR AFTER JULY 1, 1990.

**SMOKE**

THIS MODEL

0    (Grams Per Hour)    0.5

03409-13_F02.EPS

*Figure 2* EPA-certified wood stove label and hang tag.

fan that moves air over the surface of the stove to pick up additional heat.

Only properly seasoned wood should be burned in a wood-burning stove or furnace. Wood must be seasoned for several months to remove moisture. Hardwoods are preferred because they burn slower. Improperly seasoned wood (called green wood) has high moisture content. When green wood is burned, the water vapor given off combines with other combustion products to form creosote in the chimney.

### 1.1.2 Wood-Burning Furnaces

Wood-burning forced-air furnaces are used as the primary source of heat in some structures. They are typically installed in two ways: as a stand-alone heating appliance or as a supplement to an existing fossil-fuel furnace. Some wood-burning furnaces have dual-fuel capability in one cabinet. The furnace has a common circulating fan that moves air over the heat exchanger regardless of whether heat is being supplied by wood or by an integral gas or oil burner.

A wood-burning furnace used as a supplemental heater (*Figure 3*) is basically a wood stove with a sheet metal enclosure that allows air to circulate over the hot fire box. This type of furnace usually does not have a circulating air blower. Instead, the blower in the companion gas or oil furnace is used to supply the air. To accomplish this, the supply air plenum has to be modified so that the air from the gas or oil furnace blower is diverted through the cabinet of the wood-burning furnace and then into the main supply air duct (*Figure 4*). In effect, the wood-burning furnace is in series with the gas or oil furnace. There are several ways to control blower operation in the wood heat mode. When the wood-burning furnace is installed in series with the gas or oil furnace, the fan selector switch on the room thermostat subbase can be set for continuous operation. One disadvantage of this arrangement however, can be cold drafts

03409-13_F03.EPS

*Figure 3* Add-on wood-burning furnace.

## Pellet Stoves

Special stoves are available that burn wood pellets or shelled corn. The pellets are made of compressed wood waste or sawdust and come in bags. The pellets are loaded into a hopper on the side of the stove and are fed automatically into the firebox as needed. Shelled corn can often be used interchangeably with wood pellets. Both fuels are more convenient to use than traditional wood chunks and generally burn much cleaner.

03409-13_SA01.EPS

coming from the supply air register if the wood fire dies out.

Another arrangement is to install a separate fan switch in the plenum directly above the wood-fired furnace (*Figure 5*). This fan switch is connected in parallel with the fan switch in the gas or oil furnace. That way, the fan switch in the gas or oil furnace can control the fan when the gas or oil furnace is supplying heat and the fan switch on the wood-burning furnace can control the fan when wood is being burned.

Supplemental wood-burning furnaces can also be equipped with a circulating air blower and installed in parallel with an existing gas or oil furnace. By installing a blower, it can also be installed as the primary heat source (no gas or oil furnace). Supplemental wood-burning furnaces have the flexibility to adapt to a wide variety of existing forced-air heating systems. During operation, the fan switch in the supplemental furnace will keep the circulating fan operating almost constantly as

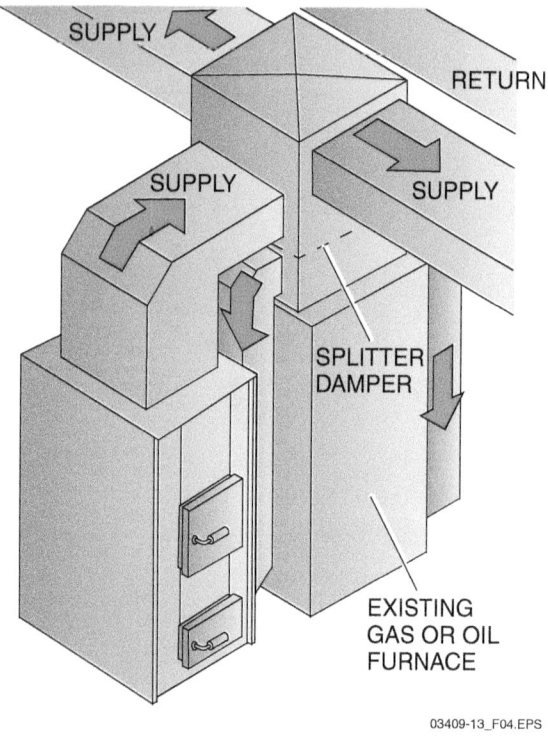

*Figure 4* Add-on furnace ducting.

long as there is fire in the firebox. As the fire dies out, the fan may begin to cycle on and off. Typically, the room thermostat of the gas or oil furnace is set low enough so that the gas or oil furnace does not operate while the wood-burning furnace is supplying heat. When the fire goes completely out, room temperature will drop low enough for the room thermostat to call for heat from the gas or oil furnace.

A dual-fuel, wood-burning furnace can best be described as a gas or oil furnace and a wood-burning furnace combined in one package. Electric heat versions are also available. Some models are also able to burn coal. A dual-fuel furnace is used as the primary heat source in a structure and the duct system is installed the same as any other forced-air furnace. The controls tend to be more sophisticated than those found on a simpler supplemental wood-burning furnace. Some even allow for the installation of air conditioning.

Two room thermostats are used in the furnace configuration shown in *Figure 6*. One controls the oil burner and the other controls combustion air for the wood. The wood burner room thermostat is set higher than the oil burner thermostat. As the fire dies down or goes out, room temperature will fall and the room thermostat controlling the oil burner will activate the burner to maintain room temperature. On a typical dual-fuel furnace, the gas or oil burner can be used to light the wood or the wood can be lit in the conventional manner.

### 1.1.3 Wood-Burning Boilers

Wood-burning boilers, like forced air furnaces, can be used as the primary source of heat in a structure. They are typically installed in two ways: as a standalone heating appliance or as a supplement to an existing boiler. Some wood-burning boilers have dual-fuel capability in one cabinet. The boiler has a common circulating pump that moves water through the boiler sections regardless of whether heat is being supplied by wood or by an integral gas or oil burner.

## EPA-Qualified Appliances

There is a significant difference between an EPA-certified appliance and one that is EPA-qualified. Wood stoves that meet the emissions standards can be EPA-certified as having met those standards. These emission guidelines are not suggestions; meeting them and acquiring the EPA certification is required to lawfully manufacture and sell the stove. However, fireplaces and outdoor, wood-burning boilers have no such EPA-required standards to meet. The EPA has established a set of voluntary emission standards for these appliances that manufacturers can choose to meet. If the standard is met, the appliance can then become EPA-qualified.

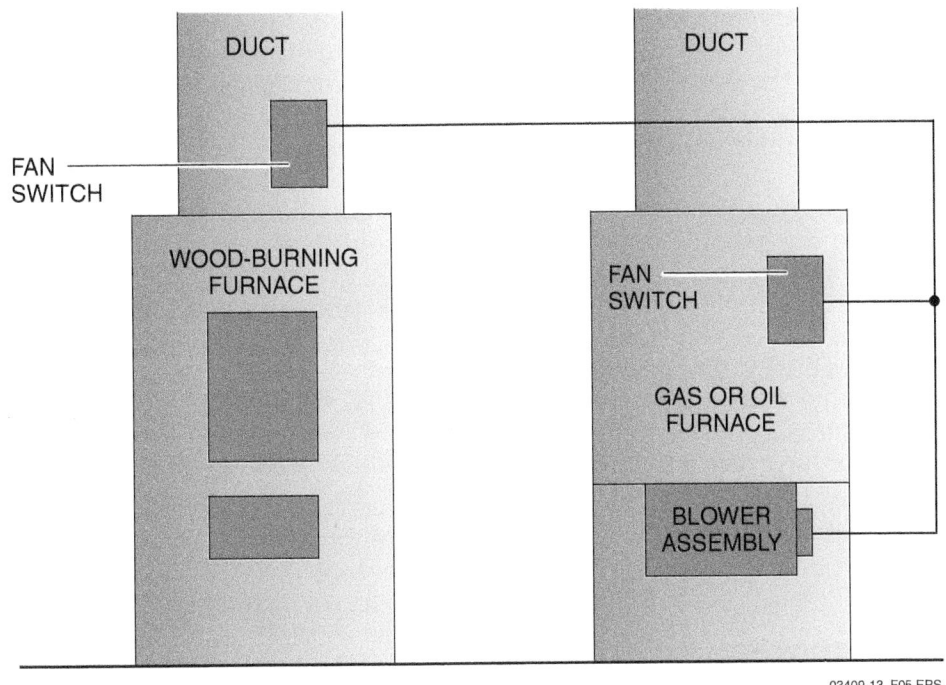

*Figure 5* Add-on furnace fan switch wiring.

**OIL OR GAS AND WOOD HEATING ONLY**

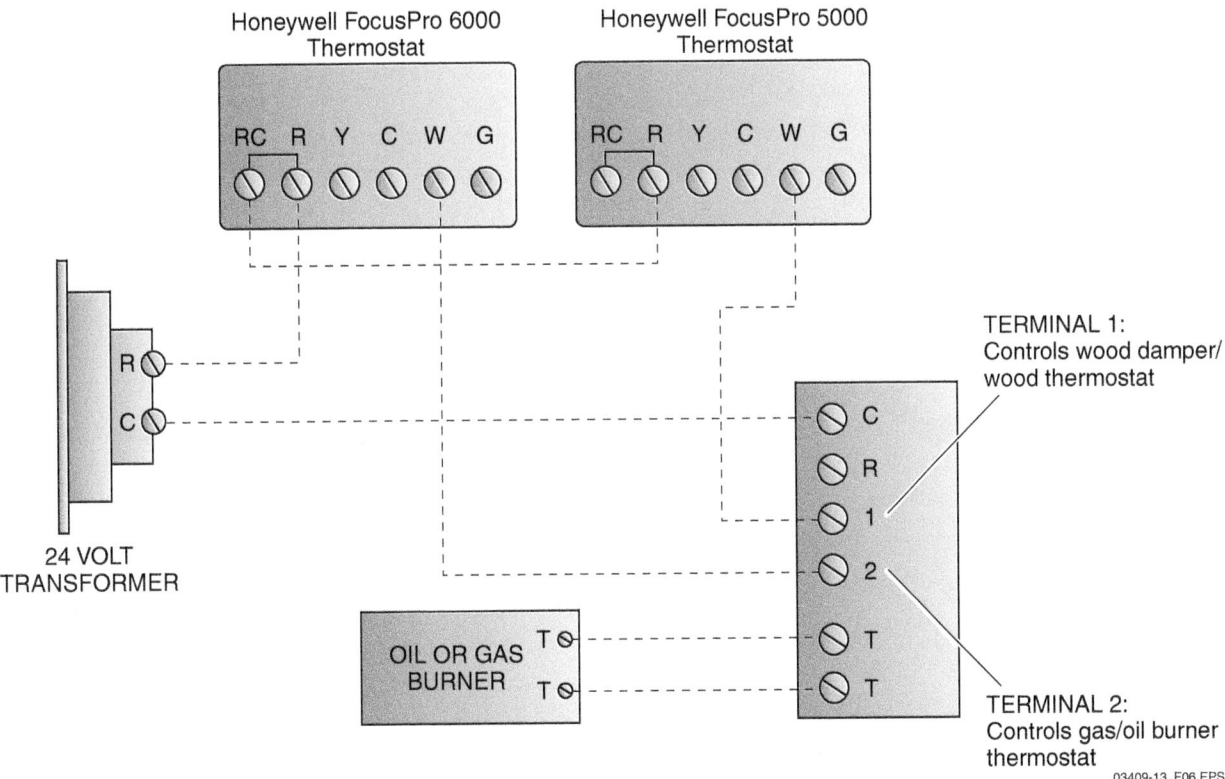

*Figure 6* Typical dual-fuel furnace thermostat wiring.

A wood-burning boiler *(Figure 7)* can provide all the heat for a building, thereby leaving the existing gas- or oil-fired boiler as a backup. The pump on the wood boiler operates continuously, circulating the hot water from the wood boiler through the existing boiler or high-efficiency exchanger. The wood-burning boiler is piped in series with the existing boiler. When there is a demand for heat, the existing boiler can remain unfired, unless the wood boiler is not refueled and the fire dies out. If the water temperature is maintained high enough, the burner on the existing boiler will not operate. If the water temperature drops further and a heat demand is present, the aquastat in the existing boiler will energize the burner (*Figure 8*).

Supplemental wood-burning boilers can be installed indoors next to the existing boiler. They can also be installed outdoors in a self-contained weatherproof structure, complete with its own chimney (*Figure 9*). The heated water is pumped through insulated pipes buried underground and connected to the existing hydronic heating system inside the structure. Advantages of this system are that it is cleaner (no ashes or wood debris inside the structure) and no new chimney or vent has to be installed in the structure.

Like its forced-air counterpart, a dual-fuel, wood-burning boiler is a gas or oil boiler and a wood-burning boiler combined in one package (*Figure 10*). It offers the same advantages as a dual-fuel furnace. A dual-fuel boiler is used as the primary heat source in a structure and the piping and radiation is installed the same as any other hydronic heating system.

### 1.1.4 Wood-Burning Appliance Installation and Maintenance

Heating appliances that burn wood or other solid fuel have unique installation and maintenance requirements, including the following:

- Clearance to combustibles
- Combustion air
- Venting
- Field wiring, piping, and ductwork
- Fuel storage
- Cleaning and maintenance

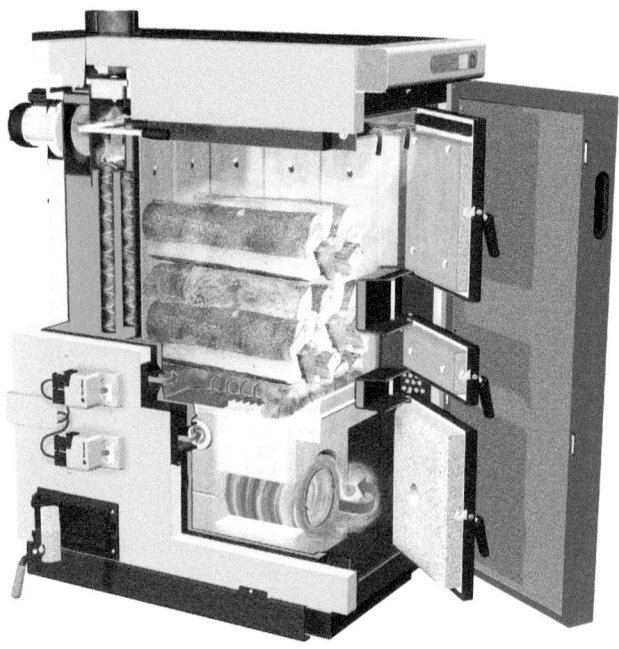

*Figure 7* Add-on wood-burning boiler.

03409-13_F09.EPS

*Figure 9* Outdoor wood-burning boiler.

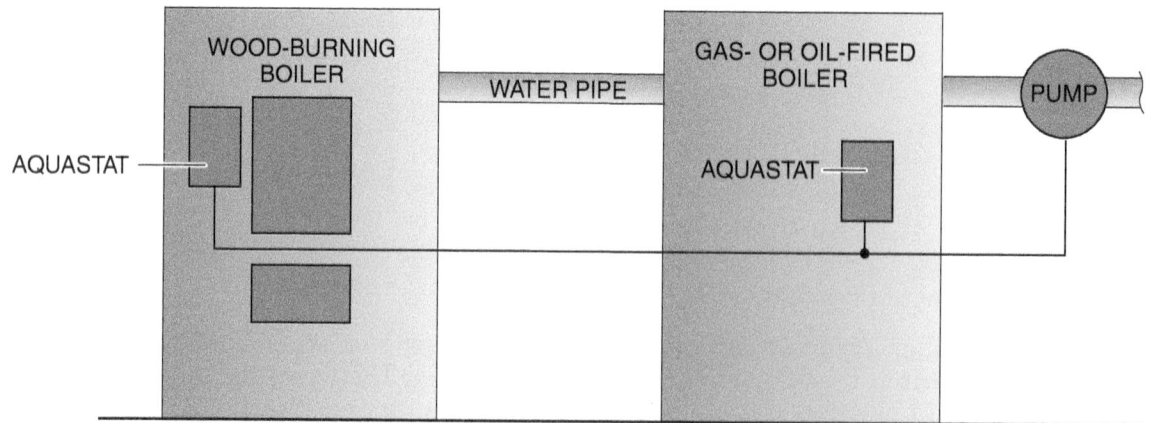

03409-13_F08.EPS

*Figure 8* Add-on boiler aquastat wiring.

03409-13_F10.EPS

*Figure 10* Dual-fuel boiler.

Before installing any wood-burning appliance, always thoroughly read the manufacturer's installation and service literature before proceeding.

Any fuel-burning appliance, such as furnaces, will always carry specifications for installation clearances from combustible materials such as floors, walls, and building structural members. Wood-burning appliances are no different. However, since wood-burning stoves and furnaces operate at higher temperatures than conventional gas or oil furnaces, those distances tend to be much greater. For example, wood stoves must be placed on an approved, non-combustible base. When installed in a room, the stove must maintain a 36" clearance from the top, sides, back, and front to any combustibles. If the stove is equipped with an approved heat shield, distances can be reduced. The closest a stovepipe can be to combustibles is three times the diameter of the pipe. In other words, a 6" stovepipe must be at least 18" away from combustibles (*Figure 11*).

Check the manufacturer's literature for specific clearance requirements when installing wood-

burning stoves, furnaces, and boilers. For the most up-to-date information on all aspects of installing wood-burning appliances, consult National Fire Protection Association (NFPA) codes.

Stoves and furnaces that burn solid fuel must be supplied with adequate air for combustion. In a loosely-built structure, air entering through normal infiltration may be adequate to supply all combustion air needs. If the structure is tightly built, outside air will have to be brought into the structure to support combustion. Outside air can be brought in through a duct into the area near the stove or furnace (*Figure 12*). As a rule of thumb, the duct size should be equal to or greater than the size of the chimney or vent of the appliance.

If the stove or furnace is installed in a confined space such as a utility room, combustion air and ventilation air grilles must be installed in a wall or door that is open to the rest of the structure. The combustion air grille is installed near the floor and the ventilation air grille is installed near the ceiling. Each grille must have a free air opening of one square inch for each 1000 Btuh of input capacity. For example, assume a combination wood/oil furnace has a maximum input capacity of 144,000 Btuh. Two grilles having at least 144 square inches of free area would be required for the utility room. Be aware that free air area is not the same as the area of the face of the grille. Consult the grille manufacturers literature for the free air area for any given grille.

Wood-burning stoves, furnaces, and boilers must be vented, either through a correctly sized, tile-lined masonry chimney or through a correctly

---

### Renewable Energy Resource

**GOING GREEN**

Wood and corn used as fuel are considered renewable energy resources because they can be constantly replenished through natural processes. Oil and natural gas are not renewable fuels; once used, they are gone forever. The use of renewable energy reduces the dependence on oil and gas and is generally easier on the environment.

---

## Comfort

Wood-burning furnaces can provide outstanding cold weather comfort due to the almost constant operation of the air-circulating blower. As long as there is a wood fire, the fan switch will keep the blower operating. However, if the outdoor temperature rises and a hot fire is maintained, the structure may overheat. Users of wood-burning furnaces learn through experience how to maintain the fire at a level that will not overheat the structure.

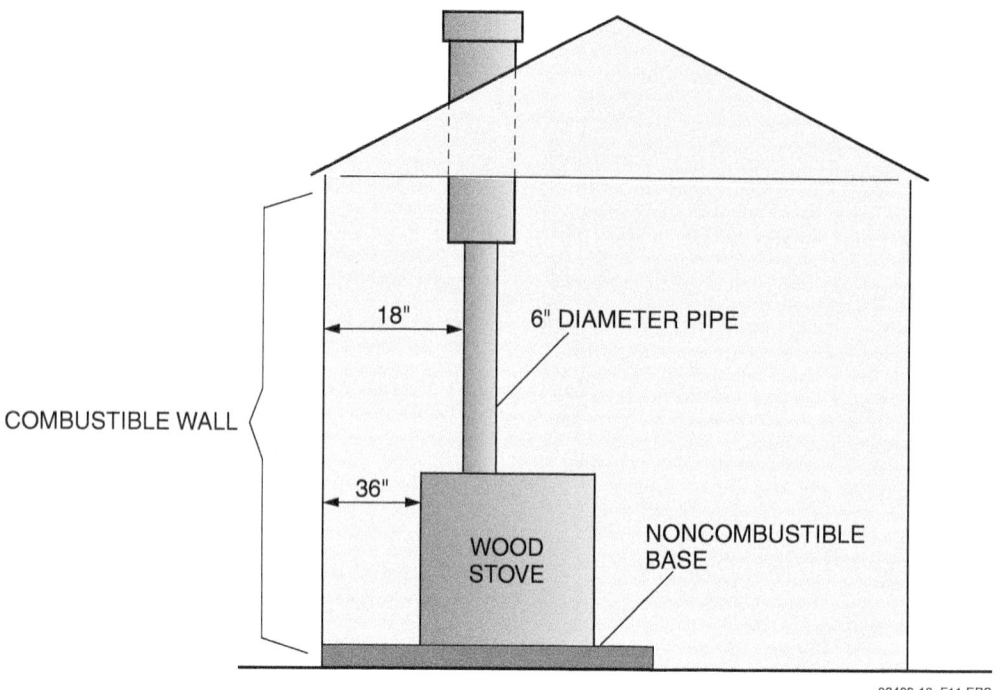

Figure 11 Wood stove installation clearances.

03409-13_F11.EPS

sized all-fuel Type HT vent (*Figure 13*). Type HT vents are rated to handle temperatures as high as 1,000°F. When installing a Type HT vent, do not mix components from a different manufacturer's Type HT vent. Mixing components may create an unsafe condition. The agency rating of the assem-

bled vent is only valid if rated components from the same manufacturer are used.

Never vent a wood-burning stove, furnace, or boiler through a chimney or vent serving a fireplace or other fuel-burning appliance such as a gas or oil furnace. When installing a Type HT

## Pellet Boilers

Like stoves, boilers that burn pellets and incorporate a self-feeding mechanism are available. The boiler shown here features automatic filling and automatic ignition. It also includes controls designed for easy integration into solar applications. Ash need only be removed every 300-500 hours of operation.

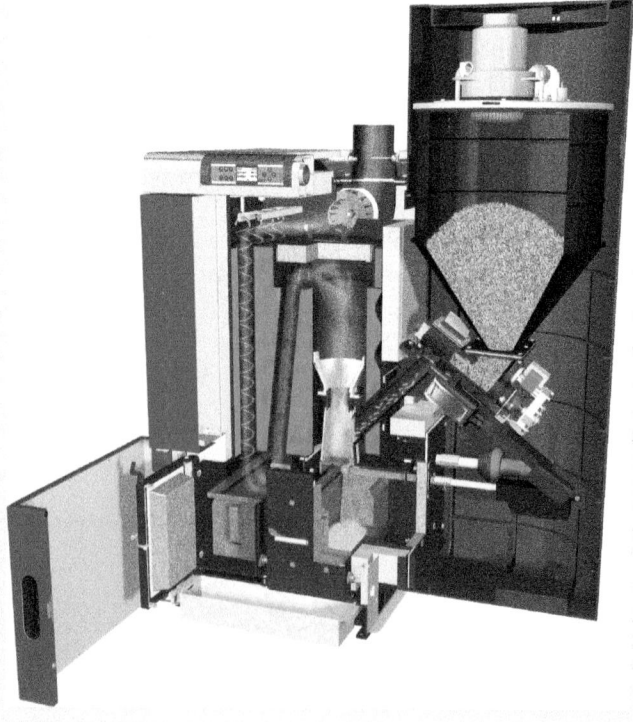

NCCER – *HVAC Level Four*   03409

vent, follow the manufacturer's installation instructions and all appropriate local and national codes. Corn and wood pellet stoves generally do not have flue temperatures that are as high as those of conventional wood stoves. For that reason, they can sometimes be directly vented through a sidewall using the vent kit supplied by the stove manufacturer. Special pellet venting pipe or Type PL vent, is also available for venting corn and pellet stoves. A corn or pellet stove may be vented through an existing vent or fireplace chimney as long as it is not being used to vent another appliance and does not violate local codes.

Dual-fuel appliances such as wood/oil or wood/gas furnaces and boilers often use a common vent. If installing such a furnace or boiler as a replacement for a conventional gas or oil furnace or boiler, the existing chimney or vent may not be adequate to handle the higher temperature of the replacement appliance. When installing a dual-fuel appliance, use a Type HT vent or a correctly sized and lined masonry chimney to ensure a safe installation.

Wood-burning appliances should be cleaned and maintained in accordance with the manufacturer's instructions. Maintenance items include oiling motors, checking belts, and replacing filters. When wood burns, it can create creosote that can build up on the internal surfaces of the stove, furnace, and chimney (*Figure 14*). Creosote is combustible, and can ignite if left to accumulate. Once ignited, creosote can burn with the intensity of a blowtorch, permanently damaging the stove and/or vent and possibly causing a structure fire. For that reason, all wood-burning appliances must be inspected and cleaned when creosote buildup is noted. Since creosote buildup varies between installations, there is no hard and fast rule for inspection intervals. It is not uncommon to have to clean a wood stove and/or chimney every few weeks.

DUCT

WOOD-
BURNING
FURNACE

FRESH AIR DUCT

03409-13_F12.EPS

*Figure 12* Wood-burning appliance combustion air supply.

03409-13_F13.EPS

*Figure 13* Type HT vent.

## Outdoor Boilers

Owners of wood-burning outdoor boilers can realize outstanding energy savings. However, outdoor boilers are not without controversy. If unseasoned wood or different fuels, such as construction debris or old tires are carelessly burned in them, obnoxious smoke pollution can result. Since the chimneys on these boilers are relatively low to the ground, the smoke pollution is released low to the ground. This has led to lawsuits and to local ordinances that ban or restrict the use of outdoor boilers. Restrictions include the size of the residential lot and minimal setback distances from lot lines.

03409-13_F14.EPS

*Figure 14* Creosote buildup in a chimney.

### 1.2.0 Waste-Oil Heaters

People today realize the importance and necessity of recycling and reusing resources to protect the environment. Heaters that burn waste oil are examples of products that use a resource that, in the past, was often discarded. Waste oil can come from a variety of sources. Automotive repair shops generate used motor oil, transmission fluid, and other petroleum-based lubricants and fluids. Restaurants and food processing facilities generate vegetable oils that were used in cooking. A waste-oil burner (*Figure 15*) is similar to a conventional gun-type oil burner that has been modified to burn various types of waste oil. The burners can be used in forced-air systems or to fire a boiler. Because of the nature of the fuel, waste-oil heaters are likely to be used in commercial and industrial applications and rarely in residential applications.

The quality and physical properties of waste oil can vary significantly. For example, used motor oil is significantly different from used french-fry oil. The fuel handling system and the burner itself must be capable of handling these variables. Filters are required to remove various particulate contaminants from the waste oil.

Although a variety of oil types can be burned, waste oil heaters need to maintain the fuel at a consistent viscosity in order to fire properly. They have a built-in heater that heats the waste oil, keeping the viscosity more uniform. Compressed air, supplied by a built-in air compressor or by an outside source, is required to help atomize the fuel. The atomized fuel is typically ignited by a high-voltage spark from the ignition transformer. An oil burner primary control using a cad cell flame detector provides a safety shutoff if a flame is not established. Installation and venting of waste oil

## Be Safe

Wood-burning appliances are safe if they are installed, operated, and maintained properly. If these factors are ignored, an unsafe condition can be created. Always install smoke detectors and carbon monoxide (CO) detectors whenever a wood-burning appliance is installed.

03409-13_F15.EPS

*Figure 15* Waste oil burner assembly.

# Chemical Cleaners

There are chemical cleaners available that claim to clean the creosote from inside wood-burning stoves and chimneys. They often take the form of powder that is sprinkled on the fire. The heat vaporizes the chemicals so that they can do their cleaning. Some chemical cleaners may be more effective than others. However, some equipment manufacturers do not recommend or support their use because the chemicals they contain can corrode the inner surfaces of the appliance.

heaters is similar to conventional oil-fired heaters. Although conventional oil burners all operate in a similar manner and use similar components, that is not the case with waste-oil burners. Different manufacturers use different engineering and different components to accomplish similar results. Always read the manufacturer's installation literature before proceeding.

## 1.2.1 Waste Oil Heating Issues

Burning waste oil can provide significant cost savings but there are negative aspects as well. The Environmental Protection Agency (EPA) has rules that cover the burning of waste oil. For example, regulations govern the conditions under which waste oil containing certain types of contaminants can be burned. Local codes may regulate or restrict the burning of waste oil.

Waste oil can contain a number of different contaminants that can affect its usefulness as a fuel. Used motor oil can contain water, gasoline, antifreeze and traces of heavy metals. Used refrigeration oil can contain refrigerant and acid, and used cooking oil can contain water and food particles. The amount of a given contaminant in used oil can significantly affect its suitability as a fuel. Burning these various fuels, with the potential contaminants they contain, usually results in decreased intervals between scheduled maintenance and more frequent equipment breakdowns. The negative aspects as well as the potential cost savings should be carefully weighed before installing a waste-oil heater.

## 1.3.0 Solar Heating Systems

Harnessing the energy of the sun to provide heat for comfort is not only energy-efficient, it is very kind to the environment. It is estimated that each square mile of the Earth receives solar energy equivalent to 5 billion kWh each year. This free source of energy is there for the taking. Solar heating systems capture that free energy. The systems fall into two categories: passive and active. Solar heating systems are popular in southwest-ern parts of the United States where winter sunshine is plentiful.

## 1.3.1 Passive Solar Heating Systems

The family cat is aware of passive solar heating. On winter days, cats can be found stretched out on the carpet in front of a south-facing window. The rays of the sun warm both the carpet and the cat. Passive solar heating systems (*Figure 16*) both capture and store solar energy. Simple passive systems require two major components: large, south-facing windows, and thermal mass to collect and store the heat. In the United States, the sun is low in the southern sky during winter. Large south-facing windows allow the sun to enter the room where it is captured in a greenhouse effect and heats the thermal mass.

The thermal mass often takes the form of dark-colored tile or masonry floors and walls. During the day, the sun's rays heat the thermal mass. After the sun goes down, the heat retained in the thermal mass is gradually released into the space.

In some schemes, the thermal mass is a wall filled with water. There are too many variations of this basic design to fully discuss them all in this module. A common feature includes thermal drapes that automatically close when the sun goes down to prevent heat loss back out through the windows. Some passive designs are laid out so that the thermal mass induces gentle convection airflow throughout the room or structure.

Passive solar hydronic heating systems are also available. In this scheme, convection currents move water from a roof-mounted solar collector to a storage tank inside the structure. A thermo-siphon system (*Figure 17*) is typical of passive systems. It operates by the different densities between the water in the storage tank and the collector. Cool water in the storage tank and hot water in the collector means rapid flow. Hot water in the storage tank and cool water in the collector means little or no flow. This condition could possibly lead to reverse thermosiphoning. A check valve must be installed to prevent this backward flow of water through the system. Thermosiphon-

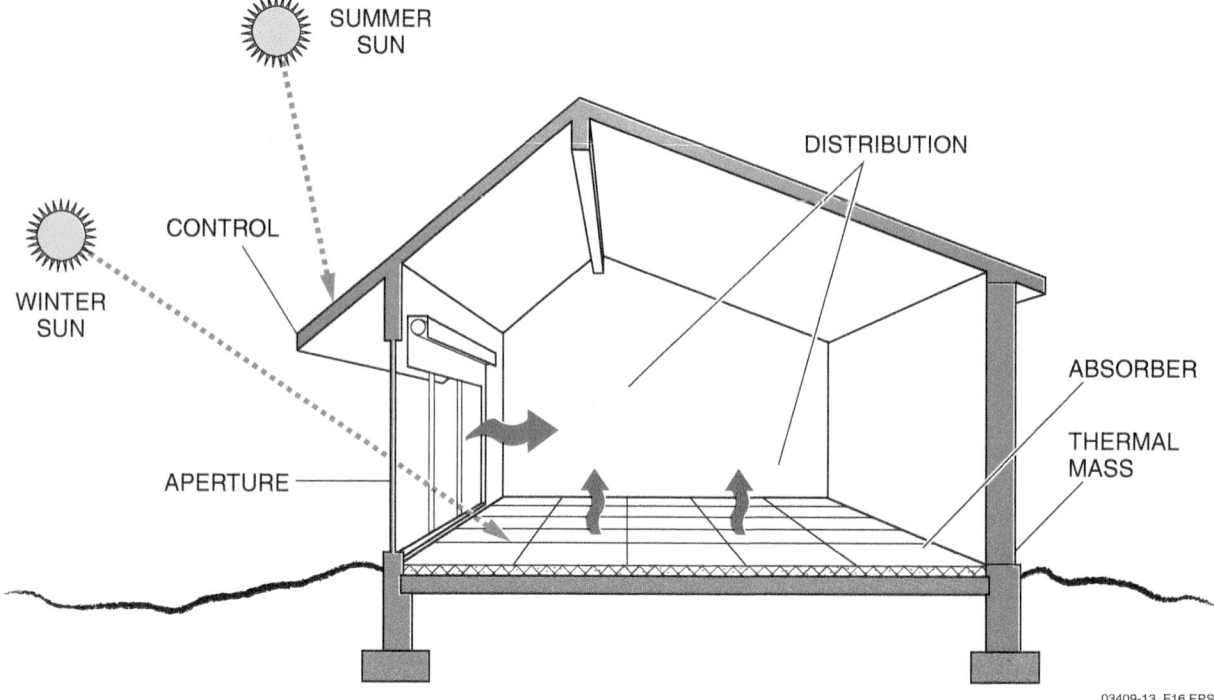

SUMMER
SUN

WINTER
SUN

CONTROL

APERTURE

DISTRIBUTION

ABSORBER

THERMAL
MASS

03409-13_F16.EPS

*Figure 16* Passive solar heating system.

ing systems do not have temperature controls, so water temperature in the system will vary with changes in the weather. The warm water in the storage tank is used to preheat water in a conventional hydronic heating system with a boiler. Passive hydronic heating systems are not as well-suited for space heating as other types of hydronic solar heating systems.

The major advantage of a passive solar heating system is simplicity. With few or no moving parts, the system heats the structure as long as the sun shines. A major disadvantage of passive systems

is that they are prone to overheat the structure on warm sunny days and provide little heat on cloudy days. A conventional heating system is often required to supplement a passive solar heating system.

### 1.3.2 *Active Solar Heating Systems*

Active solar heating systems are more complex than passive systems. They use pumps, valves, and other devices to initiate and control the flow of fluid through the solar collector and the rest of

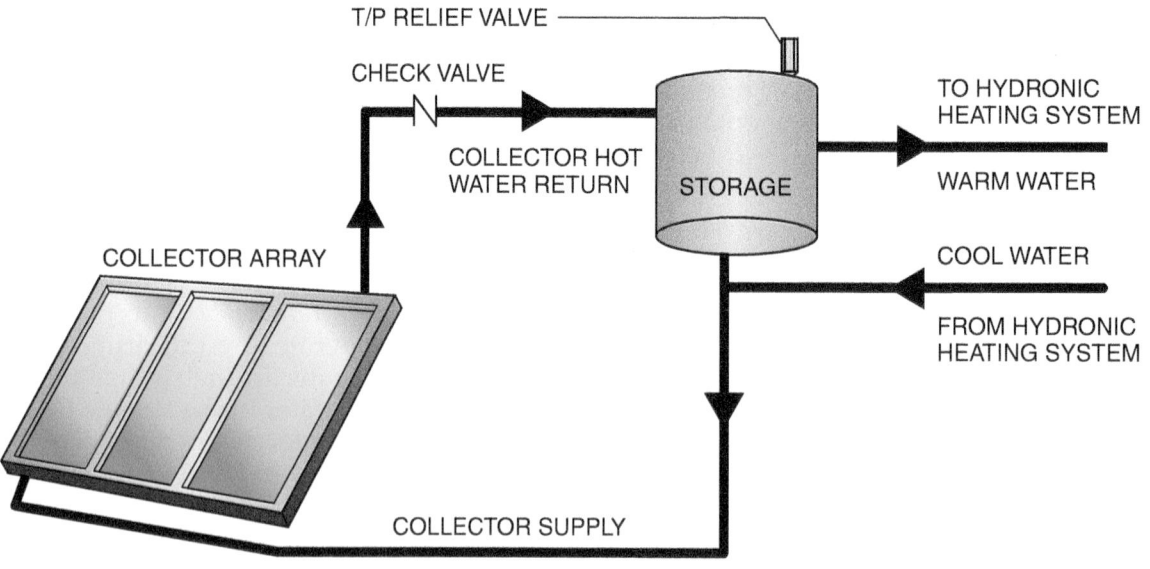

T/P RELIEF VALVE

CHECK VALVE

COLLECTOR HOT
WATER RETURN

STORAGE

TO HYDRONIC
HEATING SYSTEM

WARM WATER

COOL WATER

FROM HYDRONIC
HEATING SYSTEM

COLLECTOR ARRAY

COLLECTOR SUPPLY

03409-13_F17.EPS

*Figure 17* Passive solar thermosiphon hydronic heating system.

NCCER – *HVAC Level Four*   03409

the system. Indirect systems, often called closed loop systems, use antifreeze such as ethylene glycol as the circulating fluid in the system. The heated antifreeze solution is then passed through a heat exchanger where it gives up heat to water that is circulated through the hydronic heating system pipes and radiation. *Figure 18* shows a typical indirect solar hydronic heating system. The storage tank shown would commonly be piped in series with an existing hydronic system boiler where it acts as a preheater. The boiler would operate very little or not at all when the sun is shining. On cloudy days or at night, when little or no preheating occurs, the boiler heats the structure.

Active solar hydronic heating systems provide better temperature control due to their use of complex control systems. However, like all solar heating systems, they are ineffective on cloudy days and at night, and require some other form of backup heat.

## 1.4.0 Evaporative Coolers

Evaporative coolers, often called swamp coolers (*Figure 19*), are used in many parts of the United States for comfort cooling. They offer lower operating costs than conventional air conditioners. Evaporative coolers do not operate using the refrigeration cycle and are very simple in their construction. Instead, they operate on the principle that when water is changed from a liquid to water vapor, heat is absorbed. *Figure 19* shows models that can be roof-mounted, using a vertical dis-

charge, or mounted on the ground or outside wall with a side discharge.

An evaporative cooler consists of a louvered cabinet containing a blower assembly, water-absorbing pads, a water pump with a water distribution system, and a float valve to control the water level in the sump of the unit. In operation, the pump distributes water to the pad or pads to wet them. The blower assembly draws hot air across the wet pads. The pads used in evaporative coolers have traditionally been made of shredded aspen wood. Today, pads are available that are made of cellulose or man-made fibers that last longer.

The water on the pads absorbs heat from the air as the water evaporates. This cools and adds moisture to the air. The cooled air is then distributed throughout the structure. Evaporative coolers do not require a return duct because they take in outdoor air. To prevent the structure from being pressurized, windows or doors must be opened or a relief duct provided to relieve the pressure.

Evaporative coolers are most effective if the outdoor air has a low relative humidity. That is why they are popular in the hot, dry climate of the southwestern United States. If the relative humidity increases, evaporative coolers become less effective because the air cannot absorb additional moisture. In those situations, a conventional air conditioner is used for comfort cooling. They are also used in greenhouses to maintain a high wet-bulb temperature while lowering the temperature inside. Portable models (*Figure 20*) are also used

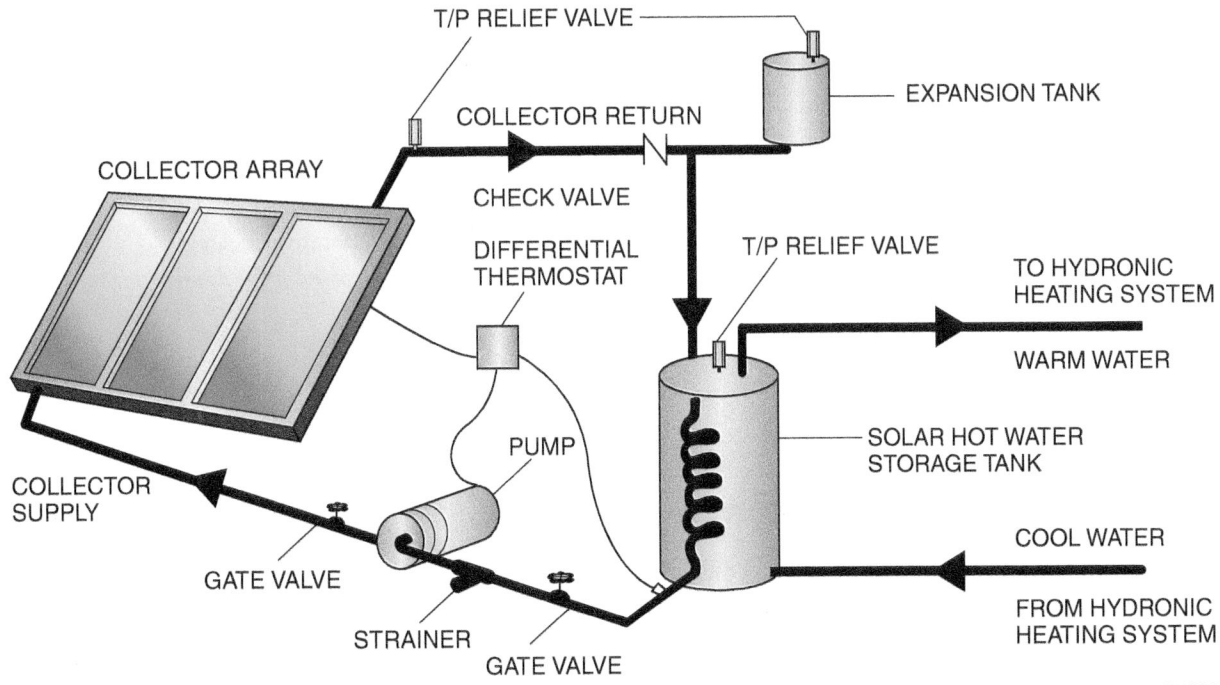

*Figure 18* Active solar indirect hydronic heating system.

SIDE DISCHARGE

VERTICAL DISCHARGE (DOWN-SHOT)

03409-13_F19.EPS

*Figure 19* Evaporative coolers.

for spot cooling applications in hot work areas. A simple garden hose and power cord is all that is required for their operation. Other applications for evaporative cooling include commercial kitchen make-up air systems.

Sometimes evaporative coolers are installed so that they share the supply duct system with a conventional air conditioner. Manual isolation panels or positive-closure dampers are used to prevent the conditioned air from one system from being bypassed through to the other. At one time, a combination heat pump/evaporative cooler was manufactured. It did not prove popular and has since been discontinued.

The major advantage of evaporative coolers is their lower installed cost and lower operating cost compared to a traditional air conditioner. The blower motor and water pump consume much less power than a compressor, condenser fan motor, and evaporator blower motor. However, they do increase water consumption. Disadvantages include poor comfort when humidity is high, as well as the increased maintenance that is often associated with water, constantly wetted surfaces, and media that require replacement.

03409-13_F20.EPS

*Figure 20* Portable evaporative cooler.

## Indirect Heat Exchangers

If an indirect solar heating system is used to heat potable water, the heat exchanger must be double-walled to prevent possible contamination of the potable water supply with ethylene glycol, a toxic substance.

## Additional Resources

*Wood Pellet Heating Systems: The Earthscan Expert Handbook of Planning, Design, and Installation.* Dilwyn Jenkins. 2010. Taylor & Francis.

*Solar Water Heating–Revised & Expanded Edition: A Comprehensive Guide to Solar Water and Space Heating Systems,* Second Edition. Bob Ramlow and Benjamin Nusz. 2010. New Society Publishers.

## 1.0.0 Section Review

1. The catalytic element of an equipped wood stove captures and collects smoke particles in a special filter that is replaced following each burn cycle.

   a. True
   b. False

2. To maintain uniform viscosity of the fuels, a waste-oil heater may be equipped with a _____.

   a. catalytic converter
   b. viscosity adapter
   c. fuel heater
   d. special filter

3. The thermal mass in a passive solar heating system is likely to be constructed of _____.

   a. tile or masonry
   b. glass
   c. wood
   d. sheet metal

4. Evaporative coolers perform well in greenhouses because they _____.

   a. maintain a low relative humidity
   b. maintain a high relative humidity
   c. reduce the wet bulb temperature
   d. increase the dry bulb temperature

### 2.0.0 UNIQUE HEATING AND COOLING SYSTEMS

#### Objectives

Identify and describe various unique heating and cooling systems and equipment.

   a. Identify and describe direct-fired make-up air units.
   b. Identify and describe computer room cooling systems and equipment.
   c. Identify and describe enclosure- and spot-cooling equipment.
   d. Identify and describe valance and chilled-beam cooling approaches.
   e. Identify and describe air turnover systems.

#### Trade Terms

**Air stratification**: The layering of air in a room based on temperature. The warmest air is concentrated at the ceiling and the coldest air is concentrated at the floor, with varying temperature layers in between.

**Chilled-beam cooling system**: A cooling system that employs radiators (chilled beams) mounted near the ceiling through which chilled water flows. Passive systems rely on convection currents for cooling. Active systems use ducted conditioned air to help induce additional airflow over the beams.

**Direct-fired make-up air unit**: An air handler that heats and replaces indoor air that is lost from a building through exhaust vents.

**Hot aisle/cold aisle configuration**: A method used to install equipment cabinets in computer rooms to manage the flow of warm and cool air out of and into the cabinets.

**Micro-channel evaporator coil**: Coil construction technology that consists of a group of very small parallel tubes bonded to enhanced heat transfer fins, replacing the single large tube common in standard evaporator and condenser coils.

**Valance cooling system**: A type of cooling system in which chilled water is circulated through finned-tube radiators located near the ceiling around the perimeter of a room. Convection currents move the cooled air instead of a blower assembly. A decorative valance conceals the system.

There are many types of HVACR systems that can be considered unique. It would require several volumes to cover them all in depth. Cascade systems, for example, are special refrigeration systems that provide extremely low temperatures. They may be used to create low temperatures in a test chamber for electronic device testing. These systems are actually two refrigeration systems in one. One system, the low-stage system, is used to provide the low evaporator temperatures required; –50°F, for example. The evaporator of the second-stage, or high-stage, system acts as a refrigerant-to-refrigerant condenser for the low-stage system. The second-stage system has a more typical air- or water-cooled condenser. However, such systems will not often be encountered.

There are systems used for unique purposes or applications that will be encountered far more often than others. Systems to be discussed in this section include:

- Direct-fired make-up air units
- Computer room cooling systems
- Enclosure and spot cooling equipment
- Valance and chilled-beam systems
- Air turnover systems

### 2.1.0 Direct-Fired Make-Up Air Units

In many commercial and industrial applications, large quantities of air must be mechanically exhausted to purge harmful contaminants, smoke, and odors. For example, air near a loading dock can be contaminated by vehicle exhaust. A second example is an industrial paint booth, which often requires massive amounts of air to be exhausted.

There are many possible industrial situations that require high volumes of exhaust air, and this exhausted air must be replaced. Further, this make-up air must often be conditioned before entering the space, in order to maintain comfort or to ensure that the industrial process environment remains consistent. Without mechanical equipment to replace exhaust air, significant negative pressures in the building result. Unconditioned outside air then enters through any available opening.

In some of these applications, cooling is not a requirement. In many locations, and in many different industrial applications, an indoor temperature that is no cooler than the outdoor temperature is acceptable. In those cases, unconditioned outdoor air is simply provided to make up the exhaust air. However, in cold climates, allowing the process and work area to become too cold is a concern.

When ensuring that the make-up air is filtered and remains above a certain temperature is the primary consideration, a direct-fired make-up air unit (*Figure 21*) can be used. This unit replaces the exhausted air, using 100 percent air from outdoors, and heats the air as it passes through. These units are typically gas-fired, but contain no heat exchanger or flue vent. The unique feature of the direct-fired unit is that the combustion process takes place directly in the primary airstream. Combustion air is drawn from the same air being supplied to the building, and all byproducts of combustion remain in the airstream entering the building as well.

Obviously, it is important that combustion byproducts be sufficiently diluted to ensure the safety of building occupants. Direct-fired units are specifically designed with this key requirement in mind, and the level of hazardous byproducts is sufficiently diluted by the high volume of air flowing through them. The velocity of the airstream across the burner is required to be in the range of 3,000 fpm to satisfy regulations. Precision static pressure switches are used to ensure that the airflow through the unit stays within the acceptable range for safety. One of the principal advantages though, is that these units achieve 100 percent combustion efficiency, since no heat is lost through a flue vent. However, they are not 100 percent thermally efficient. As a general rule, an average of 8 percent of the total thermal efficiency is lost due to the latent heat related to moisture in the combustion process.

Direct-fired units are generally controlled by a discharge-air temperature sensor arrangement,

## Indirect-Fired Make-Up Air Units

Since the discussion here focuses on direct-fired systems, you may be asking yourself "What about indirect-fired systems?"

In fact, all common furnaces are indirect-fired. A heat exchanger separates the combustion process and its byproducts from the conditioned air supply. There are indirect-fired make-up air units, as well as direct-fired models. However, they are really just large versions of common furnaces that are designed to operate on 100 percent outside air. This often requires a stainless steel heat exchanger for durability.

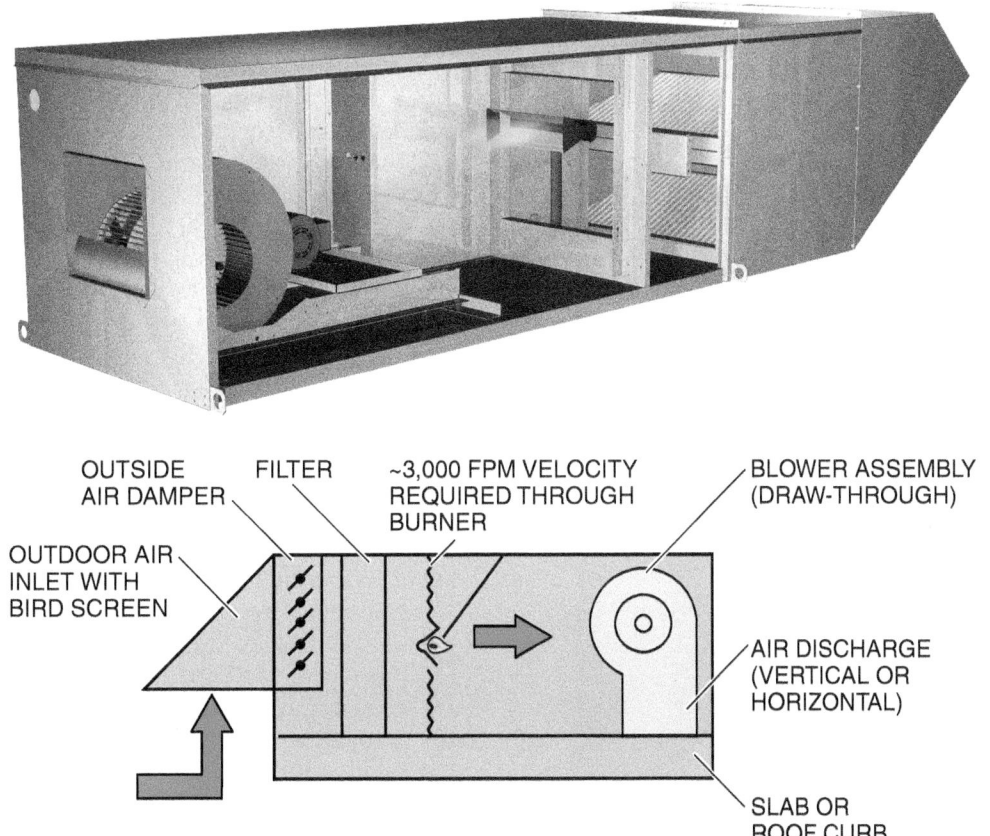

OUTSIDE AIR DAMPER  FILTER  ~3,000 FPM VELOCITY REQUIRED THROUGH BURNER  BLOWER ASSEMBLY (DRAW-THROUGH)

OUTDOOR AIR INLET WITH BIRD SCREEN  AIR DISCHARGE (VERTICAL OR HORIZONTAL)

SLAB OR ROOF CURB

03409-13_F21.EPS

*Figure 21* Direct-fired make-up air unit.

and use a modulating gas burner design. Unlike a typical residential furnace, which usually produces supply air temperatures of 110°F or more, the discharge temperature of a direct-fired unit is often set at, or slightly above, the desired room temperature. However, they are typically available with rated temperature rises ranging from 60°F to 120°F. Another traditional system feature is a control interlock between the make-up air unit and the related exhaust system. If the exhaust system is not in operation or fails, the make-up air unit is automatically shut down as well. Direct-fired units can be ducted, or designed to simply free-blow into the space they serve.

*Figure 22* shows a typical direct-fired burner assembly. Most utilize a cast-iron pipe as a manifold, drilled to incorporate the gas orifices, and stainless steel baffles. The baffles are also known as mixing plates. The burner must be designed to prevent the airstream from disturbing the combustion process, while simultaneously collecting sufficient combustion air to support the flame.

In situations that require the make-up air to be cooled and/or dehumidified as well, direct-expansion (DX) or chilled-water cooling systems can also be incorporated into the direct-fired make-up air unit. Even evaporative systems can be combined into a single piece of equipment with a direct-fired make-up air unit. These arrangements can provide a packaged system capable of maintaining reasonably precise discharge air conditions to the indoor environment utilizing either a combination of return and outdoor air, or 100 percent outdoor air on a year-round basis.

Since high entering air temperatures are typically imposed on these systems, as well as high wet bulb temperatures, evaporator coil and overall refrigerant circuit design may be somewhat

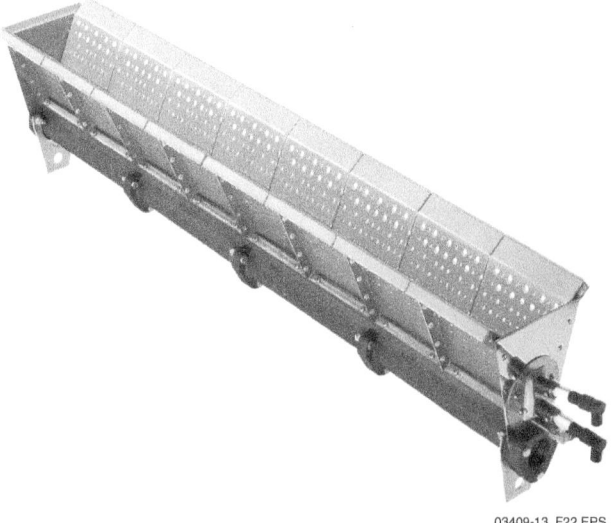

03409-13_F22.EPS

*Figure 22* Direct-fired burner assembly.

different than the typical DX system. Standard cooling systems may easily be overloaded with entering air temperatures in excess of 90°F for extended periods of time, especially when the air is also moisture-laden.

## 2.2.0 Computer Room Cooling Systems

Computers and other electronic equipment generate a great deal of heat that must be removed to prevent overheating. If allowed to overheat, computers will malfunction or fail. For security reasons, computers and banks of servers are installed in dedicated rooms and closets. The various components of the system are often stacked tightly together, allowing a great deal of heat to build up in a very small area. For this reason, special equipment is used to cool computers and similar electronic devices. The cooling equipment must be able to reliably handle a significant and constant cooling load.

In most computer and electronic cooling applications, the sensible cooling load represents an even larger amount of the total load than in common comfort applications. Since computers and other electronic devices produce only sensible heat, the only latent load is that which comes from infiltration or from moisture added by humidification systems. Maintaining proper and consistent humidity levels is also very important, because electronic systems are sensitive to damage from static electricity. For this reason, humidification accessories and dehumidification cycles are often incorporated into the cooling systems. However, dehumidification should not be needed very often if the controls are calibrated and set up properly. Dehumidification is a process that uses an excessive amount of energy, since both the refrigeration circuit and electronics are usually operating simultaneously. When computer room systems are found moving in and out of this mode regularly, it is often caused by an unnecessarily tight setting of the humidity deadband. Humidification is needed often, as it is not possible to completely stop the constant removal of moisture from the air as a direct-expansion (DX) system operates in the cooling mode. As moisture is removed through this process, it must be replenished to keep the air from becoming too dry.

One important difference between standard DX cooling systems and those used specifically for computer rooms and data centers is found at the evaporator coil. The evaporator of computer room systems is typically designed to deliver as much sensible cooling as possible, and minimal condensate removal. This is done by increasing the surface area of the coil while decreasing the thickness. As air passes through the thinner coil, it has less time in contact with the coil surface, and therefore less moisture condenses out of the air. Although condensate is not eliminated, it is minimized this way.

The cooling load from the electronics and computer system is typically consistent year-round, so computer room systems must be properly prepared or designed for operation at low ambient conditions. This includes some form of head pressure control during low outdoor temperatures, as well as crankcase heaters for compressors located outdoors. Care must also be taken to prevent refrigerant migration outdoors during the off-cycle, generally by using a liquid line solenoid valve. Another feature often found on these systems includes compressor staging and/or hot gas bypass for capacity control to maintain precise control of the environment.

Servers and other electronic equipment are often found in large racks. As a result, most of the heat generated in a computer room or data center is generated in small areas. Simply cooling the entire room from above is not as effective as delivering the air to these heat-generating locations. In addition, computer rooms and data centers are constantly growing or being rearranged. As a result, racks grow in size, using a modular approach, and are often relocated. The air distribution system must be flexible enough to accommodate the constant changes.

### 2.2.1 Rack and High-Density Cooling

Small rack systems, isolated in small rooms, represent an interesting problem of their own. The facility cooling system rarely has the capacity and air distribution system design to provide the needed cooling to these areas. In addition, the facility systems may be operating in a heating mode during the winter months, while the isolated racks continue to generate heat. In most cases, a rack system located in a small room needs a dedicated system to maintain reasonable operating conditions. One of the ways this is done is through the use of a rack cooling system. The rack unit shown in *Figure 23* provides built-in chilled water cooling. Fans move the air across chilled-water heat exchangers inside the rack, transferring the heat to the water and out to a remote chiller. The fan speed modulates depending on the cooling load. Their unique design isolates the heat exchanger from the electronics to avoid the significant damage that could occur in the event of a water leak. Rather than having an independent, dedicated chiller, the unit can be supported by the unit shown in *Figure 24*. This unit is a water-to-water heat exchanger that is tied into the facility chilled water system. It isolates the two chilled water streams from each other, and provides a single location for rack cooling systems to interface with the building chiller. Otherwise, chilled water from the main system must be piped to each rack instead. This can create complex piping and pumping problems in an existing building. This unit has its own pump, so that piping for

## Dehumidifiers

For dedicated dehumidifiers using a DX evaporator coil, the coil is designed to maximize the latent heat exchange, which results in moisture removal. Unlike computer room coils, a dehumidifier coil is much thicker than a standard evaporator coil. This maximizes the time that the passing air spends in contact with the coil surface, resulting in a greater volume of condensate.

## Controlling Multiple Computer Room Units

GOING GREEN

It is not unusual for a number of computer room units to serve a single room or data center, each operating on its own internal controls. As a result, it is also not unusual to discover that one unit is cooling, one is heating, one is adding humidity, and another is dehumidifying. Obviously, energy costs skyrocket as a result. This happens because of differences in control calibrations, and because of inconsistent unit programming. IT people are primarily concerned with the results, and rarely have an understanding of the HVAC control settings. It is usually best to use a centralized control system to coordinate the operation of the units. Such systems can pay for themselves very quickly in energy savings.

*Figure 23* Water-cooled server rack.

*Figure 24* Water-to-water heat exchanger system with integral pump.

rack systems becomes completely independent of the facility system.

Refrigerant-based systems can be used instead of chilled water. Indeed, many IT professionals prefer not to use the words *electronics* and *water* in the same sentence. *Figure 25* shows two cooling module styles used for smaller rack systems in high-density rooms. Liquid refrigerant at a low pressure and above the dewpoint temperature of the room is pumped to these units, eliminating the hazard of water in the electronics environment. Since it is above the dewpoint, condensate does not form on the refrigerant lines or on the heat exchanger. The liquid absorbs heat and

changes state to a vapor, using this latent process to cool the air. No expansion device is needed at the cooling module to reduce the pressure and temperature of the refrigerant. Since the refrigerant is above the dewpoint, it means that there is a limited temperature difference (TD) between the air and refrigerant. This approach works best by exposing the air to the refrigerant using a highly efficient micro-channel evaporator coil.

Other styles in addition to the two shown in *Figure 25* are also available. Special piping systems (*Figure 26*) that offer the flexibility needed for systems that are growing and changing are installed

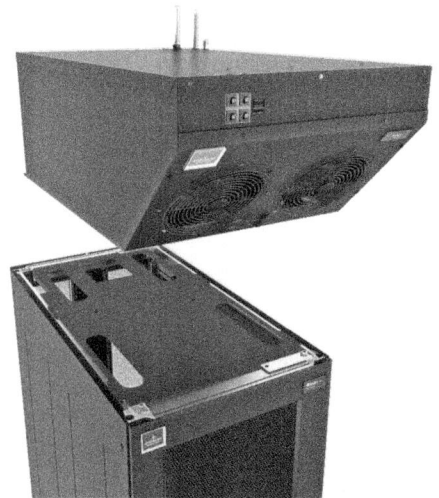

RACK-TOP MODULE

OVERHEAD MODULE

*Figure 25* Refrigerant-based cooling modules for rack systems.

Figure 26 Modular refrigerant piping system.

to allow for the easy attachment of future cooling modules. The connection from the piping to the module is flexible. The low-pressure liquid refrigerant is not provided by a traditional condensing unit. Instead, a chilled water-to-refrigerant heat exchanger is used, using the chilled water system of the facility to condense the vaporized refrigerant for circulation back to the cooling modules.

### 2.2.2 Open Data Center Systems

Large, more expansive data centers (*Figure 27*) usually require a different strategy than systems designed for individual racks. Although one would think an air distribution system for this type of facility would need to be complex, it is actually fairly simple. The answer often lies in a raised floor.

Not only does a raised floor provide a means of delivering air throughout the room, it also allows the air delivery to be very flexible. Looking again at *Figure 27*, note the tiled appearance of the floor. The floor is raised on legs, and constructed of 24" × 24" heavy-duty tiles supported by the legs (*Figure 28*). The floor provides an area for piping, wiring, and air distribution. The height of the floor varies depending upon the needs and use of the room.

Looking once again at *Figure 27*, note the perforated floor panels in the foreground. These perforated tiles are the supply air diffusers for the system. They can be placed in the desired locations so that cooled air is provided to one side of the racks.

Figure 27 Data center on raised floor.

The racks or cabinets are arranged in what is known as a hot aisle/cold aisle configuration (*Figure 29*). Most racks, especially those that are enclosed, have small fans that draw air in through the front and discharge it through the back. Some electronic components also have their own fans built into the housing. Conditioned air is available in the cold aisle and heated air from the cabinets is dumped into the hot aisles. The perforated floor tiles, acting as diffusers, are placed along the cold aisle to provide the conditioned air. In *Figure 29*, a ceiling has been installed and is being used as a return air plenum; the computer room environmental unit (CRU) is placed outside of the room. In many rooms, the unit is simply placed inside the room with the racks. The top portion of the CRU remains open and free of duct, drawing its return air from the room itself (*Figure 30*).

### 2.2.3 Computer Room Equipment

Computer room units such as the one shown in *Figure 31* are true environmental units. They are not simple indoor air conditioning units. Sophisticated CRUs are typically equipped with the following:

- Precision temperature and humidity controls
- User control panel with LCD/LED display, or a graphical user interface (GUI), built into each unit
- Belt-driven blowers for flexibility in air volume adjustments. Newer models may be equipped with one or more ECM-driven blowers that modulate the airflow based on the load and operating mode.
- Humidifiers. To achieve a significant amount of humidification capacity, steam generating canisters or infrared humidifiers are used.
- Multiple stages of cooling, both through the use of multiple compressors and compressor capacity control. For very precise control, hot gas bypass systems may also be used.
- Expansion valve metering devices. Newer models may be equipped with electronic expansion valves for even greater precision and energy efficiency.
- Electric heat. Although it is rare for a computer room to need heat, electric reheat systems are needed to maintain the room temperature setting when the unit operates in the dehumidification mode.
- Substantial condensate pumps.

The unit in *Figure 31* is equipped with three ECM-powered centrifugal blowers, seen suspended under the unit. The unit is also shown on its stand, prepared for installation in a raised floor application. Return air is either ducted into the top, or returns freely into the top from the room. A filter rack, typically holding either 2" or 4" thick pleated filters, is provided just inside the return air opening.

Through the user control panel, the operating parameters are easily programmed. Temperature

03409-13_F28.EPS

*Figure 28* Data center raised floor concept.

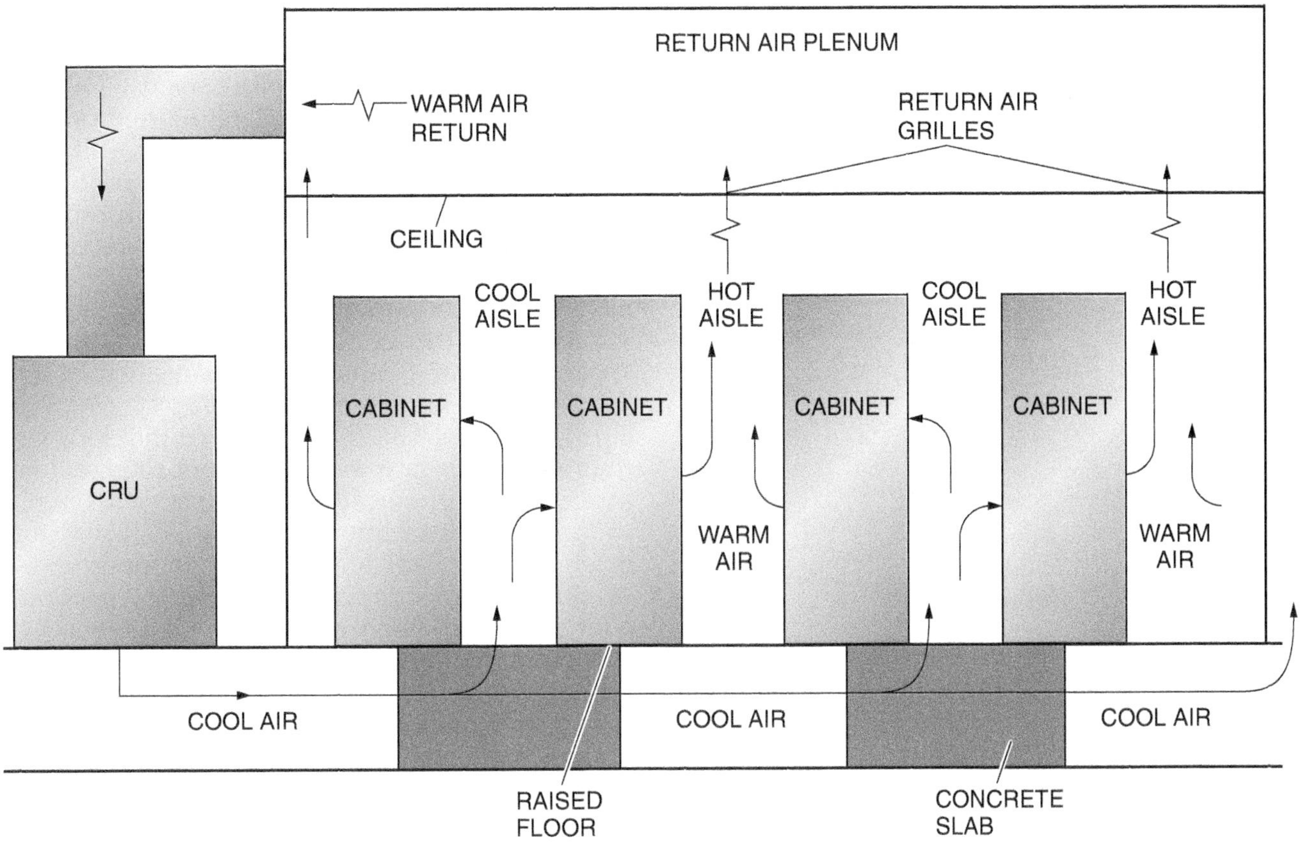

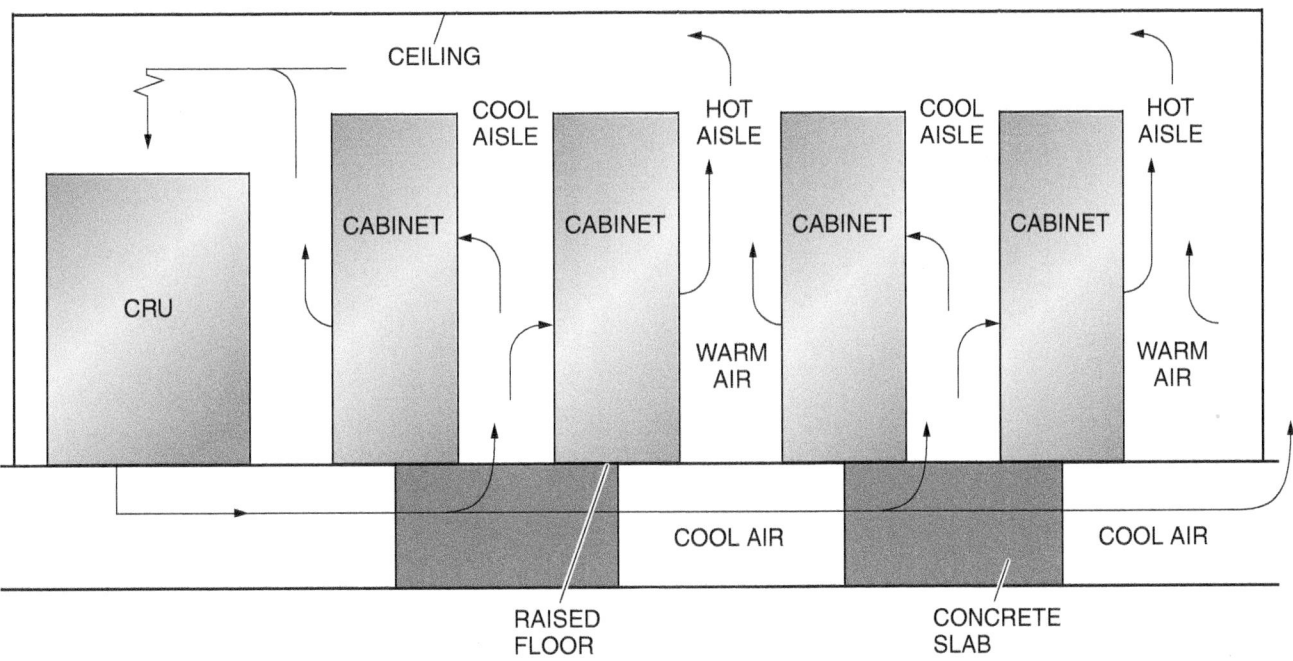

RETURN AIR PLENUM

WARM AIR RETURN

RETURN AIR GRILLES

CEILING

COOL AISLE

HOT AISLE

COOL AISLE

HOT AISLE

CRU

CABINET

CABINET

CABINET

CABINET

WARM AIR

WARM AIR

COOL AIR

COOL AIR

COOL AIR

RAISED FLOOR

CONCRETE SLAB

03409-13_F29.EPS

*Figure 29* Hot-aisle/cold-aisle cabinet configuration.

and humidity setpoints are programmed, along with the range and/or sensitivity level of the setpoints. A number of alarm setpoints can also be programmed. Of course, CRUs can also be net-

worked together and controlled as a group from a local or remote location.

Humidifiers for larger units are usually of the infrared type (*Figure 32*) or a steam-generating

CEILING

COOL AISLE

HOT AISLE

COOL AISLE

HOT AISLE

CRU

CABINET

CABINET

CABINET

CABINET

WARM AIR

WARM AIR

COOL AIR

COOL AIR

RAISED FLOOR

CONCRETE SLAB

03409-13_F30.EPS

*Figure 30* CRU placed within the room.

*Figure 31* Computer room environmental unit; shown with direct-drive ECM blowers.

canister. Infrared humidifiers use long lamps positioned directly above a relatively shallow pan of water. The intense heat of the lamps evaporates the water into the passing airstream. Ultrasonic humidification is also an option (*Figure 33*). These units use high-frequency sound waves to excite water molecules and turn the water into an extremely fine mist. They typically use less energy and are very fast, since the water does not need to become hot before moisture is added to the airstream.

INFRARED
LAMPS

WATER PAN

*Figure 32* Infrared humidifier.

With infrared and steam generating humidifiers, all the minerals and other particles in the water supply remain in the pan or canister. When the humidifiers are very active, a great deal of mineral deposits is left behind. As a result, regular maintenance of the pan is required. Steam generating canisters also require periodic replacement. The action of ultrasonic humidifiers causes minerals to be caught up in the mist and carried along into the airstream. Users may complain of white dust settling throughout the conditioned space. However, this issue can be avoided by using a purified water supply.

It is rare that CRUs are installed in a location that allows for gravity drainage of the condensate. As a result, most are equipped with high-capacity condensate pumps (*Figure 34*). Due to the extended operating hours and the humidity levels that are maintained, these units can produce a lot of condensate. A pump is needed that is capable of pumping against a significant head while moving a considerable volume of water.

For obvious reasons, water is a great concern in the computer room and/or data center environment. Condensate pumps can be fitted with switches that are interlocked with the unit controls, shutting the unit down in the event of a problem. However, this may not be sufficient pro-

*Figure 33* Ultrasonic humidifier.

tection. The under-floor area under CRUs is often fitted with additional water protection. The use of water-sensing cables along the floor, capable of sensing even a few drops of liquid, is considered to be a more comprehensive approach. This type of sensor may be placed in a number of locations, including along the perimeter. It then also provides protection against water seeping into the area through the walls or slab. A number of sensors can be networked together and connected to a network for instant alarm activation when a problem is sensed.

## 2.3.0 Enclosure and Spot-Cooling Equipment

Many different heat-producing components are housed in enclosures, and for a variety of reasons. These components include contactors, relays, fuses, circuit breakers, and power disconnect switches. Transformers are also commonly found here, as well as any number of electronics devices. These cabinets, especially those that handle a considerable amount of power, can become extremely

hot. To ensure the reliability and longevity of the components, they must be cooled.

Spot-cooling equipment can be used in permanent applications for a problem area. However, they are more often used as emergency back-up systems to maintain reasonable conditions when a primary unit fails. Both enclosure and spot-cooling systems are introduced here.

### 2.3.1 Enclosure Cooling Equipment

Switchgear and similar enclosures can be cooled by self-contained air conditioning units (*Figure 35*). Electronic cabinets related to cellular phone transmission sites are also candidates for enclosure cooling. These applications may be indoors or outdoors.

These units must often operate in harsh environments that are dusty and hot, and may even be

03409-13_F34.EPS

*Figure 34* High-capacity condensate pump.

03409-13_F35.EPS

*Figure 35* Enclosure cooling units.

## Computer Cabinets

Computer rooms are typically kept at a temperature of 68°F to 72°F, with a relative humidity between 40 and 55 percent. Important as these values are, it is more important to maintain even and stable conditions within the cabinets that contain the individual electronic components. The room temperature often is much lower than the temperature inside the individual cabinets. A variety of specialty equipment manufacturers produce cabinets designed to maintain good air circulation and even temperatures within the cabinet. When racks and cabinets are well ventilated, higher temperatures can usually be maintained in the room, saving energy. Maintaining sensitive electronic equipment at cooler temperatures is proven to dramatically increase the reliability and longevity of the equipment.

needed where explosive atmospheres could exist. To function in such environments, a number of options are available, such as the following:

- Electrical wiring that satisfies the requirements of National Electrical Manufacturers Association (NEMA) Standard 1 (indoor use only); NEMA 12 (indoor, but protected against light splashing of liquids and dust); NEMA 3R (rain-tight, indoor/outdoor use); NEMA 4 (same as 3R, but also protected against dust and heavy wash-downs); and NEMA 4X (same as NEMA 4, but also protected against corrosive agents).
- Epoxy-coated coils for protection against corrosives.
- Stainless steel chassis and/or cabinets.
- Built-in condensate evaporator; collects condensate and re-evaporates it back to the atmosphere outside of the cabinet.
- Head pressure control, usually through condenser fan cycling or speed modulation.
- Lead-lag controllers, which coordinate the operation of more than one unit.

The cooling capacity of enclosure coolers typically ranges from 1,000 Btuh up to as much as 26,000 Btuh. Models designed for hazardous locations must be explosion-proof. Electrical components which do (or could potentially) create an electrical arc must be located inside of special enclosures to keep them isolated from the ambient air. Since this type of unit is rarely needed, few are manufactured and they are quite expensive.

Although most enclosure coolers have compressors and a typical DX refrigerant circuit, they can also be chilled water-based. However, this requires that chilled water piping be provided.

### 2.3.2 Spot Coolers

Spot coolers (*Figure 36*) are portable packaged air conditioners, often on wheels, that can be moved to an area to provide temporary or supplemental localized cooling. Fully portable models must be provided with flexible or rigid round duct routed to the outdoors or above a ceiling to discharge the hot condenser air. Most of these units use air from inside the space as condenser air, which must then be discharged outside of the conditioned space. This is rather inefficient, as the unit is also trying to cool the same space. Some users may expect higher performance at times, but users must remember that any air exhausted from the space as condenser air must be replaced by air outside the space. This can increase the load considerably if the air moving in is unconditioned. Larger spot coolers can be left outdoors and ducted into the space, avoiding this issue altogether. However, for temporary use, a spot cooler often saves the day. Such units can be purchased or rented in times of emergency.

The equipment cooling unit shown in *Figure 36* is designed to discharge the condenser air into the same space through the grille on the top panel. However, it can be fitted with a round duct

## Wall-Mounted Packaged Units

Wall-mounted packaged air conditioners and heat pumps are large versions of enclosure coolers. Of course, enclosure units could be considered small versions of these units as well. These units are popular for use with small mobile offices, portable classrooms, and similar structures. They are complete, air-cooled packages that are mounted on an outside wall. Supply and return air openings are cut in the wall. The return air opening is usually covered with a return air filter grille. The supply air opening may also be covered with a simple register, or connected to supply air ductwork above the ceiling.

03409-13_SA03.EPS

take-off to duct the condenser discharge air out of the space as well. The flexible nozzles allow users to direct the cooled air onto a piece of critical equipment. The office model can be fitted with an optional nozzle kit as well.

The ceiling-mounted model is designed for a more permanent installation above a ceiling. It is also a packaged unit, and discharges its condenser air into the ceiling plenum. When the ceiling space is used as a return air plenum for a facility cooling system, the condenser discharge of this unit then becomes an added load for that system. Essentially, this unit transfers its heat load to another system.

Spot coolers may also be water-cooled. Water-cooled spot coolers require hoses to bring in and exhaust water for the condenser.

Condensate is collected in a tank at the base of the unit. The tank is fitted with a float-type or similar switch to shut the unit down once the tank has become full. The tank must then be emptied manually and reinstalled to restore operation. One option is to hook a drain hose to the tank for gravity drainage to an appropriate spot. Another

(A) EQUIPMENT COOLING –
WITH FLEXIBLE NOZZLES

(B) OFFICE MODEL

(C) CEILING MOUNTED

03409-13_F36.EPS

*Figure 36* Spot coolers.

option is to install a condensate pump kit. The kit can pump the condensate up to a reasonable height when a gravity drain is not an option.

## 2.4.0 Valance and Chilled-Beam Cooling

Valance and chilled-beam cooling approaches are both rather unique answers to unusual situations. Valance systems, for example, are installed in applications where a conventional ducted air distribution system is impractical. Chilled-beam systems also offer a unique way to avoid an air distribution system.

### 2.4.1 Valance Systems

A valance cooling system is a chilled-water system that uses finned-tube radiation installed around the perimeter of a room, just below the ceiling. It gets its name because the radiation and piping are concealed beneath a decorative valance.

Each valance cooling unit consists of a section or sections of finned-tube radiation (*Figure 37*).

Each unit is equipped with an insulated chilled-water supply and return line. A zone valve may be installed in the supply line. Beneath the finned-tube section is a pan to catch the condensate and drain it away for disposal. The condensate pan is often a part of the decorative valance. Since the units are positioned around the perimeter of a room, access to the wall to route condensate drain piping is fairly easy. In operation, chilled water flows through the coil on a call-for-cooling from the low-voltage room thermostat. The signal energizes the circulator pump and the zone valve, if so equipped. As the radiation cools, it cools the air surrounding the coil. As it cools, the air becomes denser and begins to fall toward the floor. This sets up natural air currents in the room, with the dense falling air drawing warm air from the ceiling area into contact with the coil. No circulating fans are needed or used in the system. A separate system is normally used to provide ventilation and air filtration where required.

Valance units can also be used for heating. When hot water is circulated through the coil,

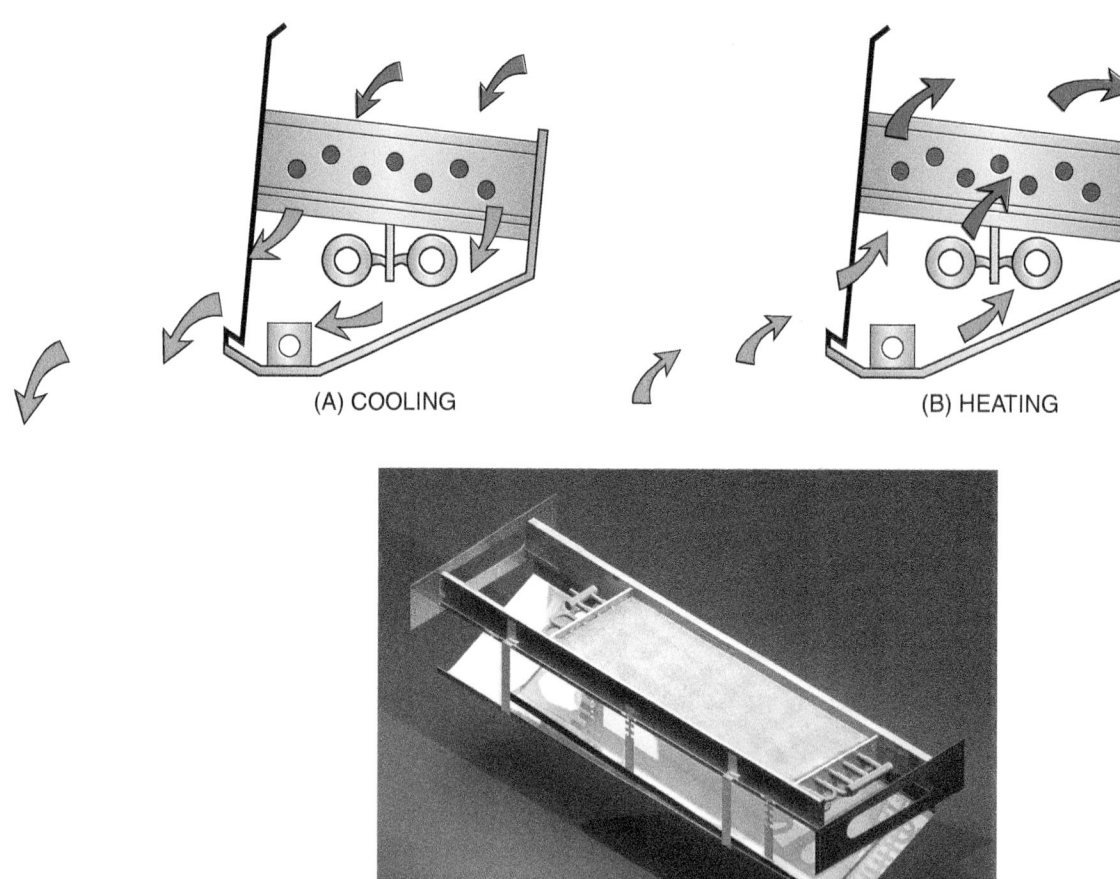

(A) COOLING

(B) HEATING

(C)

03409-13_F37.EPS

*Figure 37* Valance heating/cooling unit.

it has the opposite effect on airflow. Warm air in contact with the coils begins to rise, drawing cooler air into the bottom of the fixture. However, they are somewhat less effective for heating, as the heated air tends to remain close to the ceiling.

Because valance systems do not use a circulating fan, the systems are very quiet, do not create uncomfortable drafts, and dust is kept to a minimum. They are more energy-efficient because the circulating pump consumes much less power than a blower motor. As with other types of hydronic systems, the initial installed cost of a valance cooling system is generally higher than those of conventional forced-air cooling systems.

### 2.4.2 Chilled-Beam Systems

Chilled-beam cooling systems have been used for a number of years in Europe and Australia and are now being introduced in the United States. Chilled-beam systems are similar in some respects to valance cooling systems. The name seems to imply that structural members of the building are cooled. In fact, the units used to cool the room are long, finned-tube radiators in an enclosure. They resemble beams suspended from the ceiling, thus the name *chilled beam*. There are two main types of chilled-beam cooling systems: passive and active.

Passive chilled-beam systems (*Figure 38*) resemble valance cooling units in that they both employ finned-tube elements located near the ceiling of a room. Chilled-beam units can be located near outside walls or can be located in other areas of the ceiling. The units may be exposed or concealed behind metal grilles in a suspended ceiling. Pas-

sive systems rely on natural air currents to cool the room. Air in contact with the coil cools and becomes denser, falling gently to the floor. Warm air is then naturally drawn in through the top to replace it.

In passive systems, the chilled water supplied to the finned elements has to be slightly warmer than the dew point of the room air to prevent condensation from forming and dripping from the cooling units on the ceiling. This is the most significant difference between valance cooling and the chilled-beam approach. Since the chilled-beam can be suspended mid-room, it cannot handle and dispose of condensate. Since the TD between the chilled-water and the air is much smaller, passive chilled-beam systems require far more coil area to provide the same capacity as a valance system.

Active chilled-beam systems (*Figure 39*) use ducted, partially conditioned air to help induce additional airflow past the chilled-beam units. This air, known as the primary air, is blown into the chilled-beam fixture at a low pressure and is discharged down through jets or nozzles on the sides. This creates a slight negative pressure in the center of the assembly, drawing in warm air through the center and across the chilled water coils. The warm air from the room is referred to as the secondary air.

The primary air is partially cooled and thoroughly dehumidified before it is sent to the chilled beams. This takes care of any latent load and prevents any condensation from occurring on the chilled-beam coils. It also accounts for some of the sensible load. The function of the chilled beam is to satisfy the remainder of the room's sensible

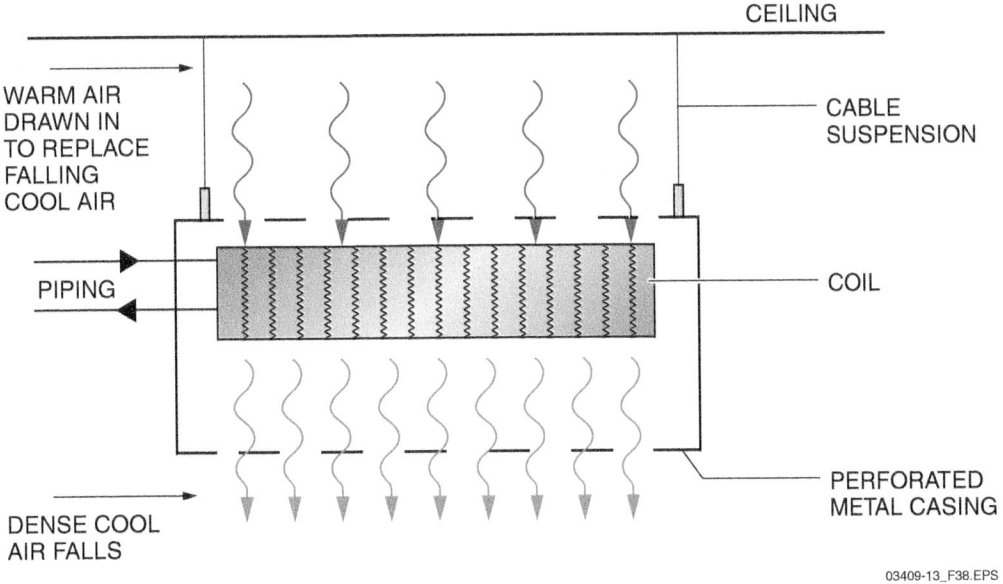

CEILING

WARM AIR DRAWN IN TO REPLACE FALLING COOL AIR

CABLE SUSPENSION

PIPING

COIL

DENSE COOL AIR FALLS

PERFORATED METAL CASING

03409-13_F38.EPS

*Figure 38* Passive chilled-beam cooling system.

load only. Again, the chilled water temperature is maintained above the room's design dew point, preventing condensation on the coil.

As shown in *Figure 39*, the ratio of ventilation air to induced air is usually between 1:2.5 and 1:4. The fin spacing used on the finned tubing has a large effect on the ratio. Tightly spaced fins, which offer greater heat transfer, are also more resistant to air movement.

Active chilled-beam fixtures can be suspended from the ceiling, or tucked into the ceiling and flush-mounted. Refer to *Figure 40* for various chilled-beam unit styles. Note that heating can be provided through them as well. Their cooling operation is the most challenging from an engineering point of view, so cooling dominates this discussion.

Chilled-beam systems are noted for their energy efficiency. They also offer quiet operation, excellent indoor air quality, and low maintenance due to their small number of moving parts. Active chilled-beam systems are appealing for these reasons, and the fact that ventilation requirements

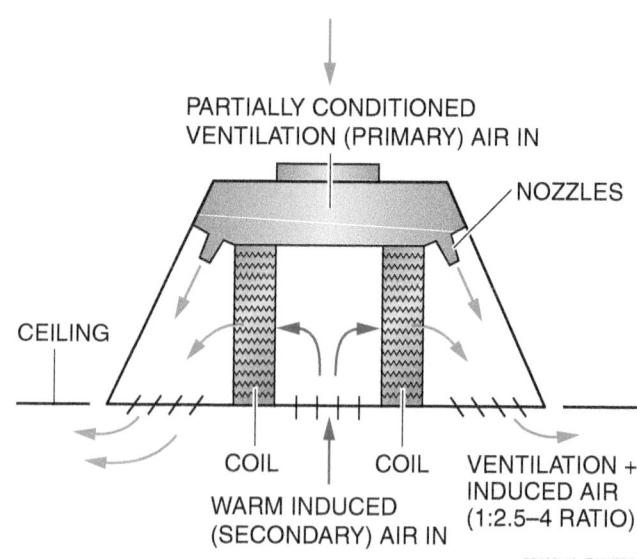

03409-13_F39.EPS

*Figure 39* Active chilled-beam cooling system.

can be addressed and provided within the chilled-beam system design. Disadvantages include high installation costs compared to more conventional systems. However, the efficiency and other fac-

(A) FLUSH-MOUNTED

(B) SUSPENDED

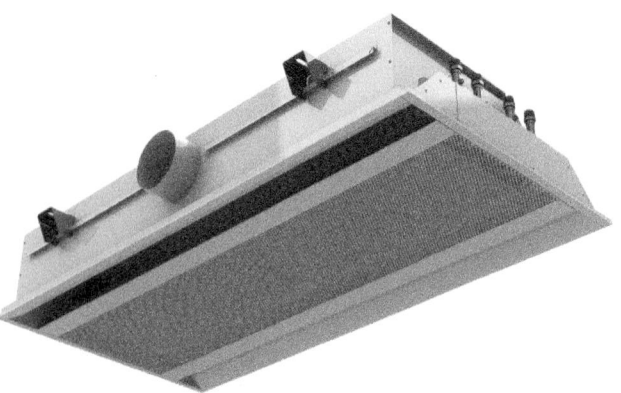

(C) CASSETTE FOR DROP CEILING

(D) CIRCULAR

03409-13_F40.EPS

*Figure 40* Active chilled-beam unit styles.

tors make them attractive to building owners and developers seeking LEED certification. Because this technology is new to the United States, many HVAC contractors may be unfamiliar with it.

### 2.5.0 Air Turnover Systems

Air stratification is a characteristic of all rooms with high ceilings. Warm, lighter air rises and is retained near the ceiling. Heavier cold air settles near the floor. This creates different temperature zones between the floor and the ceiling, which can create discomfort. By preventing the air from stratifying, a more even temperature can be maintained, resulting in lower energy costs.

Although this section focuses on systems used to prevent stratification, it should also be noted that air stratification is sometimes employed as a load-reduction strategy. For example, a light manufacturing operation may be located in a facility with a roof height of 28'. If there is no use of the space above the 12' height, the air distribution system—both supply and return ducts—can be installed from the roof down to the 12' height. This effectively leaves the hot air, which rises toward the roof, unconditioned and undisturbed. This reduces the load on the cooling system, often reduces the capacity of the equipment to be installed, and reduces operating costs. However, this strategy applies only to ducted systems. Unfortunately, routing the ductwork in this manner is usually an expensive approach. Air turnover systems represent a different concept.

One of the simplest ways to prevent air stratification is to use ceiling-mounted paddle fans. This method has been used for over 100 years and is still used today in residential, commercial, and industrial buildings. However, paddle fans may be ineffective in high-bay industrial and commercial settings. For these applications, and when the area is large, an air turnover unit is preferred.

Air turnover units are tall, vertical air handlers that often contain both heating and cooling sections. In some applications, they may provide heating only. The fans in air turnover units gently circulate the air, drawing air in at the floor level and discharging it at the top (*Figure 41*). Discharging the air at this height typically carries it over obstructions such as manufacturing operations or warehouse storage racks. Propeller fans are often used instead of centrifugal blowers, since there is no duct to add resistance. Air turnover units typically move large volumes of air. The air volume is part of the secret to their effective use—moving all of the air through a facility, leaving no stagnant areas.

As the air passes through the air handler, it is heated or cooled. Cooling can be achieved through DX coils or chilled-water coils. Heating is usually accomplished with gas burners, hot-water coils, or steam coils. Gas-fired systems are not direct-fired, since there is rarely any outside air

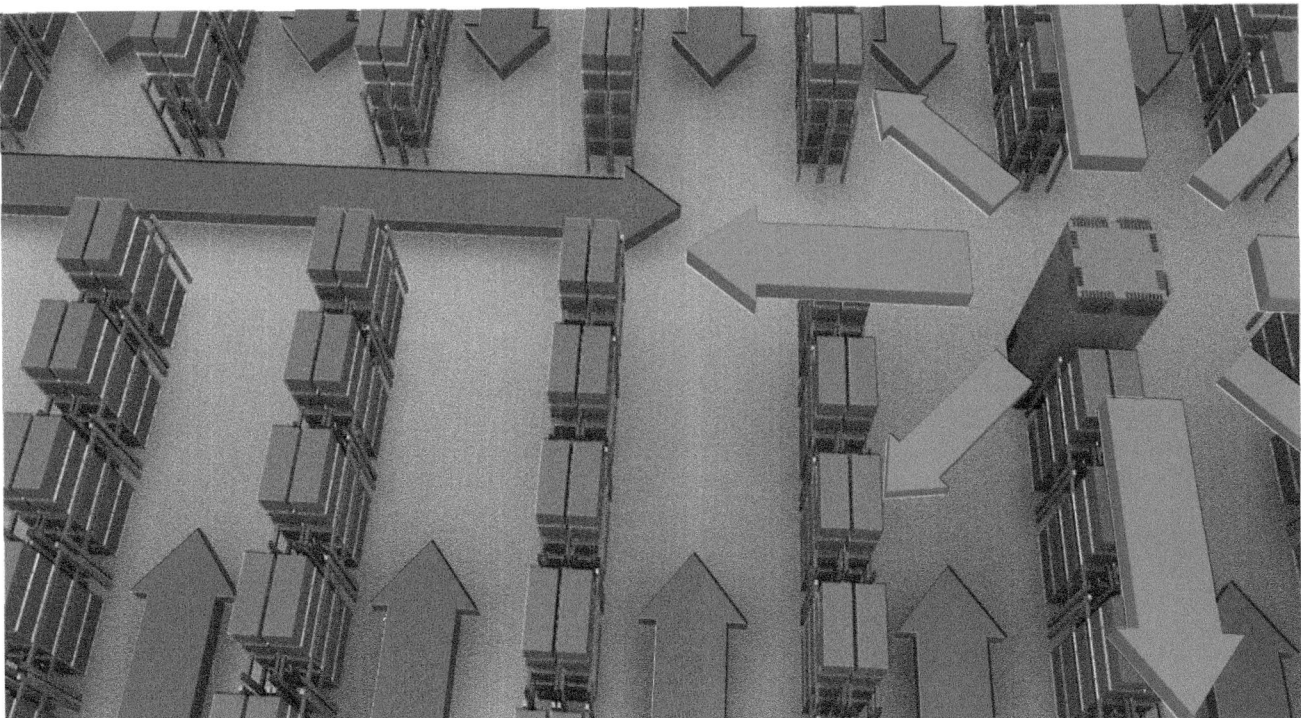

03409-13_F41.EPS

*Figure 41* Air turnover concept.

moved through the equipment itself. They almost always operate on 100 percent indoor air. In most applications, there is already plenty of outdoor air introduced through infiltration due to large rolling doors and similar openings.

Applications vary for the use of an air turnover unit. Even a large, centralized mechanical room (*Figure 42*) can be a good application. Warehouses, manufacturing areas, aircraft hangars, and gymnasiums are other potential sites. A single large air turnover unit can replace a number of traditional rooftop units, with better performance as well. In addition, eliminating the physical load of many rooftop units from the roof structure can save substantially on construction costs. In spite of their apparent simplicity, they can provide reasonably precise control over the conditions across a large area.

03409-13_F42.EPS

*Figure 42* Air turnover unit in a mechanical room.

## Taking Good Care of the Wine

The proper storage of wine as it ages is a science in itself. A certain amount of wine stored in barrels slowly evaporates over time; as much as 12 percent of the contents is often lost. To limit the loss, wine in barrels needs to be stored in a very humid but cool environment. An area at 55°F and 80 percent RH can be considered desirable. However, cooling processes generally result in a great deal of dehumidification.

03409-13_SA04.EPS

A famous California winery had literally hundreds of storage units used for aging. Rooftop-mounted cooling units were being used to maintain the proper conditions, but with limited success. Over $500,000.00 per year was being lost in evaporation losses per storage unit. Johnson Air Rotation® proposed that air turnover units be used and guaranteed the results. The result: Wine evaporation was reduced by 75 percent, saving over $375,000.00 per year in evaporation losses at each storage unit. Since there is no duct involved, the air is also being moved by half the previous amount of horsepower.

## Additional Resources

*ASHRAE Handbook – HVAC Applications,* Latest Edition. Atlanta, GA: American Society of Heating, Refrigerating, and Air Conditioning Engineers, Inc.

*ASHRAE Handbook – HVAC Systems and Equipment,* Latest Edition. Atlanta, GA: American Society of Heating, Refrigerating, and Air Conditioning Engineers, Inc.

*Design Considerations for Active Chilled Beams.* ASHRAE Journal, September, 2008. Darren Alexander and Mike O'Rourke. Atlanta, GA: American Society of Heating, Refrigerating, and Air Conditioning Engineers, Inc.

## 2.0.0 Section Review

1. Which of the following statements about direct-fired make-up air units is true?

    a. Most direct-fired units use electric resistance heat.
    b. The byproducts of combustion remain in the airstream, rather than being exhausted through a flue pipe or chimney.
    c. Fuel oil is the most popular fuel for direct-fired units.
    d. The heat exchanger of a direct-fired unit is very complex and offers a high resistance to airflow.

2. Some rack-cooling systems in small computer rooms use pumped liquid refrigerant that is at a low pressure and _____.

    a. at a temperature above the dew point
    b. at a temperature below the dew point
    c. is superheated
    d. never changes its state

3. Equipment built to a NEMA 1 standard can be installed and operated in any location.

    a. True
    b. False

4. A valance cooling system _____.

    a. is often installed in the center of the room
    b. has no means of handling condensate collection or disposal
    c. is located above a drop ceiling
    d. is suspended around the perimeter of a room

5. One common characteristic of an air turnover unit is that it needs _____.

    a. both a supply and return duct
    b. only a supply duct
    c. only a return duct
    d. no ductwork at all

# Summary

In addition to conventional heating and cooling systems, there are several alternative systems available that can heat or cool a structure. These alternative systems are used because the application is not suited to a conventional system, or the availability of a low-cost fuel or energy source makes the alternative system attractive. Solid fuels, such as wood, wood pellets, shelled corn, and coal, if they are readily available, are often cheaper than heating fuels such as natural gas, propane, and fuel oil. To be burned as fuel, these solid fuels require special appliances. Solar energy is also an available resource, but different and often unconventional methods are required to capture and use that type of heat. However, once the system is in place, the fuel is free. Evaporative coolers do require the use of a little power and some water, but they can provide comfort in certain environments.

There are also systems based on conventional fuels that are specifically designed to solve a unique problem. Such systems include direct-fired make-up air units. These are applied in many industrial environments when a large volume of air must be exhausted. Computer rooms and data centers also require special handling to ensure that crucial electronics remain reliable and long-lived. On a smaller scale, enclosure and spot-cooling equipment fill unique needs of their own.

Valance and chilled-beam cooling systems have been popular in other parts of the world for some time, but are now gaining more widespread acceptance in the US. Although they do have some drawbacks and require careful engineering, they are known for their energy efficiency and low maintenance requirements. Air turnover systems provide heating and cooling to large open areas that would otherwise require a number of systems working together to provide a desirable environment.

Alternative and specialized HVAC systems are both important parts of the complete HVAC picture. There is no doubt that both areas will continue to grow as energy costs and supplies continue in their present direction, and more unique applications requiring an innovative approach present themselves.

1. Wood stoves equipped with a catalytic element cannot have smoke emissions that exceed _____.

   a. 0.4 grams per hour
   b. 4.1 grams per hour
   c. 7.5 grams per hour
   d. 10 grams per hour

2. A dual-fuel wood-burning furnace requires a separate blower for the wood-burning portion of the system.

   a. True
   b. False

3. When installing a wood-burning furnace, the clearance to combustibles requirement is generally the same as that of a gas-fired furnace.

   a. True
   b. False

4. Which of the following vent types would be used to vent a combination gas/wood furnace?

   a. Type B
   b. Type PL
   c. PVC
   d. Type HT

5. For proper operation, waste-oil heaters must burn oil of a uniform _____.

   a. color
   b. temperature
   c. viscosity
   d. molecular weight

6. A thermal mass is an integral part of a(n) _____.

   a. passive solar heating system
   b. active solar heating system
   c. evaporative cooler
   d. waste oil heater

7. Evaporative coolers are less effective when _____.

   a. it is hot outdoors
   b. it is humid outdoors
   c. the wet bulb temperature is very low
   d. the water supply is warm

8. In a cascade system, the _____.

   a. low-stage system cools a falling stream of water
   b. high-stage system cools a falling water stream
   c. high-stage evaporator acts as the condenser for the low-stage system
   d. low-stage evaporator acts as the condenser for the high-stage system

9. A direct-fired make-up air system typically uses _____.

   a. 35 percent outside air
   b. 50 percent outside air
   c. 75 percent outside air
   d. 100 percent outside air

10. Direct-fired make-up air units can be ducted or designed to free-blow into the space.

    a. True
    b. False

11. The evaporator coil of most computer room systems is designed to maximize the transfer of latent heat.

    a. True
    b. False

12. When computer room systems are found cycling in and out of their dehumidification mode regularly, it is often caused by _____.

    a. moisture coming from operating electronics
    b. under-floor water leaks
    c. an unnecessarily tight setting of the humidity deadband
    d. a component failure in the refrigeration circuit

13. Dedicated computer room units are typically equipped with _____.

    a. direct-drive blower motors
    b. fixed-restrictor metering devices
    c. bypass humidifiers
    d. electric heat

14. Valance cooling system modules use a small fan to aid in air circulation.

    a. True

    b. False

15. A simple way to prevent air stratification in a building with a high ceiling is to use _____.

    a. window fans

    b. valance coolers

    c. paddle fans

    d. evaporative coolers

**Active solar heating system**: A type of solar heating system that uses fluids pumped through collectors to gather solar heat. It is the most complex solar heating system and provides the best temperature control.

**Air stratification**: The layering of air in a room based on temperature. The warmest air is concentrated at the ceiling and the coldest air is concentrated at the floor, with varying temperature layers in between.

**Catalytic element**: A device used in wood-burning stoves that helps reduce smoke emissions.

**Chilled-beam cooling system**: A cooling system that employs radiators (chilled beams) mounted near the ceiling through which chilled water flows. Passive systems rely on convection currents for cooling. Active systems use ducted conditioned air to help induce additional airflow over the beams.

**Creosote**: A black, sticky combustible byproduct created when wood is burned in a stove. It must be periodically cleaned from within the stove and chimney before it can build up to the point where it can cause a fire.

**Direct-fired make-up air unit**: An air handler that heats and replaces indoor air that is lost from a building through exhaust vents.

**Evaporative cooler**: A comfort cooling device that cools air by evaporating water. It is commonly used in hot, dry climates. Cooling effectiveness drops as the relative humidity of the outdoor air increases.

**Hot aisle/cold aisle configuration**: A method used to install equipment cabinets in computer rooms to manage the flow of warm and cool air out of and into the cabinets.

**Indirect solar hydronic heating system**: A type of active solar heating system in which a double-walled heat exchanger is used to prevent toxic antifreeze in the solar collectors from contaminating the potable water system.

**Micro-channel evaporator coil**: Coil construction technology that consists of a group of very small parallel tubes bonded to enhanced heat transfer fins, replacing the single large tube common in standard evaporator coils.

**Passive solar heating system**: A type of solar heating system characterized by a lack of moving parts and controls. The sun shining through a window on a winter day is an example of passive solar heat.

**Thermosiphon system**: A type of passive solar heating system in which the difference in fluid temperature in different parts of the system causes the fluids to flow through the system.

**Type HT vent**: A metal vent capable of withstanding temperatures up to 1,000°F. It is commonly used to vent wood-burning stoves and furnaces.

**Type PL vent**: A type of metal vent specifically designed for stoves that burn wood pellets or corn.

**Valance cooling system**: A type of cooling system in which chilled water is circulated through finned-tube radiators located near the ceiling around the perimeter of a room. Convection currents move the cooled air instead of a blower assembly. A decorative valance conceals the system.

# Additional Resources

This module presents thorough resources for task training. The following resource material is suggested for further study.

*ASHRAE Handbook – HVAC Applications,* Latest Edition. Atlanta, GA: American Society of Heating, Refrigerating, and Air Conditioning Engineers, Inc.

*ASHRAE Handbook – HVAC Systems and Equipment,* Latest Edition. Atlanta, GA: American Society of Heating, Refrigerating, and Air Conditioning Engineers, Inc.

*Design Considerations for Active Chilled Beams.* ASHRAE Journal, September, 2008. Darren Alexander and Mike O'Rourke. Atlanta, GA: American Society of Heating, Refrigerating, and Air Conditioning Engineers, Inc.

*Wood Pellet Heating Systems: The Earthscan Expert Handbook of Planning, Design, and Installation.* Dilwyn Jenkins. 2010. Taylor & Francis.

*Solar Water Heating–Revised & Expanded Edition: A Comprehensive Guide to Solar Water and Space Heating Systems,* Second Edition. Bob Ramlow and Benjamin Nusz. 2010. New Society Publishers.

# Figure Credits

# Section Review Answer Key

| Answer | Section Reference | Objective |
|---|---|---|
| **Section One** | | |
| 1.b | 1.1.1 | 1a |
| 2.c | 1.2.0 | 1b |
| 3.a | 1.3.1 | 1c |
| 4.b | 1.4.0 | 1d |
| **Section Two** | | |
| 1.b | 2.1.0 | 2a |
| 2.a | 2.2.1 | 2b |
| 3.b | 2.3.1 | 2c |
| 4.d | 2.4.1 | 2d |
| 5.d | 2.5.0 | 2e |

# NCCER CURRICULA — USER UPDATE

NCCER makes every effort to keep its textbooks up-to-date and free of technical errors. We appreciate your help in this process. If you find an error, a typographical mistake, or an inaccuracy in NCCER's curricula, please fill out this form (or a photocopy), or complete the online form at **www.nccer.org/olf**. Be sure to include the exact module ID number, page number, a detailed description, and your recommended correction. Your input will be brought to the attention of the Authoring Team. Thank you for your assistance.

*Instructors* – If you have an idea for improving this textbook, or have found that additional materials were necessary to teach this module effectively, please let us know so that we may present your suggestions to the Authoring Team.

**NCCER Product Development and Revision**

13614 Progress Blvd., Alachua, FL 32615

**Email:**  curriculum@nccer.org
**Online:**  www.nccer.org/olf

❏ Trainee Guide      ❏ Lesson Plans      ❏ Exam      ❏ PowerPoints      Other _____

Craft / Level: _____          Copyright Date: _____

Module ID Number / Title: _____

Section Number(s): _____

Description: _____

_____

_____

_____

Recommended Correction: _____

_____

_____

_____

Your Name: _____

Address: _____

Email: _____          Phone: _____

# Fundamentals of Crew Leadership

## OVERVIEW

When a crew is assembled to complete a job, one person is appointed the leader. This person is usually an experienced craft professional who has demonstrated leadership qualities. While having natural leadership qualities helps in becoming an effective leader, it is more true that "leaders are made, not born." Whether you are a crew leader or want to become one, this module will help you learn more about the requirements and skills needed to succeed.

## Module 46101

**NCCER**

*President:* Don Whyte
*Vice President:* Steve Greene
*Chief Operations Officer:* Katrina Kersch
*Fundamentals of Crew Leadership Project Manager:* Mark Thomas
*Senior Development Manager:* Mark Thomas

*Senior Production Manager:* Tim Davis
*Quality Assurance Coordinator:* Karyn Payne
*Desktop Publishing Coordinator:* James McKay
*Permissions Specialist:* Adrienne Payne
*Production Specialist:* Adrienne Payne
*Editor:* Graham Hack

**Writing and development services provided by Topaz Publications, Liverpool, NY**

*Lead Writer/Project Manager:* Thomas Burke
*Desktop Publisher:* Joanne Hart
*Art Director:* Alison Richmond

*Permissions Editor:* Andrea LaBarge
*Writer:* Thomas Burke

**Pearson**

*Director of Alliance/Partnership Management:* Andrew Taylor
*Editorial Assistant:* Collin Lamothe
*Program Manager:* Alexandrina B. Wolf
*Assistant Content Producer:* Alma Dabral
*Digital Content Producer:* Jose Carchi
*Director of Marketing:* Leigh Ann Simms

*Senior Marketing Manager:* Brian Hoehl
*Composition:* NCCER
*Printer/Binder:* RR Donnelley
*Cover Printer:* RR Donnelley
*Text Fonts:* Palatino, Minion Pro, and Univers

Trainees with successful module completions may be eligible for credentialing through the NCCER Registry. To learn more, go to **www.nccer.org** or contact us at 1.888.622.3720. Our website has information on the latest product releases and training, as well as online versions of our *Cornerstone* magazine and Pearson's product catalog.

Your feedback is welcome. You may email your comments to **curriculum@nccer.org**, send general comments and inquiries to **info@nccer.org**, or fill in the User Update form at the back of this module.

This information is general in nature and intended for training purposes only. Actual performance of activities described in this manual requires compliance with all applicable operating, service, maintenance, and safety procedures under the direction of qualified personnel. References in this manual to patented or proprietary devices don't constitute a recommendation of their use.

## Objectives

When you have completed this module, you will be able to do the following:

1. Describe current issues and organizational structures in industry today.
   a. Describe the leadership issues facing the construction industry.
   b. Explain how gender and cultural issues affect the construction industry.
   c. Explain the organization of construction businesses and the need for policies and procedures.
2. Explain how to incorporate leadership skills into work habits, including communications, motivation, team-building, problem-solving, and decision-making skills.
   a. Describe the role of a leader on a construction crew.
   b. Explain the importance of written and oral communication skills.
   c. Describe methods for motivating team members.
   d. Explain the importance of teamwork to a construction project.
   e. Identify effective problem-solving and decision-making methods.
3. Identify a crew leader's typical safety responsibilities with respect to common safety issues, including awareness of safety regulations and the cost of accidents.
   a. Explain how a strong safety program can enhance a company's success.
   b. Explain the purpose of OSHA and describe the role of OSHA in administering worker safety.
   c. Describe the role of employers in establishing and administering safety programs.
   d. Explain how crew leaders are involved in administering safety policies and procedures.
4. Demonstrate a basic understanding of the planning process, scheduling, and cost and resource control.
   a. Describe how construction contracts are structured.
   b. Describe the project planning and scheduling processes.
   c. Explain how to implement cost controls on a construction project.
   d. Explain the crew leader's role in controlling project resources and productivity.

## Performance Tasks

Under the supervision of your instructor, you should be able to do the following:

1. Develop and present a look-ahead schedule.
2. Develop an estimate for a given work activity.

## Trade Terms

| | |
|---|---|
| Autonomy | Look-ahead schedule |
| Bias | Negligence |
| Cloud-based applications | Organizational chart |
| Craft professionals | Paraphrase |
| Crew leader | Pragmatic |
| Critical path | Proactive |
| Demographics | Project manager |
| Ethics | Return on investment (ROI) |
| Infer | Safety data sheets (SDS) |
| Intangible | Sexual harassment |
| Job description | Smartphone |
| Job diary | Superintendent |
| Legend | Synergy |
| Lethargy | Textspeak |
| Letter of instruction (LOI) | Wide area networks (WAN) |
| Local area networks (LAN) | Wi-Fi |
| Lockout/tagout (LOTO) | Work breakdown structure (WBS) |

## Industry Recognized Credentials

If you are training through an NCCER-accredited sponsor, you may be eligible for credentials from NCCER's Registry. The ID number for this module is 46101. Note that this module may have been used in other NCCER curricula and may apply to other level completions. Contact NCCER's Registry at 888.622.3720 or go to **www.nccer.org** for more information.

# Contents

## Figures and Tables

## 1.0.0 BUSINESS STRUCTURES AND ISSUES IN THE INDUSTRY

### Objective

Describe current issues and organizational structures in industry today.

  a. Describe the leadership issues facing the construction industry.
  b. Explain how gender and cultural issues affect the construction industry.
  c. Explain the organization of construction businesses and the need for policies and procedures.

### Trade Terms

**Cloud-based applications:** Mobile and desktop digital programs that can connect to files and data located in distributed storage locations on the Internet ("the Cloud"). Such applications make it possible for authorized users to create, edit, and distribute content from any location where Internet access is possible.

**Craft professionals:** Workers who are properly trained and work in a particular construction trade or craft.

**Crew leader:** The immediate supervisor of a crew or team of craft professionals and other assigned persons.

**Demographics:** Social characteristics and other factors, such as language, economics, education, culture, and age, that define a statistical group of individuals. An individual can be a member of more than one demographic.

**Job description:** A description of the scope and responsibilities of a worker's job so that the individual and others understand what the job entails.

**Letter of instruction (LOI):** A written communication from a supervisor to a subordinate that informs the latter of some inadequacy in the individual's performance, and provides a list of actions that the individual must satisfactorily complete to remediate the problem. Usually a key step in a series of disciplinary actions.

**Local area networks (LAN):** Communication networks that link computers, printers, and servers within a small, defined location, such as a building or office, via hard-wired or wireless connections.

**Organizational chart:** A diagram that shows how the various management and operational responsibilities relate to each other within an organization. Named positions appear in ranked levels, top to bottom, from those functional units with the most and broadest authority to those with the least, with lines connecting positions indicating chains of authority and other relationships.

**Paraphrase:** Rewording a written or verbal statement in one's own words in such a way that the intent of the original statement is retained.

**Pragmatic:** Sensible, practical, and realistic.

**Sexual harassment:** Any unwelcome verbal or nonverbal form of communication or action construed by an individual to be of a sexual or gender-related nature.

**Smartphone:** The most common type of cellular telephone that combines many other digital functions within a single mobile device. Most important of these, besides the wireless telephone, are high-resolution still and video cameras, text and email messaging, and GPS-enabled features.

**Synergy:** Any type of cooperation between organizations, individuals, or other entities where the combined effect is greater than the sum of the individual efforts.

**Wide area networks (WAN):** Dedicated communication networks of computers and related hardware that serve a given geographic area, such as a work site, campus, city, or a larger but distinct area. Connectivity is by wired and wireless means, and may use the Internet as well.

**Wi-Fi:** The technology allowing communications via radio signals over a LAN or WAN equipped with a wireless access point or the Internet. (Wi-Fi stands for wireless fidelity.) Many types of mobile, portable, and desktop devices can communicate via Wi-Fi connections.

Today's managers, supervisors, and lower-level managers face challenges different from those of previous generations. To be a crew leader today, it is essential to be well prepared. Crew leaders must understand how to use various types of new technology. In addition, they must have the knowledge and skills needed to manage, train, and communicate with a culturally-diverse workforce whose attitudes toward work and job expectations may differ from those of earlier generations and cultures.

A summary of the changes in the workforce, in the work environment, and of industry needs include the following:

- A shrinking workforce
- The growth of construction, communication, and scheduling technology
- Changes in the attitudes and values of craft professionals
- The rapidly-changing demographics of gender, gender-identity, and foreign-born workers
- Increased emphasis on workplace safety and health
- Greater need for more education and training

### 1.1.0 Leadership Issues and Training Strategies

Effective craft training programs are necessary if the industry is to meet the forecasted worker demands. Many skilled, knowledgeable craft professionals, crew leaders, and managers have reached retirement age. In 2015, the generation of workers called *"Baby Boomers,"* who were born between 1946 and 1964, represented 50 percent of the workforce. Their departure creates a large demand for craftworkers across the industry. The US Department of Labor (DOL) concludes that the best way for industry to reduce shortages of skilled workers is to create more education and training opportunities. The DOL suggests that companies and community groups form partnerships and create apprenticeship programs. Such programs could provide younger workers, including women and minorities, with the opportunity to develop job skills by giving them hands-on experience.

When training workers, it is important to understand that people learn in different ways. Some people learn by doing, some people learn by watching or reading, and others need step-by-step instructions as they are shown the process. Most people learn best through a combination of styles. While you may have the tendency to teach in the style that you learn best, you must always stay in tune with what kind of learner you are teaching. Have you ever tried to teach somebody and failed, and then another person successfully teaches the same thing in a different way? A person who acts as a mentor or trainer needs to be able to determine what kind of learner they are addressing, and teach according to those needs.

*"The mediocre teacher tells.*
*The good teacher explains.*
*The superior teacher demonstrates.*
*The great teacher inspires."*
*—William Arthur Ward (1921)*

The need for training isn't limited to craft professionals. There must be supervisory training to ensure there are qualified leaders in the industry to supervise the craft professionals.

### 1.1.1 Motivation

As a supervisor or crew leader, it is important to understand what motivates your crew. Money is often considered a good motivator, but it is sometimes only a temporary solution. Once a person has reached a level of financial security, other factors come into play. Studies show that environment and conditions motivate many people. For those people, a great workplace may mean more to them than better pay.

If you give someone a raise, they tend to work harder for a period of time. Then the satisfaction wanes and they may want another raise. A sense of accomplishment is what motivates most people. That is why setting and working toward recognizable goals tends to make employees more productive. A person with a feeling of involvement or a sense of achievement is likely to be better motivated and help to motivate others.

### 1.1.2 Understanding Workers

Many older workers grew up in an environment where they learned to work hard with the same employer until retirement. They expected to stay with a company for a long time, and the structure of the companies created a family-type environment.

Times have changed. Younger workers have grown up in a highly mobile society and expect frequent rewards and rapid advancement. Some might perceive this generation of workers as lazy, narcissistic, and unmotivated, but in reality, they simply have a different perspective on life, work, and priorities. For such workers, it may be better to give them small projects or break up large projects into smaller pieces so that they are rewarded more often by successfully achieving short-term goals. The following strategies are important for keeping young workers motivated and engaged:

- *Goal setting* – Set short-term and long-term goals, including tasks to be done and expected time frames. Things can happen to change, delay, or upset the short-term goals. This is one reason to set long-term goals as well. Don't set workers up for failure, as this leads to frustration, and frustration can lead to reduced productivity.
- *Feedback* – Timely feedback is important. For example, telling someone they did a good job last year, or criticizing them for a job they did a month ago, is meaningless. Simple recognition isn't always enough. Some type of reward should accompany positive feedback, even if it is simply recognizing the employee in a public way. Constructive criticism or reprimands should always be given in private. You can also provide some positive action, such as one-on-one training, to correct a problem.

### 1.1.3 Craft Training

Craft training is often informal, taking place on the job site, outside of a traditional training classroom. According to the American Society for Training and Development (ASTD), a qualified co-worker or a supervisor conducts craft training generally through on-the-job instruction. The Society of Human Resources Management (SHRM) offers the following tips to supervisors in charge of training their employees:

- *Help crew members establish career goals* – Once career goals are established, you can readily identify the training required to meet the goals.
- *Determine what kind of training to give* – Training can be on the job under the supervision of a co-worker. It can be one-on-one with the supervisor. It can involve cross training to teach a new trade or skill, or it can involve delegating new or additional responsibilities.
- *Determine the trainee's preferred method of learning* – Some people learn best by watching, others from verbal instructions, and others by doing. These are categorized as visual learners, auditory learners, and tactile learners, respectively. When training more than one person at a time, try to use a mix of all three methods.

Communication is a critical component of training employees. SHRM advises that supervisors do the following when training their employees:

- Explain the task, why it is important, and how to do it. Confirm that the worker trainees understand these three areas by asking questions. Allow them to ask questions as well.

- Demonstrate the task. Break the task down into manageable parts and cover one part at a time.
- Ask your trainees to do the task while you observe them. Try not to interrupt them while they are doing the task unless they are doing something that is unsafe or potentially harmful.
- Give the trainees feedback. Be specific about what they did and mention any areas where they need to improve.

### 1.1.4 Supervisory Training

Because of the need for skilled craft professionals and qualified supervisory personnel, some companies offer training to their employees through in-house classes, or by subsidizing outside training programs. However, for a variety of reasons, many contractors don't offer training at all. Some common reasons include the following:

- Lack of money to train
- Lack of time to train
- Lack of knowledge about the benefits of training programs
- High rate of employee turnover
- Workforce is too small
- Past training involvement was ineffective
- The company hires only trained workers
- Lack of interest from workers

For craft professionals to move up into supervisory and managerial positions, they must continue their education and training. Those who are willing to acquire and develop new skills have the best chance of finding stable employment. This makes it critical for them to take advantage of training opportunities, and for companies to incorporate training into their business culture.

If your company has recognized the need for training, your participation in a leadership training program such as this will begin to fill the gap between craft and supervisory training.

### 1.1.5 Impact of Technology

Many industries, including the construction industry, have embraced technology to remain competitive. Benefits of technology include increased productivity and speed, improved quality of documents, greater access to common data, and better financial controls and communication. As technology becomes a greater part of supervision, crew leaders must be able to use it properly.

Cellphones (in particular the smartphone) have made it easy to keep in touch through numerous forms of media. As of 2016, 95 percent of all Americans own a cellphone of some kind

# Learning Strategies

There is a lot more to training than simply telling a group of people what you know about a subject or talking your way through a series of slides. Studies have shown that the information and understanding that learners retain depends on how they receive the information and use it. These learning processes are as follows, listed from least effective to most effective:

1. Reading about a process
2. Hearing a description of a process
3. Observing a process
4. Observing and hearing a description of a process
5. Observing, hearing, and responding to instruction
6. Observing, hearing, and doing a process

Although some companies may attempt to assign percentage effectiveness to these instructional approaches, the results are very arbitrary and inconclusive. It is more important to understand which are the most effective methods for teaching and training.

and 77 percent own a smartphone, according to the Pew Research Center. They are particularly useful communication tools for contractors or crew leaders who are on a job site, away from their offices, or constantly on the go. Workers use smartphones at any time for phone calls, emails, text messages, and voicemail, as well as to share photos and videos. The hundreds of thousands of mobile computing apps available allow smartphones to perform numerous other functions on the go. However, the number of accidents due to inattention while focusing on a smartphone or other mobile device is also rapidly rising. Always check the company's policy regarding cell phone use on the job.

As a crew leader, you should be aware that smartphones and tablets (*Figure 1*) allow supervisors to plan their calendars, schedule meetings, manage projects, and access their company email from remote locations. These devices are far more powerful than the computers that took us to the moon, and they can hold years of information from various projects. Cloud-based applications now permit remote access to and updating of files, plans, and data from any place with wireless connectivity. In fact, it is becoming common for work sites to set up local area networks (LAN), wide area networks (WAN), and dedicated Wi-Fi services to support mobile communications among workers on the job.

In all forms of electronic communication (verbal, written, and visual), it is important to keep messages brief, factual, and legal. Text-based communications can be easily misunderstood because there are no visual or auditory cues to indicate the sender's intent. In other words, it is more difficult to tell if someone is just joking via email because you can't see the sender's expression or hear their tone of voice.

## 1.2.0 Gender and Cultural Issues

In the past several years, the construction industry in the United States has experienced a shift in worker expectations and diversity. These two issues are converging at a rapid pace.

The generation of learners is also a factor in the learning process and in the workplace. The various generations include Baby Boomers, Generation X, Millennials (Generation Y), and Generation Z. The ranges of birth years for these generations are as follows (note that there may be

*Figure 1* Digital tablets are a convenient management and scheduling tool for supervisors.

some overlap, and opinions on the generational names as well as the range of years for each generation may vary):

- *Baby Boomers* – born 1946 through 1964
- *Generation X* – born 1965 through 1979
- *Millennials* (*Generation Y*) – born 1980 through 2000
- *Generation Z* – born 2000 through 2015

Each generation has been studied to some degree to determine their different interpretations of and approaches to learning and training. Family norms, religious values and morals, educational methods, music, movies, politics, and global events are all influential factors that define a generation. Remember that individuals within a generation can vary widely in their expectations and behaviors.

This trend, combined with industry diversity initiatives, has created a climate in which companies recognize the need to embrace a diverse workforce that crosses generational, gender, and ethnic boundaries (*Figure 2*). To do this effectively, they are using their own resources, as well as relying on associations with the government and trade organizations.

All current research indicates that industry will be more dependent on the critical skills of a diverse workforce—but a workforce that must be both culturally and ethnically fused. Across the United States, construction and other industries are aggressively seeking to bring new workers into their ranks, including women, racial, and ethnic minorities. Social and political issues are no longer the main factors driving workplace diversity, but by consumers and citizens who need more hospitals, malls, bridges, power plants, refineries, and many other commercial and residential structures. The construction industry needs more workers.

*Figure 2* The modern workforce is diverse.

There are some potential issues relating to a diverse workforce that may be encountered on the job site. These issues include different communication styles of men and women, language barriers associated with cultural differences, sexual harassment, and gender or racial discrimination.

### 1.2.1 Communication Styles of Men and Women

As more and more women enter construction workforce, it becomes increasingly important to break down communication barriers between men and women and to understand differences in behaviors so that men and women can work together more effectively. The Jamestown, Area Labor Management Committee (JALMC) in New York offers the following explanations and tips (*emphasized text* is from *Exchange and Deception: A Feminist Perspective*, ed. by Caroline Gerschlager, Monika Mokre, 2002. Springer Science & Business Media):

- *Women tend to ask more questions than men do* – Men are more likely to proceed with a job and figure it out as they go along, while women are more likely to ask questions first.
- *Men tend to offer solutions before empathy; women tend to do the opposite* – Both men and women should say what they want up front, whether it's the solution to a problem, or simply a sympathetic ear. That way, both genders will feel understood and supported.
- *Women are more likely to ask for help when they need it* – Women are generally more pragmatic when it comes to completing a task. If they need help, they will ask for it. Men are more likely to attempt to complete a task by themselves, even when they need assistance.
- *Men tend to communicate more competitively, and women tend to communicate more cooperatively* – Both parties need to hear one another out without interruption.

This doesn't mean that any one method is better or worse than the other; it simply means that men and women inherently use different approaches to achieve the same result. It's either genetic or cultural, but awareness can overcome these tendencies.

### 1.2.2 Language Barriers

Language barriers are a real workplace challenge for crew leaders. Millions of American workers speak languages other than English. Bilingual job sites are increasingly common. As the makeup of the immigrant population continues to change with the influx of refugees from around the world, the number of non-English speakers in the American workforce is rising and diversifying dramatically.

In addition, there are many different dialects of the English language in America, which can present some communication challenges. For example, some workers may speak a dialect commonly called *Ebonics* (referred to by linguists as *African-American Vernacular English*). The dialect called *Rural White Southern English* can also be difficult to understand by those not born in the American South. Many find the American New England accent and dialect difficult to comprehend. Some sources list as many as 16 distinct American dialects.

Companies have the following options to overcome the language challenge, mainly for foreign or English-as-a-second-language (ESL) workers:

- Offer English classes either at the work site or through school districts and community colleges.
- Offer incentives for workers to learn English.

Communication with ESL workers, and varying dialects in general, becomes even more critical as the workforce grows more diverse. The following tips will help when communicating across language barriers:

- Be patient. Give workers time to process the information in a way they can comprehend.
- Avoid humor. Humor is easily misunderstood. The worker may misinterpret what you say as a joke at the worker's expense.
- Don't assume workers are unintelligent simply because they don't understand what you are saying. Explaining something in multiple ways, and having trainees paraphrase what they heard, is an excellent way to ensure mutual understanding and prevent miscommunication.
- If a worker is not fluent in English, ask the worker to demonstrate his or her understanding through other means.
- Speak slowly and clearly, and avoid the tendency to raise your voice.
- Use face-to-face communication whenever possible. Over-the-phone communication is often more difficult when a language barrier is involved.
- Use pictures or sketches to get your point across.

### 1.2.3 Cultural Differences

As workers from a multitude of backgrounds and cultures work together, there are bound to be differences and conflicts in the workplace. To overcome cultural conflicts, the SHRM suggests the following approach to resolving cultural conflicts between individuals:

- *Define the problem from both points of view* – How does each person involved view the conflict? What does each person think is wrong? This involves moving beyond traditional thought processes to consider alternate ways of thinking.
- *Uncover cultural interpretations* – What assumptions may the parties involved be making based on cultural programming? This is particularly true for certain gestures, symbols, and words that mean different things in different cultures. By doing this, the supervisor may realize what motivated an employee to act in a specific way.
- *Create cultural* synergy – Devise a solution that works for both parties involved. The purpose is to recognize and respect other's cultural values, and work out mutually acceptable alternatives.

### 1.2.4 Sexual Harassment

In today's business world, men and women are working side-by-side in careers of all kinds, creating increased opportunity for sexual harassment to occur. Sexual harassment can be defined as any unwelcome behavior that makes someone feel uncomfortable in the workplace by focusing attention on their gender or gender identity. Activities that might qualify as sexual harassment include, but are not limited to, the following:

- Telling an offensive, sexually-oriented joke
- Displaying a poster of a man or woman in a revealing swimsuit
- Wearing a patch or article of clothing blatantly promoting or degrading a specific gender
- Making verbal or physical advances (*Figure 3*)
- Speaking abusively about a specific gender

Historically, many have thought of sexual harassment as an act perpetrated by men against women, especially those in subordinate positions. However, the nature of sexual harassment cases over the years have shown that the perpetrator can be of any gender.

*Figure 3* Any unwelcome physical contact can be considered harassment.

Sexual harassment can occur in various circumstances, including the following:

- The victim and harasser may be either male or female. The victim does not have to be of the opposite gender.
- The harasser can be the victim's supervisor, an agent of the employer, a supervisor in another area, a co-worker, a subordinate, or a nonemployee.
- Legally, the victim doesn't have to be the person harassed, but could be anyone offended by the relevant conduct.
- Unlawful sexual harassment may occur without economic injury to or the discharge of the victim.
- The harasser's conduct must be unwelcome.

The Equal Employment Opportunity Commission (EEOC) enforces sexual harassment laws within industries. When investigating allegations of sexual harassment, the EEOC looks at the whole record, including the circumstances and the context in which the alleged incidents occurred. A decision on the allegations is made from the facts on a case-by-case basis. Supervision can hold the crew leader responsible who is aware of sexual harassment and does nothing to stop it. The crew leader therefore should not only take action to stop sexual harassment, but should serve as a good example for the rest of the crew.

Prevention is the best tool to eliminate sexual harassment in the workplace. The EEOC encourages employers to take steps to prevent sexual harassment from occurring. Employers should clearly communicate to employees that they and their company won't tolerate sexual harassment. They do so by developing a policy on sexual harassment, establishing an effective complaint or grievance process, and taking immediate and appropriate action when an employee complains.

Controversial or sexually-related remarks, jokes, and swearing are not only offensive to co-workers, but also tarnish a worker's character. Crew leaders need to emphasize that abrasive or crude behavior may affect opportunities for advancement. If disciplinary action becomes necessary, written company policies should spell out the type of actions warranted by such behavior. A typical approach is a three-step process in which the perpetrator is first given a verbal reprimand. In the event of further violations, a written reprimand, often including a letter of instruction (LOI) and a warning are given. Dismissal typically accompanies any subsequent violations.

### 1.2.5 Gender and Minority Discrimination

Employers are giving more attention to fair recruitment, equal pay for equal work, and promotions for women and minorities in the workplace. Consequently, many businesses are analyzing their practices for equity, including the treatment of employees, the organization's hiring and promotional practices, and compensation.

The construction industry, which was once male-dominated, is moving away from this image and actively recruiting and training women, younger workers, people from other cultures, and disabled workers. This means that organizations hire the best person for the job, without regard for race, sex, religion, age, etc.

---

**Did You Know?**

# Respectful Workplace Training

Some companies employ a tool called *sensitivity training* in cases where individuals or groups have trouble adapting to a multi-cultural, multi-gender workforce. Sensitivity training is a psychological technique using group discussion, role playing, and other methods to allow participants to develop an awareness of themselves and how they interact with others. Critics of earlier forms of sensitivity training noted that the methods used were not very different from brainwashing. Modern forms of cultural-awareness and gender-issue training are more ethical, and go by the term *respectful workplace training*.

---

To prevent discrimination cases, employers must have valid job-related criteria for hiring, compensation, and promotion. They must apply these measures consistently for every applicant interview, employee performance appraisal, and hiring or promotion decision. Therefore, employers must train all workers responsible for recruitment, selection, supervision of employees, and evaluating job performance on how to use the job-related criteria legally and effectively.

## 1.3.0 Business Organization

An organization is the relationship among the people within the company or project. The crew leader needs to be aware of two types of organizations: formal organizations and informal organizations.

A formal organization exists when the people within a work group led by someone direct their activities toward achieving a common goal. An example of a formal organization is a work crew consisting of four carpenters and two laborers led by a crew leader, all working together to accomplish a project.

An organizational chart typically illustrates a formal organization. It outlines all the positions that make up an organization and shows how those positions are related. Some organizational charts even identify the person within each position and the person to whom that person reports, as well as the people that the person supervises. *Figure 4* and *Figure 5* show examples of what an organization chart might look like for a construction company and an industrial company. Note that each organizational position represents an opportunity for advancement in the construction industry that a crew leader can eventually achieve.

An informal organization allows for communication among its members so they can perform as a group. It also establishes patterns of behavior that help them to work as a group, such as agreeing to use a specific training or certification program.

An example of an informal organization is a trade association such as Associated Builders and Contractors (ABC), Associated General Contractors (AGC), and the National Association of Women in Construction (NAWIC). Those, along with the thousands of other trade associations in the United States, provide forums in which members with common concerns can share information, work on issues, and develop standards for their industry.

Both formal and informal organizations establish the foundation for how communication flows. The formal structure is the means used to delegate authority and responsibility and to exchange information. The informal structure provides means for the exchange of information.

Members in an organization perform best when they can do the following:

- Know the job and how it will be done
- Communicate effectively with others in the group
- Understand their role in the organization
- Recognize who has the authority and responsibility

*"Leadership is not about titles, positions, or flow charts. It is about one life influencing another."*
*—John C. Maxwell*

### 1.3.1 Division of Responsibility

The conduct of a business involves certain functions. In a small organization, one or two people may share responsibilities. However, in a larger organization with many different and complex activities, distinct activity groups may lie under the responsibility of separate department managers. In either case, the following major department functions exist in most companies:

- *Executive* – This office represents top management. It is responsible for the success of the company through short-range and long-range planning.
- *Human Resources (HR)* – This office is responsible for recruiting and screening prospective employees, managing employee benefits programs, advising management on pay and benefits, and developing and enforcing procedures related to hiring practices.
- *Accounting* – This office is responsible for all recordkeeping and financial transactions, including payroll, taxes, insurance, and audits.
- *Contract Administration* – This office prepares and executes contractual documents with owners, subcontractors, and suppliers.
- *Purchasing* – This office obtains material prices and then issues purchase orders. The purchasing office also obtains rental and leasing rates on equipment and tools.
- *Estimating* – This office is responsible for recording the quantity of material on the jobs, the takeoff, pricing labor and material, analyzing subcontractor bids, and bidding on projects.
- *Operations* – This office plans, controls, and supervises all project-related activities.

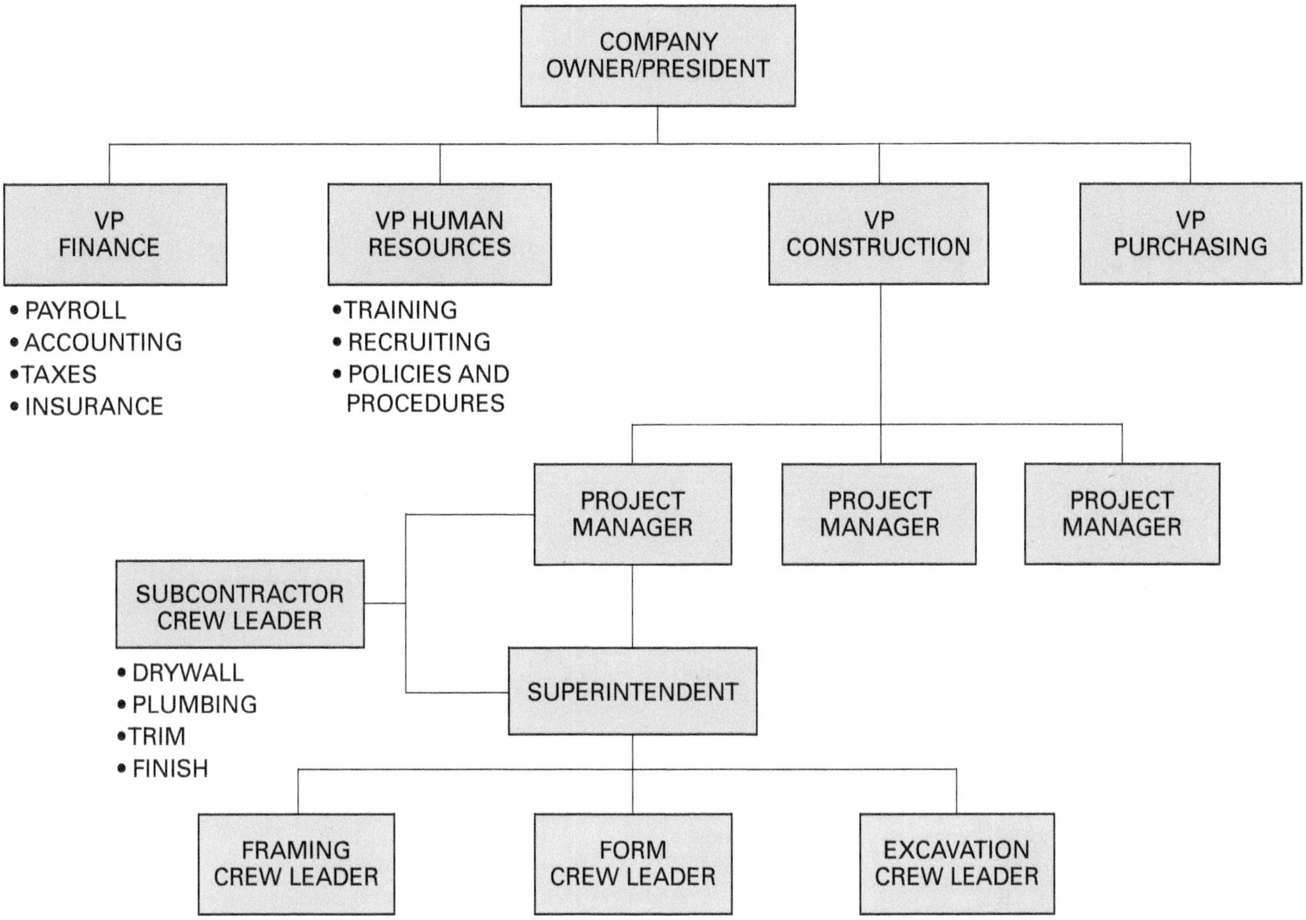

```
                          ┌─────────────────┐
                          │    COMPANY      │
                          │ OWNER/PRESIDENT │
                          └─────────────────┘
```

| VP FINANCE | VP HUMAN RESOURCES | VP CONSTRUCTION | VP PURCHASING |
|---|---|---|---|

- PAYROLL
- ACCOUNTING
- TAXES
- INSURANCE

- TRAINING
- RECRUITING
- POLICIES AND PROCEDURES

PROJECT MANAGER    PROJECT MANAGER    PROJECT MANAGER

SUBCONTRACTOR CREW LEADER

- DRYWALL
- PLUMBING
- TRIM
- FINISH

SUPERINTENDENT

FRAMING CREW LEADER    FORM CREW LEADER    EXCAVATION CREW LEADER

*Figure 4* Sample organization chart for a construction company.

Other divisions of responsibility a company may create involve architectural and engineering design functions. These divisions usually become separate departments.

### 1.3.2 Responsibility, Authority, and Accountability

As an organization grows, the manager must ask others to perform many duties so that the manager can concentrate on management tasks. Managers typically assign (delegate) activities to their subordinates. When delegating activities, the crew leader assigns others the responsibility to perform the designated tasks.

*Responsibility* is the obligation to perform the duties. Along with responsibility comes authority. *Authority* is the right to act or make decisions in carrying out an assignment. The type and amount of authority a supervisor or worker has depends on the employee's company. There must be a balance between the authority and responsibility workers have so that they can carry out their assigned tasks. In addition, the crew leader must delegate sufficient authority to make a worker accountable to the crew leader for the results.

*Accountability* is holding an employee responsible for completing the assigned activities. Even though the crew leader may delegate authority and responsibility to crew members, the overall responsibility—the accountability—for the tasks assigned to the crew always rests with the crew leader.

### 1.3.3 Job Descriptions

Many companies furnish each employee with a written job description that explains the job in detail. Job descriptions set a standard for the employee. They make judging performance easier, clarify the tasks each person should handle, and simplify the training of new employees.

Each new employee should understand all the duties and responsibilities of the job after reviewing the job description. Thus, the time is shorter for the employee to make the transition from being a new and uninformed employee to a more experienced member of a crew.

A job description need not be long, but it should be detailed enough to ensure there is no misunderstanding of the duties and responsibilities of the position. The job description should

---

 46101 Fundamentals of Crew Leadership

Module Eleven   9

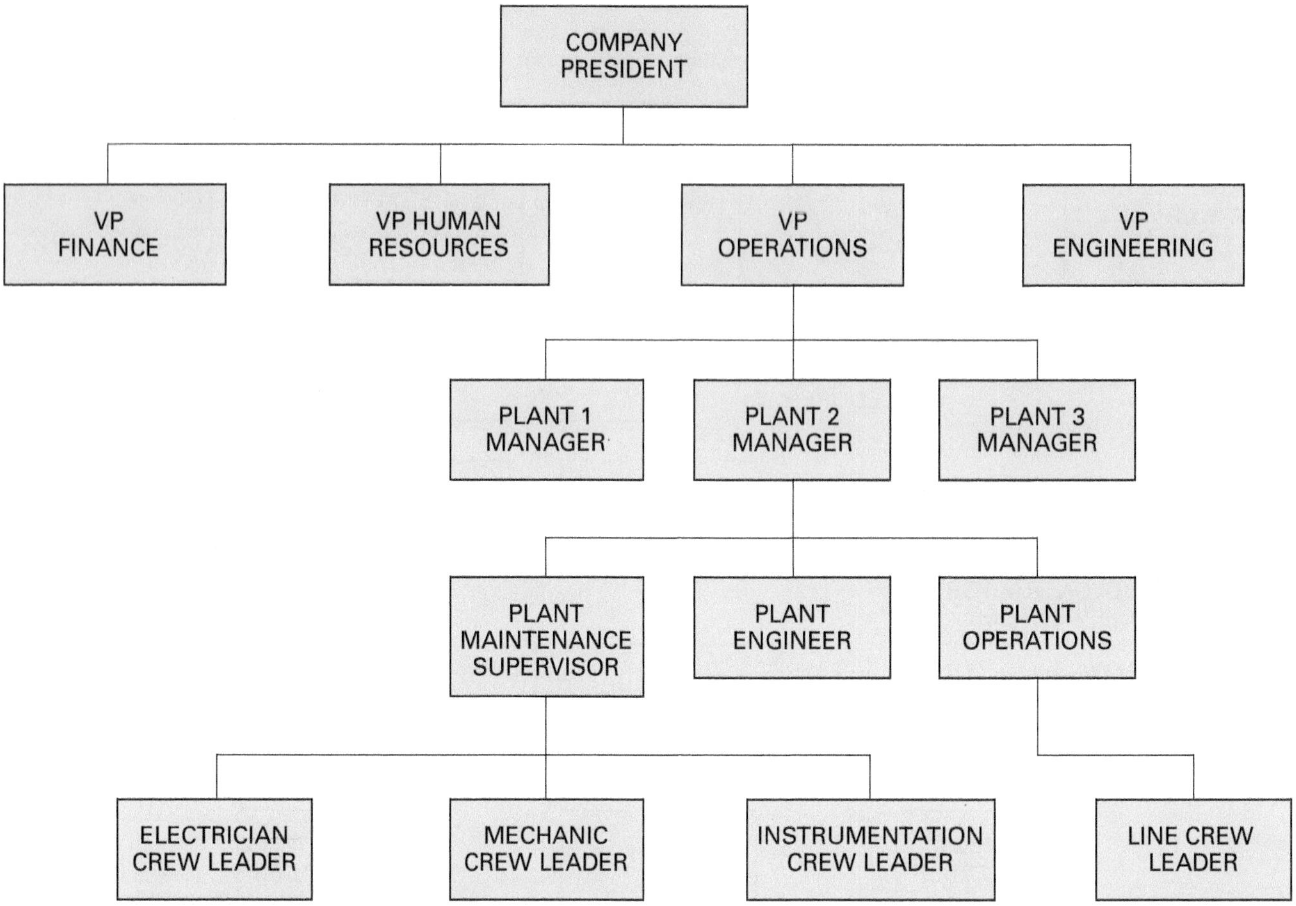

```
                          ┌─────────────┐
                          │  COMPANY    │
                          │  PRESIDENT  │
                          └─────────────┘
        ┌──────────────┬─────────┴─────────┬──────────────┐
  ┌──────────┐  ┌──────────────┐  ┌──────────────┐  ┌──────────────┐
  │    VP    │  │  VP HUMAN    │  │     VP       │  │     VP       │
  │ FINANCE  │  │  RESOURCES   │  │  OPERATIONS  │  │  ENGINEERING │
  └──────────┘  └──────────────┘  └──────────────┘  └──────────────┘
                              ┌──────────┼──────────┐
                       ┌──────────┐ ┌──────────┐ ┌──────────┐
                       │ PLANT 1  │ │ PLANT 2  │ │ PLANT 3  │
                       │ MANAGER  │ │ MANAGER  │ │ MANAGER  │
                       └──────────┘ └──────────┘ └──────────┘
                              ┌──────────┼──────────┐
                   ┌──────────────┐ ┌──────────┐ ┌──────────────┐
                   │    PLANT     │ │  PLANT   │ │    PLANT     │
                   │ MAINTENANCE  │ │ ENGINEER │ │ OPERATIONS   │
                   │ SUPERVISOR   │ │          │ │              │
                   └──────────────┘ └──────────┘ └──────────────┘
            ┌──────────────┬──────────────┬──────────────┐
   ┌──────────────┐ ┌──────────────┐ ┌──────────────────┐ ┌──────────────┐
   │  ELECTRICIAN │ │   MECHANIC   │ │  INSTRUMENTATION │ │  LINE CREW   │
   │ CREW LEADER  │ │ CREW LEADER  │ │   CREW LEADER    │ │   LEADER     │
   └──────────────┘ └──────────────┘ └──────────────────┘ └──────────────┘
```

*Figure 5* Sample organization chart for an industrial company.

contain all the information necessary to evaluate the employee's performance and hold the employee accountable.

A job description should contain, at minimum, the following:

- Job title
- General description of the position
- Minimum qualifications for the job
- Specific duties and responsibilities
- The supervisor to whom the position reports
- Other requirements, such as qualifications, certifications, and licenses

*Figure 6* is an example of a job description.

### 1.3.4 Policies and Procedures

Most companies have formal policies and procedures established to help crew leaders carry out their duties. A policy is a general statement estab-lishing guidelines for a specific activity. Examples include policies on vacations, breaks, workplace safety, and checking out tools. Procedures are formal instructions to carry out and meet policies. For example, a procedure written to implement a policy on workplace safety would include guidelines expected of all employees for reporting accidents and to follow general safety practices.

A crew leader must be familiar with the company policies and procedures, especially regarding safety practices. When OSHA inspectors visit a jobsite, they often question employees and crew leaders about the company policies related to safety. If they are investigating an accident, they will want to verify that the responsible crew leader knew the applicable company policy and followed it.

**Position:**
Crew Leader

**General Summary:**
First line of supervision on a construction crew installing concrete formwork.

**Reports To:**
Job Superintendent

**Physical and Mental Responsibilities:**
- Ability to stand for long periods
- Ability to solve basic math and geometry problems

**Duties and Responsibilities:**
- Oversee crew
- Provide instruction and training in construction tasks as needed
- Make sure proper materials and tools are on the site to accomplish tasks
- Keep project on schedule
- Enforce safety policies and procedures

**Knowledge, Skills, and Experience Required:**
- Extensive travel throughout the Eastern United States, home base in Atlanta
- Ability to operate a backhoe and trencher
- Valid commercial driver's license with no DUI violations
- Ability to work under deadlines with the knowledge and ability to foresee problem areas and develop a plan of action to solve the situation

*Figure 6* A sample job description.

**Did You Know?**

# The ADA

The Americans with Disabilities Act (ADA) is a law that prohibits employers from discriminating against disabled persons in their hiring practices. While the ADA doesn't require written job descriptions, investigators use existing job descriptions as evidence in determining whether discrimination occurred. This means that job descriptions must specifically define the job duties and the abilities needed to perform them. In making a determination of discrimination under the ADA, investigators give consideration to whether a reasonable accommodation could have qualified the disabled person to fill a job.

## Additional Resources

*Construction Workforce Development Professional*, NCCER. 2016. New York, NY: Pearson Education, Inc.

*Mentoring for Craft Professionals*, NCCER. 2016. New York, NY: Pearson Education, Inc.

*Generational Cohorts and their Attitudes Toward Work Related Issues in Central Kentucky*, Frank Fletcher, et al. 2009. Midway College, Midway, KY. **www.kentucky.com**

*The Young Person's Guide to Wisdom, Power, and Life Success: Making Smart Choices*. Brian Gahran, PhD. 2014. San Diego, CA: Young Persons Press. **www.WPGBlog.com**

The following websites offer resources for products and training:

Aging Workforce News, **www.agingworkforcenews.com**

American Society for Training and Development (ASTD), **www.astd.org**

Equal Employment Opportunity Commission (EEOC), **www.eeoc.gov**

National Association of Women in Construction (NAWIC), **www.nawic.org**

Society for Human Resources Management (SHRM), **www.shrm.org**

United States Census Bureau, **www.census.gov**

United States Department of Labor, **www.dol.gov**

Wi-Fi® is a registered trademark of the Wi-Fi Alliance, **www.wi-fi.org**

## 1.0.0   Section Review

1. According to the US Department of Labor, the best way for the construction industry to reduce skilled-worker shortages is to _____.
   a. create training opportunities
   b. avoid discrimination lawsuits
   c. update the skills of older workers who are retiring at a later age than they previously did
   d. implement better policies and procedures

2. Which of the following is *not* helpful when dealing with language diversity in the workplace?
   a. Speaking slowly and clearly in an even tone
   b. Using humor when communicating
   c. Using sketches or diagrams to explain what needs to be done.
   d. Being patient and giving bilingual workers time to process your instructions.

3. Members tend to function best within an organization when they _____.
   a. are allowed to select their own uniform for each project
   b. understand their role within the organization
   c. don't disagree with the statements of other workers or supervisors
   d. are able to work without supervision

# SECTION TWO

## 2.0.0 LEADERSHIP SKILLS

### Objective

Explain how to incorporate leadership skills into work habits, including communications, motivation, team-building, problem-solving, and decision-making skills.

a. Describe the role of a leader on a construction crew.
b. Explain the importance of written and oral communication skills.
c. Describe methods for motivating team members.
d. Explain the importance of teamwork to a construction project.
e. Identify effective problem-solving and decision-making methods.

### Trade Terms

**Autonomy:** The condition of having complete control over one's actions, and being free from the control of another. To be independent.

**Bias:** A preconceived inclination against or in favor of something.

**Ethics:** The moral principles that guides an individual's or organization's actions when dealing with others. Also refers to the study of moral principles.

**Infer:** To reach a conclusion using a method of reasoning that starts with an assumption and considers a set of logically-related events, conditions, or statements.

**Legend:** In maps, plans, and diagrams, an explanatory table defining all symbolic information contained in the document.

**Lockout/tagout (LOTO):** A system of safety procedures for securing electrical and mechanical equipment during repairs or construction, consisting of warning tags and physical locking or restraining devices applied to controls to prevent the accidental or purposeful operation of the equipment, endangering workers, equipment, or facilities.

**Proactive:** To anticipate and take action in the present to deal with potential future events or outcomes based on what one knows about current events or conditions.

**Project manager:** In construction, the individual who has overall responsibility for one or more construction projects; also called the general superintendent.

**Superintendent:** In construction, the individual who is the on-site supervisor in charge of a given construction project.

**Textspeak:** A form of written language characteristic of text messaging on mobile devices and text-based social media, which usually consists of acronyms, abbreviations, and minimal punctuation.

It is important to define some of the supervisory positions discussed throughout this module. You are already familiar with the roles of the craft professional and crew leader. A superintendent is essentially an on-site supervisor who is responsible for one or more crew leaders or front-line supervisors. A project manager or general superintendent may be responsible for managing one or more projects. This training will concentrate primarily on the supervisory role of the crew leader.

Craftworkers and crew leaders differ in that the crew leader manages the activities that the craft professionals perform. To manage a crew of craft professionals, a crew leader must have first-hand knowledge and experience in their activities. Additionally, the crew leader must be able to act directly in organizing and directing the activities of the various crew members.

This section explains the importance of developing leadership skills as a new crew leader. It will cover effective ways to communicate with co-workers and employees at all levels, build teams, motivate crew members, make decisions, and resolve problems.

Crew leaders are generally promoted up from a work crew. A worker's ability to accomplish tasks, get along with others, meet schedules, and stay within the budget have a significant influence on the selection process. The crew leader must lead the team to work safely and provide a quality product.

Making the transition from crew member to a crew leader can be difficult, especially when the new position involves overseeing a group of former peers. For example, some of the crew may try to take advantage of their friendship by seeking special favors. They may also want to be privy to supervisory information that is normally closely held. When you become a crew leader, you are no longer responsible for your work alone. Crew leaders are accountable for the work of an entire crew of people with varying skill levels, personalities, work styles, and cultural and educational backgrounds.

Crew leaders must learn to put personal relationships aside and work for the common goals of the entire crew. The crew leader can overcome these problems by working with the crew to set mutual performance goals and by freely communicating with them within permitted limits. Use their knowledge and strengths along with your own so that they feel like they are key players on the team.

As employees move from being a craftworker into the role of a crew leader and above, they will begin to spend more hours supervising the work of others (supervisory work) than practicing their own craft skills (technical work). *Figure 7* illustrates the relative amounts of time craft professionals, crew leaders, superintendents, and project managers spend on technical and supervisory work as their management responsibilities increase.

There are many ways to define a leader. One simple definition of a leader is a person who influences other people to achieve a goal. Some people may have innate leadership qualities that developed during their upbringing, or they may have worked to develop the traits that motivate others to follow and perform. Research shows that people who possess such talents are likely to succeed as leaders.

*"Leadership is the art of getting someone else to do something you want done because he wants to do it."*
*—Dwight D. Eisenhower*

## 2.1.0 The Qualities and Role of a Leader

Leadership traits are similar to the skills that a crew leader needs to be effective. Although the characteristics of leadership are many, there are some definite commonalities among effective leaders.

First and foremost, effective leaders lead by example. They work and live by the standards that they establish for their crew members or followers, making sure they set a positive example.

Effective leaders also tend to have a high level of drive and determination, as well as a persistent attitude. When faced with obstacles, effective leaders don't get discouraged. Instead, they identify the potential problems, make plans to overcome them, and work toward achieving the intended goal. In the event of failure, effective leaders learn from their mistakes and apply that knowledge to future situations. They also learn from their successes.

Effective leaders are typically good communicators who clearly lay out the goals of a project to their crew members. Accomplishing this may require that the leader overcome issues such as language barriers, gender bias, or differences in personalities to ensure that each member of the crew understands the established goals of the project.

Effective leaders can motivate their crew members to work to their full potential and become useful members of the team. Crew leaders try to develop crew-member skills and encourage them to improve and learn so they can contribute more to the team effort. Effective leaders strive for excellence from themselves and their team, so they work hard to provide the skills and leadership necessary to do so.

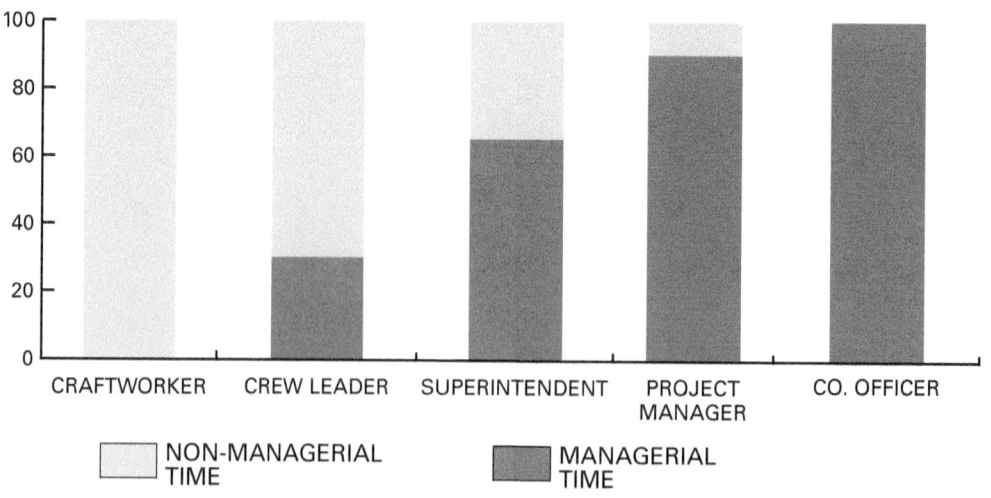

*Figure 7* Percentages of technical and supervisory work by role.

In addition, effective leaders must possess organizational skills. They know what needs to be accomplished, and they use their resources to make it happen. Because they can't do it alone, leaders require the help of their team members to share in the workload. Effective leaders delegate work to their crew members, and they implement company policies and procedures to complete the work safely, effectively, and efficiently.

Finally, effective leaders have the experience, authority, and self-confidence that allows them to make decisions and solve problems. To accomplish their goals, leaders must be able to calculate risks, take in and interpret information, assess courses of action, make decisions, and assume the responsibility for those decisions.

### 2.1.1 Functions of a Leader

The functions of a leader will vary with the environment, the group of individuals they lead, and the tasks they must perform. However, there are certain functions common to all situations that the leader must fulfill. Some of the major functions are as follows:

- Accept responsibility for the successes and failures of the group's performance
- Be sensitive to the differences of a diverse workforce
- Ensure that all group members understand and abide by company policies and procedures
- Give group members the confidence to make decisions and take responsibility for their work
- Maintain a cohesive group by resolving tensions and differences among its members and between the group and those outside the group
- Organize, plan, staff, direct, and control work
- Represent the group

### 2.1.2 Leadership Traits

There are many traits and skills that help create an effective leader. Some of these include the following:

- Ability to advocate an idea
- Ability to motivate
- Ability to plan and organize
- Ability to teach others
- Enthusiasm
- Fairness
- Good communication skills
- Initiative
- Loyalty to their company and crew
- Willingness to learn from others

### 2.1.3 Expected Leadership Behavior

Followers have expectations of their leaders. They look to their leaders to do the following:

- Abide by company policies and procedures
- Be a loyal member of the team
- Communicate effectively
- Have the necessary technical knowledge
- Lead by example
- Make decisions and assume responsibility
- Plan and organize the work
- Suggest and direct
- Trust the team members

### 2.1.4 Leadership Styles

Through history, there have been many terms used to describe different leadership styles and the amount of autonomy they permit. The amount of crew autonomy directly relates to the leadership style used by the crew leader. *Figure 8* illustrates the linear relationship between leader authority and worker autonomy.

There are three main styles of leadership. At one extreme is the *controller style* of leadership, where the crew leader makes all of the decisions independently, without seeking the opinions of crew members. (Very little autonomy exists for workers under this type of leader.) At the other extreme is the *advisor style*, where the crew leader empowers the employees to make decisions (high autonomy). In between these extremes is the *directive style*, where the crew leader seeks crew member opinions and makes the appropriate decisions based on their input.

The following are some characteristics of each of the three leadership styles:

*Controller style (high authority, low autonomy):*

- Expect crew members to work without questioning procedures
- Seldom seek advice from crew members
- Insist on solving problems alone
- Seldom permit crew members to assist each other
- Praise and criticize on a personal basis
- Have no sincere interest in creatively improving methods of operation or production

*Directive style (equal authority and autonomy):*

- Discuss problems with their crew members
- Listen to suggestions from crew members
- Explain and instruct
- Give crew members a feeling of accomplishment by commending them when they do a job well

- Are friendly and available to discuss personal and job-related problems

*Advisor style (low authority, high autonomy):*

- Believe no supervision is best
- Rarely give orders
- Worry about whether they are liked by their crew members

Effective leadership takes many forms. The correct style for a particular situation or operation depends on the nature of the crew as well as the work it has to accomplish. For example, if the crew does not have enough experience for the job ahead, then a controller style may be appropriate. The controller style of leadership is also effective when jobs involve repetitive operations that require little decision-making.

However, if a worker's attitude is an issue, a directive style may be appropriate. In this case, providing the missing motivational factors may increase performance and result in the improvement of the worker's attitude. The directive style of leadership is also used when the work is of a creative nature, because brainstorming and exchanging ideas with such crew members can be beneficial.

The advisor style is effective with an experienced crew on a well-defined project. The company must give a crew leader sufficient authority to do the job. This authority must be commensurate with responsibility, and it must be made known to crew members when they are hired so that they understand who is in charge.

A crew leader must have an expert knowledge of the activities to be supervised in order to be effective. This is important because the crew members need to know that they have someone to turn to when they have a question or a problem, when they need some guidance, or when modifications or changes are warranted by the job.

Respect is probably the most useful element of authority. Respect usually derives from being fair to employees by listening to their complaints and suggestions, and by using incentives and rewards appropriately to motivate crew members. In addition, crew leaders who have a positive attitude and a favorable personality tend to gain the respect of their crew members as well as their peers. Along with respect comes a positive attitude from the crew members.

### 2.1.5 Ethics in Leadership

Crew leaders should maintain the highest standards of honesty and legality. Every day, the crew leader must make decisions that may have fair-

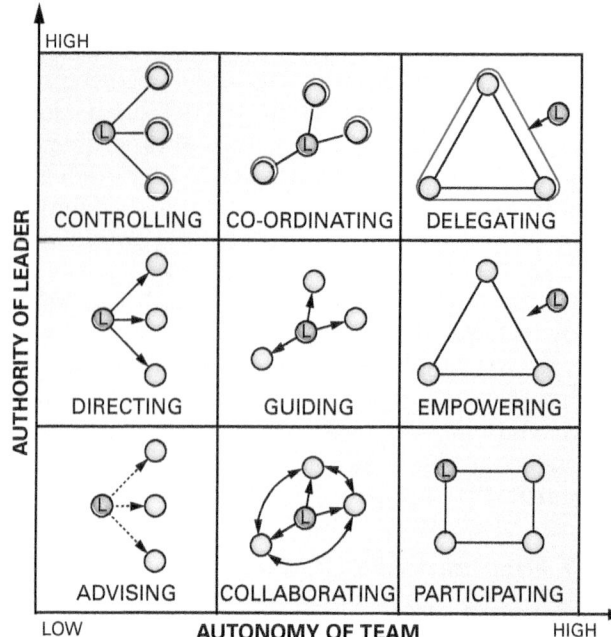

*Figure 8* The authority-autonomy leadership style matrix.

ness and moral implications. When you make a dishonorable or unethical decision, it not only reflects on you, but also impacts other workers, peers, and the company as a whole.

There are three basic types of ethics:

- *Business or legal ethics* – Business, or legal, ethics concerns adhering to all laws and published regulations related to business relationships or activities.
- *Professional/balanced ethics* – Professional, or balanced, ethics relates to carrying out all activities in such a manner as to be honest and fair to everyone under one's authority.
- *Situational ethics* – Situational ethics pertains to specific activities or events that may initially appear to be a gray area. For example, you may ask yourself, "How will I feel about myself if my actions were going to be published in the newspaper or if I need to justify my actions to my family, friends, and colleagues? Would I still do the same thing?"

You will often find yourself in a situation where you will need to assess the ethical consequences of an impending decision. For example, if one of your crew members is showing symptoms of heat exhaustion, should you keep him working just because the superintendent says the project is behind schedule? If you are the only one aware that your crew did not properly erect the reinforcing steel, should you stop the pour and correct the situation? If supervision ever asks a crew leader to carry through on an unethical decision, it is up to that individual to inform the next-higher level of authority of the unethical nature of the issue. It is a sign of good character to refuse to carry out an unethical act.

*"The right way is not always the popular and easy way. Standing for right when it is unpopular is a true test of character."*
*—Margaret Chase Smith*

> **NOTE**
> Anyone who is aware of or knowingly a party to an illegal activity can face legal consequences simply due to inactivity, even if the activities are instigated by one's supervisor. However, a person of good character will never consider weighing the legal consequences against the ethics of one's actions.

## 2.2.0 Communication

Successful crew leaders learn to communicate effectively with people at all levels of the organization. In doing so, they develop an understanding of human behavior and acquire communication skills that enable them to understand and influence others.

There are many definitions of communication. Communication is the act of accurately and effectively conveying to or exchanging facts, feelings, and opinions with another person. Simply stated, communication is the process of exchanging information and ideas. Just as there are many definitions of communication, it also comes in many forms, including verbal, written, and nonverbal.

### 2.2.1 The Communication Process and Verbal Communication

There are two basic steps to clear communication, as illustrated in *Figure 9*. First, a sender sends a message (either verbal or written) to a receiver. When the receiver gets the message, he or she figures out what it means by listening or reading carefully. If anything is unclear, the receiver gives

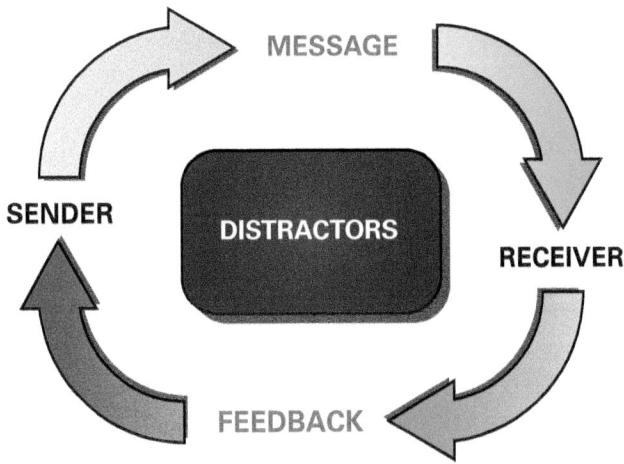

*Figure 9* The communication process.

the sender feedback by asking the sender for more information.

This process is called *two-way communication*, and it is the most effective way to make sure that everyone understands what's going on. This process sounds simple, so you may ask why good communication is so hard to achieve? When we try to communicate, a lot of things can get in the way. These communication obstacles and distractors are called *noise*.

The communication process illustrated in *Figure 9* consists of the following major components:

- *The Sender* – The sender is the person who creates and transmits the message. In verbal communication, the sender speaks the message aloud to the person(s) receiving the message. The sender must be sure to speak in a clear and concise manner easily understood by others. This isn't a natural skill; it takes practice. Some basic speaking tips are as follows:

  - Avoid talking with anything in your mouth (food, gum, etc.).
  - Avoid swearing, or other crude language, and acronyms.
  - Don't speak too quickly or too slowly. In extreme cases, people tend to focus on the rate of speech rather than the words themselves.

### Did You Know?
## Supervisor's Communication Breakdown

No, this isn't about failure of communication, but how a supervisor's daily communications break down into the various types, and how much time the supervisor spends doing each. Research shows that about 80 percent of the typical supervisor's day is spent communicating through writing, speaking, listening, or using body language. Of that time, studies suggest that approximately 20 percent of communication is in the written form, and 80 percent involves speaking or listening.

- Pronounce words carefully to prevent misunderstandings.
- Speak pleasantly and with enthusiasm. Avoid speaking in a harsh voice or in a monotone.

• *The Message* – The message is what the sender is attempting to communicate to the receiver(s). A message can be a set of directions, an opinion, or dealing with a personnel matter (praise, reprimand, etc.). Whatever its function, a message is an idea or fact that the sender wants the receiver to know.

Before speaking, determine what must be communicated, and then take the time to organize what to say. Ensure that the message is logical and complete. Take the time to clarify your thoughts to avoid confusing the receiver. This also permits the sender to get to the point quickly.

In delivering the message, the sender should assess the audience. It is important not to talk down to them. Remember that everyone, whether in a senior or junior position, deserves respect and courtesy. Therefore, the sender should use words and phrases that the receiver can understand and avoid technical language or slang if the receiver is unfamiliar with such terms. In addition, the sender should use short sentences, which gives the audience time to understand and digest one point or fact at a time.

• *The Receiver* – The receiver is the person who takes in the message. For the verbal communication process to be successful, it is important that the receiver understands the message as the speaker intended. Therefore, the receiver must avoid things that interfere with the delivery of the message. There are many barriers to effective listening, particularly on a busy construction job site. Some of these obstacles include the following:

- Noise, visitors, cell phones, or other distractions
- Preoccupation, being under pressure, or daydreaming
- Reacting emotionally to what is being communicated
- Thinking about how to respond instead of listening
- Giving an answer before the message is complete
- Personal bias/prejudice against the sender or the sender's communication style
- Finishing the sender's sentence

Some tips for overcoming these barriers include the following:

- Take steps to minimize or remove distractions; learn to tune out your surroundings
- Listen for key points
- Take notes
- Be aware of your personal biases, and try to stay open-minded; focus on the message, not the speaker
- Allow yourself time to process your thoughts before responding
- Let the sender communicate the message without interruption

There are many ways for a receiver to show that he or she is actively listening to the message. The receiver can even accomplish this without saying a word. Examples include maintaining eye contact (*Figure 10*), nodding your head, and taking notes. Feedback can also provide an important type of response to a message.

• *Feedback* – Feedback is the receiver's communication back to the sender in response to the message. Feedback is a very important part of the communication process because it shows the sender how the receiver interpreted the message and that the receiver understood it as intended. In other words, feedback is a checkpoint to make sure the receiver and sender are on the same page.

The receiver can paraphrase the message as a form of feedback to the sender. When paraphrasing, you use your own words to repeat the message. That way, you can show the sender that you interpreted the message correctly and could explain it to others if needed.

*Figure 10* Eye contact is important when communicating face-to-face.

When providing feedback, the receiver can also request clarification or additional information, generally by asking questions.

One opportunity a crew leader can take to provide feedback is in the performance of crew evaluations. Many companies use formal evaluations on a yearly basis to assess workers' performance for pay increases and eligibility for advancement. These evaluations should not come as a once-a-year surprise to workers. An effective crew leader provides frequent performance feedback, and should ultimately summarize these communications in the annual performance evaluation. It is also important to stress the importance of self-evaluation with your crew.

### 2.2.2 Written or Visual Communications

Some communication must be written or visual. Written or visual communication includes messages or information documented on paper, or transmitted electronically using words, sketches, or other types of images.

Many messages on a job are in text form. Examples include weekly reports, requests for changes, purchase orders, and correspondence on a specific subject. These items are in writing because they can form a permanent record for business and historical purposes. Some communications on the job must be in the form of images. People are visual beings, and sketches, diagrams, graphs, and photos or videos can more effectively communicate some things. Examples include the plans or drawings used on a job, or a video of a work site accident scene.

*"One picture is worth a thousand words."*
*—San Antonio Express-News (1918)*

When writing or creating a visual message, it is best to assess the receiver (the reader) or the audience before beginning. The receiver must be able to read the message and understand the content; otherwise, the communication process will be unsuccessful. Therefore, the sender (the writer) should consider the actual meaning of words or diagrams and how others might interpret them. In addition, the writer should make sure that handwriting, if applicable, is legible.

Here are some basic tips for writing:

- Avoid emotion-packed words or phrases.
- Avoid making judgments unless asked to do so.
- Avoid using unfamiliar acronyms and text-speak (*Figure 11*).

*Figure 11* Avoid textspeak for on-the-job written communications.

- Avoid using technical language or jargon unless the writer knows the receiver is familiar with the terms used.
- Be positive whenever possible.
- Be prepared to provide a verbal or visual explanation, if needed.
- Make sure that the document is legible.
- Present the information in a logical manner.
- Proofread your work; check for spelling, typographical, and grammatical errors—especially if relying on a spell-checker or an autocorrect typing feature.
- Provide an adequate level of detail.
- State the purpose of the message clearly.
- Stick to the facts.

Entire books are available that explain how to create effective diagrams and graphs. There are many factors that contribute to clear, concise, unambiguous images. The following are some basic tips for creating effective visual communications:

- Avoid making complex graphics; simplicity is better.
- Be prepared to provide a written or verbal explanation of the diagram, if needed.
- Ensure that the diagram is large enough to be useful.
- Graphs are better for showing numerical relationships than columns of numbers.
- Present the information in a logical order.
- Provide a legend if you include symbolic information.
- Provide an adequate level of detail and accuracy.

### 2.2.3 Nonverbal Communication

Unlike verbal or written communication, nonverbal communication doesn't involve spoken or written words, or images. Rather, nonverbal communication refers to things the speaker and the receiver see and hear in each other while communicating face to face. Examples include facial expressions, body movements, hand gestures, tone of voice, and eye contact.

Nonverbal communication can provide an external signal of an individual's inner emotions. It occurs at the same time as verbal communication. In most cases, the transmission of the sender and receiver nonverbal cues is totally unconscious.

People observe physical, nonverbal cues when communicating, making them just as important as words. Often, nonverbal signals influence people more than spoken words. Therefore, it's important to be conscious of nonverbal cues to avoid miscommunication based on your posture or expression. After all, these things may have nothing to do with the communication exchange; instead, they may be carrying over from something else going on in your day.

### 2.2.4 Communication Issues

It is important to note that everyone communicates differently; that is what makes us unique as individuals. As the diversity of the workforce changes, communication becomes even more challenging because the audience may include individuals from many different backgrounds. Therefore, it is necessary to assess the audience to determine how to communicate effectively with each individual.

The key to effective communication is to acknowledge that people are different, and to be able to adjust the communication style to meet the needs of the audience or the person on the receiving end of your message. This involves relaying the message in the simplest way possible, and avoiding the use of words that people may find confusing. Be aware of how you use technical language, slang, jargon, and words that have multiple meanings. Present the information in a clear, concise manner. Avoid rambling and always speak clearly, using good grammar.

> **NOTE**
> When working with bilingual workers, it is sometimes helpful to learn words and phrases commonly used in their language(s). They will respect you more for attempting to learn their language as they learn yours, and the effort will generate a closer comradery.

In addition, be prepared to communicate the message in multiple ways or adjust your level of detail or terminology to ensure that everyone understands the meaning as intended. For instance, a visual learner may need a map to comprehend directions. It may be necessary to overcome language barriers on the job site by using graphics or visual aids to relay the message. *Figure 12* shows how to tailor the message to the audience.

### 2.3.0 Motivation

The ability to motivate others is a key skill that leaders must develop. To motivate is to influence others to put forth more effort to accomplish something. For example, a crew member who skips breaks and lunch to complete a job on time is thought to be highly motivated, but crew members who do the bare minimum or just enough to keep their jobs are considered unmotivated.

A leader can *infer* an employee's motivation by observing various factors in the employee's performance. Examples of factors that may indicate a crew's motivation include the level or rate of unexcused absenteeism, the percentage of employee turnover, and the number of complaints, as well as the quality and quantity of work produced.

Different things motivate different people in different ways. Consequently, there is no single best approach to motivating crew members. It is important to recognize that what motivates one crew member may not motivate another. In addition, what works to motivate a crew member once may not motivate that same person again in the future.

Often, the needs and factors that motivate individuals are the same as those that create job satisfaction. These include the following:

- Accomplishment
- Change
- Job importance
- Opportunity for advancement
- Personal growth
- Recognition and praise
- Rewards

A crew leader's ability to satisfy these needs increases the likelihood of high morale within a crew. Morale refers to an individual's emotional outlook toward work and the level of satisfaction gained while performing the jobs assigned. High morale means that employees will be motivated to work hard, and they will have a positive attitude about coming to work and doing their jobs.

### 2.3.1 Accomplishment

Accomplishment refers to a worker's need to set challenging goals and achieve them. There is nothing quite like the feeling of achieving a goal, particularly a goal one never expected to accomplish in the first place.

Read the following verbal conversations, and identify any problems:

### Conversation I:

*Judy:* Hey, José...

*José:* What's up?

*Judy:* Has the site been prepared for the job trailer yet?

*José:* Job trailer?

*Judy:* The job trailer—it's coming in today. What time will the job site be prepared?

*José:* The trailer will be here about 1:00 PM.

*Judy:* The job site! What time will the job site be prepared?

_____

_____

_____

_____

_____

### Conversation II:

*Jimar:* Hey, Mike, I need your help.

*Mike:* What is it?

*Jimar:* You and Miguel go over and help Al's crew finish laying out the site.

*Mike:* Why me? I can't work with Miguel. He can't understand a word I say.

*Jimar:* Al's crew needs some help, and you and Miguel are the most qualified to do the job.

*Mike:* I told you, Jimar, I can't work with Miguel.

_____

_____

_____

_____

_____

### Conversation III:

*Hiro:* Hey, Jill.

*Jill:* Sir?

*Hiro:* Have you received the latest DOL, EEO requirement to be sure the OFCP administrator finds our records up to date when he reviews them in August?

*Jill:* DOL, EEO, and OFCP?

*Hiro:* Oh, and don't forget the MSHA, OSHA, and EPA reports are due this afternoon.

*Jill:* MSHA, OSHA, and EPA?

_____

_____

_____

_____

_____

_____

_____

### Conversation IV:

*Lakeisha:* Good morning, Roberto, would you do me a favor?

*Roberto:* Okay, Lakeisha. What is it?

*Lakeisha:* I was reading the concrete inspection report and found the concrete in Bays 4A, 3B, 6C, and 5D didn't meet the 3,000 psi strength requirements. Also, the concrete inspector on the job told me the two batches that came in today had to be refused because they didn't meet the slump requirements as noted on page 16 of the spec. I need to know if any placement problems happened on those bays, how long the ready mix trucks were waiting today, and what we plan to do to stop these problems in the future.

_____

_____

_____

_____

Read the following written memos, and identify any problems:

**Memo I:**

Let's start with the transformer vault $285.00 due. For what you ask? Answer: practically nothing I admit, but here is the story. Paul the superintendent decided it was not the way good ole Comm Ed wanted it, we took out the ladder and part of the grading (as Paul instructed us to do) we brought it back here to change it. When Comm Ed the architect or DOE found out that everything would still work the way it was, Paul instructed us to reinstall the work. That is the whole story there is please add the $285.00 to my next payout.

_____

_____

_____

_____

_____

_____

_____

**Memo II:**

Let's take rooms C 307-C-312 and C-313 we made the light track supports and took them to the job to erect them when we tried to put them in we found direct work in the way, my men spent all day trying to find out what to do so ask your Superintendent (Frank) he will verify seven hours pay for these men as he went back and forth while my men waited. Now the Architect has changed the system of hanging and has the gall to say that he has made my work easier, I can't see how. Anyway, we want an extra two (2) men for seven (7) hours for April 21 at $55.00 per hour or $385.00 on April 28th DOE Reference 197 finally resolved this problem. We will have no additional charges on DOE Reference 197, please note.

_____

_____

_____

_____

_____

Crew leaders can help their crew members attain a sense of accomplishment by encouraging them to develop performance plans, such as goals for the year, the attaining of which the crew leader will consider later in performance evaluations. In addition, crew leaders can provide the support and tools (such as training and coaching) necessary to help their crew members achieve these goals.

### 2.3.2 Change

Change refers to an employee's need to have variety in work assignments. Change can keep things interesting or challenging, and prevent the boredom that results from doing the same task day after day with no variety. However, frequent or significant changes in work can actually have a negative impact on morale, since people often prefer some consistency and predictability in their lives.

### 2.3.3 Job Importance

Job importance refers to an employee's need to feel that his or her skills and abilities are valued and make a difference. Employees who don't feel valued tend to have performance and attendance issues. Crew leaders should attempt to make every crew member feel like an important part of the team, as if the job wouldn't be possible without their help.

### 2.3.4 Opportunity for Advancement

Opportunity for advancement refers to an employee's need to gain additional responsibility and develop new skills and abilities. It is important that employees know that they aren't limited to their current jobs. Let them know that they have a chance to grow with the company and to be promoted as recognition for excelling in their work when such opportunities occur.

| VERBAL INSTRUCTIONS (EXPERIENCED CREW) | VERBAL INSTRUCTIONS (INEXPERIENCED CREW) | WRITTEN INSTRUCTIONS | VISUAL INSTRUCTIONS (DIAGRAM/MAP) |
|---|---|---|---|
| "Please drive to the supply shop to pick up our order." | "Please drive to the supply shop. Turn right here and left at Route 1. It's at 75th Street and Route 1. Tell them the company name and that you're there to pick up our order." | 1. Turn right at exit.<br>2. Drive 2 miles to Route 1. Turn LEFT.<br>3. Drive 1 mile (pass the tire shop) to 75th Street.<br>4. Look for supply store on right... | 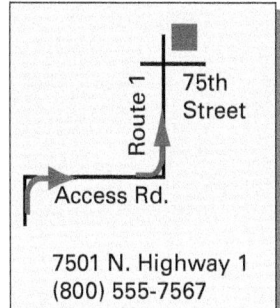<br>7501 N. Highway 1<br>(800) 555-7567 |

Different people learn in different ways. Be sure to communicate so you can be understood.

*Figure 12* Tailor your message.

Effective leaders encourage each of their crew members to work to his or her full potential. In addition, they share information and skills with their employees to help them advance within the organization.

### 2.3.5 Recognition

Recognition and praise refer to the need to have good work appreciated, applauded, and acknowledged by others. You can accomplish this by simply thanking employees for helping on a project, or it can entail more formal praise, such as an award for Employee of the Month, or tangible or monetary awards. Some tips for giving recognition and praise include the following:

- Be available on the job site so that you have the opportunity to witness good work.
- Look for and recognize good work, and look for ways to praise it.
- Give recognition and praise only when truly deserved; people can quickly recognize insincere or artificial praise, and you will lose respect as a result.
- Acknowledge satisfactory performance, and encourage improvement by showing confidence in the ability of the crew members to do above-average work.

### 2.3.6 Personal Growth

Personal growth refers to an employee's need to learn new skills, enhance abilities, and grow as a person. It can be very rewarding to master a new competency on the job. Personal growth prevents the boredom associated with contemplation of working the same job indefinitely with no increase of technical knowledge, skills, or responsibility.

Crew leaders should encourage the personal growth of their employees as well as themselves. Learning should be a two-way street on the job site; crew leaders should teach their crew members and learn from them as well. In addition, crew members should be encouraged to learn from each other

Personal growth also takes place outside the workplace. Crew leaders should encourage their employees to acquire formal education applicable to their vocation as well as in subjects that will expand their understanding of the larger world. Encourage them to develop outside interests and ways to express their innate creativity.

### 2.3.7 Rewards

Rewards are additional compensation for hard work. Rewards can include an increase in a crew member's wages, or go beyond that to include bo-

**Did You Know?**

# The Importance of Continuing Education

Education doesn't stop the day a person receives a diploma or certificate. It is a lifelong activity. Employers have long recognized and promoted continuing education as a factor in advancement, but it is essential to simply remaining in place as well. Regardless of what you do, new construction materials, methods, standards, and processes are constantly emerging. Those who don't make the effort to keep up will fall behind.

Also, studies have shown that a worker's lifetime income improved by 7–10 percent per year of community college. Consider continuing and vocational technical education (CTE and VTE) through local colleges.

nuses or other incentives. They can be monetary in nature (salary raises, holiday bonuses, etc.), or they can be nonmonetary, such as free merchandise (shirts, coffee mugs, jackets, merchant gift cards, etc.), or other prizes. Attendance at costly certification or training courses can be another form of reward.

### 2.3.8 Motivating Employees

To increase motivation in the workplace, crew leaders must individualize how they motivate different crew members. It is important that crew leaders get to know their crew members and determine what motivates them as individuals. Once again, as diversity increases in the workforce, this becomes even more challenging; therefore, effective communication skills are essential. Some tips for motivating employees include the following:

- Keep jobs challenging and interesting.
- Communicate your expectations. People need clear goals in order to feel a sense of accomplishment when they achieve them.
- Involve the employees. Feeling that their opinions are valued leads to pride in ownership and active participation.
- Provide sufficient training. Give employees the skills and abilities they need to be motivated to perform.
- Mentor the employees. Coaching and supporting employees boosts their self-esteem, their self-confidence, and ultimately their motivation.
- Lead by example. Employees will be far more motivated to follow someone who is willing to do the same things they are required to do.
- Treat employees well. Be considerate, kind, caring, and respectful; treat employees the way that you want to be treated.
- Avoid using scare tactics. Approaching your leadership responsibilities by threatening employees with negative consequences can result in higher employee turnover instead of motivation.
- Reward your crew for doing their best by giving them easier tasks from time to time. It is tempting to give your best employees the hardest or dirtiest jobs because you know they will do the jobs correctly.
- Reward employees for a job well done.
- Maintain a sense of humor, especially toward your own failings as a human being. No one is perfect. Your employees will appreciate that and be more motivated to be an encouragement to you.

### 2.4.0 Team Building

Organizations are making the shift from the traditional boss-worker mentality to one that promotes teamwork. The manager becomes the team leader, and the workers become team members. They all work together to achieve the common goals of the team.

There are several benefits associated with teamwork. These include the ability to complete complex projects more quickly and effectively, higher employee satisfaction, and reduced turnover.

### 2.4.1 Successful Teams

Successful teams consist of individuals who are willing to share their time and talents to reach a common goal—the goal of the team. Members of successful teams possess an "Us" or "We" attitude rather than an "I" and "You" attitude; they consider what's best for the team and put their egos aside.

Some characteristics of successful teams include the following:

- Everyone participates and every team member counts.
- All team members understand the goals of the team's work and are committed to achieve those goals.
- There is a sense of mutual trust and interdependence.
- The organization gives team members the confidence and means to succeed.
- They communicate.
- They are creative and willing to take risks.
- The team leader has strong people skills and is committed to the team.

### 2.4.2 Building Successful Teams

To be successful in the team leadership role, the crew leader should contribute to a positive attitude within the team. There are several ways in which the team leader can accomplish this. First, he or she can work with the team members to create a vision or purpose of what the team is to achieve. It is important that every team member is committed to the purpose of the team, and the team leader is instrumental in making this happen.

Within the construction industry, the company typically assigns a crew to a crew leader. However, it can be beneficial for the team leader to be involved in selecting the team members. The willingness of people to work on the team and the resources that they bring should be the key factors for selecting them for the team.

You are the crew leader of a masonry crew. Sam Williams is the person whom the company holds responsible for ensuring that equipment is operable and distributed tothe jobs in a timely manner.

Occasionally, disagreements with Sam have resulted in tools and equipment arriving late. Sam, who has been with the company 15 years, resents having been placed in the job and feels that he outranks all the crew leaders.

Sam figured it was about time he talked with someone about the abuse certain tools and other items of equipment were receiving on some of the jobs. Saws were coming back with guards broken and blades chewed up, bits were being sheared in half, motor housings were bent or cracked, and a large number of tools were being returned covered with mud. Sam was out on your job when he observed a mason carrying a portable saw by the cord. As he watched, he saw the mason bump the swinging saw into a steel column. When the man arrived at his workstation, he dropped the saw into the mud.

You are the worker's crew leader. Sam approached as you were coming out of the work trailer. He described the incident. He insisted, as crew leader, you are responsible for both the work of the crew and how its members use company property. Sam concluded, "You'd better take care of this issue as soon as possible! The company is sick and tired of having your people mess up all the tools!"

You are aware that some members of your crew have been mistreating the company equipment.

1. How would you respond to Sam's accusations?

_____

_____

_____

_____

_____

2. What action would you take regarding the misuse of the tools?

_____

_____

_____

_____

_____

3. How can you motivate the crew to take better care of their tools? Explain.

_____

_____

_____

_____

_____

When forming a new team, team leaders should do the following:

- Explain the purpose of the team. Team members need to know what they will be doing, how long they will be doing it (if they are temporary or permanent), and why they are needed.
- Help the team establish goals or targets. Teams need a purpose, and they need to know what it is they are responsible for accomplishing.
- Define team member roles and expectations. Team members need to know how they fit into the team and what they can expect to accomplish as members of the team.
- Plan to transfer a sense of responsibility to the team as appropriate. Teams should feel responsible for their assigned tasks. However, never forget that management will still hold the crew leader responsible for the crew's work.

*"My model for business is The Beatles. They were four guys who kept each other's kind of negative tendencies in check. They balanced each other and the total was greater than the sum of the parts. That's how I see business: Great things in business are never done by one person, they're done by a team of people.*
—*Steve Jobs*

### 2.4.3 Delegating

Once the various activities that make up the job have been determined, the crew leader must identify the person or persons who will be responsible for completing each activity. This requires that the crew leader be aware of the skills and abilities of the people on the crew. Then, the crew leader must put this knowledge to work in matching the crew's skills and abilities to accomplish the specific tasks needed to complete the job.

After matching crew members to specific activities, the crew leader must then delegate the assignments to the responsible person(s). Generally, when delegating responsibilities, the crew leader verbally communicates directly with the person who will perform or complete the activity.

When delegating work, remember to do the following:

- Delegate work to a crew member who can do the job properly. If it becomes evident that the worker doesn't perform to the desired standard, either teach the crew member to do the work correctly or turn it over to someone else who can (without making a public spectacle of the transfer).
- Make sure crew members understand what to do and the level of responsibility. Be clear about the desired results, specify the boundaries and deadlines for accomplishing the results, and note the available resources.
- Identify the standards and methods of measurement for progress and accomplishment, along with the consequences of not achieving the desired results. Discuss the task with the crew member and check for understanding by asking questions. Allow the crew member to contribute feedback or make suggestions about how to perform the task in a safe and quality manner.
- Give the crew member the time and freedom to get started without feeling the pressure of too much supervision. When making the work assignment, be sure to tell the crew member how much time there is to complete it, and confirm that this time is consistent with the job schedule.
- Examine and evaluate the result once a task is complete. Then, give the crew member some feedback as to how well the worker did the task. Get the crew member's comments. The information obtained from this is valuable and will enable the crew leader to know what kind of work to assign that crew member in the future. It will also provide a means of measuring the crew leader's own effectiveness in delegating work.

Be aware that there may be times when someone else in the company will issue written or verbal instructions to a crew member or to the crew without going through the crew leader. This kind of situation requires an extra measure of maturity and discretion on the part of the crew leader to first understand the circumstances, then establish an understanding with the responsible individual, if possible, to avoid work orders that circumvent the crew leader in the future.

### 2.4.4 Implementing Policies and Procedures

Every company establishes policies and procedures that crew leaders are expected to implement, and employees are expected to follow. Company policies and procedures are essentially guidelines for how the organization does business. They can also reflect organizational philosophies, such as putting safety first or making the customer the top priority. Examples of policies and procedures include safety guidelines, credit standards, and billing processes.

The following tips can help you effectively implement policies and procedures:

- Learn the purpose of each policy. This will help you follow the policy and apply it appropriately and fairly.
- If you're not sure how to apply a company policy or procedure, check the company manual or ask your supervisor.

> **NOTE**
> Try to obtain a supervisor's policy interpretation in writing or print out the email response so that you can append the decision to your copy of the company manual for future reference.

- Always follow company policies and procedures. Remember that they combine what's best for the customer and the company. In addition, they provide direction on how to handle specific situations and answer questions.

Crew leaders may need to issue orders to their crew members. An order is a form of communication that initiates, changes, or stops an activity. Orders may be general or specific, written or oral, and formal or informal. The decision of how an order will be issued is up to the crew leader, but the policies and procedures of the company may govern the choice.

When issuing orders, do the following:

- Make them as specific as possible. Avoid being general or vague unless it is impossible to foresee all the circumstances that could occur in carrying out the order.
- Recognize that it isn't necessary to write orders for simple tasks unless the company requires that supervisors write all orders.
- Write orders for more complex tasks, tasks that will take considerable time to complete, or that are permanent (standing) orders.
- Consider what is being said, the audience to whom it applies, and the situation under which it will be implemented to determine the appropriate level of formality for the order.

## 2.5.0 Making Decisions and Solving Problems

Decision making and problem solving and are a large part of every crew leader's daily work. They are a part of life for all supervisors, especially in fast-paced, deadlineoriented industries.

### 2.5.1 Decision Making Versus Problem Solving

Sometimes, the difference between decision making and problem solving isn't clear. Decision making refers to simply initiating an action, stopping one, or choosing an alternative course, as appropriate for the situation. Problem solving involves recognizing the difference between the way things are and the way things should be, then taking action to move toward the desired condition. The two activities are related because, to make a decision, you may have to use problem-solving techniques, just as solving problems requires making decisions.

### 2.5.2 Types of Decisions

Some decisions are routine or simple, and can be made based on past experiences. An example would be deciding how to get to work. If you've worked at the same place for a long time, you are already aware of the options for traveling to work (take the bus, drive a car, carpool with a co-worker, take a taxi, etc.). Based on past experiences with the options identified, you can make a decision about how best to get to work.

Other decisions are more difficult to make. These decisions require more careful thought about how to carry out an activity by using problem-solving techniques. An example is planning a trip to a new vacation spot. If you're not sure how to get there, where to stay, what to see, etc., one option is to research the area to determine the possible routes, hotel accommodations, and attractions. Then, you can make a decision about which route to take, what hotel to choose, and what sites to visit, without the benefit of direct experience. The Internet makes these tasks much easier, but the research still requires prioritizing what is most important to you.

### 2.5.3 Problem Solving

The ability to solve problems is an important skill in any workplace. It's especially important for craft professionals, whose workday is often not predictable or routine. This section provides a five-step process for solving problems, which you can apply to both workplace and personal issues.

Review the following steps and then see how you can be apply them to a job-related problem. Keep in mind that you can't solve a problem until everyone involved acknowledges the problem.

**Step 1**  *Define the problem.* This isn't as easy as it sounds. Thinking through the problem often uncovers additional problems. Also, drilling down to the facts of the problem may mean setting aside your own biases or presumptions toward the situation or the individuals involved.

**Step 2**  *Think about different ways to solve the problem.* There is often more than one solution to a problem, so you must think through each possible solution and pick the best one. The best solution might be taking parts of two different solutions and combining them to create a new solution.

**Step 3**  *Choose the solution that seems best, and make an action plan.* It is best to receive input both from those most affected by the problem, those who must correct the problem, and from those who will be most affected by any potential solution.

**Step 4**  *Test the solution to determine whether it actually works.* Many solutions sound great in theory but in practice don't turn out to be effective. On the other hand, you might discover from trying to apply a solution that it is acceptable with a little modification. If a solution doesn't work, think about how you could improve it, and then test your new plan.

**Step 5**  *Evaluate the process.* Review the steps you took to discover and implement the solution. Could you have done anything better? If the solution turns out to be satisfactory, you can add the solution to your knowledge base.

These five steps can be applied in specific situations. Read the following example situation, and apply the five-step problem-solving process to come up with a solution.

*Example:*
You are part of a team of workers assigned to a new shopping mall project. The project will take about 18 months to complete. The only available parking is half a mile from the job site. The crew must carry heavy toolboxes and safety equipment from their cars and trucks to the work area at the start of the day, and then carry them back at the end of their shifts. The five-step problem-solving process can be applied as follows:

*Step 1*  *Define the problem.* Workers are wasting time and energy hauling all their equipment to and from the work site.

*Step 2*  *Think about different ways to solve the problem.* Several workers have proposed solutions:

- Install lockers for tools and equipment closer to the work site.
- Have workers drive up to the work site to drop off their tools and equipment before parking.
- Bring in another construction trailer where workers can store their tools and equipment for the duration of the project.
- Provide a round-trip shuttle service to ferry workers and their tools.

> **NOTE**
>
> Each solution will have pros and cons, so it's important to receive input from the workers affected by the problem. For example, workers will probably object to any plan (like the drop-off plan) that leaves their tools vulnerable to theft.

*Step 3*  *Choose the solution that seems best, and make an action plan.* The work site superintendent doesn't want an additional trailer in your crew's area. The workers decide that the shuttle service makes the most sense. It should solve the time and energy problem, and workers can keep their tools with them. To put the plan into effect, the project supervisor arranges for a large van and driver to provide the shuttle service.

*Step 4*  *Test the solution to determine whether it actually works.* The solution works, but there is another problem. The workers' schedule has them all starting and leaving at the same time. There isn't enough room in the van for all the workers and their equipment. To solve this problem, the supervisor schedules trips spaced 15 minutes apart. The supervisor also adjusts worker schedules to correspond with the trips. That way, all the workers won't try to get on the shuttle at the same time.

*Step 5*  *Evaluate the process.* This process gave both management and workers a chance to express an opinion and discuss the various solutions. Everyone feels pleased with the process and the solution.

### 2.5.4 Special Leadership Problems

Because they are responsible for leading others, it is inevitable that crew leaders will encounter problems and be forced to make decisions about how to respond to the problem. Some problems will be relatively simple to resolve, like covering for a sick crew member who has taken a day off from work. Other problems will be complex and much more difficult to handle.

Some complex problems that are relatively common include the following:

- Inability to work with others
- Absenteeism and turnover
- Failure to comply with company policies and procedures

*Inability to Work with Others* – Crew leaders will sometimes encounter situations where an employee has a difficult time working with others on the crew. This could be a result of personality differences, gender or gender-identity prejudices, an inability to communicate, or some other cause. Whatever the reason, the crew leader must address the issue and get the crew working as a team.

The best way to determine the reason for why individuals don't get along or work well together is to talk to the parties involved. The crew leader should speak openly with the employee, as well as the other individual(s) to uncover the source of the problem and discuss its resolution.

After uncovering the reason for the conflict, the crew leader can determine how to respond. There may be a way to resolve the problem and get the workers communicating and working as a team again. On the other hand, there may be nothing the crew leader can do that will lead to a harmonious solution. In this case, the crew leader would need to either transfer one of the involved employees to another crew or have the problem crew member terminated. Resorting to this latter option should be the last measure and taken after discussing the matter with one's immediate supervisor or the company's Human Resources Department.

*Absenteeism and Turnover* – Absenteeism and turnover in the industry can delay jobs and cause companies to lose money. Absenteeism refers to workers missing their scheduled work time on a job. It has many causes, some of which cannot be helped. Sickness, family emergencies, and funerals are examples of unavoidable causes of worker absence. However, there are some causes of unexcused absenteeism that crew leaders can prevent.

The most effective way to control absenteeism is to make the company's policy clear to all employees. The policy should be explained to all new employees. This explanation should include the number of absences allowed and acceptable reasons for taking sick or personal days. In addition, all workers should know how to inform their crew leaders when they miss work and understand the consequences of exceeding the number of sick or personal days allowed.

Once crew leaders explain the policy on absenteeism to employees, they must be sure to implement it consistently and fairly. This makes employees more likely to follow it. However, if enforcement of the policy is inconsistent and the crew leader gives some employees exceptions, it won't be effective. Thus, the rate of absenteeism is likely to increase.

Despite having a policy on absenteeism, there will always be employees who are chronically late or miss work. In cases where an employee abuses the absenteeism policy, the crew leader should discuss the situation directly with the employee, ensure that they understand the policy, and insist that they comply with it. If the employee's behavior does not improve, disciplinary action will be in order.

Turnover refers to the rate at which workers leave a company and are replaced by others. Like absenteeism, there are some causes of turnover that cannot be prevented and others that can. For instance, an employee may find a job elsewhere earning twice as much money. However, crew leaders can prevent some employee turnover situations. They can work to ensure safe working conditions for their crew, treat their workers fairly and consistently, and help promote good working conditions. The key is communication and promoting the motivational factors discussed earlier. Crew leaders need to know the problems if they are going to be able to successfully resolve them.

Some major causes of employee turnover include the following:

- Unfair/inconsistent treatment by the immediate supervisor
- Unsafe project sites
- Lack of job security or opportunities for advancement

For the most part, the actions described for absenteeism are also effective for reducing turnover. Past studies have shown that maintaining harmonious relationships on the job site goes a long way in reducing both turnover and absenteeism. This requires effective and proactive leadership on the part of the crew leader.

*Failure to Comply with Company Policies and Procedures* – Policies are rules that define the relationship between the company, its employees, its clients, and its subcontractors. Procedures include the instructions for carrying out the policies. Some companies have policies that dictate dress codes. The dress code may be designed partly to ensure safety, and partly to define the image a company wants to project to the outside world.

Companies develop procedures to ensure that everyone who performs a task does it safely and efficiently. Many procedures directly relate to safety. A lockout/tagout (LOTO) procedure is an example. In this procedure, the company defines who may perform a LOTO, how to properly complete and remove a LOTO, and who has the authority to remove it. Workers who fail to follow the procedure endanger themselves, as well as their co-workers.

Companies typically have a policy on disciplinary action, which defines steps to take if an employee violates company policies or procedures. The steps range from counseling by a supervisor for the first offense, to a written warning and/or LOI, to dismissal for repeat offenses. This will vary from one company to another. For example, some companies will fire an employee for the first violation of a safety procedure with potential for loss of life or serious injury.

The crew leader has the first-line responsibility for enforcing company policies and procedures. The crew leader should take the time with a new crew member to discuss the policies and procedures and show the crew member how to access them. If a crew member shows a tendency to neglect a policy or procedure, it is up to the crew leader to counsel that individual. If the crew member continues to violate a policy or procedure, the crew leader has no choice but to refer that individual to the appropriate authority within the company for disciplinary action.

> **NOTE**
> When a crew shows a pattern of consistent policy violations, management will scrutinize the crew leader and potentially take disciplinary action if the crew leader does not handle the violations appropriately. The crew leader is responsible for all aspects of the crew, not just its work accomplishment on the job.

## Case I:

On the way over to the job trailer, you look up and see a piece of falling scrap heading for one of the laborers. Before you can say anything, the scrap material hits the ground about five feet in front of the worker. You notice the scrap is a piece of conduit. You quickly pick it up, assuring the worker you will take care of this matter.

Looking up, you see your crew on the third floor in the area from which the material fell. You decide to have a talk with them. Once on the deck, you ask the crew if any of them dropped the scrap. The men look over at Bob, one of the electricians in your crew. Bob replies, "I guess it was mine. It slipped out of my hand."

It is a known fact that the Occupational Safety and Health Administration (OSHA) regulations state that an enclosed chute of wood shall be used for material waste transportation from heights of 20 feet or more. It is also known that Bob and the laborer who was almost hit have been seen arguing lately.

1. Assuming Bob's action was deliberate, what action would you take?

_____

_____

_____

2. Assuming the conduit accidentally slipped from Bob's hand, how can you motivate him to be more careful?

_____

_____

_____

3. What follow-up actions, if any, should be taken relative to the laborer who was almost hit?

_____

_____

_____

4. Should you discuss the apparent OSHA violation with the crew? Why or why not?

_____

_____

_____

5. What acts of leadership would be effective in this case? To what leadership traits are they related?

_____

_____

_____

## Case II:

The company just appointed Antonio crew leader of a tile-setting crew. Before his promotion into management, he had been a tile setter for five years. His work had been consistently of superior quality.

Except for a little good-natured kidding, Antonio's co-workers had wished him well in his new job. During the first two weeks, most of them had been cooperative while Antonio was adjusting to his supervisory role.

At the end of the second week, a disturbing incident took place. Having just completed some of his duties, Antonio stopped by the job-site wash station. There he saw Steve and Ron, two of his old friends who were also in his crew, washing.

"Hey, Ron, Steve, you shouldn't be cleaning up this soon. It's at least another thirty minutes until quitting time," said Antonio. "Get back to your work station, and I'll forget I saw you here."

"Come off it, Antonio," said Steve. "You used to slip up here early on Fridays. Just because you have a little rank now, don't think you can get tough with us." To this Antonio replied, "Things are different now. Both of you get back to work, or I'll make trouble." Steve and Ron said nothing more, and they both returned to their work stations.

From that time on, Antonio began to have trouble as a crew leader. Steve and Ron gave him the silent treatment. Antonio's crew seemed to forget how to do the most basic activities. The amount of rework for the crew seemed to be increasing. By the end of the month, Antonio's crew was behind schedule.

1. How do you think Antonio should've handled the confrontation with Ron and Steve?

_____

_____

2. What do you suggest Antonio could do about the silent treatment he got from Steve and Ron?

_____

_____

3. If you were Antonio, what would you do to get your crew back on schedule?

_____

_____

4. What acts of leadership could be used to get the crew's willing cooperation?

_____

_____

5. To which leadership traits do they correspond?

_____

_____

## Additional Resources

*Construction Workforce Development Professional*, NCCER. 2016. New York, NY: Pearson Education, Inc.

*Mentoring for Craft Professionals*, NCCER. 2016. New York, NY: Pearson Education, Inc.

*It's Your Ship: Management Techniques from the Best Damn Ship in the Navy*, Captain D. Michael Abrashoff, USN. 2012. New York City, NY: Grand Central Publishing.

*Survival of the Fittest,* Mark Breslin. 2005. McNally International Press.

*The Definitive Book of Body Language: The Hidden Meaning Behind People's Gestures and Expressions,* Barbara Pease and Allan Pease. 2006. New York City, NY: Random House / Bantam Books.

## 2.0.0  Section Review

1. A crew leader differs from a craftworker in that a crew leader _____.

   a. does not need direct experience in the job duties a craft professional typically performs
   b. can expect to oversee one or more workers in addition to performing typical craft duties
   c. is exclusively in charge of overseeing, since performing technical work isn't part of this role
   d. has no responsibility to be present on the job site

2. Feedback is important in verbal communication because it _____.

   a. requires the sender to repeat the message
   b. involves the receiver repeating back the message word for word
   c. informs the sender of how the message was received
   d. consists of a written analysis of the message

3. Once achieved, setting challenging goals for workers gives them a sense of _____.

   a. accomplishment
   b. entitlement
   c. persecution
   d. failure

4. Which of the following is *not* a characteristic of a successful team?

   a. Everyone participates and everyone counts.
   b. There is a sense of mutual trust.
   c. Members minimize communication with each other.
   d. The team leader is committed to the team.

5. Problem solving differs from decision making because it _____.

   a. involves finding an answer
   b. expresses an opinion
   c. involves starting or stopping an action
   d. separates facts from non-facts

# SECTION THREE

## 3.0.0 SAFETY AND SAFETY LEADERSHIP

### Objective

Identify a crew leader's typical safety responsibilities with respect to common safety issues, including awareness of safety regulations and the cost of accidents.

  a. Explain how a strong safety program can enhance a company's success.
  b. Explain the purpose of OSHA and describe the role of OSHA in administering worker safety.
  c. Describe the role of employers in establishing and administering safety programs.
  d. Explain how crew leaders are involved in administering safety policies and procedures.

### Trade Terms

**Intangible:** Not touchable, material, or measureable; lacking a physical presence.

**Lethargy:** Sluggishness, slow motion, lack of activity, or a lack of enthusiasm.

**Negligence:** Lack of appropriate care when doing something, or the failure to do something, usually resulting in injury to an individual or damage to equipment.

**Safety data sheets (SDS):** Documents listing information about a material or substance that includes common and proper names, chemical composition, physical forms and properties, hazards, flammability, handling, and emergency response in accordance with national and international hazard communication standards. Also called material safety data sheets (MSDS).

Businesses lose millions of dollars every year because of on-the-job accidents. Work-related injuries, sickness, and fatalities have caused untold suffering for workers and their families. Resulting project delays and budget overruns can cause huge losses for employers, and work-site accidents damage the overall morale of the crew.

Craft professionals routinely face hazards. Examples of these hazards include falls from heights, working on scaffolds, using cranes in the presence of power lines, operating heavy

## Did You Know?

### The Fatal Four

When OSHA inspects a job site, they focus on the types of safety hazards that are most likely to cause serious and fatal injuries. These hazards result in the following most common categories of injuries:

- Falls from elevations
- Struck-by hazards
- Caught-in/between hazards
- Electrical-shock hazards

machinery, and working on electrically-powered or pressurized equipment. Despite these hazards, experts believe that applying preventive safety measures could drastically reduce the number of accidents.

As a crew leader, one of your most important tasks is to enforce the company's safety program and make sure that all workers are performing their tasks safely. To be successful, the crew leader should do the following:

- Be aware of the human and monetary costs of accidents.
- Understand all federal, state, and local governmental safety regulations applicable to your work.
- Be the most visible example of the best safe work practices.
- Be involved in training workers in safe work methods.
- Conduct training sessions.
- Get involved in safety inspections, accident investigations, and fire protection and prevention.

*"Example is not the main thing in influencing others. It is the only thing."*
—Albert Schweitzer

Crew leaders are in the best position to ensure that their crew members perform all jobs safely. Providing employees with a safe working environment by preventing accidents and enforcing safety standards will go a long way towards maintaining the job schedule and enabling a job's completion on time and within budget.

### 3.1.0 The Impact of Accidents

Each day, workers in construction and industrial occupations face the risk of falls, machinery accidents, electrocutions, and other potentially fatal hazards. The National Institute of Occupational Safety and Health (NIOSH) statistics show that roughly 1,000 construction workers are killed on the job each year, which is more than any other industry. Falls are the leading cause of deaths in the construction industry through accident or negligence, accounting for over 60 percent of the fatalities in recent years. Nearly half of the fatal falls occurred from roofs, scaffolds, or ladders. Roofers, structural metal workers, and painters are at the greatest risk of fall fatalities (*Figure 13*).

In addition to the number of fatalities that occur each year, there are a staggering number of work-related injuries. In 2015, for example, almost 200,000 job-related injuries occurred in the construction industry. NIOSH estimates that the total cost of fatal and non-fatal injuries in the construction industry represents about 15 percent of the costs for all private industry. The main causes of injuries on construction sites include falls, electrocution, fires, and mishandling of machinery or equipment. According to NIOSH, back injuries are the leading health-related problem in workplaces.

### 3.1.1 Cost of Accidents

Occupational accidents cost roughly $250 billion or more every year. These costs affect the individual employee, the company, and the construction industry as a whole.

Organizations encounter both direct and indirect costs associated with workplace accidents. Direct costs are the money companies must pay out to workers' compensation claims and sick pay; indirect costs are all the other tangible and intangible things and costs a company must account for as the result of a worker's injury or death. To compete and survive, companies must control these as well as all other employment-related

*Figure 13* Falls are the leading cause of deaths and injuries in construction.

costs. There are many costs involved with workplace accidents. A company can insure some of these costs, but not others.

*Insured costs* – Insured costs are those costs either paid directly or reimbursed by insurance carriers. Insured costs related to injuries or deaths include the following:

- Compensation for lost earnings (known as *worker's comp*)
- Funeral charges
- Medical and hospital costs
- Monetary awards for permanent disabilities
- Pensions for dependents
- Rehabilitation costs

Insurance premiums or charges related to property damages include the following:

- Fire or other safety-related peril
- Structural loss; material and equipment loss or damage
- Loss of business use and occupancy
- Public liability
- Replacement cost of equipment, material, and structures

*Uninsured costs* – The relative direct and indirect costs of accidents are comparable to the visible and hidden portions of an iceberg, as shown in *Figure 14*. The tip of the iceberg represents direct costs, which are the visible costs. Not all of these are covered by insurance. The more numerous indirect costs aren't readily measurable, but they can represent a greater financial burden than the direct costs.

Uninsured costs from injuries or deaths include the following:

- First aid expenses
- Transportation costs
- Costs of investigations
- Costs of processing reports
- Down time on the job site
- Costs to train replacement workers

**Did You Know?**

# The Costs of Negligence

If you receive an injury as the result of a workplace accident, and a completed investigation shows that your injuries were due to your own negligence, you can't sue your employer for damages. In addition, workers' compensation insurance companies may decline to pay claims for negligent employees who are injured or killed.

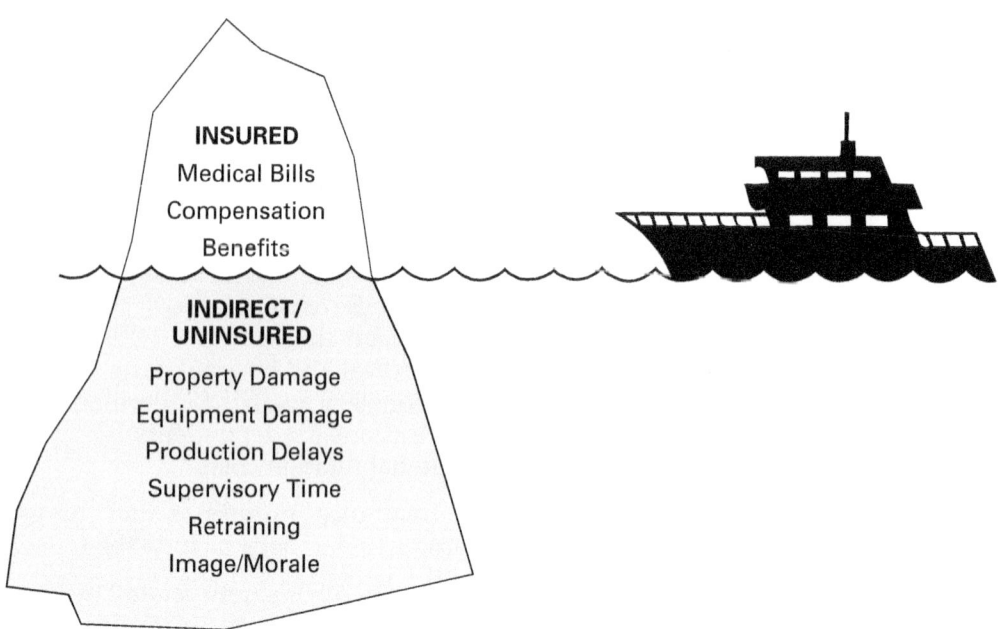

*Figure 14* Costs associated with accidents.

Uninsured costs related to wage losses include the following:

- Idle time of workers whose work is interrupted
- Time spent cleaning the accident area
- Time spent repairing damaged equipment
- Time lost by workers receiving first aid
- Costs of training injured workers in a new career

Uninsured costs related to production losses include the following:

- Product spoiled by accident
- Loss of skill and experience; worker replacement
- Lowered production capacity
- Idle machine time due to lack of qualified operators

Associated costs may include the following:

- Difference between actual losses and amount recovered
- Costs of rental equipment used to replace damaged equipment
- Costs of inexperienced temp or permanent new workers used to replace injured workers
- Wages or other benefits paid to disabled workers
- Overhead costs while production is stopped
- Impact on schedule
- Loss of client bonus or payment of forfeiture for delays

Uninsured costs related to off-the-job activities include the following:

- Time spent on ensuring injured workers' welfare
- Loss of skill and experience of injured workers
- Costs of training replacement workers

Uninsured costs related to intangible factors include the following:

- Increased labor conflict
- Loss of bid opportunities because of poor safety records
- Loss of client goodwill
- Lowered employee morale
- Unfavorable public relations

### 3.2.0 OSHA

To reduce safety and health risks and the number of injuries and fatalities on the job, the federal government has enacted laws and regulations, including the Occupational Safety and Health Act of 1970 (OSH Act of 1970). This law created the Occupational Safety and Health Administration (OSHA), which is part of the US Department of Labor. OSHA also provides education and training for employers and workers. Through the administration of OSHA, the US Congress seeks "to assure so far as possible every working man and woman in the Nation safe and healthful working conditions and to preserve our human resources…" (OSH Act of 1970, Section 2[b]).

To promote a safe and healthy work environment, OSHA issues standards and rules for working conditions, facilities, equipment, tools, and work processes. It does extensive research into occupational accidents, illnesses, injuries, and deaths to reduce the number of occurrences and adverse effects. In addition, OSHA regulatory agencies conduct workplace inspections to ensure that companies follow the standards and rules.

To enforce OSHA regulations, the government has granted regulatory agencies the right to enter public and private properties to conduct workplace safety investigations. The agencies also have the right to take legal action if companies are not in compliance with the Act. These regulatory agencies employ OSHA Compliance Safety and Health Officers (CSHO), who are experts in the occupational safety and health field. The CSHOs are thoroughly familiar with OSHA standards and recognize safety and health hazards.

States with their own occupational safety and health programs conduct inspections by enlisting the services of qualified state CSHOs.

Companies are inspected for a multitude of reasons. They may be randomly selected, or they may be chosen as a result of employee complaints, a report of an imminent danger, or major accidents/fatalities that have occurred.

OSHA has established significant monetary fines for the violation of its regulations. *Table 1* lists the penalties as of 2016. In some cases, OSHA will hold a superintendents or crew leaders personally liable for repeat violations as well. In addition to the fines, there are possible criminal charges for willful violations resulting in death or serious injury. The attitude of the employer and their safety history can have a significant effect on the outcome of a case.

### 3.3.0 Employer Safety Responsibilities

Each employer must set up a safety and health program to manage workplace safety and health and to reduce work-related injuries, illnesses, and fatalities. The program must be appropriate for the conditions of the workplace. It should consider the number of workers employed and the hazards they face while at work.

To be successful, the safety and health program must have management, leadership, and employee participation. In addition, training and

informational meetings play an important part in effective programs. Being consistent with safety policies is the key. Regardless of the employer's responsibility, however, the individual worker is ultimately responsible for his or her own safety.

### 3.3.1 Safety Program

The crew leader plays a key role in the successful implementation of the safety program. The crew leader's attitude toward the program sets the standard for how crew members view safety. Therefore, the crew leader should follow all program guidelines and require crew members to do the same.

Safety programs should consist of the following:

- Safety policies and procedures
- Safety information and training
- Posting of safety notices
- Hazard identification, reporting, and assessment
- Safety record system
- Accident reporting and investigation procedures
- Appropriate discipline for not following safety procedures

### 3.3.2 Safety Policies and Procedures

Employers are responsible for following OSHA and state safety standards. Usually, they incorporate federal and state OSHA regulations into a safety policies and procedures manual. Employees receive such a manual when they are hired.

During orientation, appropriate company staff should guide the new employees through the general sections of the safety manual and the sections that have the greatest relevance to their job.

**Table 1** OSHA Penalties for violations established in 2016

| Type of Violation | Maximum Penalty |
| --- | --- |
| Serious | |
| Other-Than-Serious | $12,471 per violation |
| Posting Requirements | |
| Failure to Abate | $12,471 per day |
| Willful or Repeated Violation | $124,709 per violation |

If the employee can't read, the employer should have someone read it to the employee and answer any questions that arise. The employee should then sign a form stating understanding of the information.

It isn't enough to tell employees about safety policies and procedures on the day they are hired and then never mention them again. Rather, crew leaders should constantly emphasize and reinforce the importance of adhering to all safety policies and procedures. In addition, employees should play an active role in determining job safety hazards and find ways to prevent and control hazards.

### 3.3.3 Hazard Identification and Assessment

Safety policies and procedures should be specific to the company. They should clearly present the hazards of the job and provide the means to report hazards to the proper level of management without prejudice to the individual doing the reporting. Crew leaders should also identify and assess hazards to which employees are exposed. They must also assess compliance with federal and state OSHA standards.

To identify and assess hazards, OSHA recommends that employers conduct periodic and random inspections of the workplace, monitor safety and health information logs, and evaluate new equipment, materials, and processes for potential hazards before they are used.

*"You get what you inspect, not what you expect."*
*—Anonymous*

Crew leaders and workers play important roles in identifying and reporting hazards. It is the crew leader's responsibility to determine what working conditions are unsafe and to inform employees of hazards and their locations. In addition, they should encourage their crew members to tell them about hazardous conditions. To accomplish this, crew leaders must be present and available on the job site.

The crew leader also needs to help the employee be aware of and avoid the built-in hazards to which craft professionals are exposed. Examples include working at elevations, working in confined spaces such as tunnels and underground vaults, on caissons, in excavations with earthen walls, and other naturally-dangerous projects. In addition, the crew leader can take safety measures, such as installing protective railings to prevent workers from falling from buildings, as well as scaffolds, platforms, and shoring.

### 3.3.4 Safety Information and Training

The employer must provide periodic information and training to new and long-term employees. This happens as often as necessary so that all employees receive adequate training. When safety and health information changes or workplace conditions create new hazards, the company must then provide special training and informational sessions. It is important to note that the company must present safety-related information in a manner that each employee will understand.

When a crew leader assigns an inexperienced employee a new task, the crew leader must ensure that the employee can do the work in a safe manner. The crew leader can accomplish this by providing safety information or training for groups or individuals.

When assigning an inexperienced employee a new task, do the following:

- Define the task.
- Explain how to do the task safely.
- Explain what tools and equipment to use and how to use them safely.
- Identify the necessary personal protective equipment and train the employee in its use.
- Explain the nature of the hazards in the work and how to recognize them.
- Stress the importance of personal safety and the safety of others.
- Hold regular safety training sessions with the crew's input.
- Review safety data sheets (SDS) that may be applicable.

### 3.3.5 Safety Record System

OSHA regulations (29 *CFR* 1904) require that employers keep records of hazards identified and document the severity of the hazard. The information should include the likelihood of employee exposure to the hazard, the seriousness of the harm associated with the hazard, and the number of exposed employees.

In addition, the employer must document the actions taken or plans for action to control the hazards. While it is best to take corrective action immediately, it is sometimes necessary to develop a plan to set priorities and deadlines and track progress in controlling hazards.

Employers who are subject to the recordkeeping requirements of the Occupational Safety and Health Act of 1970 must maintain records of all recordable occupational injuries and illnesses. The following are some OSHA forms that should be used for this recordkeeping:

- OSHA Form 300, *Log of Work-Related Injuries and Illnesses*
- OSHA Form 300A, *Summary of Work-Related Injuries and Illnesses*
- OSHA Form 301, *Injury and Illness Incident Report*

These three OSHA forms are included in the *Appendix* at the end of this module. Note that crew leaders directly handle the OSHA Form 301.

An SDS provides both workers and emergency personnel with the proper procedures for handling or working with a substance that may be dangerous. The document will include information such as physical data (melting point, boiling point, flash point, etc.), toxicity, health effects, first aid, reactivity, storage, disposal, protective equipment required for handling, and spill/leak procedures. These sheets are of particular use if a spill, fire, or other accident occurs.

Companies not exempted by OSHA must maintain required safety logs and retain them for 5 years following the end of the calendar year to which they relate. Logs must be available (normally at the company offices) for inspection and copying by representatives of the Department of Labor, the Department of Health and Human Services, or states given jurisdiction under the Act. Employees, former employees, and their representatives may also review these logs.

### 3.3.6 Accident Investigation

Employees must know from their training to immediately report any unusual event, accident, or injury. Policies and definitions of what these incidents consist of should be included in safety manuals. In the event of an accident, the employer is required to investigate the cause of the accident and determine how to avoid it in the future.

According to OSHA regulations, the employer must investigate each work-related death, serious injury or illness, or incident having the potential to cause death or serious physical harm. The employer should document any findings from the investigation, as well as the action plan to

## Summaries of Work-Related Injuries and Illnesses

Most companies with 11 or more employees must post an OSHA Form 300A, *Summary of Work-Related Injuries and Illnesses*, between February 1 and April 30 of each year. Employees have the right to review this form. Check your company's policies regarding this and the related OSHA forms.

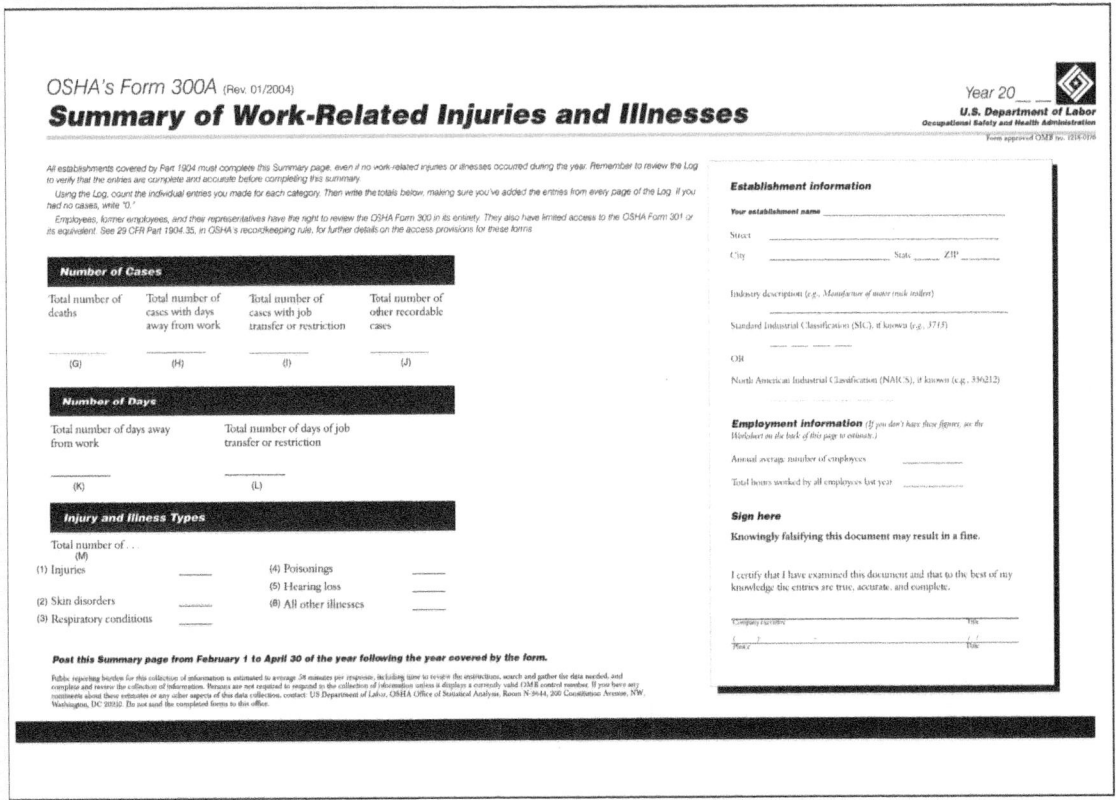

prevent future occurrences. The company should complete these actions immediately, with photos or video if possible. It's important that the investigation uncover the root cause of the accident to avoid similar incidents in the future. In many cases, the root cause was a flaw in the system that failed to recognize the unsafe condition or the potential for an unsafe act (*Figure 15*).

## 3.4.0 Leader Involvement in Safety

To be an effective, you must be actively involved in your company's safety program. Crew leader involvement includes conducting frequent safety training sessions and inspections, promoting first aid, and fire protection and prevention, preventing substance abuse on the job, and investigating accidents. Most importantly, crew leaders must practice safety at all times.

### 3.4.1 Safety Training Sessions

A safety training session may be a brief, informal gathering of a few employees or a formal meeting with instructional videos and talks by guest speakers. The size of the audience and the topics addressed determine the format of the meeting. You should plan to conduct small, informal safety sessions weekly.

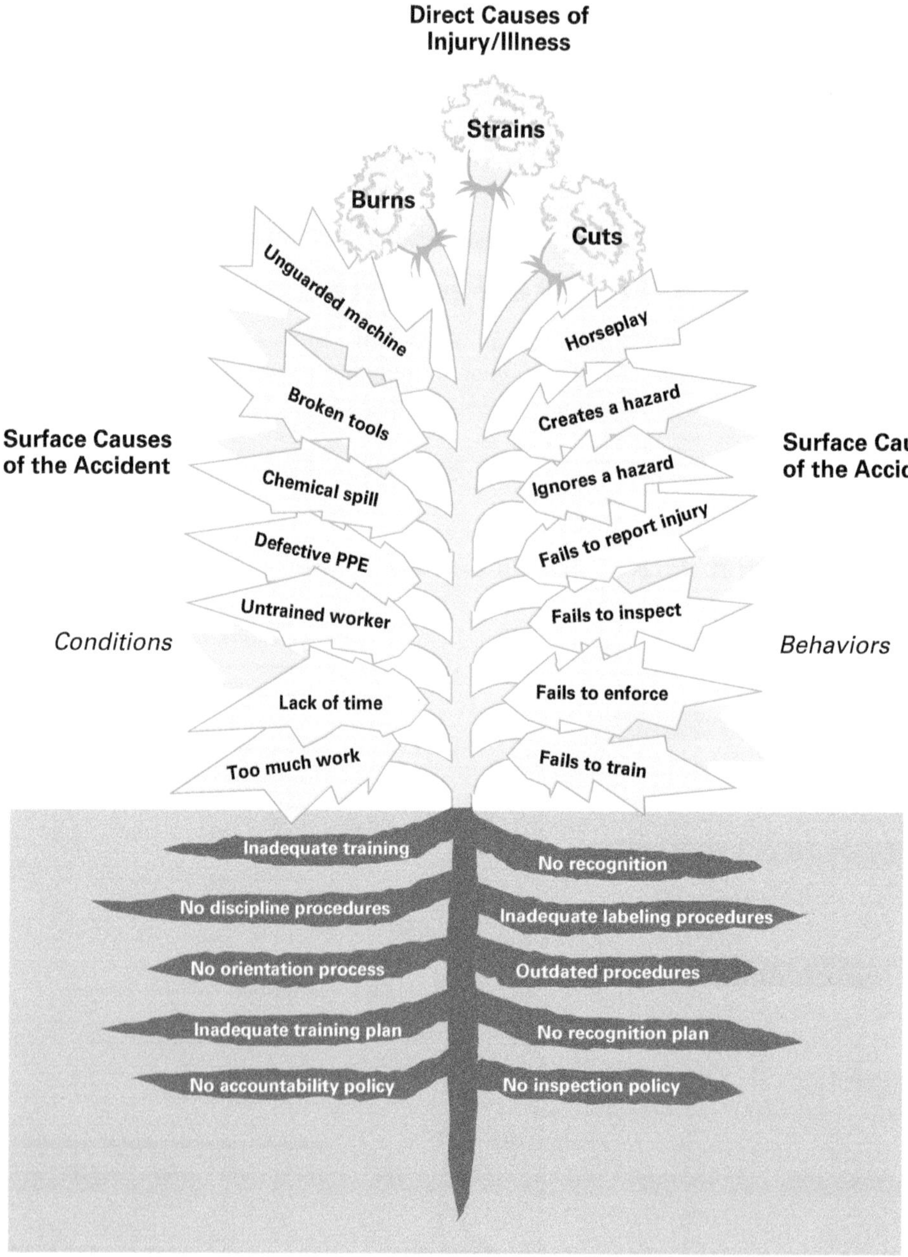

**Direct Causes of Injury/Illness**

Strains

Burns

Cuts

Unguarded machine

Horseplay

**Surface Causes of the Accident**

Broken tools

Creates a hazard

Chemical spill

Ignores a hazard

Defective PPE

Fails to report injury

Untrained worker

Fails to inspect

**Surface Causes of the Accident**

*Conditions*

Lack of time

Fails to enforce

*Behaviors*

Too much work

Fails to train

Inadequate training

No recognition

No discipline procedures

Inadequate labeling procedures

No orientation process

Outdated procedures

Inadequate training plan

No recognition plan

No accountability policy

No inspection policy

*Figure 15* Root causes of accidents.

# OSHA Accident Notification Requirements

There are urgent reporting requirements by the employer to OSHA for the following cases:

1. Within 8 hours: A work-related fatality.
2. Within 24 hours:
   - Work-related accident that resulted in an in-patient admission of one or more employees
   - Work-related amputation
   - Work-related loss of an eye

You should also plan safety training sessions in advance, and you should communicate the information to all affected employees. In addition, the topics covered in these training sessions should be timely and practical. Keep a log of each safety session signed by all attendees. It must be maintained as a record and available for inspection. It is advisable to attach a copy of a summary of the training session to the attendance record so you can keep track of what topics you covered and when.

## 3.4.2 Inspections

Crew leaders must make routine, frequent inspections to prevent accidents from happening. They must also take steps to avoid accidents. For that purpose, they need to inspect the job sites where their workers perform tasks. It's advisable to do these inspections before the start of work each day and during the day at random times.

Crew leaders must protect workers from existing or potential hazards in their work areas. Crews are sometimes required to work in areas controlled by other contractors. In these situations, the crew leader must maintain control over the safety exposure of the crew. If hazards exist, the crew leader should immediately bring the hazards to the attention of the contractor at fault, their superior, and the person responsible for the job site.

Crew leader inspections are only valuable if follow-up action corrects potential hazards. Therefore, crew leaders must be alert for unsafe and negligent acts on their work sites. When an employee performs an unsafe action, the crew leader must explain to the employee why the act was unsafe, tell that employee to not do it again, and request cooperation in promoting a safe working environment. The crew leader must document what happened and what the employee was told to do to correct the situation.

It is then important that crew leaders follow up to make certain the employee is complying with the safety procedures. Never allow a safety violation to go uncorrected. There are three courses of action that you, as a crew leader, can take in an unsafe work site situation:

- Get the appropriate party to correct the problem.
- Fix the problem yourself.
- Refuse to have the crew work in the area until the responsible party corrects the problem.

*"As soon as you see a mistake and don't fix it, it becomes your mistake."*
—*Anonymous*

## 3.4.3 First Aid

The primary purpose of first aid is to provide immediate and temporary medical care to employees involved in accidents, as well as employees experiencing non-work-related health emergencies, such as chest pain or breathing difficulty. To meet this objective, every crew leader should be aware of the location and contents of first aid kits available on the job site. Emergency numbers should be posted in the job trailer. In addition, OSHA requires that at least one person trained in first aid be present at the job site at all times. It's also advisable, but not required, that someone on site should have cardiopulmonary resuscitation (CPR) training.

> **NOTE**
> CPR certifications must be renewed every two years.

The victim of an accident or sudden illness at a job site may be harder to aid than elsewhere since the worker may be at a remote location. The site may be far from a rescue squad, fire department, or hospital, presenting a problem in the rescue and transportation of the victim to a hospital. The worker may also have received an injury from falling rock or other materials, so immediate special rescue equipment or first-aid techniques are often needed.

Employer benefits of having personnel trained in first aid at job sites include the following:

- The immediate and proper treatment of minor injuries may prevent them from developing into more serious conditions. Thus, these precautions can eliminate or reduce medical expenses, lost work time, and sick pay.

- It may be possible to determine if the injured person requires professional medical attention.
- Trained individuals can save valuable time preparing the inured for treatment for when professional medical care arrives. This service increases the likelihood of saving a life.

The American Red Cross, Medic First Aid, and the United States Bureau of Mines provide basic and advanced first aid courses at nominal costs. These courses include both first aid and CPR. The local area offices of these organizations can provide further details regarding the training available.

### 3.4.4 Fire Protection and Prevention

Fires and explosions kill and injure many workers each year, so it is important that crew leaders understand and practice fire-prevention techniques as required by company policy.

The need for protection and prevention is increasing as manufacturers introduce new building materials. Some building materials are highly flammable. They produce great amounts of smoke and gases, which cause difficulties for fire fighters, and can quickly overcome anyone present. Other materials melt when they burn and may puddle over floors, preventing fire-fighting personnel from entering areas where this occurs.

OSHA has specific standards for fire safety. Employers are required to provide proper exits, fire-fighting equipment, and employee training on fire prevention and safety. For more information, consult OSHA guidelines (available at **www.osha.gov**).

### 3.4.5 Substance Abuse

Substance abuse is a continuing problem in the workplace. Substance abuse is the inappropriate overuse of drugs and chemicals, whether they are legal or illegal. All substance abuse results in some form of mental, sensory, or physical impairment. Some people use illegal "street drugs", such as cocaine or crystal meth. Others use legal prescription drugs incorrectly by taking too many (or too few) pills, using other people's medications, or self-medicating. Alcohol can also be abused by consuming to the point of intoxication. Other substances that are legal in some states (e.g., marijuana) can cause prolonged impairment with heavy usage.

It is essential that crew leaders enforce company policies and procedures regarding substance abuse. Crew leaders must work with management to deal with suspected drug and alcohol abuse and should not deal with these situations themselves. The Human Resources department or a designated manager usually handles these cases.

There are legal consequences of substance abuse and the associated safety implications. If you observe an employee showing impaired behavior for any reason, immediately contact your supervisor and/or Human Resources department for assistance. You protect the business, and the employee's and other workers' safety by taking these actions. It is the crew leader's responsibility to maintain safe working conditions at all times. This may include removing workers from a work site where they may be endangering themselves or others.

For example, suppose several crew members go out and smoke marijuana or drink during lunch. Then, they return to work to erect scaffolding for a concrete pour in the afternoon. If you can smell marijuana on the crew member's clothing or alcohol on their breath, you must step in and take action. Otherwise, they might cause an accident that could delay the project, or cause serious injury or death to themselves or others.

> *"Concern for man himself and his safety must always form the chief interest of all technical endeavors."*
> —*Albert Einstein*

It is often difficult to detect drug and alcohol abuse because the effects can be subtle. The best way is to look for identifiable effects, such as those mentioned above, or sudden changes in behavior that aren't typical of the employee. Some examples of such behaviors include the following:

- Unscheduled absences; failure to report to work on time
- Significant changes in the quality of work
- Unusual activity or lethargy
- Sudden and irrational temper flare-ups
- Significant changes in personal appearance, cleanliness, or health

There are other more specific signs that should arouse suspicion, especially if more than one is visible:

- Slurring of speech or an inability to communicate effectively
- Shiftiness or sneaky behavior, such as an employee disappearing to wooded areas, storage areas, or other private locations
- Wearing sunglasses indoors or on overcast days to hide dilated or constricted pupils, conditions which impair vision (*Figure 16*)

*Figure 16* An impaired worker is a dangerous one.

- Wearing long-sleeved garments, particularly on hot days, to cover marks from needles used to inject drugs
- Attempting to borrow money from co-workers
- The loss of an employee's tools or company equipment

### 3.4.6 Job-Related Accident Investigations

Crew leaders are sometimes involved with an accident investigation. When an accident, injury, or report of work-connected illness takes place. If present on site, the crew leader should proceed immediately to the accident location to ensure that the victim receives proper first aid. The crew leader will also want to make sure that responsible individuals take other safety and operational measures to prevent another incident.

If required by company policy, the crew leader will also need to make a formal investigation and submit a report after an incident (including the completion of an *OSHA Form 301* report). An investigation looks for the causes of the accident by examining the circumstances under which it occurred and talking to the people involved. Investigations are perhaps the most useful tool in the prevention of future accidents.

The following are four major parts to an accident investigation:

- Describing the accident and related events leading to the accident
- Determining the cause(s) of the accident
- Identifying the persons or things involved and the part played by each
- Determining how to prevent reoccurrences

### 3.4.7 Promoting Safety

The best way for crew leaders to encourage safety is through example. Crew leaders should be aware that their behavior sets standards for their crew members. If a crew leader cuts corners on safety, then the crew members may think that it is okay to do so as well.

> CAUTION
>
> Workers often "follow the leader" when it comes to unsafe work practices. It is common for supervisory personnel to engage in unsafe practices and take more risks because they are more experienced. However, inexperienced or careless workers who take the same risks won't be as successful avoiding injury. As a leader, you must follow all safety practices to encourage your crew to do the same.

*"I cannot trust a man to control others who cannot control himself."*
—Robert E. Lee

The key to effectively promote safety is good communication. It is important to plan and coordinate activities and to follow through with safety programs. The most successful safety promotions occur when employees actively participate in planning and carrying out activities.

Some activities used by organizations to help motivate employees on safety and help promote safety awareness include the following:

- Safety training sessions
- Contests
- Recognition and awards
- Publicity

Safety training sessions can help keep workers focused on safety and give them the opportunity to discuss safety concerns with the crew. A previous section addressed this topic.

> **Did You Know?**
> # Substance Abuse
> An employee who is involved in an accident while under the influence of drugs or alcohol may be denied workers compensation insurance benefits.

# Case Study

For years, a prominent safety engineer was confused as to why sheet-metal workers fractured their toes frequently. The crew leader had not performed thorough accident investigations, and the injured workers were embarrassed to admit how the accidents really occurred. Further investigation discovered they used the metal-reinforced cap on their safety shoes as a "third hand" to hold the sheet metal vertically in place when they fastened it. The rigid and heavy metal sheet was inclined to slip and fall behind the safety cap onto the toes, causing fractures. The crew leader could have prevented several injuries by performing a proper investigation after the first accident.

### 3.4.8 Safety Contests

Contests are a great way to promote safety in the workplace. Examples of safety-related contests include the following:

- Sponsoring housekeeping contests for the cleanest job site or work area
- Challenging employees to come up with a safety slogan for the company or department
- Having a poster contest that involves employees or their children creating safety-related posters
- Recording the number of accident-free workdays or worker-hours
- Giving safety awards (hats, T-shirts, other promotional items or prizes)

One of the positive aspects of safety contests is their ability to encourage employee participation. It is important, however, to ensure that the contest has a valid purpose. For example, workers can display the posters or slogans created in a poster contest throughout the organization as safety reminders.

> **CAUTION**
>
> One mistake that some companies make when offering safety contests is providing tangible or monetary awards to departments or teams specifically for the lowest number of reported accidents or accident-free work hours. While well-intentioned, this approach appeals to the tendency of people to inflate their performance to win. Consequently, history shows this has the negative effect of encouraging the under-reporting of accidents and injuries, which defeats the purposes of safety contests.

### 3.4.9 Incentives and Rewards

Incentives and awards serve several purposes. Among them are acknowledging and encouraging good performance, building goodwill, reminding employees of safety issues, and publicizing the importance of practicing safety standards. There are countless ways to recognize and award safety. Examples include the following:

- Supplying food at the job site when a certain goal is achieved

- Providing a reserved parking space to acknowledge someone for a special achievement
- Giving gift items such as T-shirts or gift certificates to reward employees
- Giving awards to a department or an individual (*Figure 17*)
- Sending a letter of appreciation
- Publicly honoring a department or an individual for a job well done

You can use creativity to determine how to recognize and award good safety on the work site. The only precautionary measure is that the award should be meaningful and not perceived as a bribe. It should be representative of the accomplishment.

### 3.4.10 Publicity

Publicizing safety is the best way to get the message out to employees. An important aspect of publicity is to keep the message accurate and current. Safety posters that hang for years on end tend to lose effectiveness. It is important to keep ideas fresh.

Examples of promotional activities include posters or banners, advertisements or information on bulletin boards, payroll mailing stuffers, and employee newsletters. In addition, the company can purchase merchandise that promotes safety, including buttons, hats, T-shirts, and mugs.

*Figure 17* An example of a safety award.

Described here are three scenarios that reflect unsafe practices by craft workers. For each of these scenarios write down how you would deal with the situation, first as the crew leader of the craft worker, and then as the leader of another crew.

1. You observe a worker wearing his hard hat backwards and his safety glasses hanging around his neck. He is using a concrete saw.

_____

_____

_____

_____

_____

_____

2. As you are supervising your crew on the roof deck of a building under construction, you notice that a section of guard rail has been removed. Another contractor was responsible for installing the guard rail.

_____

_____

_____

_____

_____

_____

3. Your crew is part of plant shutdown at a power station. You observe that a worker is welding without a welding screen in an area where there are other workers.

_____

_____

_____

_____

_____

_____

## Additional Resources

*Construction Workforce Development Professional*, NCCER. 2016. New York, NY: Pearson Education, Inc.

*Mentoring for Craft Professionals*, NCCER. 2016. New York, NY: Pearson Education, Inc.

The following websites offer resources for products and training:

National Census of Fatal Occupational Injuries (NCFOI), **www.bls.gov**

National Institute of Occupational Safety and Health (NIOSH), **www.cdc.gov/niosh**

National Safety Council, **www.nsc.org**.

Occupational Safety and Health Administration (OSHA), **www.osha.gov**

## 3.0.0 Section Review

1. One of a crew leader's most important responsibilities to the employer is to _____.
   a. enforce company safety policies
   b. estimate material costs for a project
   c. make recommendations for setting up a crew
   d. provide input for fixed-price contracts

2. What amount can OSHA fine a company for willfully committing a violation of an OSHA safety standard?
   a. $1,000
   b. $7,000
   c. $12,471
   d. $124,709

3. Who is ultimately responsible for a worker's safety?
   a. The individual worker
   b. The crew leader
   c. The project superintendent
   d. The company HR department head

4. A crew leader's safety responsibilities include all of the following *except* _____.
   a. conducting safety training sessions
   b. developing a company safety program
   c. performing safety inspections
   d. participating in accident investigations

## 4.0.0 PROJECT PLANNING

### Objective

Demonstrate a basic understanding of the planning process, scheduling, and cost and resource control.

a. Describe how construction contracts are structured.
b. Describe the project planning and scheduling processes.
c. Explain how to implement cost controls on a construction project.
d. Explain the crew leader's role in controlling project resources and productivity.

### Performance Tasks

1. Develop and present a look-ahead schedule.
2. Develop an estimate for a given work activity.

### Trade Terms

**Critical path:** In manufacturing, construction, and other types of creative processes, the required sequence of tasks that directly controls the ultimate completion date of the project.

**Job diary:** A written record that a supervisor maintains periodically (usually daily) of the events, communications, observations, and decisions made during the course of a project.

**Look-ahead schedule:** A manual- or software-scheduling tool that looks several weeks into the future at the planned project events; used for anticipating material, labor, tool, and other resource requirements, as well as identifying potential schedule conflicts or other problems.

**Return on investment (ROI):** A measure of the gain or loss of money resulting from an investment; normally measured as a percentage of the original investment.

**Work breakdown structure (WBS):** A diagrammatic and conceptual method for subdividing a complex project, concept, or other thing into its various functional and organizational parts so that planners can analyze and plan for each part in detail.

This section describes methods of efficient project control. It examines estimating, planning and scheduling, and resource and cost control. All workers who participate in a job are responsible at some level for controlling cost and schedule performance, and for ensuring that they complete the project according to plans and specifications.

> **NOTE**
>
> This section mainly pertains to building-construction projects, but the project control principles described here apply generally to all types of projects.

The contractor, project manager, superintendent, and crew leader each have management responsibilities for their assigned jobs. For example, the contractor's responsibility begins with obtaining the contract, and it doesn't end until the client takes ownership of the project. The project manager is generally the person with overall responsibility for coordinating the project. Finally, the superintendent and crew leader are responsible for coordinating the work of one or more workers, one or more crews of workers within the company or, on occasion, one or more crews of subcontractors. The crew leader directs a crew in the performance of work tasks.

### 4.1.0 Construction Project Phases, Contracts, and Budgeting

Construction projects consist of three phases: the *development phase*, the *planning phase*, and the *construction phase*. Throughout these phases, the property owner must work directly with engineers and architects to gather data necessary for persuading a contractor to accept the job. Once a job is in progress, the crew leader fills an active role in maintaining the planned budget for the project.

#### 4.1.1 Development Phase

A new building project begins when an owner has decided to build a new facility or add to an existing facility. The development process is the first stage of planning for a new building project. This process involves land research and feasibility studies to ensure that the project has merit. Architects or engineers develop the conceptual drawings that define the project graphically. They then provide the owner with sketches of room layouts and elevations and make suggestions about what construction materials to use.

During the development phase, architects, engineers, and/or the owner develop an estimate for the proposed project and establish a preliminary budget. Once that budget is established, the financing of the project with lending institutions begins. The development team begins preliminary reviews with government agencies. These reviews include zoning, building restrictions, landscape requirements, and environmental impact studies.

The owner must analyze the project's cost and potential return on investment (ROI) to ensure that its costs won't exceed its market value and that the project provides a reasonable profit during its existence. If the project passes this test, the responsible architects/engineers will proceed to the planning phase.

### 4.1.2 Planning Phase

When the architects/engineers begin to develop the project drawings and specifications, they consult with other design professionals such as structural, mechanical, and electrical engineers. They perform the calculations, make a detailed technical analysis, and check details of the project for accuracy.

The design professionals create drawings and specifications. They use these drawings and specifications to communicate the necessary information to the contractors, subcontractors, suppliers, and workers that contribute to a project.

During the planning phase, the owners hold many meetings (*Figure 18*) to refine estimates, adjust plans to conform to regulations, and secure a construction loan. If the project is a condominium, an office building, or a shopping center, then a marketing firm develops a marketing program. In such cases, the selling of the project often starts before actual construction begins.

*Figure 18* Architects and clients meet to refine plans.

Next, the design team produce a complete set of drawings, specifications, and bid documents. Then the owner will select the method to obtain contractors. The owner may choose to negotiate with several contractors or select one through competitive bidding. Everyone concerned must also consider safety as part of the planning process. A safety crew leader may walk through the site as part of the pre-bid process.

Contracts for construction projects can take many forms. All types of contracts fall under three basic categories: *firm-fixed-price*, *cost reimbursable*, and *guaranteed maximum price*.

*Firm-fixed-price* – In this type of contract, the buyer generally provides detailed drawings and specifications, which the contractor uses to calculate the cost of materials and labor. To these costs, the contractor adds a percentage representing company overhead expenses such as office rent, insurance, and accounting/payroll costs. At the end, the contractor adds a profit factor.

When submitting the bid, the contractor will state very specifically the conditions and assumptions on which the company based the bid. These conditions and assumptions also form the basis from which parties can price allowable changes to the contract. Because contracting parties establish the price in advance, any changes in the job requirements once the job is started will impact the contractor's profit margin.

This is where the crew leader can play an important role by identifying problems that increase the amount of planned labor or materials. By passing this information up the chain of command, the crew leader allows the company to determine if the change is outside the scope of the bid. If so, they can submit a change order request to cover the added cost.

*Cost reimbursable contract* – In this type of contract, the buyer reimburses the contractor for labor, materials, and other costs encountered in the performance of the contract. Typically, the contractor and buyer agree in advance on hourly or daily labor rates for different categories of worker. These rates include an amount representing the contractor's overhead expense. The buyer also reimburses the contractor for the cost of materials and equipment used on the job.

The buyer and contractor also negotiate a profit margin. On this type of contract, the profit margin is likely to be lower than that of a fixed-price contract because of the significantly-reduced contractor's cost risk. The profit margin is often subject to incentive or penalty clauses that make the amount of profit awarded subject to performance by the contractor. The contract usually ties performance to project schedule milestones.

*Guaranteed maximum price (GMP) contract* – This form of contract, also called a *not-to-exceed contract*, is common on projects negotiated mainly with the owner. Owner's involvement in the process usually includes preconstruction, and the entire team develops the parameters that define the basis for the work.

With a GMP contract, the owner reimburses the contractor for the actual costs incurred by the contractor. The contract also includes a payment of a fixed fee up to the maximum price allowed in the contract. The contractor bears any cost overruns.

The advantages of the GMP contract vehicle may include the following:

- Reduced design time
- Allows for phased construction
- Uses a team approach to a project
- Reduction in changes related to incomplete drawings

### 4.1.3 Construction Phase

The designated contractor enlists the help of mechanical, electrical, elevator, and other specialty subcontractors to complete the construction phase. The contractor may perform one or more parts of the construction, and rely on subcontractors for the remainder of the work. However, the general contractor is responsible for managing all the trades necessary to complete the project. *Figure 19* shows the flow of a typical project from beginning to end.

As construction nears completion, the architect/engineer, owner, and government agencies start their final inspections and acceptance of the project. If the general contractor has managed the project, the subcontractors have performed their work, and the architects/engineers have regularly inspected the project to ensure it satisfied the local code, then the inspection process can finish up in a timely manner. This results in a satisfied client and a profitable project for all.

On the other hand, if the inspection reveals faulty workmanship, poor design, incorrect use of materials, or violation of codes, then the inspection and acceptance will become a lengthy battle and may result in a dissatisfied client and an unprofitable project.

The initial set of drawings for a construction project reflects the completed project as conceived by the architect and engineers. During construction, changes are usually necessary because of factors unforeseen during the design phase. For example, when electricians must reroute cabling or conduit, or the installed equipment location is different than shown on the original drawing, such changes must be marked on the drawings. Without this record, technicians called to perform maintenance or modify the equipment later will have trouble locating all the cabling and equipment.

Project supervision must document any changes made during construction or installation on the drawings as the changes occur. Architects usually note changes on hard-copy drawings using a colored pen or pencil, so users can readily spot the change. These marked-up versions are commonly called redline drawings. With mobile digital technology, architects and engineers can revise and promulgate the latest drawing versions almost instantly. After the drawings have been revised to reflect the redline changes, the final drawings are called as-built drawings, and are so marked. These become the drawings of record for the project.

### 4.1.4 Project-Delivery Systems

Project-delivery systems are processes for constructing projects, from development through construction. Project delivery systems focus on the following three primary systems, shown in *Figure 20*:

- *General contracting* – The traditional project delivery system uses a general contractor. In this type of project, the owner determines the design of the project, and then solicits proposals from general contractors. After selecting a general contractor, the owner contracts directly with that contractor, who builds the project as the prime, or controlling, contractor.
- *Design-build* – In the design-build system, a single entity manages both the design and construction of a project. Design-build delivery commonly use GMP contracts.
- *Construction management* – The construction management project delivery system uses a construction manager to facilitate the design and construction of a project. Construction managers are very involved in project control; their main concerns are controlling time, cost, and the quality of the project.

### 4.1.5 Cost Estimating and Budgeting

Before building a project, an estimate must be prepared. Estimating is the process of calculating the cost of a project. There are two types of costs to consider, including direct and indirect costs. Direct costs, also known as general conditions, are those that planners can clearly assigned to a

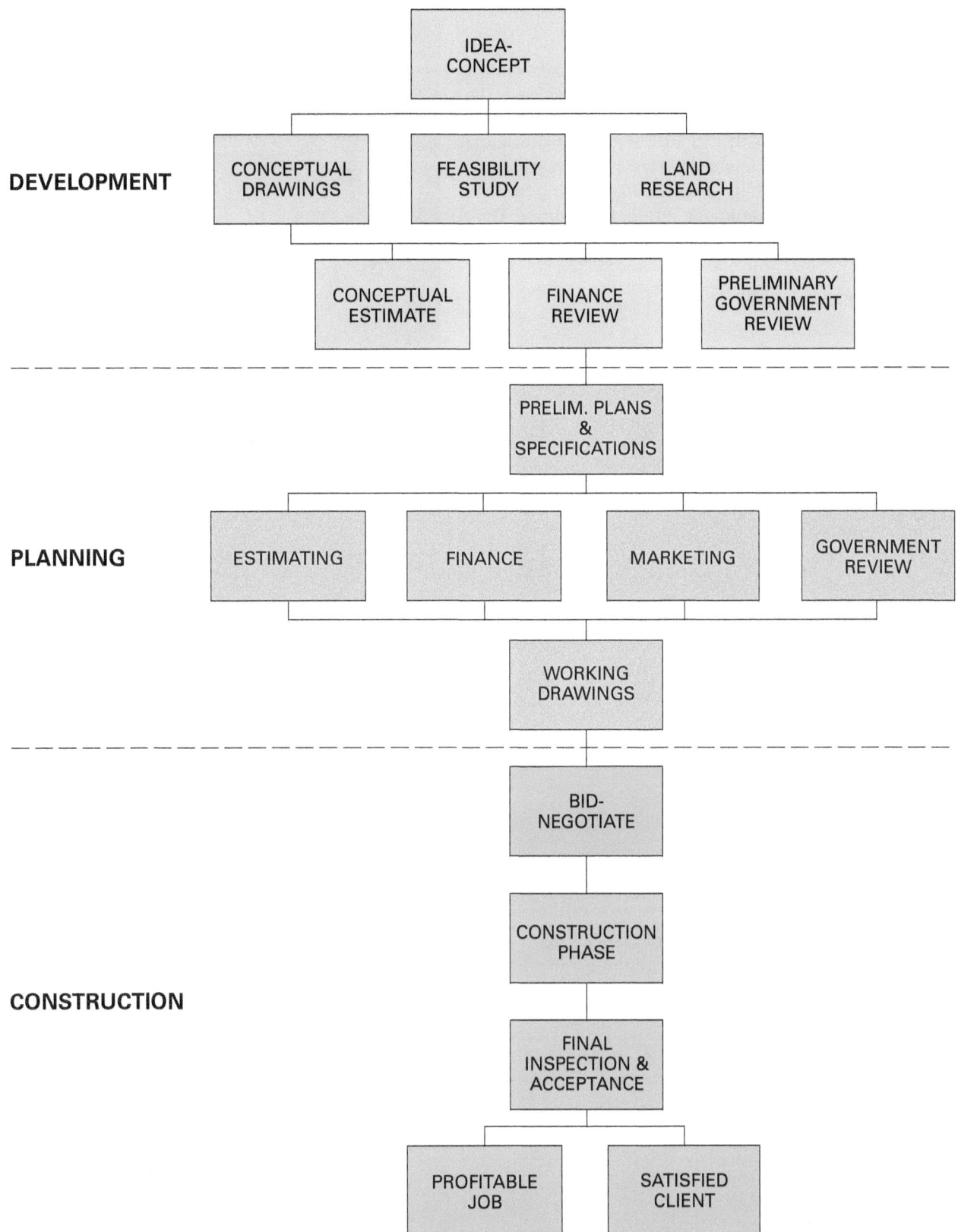

*Figure 19* Project flow diagram.

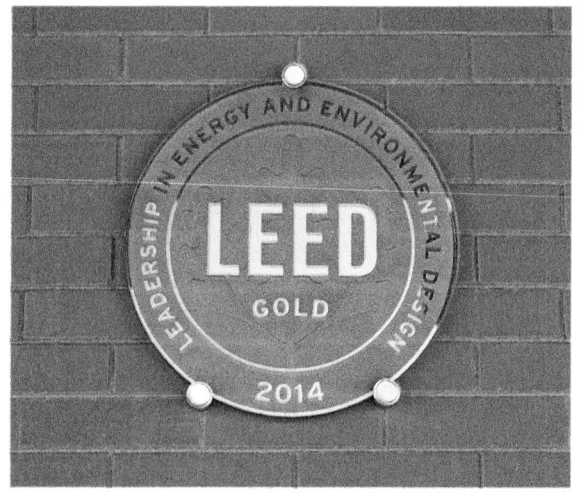
# LEED

LEED stands for *Leadership in Energy and Environmental Design*, which is an initiative started by the US Green Building Council (USGBC) to encourage and accelerate the adoption of sustainable construction standards worldwide through its Green Building Rating System. USGBC is a non-government, not-for-profit group. Their rating system addresses six categories:

1. Sustainable Sites (SS)
2. Water Efficiency (WE)
3. Energy and Atmosphere (EA)
4. Materials and Resources (MR)
5. Indoor Environmental Quality (EQ)
6. Innovation in Design (ID)

Building owners are the driving force behind the LEED voluntary program. Construction crew leaders may not have input into the decision to seek LEED certification for a project, or what materials are used in the project's construction. However, crew leaders can help to minimize material waste and support recycling efforts, both of which are factors in obtaining LEED certification.

An important question to ask is whether your project is seeking LEED certification. If the project is seeking certification, the next step is to ask what your role will be in getting the certification. If you are procuring materials, what information does the project need and who should receive it? What specifications and requirements do the materials need to meet? If you are working outside the building or inside in a protected area, what do you need to do to protect the work area? How does your crew manage waste? Are there any other special requirements that will be your responsibility? Do you see any opportunities for improvement? LEED principles are described in more detail in NCCER's *Your Role in the Green Environment* (Module ID 70101-15) and *Sustainable Construction Supervisor* (Module ID 70201-11).

## DELIVERY SYSTEMS

| RESPONSIBLE PARTIES | | GENERAL CONTRACTING | DESIGN-BUILD | CONSTRUCTION MANAGEMENT |
|---|---|---|---|---|
| | OWNER | Designs project (or hires architect) | Hires general contractor | Hires construction management company |
| | GENERAL CONTRACTOR | Builds project (with owner's design) | Involved in project design, builds project | Builds, may design (hired by construction management company) |
| | CONSTRUCTION MANAGEMENT COMPANY | | | Hires and manages general contractor and architect |

*Figure 20* Project-delivery systems.

budget. Indirect costs are overhead costs shared by all projects. Planners generally calculate these costs as an overhead percentage to labor and material costs.

Direct costs include the following:

- Materials
- Labor
- Tools
- Equipment

Indirect costs refer to overhead items such as the following:

- Office rent
- Utilities
- Telecommunications
- Accounting
- Office supplies, signs

The bid price includes the estimated cost of the project as well as the profit. Profit refers to the amount of money that the contractor will make after paying all the direct and indirect costs. If the direct and indirect costs exceed the estimate for the job, the difference between the actual and estimated costs must come out of the company's profit. This reduces what the contractor makes on the job.

Profit is the fuel that powers a business. It allows the business to invest in new equipment and facilities, provide training, and to maintain a reserve fund for times when business is slow. In large companies, profitability attracts investors who provide the capital necessary for the business to grow. For these reasons, contractors can't afford to consistently lose money on projects. If they can't operate profitably, they are forced out of business. Crew leaders can help their companies remain profitable by managing budget, schedule, quality, and safety adhering to the drawings, specifications, and project schedule.

The cost estimate must consider many factors. Many companies employ professional cost estimators to do this work. They also maintain performance data for previous projects. They use this data as a guide in estimating new projects. Development of a complete estimate generally proceeds as follows:

*Step 1* Using the drawings and specifications, an estimator records the quantity of the materials needed to construct the job. Construction companies call this step the *quantity takeoff*. The estimator enters the information on a hard-copy or digital takeoff sheet like the one shown in *Figure 21*.

*Step 2* The estimator uses the company's productivity rates to calculate the amount of labor required to complete the project. Most companies keep records of these rates for the type and size of the jobs that they perform. The company's estimating department maintains and updates these records.

*Step 3* The estimator calculates the labor hours required by dividing the estimated amount of work by the productivity rate.
- For example, if the productivity rate for concrete finishing is 40 square feet per hour, and there are 10,000 square feet of concrete to be finished, then 250 hours of concrete finishing labor is required (10,000 ÷ 4 = 250).
- The estimator multiplies this number by the hourly rate for concrete finishing to determine the cost of that labor category.
- If this work is subcontracted, then the estimator uses the subcontractor's cost estimate, raised by an overhead factor, in place of direct-labor cost.

*Step 4* The estimator transfers the total material quantities from the quantity takeoff sheet to a summary or pricing sheet (*Figure 22*). The total cost of materials is calculated after obtaining material prices from local suppliers.

*Step 5* Next, the estimator determines the cost of equipment needed for the project. This number could reflect rental cost or a factor applied by the company when they plan to use their own equipment.

*Step 6* The estimator totals the cost of all resources on the summary sheet—materials, equipment, tools, and labor. The estimator can also calculate the material unit cost—the total cost divided by the total number of units of listed materials.

*Step 7* The estimator adds the cost of taxes, bonds, insurance, subcontractor work, and other indirect costs to the direct costs of the materials, equipment, tools, and labor.

*Step 8* A sum of direct and indirect costs yields the total project cost. The contractor adds the expected profit to that total.

**WORKSHEET**

PAGE #

Takeoff By:

Checked By:

DATE _____

SHEET ___ of ___

PROJECT _____

ARCHITECT _____

| REF. | DESCRIPTION | NO | DIMENSIONS | | | EXTENSION | | QUANTITY | UNIT | TOTAL | | REMARKS |
|---|---|---|---|---|---|---|---|---|---|---|---|---|
| | | | LENGTH | WIDTH | HEIGHT | | | | | QUANTITY | UNIT | |
| | | | | | | | | | | | | |
| | | | | | | | | | | | | |
| | | | | | | | | | | | | |
| | | | | | | | | | | | | |
| | | | | | | | | | | | | |
| | | | | | | | | | | | | |
| | | | | | | | | | | | | |
| | | | | | | | | | | | | |
| | | | | | | | | | | | | |
| | | | | | | | | | | | | |
| | | | | | | | | | | | | |
| | | | | | | | | | | | | |
| | | | | | | | | | | | | |
| | | | | | | | | | | | | |

*Figure 21* Quantity takeoff sheet.

**SUMMARY SHEET**

DATE _____

SHEET _____ of _____

By:

PROJECT _____

WORK ORDER # _____

TITLE: _____

PAGE #

| DESCRIPTION | QUANTITY | | MATERIAL COST | | LABOR MAN HOURS FACTORS | | | | | LABOR COST | | ITEM COST | |
|---|---|---|---|---|---|---|---|---|---|---|---|---|---|
| | TOTAL | UT | PER UNIT | TOTAL | CRAFT | PR UNIT | TOTAL | RATE | COST PR | PER | TOTAL | TOTAL | PER UNIT |
| | | | | | | | | | | | | | |
| | | | | | | | | | | | | | |

MATERIAL | LABOR | TOTAL

*Figure 22* Summary sheet.

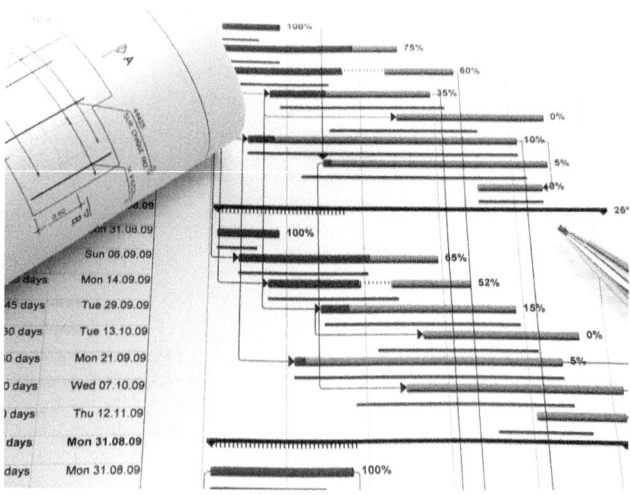

*Figure 23* Proper prior planning prevents poor performance.

As a crew leader, you may be required to estimate quantities of materials. You will need a set of construction drawings and specifications to estimate the amount of a certain type of material required to perform a job. You should carefully review the appropriate section of the technical specifications and page(s) of drawings to determine the types and quantities of materials required. Then enter the quantities on the worksheet. For example, you should review the specification section on finished carpentry along with the appropriate pages of drawings before taking off the linear feet of door and window trim.

If insufficient materials are available to complete the job and an estimate is required, the estimator must determine how much more construction work is necessary. Knowing this, the crew leader can then determine the materials needed. You must also reference the construction drawings in this process.

## 4.2.0 Planning

One definition of planning is determining the methods used and their sequence to carry out the different tasks to complete a project (*Figure 23*). Planning involves the following:

- Determining the best method for performing the job
- Identifying the responsibilities of each person on the work crew
- Determining the duration and sequence of each activity
- Identifying the tools and equipment needed to complete a job
- Ensuring that the required materials are at the work site when needed
- Making sure that heavy construction equipment is available when required

- Working with other contractors in such a way as to avoid interruptions and delays

Some important reasons for crew-leader planning include the following:

- Controlling the job in a safe manner so that it is built on time and within cost
- Lowering job costs through improved productivity
- Preparing for bad weather or unexpected occurrences
- Promoting and maintaining favorable employee morale
- Determining the best and safest methods for performing the job

With a plan, a crew leader can direct work efforts efficiently and can use resources such as personnel, materials, tools, equipment, work area, and work methods to their full potential.

A proactive crew leader will always have a backup plan in case circumstances prevent the original plan from working. There are many circumstances that can cause a plan to go awry, including adverse weather, equipment failure, absent crew members, and schedule slippage by other crafts.

Project planners establish time and cost limits for the project; the crew leader's planning must fit within those constraints. Therefore, it is important to consider the following factors that may affect the outcome:

- Site and local conditions, such as soil types, accessibility, or available staging areas
- Climate conditions that should be anticipated during the project
- Timing of all phases of work
- Types of materials to be installed and their availability

Assume you are the leader of a crew building footing formwork for the construction shown in the figure below. You have used all of the materials provided for the job, yet you have not completed it. You study the drawings and see that the formwork consists of two side forms, each 12" high. The total length of footing for the entire project is 115'-0". You have completed 88'-0" to date; therefore, you have 27'-0" remaining (115' – 88' = 27').

Your task is to prepare an estimate of materials that you will need to complete the job. In this case, you will estimate only the side form materials (do not consider the miscellaneous materials here).

- Footing length to complete: 27'-0"
- Footing height: 1'-0"

Refer to the worksheet on the next page for a final tabulation of the side forms needed to complete the job.

1. Using the same footing as described in the example above, calculate the quantity (square feet) of formwork needed to finish 203 linear feet of the footing. Place this information directly on the worksheet.
2. You are the crew leader of a carpentry crew whose task is to side a warehouse with plywood sheathing. The wall height is 16', and there is a total of 480 linear feet of wall to side. You have done 360 linear feet of wall and have run out of materials. Calculate how many more feet of plywood you will need to complete the job. If you are using 4 × 8 plywood panels, how many will you need to order to cover the additional work? Write your estimate on the worksheet.

Show your calculations to the instructor.

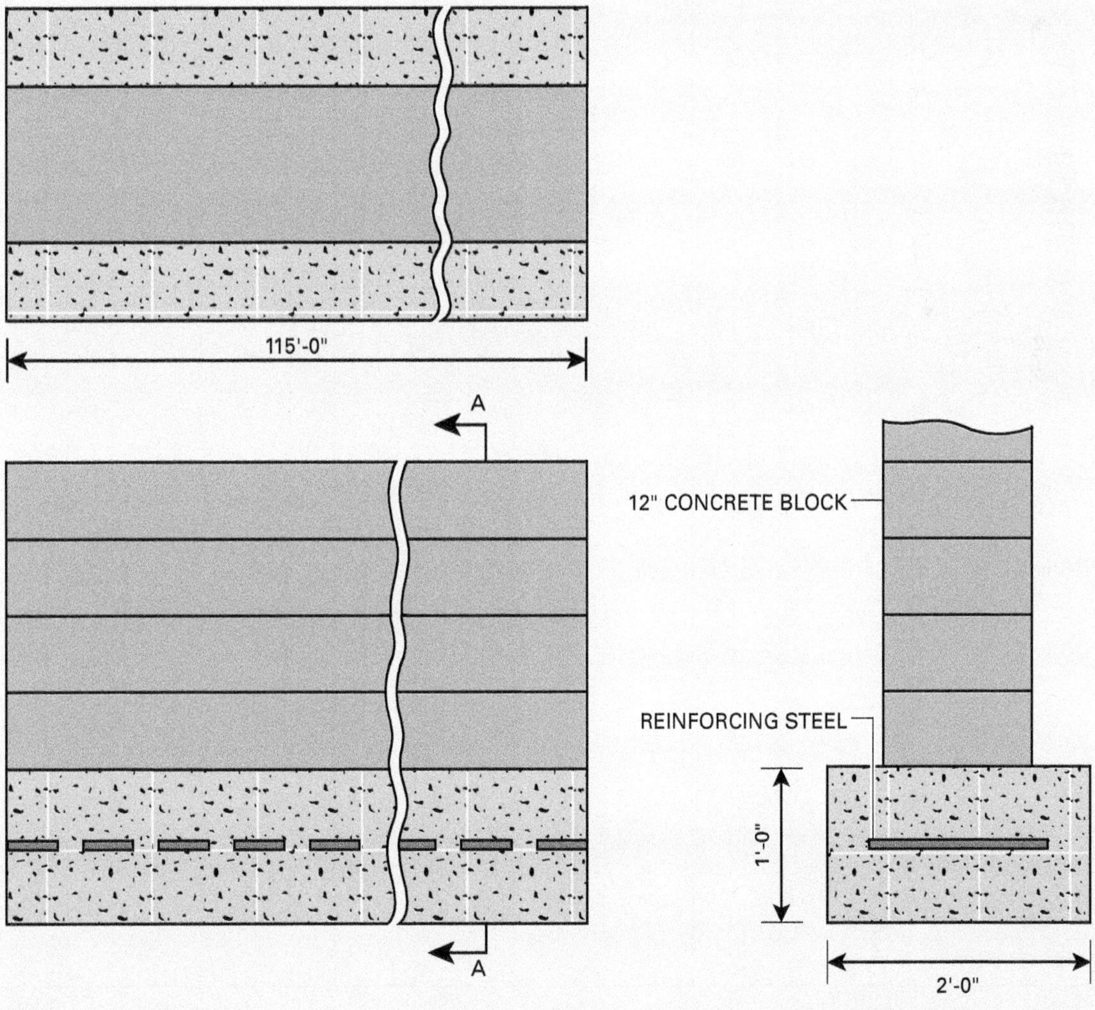

115'-0"

A

A

12" CONCRETE BLOCK

REINFORCING STEEL

1'-0"

2'-0"

## WORKSHEET

Takeoff By: RWH

Checked By:

DATE __2/1/17__

SHEET __01__ of __01__

PROJECT __Sam's Diner__

ARCHITECT __654b__

PAGE #1

| REF. | DESCRIPTION | NO | DIMENSIONS LENGTH | DIMENSIONS WIDTH | DIMENSIONS HEIGHT | EXTENSION | QUANTITY | UNIT | TOTAL QUANTITY | TOTAL UNIT | REMARKS |
|------|-------------|-----|--------|-------|--------|-----------|----------|------|----------|------|---------|
|  | Footing Side Forms | 2 | 27'0" |  | 1'0" | 2x27x1 | 54 | SF | 54 | SF |  |

- Equipment and tools required and their availability
- Personnel requirements and availability
- Relationships with the other contractors and their representatives on the job

On a simple job, crews can handle these items almost automatically. However, larger or more complex jobs require the planner to give these factors more formal consideration and study.

### 4.2.1 Stages of Planning

Formal planning for a construction job occurs at specific times in the project process. The two most important stages of planning occur in the preconstruction phase and during the construction work.

*Preconstruction planning* – The preconstruction stage of planning occurs before the start of construction. The preconstruction planning process doesn't always involve the crew leader, but it's important to understand what it consists of.

There are two phases of preconstruction planning. The first is developing the proposal, bid, or negotiated price for the job. This is when the estimator, the project manager, and the field superintendent develop a preliminary plan for completing the work. They apply their experience and knowledge from previous projects to develop the plan. The process involves determining what methods, personnel, tools, and equipment the work will require and what level of productivity they can achieve.

The second phase of preconstruction planning occurs after the client awards the contract. This phase requires a thorough knowledge of all project drawings and specifications. During this process, planners select the actual work methods and resources needed to perform the work. Here, crew leaders might get involved, but their planning must adhere to work methods, production rates, and resources that fit within the estimate prepared during contract negotiations. If the project requires a method of construction different

---

### Participant Exercise F

1. In your own words, define planning, and describe how a job can be done better if it is planned. Give an example.

   _____

   _____

   _____

   _____

2. Consider a job that you recently worked on to answer the following:
   a. List the material(s) used.
   b. List each member of the crew with whom you worked and what each person did.
   c. List the kinds of equipment used.

   _____

   _____

   _____

   _____

   _____

3. List some suggestions for how the job could have been done better, and describe how you would plan for each of the suggestions.

   _____

   _____

   _____

   _____

---

from what is normal, planners will usually inform the crew leader of what method to use.

*Construction planning* – During construction, the crew leader is directly involved in daily planning. This stage of planning consists of selecting methods for completing tasks before beginning work. Effective planning exposes likely difficulties, and enables the crew leader to minimize the unproductive use of personnel and equipment. Proper planning also provides a gauge to measure job progress.

Effective crew leaders develop a tool known as the look-ahead schedule. These schedules consider actual circumstances as well as projections two-to-three weeks into the future. Developing a look-ahead schedule helps ensure that all resources are available on the project when needed. Most scheduling apps and programs include this feature that automatically help you focus on this period in the job.

### 4.2.2 The Planning Process

The planning process consists of the following five steps:

*Step 1*   Establish a goal.

*Step 2*   Identify the completed work activities required to achieve the goal.

*Step 3*   Identify the required tasks to accomplish those activities.

*Step 4*   Communicate responsibilities.

*Step 5*   Follow up to verify the goal achievement.

*Establishing a goal* – The term *goal* has different meanings for different people. In general, a goal is a specific outcome that one works toward. For example, the project superintendent of a home construction project could establish the goal to have a house dried-in by a certain date. (The term *dried-in* means ready for the application of roofing and siding.) To meet that goal, the leader of the framing crew and the superintendent would need to agree to a goal to have the framing completed by a given date. The crew leader would then establish sub-goals (objectives) for the crew to complete each element of the framing (floors, walls, roof) by a set time. The superintendent would need to set similar goals with the crews that install sheathing, building wrap, windows, and exterior doors. However, if the framing crew doesn't meet its goal, that will delay the other crews.

*Identifying the required work* – The second step in planning is to identify the necessary work to achieve the goal as a series of activities in a certain sequence. You will learn how to break down a job into activities later in this section. At this point, the crew leader should know that, for each activity, one or more objectives must be set.

An objective is a statement of a condition the plan requires to exist or occur at a specific time. An objective must:

• Mean the same thing to everyone involved
• Be measurable, so that everyone knows when it has been reached
• Be achievable with the resources available
• Have everyone's full support

Examples of practical objectives include the following:

• "By 4:30 p.m. today, the crew will have completed installation of the floor joists."
• "By closing time Friday, the roof framing will be complete."

Notice that both examples meet the first three requirements of an objective. Planners assume that everyone involved in completing the task is committed to achieving the objective. The advantage in developing objectives for each work activity is that it allows the crew leader to determine if the crew is following the plan. In addition, objectives serve as sub-goals that are usually under the crew leader's control.

Some construction work activities, such as installing 12" footing forms, are done so often that they require little planning. However, other jobs, such as placing a new type of mechanical equipment, require substantial planning. This type of job demands that the crew leader set specific objectives.

Whenever faced with a new or complex activity, take the time to establish objectives that will serve as guides for accomplishing the job. You can use these guides in the current situation, as well as for similar work in the future.

*Identifying the required tasks* – To plan effectively, the crew leader must be able to break a work activity assignment down into smaller tasks. Large jobs include a greater number of tasks than small ones, but all jobs can be broken down into manageable components.

When breaking down an assignment into tasks, make each task identifiable and definable. A task is identifiable when one knows the types and amounts of resources it requires. A task is definable if it has a specific duration. For purposes of efficiency, the job breakdown should not be too detailed or complex, unless the job has never been done before or must be performed with strictest efficiency.

For example, a suitable breakdown for the work activity to install square vinyl floor tiles in a cafeteria might be the following:

**Step 1** Prepare the floor.

**Step 2** Stage the tiles.

**Step 3** Spread the adhesive.

**Step 4** Lay the tiles.

**Step 5** Clean the tiles.

**Step 6** Wax the floor.

The crew leader could create even more detail by breaking down any one of the tasks into subtasks. In this case, however, that much detail is unnecessary and wastes the crew leader's time and the project's money.

Planners can divide every work activity into three general parts:

- Preparing
- Performing
- Cleaning up

Some of the most frequent mistakes made in the planning process are forgetting to prepare and to clean up. The crew leader must not overlook preparation and cleanup.

After identifying the various tasks that make up the job and developing an objective for each task, the crew leader must determine what resources the job requires. Resources include labor, equipment, materials, and tools. In most jobs, the job estimate identifies these resources. The crew leader must make sure that these resources are available on the site when needed.

*"By failing to prepare, you are preparing to fail."*
*—Benjamin Franklin*

*Communicating responsibilities* – No supervisor can complete all the activities within a job alone. The crew leader must rely on other people to get everything done. Therefore, most jobs have a crew of people with various experiences and skill levels to assist in the work. The crew leader's job is to draw from this expertise to get the job done well and in a safe and timely manner.

Once the various activities that make up the job have been determined, the crew leader must identify the person or persons responsible for completing each activity. This requires that the crew leader be aware of the skills and abilities of the people on the crew. Then, the crew leader

must put this knowledge to work in matching the crew's skills and abilities to specific tasks required to complete the job.

After matching crew members to specific activities, the crew leader must then assign work to the crew. The crew leader normally communicates responsibilities verbally; the crew leader often talks directly to the responsible person for the activity. There may be times when the crew leader assigns work through written instructions or indirectly through someone other than the crew leader. Either way, crew members should know what it is they are responsible for accomplishing on the job.

*Following up* – Once the crew leader has delegated the activities to the appropriate crew members, there must be follow-up to make sure that the crew has completed them correctly and efficiently. Task follow-up involves being present on the job site to make sure all the resources are available to complete the work, ensuring that the crew members are working on their assigned activities, answering any questions, and helping to resolve any problems that occur while the work is being done. In short, follow-up activity means that the crew leader is aware of what's going on at the job site and is doing whatever is necessary to make sure that the crew completes the work on schedule. *Figure 24* reviews the planning steps.

The crew leader should carry a small note pad or electronic device for planning and taking notes. That way, you can record thoughts about the project as they occur, and you won't forget pertinent details. The crew leader may also choose to use a manual planning form such as the one illustrated in *Figure 25*.

As the job progresses, refer to these resources to see that the tasks are being done according to plan. This is job analysis. Construction projects that don't proceed according to work plans usually end up costing more and taking longer. Therefore, it is important that crew leaders refer to the planning documents periodically.

The crew leader is involved with many activities on a day-to-day basis. Thus, it is easy to forget important events if they aren't recorded. To help keep track of events such as job changes, interruptions, and visits, the crew leader should keep a job diary. A job diary is a notebook in which the crew leader records activities or events that take place on the

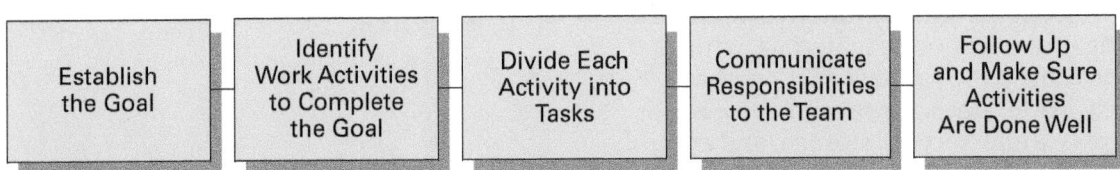

*Figure 24* Steps to effective planning.

## DAILY WORK PLAN

*"PLAN YOUR WORK AND WORK YOUR PLAN = EFFICIENCY"*

Plan of _____    Date _____

| PRIORITY | DESCRIPTION | ✓ When Completed<br>✗ Carried Forward |
|---|---|---|
| | | |
| | | |
| | | |
| | | |
| | | |

*Figure 25* Planning form.

job site that may be important later. When making entries in a job diary, make sure that the information is accurate, factual, complete, consistent, organized, and up-to-date. This is especially true if documenting personnel problems. If the company requires maintaining a job diary, follow company policy in determining which events and what details you should record. However, if there is a doubt about what to include, it is better to have too much information than too little. *Figure 26* shows a sample page from a job diary.

### 4.2.3 Planning Resources

Once a job has been broken down into its tasks or activities, the various resources needed to perform them must be determined and accounted for. Resource planning includes the following specific considerations:

- *Safety planning* – Using the company safety manual as a guide, the crew leader must assess the safety issues associated with the job and take necessary measures to minimize any risk to the crew. This may involve working with the company or site safety officer and may require a formal hazard analysis.
- *Materials planning* – Preconstruction planning identifies the materials required for the job and lists them on the job estimate. Companies usually order the materials from suppliers who have previously provided quality materials on schedule and within estimated cost.

The crew leader is usually not involved in the planning and selection of materials, which happens during the preconstruction phase. The crew leader does, however, have a major role to play in the receipt, storage, and control of the materials after they reach the job site.

- The crew leader is also involved in planning materials for tasks such as job-built formwork and scaffolding. In addition, the crew leader may run out of a specific material, such as fasteners, and need to order more. In such cases, be sure to consult the appropriate supervisor, since most companies have specific purchasing policies and procedures.

July 8, 2017

Weather: Hot and Humid

Project: Company XYZ Building

- The paving contractor crew arrived late (10 am).
- The owner representative inspected the footing foundation at approximately 1 pm.
- The concrete slump test did not pass. Two trucks had to be ordered to return to the plant, causing a delay.
- John Smith had an accident on the second floor. I sent him to the doctor for medical treatment. The cause of the accident is being investigated.

*Figure 26* Sample page from a job diary.

- *Site planning* – There are many planning elements involved in site work. The following are some of the key elements:

  - Access roads
  - Emergency procedures
  - Material and equipment storage
  - Material staging
  - Parking
  - Sedimentation control
  - Site security
  - Storm water runoff

- *Equipment planning* – The preconstruction phase addresses much of the planning for use of construction equipment. This planning includes the types of equipment needed, the use of the equipment, and the length of time it will be on the site. The crew leader must work with the main office to make certain that the equipment reaches the job site on time. The crew leader must also ensure that crew equipment operators are properly trained.

  Coordinating the use of the equipment is also very important. Some equipment operates in combination with other equipment. For example, dump trucks are generally required when loaders and excavators are used. The crew leader should also coordinate equipment with other contractors on the job. Sharing equipment can save time and money and avoid duplication of effort.

  The crew leader must reserve time for equipment maintenance to prevent equipment failure. In the event of an equipment failure, the crew leader must know who to contact to resolve the problem. An alternate plan must be ready in case one piece of equipment breaks down, so that the other equipment doesn't sit idle. The crew leader should coordinate these contingency plans with the main office or the crew leader's immediate superior.

- *Tool planning* – A crew leader is responsible for planning tool usage for a job. This task includes the following:

  - Determining the tools required
  - Informing the workers who will provide the tools (company or worker)
  - Making sure the workers are qualified to use the tools safely and effectively
  - Determining what controls to establish for tools

- *Labor planning* – All jobs require some sort of labor because the crew leader can't complete all the work alone. When planning for labor, the crew leader must do the following:

  - Identify the skills needed to perform the work.

  - Determine the number of people having those specific skills that are required.
  - Decide who will be on the crew.

In many companies, the project manager or job superintendent determines the size and make-up of the crew. Then, supervision expects the crew leader to accomplish the goals and objectives with the crew provided. Even though the crew leader may not have any involvement in staffing the crew, the crew leader is responsible for training the crew members to ensure that they have the skills needed to do the job. In addition, the crew leader is responsible for keeping the crew adequately staffed at all times to avoid job delays. This involves dealing with absenteeism and turnover, two common problems discussed in earlier sections.

### 4.2.4 Scheduling

Planning and scheduling are closely related and are both very important to a successful job. Planning identifies the required activities and how and in what order to complete them. Scheduling involves establishing start and finish times/dates for each activity.

A schedule for a project typically shows the following:

- Operations listed in sequential order
- Units of construction
- Duration of activities
- Estimated date to start and complete each activity
- Quantity of materials to be installed

There are different types of schedules used today. They include the bar chart, the network schedule, also called the critical path method (CPM) or precedence diagram, and the short-term, or look-ahead schedule.

The following is a summary of the steps a crew leader must complete to develop a schedule.

Step 1  Make a list of all the activities that the plan requires to complete the job, including individual work activities and special tasks, such as inspections or the delivery of materials. At this point, the crew leader should just be concerned with generating a list, not with determining how to accomplish the activities, who will perform them, how long they will take, or the necessary sequence to complete them.

Step 2  Use the list of activities created in Step 1 to reorganize the work activities into a logical sequence. When doing this, keep in mind that certain steps can't happen before the completion of others. For example, footing excavation must occur before concrete emplacement.

*Step 3* Assign a duration or length of time that it will take to complete each activity and determine the start time for each. Then place each activity into a schedule format. This step is important because it helps the crew leader compare the task time estimates to the scheduled completion date or time.

The crew leader must be able to read and interpret the job schedule. On some jobs, the form provides the beginning and expected end date for each activity, along with the expected crew or worker's production rate. The crew leader can use this information to plan work more effectively, set realistic goals, and compare the starts and completions of tasks to those on the schedule.

Before starting a job, the crew leader must do the following:

- Determine the materials, tools, equipment, and labor needed to complete the job.
- Determine when the various resources are needed.
- Follow up to ensure that the resources are available on the job site when needed.

The crew leader should verify the availability of needed resources three to four working days before the start of the job. This should occur even earlier for larger jobs. Advance preparation will help avoid situations that could potentially delay starting the job or cause it to fall behind schedule.

Supervisors can use bar chart schedules, also known as *Gantt charts*, for both short-term and long-term jobs. However, they are especially helpful for jobs of short duration.

Bar charts provide management with the following:

- A visual presentation of the overall time required to complete the job using a logical method rather than a calculated guess
- A means to review the start and duration of each part of the job
- Timely coordination requirements between crafts
- Alternative sequences of performing the work

A bar chart works as a control device to see whether the job is on schedule. If the job isn't on schedule, supervision can take immediate action in the office and the field to correct the problem and increases the likelihood of completing the activity on schedule. *Figure 27* illustrates a bar or Gantt chart.

Another type of schedule is the network schedule, which shows dependent (critical path) activities and other activities completed in parallel with but not part of the critical path. In *Figure 28*, for example, reinforcing steel can't be set until the concrete forms have been built and placed. Other activities are happening in parallel, but the forms are in the critical path.

When building a house, builders can't install and finish drywall until wiring, plumbing, and HVAC ductwork have been roughed-in. Because other activities, such as painting and trim work, depend on drywall completion, the drywall work is a critical-path operation. In other words, until it is complete, workers can't start the other tasks, and the project itself will likely experience delay by the amount of delay starting any dependent

**Did You Know?**

## Recycling Materials

Habitat for Humanity, a charitable organization that builds homes for disadvantaged families, accepts surplus building materials for use in their projects. In some cities, they have stores called ReStores, which serve as retail outlets for such materials. They may also accept materials salvaged during demolition of a structure. Companies can obtain LEED credits through practices such as salvaging building materials, segregating scrap materials for recycling, and taking steps to minimize waste.

**Figure Credit:** Sushil Shenoy/Virginia Tech

activity. Likewise, workers can't even start the drywall installation until the rough-ins are complete. Therefore, the project superintendent is likely to focus on those activities when evaluating schedule performance.

The advantage of a network schedule is that it allows project leaders to see how a schedule change with one activity is likely to affect other activities and the project in general. Planners lay out a network schedule on a timeline and usually show the estimated duration for each activity. Planners use network schedules for complex jobs that take a long time to complete. The PERT (program evaluation and review technique) schedule is a form of network schedule.

Since the crew needs to hold to the job schedule, the crew leader needs to be able to plan daily production. As discussed earlier, short-term scheduling is a method used to do this. *Figure 29* displays an example of a short-term, look-ahead schedule.

The information to support short-term scheduling comes from the estimate or cost breakdown. The schedule helps to translate estimate data and the various job plans into a day-to-day schedule of events. The short-term schedule provides the crew leader with visibility over the immediate future of the project. If actual production begins to slip behind estimated production, the schedule will warn the crew leader that a problem lies ahead and that a schedule slippage is developing.

Crew leaders should use short-term scheduling to set production goals. Generally, workers can improve production when they:

- Know the amount of work to be accomplished
- Know the time available to complete the work
- Can provide input when setting goals

*Example:*

A carpentry crew on a retaining wall project is about to form and pour catch basins and put up wall forms. The crew has put in several catch basins, so the crew leader is sure that they can perform the work within the estimate. However, the crew leader is concerned about their production of the wall forms. The crew will work on both the basins and the wall forms at the same time. The scheduling process in this scenario could be handled as follows:

1. The crew leader notices the following in the estimate or cost breakdown:
   a. Production factor for wall forms = 16 worker-hours (w-h) per 100 ft$^2$
   b. Work to be done by measurement = 800 ft$^2$
   c. Total time: (800 ft$^2$ × 16 w-h/100 ft$^2$) = 128 w-h

2. The carpenter crew consists of the following:
   a. One carpenter crew leader
   b. Four carpenters
   c. One laborer

3. The crew leader determines the goal for the job should be set at 128 w-h (from the cost breakdown).

4. If the crew remains the same (six workers), the work should be completed in about 21 crew-hours (128 w-h ÷ 6 workers/crew = 21.3 crew-hours).

5. The crew leader then discusses the production goal (completing 800 ft$^2$ in 21 crew-hours) with the crew and encourages them to work together to meet the goal of getting the forms erected within the estimated time.

In this example, the crew leader used the short-term schedule to translate production into work-hours or crew-hours and to schedule work so that the crew can accomplish it within the estimate. In addition, setting production targets provides the motivation to produce more than the estimate requires.

No matter what type of schedule is used, supervision must keep it up to date to be useful to the crew leader. Inaccurate schedules are of no value. The person responsible for scheduling in the office handles the updates. This person uses information gathered from job field reports to do the updates.

The crew leader is usually not directly involved in updating schedules. However, completing field or progress reports used by the company may be a daily responsibility to keep the schedule up to date. It's critical that the crew leader fill out any required forms or reports completely and accurately.

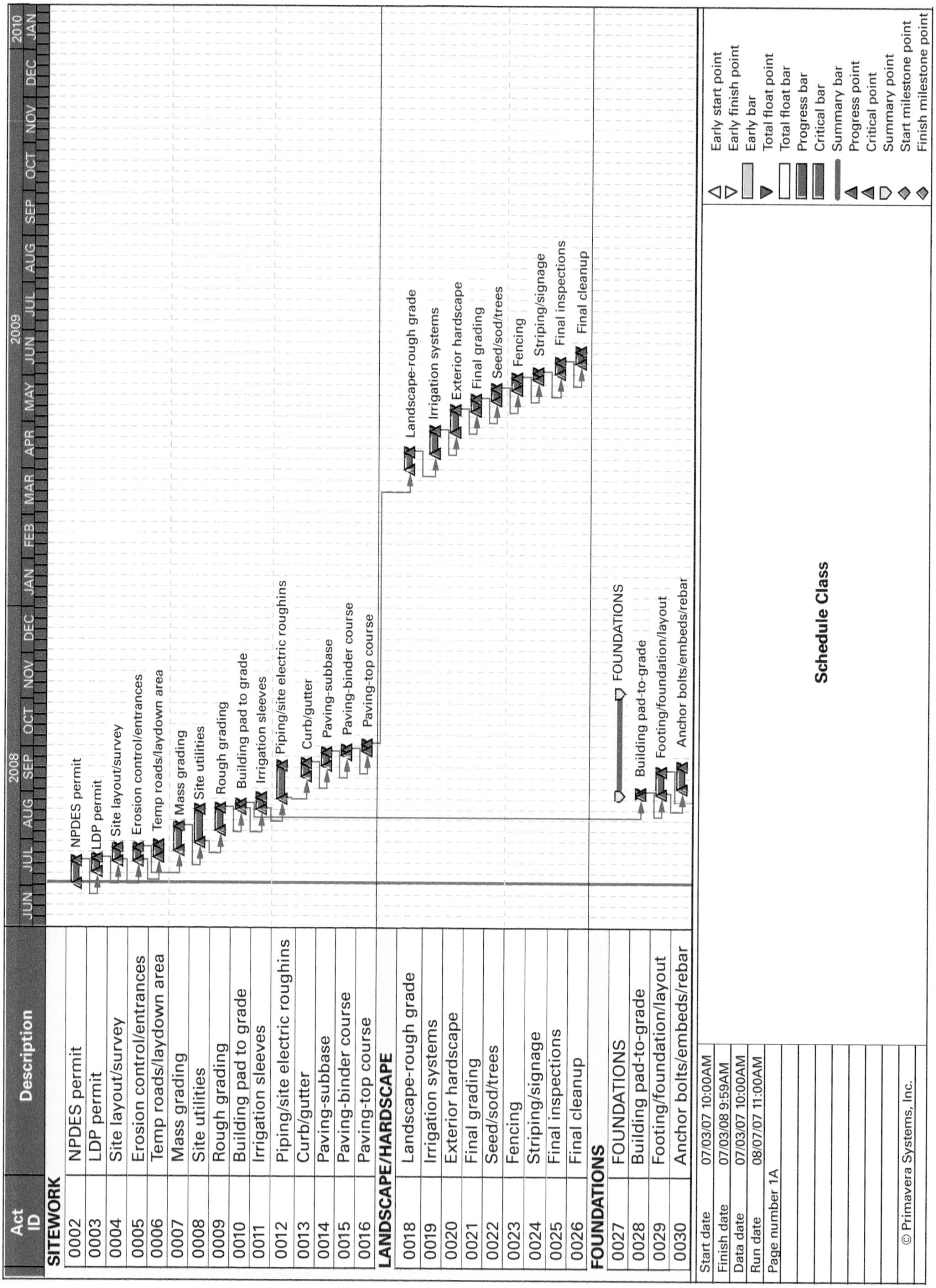
Act ID
Description
2008
JUN JUL AUG SEP OCT NOV DEC
2009
JAN FEB MAR APR MAY JUN JUL AUG SEP OCT NOV DEC
2010
JAN
SITEWORK
0002 NPDES permit
0003 LDP permit
0004 Site layout/survey
0005 Erosion control/entrances
0006 Temp roads/laydown area
0007 Mass grading
0008 Site utilities
0009 Rough grading
0010 Building pad to grade
0011 Irrigation sleeves
0012 Piping/site electric roughins
0013 Curb/gutter
0014 Paving-subbase
0015 Paving-binder course
0016 Paving-top course
LANDSCAPE/HARDSCAPE
0018 Landscape-rough grade
0019 Irrigation systems
0020 Exterior hardscape
0021 Final grading
0022 Seed/sod/trees
0023 Fencing
0024 Striping/signage
0025 Final inspections
0026 Final cleanup
FOUNDATIONS
0027 FOUNDATIONS
0028 Building pad-to-grade
0029 Footing/foundation/layout
0030 Anchor bolts/embeds/rebar
Start date 07/03/07 10:00AM
Finish date 07/03/08 9:59AM
Data date 07/03/07 10:00AM
Run date 08/07/07 11:00AM
Page number 1A
© Primavera Systems, Inc.
Schedule Class
Early start point
Early finish point
Early bar
Total float point
Total float bar
Progress bar
Critical bar
Summary bar
Progress point
Critical point
Summary point
Start milestone point
Finish milestone point

*Figure 27* Example of a bar-chart schedule.

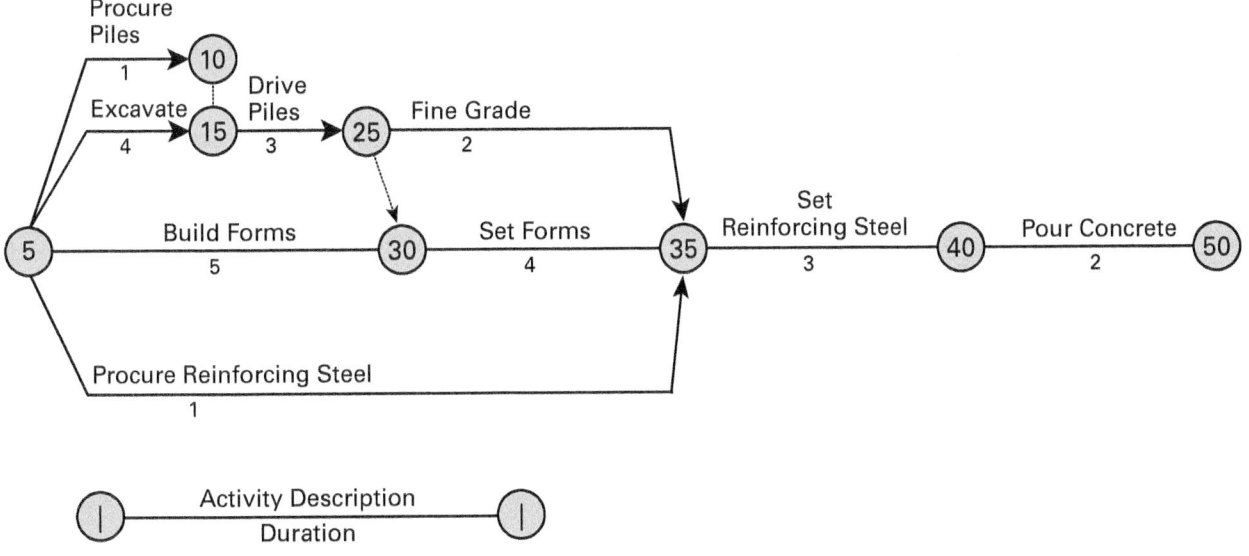

*Figure 28* Example of a network schedule.

## 4.3.0 Cost Control

Being aware of costs and controlling them is the responsibility of every employee on the job. It's the crew leader's job to ensure that employees uphold this responsibility. Control refers to the comparison of estimated performance against actual performance and following up with any needed corrective action. Crew leaders who use cost-control practices are more valuable to the company than those who do not.

On a typical job, many activities are going on at the same time, even within a given crew. This can make it difficult to control the activities involved. The crew leader must be constantly aware of the costs of a project and effectively control the various resources used on the job.

When resources aren't controlled, the cost of the job increases. For example, a plumbing crew of four people is installing soil pipe and runs out of fittings. Three crew members wait (*Figure 30*) while one crew member goes to the plumbing-supply dealer for a part that costs only a few dollars. It takes the crew member an hour to get the part, so four worker-hours of production have been lost. In addition, the total cost of the delay must include the travel costs for retrieving the supplies.

### 4.3.1 Assessing Cost Performance

Cost performance on a project is determined by comparing actual costs to estimated costs. Regardless of whether the job is a contract-bid project or an in-house project, the company must first establish a budget. In the case of a contract bid, the budget is generally the cost estimate used to bid the job. For an in-house job, participants will submit labor and material forecasts, and someone in authority will authorize a project budget.

It is common to estimate cost by either breaking the job into funded tasks or by forecasting labor and materials expenditures on a timeline. Many companies create a work breakdown structure (WBS) for each project. Within the WBS, planners assign each major task a discrete charge number. Anyone working on that task charges that number on their time sheet, so that project managers can readily track cost performance. However, knowing how much money the company is spending doesn't necessarily determine cost performance.

Although financial reports can show that actual expenses are tracking with forecast expenses, they don't show if the work itself is occurring at the required rate. Thus, it is possible to have spent half the budget, but have less than half of the work compete. When the project is broken down into funded tasks related to schedule activities and events, there is far greater control over cost performance.

### 4.3.2 Field Reporting System

The total estimated cost comes from the job estimates, but managers obtain the actual cost of doing the work from an effective field-reporting system. A field-reporting system consists of a series of forms, which are completed by the crew leader and others. Each company has its own forms and methods for obtaining information. The following paragraphs describe the general information and the process of how they are used. First, records must document the number of hours each person worked on each task. This

| ACTIVITY DESCRIPTION | 7/1 | 7/2 | 7/3 | 7/7 | 7/8 | 7/9 | 7/10 | 7/11 | 7/14 | 7/15 | 7/16 | 7/17 | 7/18 | 7/21 | 7/22 | 7/23 |
|---|---|---|---|---|---|---|---|---|---|---|---|---|---|---|---|---|
| **Work Days** | 1 | 2 | 3 | 4 | 5 | 6 | 7 | 8 | 9 | 10 | 11 | 12 | 13 | 14 | 15 | 16 |
| Process Piles | ▓ | | | | | | | | | | | | | | | |
| Excavate | ▓ | ▓ | ▓ | ▓ | | | | | | | | | | | | |
| Build Forms | ▓ | ▓ | ▓ | ▓ | ▓ | | | | | | | | | | | |
| Process Reinforcing Steel | ▓ | | | | | | | | | | | | | | | |
| Drive Piles | | | | | ▓ | ▓ | ▓ | | | | | | | | | |
| Fine Grade | | | | | | | | ▓ | ▓ | | | | | | | |
| Set Forms | | | | | | | | ▓ | ▓ | ▓ | ▓ | | | | | |
| Set Reinforced Steel | | | | | | | | | | | | ▓ | ▓ | ▓ | | |
| Pour Concrete | | | | | | | | | | | | | | | ▓ | ▓ |

NOTES: The project start date is July 1st, which is a Tuesday.
Time placement of activity and duration may be done anytime through shaded portion.
Bottom portion of line available to show progress as activities are completed.

ADDITIONAL TASKS:

| | 1 | 2 | 3 | 4 | 5 | 6 | 7 | 8 | 9 | 10 | 11 | 12 | 13 | 14 | 15 | 16 |
|---|---|---|---|---|---|---|---|---|---|---|---|---|---|---|---|---|
| | | | | | ▓ | | | | | | | | | | | |
| | | | | | | | | | | | | | | | | |
| | | | | | | | | | | | | | | | | |
| | | | | | | | | | | | | | | | | |

*Figure 29* Short-term schedule.

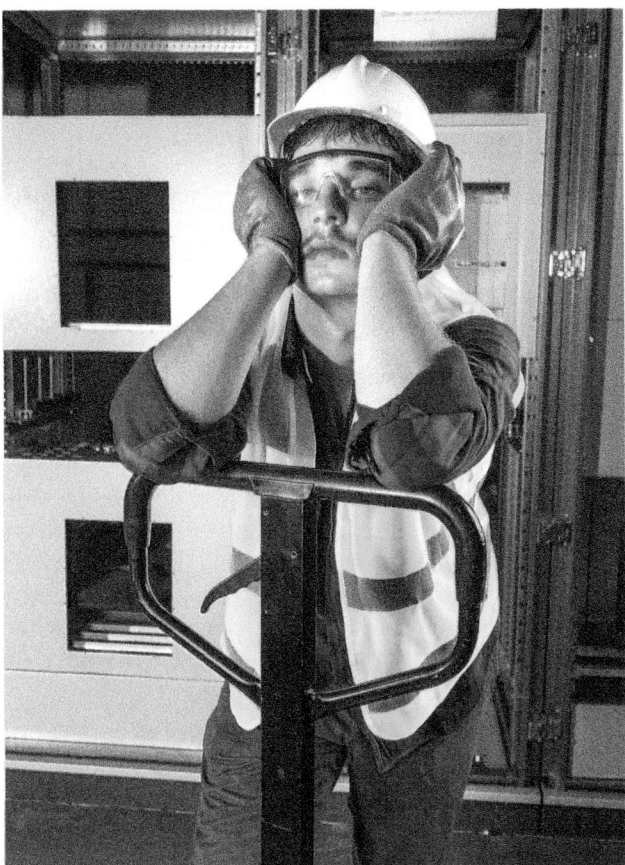

*Figure 30* Idle workers cost money.

information comes from daily time cards. Once the accounting department knows how many hours each employee worked on an activity, it can calculate the total cost of the labor by multiplying the number of hours worked by the wage rate for each worker. Managers can then calculate the cost for the labor to do each task as the job progresses. They compare this cost and the estimated costs during the project and at its completion. Managers use a similar process to determine if the costs to operate equipment are comparable to the estimated equipment operating costs.

As the crew places materials, a designated person will measure the quantities from time to time, and send this information to the company office and, possibly, the crew leader. This information, along with the actual cost of the material and the number of hours it took the workers to install it, is compared to the estimated cost. If the cost is greater than the estimate, management and the crew leader must take action to reduce the cost.

For this comparison process to be of use, the information obtained from field personnel must be correct. It is important that the crew leader be accurate in reporting. This is another reason to maintain a daily job diary as discussed earlier.

In the event of a legal/contractual conflict with the client, courts can use such diaries as evidence in legal proceedings, and they can be helpful in reaching a settlement.

*Example:*

You are running a crew of five concrete finishers for a subcontractor. When you and your crew show up to finish a slab, the general contractor (GC) informs you, "We're a day behind on setting the forms, so I need you and your crew to stand down until tomorrow." What do you do?

*Solution:*

You should first call your office to let them know about the delay. Then, immediately record it in your job diary. A six-man crew for one day represents 48 worker-hours. If your company charges $30 an hour, that's a potential loss of $1,440, which the company would want to recover from the GC. If there is a dispute, your entry in the job diary could result in a favorable decision for your employer.

### 4.3.3 Crew Leader's Role in Cost Control

The crew leader is often the company's representative in the field, where the work takes place. Therefore, the crew leader contributes a great deal to determining job costs. When supervision assigns work to a crew, the crew leader should receive a budget and schedule for completing the job. It is then up to the crew leader to make sure the crew finishes the job on time and stays within budget. The crew leader achieves this by actively managing the use of labor, materials, tools, and equipment.

*"Beware of little expenses;*
*a small leak will sink a great ship."*
*—Benjamin Franklin*

If the actual costs are at or below the estimated costs, the job is progressing as planned and scheduled, then the company will realize the expected profit. However, if the actual costs exceed the estimated costs, one or more problems may result in the company losing its expected profit, and maybe more. No company can remain in business if it continually loses money.

One of the factors that can increase cost is client-related changes (often called scope creep). The crew leaders must be able to assess the potential impact of such changes and, if necessary, confer with the employer to determine the course of action. If contractor-related losses are occurring, the crew leader and superintendent will need to work together to get the costs back in line.

*"There is no such thing as scope creep, only scope gallop."*
—Cornelius Fitchner

The following are some causes that can make actual costs exceed estimated costs, along with some potential actions the crew leader can take to bring the costs under control:

> **NOTE**
>
> Before starting any action, the crew leader should check with company supervision to see that the action proposed is acceptable and within the company's policies and procedures.
>
> - *Late delivery of materials, tools, and/or equipment* – Plan ahead to ensure that job resources will be available when needed
> - *Inclement weather* – Work with the superintendent and have alternate plans ready
> - *Unmotivated workers* – Counsel the workers
> - *Accidents* – Enforce the existing safety program (or evaluate the program for adequacy)

There are many other methods to get the job done on time if it gets off schedule. Examples include working overtime, increasing the size of the crew, prefabricating assemblies, or working staggered shifts. However, these examples may increase the cost of the job, so the crew leader should not do them without the approval of the project manager.

## 4.4.0 Resource Control

The crew leader's job is to complete assigned tasks safely according to the plans and specifications, on schedule, and within the scope of the estimate. To accomplish this, the crew leader must closely control how resources of materials, equipment, tools, and labor are used. The crew must minimize waste whenever possible.

Control involves measuring performance and correcting departures from plans and specifications to accomplish objectives. Control anticipates predictable deviations from plans and specifications based on experience and takes measures to prevent them from occurring.

An effective control process can be broken down into the following steps:

*Step 1*   Establish standards and divide them into measurable units. For example, a baseline can be created using experience gained on a typical job, where 2,000 linear feet (LF) of 1¼" copper water tube was installed in

five days. Thus, dividing the total LF installed by the number of installation days gives the average installation rate. In this case, for 1¼" copper water tube, the standard rate was 400 LF/day.

*Step 2*   Measure performance against a standard. On another job, a crew placed 300 LF of the same tube during a given day. Thus, this actual production of 300 LF/day did not meet the standard rate of 400 LF/day.

*Step 3*   Adjust operations to meet the standard. In Step 2, if the plan scheduled the job for five days, the crew leader would have to take action to meet this goal. If 300 LF/day is the actual average daily production rate for that job, the crew must increase their placement rate by 100 LF/day to meet the standard. This increase may come by making the workers place piping faster, adding workers to the crew, or receiving authorization for overtime.

The crew leader's responsibility in materials control depends on the policies and procedures of the company. In general, the crew leader is responsible for ensuring on-time delivery, preventing waste, controlling delivery and storage, and preventing theft of materials.

### 4.4.1 Ensuring On-Time Delivery

It's essential that the materials required for each day's work be on the job site when needed. The crew leader should confirm in advance the placement of orders for all materials and that they will arrive on schedule. A week or so before the delivery date, follow up to make sure there will be no delayed deliveries or items on backorder.

If other people are responsible for providing the materials for a job, the crew leader must follow up to make sure that the materials are available when needed. Otherwise, delays occur as crew members stand around waiting for the delivery of materials.

### 4.4.2 Preventing Waste

Waste in construction can add up to loss of critical and costly materials and may result in job delays. The crew leader needs to ensure that every crew member knows how to use the materials efficiently. Crew leaders should monitor their crews to make certain that no materials are wasted.

An example of waste is a carpenter who saws off a piece of lumber from a full-sized piece, when workers could have found the length needed in the lumber scrap pile. Another example of waste

# Just in Time (JIT)

Just-in-time (JIT) delivery is a strategy in which materials arrive at the job site when needed. This means that a crew may install the materials right off the truck. This method reduces the need for on-site storage and staging areas. It also reduces the risk of loss or damage while moving products about the site. Other modern material management methods include the use of radiofrequency identification (RFID) tags that make it easy to inventory and locate material in crowded staging areas. Using an RFID inventory control system requires special scanners.

involves rework due to bad technique or a conflict in craft scheduling requiring removal and reinstallation of finished work. The waste in time spent and replacement material costs occur when redoing the task a second time.

Under LEED, waste control is very important. Companies receive LEED credits for finding ways to reduce waste and for recycling construction waste. Workers should segregate waste materials by type for recycling, if feasible (*Figure 31*).

### 4.4.3 Verifying Material Delivery

A crew leader may be responsible for the receipt of materials delivered to the work site. When this happens, the crew leader should require a copy of the shipping invoice or similar document and check each item on the invoice against the actual materials to verify delivery of the correct amounts.

The crew leader should also check the condition of the materials to verify that nothing is defective before signing the invoice. This can be difficult and time consuming because it requires the crew to open cartons and examine their contents. However, this step cannot be overlooked, because a signed invoice indicates that the recipient accepted all listed materials and in an undamaged state. If the crew leader signs for the materials without checking them, and then finds damage, no one will be able to prove that the materials came to the site in that condition.

After checking and signing the shipping invoice, the crew leader should give the original or a copy to the superintendent or project manager. The company then files the invoice for future reference because it serves as the only record the company has to check bills received from the supplier company.

### 4.4.4 Controlling Delivery and Storage

Another essential element of materials control is the selection of where the materials will be stored on the job site. There are two factors in determining the appropriate storage location. The first is

*Figure 31* Waste materials separated for recycling.

convenience. If possible, the materials should be stored near where the crew will use them. The time and effort saved by not having to carry the materials long distances will greatly reduce the installation costs.

Next, the materials must be stored in a secure area to avoid damage. It is important that the storage area suit the materials being stored. For instance, materials that are sensitive to temperature, such as chemicals or paints, should be stored in climate-controlled areas to prevent waste.

### 4.4.5 Preventing Theft and Vandalism

Theft and vandalism of construction materials increase costs. The direct and indirect costs of lost materials include obtaining replacement materials, production time lost while the needed materials are missing, and the costs of additional security precautions can add significantly to the overall cost to the company. In addition, the insurance that the contractor purchases will increase in cost as the theft and vandalism rate grows.

The best way to avoid theft and vandalism is a secure job site. At the end of each work day, store unused materials and tools in a secure location, such as a locked construction trailer. If the job site is fenced or the building has lockable accesses, the materials can be stored within. Many sites have security cameras and/or intrusion alarms to help minimize theft and vandalism.

### 4.4.6 Equipment Control

The crew leader may not be responsible for long-term equipment control. However, the equipment required for a specific job is often the crew leader's responsibility. The first step is to schedule when a crew member must transport the required equipment from the shop or rental yard. The crew leader is responsible for informing the shop of the location of its use and seeing a worker returns it to the shop when finished with it.

It is common for equipment to lay idle at a job site because of a lack of proper planning for the job and the equipment arrived early. For example, if wire-pulling equipment arrives at a job site before the conduit is in place, this equipment will be out of service while awaiting the conduit installation. In addition to the wasted rental cost, damage, loss, or theft can also occur with idle equipment while awaiting use.

The crew leader needs to control equipment use, ensure that the crew operates the equipment according to its instructions, and within time and cost guidelines. The crew leader must also assign or coordinate responsibility for maintaining and repairing equipment as indicated by the applicable preventive maintenance schedule. Delaying maintenance and repairs can lead to costly equipment failures. The crew leader must also ensure that the equipment operators have the necessary credentials to operate the equipment, including applicable licenses.

The crew leader is responsible for the proper operation of all other equipment resources, including cars and trucks. Reckless or unsafe operation of vehicles will likely result in damaged equipment, leading to disciplinary action in the case of the workers, fines and repair costs levied by the vehicle vendors, and a loss of confidence in the crew leader's ability to govern crew-members' actions.

The crew leader should also arrange to secure all equipment at the close of each day's work to prevent theft. If continued use of the equipment is necessary for the job, the crew leader should make sure to lock it in a safe place; otherwise, return it to the shop.

### 4.4.7 Tool Control

Among companies, various policies govern who provides hand and power tools to employees. Some companies provide all the tools, while others furnish only the larger power tools. The crew leader should be familiar with and enforce any company policies related to tools.

Tool control has two aspects. First, the crew leader must control the issue, use, and maintenance of all tools provided by the company. Second, the crew leader must control how crew members use the tools to do the job. This applies to tools supplied by the company as well as tools that belong to the workers.

Using the proper tools correctly saves time and energy. In addition, proper tool use reduces the chance of damage to the tool during use, as well as injury to the user and to nearby workers. This is especially true of edge tools and those with any form of fixed or rotating blade. A dull tool is far more dangerous than a sharp one.

Tools must be adequately maintained and properly stored. Making sure that tools are cleaned, dried, and lubricated prevents rust and ensures that the tools are in the proper working order. If a tool is damaged, it is essential to repair or replace it promptly. Otherwise, an accident or injury could result.

> **NOTE**
>
> Regardless of whether a worker or the company owns a tool, OSHA holds companies responsible for the consequences of using a tool on a job site. The company is accountable if an employee receives an injury from a defective tool. Therefore, the crew leader needs to be aware of any defects in the tools the crew members are using.

Users should take care of company-issued tools as if they were their own. Workers should not abuse tools simply because they don't their property.

One of the major sources of low productivity on a job is the time spent searching for a tool. To prevent this from occurring, supervision should establish a storage location for company-issued tools and equipment. The crew leader should make sure that crew members return all company-issued tools and equipment to this designated location after use. Similarly, workers should organize their personal toolboxes so that they can readily find the appropriate tools and return their tools to their toolboxes when they are finished using them. This is a matter of professionalism that crew leaders are in an excellent position to develop.

Studies have shown that some keys to an effective tool control system are:

- Limiting the number of people allowed access to stored tools
- Limiting the number of people held responsible for tools and holding them accountable
- Controlling the ways in which a tool can be returned to storage
- Making sure tools are available when needed

### 4.4.8 Labor Control

Labor typically represents more than half the cost of a project, and therefore has an enormous impact on profitability. For that reason, it's essential to manage a crew and their work environment in a way that maximizes their productivity. One of the ways to do that is to minimize the time spent on unproductive activities such as the following:

- Engaging in bull sessions
- Correcting drawing errors
- Retrieving tools, equipment, and materials
- Waiting for other workers to finish

If crew members are habitually goofing off, it is up to the crew leader to counsel those workers. The crew leader's daily diary should document the counseling. The crew leader will need to refer repeated violations to the attention of higher management as guided by company policy.

All sorts of errors will occur during a project that need correction. Some errors, such as mistakes on drawings, may be outside of the crew leader's control. However, before a drawing error results in performing an action that will need to be corrected later, a careful examination of the drawings by the crew leader may discover the error before work begins, saving time and materials. If crew members are making mistakes due to inexperience, the crew leader can help avoid these errors by providing on-the-spot training and by checking on inexperienced workers more often.

The availability and location of tools, equipment, and materials to a crew can have a profound effect on their productivity. As discussed in the previous section, the key to minimizing such problems is proactive management of these resources.

Delays caused by crews or contractors can be minimized or avoided by carefully tracking the project schedule. In doing so, crew leaders can anticipate delays that will affect the work and either take action to prevent the delay or redirect the crew to another task.

### 4.4.9 Production and Productivity

Production is the amount of construction materials put in place. It is the quantity of materials installed on a job, such as 1,000 linear feet of waste pipe installed for a task. On the other hand, productivity is a rate, and depends on the level of efficiency of the work. It is the amount of work done per hour or day by one worker or a crew.

Production levels are set during the estimating stage. The estimator determines the total amount of materials placed from the plans and specifications. After the job is complete, supervision can assess the actual amount of materials installed, and can compare the actual production to the estimated production.

Productivity relates to the amount of materials put in place by the crew over a certain period of time. The estimator uses company records during the estimating stage to determine how much time and labor it will take to place a certain quantity of materials. From this information, the estimator calculates the productivity necessary to complete the job on time.

1. List the methods your company uses to minimize waste.

_____

_____

_____

_____

_____

2. List the methods your company uses to control small tools on the job.

_____

_____

_____

_____

_____

3. List five ways that you feel your company could control labor to maximize productivity.

_____

_____

_____

_____

_____

For example, it might take a crew of two people ten days to paint 5,000 square feet. To calculate the required average productivity, divide 5,000 ft² by 10 days. The result is 500 ft²/day. The crew leader can compare the daily production of any crew of two painters doing similar work with this average, as discussed previously.

Planning is essential to productivity. The crew must be available to perform the work and have all the required materials, tools, and equipment in place when the job begins.

The time on the job should be for business, not for taking care of personal problems. Anything not work-related should be handled after hours, away from the job site, if possible. Even the planning of after-work activities, arranging social functions, or running personal errands should occur after work or during breaks. Only very limited and necessary exceptions to these rules should be permitted (e.g., making or meeting medical appointments), and these should be spelled out in company or crew policies.

Organizing field work can save time. The key to effectively using time is to work smarter, not necessarily harder. Working smart can save your crew from doing unnecessary work. For example, most construction projects require that the contractor submit a set of as-built plans at the completion of the work. These plans describe the actual structure and installation of materials. The best way to prepare these plans is to mark up a set of working plans as the work is in progress as described earlier. That way, workers and supervision won't forget pertinent details and waste time trying to reconstruct how the company did the work.

*"Knowledge is power is time is money."*
—*Robert Thier, Storm and Silence*

The amount of material used should not exceed the estimated amount. If it does, either the estimator has made a mistake, undocumented changes have occurred, or rework has caused the need for additional materials. Whatever the case, the crew leader should use effective control techniques to ensure the efficient use of materials.

When bidding a job, most companies calculate the cost per worker-hour. For example, a ten-day job might convert to 160 w-h (two painters for ten days at eight hours per day). If the company charges a labor rate of $30/hour, the labor cost would be $4,800. The estimator then adds the cost of materials, equipment, and tools, along with overhead costs and a profit factor, to determine the price of the job.

After completion of a job, information gathered through field reporting allows the company office to calculate actual productivity and compare it to the estimated figures. This helps to identify productivity issues and improves the accuracy of future estimates.

The following labor-related practices can help to ensure productivity:

- Ensure that all workers have the required resources when needed.
- Ensure that all personnel know where to go and what to do after each task is completed.
- Make reassignments as needed.
- Ensure that all workers have completed their work properly.

<br/>

## Additional Resources

*Construction Workforce Development Professional*, NCCER. 2016. New York, NY: Pearson Education, Inc.

*Mentoring for Craft Professionals*, NCCER. 2016. New York, NY: Pearson Education, Inc.

*Blueprint Reading for Construction*, James A. S. Fatzinger. 2003. New York, NY: Pearson Education, Inc.

*Construction Leadership from A to Z: 26 Words to Lead By*, Wally Adamchik. 2011. Live Oak Book Company.

The following websites offer resources for products and training:

Architecture, Engineering, and Construction Industry (AEC), **www.aecinfo.com**

US Green Building Council (USGBC), **www.usgbc.org/leed**

### 4.0.0  Section Review

1. Which of these activities occurs during the development phase of a project?

   a. Architect/engineer sketches are prepared and a preliminary budget is developed.
   b. Government agencies give a final inspection of the design, check for adherence to codes, and inspect materials used.
   c. Detailed project drawings and specifications are prepared.
   d. Contracts for the project are awarded.

2. A job diary should typically record any of the following *except* _____.

   a. items such as job interruptions and visits
   b. the exact times of scheduled lunch breaks
   c. changes needed to project drawings
   d. the actual time each job task related to a particular project took to complete

3. Which of the following is a correct statement regarding project cost?

   a. Cost is handled by the accounting department and isn't a concern of the crew leader.
   b. The difference between the estimated cost and the actual cost affects a company's profit.
   c. Wasted material is factored into the estimate and is never a concern.
   d. The contractor's overhead costs aren't included in the cost estimate.

4. To prevent job delays due to late delivery of materials, the crew leader should _____.

   a. demand a discount from the supplier to compensate for the delay
   b. tell a crew member to go look for the delivery truck
   c. refuse the late delivery and re-order the materials from another supplier
   d. check with the supplier in advance of the scheduled delivery

# SUMMARY

In construction, the crew leader is the company's supervisor on the ground at the scene where the actual work takes place. The modern crew leader must be prepared to deal with a host of challenges ranging from differences in age, culture, race, gender, language, and educational backgrounds to attitudes regarding authority and personal integrity. Rather than being overwhelmed, the effective and successful crew leader will use these challenges to broaden his or her understanding of people and to grow in character.

Crew leaders must develop an understanding of the company, its policies, and how to become an effective and valuable member of the supervisory team. They must also be able to effectively communicate up and down the chain, to motivate the crew's workers, and develop a sense of ownership within the crew so that they have the confidence and will to do the jobs.

Safety is an overriding concern for the company, and crew leaders need to make safe work practices and conditions the number-one priority. Once a worker has lost an eye, a limb, or life itself, it's too late to reconsider how the task might have been done safer. Workers are a company's most valuable asset and the crew leader is in the position of being the best example of safety to the crew.

While crew leaders view their role as mainly people and work managers, these tasks gain a greater sense of importance when crew leaders understand their responsibilities to the company for whom they work. Effectively controlling worker productivity, resources, schedules, and other aspects of the project directly contribute to cost control, which reflects in greater profits for the company. Companies recognize effective crew leaders and they can expect long and successful careers as a result.

1. Younger-generation workers may have a better perception of success if _____.

    a. crew leaders assign their projects in smaller, well-defined units
    b. they receive closer supervision
    c. crew leaders leave them alone
    d. they are given the perception that they are in charge

2. Companies can help prevent sexual harassment in the workplace by _____.

    a. requiring employee training that avoids the potentially offensive subject of stereotypes
    b. developing a consistent policy with appropriate consequences for engaging in sexual harassment
    c. communicating to workers that, legally, the victim of sexual harassment is only the one being directly harassed
    d. establishing an effective complaint or grievance process for victims of harassment

3. Which of the following statements about a formal organization is *true*?

    a. It is a group of members created mainly to share professional information.
    b. It is a group established for developing industry standards.
    c. It is a group of individuals led by someone all working toward the same goal.
    d. It is a group that is identified by its overall function rather than by key positions.

4. Which of these departments in a large company would likely arrange equipment rentals?

    a. Accounting
    b. Estimating
    c. Human Resources
    d. Purchasing

5. When a person is promoted to crew leader, the amount of time they spend on craft work _____.

    a. increases
    b. decreases
    c. stays the same
    d. doubles

6. Which of the following is a *correct* statement regarding annual crew evaluations?

    a. Their results should never come as a surprise to the crew member.
    b. Formal evaluations should be conducted every three months.
    c. Comments on performance should be withheld until the formal evaluation.
    d. Advise the crew members to improve their performance just before the formal evaluation is scheduled.

7. As a crew leader, what should you avoid when giving recognition to crew members?

    a. Looking for good work and knowing what good work is
    b. Being available at the job site to witness good work
    c. Giving praise for poor work to make the worker feel better
    d. Encouraging work improvement through your confidence in a worker

8. Within the construction industry, who normally assigns a crew to the crew leader?

    a. The US Department of Labor
    b. The company
    c. The crew leader
    d. The crew members

9. Some decisions are difficult to make because you _____.

    a. have no experience dealing with the circumstances requiring a decision
    b. don't want to make a decision
    c. prefer to think it's someone else's responsibility to make a decision
    d. don't recognize the need for a decision

10. The first step in problem solving should be _____.

    a. choosing a solution to the problem
    b. evaluating the solution to the problem
    c. defining the problem
    d. testing the solution to the problem

11. The leading cause of fatalities in the construction industry is _____.
    a. electrocution
    b. asphyxiation
    c. struck by something
    d. falls

12. OSHA inspection of a business or job site _____.
    a. can be done only by invitation
    b. can be conducted at random
    c. is done only after an accident
    d. is conducted only if a safety violation occurs

13. To proactively identify and assess hazards, it is essential that employers _____.
    a. ensure that crew leaders are present and available on the job site
    b. conduct safety inspections of contractor offices
    c. design safety logs to document safety violations
    d. conduct accident investigations

14. Employers who are subject to OSHA record-keeping requirements must maintain a log of recordable occupational injuries and illnesses for _____.
    a. three years
    b. five years
    c. seven years
    d. as long as the company is in business

15. Which of the following statements regarding safety training sessions is *correct*?
    a. The project manager usually holds them.
    b. They are held only for new employees.
    c. They should be conducted frequently by the crew leader.
    d. They are required only after an accident has occurred.

16. The type of contract in which the client pays the contractor for their actual labor and material expenses they incur is known as a _____.
    a. firm fixed-price contract
    b. time-spent contract
    c. performance-based contract
    d. cost-reimbursable contract

17. On a design-build project, _____.
    a. the same contractor is responsible for both design and construction
    b. the owner is responsible for providing the design
    c. the architect does the design and the general contractor builds the project
    d. a construction manager is hired to oversee the project

18. When planning labor for onsite work, the crew leader must _____.
    a. factor in the absenteeism and turnover rates
    b. factor in time for training classes
    c. select the material to be used by each crew member
    d. identify the skills needed to perform the work

19. Which of the following is *not* a source of data in the field-reporting system?
    a. The cost-estimate for the project
    b. Crew leader's daily diary
    c. Onsite inventory of placed materials
    d. Daily time cards

20. The following duties are important in work site equipment control *except* _____.
    a. scheduling the pickup and return of a piece of equipment
    b. ensuring the equipment is operated and maintained correctly
    c. monitoring safe and responsible operation of vehicles
    d. selection and purchase of equipment to be stocked at the work site equipment room

# Trade Terms Introduced in This Module

**Autonomy:** The condition of having complete control over one's actions, and being free from the control of another. To be independent.

**Bias:** A preconceived inclination against or in favor of something.

**Cloud-based applications:** Mobile and desktop digital programs that can connect to files and data located in distributed storage locations on the Internet ("the Cloud"). Such applications make it possible for authorized users to create, edit, and distribute content from any location where Internet access is possible.

**Craft professionals:** Workers who are properly trained and work in a particular construction trade or craft.

**Crew leader:** The immediate supervisor of a crew or team of craft professionals and other assigned persons.

**Critical path:** In manufacturing, construction, and other types of creative processes, the required sequence of tasks that directly controls the ultimate completion date of the project.

**Demographics:** Social characteristics and other factors, such as language, economics, education, culture, and age, that define a statistical group of individuals. An individual can be a member of more than one demographic.

**Ethics:** The moral principles that guides an individual's or organization's actions when dealing with others. Also refers to the study of moral principles.

**Infer:** To reach a conclusion using a method of reasoning that starts with an assumption and considers a set of logically-related events, conditions, or statements.

**Intangible:** Not touchable, material, or measureable; lacking a physical presence.

**Job description:** A description of the scope and responsibilities of a worker's job so that the individual and others understand what the job entails.

**Job diary:** A written record that a supervisor maintains periodically (usually daily) of the events, communications, observations, and decisions made during the course of a project.

**Legend:** In maps, plans, and diagrams, an explanatory table defining all symbolic information contained in the document.

**Lethargy:** Sluggishness, slow motion, lack of activity, or a lack of enthusiasm.

**Letter of instruction (LOI):** A written communication from a supervisor to a subordinate that informs the latter of some inadequacy in the individual's performance, and provides a list of actions that the individual must satisfactorily complete to remediate the problem. Usually a key step in a series of disciplinary actions.

**Local area networks (LAN):** Communication networks that link computers, printers, and servers within a small, defined location, such as a building or office, via hard-wired or wireless connections.

**Lockout/tagout (LOTO):** A system of safety procedures for securing electrical and mechanical equipment during repairs or construction, consisting of warning tags and physical locking or restraining devices applied to controls to prevent the accidental or purposeful operation of the equipment, endangering workers, equipment, or facilities.

**Look-ahead schedule:** A manual- or software-scheduling tool that looks several weeks into the future at the planned project events; used for anticipating material, labor, tool, and other resource requirements, as well as identifying potential schedule conflicts or other problems.

**Negligence:** Lack of appropriate care when doing something, or the failure to do something, usually resulting in injury to an individual or damage to equipment.

**Organizational chart:** A diagram that shows how the various management and operational responsibilities relate to each other within an organization. Named positions appear in ranked levels, top to bottom, from those functional units with the most and broadest authority to those with the least, with lines connecting positions indicating chains of authority and other relationships.

**Paraphrase:** Rewording a written or verbal statement in one's own words in such a way that the intent of the original statement is retained.

**Pragmatic:** Sensible, practical, and realistic.

**Proactive:** To anticipate and take action in the present to deal with potential future events or outcomes based on what one knows about current events or conditions.

**Project manager:** In construction, the individual who has overall responsibility for one or more construction projects; also called the general superintendent.

**Return on investment (ROI):** A measure of the gain or loss of money resulting from an investment; normally measured as a percentage of the original investment.

**Safety data sheets (SDS):** Documents listing information about a material or substance that includes common and proper names, chemical composition, physical forms and properties, hazards, flammability, handling, and emergency response in accordance with national and international hazard communication standards. Also called material safety data sheets (MSDS).

**Sexual harassment:** Any unwelcome verbal or nonverbal form of communication or action construed by an individual to be of a sexual or gender-related nature.

**Smartphone:** The most common type of cellular telephone that combines many other digital functions within a single mobile device. Most important of these, besides the wireless telephone, are high-resolution still and video cameras, text and email messaging, and GPS-enabled features.

**Superintendent:** In construction, the individual who is the on-site supervisor in charge of a given construction project.

**Synergy:** Any type of cooperation between organizations, individuals, or other entities where the combined effect is greater than the sum of the individual efforts.

**Textspeak:** A form of written language characteristic of text messaging on mobile devices and text-based social media, which usually consists of acronyms, abbreviations, and minimal punctuation.

**Wide area networks (WAN):** Dedicated communication networks of computers and related hardware that serve a given geographic area, such as a work site, campus, city, or a larger but distinct area. Connectivity is by wired and wireless means, and may use the Internet as well.

**Wi-Fi:** The technology allowing communications via radio signals over a LAN or WAN equipped with a wireless access point or the Internet. (Wi-Fi stands for wireless fidelity.) Many types of mobile, portable, and desktop devices can communicate via Wi-Fi connections.

**Work breakdown structure (WBS):** A diagrammatic and conceptual method for subdividing a complex project, concept, or other thing into its various functional and organizational parts so that planners can analyze and plan for each part in detail.

## OSHA Forms for Safety Records

Employers who are subject to the recordkeeping requirements of the Occupational Safety and Health Act of 1970 must maintain records of all recordable occupational injuries and illnesses. The following are three important OSHA forms used for this recordkeeping (these forms can be accessed at **www.osha.gov**):

- OSHA Form 300, *Log of Work-Related Injuries and Illnesses*
- OSHA Form 300A, *Summary of Work-Related Injuries and Illnesses*
- OSHA Form 301, *Injury and Illness Incident Report*

# OSHA's Form 300 (Rev. 01/2004)
## Log of Work-Related Injuries and Illnesses

**Attention:** This form contains information relating to employee health and must be used in a manner that protects the confidentiality of employees to the extent possible while the information is being used for occupational safety and health purposes.

U.S. Department of Labor
Occupational Safety and Health Administration

Form approved OMB no. 1218-0176

Year _____

You must record information about every work-related injury or illness that involves loss of consciousness, restricted work activity or job transfer, days away from work, or medical treatment beyond first aid. You must also record significant work-related injuries and illnesses that are diagnosed by a physician or licensed health care professional. You must also record work-related injuries and illnesses that meet any of the specific recording criteria listed in 29 CFR 1904.8 through 1904.12. Feel free to use two lines for a single case if you need to. You must complete an injury and illness incident report (OSHA Form 301) or equivalent form for each injury or illness recorded on this form. If you're not sure whether a case is recordable, call your local OSHA office for help.

Establishment name _____

City _____   State _____

| Identify the person | | | Describe the case | | Classify the case | | | | | | | | | | |
|---|---|---|---|---|---|---|---|---|---|---|---|---|---|---|---|
| (A) Case No. | (B) Employee's Name | (C) Job Title (e.g., Welder) | (D) Date of injury or onset of illness (mo./day) | (E) Where the event occurred (e.g. Loading dock north end) | (F) Describe injury or illness, parts of body affected, and object/substance that directly injured or made person ill (e.g. Second degree burns on right forearm from acetylene torch) | CHECK ONLY ONE box for each case based on the most serious outcome for that case: | | | | Enter the number of days the injured or ill worker was: | | Check the "injury" column or choose one type of illness: (M) | | | | | |
| | | | | | | Days away from work (H) | Remained at work | | Other record-able cases (J) | Away From Work (days) (K) | On job transfer or restriction (days) (L) | Injury (1) | Skin Disorder (2) | Respiratory Condition (3) | Poisoning (4) | Hearing Loss (5) | All other illnesses (6) |
| | | | | | | | Job transfer or restriction (I) | | | | | | | | | | |
| | | | | | | | | | | | | | | | | | |
| | | | | | | | | | | | | | | | | | |
| **Page totals** | | | | | | 0 | 0 | | 0 | 0 | 0 | 0 | 0 | 0 | 0 | 0 | 0 |

Be sure to transfer these totals to the Summary page (Form 300A) before you post it.

Page _____ 1 of 1

*Figure A01* OSHA Form 300

**OSHA's Form 300A** (Rev. 01/2004)
# Summary of Work-Related Injuries and Illnesses

Year _____

**U.S. Department of Labor**
Occupational Safety and Health Administration

Form approved OMB no. 1218-0176

*All establishments covered by Part 1904 must complete this Summary page, even if no injuries or illnesses occurred during the year. Remember to review the Log to verify that the entries are complete and accurate before you've added the entries from every page of the log. If you had no cases write "0."*

*Using the Log, count the individual entries you made for each category. Then write the totals below, making sure you've added the entries from every page of the log. If you had no cases write "0."*

*Employees former employees, and their representatives have the right to review the OSHA Form 300 in its entirety. They also have limited access to the OSHA Form 301 or its equivalent. See 29 CFR 1904.35, in OSHA's Recordkeeping rule, for further details on the access provisions for these forms.*

**Establishment information**

Your establishment name _____

Street _____

City _____ State _____ Zip _____

Industry description (e.g., Manufacture of motor truck trailers) _____

Standard Industrial Classification (SIC), if known (e.g., SIC 3715) ____ ____ ____ ____

OR  North American Industrial Classification (NAICS), if known (e.g., 336212) ____ ____ ____ ____ ____ ____

**Number of Cases**

Total number of deaths

0

(G)

Total number of cases with days away from work

0

(H)

Total number of cases with job transfer or restriction

0

(I)

Total number of other recordable cases

0

(J)

**Number of Days**

Total number of days away from work

0

(K)

Total number of days of job transfer or restriction

0

(L)

**Injury and Illness Types**

Total number of...
(M)

(1) Injury                    0
(2) Skin Disorder             0
(3) Respiratory Condition     0
(4) Poisoning                 0
(5) Hearing Loss              0
(6) All Other Illnesses       0

**Employment information**

Annual average number of employees _____

Total hours worked by all employees last year _____

**Sign here**

Knowingly falsifying this document may result in a fine.

I certify that I have examined this document and that to the best of my knowledge the entries are true. accurate. and complete.

_____    _____
Company executive        Title

_____    _____
Phone                    Date

**Post this Summary page from February 1 to April 30 of the year following the year covered by the form**

Public reporting burden for this collection of information is estimated to average 58 minutes per response, including time to review the instruction, search and gather the data needed, and complete and review the collection of information. Persons are not required to respond to the collection of information unless it displays a currently valid OMB control number. If you have any comments about these estimates or any aspects of this data collection, contact: US Department of Labor, OSHA Office of Statistics, Room N-3644, 200 Constitution Ave, NW, Washington, DC 20210. Do not send the completed forms to this office.

*Figure A02*  OSHA Form 300A

**OSHA's Form 301**
# Injuries and Illnesses Incident Report

**U.S. Department of Labor**

Occupational Safety and Health Administration

Form approved OMB no. 1218-0176

**Attention:** This form contains information relating to employee health and must be used in a manner that protects the confidentiality of employees to the extent possible while the information is being used for occupational safety and health purposes.

This *Injury and Illness Incident Report* is one of the first forms you must fill out when a recordable work-related injury or illness has occurred. Together with the *Log of Work-Related Injuries and Illnesses* and the accompanying *Summary*, these forms help the employer and OSHA develop a picture of the extent and severity of work-related incidents.

Within 7 calendar days after you receive information that a recordable work-related injury or illness has occurred, you must fill out this form or an equivalent. Some state workers' compensation, insurance, or other reports may be acceptable substitutes. To be considered an equivalent form, any substitute must contain all the information asked for on this form.

According to Public Law 91-596 and 29 CFR 1904, OSHA's recordkeeping rule, you must keep this form on file for 5 years following the year to which it pertains.

If you need additional copies of this form, you may photocopy and use as many as you need.

Completed by _____

Title _____

Phone _____ Date _____

## Information about the employee

1) Full Name _____

2) Street _____

   City _____ State _____ Zip _____

3) Date of birth _____

4) Date hired _____

5) ☐ Male
   ☐ Female

## Information about the physician or other health care professional

6) Name of physician or other health care professional

   _____

7) If treatment was given away from the worksite, where was it given?

   Facility _____

   Street _____

   City _____ State _____ Zip _____

8) Was employee treated in an emergency room?
   ☐ Yes
   ☐ No

9) Was employee hospitalized overnight as an in-patient?
   ☐ Yes
   ☐ No

## Information about the case

10) Case number from the Log _____ *(Transfer the case number from the Log after you record the case.)*

11) Date of injury or illness _____

12) Time employee began work _____ AM/PM

13) Time of event _____ AM/PM ☐ Check if time cannot be determined

14) **What was the employee doing just before the incident occurred?** Describe the activity, as well as the tools, equipment or material the employee was using. Be specific. Examples: "climbing a ladder while carrying roofing materials"; "spraying chlorine from hand sprayer"; "daily computer key-entry."

15) **What happened?** Tell us how the injury occurred. Examples: "When ladder slipped on wet floor, worker fell 20 feet"; "Worker was sprayed with chlorine when gasket broke during replacement"; "Worker developed soreness in wrist over time."

16) **What was the injury or illness?** Tell us the part of the body that was affected and how it was affected; be more specific than "hurt", "pain", or "sore." Examples: "strained back"; "chemical burn, hand"; "carpal tunnel syndrome."

17) **What object or substance directly harmed the employee?** Examples: "concrete floor"; "chlorine"; "radial arm saw." If this question does not apply to the incident, leave it blank.

18) **If the employee died, when did death occur?** Date of death _____

Public reporting burden for this collection of information is estimated to average 22 minutes per response, including time for reviewing instructions, searching existing data sources, gathering and maintaining the data needed, and completing and reviewing the collection of information. Persons are not required to respond to the collection of information unless it displays a current valid OMB control number. If you have any comments about this estimate or any other aspects of this data collection, including suggestions for reducing this burden, contact: US Department of Labor, OSHA Office of Statistics, Room N-3644, 200 Constitution Ave, NW, Washington, DC 20210. Do not send the completed forms to this office.

*Figure A03* OSHA Form 301

# Additional Resources

This module presents thorough resources for task training. The following reference material is recommended for further study.

*Construction Workforce Development Professional*, NCCER. 2016. New York, NY: Pearson Education, Inc.

*Mentoring for Craft Professionals*, NCCER. 2016. New York, NY: Pearson Education, Inc.

*Blueprint Reading for Construction*, James A. S. Fatzinger. 2003. New York, NY: Pearson Education, Inc.

*Construction Leadership from A to Z: 26 Words to Lead By*, Wally Adamchik. 2011. Live Oak Book Company.

*Generational Cohorts and their Attitudes Toward Work Related Issues in Central Kentucky*, Frank Fletcher, et al. 2009. Midway College, Midway, KY. **www.kentucky.com.**

*It's Your Ship: Management Techniques from the Best Damn Ship in the Navy*, Captain D. Michael Abrashoff, USN. 2012. New York City, NY: Grand Central Publishing.

*Survival of the Fittest*, Mark Breslin. 2005. McNally International Press.

*The Definitive Book of Body Language: The Hidden Meaning Behind People's Gestures and Expressions*, Barbara Pease and Allan Pease. 2006. New York City, NY: Random House / Bantam Books.

*The Young Person's Guide to Wisdom, Power, and Life Success: Making Smart Choices*, Brian Gahran, PhD. 2014. San Diego, CA: Young Persons Press. **www.WPGBlog.com.**

The following websites offer resources for products and training:

Aging Workforce News, **www.agingworkforcenews.com.**

American Society for Training and Development (ASTD), **www.astd.org.**

Architecture, Engineering, and Construction Industry (AEC), **www.aecinfo.com.**

Equal Employment Opportunity Commission (EEOC), **www.eeoc.gov.**

National Association of Women in Construction (NAWIC), **www.nawic.org.**

National Census of Fatal Occupational Injuries (NCFOI), **www.bls.gov.**

National Institute of Occupational Safety and Health (NIOSH), **www.cdc.gov/niosh.**

National Safety Council, **www.nsc.org.**

Occupational Safety and Health Administration (OSHA), **www.osha.gov.**

Society for Human Resources Management (SHRM), **www.shrm.org.**

United States Census Bureau, **www.census.gov.**

United States Department of Labor, **www.dol.gov.**

US Green Building Council (USGBC), **www.usgbc.org/leed.**

Wi-Fi® is a registered trademark of the Wi-Fi Alliance, **www.wi-fi.org**

# Figure Credits

© Photographerlondon/Dreamstime.com, Module Opener

© iStockphoto.com/shironosov, Figure 1

© iStockphoto.com/kali9, Figures 2, 10

© iStockphoto.com/nanmulti, Figure 3

© iStockphoto.com/Halfpoint, Figure 13

© iStock.com/vejaa, Figure 16

DIYawards, Figure 17

© iStockphoto.com/Tashi-Delek, Figure 18

© iStockphoto.com/kemaltaner, Figure 23

Sushil Shenoy/Virginia Tech, Figure 31

John Ambrosia, Figure 27

© iStockphoto.com/izustun, Figure 30

| Answer | Section Reference | Objective |
|---|---|---|
| **Section One** | | |
| 1. a | 1.1.0 | 1a |
| 2. b | 1.2.2 | 1b |
| 3. b | 1.3.0 | 1c |
| **Section Two** | | |
| 1. b | 2.0.0 | 2a |
| 2. c | 2.2.1 | 2b |
| 3. a | 2.3.1 | 2c |
| 4. c | 2.4.1 | 2d |
| 5. a | 2.5.1 | 2e |
| **Section Three** | | |
| 1. a | 3.0.0 | 3a |
| 2. d | 3.2.0; Table 1 | 3b |
| 3. a | 3.3.0 | 3c |
| 4. b | 3.4.0 | 3d |
| **Section Four** | | |
| 1. a | 4.1.1 | 4a |
| 2. b | 4.2.2 | 4b |
| 3. b | 4.3.3 | 4c |
| 4. d | 4.4.1 | 4d |

# NCCER CURRICULA — USER UPDATE

NCCER makes every effort to keep its textbooks up-to-date and free of technical errors. We appreciate your help in this process. If you find an error, a typographical mistake, or an inaccuracy in NCCER's curricula, please fill out this form (or a photocopy), or complete the online form at **www.nccer.org/olf**. Be sure to include the exact module ID number, page number, a detailed description, and your recommended correction. Your input will be brought to the attention of the Authoring Team. Thank you for your assistance.

*Instructors* – If you have an idea for improving this textbook, or have found that additional materials were necessary to teach this module effectively, please let us know so that we may present your suggestions to the Authoring Team.

**NCCER Product Development and Revision**

13614 Progress Blvd., Alachua, FL 32615

**Email:** curriculum@nccer.org
**Online:** www.nccer.org/olf

❏ Trainee Guide    ❏ Lesson Plans    ❏ Exam    ❏ PowerPoints    Other _____

Craft / Level: _____ Copyright Date: _____

Module ID Number / Title: _____

Section Number(s): _____

Description: _____

_____

_____

_____

Recommended Correction: _____

_____

_____

_____

Your Name: _____

Address: _____

_____

Email: _____ Phone: _____

# Glossary

**Active solar heating system:** A type of solar heating system that uses fluids pumped through collectors to gather solar heat. It is the most complex solar heating system and provides the best temperature control.

**Air stratification:** The layering of air in a room based on temperature. The warmest air is concentrated at the ceiling and the coldest air is concentrated at the floor, with varying temperature layers in between.

**A$_k$ factor:** The actual free area of an air distribution outlet or an inlet stated in square feet. The terminal device manufacturer specifies this value.

**Algorithm:** A mathematical equation consisting of a series of logic statements used in a computer or microprocessor to solve a specific kind of problem. In HVAC applications, algorithms are typically used in microprocessor-controlled equipment to control a wide range of control function operations based on the status of various system sensor input signals.

**Alkalinity:** A water quality parameter in which the pH is higher than 7. It is also a measure of the water's capacity to neutralize strong acids.

**Anhydrous:** Containing no water.

**Application-specific controller:** A digital controller installed by a manufacturer on a specific product at the factory.

**Arrestance efficiency:** The percentage of dust that is removed by an air filter. It is based on a test where a known amount of synthetic dust is passed through the filter at a controlled rate, then the weight of the concentration of dust in the air leaving the filter is measured.

**Atmospheric pressure:** The pressure exerted on all things on the surface of the earth as the result of the weight of the atmosphere.

**Backwashing:** A procedure that reverses the direction of water flow through a multimedia-type filter by forcing the water into the bottom of the filter tank and out the top. Backwashing is performed on a regular basis to prevent accumulated particles from clogging the filter.

**Baud rate:** The rate at which information is transmitted across communication lines.

**Biological contaminants:** Airborne agents such as bacteria, fungi, viruses, algae, insect parts, pollen, and dust. Sources include wet or moist walls, duct, duct liner, fiberboard, carpet, and furniture. Other sources include poorly-maintained humidifiers, dehumidifiers, cooling towers, condensate drain pans, evaporative coolers, showers, and drinking fountains. Also given the terms *microbial* or *microbiological contaminants*.

**Bit:** Short for binary digit. The smallest element of data that a computer can handle. It represents an off or on state (zero or one) in a binary system.

**Bleed-off:** A method used to help control corrosion and scaling in a cooling tower. A metered amount of water is consistently drained from the water circuit to help carry away minerals and other impurities that remain as water evaporates from the cooling tower.

**Bleed-off:** A method used to help control corrosion and scaling in a water system. It involves the periodic draining and disposal of a small amount of the water circulating in a system. Bleed-off aids in limiting the buildup of impurities caused by the continuous addition of makeup water to a system.

**Boyle's law:** With a constant temperature, the pressure on a given quantity of confined gas varies inversely with the volume of the gas. Similarly, at a constant temperature, the volume of a given quantity of confined gas varies inversely with the applied pressure.

**Brake horsepower:** The actual total power needed to drive a fan to deliver the required volume of air through a duct system. It is greater than the expected power needed to deliver the air because it includes losses due to turbulence, inefficiencies in the fan, and bearing losses.

**Building management system (BMS):** A centralized, computer-controlled system for managing the various systems in a building. Also known as a *building automation system* (BAS).

**Building-related illness (BRI):** A situation in which the symptoms of a specific illness can be traced directly to airborne building contaminants.

**Bus:** A multi-wire communication cable that links all the components in a hard-wired computer network.

**Catalytic element:** A device used in wood-burning stoves that helps reduce smoke emissions.

**Change order:** the documentation of a change management process, where changes in the Scope of Work that were originally agreed to by owner(s), contractor(s), and architects or engineer are recorded. Typically, the order is related to additions or deletions from the Scope of Work, and alters the completion date and/or the contract value.

**Charles' law:** With a constant pressure, the volume for a given quantity of confined gas varies directly with its absolute temperature. Similarly, with a constant volume of gas, the pressure varies directly with its temperature.

**Chemically-inert:** The property of a substance that does not readily react chemically with other substances, even under high temperatures and pressures, which usually accelerate chemical changes.

**Chilled-beam cooling system:** A cooling system that employs radiators (chilled beams) mounted near the ceiling through which chilled water flows. Passive systems rely on convection currents for cooling. Active systems use ducted conditioned air to help induce additional airflow over the beams.

**Colloidal substance:** A jelly-like material made up of very small, insoluble, nondiffusible particles larger than molecules, but small enough to remain suspended in a fluid without settling to the bottom.

**Commodities:** Commercial items such as merchandise, wares, goods, and produce.

**Concentration:** Indicates the strength or relative amount of an element present in a water solution.

**Control point:** The name for each input and output device wired to a digital controller.

**Cooler:** A refrigerated storage device that protects commodities at temperatures above 32°F.

**Coordination drawings:** Elevation, location, and other drawings produced for a project by the individual contractors for each trade to prevent a conflict between the trades regarding the installation of their materials and equipment. Development of these drawings evolves through a series of review and coordination meetings held by the various contractors.

**Creosote:** A black, sticky combustible byproduct created when wood is burned in a stove. It must be periodically cleaned from within the stove and chimney before it can build up to the point where it can cause a fire.

**Cryogenic fluid:** A substance that exists as a liquid or gas at ultra-low temperatures of –250°F and below.

**Cryogenics:** Refrigeration that deals with producing temperatures of –250°F and below.

**Cut list:** An information sheet that is derived from shop drawings. It is the shop guide for fabricating duct runs and fittings.

**Cycles of concentration:** A measurement of the ratio of dissolved solids contained in a heating/cooling system's water to the quantity of dissolved solids contained in the related makeup water supply. The term *cycles* indicates when the concentration of an element in the water system has risen above the concentration contained in the makeup water. For instance, if the hardness in the system water is determined to be two times as great as the hardness in the makeup water, then the system water is said to have two cycles of hardness.

**Dalton's law:** The total pressure of a mixture of confined gases is equal to the sum of the partial pressures of the individual component gases. The partial pressure is the pressure that each gas would exert if it alone occupied the volume of the mixture at the same temperature.

**Data collection:** The collection of trend, run-time, and consumable data from the digital controllers in a building.

**Deadband:** In a chiller, the tolerance on the chilled-water temperature control point. For example, a 1°F deadband controls the water temperature to within ±0.5°F of the control point temperature (0.5°F + 0.5°F = 1°F deadband).

**Deck:** A refrigeration industry trade term that refers to a shelf, pan, or rack that supports the refrigerated items stored in coolers and display cases.

**Desiccant:** A moisture-absorbing material.

**Desiccant:** A substance that has a high affinity to water vapor, creating and maintaining a state of dryness near it. Desiccants for HVAC purposes have low or no chemical reactivity.

**Detail drawing:** A drawing of a feature that provides more elaborate information than is available on a plan.

**Dew point:** The temperature at which water vapor in the air becomes saturated and starts to condense into water droplets.

**Digital controller:** A digital device that uses an input module, a microprocessor, and an output module to perform control functions.

**Direct digital control (DDC):** The use of a digital controller is usually referred to as *direct digital control system* or *DDC system*.

**Direct-fired make-up air unit:** An air handler that heats and replaces indoor air that is lost from a building through exhaust vents.

**Dissolved solids:** The dissolved amounts of substances such as calcium, magnesium, chloride, and sulfate contained in water. Dissolved solids can contribute to the corrosion and scale formation in a system.

**Drop:** The vertical distance that the lower edge of a horizontally projected airstream falls or rises.

**Dust spot efficiency:** The percentage of dust that is removed by an air filter. It is the number that is normally referenced in the manufacturer's literature, filter labeling, and specifications. The atmospheric dust spot efficiency of a filter is based on a test where atmospheric dust is passed through a filter, then the discoloration effect of the cleaned air is compared with that of the incoming air.

**Dust spot efficiency:** The percentage of dust that is removed by an air filter. It is the number that is normally referenced in the manufacturer's literature, filter labeling, and specifications. The atmospheric dust spot efficiency of a filter is based on a test where atmospheric dust is passed through a filter, then the discoloration effect of the cleaned air is compared with that of the incoming air.

**Electret:** A material or object that has the property of having both a positive and a negative electrical pole and generating an electrical field between them. The electrical equivalent of a magnet.

**Electrolysis:** The process of changing the chemical composition of a material (called the *electrolyte*) by passing electrical current through it.

**Electrolyte:** A substance in which conduction of electricity is accompanied by chemical action.

**Elevation view:** A view that depicts a vertical side of a building, usually designated by the direction that side is facing; for example, right, left, east, or west elevation.

**Enthalpy:** When used in HVACR technology, the total amount of thermal energy contained by a fluid; the sum of its thermal energy based on heat capacity, and its latent heats of fusion and vaporization, as applicable.

**Entrained air:** Also known as *secondary air*; the induced flow of room air by the primary air from an outlet.

**Environmental tobacco smoke (ETS):** A combination of side-stream smoke from the burning end of a cigarette, cigar, or pipe and the exhaled mainstream smoke from the smoker.

**Equal-friction method:** A method of sizing air distribution systems by designing for a consistent amount of pressure loss per unit length of duct (usually 100 feet).

**Ethernet:** A family of frame-based computer networking technologies for local area networks (LANs). It defines a number of wiring and signaling standards.

**Evaporative cooler:** A comfort cooling device that cools air by evaporating water. It is commonly used in hot, dry climates. Cooling effectiveness drops as the relative humidity of the outdoor air increases.

**Firmware:** Computer programs that are permanently stored on the computer's memory during a manufacturing process.

**Floor plan:** A building drawing indicating a plan view of a horizontal section at some distance above the floor, usually midway between the ceiling and the floor.

**Formaldehyde:** A colorless, pungent byproduct of synthetic and natural biological processes that can cause irritation of the eyes and upper air passages. A known human carcinogen.

**Fouling:** A term used for problems caused by suspended solid matter that accumulates and clogs nozzles and pipes, which restricts circulation or otherwise reduces the transfer of heat in a system.

**Friable:** The condition in which brittle materials can easily fragment, releasing particulates into the air.

**Gateway:** A link between two computer programs allowing them to share information by translating between protocols.

**Grains per gallon (gpg):** An alternate unit of measure sometimes used to describe the amounts of dissolved material in a sample of water. Grains per gallon can be converted to ppm by multiplying the value of gpg by 17.

**Half-life:** The time required for half of a number of atoms to undergo radioactive decay.

**Hardness:** A measure of the amount of calcium and magnesium contained in water. It is one of the main factors affecting the formation of scale in a system. As hardness increases, the potential for scaling also increases.

**High-efficiency particulate air (HEPA) filter:** An extended media, dry-type filter mounted in a rigid frame. It has a minimum efficiency of 99.97 percent for 0.3-micron particles when a clean filter is tested at its rated airflow capacity.

**Homogeneous:** A mixture or material that is uniform in its composition throughout.

**Hot aisle/cold aisle configuration:** A method used to install equipment cabinets in computer rooms to manage the flow of warm and cool air out of and into the cabinets.

**Hypertext transfer protocol (http):** The base protocol used by the worldwide web.

**Immiscible:** A condition in which a refrigerant does not dissolve in oil and vice versa.

**Indirect solar hydronic heating system:** A type of active solar heating system in which a double-walled heat exchanger is used to prevent toxic antifreeze in the solar collectors from contaminating the potable water system.

**Induction unit system:** An air conditioning system that uses heating/cooling terminals with circulation provided by a central primary air system that handles part of the load, instead of a blower in each cabinet. High-pressure air (primary air) from the central system flows through nozzles arranged to induce the flow of room air (secondary air) through the unit's coil. The room air is either cooled or heated at the coil, depending on the season. Mixed primary and room air is then discharged from the unit.

**Inhibitor:** A chemical substance that reduces the rate of corrosion, scale formation, or slime production.

**Internet protocol (IP):** A data-oriented protocol used for communicating data across a packet-switched internetwork.

**Internet protocol address (IP address):** A unique address (computer address) that certain electronic devices use in order to identify and communicate with each other on a computer network utilizing the internet protocol (IP) standard.

**Interoperability:** The ability of digital controllers with different protocols to function together accurately.

**King valve:** The manual refrigerant shut-off valve located at the outlet of the receiver.

**Latent-heat defrost:** A method used in hot-gas defrost applications that uses the heat added in one or more active evaporators as a source of heat to defrost another evaporator coil.

**Layup:** An industry term referring to the period of time a boiler is shut down.

**Load-shedding:** Systematically switching loads out of a system to reduce energy consumption.

**Local area network (LAN):** A server/client computer network connecting multiple computers within a building or building complex.

**Local interface device:** A keypad with an alphanumeric data display and keys for data entry. It is connected to the digital controller with a phone-type cable that provides power and communication.

**Longitudinal section:** A section drawing in which the cut is made along the long dimension of a building.

**Lyophilization:** Also known as freeze-drying. A dehydration process which first incorporates freezing of a material, then a reduction of the surrounding pressure with simultaneous addition of heat to remove moisture. Moisture leaves the material through sublimation.

**Micro-channel evaporator coil:** Coil construction technology that consists of a group of very small parallel tubes bonded to enhanced heat transfer fins, replacing the single large tube common in standard evaporator coils.

**Milligrams per liter (mg/l):** Metric unit of measure used to specify exactly how much of a certain material or element is dissolved in a sample of water. One mg/l is equivalent to one ppm.

**Monel®:** An alloy made of nickel, copper, iron, manganese, silicon, and carbon that is very resistant to corrosion.

**Multiple-chemical sensitivity (MCS):** A condition found in some individuals who believe they are vulnerable to exposure to certain chemicals and/or combinations of chemicals. Currently, there is some debate as to whether or not MCS really exists.

**Network:** A means of linking devices in a computer-controlled system and controlling the flow of information among these devices.

**New building syndrome:** A condition that refers to indoor air quality problems in new buildings. The symptoms are the same as those for sick building syndrome.

**Off-gassing:** The process by which furniture and other materials release chemicals and other volatile organic compounds (VOC) into the air.

**Ozone:** An unstable form of oxygen that has a sharp, pungent odor like chlorine bleach, formed in nature in the presence of electric discharges or exposure to ultraviolet light. It is irritating to the mucous membranes and lungs in animals and to plant tissues. It is a strong oxidizer and forms a hazardous air pollutant near ground level.

**Parts per million (ppm):** Unit of measure used to specify exactly how much of a certain material or element is dissolved in a sample of water. For example, one ounce of a contaminant mixed with 7,500 gallons of water equals a concentration of about one ppm.

**Passive solar heating system:** A type of solar heating system characterized by a lack of moving parts and controls. The sun shining through a window on a winter day is an example of passive solar heat.

**pH:** A measure of alkalinity or acidity of water. The pH scale ranges from 0 (extremely acidic) to 14 (extremely alkaline), with the pH of 7 being neutral. Specifically, pH defines the relative concentration of hydrogen ions and hydroxide ions. As pH increases, the concentration of hydroxide ions increases.

**Plan view:** The overhead view of an object or structure.

**Pontiac fever:** A mild form of Legionnaires' disease.

**Primary air:** In the context of air distribution and balancing, it is the air delivered to the room or conditioned space from the supply duct.

**Product-integrated controller (PIC):** A digital controller installed by a manufacturer on a product at the factory.

**Product-specific controller:** A digital controller designed to control specific equipment that may be installed by the equipment manufacturer. The controllers for chillers or boilers are examples of product-specific controllers. Also known as a product-integrated controller.

**Psychrometrics:** The study of air and its properties.

**Queen valve:** The manual refrigerant shut-off valve located at the inlet of the receiver.

**R-value:** The thermal resistance of a given thickness of insulating material.

**Radon:** A colorless, odorless, radioactive, and chemically inert gas that is formed by the natural breakdown of uranium in soil and groundwater. Radon exposure over an extended period of time can increase the risk of lung cancer.

**Recycle shutdown mode:** A chiller mode of operation in which automatic shutdown occurs when the compressor is operating at minimum capacity and the chilled-water temperature has dropped below the chilled-water temperature setpoint. In this mode, the chilled-water pump remains running so that the chilled-water temperature can be monitored.

**Riser diagram:** A one-line schematic depicting the layout, components, and connections of a piping system or electrical system.

**Runaround loop:** A closed-loop energy recovery system in which finned-tube water coils are installed in the supply and exhaust airstreams and connected by counterflow piping.

**Satellite compressor:** A separate compressor that uses the same source of refrigerant as those used by a related group of parallel compressors. However, the satellite compressor functions as an independent compressor connected to a cooling area that requires a lower temperature level than that being maintained by the related compressors.

**Scale:** A dense coating of mineral matter that precipitates and settles on internal surfaces of equipment as a result of falling or rising temperatures.

**Schedules:** Tables that describe and specify the types and sizes of items required for the construction of a building.

**Scintillation detectors:** Radiation detectors that contain a substance which emits light when it absorbs a fast-moving nuclear particle (proton, neutron, or alpha). The sparkling effect caused by these emissions is called scintillation. The detector includes features that amplify the light flashes and display the rate of detections.

**Secondary coolant:** Any cooling liquid that is used as a heat transfer fluid. It changes temperature without changing state as it gains or loses heat.

**Section drawing:** A drawing that depicts a feature of a building as if there were a cut made through the middle of it.

**Sensible heat:** The heat exchange that simply raises or lowers the temperature of a fluid without causing or resulting from a phase change (condensation or evaporation). Compare with the term *total heat*.

**Shop drawing:** A drawing that indicates how to fabricate and install individual components of a construction project. A shop drawing may be drafted from the construction drawings of a project building or provided by the manufacturer.

**Sick building syndrome (SBS):** A condition that exists when more than 20 percent of a building's occupants complain during a two-week period of a set of symptoms, including headaches, fatigue, nausea, eye irritation, and throat irritation, that are alleviated by leaving the building and are not known to be caused by any specific contaminant.

**Site plan:** A construction drawing that indicates the location of a building on a land site.

**Sorbent:** A substance whose main purpose is to absorb something else by forming a mixture of the two substances.

**Specific density:** The weight of one pound of air. At 70°F at sea level, one pound of dry air weighs 0.075 pound per cubic foot.

**Specific volume:** The space one pound of dry air occupies. At 70°F at sea level, one pound of dry air occupies a volume of 13.33 cubic feet.

**Spread:** The horizontal divergence of an airstream after it leaves the outlet.

**Static-regain method:** A method of sizing ducts such that the regain in static pressure due to a decreasing velocity between two points fully or partially makes up for the frictional resistance between the two points.

**Sublimation:** The change in state directly from a solid to a gas, such as the changing of ice to water vapor, without passing through the liquid state at any point.

**Suspended solids:** The amount of visible, individual particles in water or those particles that give water a cloudy appearance. They can include silt, clay, decayed organisms, iron, manganese, sulfur, and microorganisms. Suspended solids can clog treatment devices or shield microorganisms from disinfection.

**Takeoff:** The process of surveying, measuring, itemizing, and counting all materials and equipment needed for a construction project, as indicated by the drawings.

**Thermal conductivity:** the heat flow per hour (Btuh) through one square foot of one-inch thick homogeneous material for every 1°F of temperature difference between the two surfaces.

**Thermosiphon system:** A type of passive solar heating system in which the difference in fluid temperature in different parts of the system causes the fluids to flow through the system.

**Throw:** The throw of an outlet is measured in feet and is the distance from the center of the outlet to a point in the mixed airstream where the highest sustained velocity (fpm) has been reduced to a specified level, usually 50 fpm.

**Total heat rejection (THR) value:** A value used to rate condensers. It represents the total heat removed in desuperheating, condensing, and subcooling a refrigerant as it flows through the condenser.

**Total heat:** Sensible heat plus latent heat.

**Total heat:** The heat exchange that not only raises or lowers the temperature of a fluid, but also causes or results from a phase change (condensation or evaporation).

**Total pressure:** The sum of the static pressure and the velocity pressure in an air duct. It is the pressure produced by the fan or blower.

**Total solids:** The total amount of both dissolved and suspended solids contained in water.

**Transverse section:** A section drawing in which the cut is made through the short dimension of the building.

**Traverse readings:** A series of velocity readings taken at several points over the cross-sectional area of a duct or grille.

**Type HT vent:** A metal vent capable of withstanding temperatures up to 1,000°F. It is commonly used to vent wood-burning stoves and furnaces.

**Type PL vent:** A type of metal vent specifically designed for stoves that burn wood pellets or corn.

**U-factor:** The unit of measure for thermal conductivity.

**Unit cooler:** A packaged refrigeration system assembly containing the evaporator, expansion device, and fans. It is commonly used in chill rooms and walk-in coolers.

**Valance cooling system:** A type of cooling system in which chilled water is circulated through finned-tube radiators located near the ceiling around the perimeter of a room. Convection currents move the cooled air instead of a blower assembly. A decorative valance conceals the system.

**Velocity pressure:** The pressure in a duct that results from the movement of the air. It is the difference between the total pressure and the static pressure.

**Velocity-reduction method:** A method of sizing ducts such that the desired velocities occur in specific duct lengths.

**Volatile organic compounds (VOC):** A wide variety of compounds and chemicals found in such things as solvents, paints, and adhesives, which easily evaporate at room temperature.

**Zeolite:** An aluminosilicate compound that has a high affinity for water and water vapor, which can easily be dehydrated and rehydrated by heating and cooling. A compound commonly used in certain humidity-control systems.

# Index

Anemometers, (03402):21–23
Anhydrous, (03408):51, 59
Antifoaming agents, (03308):26
AO. *See* Analog output (AO) control point; Analog output (AO) devices
Appliances, energy-efficient. *See* Super-Efficient Refrigerator Program (SERP)
Application-specific controller, (03405):8, 11, 43
Apprenticeship programs, (46101):2
Apps
    building management systems, (03405):36
    construction calculators, (03401):43
Architectural plans, (03401):5
Arrestance efficiency, (03403):26, 29, 48
Asbestos, (03403):7
As-built drawings, (03401):28–29, 33, (46101):48
As-built drawings, using for duct cleaning, (03403):39
ASHRAE. *See* American Society of Heating, Refrigeration, and Air Conditioning Engineers (ASHRAE)
Aspect ratio, (03402):18
Associated Builders and Contractors (ABC), (46101):8
Associated General Contractors (AGC), (46101):8
ASTD. *See* American Society for Training and Development (ASTD)
ASTM International, (03401):36
Atmosphere, (03402):1
Atmospheric pressure, (03402):1–4, 47
Attitude, worker, (46101):16
Authority
    defined, (46101):9
    delegating, (46101):9
    leadership and, (46101):15–16
    respect and, (46101):15–16
Automated building management systems, (03403):26–27
Autonomy, (46101):13, 15, 78
Awards, safety-related, (46101):42–43

# B

Baby Boomers, (46101):2, 5
Backwashing, (03308):10, 11, 33
BACnet, (03405):34–35
Bacterial slimes, (03308):5
Bag filter, (03403):30
Bag-type water filters, (03308):12, 14
Bar chart schedule, (46101):62, 64, 66
Barometer, (03402):1–2
Basket strainers, (03308):10
Batch freezing, (03408):2–3
Baud rate, (03405):8, 10, 43
Becquerel (Bq), (03403):13
Belt-drive blowers, (03402):23–24
Belt freezers, (03408):3, 4
Bias, (46101):13, 14, 78
Bid price, (46101):51
Biocides, (03403):39
Biological contaminants
    defined, (03403):5, 48
    growth of, (03308):5, 22, (03403):8, 36
    treating, (03308):24
Bit, (03405):8, 10, 43
Bleed-off
    controlled, (03308):16
    defined, (03308):1, 33, (03406):12, 25, 40
    function, (03406):25
    highlighted in text, (03308):2
    procedure, (03308):21
    for water control, (03308):22, (03406):25
Blowdown controllers and separators, (03308):18

Blowdowns, (03308):25
Blowdown valves, (03308):27
Blowdown water and heat recovery system, (03404):20–21
Blueprints, (03401):2–3
BMS. *See* Building management systems (BMS)
BOCA. *See* Building Officials and Code Administrators (BOCA)
Boiler blowdown, (03404):20–21
Boilers
    corrosion, causes and treatment, (03406):1, 2
    dry storage, shutdown for, (03406):1–3
    water treatment chemicals, (03406):4
    wet storage, preparation for, (03406):3–4
    wood-burning, (03409):4–6, 7, 9
Boilers, steam/hot water
    carryover, (03308):26
    corrosion, causes and treatment, (03308):25
    dual-rated pressure vessels, (03406):8
    firetube, scale on, (03308):25
    foaming in, (03308):25, 26
    hard water treatment in, (03308):25
    priming, (03308):26
    scale deposits, (03308):25
    sludge deposits, (03308):25–26
    startup
        gas-fired steam boilers, (03406):6, 41
        hot-water boilers, (03406):7–10
        steam boilers, (03406):4–6
    water circulation rates, (03406):9–10
    water treatment equipment
        blowdown controllers and separators, (03308):18
        deaerators, (03308):15–16
        steam separator, (03308):11, 26
    water treatment guidelines, (03308):26–28
Boiler water log sheet, (03308):29
Box filter, (03403):30
Boyle's law, (03402):1, 5–6, 47
Bq. *See* Becquerel (Bq)
Brake horsepower, (03407):36, 40, 69
BRI. *See* Building-related illness (BRI)
Budgeting in construction planning, (46101):48, 51–54
Building air quality. *See* Indoor air quality (IAQ)
Building codes, (03401):30
Building evaluation and survey, (03407):3–7, 72–76
Building management systems (BMS)
    defined, (03405):1, 43
    demand response, (03405):34
    function, (03405):8
    integrated design process for, (03405):1
Building management systems (BMS) architecture
    applied HVAC systems
        chilled-water systems, (03405):11–13
        chiller plant control module, (03405):13
        general-purpose controller functions, (03405):13–14
        VAV systems, (03405):14–16
    Internet access, (03405):19
    interoperability, (03405):33–37
    overview, (03405):8–9
    peer-to-peer networks, (03405):8, 10–11
    user interfaces, (03405):17–21, 22, 36
    web browser system integration, (03405):36
Building management systems (BMS) digital controllers
    closed control loop algorithms
        proportional control, (03405):5
        proportional-integral control, (03405):5
        proportional-integral-derivative, (03405):5–6
    closed control loops, (03405):4–5
    components, (03405):2

CRUs. *See* Computer room units (CRUs)
Cryogenic fluid, (03408):1, 3, 59
Cryogenics, (03408):1, 3–4, 59
Cryomechanical freezing, (03408):3
CSC. *See* Construction Specifications Canada (CSC)
CSI. *See* Construction Specifications Institute (CSI)
Cultural issues in the workplace
    conflicts, resolving, (46101):6
    language barriers, (46101):5–6
Curie (Ci), (03403):13
Cut list, (03401):1, 21, 22, 26–27, 56
Cycles of concentration, (03308):1, 3, 33

**D**

Dalton's law, (03402):1, 5–6, 47
Dampers
    on drawings, (03401):28, 33
    types of
        fire dampers, (03401):28, 33
        fire/smoke combination dampers, (03401):28, 33
        smoke dampers, (03401):28, 33
Database management, building management systems, (03405):21
Data collection, building management systems, (03405):17, 21, 23, 43
DCV. *See* Demand-control (DCV) ventilation
DDC. *See* Direct digital control (DDC)
Deadband, (03406):12, 20, 40
Deaerators, (03308):15–16, 17, 25
Deaths, construction industry, (46101):33
Decision making
    leadership and, (46101):15
    problem solving vs., (46101):27
Decisions, types of, (46101):27
Deck, (03408):12, 27, 59
Defrosting refrigeration systems, methods of
    electric defrost, (03408):46–47
    hot-gas defrost, (03408):48–49
    latent-heat defrost, (03408):45, 48, 59
    off-cycle defrost, (03408):46
    thermal bank, (03408):48–49
Dehumidification systems, (03404):10–11, 15, 18
Dehumidifiers, (03403):31–34, (03409):18, 19
Delegating, (46101):9, 25–26
Delivery systems
    design-build, (46101):48, 50
    general contracting, (46101):47, 50
Demand-control (DCV) ventilation, (03403):19
Demand-side management (DSM) programs, (03404):24–26
Demographics, (46101):1, 2, 78
Department of Labor (DOL), (46101):2
Desiccant, (03403):26, 34, 48, (03406):2, 3, 40
Design-build delivery systems, (46101):48, 50
Dessicant, (03406):1
Detail drawings, (03401):1, 7, 11, 56
Dew point, (03402):8, 11, 13, 47
DI. *See* Discrete input (DI) control point; Discrete input (DI) devices
Differential pressure gauge, (03402):21
Diffuser, (03402):18
Digital controllers (BMS/DDC)
    closed control loop algorithms
        proportional control, (03405):5
        proportional-integral control, (03405):5
        proportional-integral-derivative, (03405):5–6
    closed control loops, (03405):4–5
    components, (03405):2
    control points, analog and discrete, (03405):2

defined, (03405):1, 43
general-purpose, (03405):2, 13–14
input devices, analog and discrete, (03405):3
limitations, (03405):8
output devices, analog and discrete, (03405):3–4
packaged units, (03405):11
peer-to-peer networks for information sharing, (03405):8, 10–11
polling networks, (03405):10
product-integrated, (03405):1, 2, 11, 43
product-specific, (03405):1, 2, 43
typical, (03405):2
Digital meters, (03404):25
Dig Safely cards, (03401):29
Direct digital control (DDC), (03405):1, 43
Direct digital control (DDC) systems digital controllers
    closed control loop algorithms
        proportional control, (03405):5
        proportional-integral control, (03405):5
        proportional-integral-derivative, (03405):5–6
    closed control loops, (03405):4–5
    components, (03405):2
    control points, analog and discrete, (03405):2
    defined, (03405):1, 43
    general-purpose, (03405):2, 13–14
    input devices, analog and discrete, (03405):3
    limitations, (03405):8
    output devices, analog and discrete, (03405):3–4
    packaged units, (03405):11
    peer-to-peer networks for information sharing, (03405):8, 10–11
    polling networks, (03405):10
    product-integrated, (03405):1, 2, 11, 43
    product-specific, (03405):1, 2, 43
    typical, (03405):2
Direct-drive blowers, (03402):23–24
Direct-expansion (DX) cooling systems, (03409):18
Direct-fired make-up air unit, (03409):16–18, 37
Directive style of leadership, (46101):15–16
Direct load control programs, (03404):25
Disability discrimination, (46101):11
Disciplinary action, (46101):29
Discrete input (DI) control point, (03405):2
Discrete input (DI) devices, (03405):3
Discrete output (DO) control point, (03405):2
Discrete output (DO) devices, (03405):4
Discrimination in the workplace, (46101):7–8, 11
Dispersed oil particulate (DOP), (03403):29
Display cases, (03408):26–28
Dissolved solids, (03308):1, 2, 3, 33
Diversity, workplace, (46101):5–8
DO. *See* Discrete output (DO) control point; Discrete output (DO) devices
DOL. *See* Department of Labor (DOL)
DOP. *See* Dispersed oil particulate (DOP)
Double-bundle condensers, (03404):9–10
Drawings
    abbreviations on, typical, (03401):59
    addenda, (03401):36
    as-built drawings, (46101):48
    computer generated, (03401):5
    construction phase, (46101):48
    dimensions on, proper use of, (03401):20
    legends, (03401):9, 11, 16, 20
    notes on, (03401):35
    plan set
        as-built drawings, (03401):28–29, 33
        categories, typical, (03401):3

Gases
  in air, (03402):1, 8
  in water
    removing, (03308):15–16, 25
    types of, (03308):2
Gas filters, (03403):29, 31
Gas-fired steam boilers startup, (03406):6, 41
Gateway, (03405):33, 35, 43
Gauge pressure, (03402):2
Gender issues in the workplace
  communication, (46101):5
  discrimination based on gender, (46101):7
  sexual harassment, (46101):6–7
General contracting delivery systems, (46101):48, 50
General-purpose (GPC) controllers, (03405):13–14
Generation X, (46101):5
Generation Y, (46101):5
Generation Z, (46101):5
Glass doors, load estimation, (03407):17
Goal setting
  motivation and, (46101):2, 3
  in planning projects, (46101):58
  successful teams and, (46101):24
GPC. *See* General-purpose (GPC) controllers
gpg. *See* Grains per gallon (gpg)
Grains of moisture scale, (03402):13
Grains per gallon (gpg), (03308):1, 2, 33
Green buildings, (03405):1
Grille, (03402):18
Groove corrosion, (03308):4
Guaranteed maximum price contracts, (46101):48

# H

$H_2$. *See* Hydrogen ($H_2$) detector
Habitat for Humanity, (46101):62
Half-life, (03403):5, 7, 48
Hardness, (03308):1, 3, 33
Hard water
  classification systems, (03308):3
  hardness testing, (03308):8
  in steam boilers, treating, (03308):25
  water treatment equipment, (03308):15
Hazard identification and assessment, (46101):36
Head pressure controls, refrigeration systems, (03408):44
Healthcare facilities, air pressure relationships, (03403):22
Health effects
  asbestos, (03403):7
  carbon monoxide (CO), (03403):9
  cleaning compounds, (03403):10
  lead paint, (03403):6
  ozone, (03403):35
  pesticides, (03403):10
  poor IAQ, (03403):3
  radon, (03403):12
  toxic mold, (03403):8
  volatile organic compounds (VOCs), (03403):6
Health hazards
  ammonia, (03408):51–54
  chemical inhibitors, (03308):20, 24
  controlled atmosphere containers, (03408):10
  open recirculating water system maintenance, (03308):22
Heat
  latent, (03404):6
  sensible, (03404):1, 2, 6, 33
  total, (03404):1, 5, 33
Heat conversion, (03404):9
Heat exchangers
  chilled-water, (03409):19–21

chilled water-to-refrigerant, (03409):21
  fixed-plate, (03404):2, 5–6
  heat-pipe, (03404):15–16
  indirect solar hydronic heating system, (03409):14
  inspecting and cleaning, (03406):28–29
  rotary air-to-air, (03404):2, 6–7, 8
  thermosiphons, (03404):15–17
  wheel, (03404):2, 6–7
Heating equipment selection, (03407):28–29
Heat-pipe heat exchangers, (03404):15–16
Heat pipes, (03403):34
Heat pumps, (03407):29–30
Heat recovery
  coil energy-recovery loops, (03404):13–14
  condenser systems
    air conditioning/refrigeration, (03404):7–9
    chilled-water, (03404):9–10
  flue-gas system, (03404):18–19
  fume hood exhaust, (03404):2
  heat exchangers
    fixed-plate, (03404):2, 5–6
    heat-pipe, (03404):15–16
    rotary air-to-air, (03404):2, 6–7, 8
    thermosiphons, (03404):15–17
    wheel, (03404):6–7
  in steam systems
    blowdown water and, (03404):20–21
    flash-steam (flash-tank), (03404):19–20
  swimming pool systems, (03404):10–13
  Tri-Stack® systems, (03404):2
  twin-tower enthalpy recovery loops, (03404):18
  ventilators
    energy recovery (ERVs), (03404):1–3, 5
    heat recovery (HRVs), (03404):1–3, 5
Heat recovery ventilators (HRVs), (03404):1–3, 5
Heat transfer
  basics, (03407):10
  conduction, (03407):12
  convection, (03407):13
  heat gain and loss through walls, (03407):13–14
  insulation and, (03407):10, 12
  radiation, (03407):12
  roof color, (03407):13
Heat transfer multipliers (HTMs), (03407):8
Heat wheel, (03404):6
HEPA. *See* High-efficiency particulate air (HEPA) filter
High-efficiency particulate air (HEPA) filter, (03403):26, 29, 30, 48
High-sidewall outlet systems, (03407):45
Homogeneous, (03407):8, 14, 69
Horizontal plate freezer, (03408):3
Hot aisle/cold aisle configuration, (03409):16, 22, 37
Hot gas bypass valves, refrigeration systems, (03408):39–41
Hot-gas defrost, (03408):48–49
Hot gas line piping, (03408):13
HRVs. *See* Heat recovery ventilators (HRVs)
HTMs. *See* Heat transfer multipliers (HTMs)
http. *See* Hypertext transfer protocol (http)
Human occupancy, IAQ and, (03403):10–11
Human Resources Department, (46101):8, 40
Humidifiers, (03403):31–34
  computer rooms, (03409):18, 23–24
  energy-saving, (03402):11
  infrared, (03409):23–24
  steam generating, (03409):23–24
  ultrasonic, (03409):24
Humidity
  building management systems for controlling, (03405):31

**W**

Walls, load estimation, (03407):16, 19
WAN. *See* Wide area network (WAN)
Warehouses, refrigerated, (03408):5–6
Warm climates duct layout, (03407):45–46
Waste-oil heaters, (03409):10–11
Waste prevention, (46101):68–69
Water
  in air, (03402):9–11
  biological growth, (03308):5, 21
  characteristics of, (03308):2–3
  corrosion, (03308):4–5
  hardness, (03308):1, 3, 8, 33
  scaling, (03308):5
  suspended solids, (03308):5
  testing
    for alkalinity, (03308):7–8
    collection process, (03308):5, 7
    data sheet, (03308):6
    for hardness, (03308):8
    laboratory analysis, (03308):5–6
    pH, (03308):7
  testing kits, (03308):5, 7
Water analysis data sheet, (03308):6
Water filtration equipment
  bag-type filters, (03308):12, 14
  cartridge filters, (03308):10–12
  centrifugal separators, (03308):12–14
  maintenance, (03308):10–12
  multimedia filters, (03308):11–12, 13
  strainers, (03308):10, 11
Water hammer, (03308):26
Water quality, equipment problems related to, (03308):3–5, 21
Water-side economizers, (03404):22
Water softeners, (03308):15, 16, 25
Water treatment
  chemical-free, (03308):19
  chemical safety precautions, (03308):8
  closed recirculating systems, (03308):24–25
  hard water, (03308):15, 16, 25
  importance of, (03308):2
  means of, (03308):2
  mechanical, (03308):10
  open recirculating systems, (03308):20–24
  process, (03308):2
  steam boilers, (03308):25
Water treatment equipment
  blowdown controllers and separators, (03308):18
  chemical feed systems, automatic, (03308):16–18
  deaerators, (03308):15–16, 17, 25
  evaporators, (03308):15
  water softeners, (03308):15, 16

WBS. *See* Work breakdown structure (WBS)
Weight of air, (03402):3
Wet-bulb temperature, (03402):13, 14–15
Wheel heat exchangers, (03404):2, 6–7
Whirlpool. *See* Super-Efficient Refrigerator Program (SERP)
Wide area network (WAN), (03405):17, 18, 44, (46101):1, 4, 79
Wi-Fi, (46101):1, 4, 79
Windows, load estimation, (03407):15–16, 17–19
Wine storage, (03409):32
Women in industry
  communication styles, (46101):5
  sexual harassment of, (46101):5, 6–7
Wood-burning appliances
  boilers, (03409):4–6, 7, 9
  creosote in, (03409):1, 3, 9–10, 11, 37
  fireplace inserts, (03409):1–3
  furnaces, (03409):3–4, 5, 7
  installation, (03409):7–9
  maintenance, (03409):9
  safety guidelines, (03409):10
  stoves, (03409):1–3, 11
Work breakdown structure (WBS), (46101):46, 75, 79
Workers compensation, (46101):33
Workforce
  common problems
    absenteeism, (46101):28–29
    inability to work with others, (46101):28
    policies and procedures, failure to comply, (46101):29
    substance abuse, (46101):40–41
    turnover, (46101):29
  disciplining, (46101):29
  diversity, (46101):5
  generational differences, (46101):2, 4–5
  motivators
    accomplishment, (46101):20, 22
    advancement opportunity, (46101):22–23
    change, (46101):22
    feedback, (46101):3
    financial, (46101):2
    goal setting, (46101):2, 3
    job importance, (46101):22
    personal growth, (46101):23
    recognition and praise, (46101):23
    rewards, (46101):23–24
    sense of achievement, (46101):2
  shortages predicted, (46101):2
Written communication, (46101):19

**Z**

Zeolite, (03308):15, (03404):1, 6, 33